The Padé approximant of a given power series is a rational function of numerator degree L and denominator degree M whose power series agrees with the given one up to degree $L + M$ inclusively. A collection of Padé approximants formed by using a suitable set of values of L and M often provides a means of obtaining information about the function outside its circle of convergence, and of more rapidly evaluating the function within its circle of convergence.

Applications of these ideas in physics, chemistry, electrical engineering, and other areas have led to a large number of generalizations of Padé approximants which are tailor-made for specific applications. Applications to statistical mechanics and critical phenomena are extensively covered, and there are newly extended sections devoted to circuit design, matrix Padé approximation, computational methods, and integral and algebraic approximants.

The book is written with a smooth progression from elementary ideas to some of the frontiers of research in approximation theory. Its main purpose is to make the various techniques described accessible to scientists, engineers, and other researchers who may wish to use them while also presenting the rigorous mathematical theory.

This second edition has been thoroughly updated, with several new sections added, including a substantial new chapter on multiseries approximants.

ENCYCLOPEDIA OF MATHEMATICS AND ITS
APPLICATIONS

EDITED BY G.-C. ROTA

Volume 59

Padé Approximants, second edition

ENCYCLOPEDIA OF MATHEMATICS AND ITS APPLICATIONS

4   W. Miller, Jr.   *Symmetry and separation of variables*
6   H. Minc   *Permanents*
11   W. B. Jones and W. J. Thron   *Continued fractions*
12   N. F. G. Martin and J. W. England   *Mathematical theory of entropy*
18   H. O. Fattorini   *The Cauchy problem*
19   G. G. Lorentz, K. Jetter, and S. D. Riemenschneider   *Birkhoff interpolation*
21   W. T. Tutte   *Graph theory*
22   J. R. Bastida   *Field extensions and Galois theory*
23   J. R. Cannon   *The one-dimensional heat equation*
25   A. Salomaa   *Computation and automata*
26   N. White (ed.)   *Theory of matroids*
27   N. H. Bingham, C. M. Goldie, and J. L. Teugels   *Regular variation*
28   P. P. Petrushev and V. A. Popov   *Rational approximation of real functions*
29   N. White (ed.)   *Combinatorial geometries*
30   M. Pohst and H. Zassenhaus   *Algorithmic algebraic number theory*
31   J. Aczel and J. Dhombres   *Functional equations containing several variables*
32   M. Kuczma, B. Choczewski, and R. Ger   *Iterative functional equations*
33   R. V. Ambartzumian   *Factorization calculus and geometric probability*
34   G. Gripenberg, S.-O. Londen, and O. Staffans   *Volterra integral and functional equations*
35   G. Gasper and M. Rahman   *Basic hypergeometric series*
36   E. Torgersen   *Comparison of statistical experiments*
37   A. Neumaier   *Interval methods for systems of equations*
38   N. Korneichuk   *Exact constants in approximation theory*
39   R. A. Brualdi and H. J. Ryser   *Combinatorial matrix theory*
40   N. White (ed.)   *Matroid applications*
41   S. Sakai   *Operator algebras in dynamical systems*
42   W. Hodges   *Model theory*
43   H. Stahl and V. Totik   *General orthogonal polynomials*
44   R. Schneider   *Convex bodies*
45   G. Da Prato and J. Zabczyk   *Stochastic equations in infinite dimensions*
46   A. Björner, M. Las Vergnas, B. Sturmfels, N. White, and G. Ziegler   *Oriented matroids*
47   G. A. Edgar and L. Sucheston   *Stopping times and directed processes*
48   C. Sims   *Computation with finitely presented groups*
49   T. Palmer   *Banach algebras and the general theory of *-algebras*
50   F. Borceux   *Handbook of Categorical Algebra I*
51   F. Borceux   *Handbook of Categorical Algebra II*
52   F. Borceux   *Handbook of Categorical Algebra III*
54   A. Katok and B. Hasselblatt   *Introduction to the modern theory of dynamical systems*

ENCYCLOPEDIA OF MATHEMATICS AND ITS APPLICATIONS

PADÉ APPROXIMANTS
Second Edition

GEORGE A. BAKER, JR.
Theoretical Division
Los Alamos National Laboratory

PETER GRAVES-MORRIS
Department of Mathematics
University of Bradford

Published by the Press Syndicate of the University of Cambridge
The Pitt Building, Trumpington Street, Cambridge CB2 1RP
40 West 20th Street, New York, NY 10011-4211, USA
10 Stamford Road, Oakleigh, Melbourne 3166, Australia

© Cambridge University Press 1996

Second edition published 1996

First edition originally published
in two volumes by Addison-Wesley, 1982

Printed in the United States of America

Library of Congress Cataloging-in-Publication Data
Baker, George A. (George Allen), 1932 –
Padé approximants / George A. Baker, Jr., Peter Graves-Morris. –
2nd ed.
p. cm. – (Encyclopedia of mathematics and its applications ;
v. 59)
Includes bibliographical references and index.
ISBN 0-521-45007-1
1. Padé approximant.   2. Mathematical physics.   I. Graves-Morris,
P. R.   II. Title.   III. Series.
QC20.7.P3B35   1995
515'.2432 – dc20   94-48506
                                                                CIP

A catalogue record for this book is available from the British Library.

ISBN 0-521-45007-1 Hardback

To our wives
Carroll Thomas *and* **Lucia Graves-Morris**
and to our families
and
to the memory of
Elizabeth Coles Baker

CONTENTS

Preface		*page* xi
Preface to the first edition		xiii
1	**Introduction and definitions**	**1**
1.1	Introduction and Notational Conventions	1
1.2	Padé Approximants to the Exponential Function	8
1.3	Sequences and Series; Obstacles	15
1.4	The Baker Definition, the C-Table, and Block Structure	20
1.5	Duality and Invariance	32
2	**Elementary developments**	**38**
2.1	Numerical Calculation of Padé Approximants	38
2.2	Decipherment of Singularities from Padé Approximants and Apparent Errors	44
2.3	Some Explicit Forms for Padé Denominators	56
2.4	Bigradients and Hadamard's Formula	62
3	**Padé approximants and numerical methods**	**67**
3.1	Aitken's Δ^2 Method as $[L/1]$ Padé Approximants	67
3.2	Acceleration and Overacceleration of Convergence	71
3.3	The ε-Algorithm and the η-Algorithm	73
3.4	Wynn's Identity and the ε-Algorithm	81
3.5	Common Identities and Recursion Formulas	85
3.6	Recursive Calculation of the Coefficients of Padé Approximants	92
3.7	Kronecker's Algorithm and Cordellier's Identity	106
3.8	The Q.D. Algorithm and the Root Problem	115
4	**Connection with continued fractions**	**122**
4.1	Definitions, Recursion Relations, and Computation	122
4.2	Continued Fractions Derived from Maclaurin Series	129
4.3	Various Representations of Continued Fractions	141
4.4	The Berlekamp–Massey Algorithm and an Application of It	153
4.5	Different Types of Continued Fractions	165
4.6	Examples of Continued Fractions Which Are Padé Approximants	173
4.7	Convergence of Continued Fractions	182

5 Stieltjes series and Pólya series — 193
- 5.1 Introduction to Stieltjes Series — 193
- 5.2 Convergence of Stieltjes Series — 201
- 5.3 Moment Problems and Orthogonal Polynomials — 213
- 5.4 Stieltjes Series Convergent in $|z| < R$ — 220
  - 5.4.1 Hausdorff Moment Problem — 233
  - 5.4.2 Integer Moment Problem — 234
- 5.5 Stieltjes Series with Zero Radius of Convergence — 236
- 5.6 Hamburger Series and the Hamburger Moment Problem — 245
- 5.7 Pólya Frequency Series — 264

6 Convergence theory — 276
- 6.1 Introduction to Convergence Theory: Rows — 276
- 6.2 de Montessus's Theorem — 280
- 6.3 Hermite's Formula and de Montessus's Theorem — 290
- 6.4 Uniqueness of Convergence — 297
- 6.5 Convergence in Measure — 305
- 6.6 Lemniscates, Capacity, and Measure — 316
- 6.7 The Padé Conjecture — 330

7 Extensions of Padé approximants — 335
- 7.1 Multipoint Padé Approximants — 335
- 7.2 Baker–Gammel Approximants — 362
- 7.3 Series Analysis — 372
- 7.4 Padé–Laurent, Padé–Fourier, and Padé–Tchebycheff Approximants — 378
- 7.5 Laurent–Padé Approximation and Toeplitz Systems — 389
- 7.6 Multivariable Approximants — 402

8 Multiseries approximants — 415
- 8.1 Simultaneous Padé Approximants — 415
- 8.2 Operator Padé Approximants — 429
- 8.3 Rectangular Matrix Padé Approximants for Minimal Partial-Realization Problems — 442
- 8.4 Vector Padé Approximants — 466
  - 8.4.1 Functional Padé Approximants — 492
- 8.5 Hermite–Padé Polynomials — 494
  - 8.5.1 Minimality Definitions and Uniqueness — 497
  - 8.5.2 Table Structure Results — 501
  - 8.5.3 Recursion Relations — 515
  - 8.5.4 Existence of Sequences and the Modified Minimality Definition — 521
- 8.6 Integral and Algebraic Approximants — 524
  - 8.6.1 Monodromy Theory — 525
  - 8.6.2 Definitions and the Accuracy-through-Order Principle — 531
  - 8.6.3 Equivalence Properties — 538
  - 8.6.4 Invariance Properties — 539
  - 8.6.5 Separation Properties — 543
  - 8.6.6 Convergence Theory — 544
  - 8.6.7 Singular Index and Amplitude Computations — 564

9 Connection with integral equations and quantum mechanics — 570
- 9.1 The General Method and Finite-Rank Kernels — 570

9.2	Padé Approximants and Integral Equations with Compact Kernels	573
9.3	Projection Techniques	578
9.4	Potential Scattering	584
9.5	Derivation of Padé Approximants from Variational Principles	596
9.6	An Error Bound on Padé Approximants from Variational Principles	606
9.7	Single-sign Potentials in Scattering Theory etc.	608
9.8	Variational Padé Approximants	616
9.9	Singular Potentials	622
10	**Connection with numerical analysis**	**628**
10.1	Acceleration of Convergence	628
10.2	Tchebycheff's Inequalities for the Density Function	633
10.3	Collocation and the τ-method	639
10.4	Crank–Nicholson and Related Methods for the Diffusion Equation	646
10.5	Inversion of the Laplace Transform	654
10.6	Connection with Rational Approximation	656
	10.6.1 The Carathéodory–Fejér Method	663
10.7	Padé Approximants for the Riccati Equation	670
11	**Connection with quantum field theory**	**674**
11.1	Perturbed Harmonic Oscillators	674
	11.1.1 The Peres Model	675
	11.1.2 The Anharmonic Oscillator	678
11.2	Pion–Pion Scattering	679
11.3	Lattice-Cutoff $\lambda\phi_n^4$ Euclidean Field Theory, or the Continuous-Spin Ising Model	684
Appendix: A FORTRAN FUNCTION		**690**
Bibliography		**695**
Index		**741**

PREFACE

We are glad that the first edition of these volumes is thought to have achieved its main aim of making mathematical techniques more available, not only to mathematicians, but also to the wider scientific and engineering community.

We have been glad to take the opportunity provided by this edition to incorporate the most salient aspects of the large body of new results which have been obtained since the publication of the original edition. The incorporation of this new material has led to the need to make several significant rearrangements of the previous material.

We wish to record our gratitude for the mathematical contributions and company of Arne Magnus and Helmut Werner, both of them friends who are missed by many of us. The influence of their work is to be found in Chapter 4.

A few infelicities which have been noticed in the original edition have been corrected.

George A. Baker, Jr.
Peter Graves-Morris

PREFACE TO THE FIRST EDITION

These two volumes are intended to serve as a basic text on one approach to the problem of assigning a value to a power series. We have attempted to present the basic results and methods in as transparent a form as possible, in line with the general objectives of the Encyclopedia. The general topic of Padé approximants, which is, among other things, a highly practical method of definition and of construction of the value of a power series, seems to have begun independently at least twice. Padé's claim for credit is based on his thesis (1892), in which he developed the approximants and organized them in a table. He paid particular attention to the exponential function. He was presumably unaware of the prior work of Jacobi (1846), who gave the determinantal representation in his paper on the simplification of Cauchy's solution to the problem of rational interpolation. Also, Padé's work was preceded by that of Frobenius (1881), who derived identities between the neighboring rational fractions of Jacobi. It is interesting to note that Anderson seems to have stumbled upon some Padé approximants for the logarithmic function in 1740. A photograph of H. Padé is to be found in *The Padé Approximant Method and Its Application to Mechanics*, edited by H. Cabannes. A copy of his autographed thesis is to be found in the Cornell University Library.

This work has been distilled from an extensive literature, and *The Essentials of Padé Approximants*, written by one of us, has been an essential reference. We use the abbreviation EPA for this book, and refer to it often for a different or fuller treatment of some of the more advanced topics. While each book is entirely self-contained, our notation is normally compatible with EPA, and to a large extent the books complement each other. An important exception is that the Padé table in EPA is reflected through its main diagonal in our present notation. The proceedings of the Canterbury Summer School and International Conference, edited by the other of us, contain diverse contributions which initiated in print the multidisciplinary view of the subject—a view we hope we have transmitted herein. The many publications which have contributed substantially to our text are listed in the bibliography. We are grateful to our numerous colleagues at Brookhaven, Canterbury, Cornell, Los Alamos, and Saclay in freely discussing so many topics which have made possible the breadth of our treatment. Especially, we thank Roy Chisholm, John Gammel, and Daniel Bessis for many conversations, and the C.E.A. at Saclay, where part of this book was written, for hospitality.

Our hardest task in writing this book was to choose a presentation which is both correct and readily comprehensible. A fully precise system based on rigorous analysis and set-theoretic language would have ensured total obscurity of the more practical techniques. Conversely, omission of all the conditions under which the theorems hold good would be absurdly misleading. We have chosen a level of presentation suitable for the topic in hand. For example, the connectivity of sets is mentioned where it is important, and otherwise it is omitted. The meaning of the order notation is clear in context. Both applications in physics and techniques recently developed are treated in a practical fashion.

Equations are referenced by a default option. Equation (I.6.5.3) is Equation (5.3) of Part I. Chapter 6; the Part and Chapter are dropped by default if they are the same as the source of the reference.

Finally, a spirit of evangelism may be detected in the text. When a review of rational approximation in 1963 can claim that Padé approximants cannot approximate on the entire range $(0, \infty)$ and be believed, a revision of view is overdue.

George A. Baker, Jr.
Peter Graves-Morris
1 October, 1980.

1
Introduction and definitions

1.1 Introduction and notational conventions

Suppose that we are given a power series $\sum_{i=0}^{\infty} c_i z^i$, representing a function $f(z)$, so that

$$f(z) = \sum_{i=0}^{\infty} c_i z^i. \tag{1.1}$$

This expansion is the fundamental starting point of any analysis using Padé approximants. Throughout this work we reserve the notation $c_i = 0, 1, 2, \ldots$, for the given set of coefficients, and $f(z)$ is the associated function. A Padé approximant is a rational fraction

$$[L/M] = \frac{a_0 + a_1 z + \cdots + a_L z^L}{b_0 + b_1 z + \cdots + b_M z^M} \tag{1.2}$$

which has a Maclaurin expansion which agrees with (1.1) as far as possible. We give a more complete and precise definition of Padé approximants in Section 1.4. Notice that in (1.2) there are $L + 1$ numerator coefficients and $M + 1$ denominator coefficients. There is a more or less irrelevant common factor between them, and for definiteness we take $b_0 = 1$. This choice turns out to be an essential part of the precise definition, and (1.2) is our conventional notation with this choice for b_0. So there are $L + 1$ independent numerator coefficients and M independent denominator coefficients, making $L + M + 1$ unknown coefficients in all. This number suggests that normally the $[L/M]$ ought to fit the power series (1.1) through the orders $1, z, z^2, \ldots, z^{L+M}$. In the notation of formal power series,

$$\sum_{i=0}^{\infty} c_i z^i = \frac{a_0 + a_1 z + \cdots + a_L z^L}{b_0 + b_1 z + \cdots + b_M z^M} + O(z^{L+M+1}). \tag{1.3}$$

Example

$$f(z) = 1 - \tfrac{1}{2}z + \tfrac{1}{3}z^2 + \cdots.$$

$$[1/0] = 1 - \tfrac{1}{2}z = f(z) + O(z^2),$$

$$[0/1] = \frac{1}{1 + \tfrac{1}{2}z} = f(z) + O(z^2),$$

$$[1/1] = \frac{1 + \tfrac{1}{6}z}{1 + \tfrac{2}{3}z} = f(z) + O(z^3).$$

Returning to (1.3) and cross-multiplying, we find that

$$(b_0 + b_1 z + \cdots + b_M z^M)(c_0 + c_1 z + \cdots) =$$
$$a_0 + a_1 z + \cdots + a_L z^L + O(z^{L+M+1}) \quad (1.4)$$

Equating the coefficients of $z^{L+1}, z^{L+2}, \ldots, z^{L+M}$, we find

$$b_M c_{L-M+1} + b_{M-1} c_{L-M+2} + \cdots + b_0 c_{L+1} = 0,$$
$$b_M c_{L-M+2} + b_{M-1} c_{L-M+3} + \cdots + b_0 c_{L+2} = 0, \quad (1.5)$$
$$\vdots$$
$$b_M c_L + b_{M-1} c_{L+1} + \cdots + b_0 c_{L+M} = 0.$$

If $j < 0$, we define $c_j = 0$ for consistency. Since $b_0 = 1$, Equations (1.5) become a set of M linear equations for the M unknown denominator coefficients:

$$\begin{bmatrix} c_{L-M+1} & c_{L-M+2} & c_{L-M+3} & \cdots & c_L \\ c_{L-M+2} & c_{L-M+3} & c_{L-M+4} & \cdots & c_{L+1} \\ c_{L-M+3} & c_{L-M+4} & c_{L-M+5} & \cdots & c_{L+2} \\ \vdots & \vdots & \vdots & & \vdots \\ c_L & c_{L+1} & c_{L+2} & \cdots & c_{L+M-1} \end{bmatrix} \begin{bmatrix} b_M \\ b_{M-1} \\ b_{M-2} \\ \vdots \\ b_1 \end{bmatrix} = - \begin{bmatrix} c_{L+1} \\ c_{L+2} \\ c_{L+3} \\ \vdots \\ c_{L+M} \end{bmatrix}, \quad (1.6)$$

from which the b_i may be found. The numerator coefficients, $a_0, a_1, \ldots, a_L$, follow immediately from (1.4) by equating the coefficients $1, z, z^2, \ldots, z^L$:

$$a_0 = c_0,$$
$$a_1 = c_1 + b_1 c_0,$$
$$a_2 = c_2 + b_1 c_1 + b_2 c_0,$$
$$\vdots$$
$$a_L = c_L + \sum_{i=1}^{\min(L,M)} b_i c_{L-i}. \quad (1.7)$$

Thus (1.6) and (1.7) normally determine the Padé numerator and denominator and are called the Padé equations; we have constructed an $[L/M]$ Padé approximant which agrees with $\sum_{i=0}^{\infty} c_i z^i$ through order z^{L+M}. Because the starting point of these manipulations is the given power series, we do not ever need to know about the existence of any function $f(z)$ with $\sum_{i=0}^{\infty} c_i z^i$ as its Maclaurin series, as in (1.1). Of course, we expect that a well-chosen sequence of Padé approximants will normally approximate a function $f(z)$ with the Maclaurin expansion $\sum_{i=0}^{\infty} c_i z^i$, but it is important to distinguish between problems of convergence of Padé approximants and problems of construction of Padé approximants. Given the power series, (1.6) and (1.7) show how the Padé approximants are constructed.

Every power series has a circle of convergence $|z| = R$. If $|z| < R$, the series converges, and if $|z| > R$, it does not. If $R = \infty$, the power series represents an analytic function (functions analytic everywhere we often call *entire*) and the series may be summed directly for any value of z to yield the function $f(z)$. If $R = 0$, the power series is undoubtedly formal. It contains information about $f(z)$, but just how this information is to be used is not immediately clear. However, if a sequence of Padé approximants of the formal power series converges to a function of $g(z)$ for $z \in \mathcal{D}$, then we may reasonably conclude that $g(z)$ is a function with the given power series. In certain circumstances (see Chapter 5) we make such statements precise and prove them. Nevertheless, in this book we will not be hampered by a lack of rigorous justification of any technique, and empirical convergence is regarded as entirely satisfactory within its limitations. If the given power series converges to the same function for $|z| < R$ with $0 < R < \infty$, then a sequence of Padé approximants may converge for $z \in \mathcal{D}$ where $\mathcal{D}$ is a domain larger than $|z| < R$. We will then have extended our domain of convergence. This is frequently a practical approach to what amounts to analytic continuation. The method of expansion and reexpansion due to Weierstrass is more suited to principle than practice. As an example of how well Padé approximants may work in their natural context, we consider an example.

Example

$$f(z) = \sqrt{\frac{1 + \frac{1}{2}z}{1 + 2z}} = 1 - \frac{3}{4}z + \frac{39}{32}z^2 - \cdots.$$

To calculate [1/1], Equation (1.6) becomes

$$(-\tfrac{3}{4})b_1 = -\tfrac{39}{32},$$

and so $b_1 = \frac{13}{8}$. Equation (1.7) gives $a_0 = 1$ and $a_1 = \frac{7}{8}$, with the check

$$(1 + \tfrac{13}{8}z)(1 - \tfrac{3}{4}z + \tfrac{39}{32}z^2) = 1 + \tfrac{7}{8}z + O(z^3).$$

Hence

$$[1/1] = \frac{1 + \frac{7}{8}z}{1 + \frac{13}{8}z},$$

and in Figure 1.1.1 we compare this with $f(z)$ for $z \geq 0$. In particular, $f(\infty) = 0.5$ and $[1/1](\infty) = \frac{7}{13} = 0.54\ldots$, giving 8% accuracy at infinity. This example shows remarkable accuracy for a function with a radius of convergence of $\frac{1}{2}$, using just three terms of the series.

There is one feature of the calculation of Padé approximations to be emphasized at the start — these calculations require more numerical accuracy than one might at first expect. The Padé approximant exploits the differences of the coefficients to do its long-range extrapolation, and so the differences must all be accurate. We consider the problem of deciding how much numerical accuracy is needed to calculate an $[L/M]$ Padé approximant in Section 2.1.

Thus far, we have assumed that Padé approximants are calculated directly from (1.6) and (1.7) without implying any particular method. If Cramer's rule is used, we may calculate $b_0:b_1:\cdots:b_M$ from (1.6) and hence the denominator of (1.2). Aside from a common factor, the result is

$$Q^{[L/M]}(z) = \begin{vmatrix} c_{L-M+1} & c_{L-M+2} & \cdots & c_L & c_{L+1} \\ c_{L-M+2} & c_{L-M+3} & \cdots & c_{L+1} & c_{L+2} \\ \vdots & \vdots & & \vdots & \vdots \\ c_{L-1} & c_L & \cdots & c_{L+M-2} & c_{L+M-1} \\ c_L & c_{L+1} & \cdots & c_{L+M-1} & c_{L+M} \\ z^M & z^{M-1} & \cdots & z & 1 \end{vmatrix}. \quad (1.8)$$

We take (1.8) to define $Q^{[L/M]}(z)$ and use this convention throughout. Again, recall that $c_j = 0$ if $j < 0$. Now consider

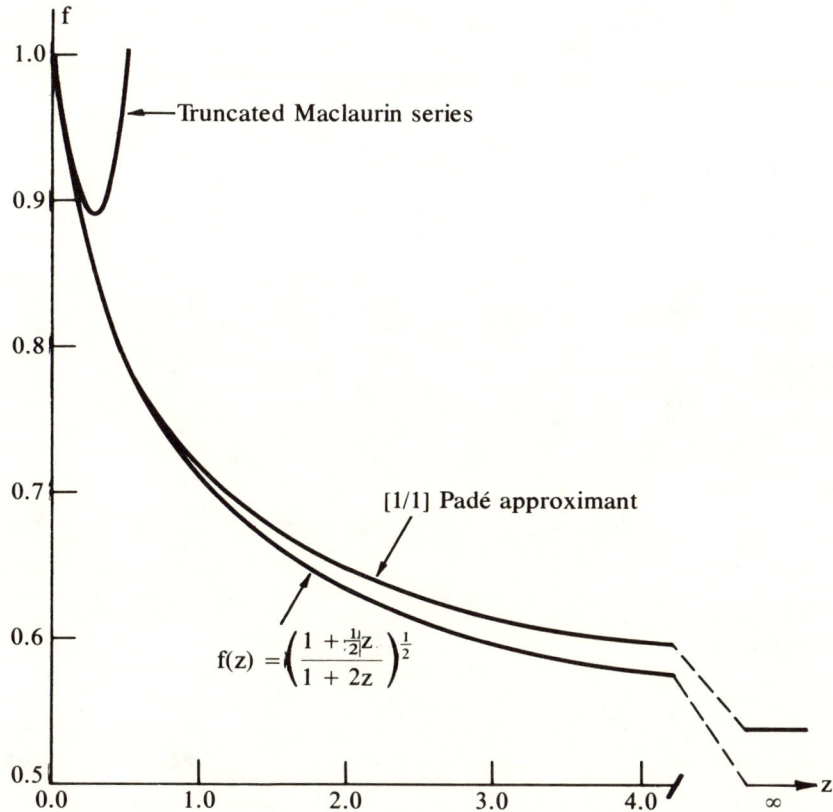

Figure 1.1.1. Values of $f(z) = \sqrt{(1 + z/2)/(1 + 2z)}$, its [1/1] Padé approximant, and its truncated Maclaurin series, $1 - 3z/4 + 39z^2/32$.

$$Q^{[L/M]}(z)\sum_{i=0}^{\infty} c_i z^i = \begin{vmatrix} c_{L-M+1} & c_{L-M+2} & \cdots & c_{L+1} \\ c_{L-M+2} & c_{L-M+3} & \cdots & c_{L+2} \\ \vdots & \vdots & & \vdots \\ c_{L-1} & c_L & \cdots & c_{L+M-1} \\ c_L & c_{L+1} & \cdots & c_{L+M} \\ \sum_{i=0}^{\infty} c_i z^{M+i} & \sum_{i=0}^{\infty} c_i z^{M+i-1} & \cdots & \sum_{i=0}^{\infty} c_i z^i \end{vmatrix}.$$

By subtracting z^{L+1} times the first row from the last, z^{L+2} times the second row from the last, etc., up to z^{L+M} times the penultimate row from the last, we reduce the series in the last row. They become lacunary series, with a gap of M terms missing. Using the initial terms of these series, we define

$$P^{[L/M]}(z) = \begin{vmatrix} c_{L-M+1} & c_{L-M+2} & \cdots & c_{L+1} \\ c_{L-M+2} & c_{L-M+3} & \cdots & c_{L+2} \\ \vdots & \vdots & & \vdots \\ c_{L-1} & c_L & \cdots & c_{L+M-1} \\ c_L & c_{L+1} & \cdots & c_{L+M} \\ \sum_{i=0}^{L-M} c_i z^{M+i} & \sum_{i=0}^{L-M+1} c_i z^{M+i-1} & \cdots & \sum_{i=0}^{L} c_i z^i \end{vmatrix}. \quad (1.9)$$

Again, (1.9) is our notational convention. We now prove our first theorem.

Theorem 1.1.1. *With the definitions* (1.8) *and* (1.9),

$$Q^{[L/M]}(z) \sum_{i=0}^{\infty} c_i z^i - P^{[L/M]}(z) = O(z^{L+M+1}). \quad (1.10)$$

Proof. We note that $\deg\{P^{[L/M]}\} \leq L$, $\deg\{Q^{[L/M]}\} \leq M$ and that the remainder is

$$Q^{[L/M]}(z) \sum_{i=0}^{\infty} c_i z^i - P^{[L/M]}(z)$$

$$= \begin{vmatrix} c_{L-M+1} & c_{L-M+2} & \cdots & c_{L+1} \\ c_{L-M+2} & c_{L-M+3} & \cdots & c_{L+2} \\ \vdots & \vdots & & \vdots \\ c_{L-1} & c_L & \cdots & c_{L+M-1} \\ c_L & c_{L+1} & \cdots & c_{L+M} \\ \sum_{i=L+1}^{\infty} c_i z^{M+i} & \sum_{i=L+2}^{\infty} c_i z^{M+i-1} & \cdots & \sum_{i=L+M+1}^{\infty} c_i z^i \end{vmatrix}$$

$$= \sum_{i=1}^{\infty} z^{L+M+i} \begin{vmatrix} c_{L-M+1} & c_{L-M+2} & \cdots & c_{L+1} \\ c_{L-M+2} & c_{L-M+3} & \cdots & c_{L+2} \\ \vdots & \vdots & & \vdots \\ c_{L-1} & c_L & \cdots & c_{L+M-1} \\ c_L & c_{L+1} & \cdots & c_{L+M} \\ c_{L+i} & c_{L+i+1} & \cdots & c_{L+M+i} \end{vmatrix}. \quad (1.11)$$

Equation (1.11) is occasionally a useful form for the error using Padé approximation. Equation (1.10) goes a long way towards satisfying (1.3). To this end, consider

$$Q^{[L/M]}(0) = \begin{vmatrix} c_{L-M+1} & c_{L-M+2} & \cdots & c_L \\ c_{L-M+2} & c_{L-M+3} & \cdots & c_{L+1} \\ \vdots & \vdots & & \vdots \\ c_{L-1} & c_L & \cdots & c_{L+M-2} \\ c_L & c_{L+1} & \cdots & c_{L+M-1} \end{vmatrix}.$$

This is called a Hankel determinant, because of the systematic way in which its rows are formed from the given coefficients c_j. Notice that if $Q^{[L/M]}(0) \neq 0$, then the linear equations (1.6) are nonsingular and the solution given by (1.8) is unambiguous. Furthermore, we may divide (1.10) by $Q^{[L/M]}(z)$, yielding

$$\sum_{i=0}^{\infty} c_i z^i - \frac{P^{[L/M]}(z)}{Q^{[L/M]}(z)} = O(z^{L+M+1}).$$

This result has proved our second theorem:

***Theorem* 1.1.2 [Jacobi, 1846].** *With the definitions* (1.8) *and* (1.9), *the* $[L/M]$ *Padé approximant of* $\sum_{i=0}^{\infty} c_i z^i$ *is given by*

$$[L/M] = \frac{P^{[L/M]}(z)}{Q^{[L/M]}(z)} \tag{1.12}$$

provided $Q^{[L/M]}(0) \neq 0$.

The only difficulties, which we defer to Section 1.4, are those occurring when $Q^{[L/M]}(0) = 0$. We extend the notation $[L/M]$ of (1.12) as $[L/M]_f$ to emphasize approximation of $f(z)$, and as $[L/M](z)$ to emphasize the z-dependence. We will thus have the various forms

$$[L/M] = [L/M]_f = [L/M](z) = [L/M]_f(z)$$

available for convenience. It is common practice to display the approximants in a table, called the Padé table, shown as Table 1.1.1. Among other things, we prove in Section 1.2 that part of the Padé table of $\exp(z)$ is given by the entries in Table 1.1.2.

Table 1.1.1. *The Padé table.*

M \ L	0	1	2	$\cdots$
0	[0/0]	[1/0]	[2/0]	$\cdots$
1	[0/1]	[1/1]	[2/1]	$\cdots$
2	[0/2]	[1/2]	[2/2]	$\cdots$
$\vdots$	$\vdots$	$\vdots$	$\vdots$	$\ddots$

Table 1.1.2. *Part of the Padé table of* $\exp(z)$ *[Padé, 1892]*.

M \ L	0	1	2
0	$\dfrac{1}{1}$	$\dfrac{1+z}{1}$	$\dfrac{2+2z+z^2}{2}$
1	$\dfrac{1}{1-z}$	$\dfrac{2+z}{2-z}$	$\dfrac{6+4z+z^2}{6-2z}$
2	$\dfrac{2}{2-2z+z^2}$	$\dfrac{6+2z}{6-4z+z^2}$	$\dfrac{12+6z+z^2}{12-6z+z^2}$

1.2 Padé approximants to the exponential function

The coefficients c_i of the Maclaurin expansion of the exponential function are sufficiently simple that explicit forms of the numerator and denominator of the Padé approximants can be found. In this section we will calculate the denominator $Q^{[L/M]}(z)$. The numerator follows by an extremely simple and elegant trick, based on the identity $\exp(-z) = 1/\exp(z)$, and this derivation is discussed in Section 1.5. Padé, in his thesis, elaborated the properties of his approximants with special emphasis on the example of the exponential function: it is a beautiful example of how the approximants work in an ideal situation. Further properties of Padé approximants of $\exp(z)$ are to be found in Section 4.6, Section 5.7, and Sections 10.3–10.4.

Our task is to calculate

$$Q^{[L/M]}(z) = \begin{vmatrix} \dfrac{1}{(L-M+1)!} & \dfrac{1}{(L-M+2)!} & \cdots & \dfrac{1}{L!} & \dfrac{1}{(L+1)!} \\ \dfrac{1}{(L-M+2)!} & \dfrac{1}{(L-M+3)!} & \cdots & \dfrac{1}{(L+1)!} & \dfrac{1}{(L+2)!} \\ \vdots & \vdots & & \vdots & \vdots \\ \dfrac{1}{L!} & \dfrac{1}{(L+1)!} & \cdots & \dfrac{1}{(L+M-1)!} & \dfrac{1}{(L+M)!} \\ z^M & z^{M-1} & \cdots & z & 1 \end{vmatrix}.$$

(2.1)

It is easier to begin with the constant term in (2.1), and so we define $C(L/M) \equiv Q^{[L/M]}(0)$, which is the coefficient of the '1' in the lower right-hand corner of (2.1),

$$C(L/M) = \begin{vmatrix} \dfrac{1}{(L-M+1)!} & \dfrac{1}{(L-M+2)!} & \cdots & \dfrac{1}{L!} \\ \dfrac{1}{(L-M+2)!} & \dfrac{1}{(L-M+3)!} & \cdots & \dfrac{1}{(L+1)!} \\ \vdots & \vdots & & \vdots \\ \dfrac{1}{L!} & \dfrac{1}{(L+1)!} & \cdots & \dfrac{1}{(L+M-1)!} \end{vmatrix}. \tag{2.2}$$

We assume that $L \geq M - 1$. If this condition does not hold, the factorial functions must be suitably reinterpreted as gamma functions for the analysis to be valid. We remove the denominators from each row, by defining

$$p = \prod_{i=1}^{M} \frac{1}{(L+i-1)!},$$

and then

$C(L/M) =$

$$p \cdot \begin{vmatrix} \dfrac{L!}{(L-M+1)!} & \dfrac{L!}{(L-M+2)!} & \cdots & L & 1 \\ \dfrac{(L+1)!}{(L-M+2)!} & \dfrac{(L+1)!}{(L-M+3)!} & \cdots & L+1 & 1 \\ \vdots & \vdots & & \vdots & \vdots \\ \dfrac{(L+M-1)!}{L!} & \dfrac{(L+M-1)!}{(L+1)!} & \cdots & L+M-1 & 1 \end{vmatrix}. \tag{2.3}$$

In (2.3), the determinant has M rows. Subtract the $(M-1)$th row from the Mth, then the $(M-2)$th row from the $(M-1)$th, etc. The identity

$$\frac{r!}{s!} - \frac{(r-1)!}{(s-1)!} = (r-s)\frac{(r-1)!}{s!} \tag{2.4}$$

is used repeatedly. In column 1 of (2.3), $r - s = M - 1$; in column 2, $r - s = M - 2$; etc., and so one finds that

$$C(L/M) = p(M-1)! \begin{vmatrix} \dfrac{L!/(M-1)}{(L-M+1)!} & \dfrac{L!/(M-2)}{(L-M+2)!} & \cdots & L & 1 \\ \dfrac{L!}{(L-M+2)!} & \dfrac{L!}{(L-M+3)!} & \cdots & 1 & 0 \\ \vdots & \vdots & & \vdots & \vdots \\ \dfrac{(L+M-2)!}{L!} & \dfrac{(L+M-2)!}{(L+1)!} & \cdots & 1 & 0 \end{vmatrix}$$

$$= p(-)^{M-1}(M-1)! \begin{vmatrix} \dfrac{L!}{(L-M+2)!} & \dfrac{L!}{(L-M+3)!} & \cdots & 1 \\ \dfrac{(L+1)!}{(L-M+3)!} & \dfrac{(L+1)!}{(L-M+4)!} & \cdots & 1 \\ \vdots & \vdots & & \vdots \\ \dfrac{(L+M-2)!}{L!} & \dfrac{(L+M-2)!}{(L+1)!} & \cdots & 1 \end{vmatrix}.$$

(2.5)

This is a $(M-1) \times (M-1)$ determinant with a form identical to (2.3) but with M replaced by $M-1$. Consequently, an obvious inductive argument shows that

$$C(L/M) = p \cdot \prod_{i=1}^{M} (-1)^{i-1}(i-1)!$$

$$= (-1)^{M(M-1)/2} \prod_{i=1}^{M} \frac{(i-1)!}{(L+i-1)!}. \qquad (2.6)$$

Thus, for the case $M = 1$,

$$C(L/1) = \frac{1}{L!},$$

and for the case $M = 2$,

$$C(L/2) = \begin{vmatrix} \dfrac{1}{(L-1)!} & \dfrac{1}{L!} \\ \dfrac{1}{L!} & \dfrac{1}{(L+1)!} \end{vmatrix} = \dfrac{-1}{L!(L+1)!}.$$

The sign pattern of (2.6) distinguishes the Polyá frequency series, to which we refer in Section 5.7. The row operations we have performed to deduce (2.6) from (2.2) are still permissible with the form (2.1), except that the situation is more complicated. We consider the coefficient of $(-z)^j$ in $Q^{[L/M]}(z)$, which is

$$(-)^j q_j^{[L/M]} = \begin{vmatrix} \dfrac{1}{(L-M+1)!} & \dfrac{1}{(L-M+2)!} & \cdots & \vdots\, \dfrac{1}{(L-j+1)!}\, \vdots & \cdots & \dfrac{1}{(L+1)!} \\ \dfrac{1}{(L-M+2)!} & \dfrac{1}{(L-M+3)!} & \cdots & \vdots\, \dfrac{1}{(L-j+2)!}\, \vdots & \cdots & \dfrac{1}{(L+2)!} \\ \vdots & \vdots & & \vdots\quad\vdots & & \vdots \\ \dfrac{1}{L!} & \dfrac{1}{(L+1)!} & & \vdots\, \dfrac{1}{(L+M-j)!}\, \vdots & \cdots & \dfrac{1}{(L+M)!} \end{vmatrix}$$
(2.7)

where the column surrounded by $\vdots\;\vdots$ is deleted. We perform a similar analysis: define

$$p' = \prod_{i=1}^{M} \dfrac{1}{(L+i)!},$$

and then

$$(-)^j q_j^{[L/M]} = p' \cdot \begin{vmatrix} \dfrac{(L+1)!}{(L-M+1)!} & \cdots & \vdots\, \dfrac{(L+1)!}{(L-j+1)!}\, \vdots & \cdots & 1 \\ \dfrac{(L+2)!}{(L-M+2)!} & \cdots & \vdots\, \dfrac{(L+2)!}{(L-j+2)!}\, \vdots & \cdots & 1 \\ \vdots & & \vdots\quad\vdots & & \vdots \\ \dfrac{(L+M)!}{L!} & \cdots & \vdots\, \dfrac{(L+M)!}{(L+M-j)!}\, \vdots & \cdots & 1 \end{vmatrix}$$
(2.8)

Subtracting rows, and using the identity (2.4), we have

$$(-)^j q_j^{[L/M]} = (-)^M p' \frac{M!}{j} \times \begin{vmatrix} \dfrac{(L+1)!}{(L-M+2)!} & \cdots & \dfrac{(L+1)!}{(L-j+2)!} & \cdots & 1 \\ \dfrac{(L+2)!}{(L-M+3)!} & \cdots & \dfrac{(L+2)!}{(L-j+3)!} & \cdots & 1 \\ \vdots & & \vdots & & \vdots \\ \dfrac{(L+M-1)!}{L!} & \cdots & \dfrac{(L+M-1)!}{(L+M-j)!} & \cdots & 1 \end{vmatrix},$$

which again is an $(M-1) \times (M-1)$ determinant with a form similar to (2.8). We make j similar reductions from (2.8) to obtain

$$(-)^j q_j^{[L/M]} = \pm \frac{p'}{j!} \prod_{i=1}^{j}(M-i+1)! \begin{vmatrix} \dfrac{(L+1)!}{(L-M+j+1)!} & \cdots & \dfrac{(L+1)!}{L!} & \cdots & 1 \\ \dfrac{(L+2)!}{(L-M+j+2)!} & \cdots & \dfrac{(L+2)!}{(L+1)!} & \cdots & 1 \\ \vdots & & \vdots & & \vdots \\ \dfrac{(L+M-j)!}{L!} & \cdots & \dfrac{(L+M-j)!}{(L+M-j-1)!} & \cdots & 1 \end{vmatrix}.$$

Removing a common factor from each row,

$$(-)^j q_j^{[L/M]} = \pm \frac{p'}{j!} \frac{(L+M-j)!}{L!} \prod_{i=1}^{j}(M-i+1)! \begin{vmatrix} \dfrac{L!}{(L-M+j+1)!} & \cdots & 1 \\ \dfrac{(L+1)!}{(L-M+j+2)!} & \cdots & 1 \\ \vdots & & \vdots \\ \dfrac{(L+M-j-1)!}{L!} & \cdots & 1 \end{vmatrix}.$$

1.2 Padé approximants to the exponential function

The analysis now follows the familiar pattern using the identity (2.4), and we deduce that

$$(-)^j q_j^{[L/M]} = \pm \left\{ \prod_{i=1}^{M} \frac{1}{(L+i)!} \right\} \frac{(L+M-j)!}{L!j!}$$

$$\times \left\{ \prod_{i=1}^{j} (M-i+1)! \right\} \prod_{i=1}^{M-j-1} i!$$

$$= \pm \frac{(L+M-j)!}{L!j!(M-j)!} \prod_{i=1}^{M} \frac{i!}{(L+i)!}. \quad (2.9)$$

The sign of the right-hand side of (2.9) is easily determined to be the same as that of (2.6), because the determinants (2.2) and (2.7) have the same dimension, and are expanded by the same top right-hand elements recursively. Hence

$$(-)^j q_j^{[L/M]} = (-)^{M(M-1)/2} \frac{(L+M-j)!}{L!j!(M-j)} \prod_{i=1}^{M} \frac{i!}{(L+i)!}. \quad (2.10)$$

Notice that (2.6) emerges as the special case with $j=0$. Consequently we have

$$q_j^{[L/M]} = (-)^j C(L/M) \frac{(L+M-j)!}{(L+M)!} \frac{M!}{(M-j)!} \frac{1}{j!}$$

and

$$Q^{[L/M]}(z) = C(L/M) \sum_{j=0}^{M} \frac{(L+M-j)!}{(L+M)!} \frac{M!}{(M-j)!} \frac{(-z)^j}{j!}$$

$$= C(L/M) \left\{ 1 + \frac{M}{L+M} \frac{(-z)}{1!} + \frac{M(M-1)}{(L+M)(L+M-1)} \frac{(-z)^2}{2!} + \cdots \right\}$$

$$= C(L/M) \left\{ 1 + \frac{-M}{-L-M} \frac{(-z)}{1!} + \frac{-M(-M+1)}{(-L-M)(-L-M+1)} \frac{(-z)^2}{2!} + \cdots \right\}$$

$$= C(L/M) {}_1F_1(-M, -L-M; -z). \quad (2.11)$$

Following the method of Section 1.6, we may deduce from (2.11) that

$$P^{[L/M]}(z) = C(L/M) {}_1F_1(-L, -L-M; z),$$

and hence the $[L/M]$ Padé approximant for $\exp(z)$ is

$$[L/M] = \frac{{}_1F_1(-L, -L-M; z)}{{}_1F_1(-M, -L-M; -z)}. \quad (2.12)$$

The way in which an $[L/M]$ Padé approximant matches a function $f(z)$ is vividly portrayed by its order star, and the exponential function provides an ideal example. An order star is a separation of the complex plane into two regions

$$A_+ = \{z: |f(z)^{-1}[L/M](z)| > 1\} \quad (2.13)$$

$$A_- = \{z: |f(z)^{-1}[L/M](z)| < 1\} \quad (2.14)$$

and A_0 denotes the common boundary [Iserles and Nørsett, 1991]. Suppose that

$$f(z) - [L/M](z) = e_{N+1}z^{N+1} + O(z^{N+2}) \quad (2.15)$$

with $e_{N+1} \neq 0$ so that the order of the error of Padé approximation is precisely $N + 1$. On the circle $z = \eta e^{i\theta}$, having some sufficiently small radius $\eta > 0$, it follows from (2.15) that

$$R(z) = f(z)^{-1}[L/M](z) = 1 + \varepsilon e^{i(N+1)\theta + i\delta} + o(\varepsilon) \quad (2.16)$$

for some $\varepsilon > 0$. Define

$$\psi = (N + 1)\theta + \delta \quad (2.17)$$

which dominates the phase of $R(z) - 1$, for small ε. Hence

$$R(z) = f(z)^{-1}[L/M](z) = (1 + \varepsilon \cos \psi) + i\varepsilon \sin \psi + o(\varepsilon). \quad (2.18)$$

From (2.13), (2.14), and (2.18), we see that z belongs to an A_- region near the origin if $\cos \psi < 0$ and to an A_+ region if $\cos \psi > 0$. Equation (2.18) implies that there are $(N + 1)$ A_+ regions and $(N + 1)$ A_- regions alternately abutting the origin, as shown in Figure 1.2.1 for $N = 2$.

From (2.13) and (2.14), it is clear that poles of the approximants belong to A_+ regions, and zeros to A_- regions.

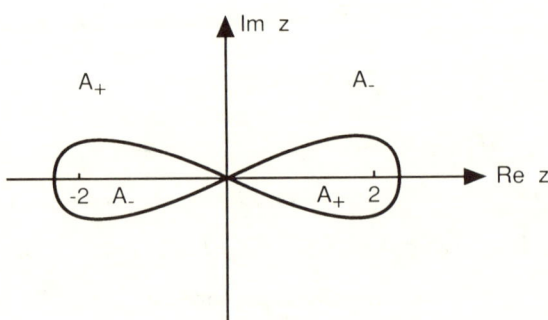

Figure 1.2.1. The order star for the $[1/1]$ Padé approximant of $\exp(z)$, showing the regions A_+, A_-.

For the case of the exponential function, e^z is larger than any rational function in modulus for Re $z > 0$ and $|z|$ sufficiently large. Hence there is an infinite A_+ region for $|\arg(z)| < \pi/2$ and $|z| \to \infty$, and an infinite A_- region for $-\pi/2 < \arg z < \pi/2$ and $|z| \to \infty$. In Section 5.7, we prove that the order star of the $[L/M]$ Padé approximant of $\exp(z)$ contains M finite A_+ regions, each containing a pole of the approximant, and L finite A_- regions, each containing a zero of the approximant; this geometry is exemplified in Figure 1.2.1.

1.3 Sequences and series; obstacles

In Section 1.1, we showed how Padé approximants are constructed from given power series, and in Section 1.2 we saw how the Padé approximants of $\exp(z)$ are obtained. In this section, we preview a few of the techniques and problems to be discussed in later chapters. The Padé method is directly applicable for the improvement of convergence of series and sequences. This application is fully discussed in Chapter 3. For the moment, assuming convergence of a sequence of approximants $[L_k/M_k](z)$, $k = 1, 2, \ldots$, at the point $z = 1$,

$$\sum_{k=0}^{\infty} c_k = \lim_{k \to \infty} [L_k/M_k](1). \tag{3.1}$$

In this sense, ordinary Padé approximants may be used to sum series. Likewise, given a sequence $\{S_n, n = 0, 1, 2, \ldots\}$, we define a series from it using forward differences,

$$c_0 = S_0,$$
$$c_{n+1} = S_{n+1} - S_n = \Delta S_n, \quad n = 0, 1, 2, \ldots,$$

and then Padé approximants may be used to extrapolate sequences. It is common practice to use the diagonal sequence of Padé approximants unless there are reasons to the contrary. We take up these points in Chapter 3. From (1.8),

$$Q^{[L/M]}(z) = \begin{vmatrix} c_{L-M+1} & c_{L-M+2} & \cdots & c_{L+1} \\ c_{L-M+2} & c_{L-M+3} & \cdots & c_{L+2} \\ \vdots & \vdots & & \vdots \\ c_L & c_{L+1} & \cdots & c_{L+M} \\ z^M & z^{M-1} & \cdots & 1 \end{vmatrix}$$

We reduce this by subtracting z times each column from the previous column, to yield

$$Q^{[L/M]}(z) = \begin{vmatrix} c_{L-M+1} - zc_{L-M+2} & \cdots & c_L - zc_{L+1} \\ c_{L-M+2} - zc_{L-M+3} & \cdots & c_{L+1} - zc_{L+2} \\ \vdots & & \vdots \\ c_L - zc_{L+1} & \cdots & c_{L+M-1} - zc_{L+M} \end{vmatrix} \quad (3.2)$$

which is a compact and symmetric form. For the numerator, we use (1.9),

$$P^{[L/M]}(z) = \begin{vmatrix} c_{L-M+1} & c_{L-M+2} & \cdots & c_{L+1} \\ c_{L-M+2} & c_{L-M+3} & \cdots & c_{L+2} \\ \vdots & \vdots & & \vdots \\ c_L & c_{L+1} & \cdots & c_{L+M} \\ \sum_{i=0}^{L-M} c_i z^{M+i} & \sum_{i=0}^{L-M+1} c_i z^{M+i-1} & \cdots & \sum_{i=0}^{L} c_i z^i \end{vmatrix}$$

and a similar reduction yields

$$P^{[L/M]}(z) = \begin{vmatrix} c_{L-M+1} - zc_{L-M+2} & \cdots & c_L - zc_{L+1} & c_{L+1} \\ c_{L-M+2} - zc_{L-M+3} & \cdots & c_{L+1} - zc_{L+2} & c_{L+2} \\ \vdots & & \vdots & \vdots \\ c_L - zc_{L+1} & \cdots & c_{L+M-1} - zc_{L+M} & c_{L+M} \\ -c_{L-M+1}z^{L+1} & \cdots & -c_L z^{L+1} & \sum_{i=0}^{L} c_i z^i \end{vmatrix}.$$

Dividing each column, except the last, by z^M, z^{M-1}, ..., z and adding to the last, we find

$$P^{[L/M]}(z) =$$

$$\begin{vmatrix} c_{L-M+1} - zc_{L-M+2} & \cdots & c_L - zc_{L+1} & c_{L-M+1}z^{-M} \\ c_{L-M+2} - zc_{L-M+3} & \cdots & c_{L+1} - zc_{L+2} & c_{L-M+2}z^{-M} \\ \vdots & & \vdots & \vdots \\ c_L - zc_{L+1} & \cdots & c_{L+M-1} - zc_{L+M} & c_L z^{-M} \\ -c_{L-M+1}z^{L+1} & \cdots & -c_L z^{L+1} & \sum_{i=0}^{L-M} c_i z^i \end{vmatrix}. \quad (3.3)$$

We are now led to define the matrices

$$W(z) = \begin{bmatrix} c_{L-M+1} - zc_{L-M+2} & c_{L-M+2} - zc_{L-M+3} & \cdots & c_L - zc_{L+1} \\ c_{L-M+2} - zc_{L-M+3} & c_{L-M+3} - zc_{L-M+4} & \cdots & c_{L+1} - zc_{L+2} \\ \vdots & \vdots & & \vdots \\ c_L - zc_{L+1} & c_{L+1} - zc_{L+2} & \cdots & c_{L+M-1} - zc_{L-M} \end{bmatrix}$$
(3.4)

1.3 Sequences and series; obstacles

and
$$\mathbf{c}^T = (c_{L-M+1}, c_{L-M+2}, \ldots, c_L). \tag{3.5}$$

With these definitions, we recognize that $Q^{[L/M]}(z) = \det W(z)$. We expand (3.3) by its last row and then by its last column, using cofactors of $W(z)$, to deduce

$$[L/M] = \sum_{i=0}^{L-M} c_i z^i + \{\mathbf{c}^T W(z)^{-1} \mathbf{c}\} z^{L-M+1}. \tag{3.6}$$

This equation is called Nuttall's compact form for a Padé approximant. If $L < M$, the polynomial term in (3.6) is understood as zero, because we use the convention that $c_j = 0$ if $j < 0$.

Reductions such as these lead to interesting forms for the Padé approximants when they are used for acceleration of convergence of sequences. Given the sequence $\{S_n, n = 0, 1, \ldots\}$, we define

$$c_0 = S_0, \quad S_j = 0 \text{ for } j < 0,$$

$$c_{k+1} = \Delta S_k = S_{k+1} - S_k \text{ for } k = 0, 1, 2, \ldots \text{ and} \tag{3.7}$$

$$\Delta^i S_k = \Delta^{i-1} S_{k+1} - \Delta^{i-1} S_k \text{ for all } k \text{ and } i = 1, 2, 3, \ldots.$$

From (1.8),

$$Q^{[L/M]}(1) = \begin{vmatrix} \Delta S_{L-M} & \Delta S_{L-M+1} & \cdots & \Delta S_L \\ \Delta S_{L-M+1} & \Delta S_{L-M+2} & \cdots & \Delta S_{L+1} \\ \vdots & \vdots & & \vdots \\ \Delta S_{L-1} & \Delta S_L & \cdots & \Delta S_{L+M-1} \\ 1 & 1 & \cdots & 1 \end{vmatrix}.$$

and after several elementary operations, we find

$$Q^{[L/M]}(1) = (-1)^M \begin{vmatrix} \Delta^2 S_{L-M} & \Delta^3 S_{L-M} & \cdots & \Delta^{M+1} S_{L-M} \\ \Delta^3 S_{L-M} & \Delta^4 S_{L-M} & \cdots & \Delta^{M+2} S_{L-M} \\ \vdots & \vdots & & \vdots \\ \Delta^{M+1} S_{L-M} & \Delta^{M+2} S_{L-M} & \cdots & \Delta^{2M} S_{L-M} \end{vmatrix},$$

which is an $M \times M$ symmetric determinant with all difference operators acting on S_{L-M}. Similarly we find that the numerator is given by the $(M+1) \times (M+1)$ symmetric determinant:

$$P^{[L/M]}(1) = (-)^M \begin{vmatrix} S_{L-M} & \Delta S_{L-M} & \cdots & \Delta^M S_{L-M} \\ \Delta S_{L-M} & \Delta^2 S_{L-M} & \cdots & \Delta^{M+1} S_{L-M} \\ \vdots & \vdots & & \vdots \\ \Delta^M S_{L-M} & \Delta^{M+1} S_{L-M} & \cdots & \Delta^{2M} S_{L-M} \end{vmatrix}.$$

The sequence $\{[M/M], M = 0, 1, 2, \ldots\}$ is called the diagonal sequence. These formulas suggest, but in no way compel, the choice of diagonal approximants for acceleration of convergence. An element of the final sequence is given by

$$[L/M](1) = \begin{vmatrix} S_{L-M} & \Delta S_{L-M} & \cdots & \Delta^M S_{L-M} \\ \Delta S_{L-M} & \Delta^2 S_{L-M} & \cdots & \Delta^{M+1} S_{L-M} \\ \vdots & \vdots & & \vdots \\ \Delta^M S_{L-M} & \Delta^{M+1} S_{L-M} & \cdots & \Delta^{2M} S_{L-M} \end{vmatrix} \div \begin{vmatrix} \Delta^2 S_{L-M} & \Delta^3 S_{L-M} & \cdots & \Delta^{M+1} S_{L-M} \\ \Delta^3 S_{L-M} & \Delta^4 S_{L-M} & \cdots & \Delta^{M+2} S_{L-M} \\ \vdots & \vdots & & \vdots \\ \Delta^{M+1} S_{L-M} & \Delta^{M+2} S_{L-M} & \cdots & \Delta^{2M} S_{L-M} \end{vmatrix}, \quad (3.8)$$

which is a remarkably elegant result [Shanks, 1955].

Finally in this section we mention obstacles. We will take note of some examples which illustrate why precision is mandatory in the treatment of formal power series.

Using the *real non-negative variable* x, $f(x) = \exp(-1/x)$ is smooth and infinitely differentiable throughout $0 \leq x < \infty$. The problem is that $f(0) = f'(0) = \cdots = f^{(n)}(0) = 0$ for all n, and so $f(x)$ has the Maclaurin expansion $f(x) = 0$. Of course, this example is contrived, and is based on a function not analytic in a neighborhood of the origin (this statement means a domain $|z| < \delta$, where δ is arbitrarily small but positive). It is clear that $\exp(-1/z)$ is not determined by its Maclaurin series, and so our theorems are always phrased so as to exclude the possibility that we are representing such functions.

Another notorious function is Euler's function. This function is a more constructive example, and in Chapter 5 we show how its Padé approximants converge. It is defined by the series

$$E(z) = 1 - 1!z + 2!z^2 - 3!z^3 + \cdots, \quad (3.9)$$

and we assume that a certain sequence of Padé approximants has been empirically found to converge. The full theory relevant to this example is explained in Sections 5.5, 5.6. The moot point is whether convergence is to $E(z)$, again begging the question of the extent to which $E(z)$ is defined by (3.9). With the information that (3.9) is an asymptotic series, an entirely satisfactory definition exists for $E(x)$ with $x \geq 0$.

To be very pragmatic, take $x = \frac{1}{10}$. The magnitudes of the terms in the series are shown in Figure 1.3.1, and in the sense of the previous definitions, a plausible value for $E(0.1)$ is reasonably well determined by

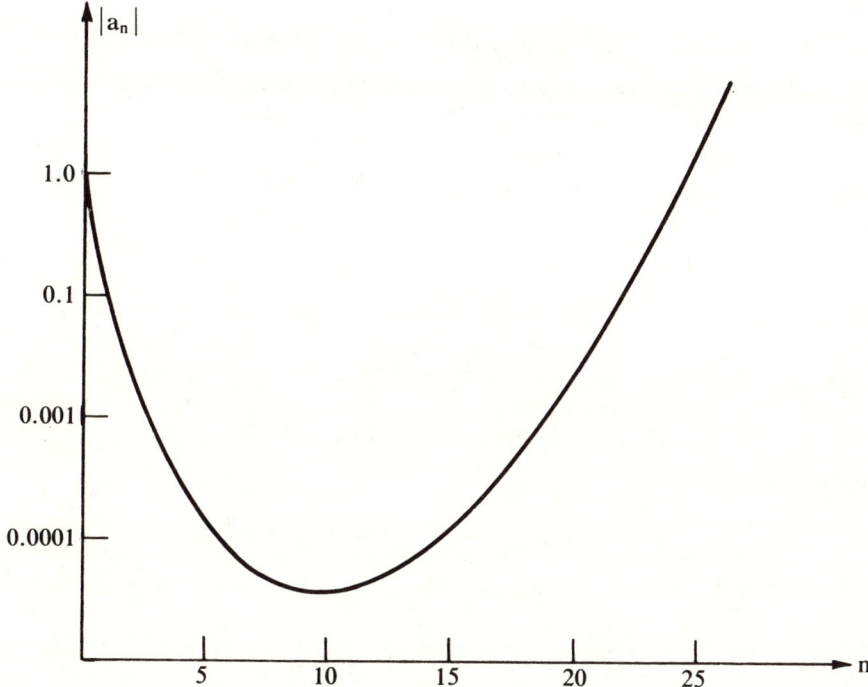

Figure 1.3.1. The magnitude of the terms $|a_n| = (0.1)^n n!$ in the hypergeometric series for $E(0.1)$.

truncation at the minimum, namely $\sum_{n=0}^{10} n!(-1/10)^n$. This procedure is much less satisfactory for large values of z, and also it would seem to work badly for z small and negative. The problem turns out to be that with the natural approach, $E(z)$ is defined with a branch cut along $-\infty < z \leq 0$, and so $E(z)$ is only determined uniquely in the sense of cut-plane analyticity. To be pedestrian, $E(-0.1)$ is really only determined by a convention about the location of the cut. This point is easy to overlook using Padé approximants, because the approximants 'choose' the cut in the natural place in a sense we describe in Section 2.2.

As a final example of mathematical perversity, we refer to Section 10.6, where we exhibit a nontrivial function which is analytic in an annulus, and which cannot be properly approximated by polynomials in z in the annulus.

Our demonstration example of Section 1.2 shows the success of low-order Padé approximants in practice for a function having cut-plane analyticity, and rapid convergence at high order may be proved using the Stieltjes series methods of Chapter 5. For the most part in this book, we are concerned primarily with how, one way or another, various natural obstacles can be overcome.

1.4 The Baker definition, the C-table, and block structure

To motivate the discussion of uniqueness and the modern definition, of which the analysis is due to Baker [1973b], we must consider what can go wrong with the basic approach discussed so far. The classic example which demonstrated the shortcoming of a simple-minded approach is the construction of [1/1] for $1 + z^2$.

We require

$$\frac{p_0 + p_1 z}{q_0 + q_1 z} = 1 + z^2 + O(z^3). \tag{4.1}$$

Hence

$$p_0 + p_1 z = q_0 + q_1 z + q_0 z^2 + O(z^3).$$

Therefore

$$p_0 = q_0, \quad p_1 = q_1, \quad \text{and} \quad q_0 = 0. \tag{4.2}$$

Consequently, the approximant is

$$\frac{0 + q_1 z}{0 + q_1 z} = 1. \tag{4.3}$$

However, $1 \neq 1 + z^2 + O(z^3)$, and we rapidly and correctly conclude that (4.1) has no solution. What we did achieve is a 'solution' of

$$p_0 + p_1 z = (q_0 + q_1 z)(1 + z^2) + O(z^3), \tag{4.4}$$

but that is not what (4.1) requires. This solution, given by (4.2), is also given by the determinantal method:

$$P^{[1/1]}(z) = \begin{vmatrix} c_1 & c_2 \\ c_0 z & c_0 + c_1 z \end{vmatrix} = \begin{vmatrix} 0 & 1 \\ z & 1 \end{vmatrix} = -z$$

and

$$Q^{[1/1]}(z) = \begin{vmatrix} c_1 & c_2 \\ z & 1 \end{vmatrix} = \begin{vmatrix} 0 & 1 \\ z & 1 \end{vmatrix} = -z.$$

For a long time, the accepted solution was to take an analogue of (4.4) as the agreed definition of Padé approximants. Specifically, the classical definition, also called the Frobenius and Padé Frobenius definition, is that if $p_{L,M}(z)$, $q_{L,M}(z)$ are polynomials of orders L, M, respectively, and if

$$q_{L,M}(z)f(z) - p_{L,M}(z) = O(z^{L+M+1}), \tag{4.5}$$

then $p_{L,M}(z)/q_{L,M}(z)$ is the (unique) Padé approximant of $f(z)$. Equation (4.5) is the general form of (4.4), and it is remarkable that polynomials

1.4 The Baker definition, the C-table, and block structure

$q_{L,M}(z)$, $p_{L,M}(z)$ of orders M, L can always be found to satisfy (4.5). However, our specific example emphasizes that if $q_{L,M}(0) = 0$, then in this case

$$f(z) \neq \frac{p_{L,M}(z)}{q_{L,M}(z)} + O(z^{L+M+1}).$$

Padé defined a deficiency index, $\omega_{L,M}$, which is the least integer for which

$$f(z) \neq \frac{p_{L,M}(z)}{q_{L,M}(z)} + O(z^{L+M-\omega_{L,M}+1}),$$

and ω_{LM} is a measure of the shortcoming of the approximation. Quite simply, the rational function $p_{L,M}(z)/q_{L,M}(z)$ does not approximate $f(z)$ through order $L + M$ in certain circumstances, and then we prefer to say that the Padé approximant does not exist.

In the general theory of rational interpolation, it is well known that there are certain unattainable values. If values $f(z_i)$ at certain points z_i are specified, the specification may be inconsistent, and in such circumstances, the rational interpolants are declared not to exist. Our approach is entirely in line with this attitude.

Because the accuracy-through-order requirement is fundamental, a definition which preserves it is essential, and we use the modern definition which was fully analyzed by Baker [1973b].

Definition (Baker). If polynomials $A^{[L/M]}(z)$, $B^{[L/M]}(z)$, of degrees L, M, respectively, can be found such that

$$\frac{A^{[L/M]}(z)}{B^{[L/M]}(z)} = f(z) + O(z^{L+M+1}) \tag{4.6}$$

with

$$B^{[L/M]}(0) = 1, \tag{4.7}$$

then we define

$$[L/M] = \frac{A^{[L/M]}(z)}{B^{[L/M]}(z)}.$$

The notation emphasizes that numerator and denominator depend on both L and M. An entirely equivalent specification of the definition is to replace (4.6) by

$$A^{[L/M]}(z) - f(z)B^{[L/M]}(z) = O(z^{L+M+1})$$

provided that (4.7) is retained. The notation of (4.6) and (4.7) is exclusively reserved for this purpose throughout the work, and without further explanation.

If, with Equation (1.8), $Q^{[L/M]}(0) \neq 0$, then the rescaling

$$A^{[L/M]}(z) = \frac{P^{[L/M]}(z)}{Q^{[L/M]}(0)}$$

and

$$B^{[L/M]}(z) = \frac{Q^{[L/M]}(z)}{Q^{[L/M]}(0)}$$

implies that the two definitions correspond up to an unimportant numerical factor. Consequently, and charitably speaking, the distinction between the definitions is sometimes taken for granted. However, if $Q^{[L/M]}(0) = 0$, precise terminology is mandatory. Because the vanishing of $Q^{[L/M]}(0)$ is so important, a special symbol is exclusively reserved for this quantity:

Definition

$$C(L/M) = Q^{[L/M]}(0) = \begin{vmatrix} c_{L-M+1} & c_{L-M+2} & \cdots & c_L \\ c_{L-M+2} & c_{L-M+3} & \cdots & c_{L+1} \\ \vdots & \vdots & & \vdots \\ c_L & c_{L+1} & \cdots & c_{L+M-1} \end{vmatrix}. \quad (4.8)$$

Summary. If $C(L/M) \neq 0$, the classical Padé–Frobenius and the Baker definitions are entirely equivalent. If $C(L/M) = 0$, it is possible that the $[L/M]$ Padé approximant does not exist. However, polynomials satisfying (4.5) exist, defining a rational fraction which is called the classical Padé approximant.

It is convenient to display $\{C(L/M), L, M = 0, 1, 2, \ldots\}$ in a table (Table 1.4.1), called the C-table. This is an array of values of determinants, and should be distinguished from the Padé table.

Example. Let

$$f(z) = \frac{1 + 2z + z^2 + z^3}{1 + z + z^3}.$$

This is given by the power series

$$f(z) = 1 + z - z^4 + z^5 - z^6 + 2z^7 - 3z^8 + 4z^9 - \cdots, \quad (4.9)$$

and the C-table of Table 1.4.2 results. The most conspicuous features of the table are the square blocks of zeros. In fact, the zero at $C(4/4)$ is the start of an infinite square block. Before proceeding with the proof of these statements, let us investigate how Table 1.4.1 was actually

1.4 The Baker definition, the C-table, and block structure

Table 1.4.1. *The C-table.*

C(0/0)	C(1/0)	C(2/0)	...
C(0/1)	C(1/1)	C(2/1)	...
C(0/2)	C(1/2)	C(2/2)	...
⋮	⋮	⋮	⋱

Table 1.4.2. *The C-table for* $(1 + 2z + z^2 + z^3)/(1 + z + z^3)$.

M \ L	0	1	2	3	4	5	6	7	8	9	...
0	1	1	1	1	1	1	1	1	1	1	...
1	1	1	0	0	−1	1	−1	2	−3	4	...
2	−1	−1	0	0	−1	0	1	−1	−1		...
3	−1	−1	1	−1	1	−1	1	−1			...
4	1	2	0	1	0	0	0	0	0	0	...
5	1	4	2	1	0	0	0	0	0	0	...
⋮	⋮	⋮	⋮	⋮	⋮	⋮	⋮	⋮	⋮	⋮	

constructed. A substantial amount of elementary algebra is needed to construct Table 1.4.2 from the basic definition (4.8) and the series (4.9); this algebra was not the method used for constructing most of the entries. A better procedure is this one:

The first row is, by definition, $C(L/0) = 1$.
The second row is $C(L/1) = c_L$.
The first column is $C(0/M) = (-1)^{M(M-1)/2} c_0^M$, from (4.8).
Most of the remaining entries were calculated from Frobenius' formula

$$C(L/M + 1) = \frac{C(L + 1/M)C(L - 1/M) - C(L/M)^2}{C(L/M - 1)}, \quad (4.10)$$

which is valid if $C(L/M - 1) \neq 0$. Equation (4.10) is called a

$$\begin{bmatrix} & * & \\ * & * & * \\ & * & \end{bmatrix}$$

star identity, showing how it relates the entries in the C-table. In order to understand fully the consequences of (4.10), it is worthwhile reconstructing Table 1.4.1 from the initializing values. We proceed with Sylvester's theorem, of which (4.10) is a corollary.

Theorem 1.4.1. Let A be a matrix, and let A_{rp} denote the matrix with row r and column p deleted. Also let $A_{rs;pq}$ denote the matrix A with rows r and s and columns p and q deleted. Provided $r < s$ and $p < q$,

$$\det A \det A_{rs;pq} = \det A_{rp} \det A_{sq} - \det A_{rq} \det A_{sp}.$$

Proof. Suppose that A is an $(n+2) \times (n+2)$ matrix, and we consider deletion of its last two rows and columns. Take $r = p = n+1$ and $s = q = n+2$. Write the matrices in block form, e.g., $A_{n+1,n+2;n+1,n+2} = M$.

$$A = \begin{bmatrix} M & h & g \\ f & e & d \\ c & b & a \end{bmatrix}.$$

$$A_{n+2,n+2} = \begin{bmatrix} M & h \\ f & e \end{bmatrix}, \quad A_{n+2,n+1} = \begin{bmatrix} M & g \\ f & d \end{bmatrix}.$$

Next we consider a $(2n+2) \times (2n+2)$ block matrix, with determinant given by

$$\begin{vmatrix} M & h & g & 0 \\ f & e & d & 0 \\ c & b & a & c \\ 0 & 0 & 0 & M \end{vmatrix} = \begin{vmatrix} M & h & g & 0 \\ f & e & d & 0 \\ c & b & a & c \\ M & h & g & M \end{vmatrix} = \begin{vmatrix} M & h & g & 0 \\ f & e & d & 0 \\ 0 & b & a & c \\ 0 & h & g & M \end{vmatrix}$$

$$= \begin{vmatrix} M & h & g & 0 \\ f & e & d & 0 \\ 0 & 0 & a & c \\ 0 & 0 & g & M \end{vmatrix} + \begin{vmatrix} M & 0 & g & 0 \\ f & 0 & d & 0 \\ 0 & b & a & c \\ 0 & h & g & M \end{vmatrix}$$

$$= \begin{vmatrix} M & h \\ f & e \end{vmatrix} \cdot \begin{vmatrix} a & c \\ g & M \end{vmatrix} - \begin{vmatrix} M & g \\ f & d \end{vmatrix} \cdot \begin{vmatrix} b & c \\ h & M \end{vmatrix},$$

which is interpreted as

$$\det A \det A_{n+1,n+2;n+1,n+2} = \det A_{n+2,n+2} \det A_{n+1,n+1}$$
$$- \det A_{n+1,n+2} \det A_{n+2,n+1}.$$

By interchanging rows r, $n-1$ and s, $n-2$ and columns p, $n-1$ and q, $n-2$, the theorem is proved.

1.4 The Baker definition, the C-table, and block structure

Corollary

$$C(L/M + 1)C(L/M - 1) = C(L + 1/M)C(L - 1/M) - C(L/M)^2.$$

(4.11)

Proof. Let $\det A = C(L/M + 1)$ as given by (4.8). With $r = p = 1$ and $s = q = M + 1$, the result and consequently (4.10), if $C(L/M - 1) \neq 0$, are proved.

This relation (4.11) is the key to the block structure of the C-table.

Theorem 1.4.2. *Zero entries occur in the C-table in square blocks which are entirely surrounded by nonzero entries (except at infinity).*

Proof. We identify the top left-hand corner of the block by requiring that

$$C(l/m) = 0, \quad C(l/m - 1) \neq 0, \quad \text{and} \quad C(l - 1/m) \neq 0. \quad (4.12a)$$

It is obvious how to redefine l or m if either of the latter two conditions does not hold. From the $l - 1/m$ star identity,

$$C(l - 1/m - 1)C(l - 1/m + 1) = C(l - 2/m)C(l/m) - C(l - 1/m)^2,$$

we deduce that

$$C(l - 1/m - 1)C(l - 1/m + 1) = -C(l - 1/m)^2.$$

Hence

$$C(l - 1/m - 1) \neq 0 \quad \text{and} \quad C(l - 1/m + 1) \neq 0.$$

Similarly, the $l/m - 1$ star identity yields $C(l + 1/m - 1) \neq 0$, and the relevant portion of the C-table is shown in Table 1.4.3. Suppose that

$$C(l/m + 1) \neq 0. \quad (4.12b)$$

Table 1.4.3. Consequences of (4.12).

$C(l - 1/m - 1)$ $\neq 0$	$C(l/m - 1)$ $\neq 0$	$C(l + 1/m - 1)$ $\neq 0$
$C(l - 1/m)$ $\neq 0$	$C(l/m)$ $= 0$	
$C(l - 1/m + 1)$ $\neq 0$		

Then the $l/m + 1$ star identity establishes that $C(l + 1/m + 1) \neq 0$, and the $l + 1/m$ star identity establishes that $C(l + 1/m) \neq 0$. Hence the theorem is proved for a unit square block. The only alternative to (4.12b) is that $C(l/m + 1) = 0$. Suppose that $C(l/m + k) = 0$ for $k = 0, 1, \ldots, \kappa - 1$ and $C(l/m + \kappa) \neq 0$. Using the $l - 1/m + k$ star identity, we establish iteratively that $C(l - 1/m + k + 1) \neq 0$ for $k = 0, 1, \ldots, \kappa - 1$. Since $C(l/m + k) \neq 0$ and $C(l - 1/m + \kappa) \neq 0$, we deduce that $C(l + 1/m + \kappa) \neq 0$. Thus we have a column of zeros bordered on top, bottom, and left by nonzero elements, as shown in Table 1.4.4. Similarly, we establish that the block is rectangular and, if finite, it is entirely bordered by nonzero elements as shown in Table 1.4.5. To establish that the blocks are square, we consider a block with r rows and s columns. First we prove that $r \geq s$, and to do this we choose a simple example which makes the general case obvious. Suppose we know that $C(2/2) = C(3/2) = C(4/2) = 0$. Then

$$\begin{vmatrix} c_1 & c_2 \\ c_2 & c_3 \end{vmatrix} = \begin{vmatrix} c_2 & c_3 \\ c_3 & c_4 \end{vmatrix} = \begin{vmatrix} c_3 & c_4 \\ c_4 & c_5 \end{vmatrix} = 0.$$

These statements are interpreted as implying that a linear combination of

$$\begin{bmatrix} c_1 \\ c_2 \end{bmatrix} \text{ and } \begin{bmatrix} c_2 \\ c_3 \end{bmatrix}$$

vanishes, and also of

$$\begin{bmatrix} c_2 \\ c_3 \end{bmatrix} \text{ and } \begin{bmatrix} c_3 \\ c_4 \end{bmatrix},$$

and also of

$$\begin{bmatrix} c_3 \\ c_4 \end{bmatrix} \text{ and } \begin{bmatrix} c_4 \\ c_5 \end{bmatrix}.$$

Using the implied multipliers, we deduce that, in general,

$$\begin{vmatrix} c_0 & c_1 & c_2 \\ c_1 & c_2 & c_3 \\ c_2 & c_3 & c_4 \end{vmatrix} =. \begin{vmatrix} x & c_1 & x \\ 0 & c_2 & 0 \\ 0 & c_3 & 0 \end{vmatrix} = 0, \qquad (4.13)$$

where the x denotes an unknown and totally immaterial entry, with the same conclusion in the exceptional cases. Likewise,

$$\begin{vmatrix} c_1 & c_2 & c_3 \\ c_2 & c_3 & c_4 \\ c_3 & c_4 & c_5 \end{vmatrix} = \begin{vmatrix} c_2 & c_3 & c_4 \\ c_3 & c_4 & c_5 \\ c_4 & c_5 & c_6 \end{vmatrix} = 0. \qquad (4.14)$$

Table 1.4.4. *The left edge of a block.*

$\neq 0$	$\neq 0$	$\neq 0$
$\neq 0$	0	
$\neq 0$	0	
$\neq 0$	0	
$\neq 0$	0	
$\neq 0$	$\neq 0$	$\neq 0$

Table 1.4.5. *A hypothetical section of the C-table.*

$\neq 0$	$\neq 0$	$\neq 0$	$\neq 0$	$\neq 0$	$\neq 0$
$\neq 0$	0	0	0	0	$\neq 0$
$\neq 0$	0	0	0	0	$\neq 0$
$\neq 0$	0	0	0	0	$\neq 0$
$\neq 0$	$\neq 0$	$\neq 0$	$\neq 0$	$\neq 0$	$\neq 0$

Finally, by interpreting (4.13) and (4.14) as asserting the existence of linearly dependent column vectors, we find that

$$C(2/4) = \begin{vmatrix} 0 & c_0 & c_1 & c_2 \\ c_0 & c_1 & c_2 & c_3 \\ c_1 & c_2 & c_3 & c_4 \\ c_2 & c_3 & c_4 & c_5 \end{vmatrix} = \begin{vmatrix} x & c_0 & c_1 & x \\ 0 & c_1 & c_2 & 0 \\ 0 & c_2 & c_3 & 0 \\ 0 & c_3 & c_4 & 0 \end{vmatrix} = 0.$$

Likewise, $C(2/4) = C(3/4) = C(4/4) = 0$. We claim that it is now obvious that if we have a block of the C-table with s columns, $L = l, l+1, \ldots, l+s-1$, then there are at least s rows. In other words, a block with r rows and s columns has $r \geq s$.

To prove the converse we refer forward to Hadamard's theorem of Section 2.4. We may always reset the problem for a function with $c_0 \neq 0$. We now assume that $c_0 \neq 0$, so that we may consider the C-table for the reciprocal function $g(z) = 1/f(z)$. Let $C'(M/L)$ be the entry for the (M/L) Hankel determinant of $g(z)$. From (6.9) we know that $C(L/M) = 0$ implies that $C'(M/L) = 0$ and vice versa. For the reciprocal function, the block with r' rows and s' columns satisfies our previous rule, $r' \geq s'$. But $r' = s$ and $s' = r$, and consequently the converse is proved. Hence $r = s$, and we have proved that all blocks of the C-table are square, and are entirely bordered by nonzero entries.

This completes the preliminary to Padé's [1892] theorem. This theorem has been modernized by Baker, but the content is essentially unchanged. The style of proof is quite different from that of the previous century.

28   Introduction and definitions

Theorem 1.4.3. *The Padé table consists of uniquely determined entries given by*

$$[L/M] = \frac{A^{[L/M]}(z)}{B^{[L/M]}(z)} = \frac{P^{[L/M]}(z)}{Q^{[L/M]}(z)}$$

following the definitions (1.8), (1.9), (4.6), *and* (4.7), *provided* $C(L/M) \neq 0$. *Otherwise, suppose that* $C(L/M) = 0$, *in each entry of an* $r \times r$ *block of the C-table. Corresponding to this, blocks of the Padé table are* $(r+1) \times (r+1)$ *blocks, for which* $C(\lambda/\mu) \neq 0$, $C(\lambda + i/\mu) \neq 0$, *and* $C(\lambda/\mu + i) \neq 0$ *for* $i = 1, 2, \ldots, r$ *and* $C(\lambda + i/\mu + j) = 0$ *for all* $i, j = 1, 2, \ldots, r$. *An* $[L/M]$ *Padé approximant in the block obeys*

either $[L/M] = [\lambda/\mu]$,   *if* $L + M \leq \lambda + \mu + r$,

or $[L/M]$ *does not exist*,   *if* $L + M > \lambda + \mu + r$.

Proof. The proof is divided into three parts.

Part 1. If $C(L/M) \neq 0$, the Padé equations are linearly independent; with $b_0 = 1$ the coefficients are uniquely determined. In particular, this observation excludes the possibility that $A^{[L/M]}(z)$ and $B^{[L/M]}$ have a common factor.

Part 2. The Padé approximants on the top edge and left-hand side of the block of the Padé table satisfy

$$[\lambda + i/\mu] = [\lambda/\mu + i] = [\lambda/\mu] \quad \text{for} \quad i = 1, 2, \ldots, r.$$

Subproof. This result follows by noting that these approximants are uniquely determined, using part 1 and the block structure of the C-table. For $i = 1$, the result follows by noting that

the coefficient of $z^{\lambda+1}$ in $P^{[\lambda+1/\mu]}(z)$ is $\pm C(\lambda + 1/\mu + 1) = 0$, and

the coefficient of $z^{\mu+1}$ in $Q^{[\lambda/\mu+1]}(z)$ is $\pm C(\lambda + 1/\mu + 1) = 0$.

In fact, by induction, it follows that because the coefficient of $z^{\lambda+i}$ in $P^{[\lambda+i/\mu]}(z)$ is $\pm C(\lambda + i/\mu + 1) = 0$,

$$[\lambda + i/\mu] = [\lambda + i - 1/\mu] \quad \text{for } i = 1, 2, \ldots, r.$$

Similarly,

$$[\lambda/\mu + i] = [\lambda/\mu + i - 1] \quad \text{for } i = 1, 2, \ldots, r,$$

and part 2 is proved.

1.4 The Baker definition, the C-table, and block structure

Part 3. $[L/M]$ Padé approximants in the relevant block of the Padé table of the theorem satisfying

$$L + M \leq \lambda + \mu + r$$

exist and equal $[\lambda/\mu]$. Otherwise

$$L + M > \lambda + \mu + r$$

and the corresponding Padé approximants in the block do not exist.

Subproof. Because the Padé approximants on the top edge, $[\lambda + i/\mu]$, $i = 0, 1, \ldots, r$, are identical, we know that the following $\mu + r$ equations for the denominator coefficients are satisfied:

$$\begin{bmatrix} c_{\lambda-\mu+1} & \cdots & c_{\lambda+1} \\ \vdots & & \vdots \\ c_{\lambda} & \cdots & c_{\lambda+\mu} \\ \vdots & & \vdots \\ c_{\lambda+r} & \cdots & c_{\lambda+\mu+r} \end{bmatrix} \begin{bmatrix} q_\mu \\ \vdots \\ q_0 \end{bmatrix} = 0, \qquad (4.15)$$

and because $[\lambda + r + 1/\mu]$ is not in this block,

$$c_{\lambda+r+1}q_\mu + c_{\lambda+r+2}q_{\mu-1} + \cdots + c_{\lambda+r+\mu+1}q_0 \neq 0. \qquad (4.16)$$

Suppose that $[L/M]$ is a Padé approximant in the block, but not on the top edge or left-hand side. Thus $\lambda + 1 \leq L \leq \lambda + r$ and $\mu + 1 \leq M \leq \mu + r$ and $B^{[L/M]}(0) = 1$. The explicit solution shows that this situation is impossible: if an explicit solution exists, $Q^{[L/M]}(0) = 0$ and $[L/M] = [L - 1/M - 1]$. Using part 2, we see that if a Padé approximant in the block exists, then it reduces to $[\lambda/\mu]$, and has denominator coefficients obeying (4.15). Hence we see that the accuracy-through-order conditions are satisfied for $L + M \leq \lambda + \mu + r$, and so the Padé approximants exist and are reducible. Also, from (4.16), if $L + M > \lambda + \mu + r$, the accuracy at order $z^{\lambda+\mu+r+1}$ is not satisfied, and so the Padé approximant does not exist. The theorem is now completely proved.

An elementary view of a set of linear equations is that either the equations are consistent and have a solution, or they are inconsistent and have no solution. This view is mirrored by the previous theorem, in which linear equations either do or do not determine Padé approximants; of course, the theorem also considers the question of uniqueness. Table 1.4.6 summarizes graphically the link between a block in the C-table and the Padé table. Proofs using the methods of Section 3.7 or 4.2 which avoid the use of determinants can also be used to prove Padé's block structure theorem.

Table 1.4.6. *Part of the C-table showing a 3 × 3 block and part of the Padé table showing the corresponding 4 × 4 block.*

(λ/μ) $\neq 0$	$\neq 0$	$\neq 0$	$\neq 0$
$\neq 0$	0	0	0
$\neq 0$	0	0	0
$\neq 0$	0	0	0

↔

$[\lambda/\mu]$	Red.	Red.	Red.
Red.	Con.	Con.	Inc.
Red.	Con.	Inc.	Inc.
Red.	Inc.	Inc.	Inc.

Red. denotes a regular Padé approximant with nonsingular equations, but which happens to reduce to order $[\lambda/\mu]$.
Con. denotes a Padé approximant with a singular Hankel determinant, but which is determined by a consistent set of equations, and also reduces to $[\lambda/\mu]$.
Inc. denotes a nonexistent Padé approximant. The equations for it are inconsistent.

Up until now, we have tended to take the question of infinite blocks of the C-table and the Padé table for granted. This casualness is for the good reason that an infinite block in the Padé table turns out to be uniquely associated with a rational function. It is certainly a consequence of Theorem 1.4.2 that if either side of any block in the C-table is of infinite length, then so is the other side of the block.

Theorem 1.4.4. *Suppose that a function $f(z)$ is analytic at the origin and is uniquely determined by its Maclaurin series: $f(z) = \sum_{i=0}^{\infty} c_i z^i$. Then the existence of an infinite block in the Padé table of $\sum_{i=0}^{\infty} c_i z^i$ is a necessary and sufficient condition for $f(z)$ to be rational; let $f(z)$ be of type $[\lambda/\mu]$. Then the corresponding block in the C-table is defined by $C(\lambda/\mu + 1) \neq 0$, $C(\lambda + 1/\mu) \neq 0$, $C(\lambda + i/\mu + j) = 0$, $i, j = 1, 2, \ldots, \infty$. This condition in turn is necessary and sufficient for the representation*

$$c_k = \sum_{i=1}^{N} \sum_{\tau=1}^{\mu_i} r_{i\tau} \binom{-\mu_i}{k} (-z_i)^{-k} \quad \text{for all } k \geq \lambda - \mu. \tag{4.17}$$

Proof. Suppose that there is an infinite block in the Padé table, containing approximants which, according to Theorem 1.4.3, all reduce to $[\lambda'/\mu']$. Then, working term by term, we find that

$$B^{[\lambda'/\mu']}(z) \sum_{i=1}^{\infty} c_i z^i - A^{[\lambda'/\mu']}(z) = 0.$$

Hence, by the hypothesis of the theorem,

$$f(z) = \frac{A^{[\lambda'/\mu']}(z)}{B^{[\lambda'/\mu']}(z)}$$

and $\lambda = \lambda'$, $\mu = \mu'$. Let

$$b(z) = B^{[\lambda/\mu]}(z) = \prod_{i=1}^{N}\left(1 - \frac{z}{z_i}\right)^{\mu_i}, \quad \text{where } \sum_{i=1}^{N}\mu_i = \mu.$$

Let

$$\pi_1(z) = \sum_{k=0}^{\lambda-\mu-1} c_k z^k \text{ if } \lambda > \mu \text{ and } \pi_1(z) = 0 \quad \text{otherwise}.$$

Then a polynomial $\pi_2(z)$, of degree μ at most, exists such that $f(z)$ decomposes into partial fractions:

$$f(z) = \pi_1(z) + \frac{\pi_2(z)}{b(z)} = \pi_1(z) + \sum_{i=1}^{N}\sum_{\tau=1}^{\mu_i} r_{i\tau}\left(1 - \frac{z}{z_i}\right)^{-\tau}.$$

Hence c_k is given by (4.17) for $k \geq \lambda - \mu$.

The converse, in which we suppose that $f(z)$ is rational, is a consequence of the uniqueness property of Padé approximants, given by Theorem 1.4.3, and the rest is obvious.

Having established that certain Padé approximants do not exist in certain circumstances, the following theorem establishes that, in every row, column, or diagonal, there are an infinite number of extant approximants.

Theorem 1.4.5. *Let $\sum_{i=0}^{\infty} c_i z^i$ be a formal power series. In its Padé table, an infinite number of Padé approximants exist*

(i) *in any row $[L/M]$, M fixed, $L \to \infty$,*
(ii) *in any column $[L/M]$, L fixed, $M \to \infty$, and*
(iii) *in any paradiagonal $[M + J/M]$, J fixed, $M \to \infty$.*

Proof. Consider the rows. In any row, there are either a finite number of blocks and consequently an infinite number of well-defined approximants, or else an infinite number of blocks. In any block, there is at least one extant Padé approximant in any row. Hence an infinite number of Padé approximants exist in any row of the Padé table. The argument is the same for columns and paradiagonals.

Note that the theorem makes no mention of convergence of these approximants, but it does mean that the discussion of convergence of rows, columns, and diagonals is more than a rhetorical exercise.

1.5 Duality and invariance

In this section, we state some algebraic properties of Padé approximants which are attractive features of a class of approximating functions. The theorems are easy to prove, and the interpretation of the theorems is important. All the theorems of the section concern algebraic properties of power series, and no convergence property is in any way directly implied.

Theorem 1.5.1 (Duality). *Let* $g(z) = \{f(z)\}^{-1}$ *and* $f(0) \neq 0$. *Then* $[L/M]_g(z) = \{[M/L]_f(z)\}^{-1}$ *provided either Padé approximant exists.*

Proof. Suppose that $[M/L]_f$ exists, and define

$$\frac{p_M(z)}{q_L(z)} = [M/L]_f(z).$$

Because $f(0) \neq 0$, $[M/L]_f(0) \neq 0$, and hence $p_M(z)$, $q_L(z)$ are polynomials of degrees at most M, L, respectively, and $p_M(0) \neq 0$. Hence

$$g(z) - \frac{q_L(z)}{p_M(z)} = \frac{p_M(z) - f(z)q_L(z)}{f(z)p_M(z)}$$
$$= O(z^{L+M+1}),$$

and so $[L/M]_g = q_L(z)/p_M(z)$ as required. The proof, given that $[L/M]_g$ exists, is similar.

The duality may be summarized by saying that the Padé approximant of the reciprocal function is the reciprocal of the Padé approximant of the function; it may be glibly restated by saying that Padé approximation and reciprocation commute.

If a class of functions and the reciprocal functions have a common property, e.g., they are meromorphic, then the duality property shows a valuable symmetry feature of the Padé approximants as approximating functions.

Theorem 1.5.2 (Homographic invariance under argument transformations). *Let* $f(z) = \sum_{i=0}^{\infty} c_i z^i$. *Define an origin-preserving linear fractional transformation of the argument*

$$w = \frac{az}{1 + bz},$$

and thereby a new function $g(w) = f(z)$. *Then*

$$[M/M]_g(w) = [M/M]_f(z)$$

provided either Padé approximant exists.

1.5 Duality and invariance

Proof. Suppose that $[M/M]_g(f)$ exists. Then

$$[M/M]_g(w) = \frac{\sum_{i=0}^{M} a_i w^i}{\sum_{i=0}^{M} b_i w^i} = g(w) + O(w^{2M+1}).$$

Let

$$A_M(z) = (1 + bz)^M \sum_{i=0}^{M} a_i \left(\frac{az}{1 + bz}\right)^i$$

and

$$B_M(z) = (1 + bz)^M \sum_{i=0}^{M} b_i \left(\frac{az}{1 + bz}\right)^i.$$

Then

$$\frac{A_M(z)}{B_M(z)} = f(z) + O(z^{2M+1}),$$

where $A_M(z)$, $B_M(z)$ are polynomials in z of order M. Hence $A_M(z)/B_M(z) = [M/M]_f(z)$.

The proof given that $[M/M]_f$ exists is similar, and the theorem is proved.

Notice that the proof is only valid for diagonal approximants, and so this homographic invariance only holds for the diagonal sequences. Theorem 1.5.2 is usually called the theorem of Baker, Gammel, and Wills [1961]; a closely related result was previously proved by Edrei [1939].

The homographic invariance theorem for argument transformations is the cornerstone of the optimistic approach to Padé approximants. This optimism is entirely validated by practical experience, but not as yet by proven theorems. The fundamental reason for this optimism is that the mapping $w = az/(1 + bz)$ shown in Figure 1.5.1 allows any circle Γ in the z-plane enclosing the origin as an interior point to be mapped onto a given circle $|w| = \rho$ centered on the origin. If a sequence of diagonal approximants may be proved to converge to $g(w)$ within $|w| \leq \rho$, then convergence of the same sequence for $f(z)$ follows in the interior of Γ (see the quasitheorem of Section 6.7). We also understand the acceleration of convergence using the sequence of *diagonal* Padé approximants, mentioned in Chapter 3, in terms of a generalized Euler transformation [Thatcher, 1974]. Other implications of this theorem are discussed in Section 6.7.

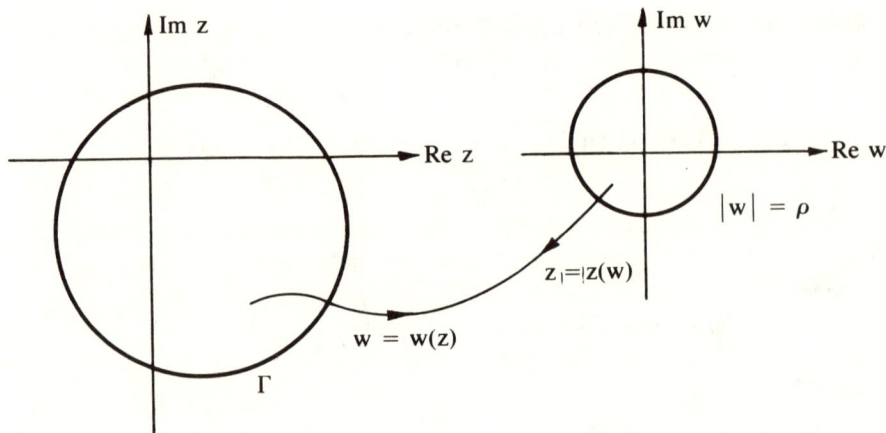

Figure 1.5.1. The mapping $w = az/(1 + bz)$ and its inverse.

Theorem 1.5.3 **(Homographic invariance of value transformations).** *Given a function* $f(z) = \sum_{i=0}^{\infty} c_i z^i$, *we define*

$$g(z) = \frac{a + bf(z)}{c + df(z)}.$$

If $c + df(0) \neq 0$, *then*

$$[M/M]_g(z) = \frac{a + b[M/M]_f(z)}{c + d[M/M]_f(z)}$$

provided $[M/M]_f(z)$ *exists*.

Proof. Because $c + df(0) \neq 0$, we may find polynomials $p_M(z)$ and $q_M(z)$ of degree M at most such that

$$\frac{p_M(z)}{q_M(z)} = \frac{a + b[M/M]_f(z)}{c + d[M/M]_f(z)}$$

with $q_M(0) \neq 0$. Then

$$\frac{p_M(z)}{q_M(z)} - g(z) = \frac{a + b[M/M]_f(z)}{c + d[M/M]_f(z)} - \frac{a + bf(z)}{c + df(z)}$$

$$= \frac{(bc - ad)\{[M/M]_f(z) - f(z)\}}{\{c + d[M/M]_f(z)\}\{c + df(z)\}}$$

$$= O(z^{2M+1}).$$

Therefore $p_M(z)/q_M(z)$ is the $[M/M]$ Padé approximant of $g(z)$.

1.5 Duality and invariance

Homographic invariance of the values, like invariance under argument transformations, is generally only valid for diagonal approximants. An interesting feature of this result is that the value ∞ of a Padé approximant is treated on a par with any other value: this is significant in the context of convergence on the Riemann sphere (see Section 6.4). The transformation $g = (a + bf)/(c + df)$ can be broken down into successive elementary transformations, each with a simple interpretation: the mapping

$$g = \frac{a + bf}{c + df}$$

is a composite of translations and inversion given by

$$g_1 = \alpha + f, \quad g_2 = \beta/g_1, \quad \text{and } g = \gamma + g_2.$$

Details of the interpretation are to be found in Baker [1975a, p. 113].

Theorem 1.5.4 (Truncation theorem). Let $f(z) = \sum_{i=0}^{\infty} c_i z^i$ and let

$$g(z) = \sum_{i=0}^{\infty} g_i z^i = \left\{ f(z) - \sum_{i=0}^{k-1} c_i z^i \right\} z^{-k}.$$

Then $[L - k/M]_g(z) = \{[L/M]_f - \sum_{i=0}^{k-1} c_i z^i\} z^{-k}$ for $k \geq 1$, $L - k \geq M - 1$, provided either Padé approximant exists.

Proof. Suppose that $[L/M]_f = A^{[L/M]}(z)/B^{[L/M]}(z)$ exists; we leave the proof for the alternative case as an exercise. Define

$$p_{L-k}(z) = \left\{ [L/M]_f - \sum_{i=0}^{k-1} c_i z^i \right\} z^{-k} B^{[L/M]}(z).$$

By construction, $p_{L-k}(z)$ is a polynomial of order $L - k$ under the conditions of the theorem. Hence

$$\frac{p_{L-k}(z)}{B^{[L/M]}(z)} - g(z) = \{[L/M]_f - f(z)\} z^{-k} = O(z^{L+M-k+1}),$$

and therefore

$$\frac{p_{L-k}(z)}{B^{[L/M]}(z)} = [L - k/M]_g(z).$$

This theorem is used repeatedly in Chapter 5, where we prove a series of results for $[M - 1/M]$ Padé approximants and generalize them to $[M + J/M]$ Padé approximants for $J \geq -1$, using a method equivalent to the truncation theorem.

***Theorem 1.5.5* (Unitarity)** [Gammel and McDonald, 1966]. *Let $f(z) = \sum_{i=0}^{\infty} c_i z^i$ be unitary, by which we mean that*

$$f(z)f^*(z) = 1.$$

If $[M/M] = [M/M]_f(z)$ is a diagonal Padé approximant of $f(z)$, then

$$[M/M][M/M]^* = 1,$$

where the asterisk denotes complex conjugation of the coefficients only.

Proof. Let

$$[M/M] = \frac{A^{[M/M]}(z)}{B^{[M/M]}(z)}.$$

Then

$$[M/M][M/M]^* = \{f(z) + O(z^{2M+1})\}\{f^*(z) + O(z^{2M+1})\}$$
$$= 1 + O(z^{2M+1}).$$

Hence

$$A^{[M/M]}(z)A^{[M/M]*}(z) - B^{[M/M]*}(z)B^{[M/M]}(z) = O(z^{2M+1}).$$

The left-hand side of this expression is a polynomial of order $2M$ at most; therefore

$$A^{[M/M]}(z)A^{[M/M]*}(z) = B^{[M/M]*}(z)B^{[M/M]}(z),$$

and hence

$$[M/M][M/M]^* = 1.$$

This theorem is summarized by saying that diagonal Padé approximants preserve unitarity. It is important in S-matrix theory (see Chapters 9, 11) because the S-matrix is unitary. A timely note of caution is that complex conjugation may alter the analytic properties of f, and it is prudent to define $g(z) = f^*(z) = 1/f(z)$ before discussing the analytic structure of $f^*(z)$.

The invariance theorems of this section have justified the value of Padé approximants as practical approximating functions. In fact, one obvious esthetic test of the merit of any generalization of Padé approximants is whether the generalizations have such useful invariance properties.

Example 1. Let $f(z)$ satisfy the conditions of Theorem 1.5.5. Define $f(z) = 1 + it(z)$, so that $t(z)$ obeys a unitarity condition

$$t(z) - t^*(z) = it(z)t^*(z).$$

Then formation of $[L/M]$ Padé approximants of $t(z)$ with $L \leq M$ preserves this unitarity property [Masson, 1967a, b].

Example 2. Let
$$g(z) = \frac{bf(z)}{c + df(z)}.$$
Assuming that $L \leq M$, $[L/M]_f$ exists, and $c + df(0) \neq 0$, we find that[†]
$$[L/M]_g = \frac{b[L/M]_f}{c + d[L/M]_f}.$$

[†] D. Bessis, private communication.

2

Elementary developments

2.1 Numerical calculation of Padé approximants

Quite probably, the first problem to be tackled by anyone interested in Padé approximants involves the calculation of values of a set of diagonal approximants using the coefficients of the Maclaurin series as data. Explicit calculation is only feasible by hand for the lowest-order approximants, and so a good numerical algorithm becomes an early requirement. In this section we discuss the methods available and the qualities which are required of a 'good' method.

We start by considering the direct method of solution based on solving (1.1) as a linear system and defer the detailed discussion of the various distinct objectives and the other methods available for dealing with 'exact' data until this discussion can be put in context. Numerical solution of the linear equations

$$\begin{bmatrix} c_{L-M+1} & c_{L-M+2} & \cdots & c_L \\ c_{L-M+2} & c_{L-M+3} & \cdots & c_{L+1} \\ \vdots & \vdots & & \vdots \\ c_L & c_{L+1} & \cdots & c_{L+M-1} \end{bmatrix} \begin{bmatrix} b_M \\ b_{M-1} \\ \vdots \\ b_1 \end{bmatrix} = \begin{bmatrix} -c_{L+1} \\ -c_{L+2} \\ \vdots \\ -c_{L+M} \end{bmatrix} \quad (1.1)$$

is the core of the problem. We assume that the data coefficients $\{c_i, i = 0, 1, 2, \ldots, L + M\}$ are real or complex numbers known to a certain accuracy and that $c_0 \neq 0$. If $c_0 = 0$ in the data, the problem of type $[L/M]$ is replaced by another problem of type $[L - \alpha/M]$ in which $c_0 \neq 0$; the factor z^α is subsequently replaced. If the coefficients c_i have pronounced geometric growth (or diminution) with increasing values of i, a compensatory implicit scaling of z is essential with fixed-point arithmetic and desirable with floating-point arithmetic. This procedure is called equilibration: see Golub and van Loan [1989] or Wilkinson [1965]. If $i < 0$ in (1.1), we take $c_i = 0$. The simplest approach to (1.1) which can be recommended [Graves-Morris, 1979] is to solve the linear system using Gaussian elimination with full pivoting. A FORTRAN routine, called FUNCTION PADE($\cdots$), is given in the Appendix. This routine, however unsophisticated, has proved robust in applications. We emphasize that most calculations do not involve high-order approximants, because accuracy is more important than efficiency. An algorithm is called *robust* if it works well in practice on a wide variety of problems. A robust but more sophisticated algorithm is in the NAG library.

For our purposes, we need a method which will recognize whether the system of equations is degenerate, which occurs if $C(L/M) = 0$. Unless the data coefficients $c_0, c_1, \ldots, c_{L+M}$ are integers (or rational numbers with an exact finite binary representation) for which symbolic methods are suitable [Geddes, 1979], it is likely that rounding errors will prevent our computer program from discovering that $C(L/M) = 0$, even if it is true. If the data coefficients correspond to a degenerate case in which $C(L/M) = 0$, any results of the calculation which we obtain are generated by rounding error and are likely to be misleading. We stress that this sort of case does arise: for example, if $f(z)$ is a geometric series, all its $[L/2]$ Padé approximants are degenerate. Our numerical algorithm must be able to recognize similar cases in which a lower order of Padé approximation is quite possibly what is required. In short, one of our requirements is an accurate method of solving the system (1.1); a method which is more accurate than one might at first sight expect is needed to discriminate against exactly degenerate equations.

Standard numerical methods for solving (1.1) are given by Wilkinson [1963, 1965] and Golub and van Loan [1989, Chap. 3]. Gaussian elimination with full pivoting and Gaussian elimination with partial pivoting supplemented by iterative refinement are methods that can be recommended because they admit a sophisticated numerical test for degeneracy.

Whatever method of solution of (1.1) is chosen, the complete algorithm must contain an error exit for use when the requested approximant is degenerate according to agreed criteria. If the algorithm includes pointwise evaluation of an approximant, the algorithm should contain an error exit in case the approximant is being evaluated at a pole.

We have emphasized that a good numerical algorithm should contain a fairly sophisticated degeneracy test. If the Padé approximant corresponding to given data does not exist, we expect good algorithms to recognize this situation and to detect the degeneracy. We call such algorithms *reliable*. We might also hope that small variations of the data coefficients within their accuracy limits should not significantly affect the computed value of the Padé approximants. However, experience shows that we must expect to face ill-conditioned (unstable) problems in Padé approximation, in which small variations of the data coefficients lead to much larger variations in the computed numerator and denominator coefficients. This comparison reflects the idea of mathematical continuity, with substantial magnification of rounding errors in the solution. Methods are called stable if they minimize this magnification as far as is possible. Beyond these considerations, we expect our algorithm to be efficient, but efficiency is not as important as reliability and stability. Experienced Padé approximators are wary of degenerate and near-degenerate cases; they prefer to use more sophisticated approximation methods (see Sections 7.2, 7.3) rather than to resort to high-order approximants. Furthermore, the actual calculation of the Padé approximants is usually the least time-consuming part of a complete calculation, in which it is the computation of the coefficients to high accuracy which is expensive in computer time. We note that solution of (1.1) by elimination requires $O(\frac{1}{3}M^3)$ operations, unless iterative refinement is necessary. We consider methods of lower order, which are more efficient when M is large, later on in the section and discuss the penalties incurred by using methods with increased efficiency.

Our concern for high accuracy of the numerical algorithm stems primarily from experience [e.g., Luke, 1982]. In the previous section, we mentioned the empirical rule of keeping M extra decimal places of precision for computation of an $[L/M]$ approximant. Of course, one cannot say a priori that M, $M - 1$, $2M$, or any other number of extra decimal places are generally required. We simply sound a clear warning about the minimum number of guarding figures likely to be required, and suggest that any calculations of $[L/M]$ approximants performed with less than M guarding figures be looked at critically. From a numerical point of view, the Padé approximant derives its capacity to extrapolate certain power series beyond their circle of convergence from using the information contained in the tails of the decimal expansion of the data: naturally, accurate data are of paramount importance. We understand some of these ideas in terms of the following example. We consider a Stieltjes series (see Chapter 5; this and other material of the section may be more suitable for a second reading of this work) in which the coefficients are given by

$$c_j = \int_0^a u^j \, d\varphi(u), \qquad j = 0, 1, 2, \ldots, \tag{1.2}$$

where $a > 0$ and $\varphi(u)$ is bounded and nondecreasing. (Our sign convention here differs from (5.1.2) in that $f(-z)$ is a Stieltjes function.) A special case of (1.2) is given by

$$c_j = \int_0^1 u^j \, du = \frac{1}{j+1}, \quad j = 0, 1, 2, \ldots. \tag{1.3}$$

For Stieltjes series, various sequences of Padé approximants are known to converge systematically, and Stieltjes functions form the major class of functions for which we have a complete convergence theory. We see that the equations (1.1) with coefficients given by (1.3) are

$$\begin{bmatrix} \frac{1}{L-M+2} & \frac{1}{L-M+3} & \cdots & \frac{1}{L+1} \\ \frac{1}{L-M+3} & \frac{1}{L-M+4} & \cdots & \frac{1}{L+2} \\ \vdots & \vdots & & \vdots \\ \frac{1}{L+1} & \frac{1}{L+2} & \cdots & \frac{1}{L+M} \end{bmatrix} \begin{bmatrix} b_M \\ b_{M-1} \\ \vdots \\ b_1 \end{bmatrix} = \begin{bmatrix} \frac{-1}{L+2} \\ \frac{-1}{L+3} \\ \vdots \\ \frac{-1}{L+M+1} \end{bmatrix} \tag{1.4}$$

for $L \geq M - 1$. The coefficient matrix of (1.4) is a Hilbert segment, which is notorious for its ill-conditioning. In fact, we have encountered ill-conditioning in an ideal case which is in no way degenerate. In this context, it is interesting to consider the condition number of the Hankel matrix

$$H_M = \begin{bmatrix} c_{L-M+1} & c_{L-M+2} & \cdots & c_L \\ c_{L-M+2} & c_{L-M+3} & \cdots & c_{L+1} \\ \vdots & \vdots & & \vdots \\ c_L & c_{L+1} & \cdots & c_{L+M-1} \end{bmatrix} \tag{1.5}$$

in which the elements of H_M are defined by (1.2). The (L_2 norm) condition number of H_M is defined by

$$\kappa(H_M) = \|H_M^{-1}\|_2 \|H_M\|_2.$$

A matrix with a large condition number corresponds to an ill-posed problem and, very crudely, one may expect that $\log_{10} \kappa(H_M)$ is the number of significant figures lost in the solution of (1.1) with rounded data. More precise statements about the loss of precision are given by Golub and van Loan [1989]. In the general case defined by (1.2) and (1.5), Taylor [1978] has shown that

$$\liminf_{M \to \infty} \kappa(H_M)^{1/M} \geq 4, \tag{1.6}$$

provided $d\varphi(a-) > 0$. Other methods are given by Wilf [1970]. For the particular case of the Hilbert segment defined by (1.3) and (1.5),

$$\liminf_{M \to \infty} \kappa(H_M)^{1/M} \geq 16. \tag{1.7}$$

Bruno and Reitich [1994] have devised a remarkably successful approach to dealing with the problems raised by the series (1.2). They modify (1.5) to take account of the value of z_0 at which the approximant is to be evaluated and use the homographic transformation (6.7.2) to distance the nearest singularity a^{-1} of $f(z)$ away from the origin, compared to $|z_0|$. Whereas Theorem 1.5.2 shows that this transformation makes no difference in principle to the calculated value of a diagonal approximant, in practice there is marked improvement both in the conditioning of the equations and precision of the estimates of $f(z_0)$.

Hopkins [1982] surveys the ill-conditioning of the Padé equations in a number of test examples, which confirm the estimates (1.7). Moreover, he notes that the matrix H_M with elements $(H_M)_{i,j} = (i+j+1)!^{-1}$ associated with the Padé approximant of type $[M/M]$ for e^z is particularly ill-conditioned; other supporting numerical examples are given by Higham [1991].

These results strongly support our rule of thumb that *approximately M decimal places of accuracy are lost in the calculation of an $[L/M]$ approximant by direct solution of the linear system*. In reality, this means that approximately M extra decimal places of precision are required of the data coefficients $c_{L-M+1}, c_{L-M+2}, \ldots, c_{L+M}$ than is expected of the solution coefficients $b_0, b_1, \ldots, b_M$. It is unusual for any further significant loss of accuracy to occur in the calculation of the numerator coefficients using (1.7).

The problem of numerical calculation of the denominator polynomial does not necessarily take the form (1.1). For classical Padé approximation, a solution of the homogeneous system

$$\begin{bmatrix} c_{L-M+1} & \cdots & c_{L+1} \\ \vdots & & \vdots \\ c_L & \cdots & c_{L+M} \end{bmatrix} \begin{bmatrix} q_M \\ \vdots \\ q_0 \end{bmatrix} = 0 \tag{1.8}$$

is required. If this system implies that $q_0 = 0$, then (1.1) is not applicable, and we have previously noted that the Padé approximant, as such, does not exist. However, the solution of such systems for the associated Padé polynomials and Padé form is useful later. The system can be solved, say, using the QR decomposition, or by Gaussian elimination with full pivoting in much the same way as the square system (1.1) until the final equation is reached. This equation is homogeneous and contains at least two variables. Decisions have to be made about which coefficients in this equation are zero within rounding error. If the equations are rank degenerate, there is a muliplicity of solutions. A good discussion of the general numerical problem of testing for rank degeneracy and on identifying a numerical zero is given by Stewart [1984]. If the equations appear numerically to have full row rank M, there is no problem and the solution is said to be essentially unique as it is unique up to an overall scale factor.

Hitherto in this section, we have tacitly assumed that our problem is the calculation of the coefficients of a particular Padé approximant with a view to calculating the approximant at a prespecified value of z. In practice, it is much more likely that our task consists of the tabulation of a particular $[L/M]$ approximant at a number of values of z, or else that it consists of pointwise evaluation of a whole sequence of Padé approximants at a prespecified value of z. Which is the best algorithm to choose for Padé approximation depends on the problem at hand. It is normal to distinguish between the coefficient problem and the value problem. If tabulation of values of a particular approximant is required, it is probably best to calculate the coefficients $a_0, a_1 \ldots, a_L, b_0, b_1, \ldots, b_M$ first, and then to evaluate the approximants. If pointwise evaluation of a whole sequence is required, it may be better to consider methods such as using the ε-algorithm in which a whole sequence is calculated recursively.

The direct method of calculating Padé approximants is reliable and as stable as possible, but may not be the most efficient such method. There are several '$O(\alpha M^2)$' methods, where α is typically 4 or 6. For the simplest of them, computational efficiency is gained at the expense of reliability, meaning that they are only suitable for use in contexts where the existence and nondegeneracy of the approximants are not in question. For those methods which have modifications for reliability, the question of stability always has to be addressed. A summary list of $O(\alpha M^2)$ methods, each with its own individual merits, for the coefficient problem is as follows:

(i) Kronecker's algorithm for the numerator and denominator coefficients. This relatively old algorithm [Kronecker, 1881] is a recursion based on an antidiagonal sequence in the Padé table. It has a straightforward modification to make it reliable, based on the Euclidean algorithm, and these matters are explained in Section 3.7. The corresponding algorithm for the antidiagonal staircase sequence was given by Baker [1973a].

(ii) The Q.D. algorithm for the corresponding continued fraction. This method consists of using the Q.D. algorithm to construct the coefficients of the corresponding continued-fraction representation, and then using these coefficients to construct numerator and denominator polynomials. It is described in Sections 4.2 and 4.3.

(iii) Hankel matrix methods for the denominator coefficients. These are based on the relationship between two successive entries on a diagonal in the Padé table, and are described in Section 3.6. The method is closely related to the Berlekamp–Massey algorithm described in Section 4.4 which is based on a diagonal staircase sequence.

2.2 Decipherment of singularities from Padé approximants and apparent errors

According to the theory of analytic continuation of functions of a complex variable, all the properties of a function, analytic at a point, are contained in its power-series expansion at that point. While in principle the function may be continued by reexpansion about other points, the central practical problem is to decipher the properties of the function from the given series coefficients. Padé approximants can be used very effectively in determining quantitative results about functions when the analytic properties are qualitatively known, and they can also be used to deduce considerable information about the singularity structure of a function from its Taylor-series coefficients. Before embarking on the proper use of Padé approximants, we review briefly the nature of the permissible singularities of an analytic function.

First, let $f(z)$ be analytic and single-valued in $|z - z_0| < R$ except at $z = z_0$. Then, for all r such that $0 < r < R$, $f(z)$ is analytic and single-valued in the annulus $r < |z - z_0| < R$. Laurent's theorem states that in these circumstances the expansion

$$f(z) = \sum_{n=0}^{\infty} a_n(z - z_0)^n + \sum_{n=1}^{\infty} b_n(z - z_0)^{-n} \tag{2.1}$$

is convergent within the annulus. Laurent's theorem follows from Cauchy's theorem, and gives a specific representation for a_n and b_n,

$$a_n = \frac{1}{2\pi i} \int_{|z-z_0|=R} \frac{f(z)\,dz}{(z - z_0)^{n+1}}, \quad b_n = \frac{1}{2\pi i} \int_{|z-z_0|=r} f(z)(z - z_0)^{n-1}\,dz. \tag{2.2}$$

Equation (2.1) permits unique categorization of singularities at $z = z_0$.

If $b_n = 0$, $n = 1, 2, \ldots$, then $z = z_0$ is a regular point and $f(z)$ is analytic at $z = z_0$.

2.2 Decipherment of singularities from Padé approximants and apparent errors

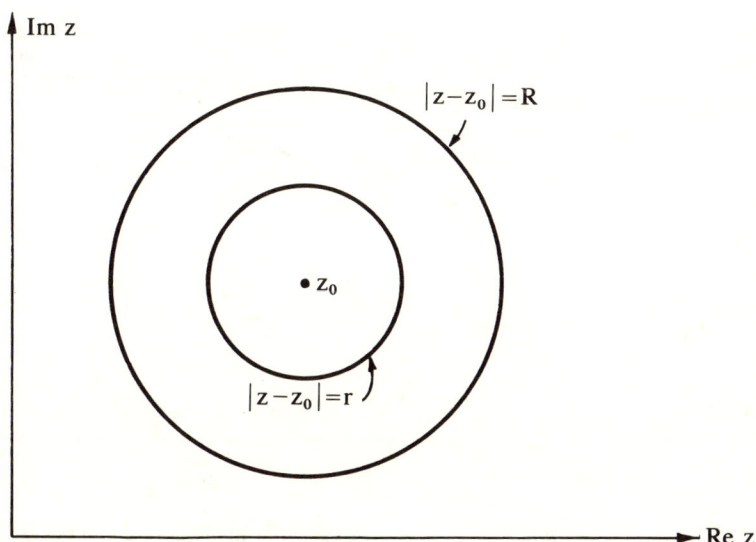

Figure 2.2.1. An annulus of analyticity.

If $b_n = 0$, $n = 2, 3, \ldots$, and $b_1 \neq 0$, then $z = z_0$ is a simple pole.

If $b_n = 0$, $n = m + 1, m + 2, \ldots$, and $b_m \neq 0$, then z_0 is an mth-order pole, and for $m > 1$, $f(z)$ is said to have a multipole.

If, for all m and some $n > m$, $b_n \neq 0$, then z_0 is an essential singularity of $f(z)$.

Provided z_0 is not an essential singularity, (2.1) suggests that $f(z)$ may be approximated by Padé approximants. However, even if z_0 is an essential singularity, provided it is not approached too closely, the essential singularity resembles a finite sum of multipoles. This is because b_n given by (2.2) decrease rapidly with n; in fact $|b_n| < r^n$ for any $r > 0$ and n sufficiently large.

Example.

$$f(z) = \exp\left(\frac{1}{1+z}\right) = \sum_{n=0}^{\infty} \frac{1}{n!}\left(\frac{1}{1+z}\right)^n. \tag{2.3}$$

$f(z)$ has an essential singularity at $z = -1$, but the coefficients $b_n = 1/n!$ decrease very rapidly. This qualitative fact can be exploited rigorously as in the proof of Theorem 6.5.4.

Secondly, $f(z)$ may have branch points. If $f(z)$ is analytically continued by expansion and reexpansion and remains a single-valued function of z, there are no branch points. However, if there is any point z_0 for which analytic continuation by a clockwise and counterclockwise path of

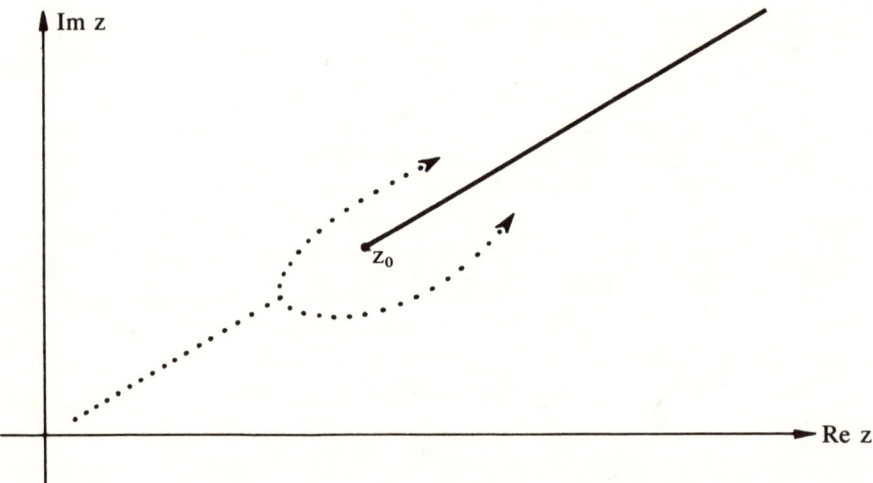

Figure 2.2.2. A branch point at z_0, an associated branch cut, and two paths of continuation.

arbitrarily small radius yields different values at the same point, then $f(z)$ is not single-valued and z_0 is a branch point. A single-valued function can be obtained for $f(z)$ in various ways. One may form the Mittag–Leffler star by drawing a straight line from each branch point to infinity along a ray from the origin. It sometimes happens that a single-valued function can be formed by connecting two or more branch points by branch cuts. For example $f(z) = \sqrt{(z-a)(z-b)}$ may be uniquely defined with a branch cut from a to b. This line is a discontinuity of the given function.

Finally, there may exist natural boundaries. For example, the lacunary series

$$f(z) = z + z^2 + z^4 + z^8 \cdots = \sum_{i=0}^{\infty} z^{(2^i)} \qquad (2.4)$$

has a natural boundary on $|z| = 1$. To prove this assertion, let us note that $f(z)$ is well defined by a convergent Maclaurin expansion in $|z| < 1$. Yet if

$$z = z_{pq} = \exp(2\pi i p/2^q), \quad \text{where } p \text{ and } q \text{ are integers,}$$

then

$$f(z) = \sum_{n=0}^{q-1} z^{(2^n)} + \sum_{n=q}^{\infty} 1, \quad \text{which diverges.}$$

Since the points $\{z_{pq}\}$ are dense on the unit circle, continuation to $|z| > 1$ is no longer straightforward. Sometimes such analytic continuation is possible, and sometimes Padé approximants converge to this continued function: standard references to Padé approximants and quasianalytic functions are Gammel and Nuttall [1973] and Gammel [1974].

In order to interpret the results of using Padé approximants on a new function, we consider how Padé approximants represent the singularities in the previous categories. If $f(z)$ has a simple pole, then a simple zero in the denominator of the Padé approximant near the pole is expected. If $f(z)$ has a multiple pole, a cluster of zeros of the Padé denominator is expected; these zeros should tend to coalesce at the multipole with increasing order of approximation. For an essential singularity, we recall *Weierstrass's theorem*: Let $f(z)$ have an essential singularity at $z = z_0$. Given $\rho > 0$, $\varepsilon > 0$, and any value v, there exists a point in the disk $|z - z_0| < \rho$ at which $|f(z) - v| < \varepsilon$ [Titchmarsh, 1939, p. 93]. In other words, this theorem shows that $f(z)$ tends to any desired limit v as $z \to z_0$ through a suitable sequence of z values. Thus we expect a clustering of poles and zeros of Padé approximants at an essential singularity. In a low order of approximation, of course, a multiple pole and an essential singularity will appear the same.

To simulate a branch cut, we expect to find a path delineated by roughly alternating poles and zeros of the Padé approximants. To appreciate this, consider

$$f(z) \equiv \int_P \frac{\rho(u)\,du}{z - u} = \frac{-1}{2\pi i}\int_\Gamma \frac{f(u)}{z - u}\,du, \qquad (2.5)$$

where P is a path from a to b in the complex plane, Γ is a contour enclosing P but not z, and the second part of (2.5) is a consequence of Cauchy's theorem. $f(z)$ is uniformly differentiable and so is analytic except on the path P. $f(z)$ may be approximated by a Riemann sum

$$f(z) \approx \sum_{i=1}^{N} \frac{\rho(u_i)\delta u_i}{z - u_i}, \qquad z \notin P,$$

which consists of a sequence of poles with residues $\rho(u_i)\delta u_i$. This behavior is what one expects the Padé approximants to reproduce.

Example

$$f(z) = \frac{1}{\sqrt{1 + z}} = \frac{1}{\pi}\int_{-\infty}^{-1} \frac{dx}{(z - x)\sqrt{-1 - x}}.$$

$f(z)$ is defined with a cut on $-\infty < z \leq -1$, and the poles of the Padé approximant are located on that cut, as shown in Figure 2.2.3.

Since there are many ways of defining the cuts to define, in turn, a single-valued function, and the Mittag–Leffler star is but one, it is

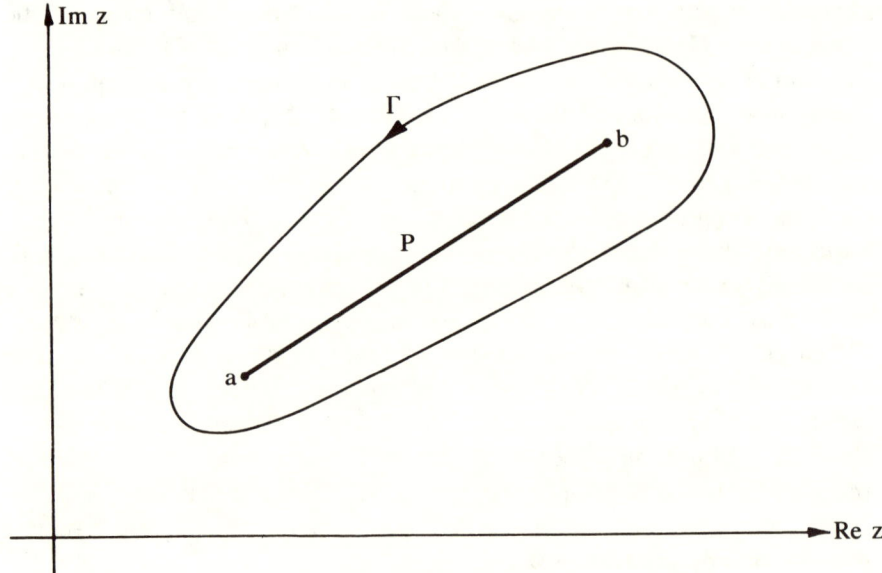

Figure 2.2.3. The path P and the contour Γ in the complex plane.

interesting to speculate that the limiting distribution of the poles and zeros of a suitable sequence of Padé approximants delineate a natural cut structure (or principal Riemann surface) from the Maclaurin series. In general terms, the cuts so defined are such as to minimize their capacity in the inverse z-plane, a point which we discuss in Section 6.7.

The foregoing account states what we expect from Padé approximants to analytic functions with various singularities. In ideal circumstances precisely these results are obtained; in practice, whether one expects them or not, *defects* occur for all but the simplest functions. A *defect* is the name given to an extraneous pole and a nearby zero. We consider, nonrigorously, a function $f(z)$ which is smooth near $z = \alpha$ and let

$$g(z) = f(z) + \frac{\varepsilon}{z - \alpha}, \quad |\varepsilon| \ll 1.$$

If ε is very small and $|z - \alpha|$ is not small, $g(z) \approx f(z)$. Only in a neighborhood of $z = \alpha$ do $g(z)$ and $f(z)$ differ appreciably. We expect $g(\alpha) = \infty$, but also we have

$$g(z) \approx f(\alpha) + \frac{\varepsilon}{z - \alpha} = 0$$

if $z - \alpha = -\varepsilon/f(\alpha)$, which is very small. Hence there is a zero close to the pole. There are real difficulties in the numerical detection of defects; we refer to Abd-Elall et al. [1970] for details. In short, the addition of an

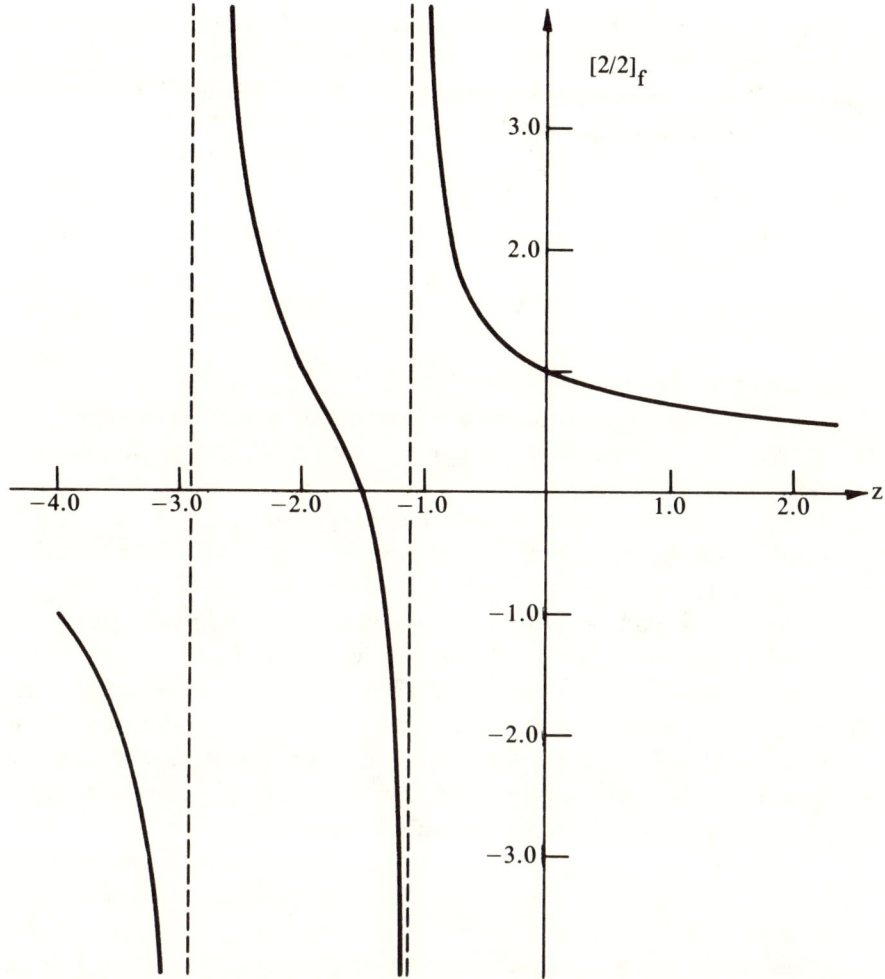

Figure 2.2.4. The [2/2] Padé approximant of $f(z) = (1-z)^{-1/2}$ showing poles at $z_{\pm} = -2 \pm 2/\sqrt{5}$ lying on the branch cut of $f(z)$, $-\infty < z \leq -1$.

'insignificant' pole to $f(z)$ produces a defect, and this problem has to be expected with Padé approximation. We regard the nearby zero of numerator and denominator as canceling approximately, which puts the defects in their proper perspective. Defects are easily recognized by their transient nature. They tend to appear and disappear as one looks at one approximant and then the next. They contrast strongly with the more stable patterns seen in conjunction with the true singularities of the function. We will return to this topic in the next section; in Section 6.5, we prove that the residue of the pole of a defect is small for high-order Padé approximation of meromorphic functions.

The following example shows why it is normally wise to ignore results derived from Padé approximants with defects close to the origin. The function $1 + z^2$ has no [1/1] Padé approximant (see Section 1.4). Consider the function [Zinn-Justin, 1970]

$$f(z) = 1 + \varepsilon z + z^2,$$

for which the [1/1] Padé approximant is

$$[1/1] = \frac{1 + (\varepsilon - 1/\varepsilon)z}{1 - z/\varepsilon}.$$

The pole of the [1/1] approximant occurs at $z = \varepsilon$, and its residue is $-\varepsilon^3$. If $|\varepsilon|$ is small, we see that the pole is close to the origin and its residue is small. We understand a Padé approximant with a defect near the origin as a nearly degenerate approximant, and we are suspicious of drawing any implications from the values of such an approximant.

Given a function defined by a formal power series as the starting point, the first step toward deciphering the information contained in the series is formation of the Padé approximants lying in at least a broad band about the central diagonal of the Padé table. Unless there are reasons to the contrary, as much of the Padé table as possible should be constructed. Normally, most computing effort goes into construction of the coefficients rather than formation of the Padé approximants, and so construction of all available Padé approximants is relatively inexpensive. The next step is the examination of the distribution of poles and zeros. Are these persistent, or do they form defects? It should then be possible to decide which poles and zeros closest to the origin represent true singularities of the function. It is sometimes possible at this stage to detect that the function has asymptotic behavior z^{J_0} for some J_0 and in some half plane by virtue of the stability of $[M + J_0/M]$ approximants, and poles receding to infinity for $L - M < J_0$.

Once the general nature of the structure of the function has been determined, either by the preceding analysis or by other qualitative information (see Section 7.3), one is in a position to make a more refined analysis. The presence of poles is normally sufficiently clear. Their influence on the function depends on their residues, their multiplicity, and their proximity to the origin. The type of structure which has been further analyzed profitably is principally the branch point and branch cut. An important strategy is the manipulation of the series to a form which can be exactly represented by Padé approximants, except for small corrections. We consider the particular case where

$$f(z) = A(z)(1 - \mu z)^{-\gamma} + B(z) \tag{2.6}$$

is expected to represent the function. $f(z)$ has a cut from $z = \mu^{-1}$ to ∞, and $A(z)$, $B(z)$ are to be analytic at $z = \mu^{-1}$ and are expected to have little structure. The following methods have been used to good effect [Baker, 1961; Baker et al., 1967; Hunter and Baker, 1973a; Baker, 1977a].

(i) Form Padé approximants to

$$F_1(z) = \frac{d}{dz} \ln f(z)$$

$$\simeq \frac{\gamma}{\mu^{-1} - z} \quad \text{near } z = \mu^{-1}. \tag{2.7}$$

The use of the appropriate Padé approximant of $F_1(z)$ determines the pole position $z = \mu^{-1}$, and $-\gamma$ is its residue. The approximants we have defined for $f(z)$ are not necessarily rational. They are commonly called D-log Padé approximants (see Section 5.3). The estimates of μ and γ are called 'unbiased' because no assumed values of μ, γ, etc. are used: μ, γ are determined directly from the series coefficients.

(ii) Form Padé approximants to

$$F_2(z) = (\mu^{-1} - z)\frac{d}{dz} \ln f(z) \approx \gamma \tag{2.8}$$

for an assumed value of μ, and obtain a biased estimate of γ by evaluating the Padé approximant at $z = \mu^{-1}$.

(iii) Form Padé approximants to

$$F_3(z) = [f(z)]^{1/\gamma} \approx \frac{[A(\mu^{-1})]^{1/\gamma}}{1 - \mu z} \tag{2.9}$$

by assuming a value for γ, and obtain biased estimates of $A(\mu^{-1})$ and μ^{-1} from the roots and residues of the Padé approximants.

(iv) Form Padé approximants to

$$F_4(z) = (1 - \mu z)^\gamma f(z) \approx A(\mu^{-1}) \tag{2.10}$$

by assuming values for μ, γ, and obtain a biased estimate of $A(\mu^{-1})$ by evaluating the Padé approximant at $z = \mu^{-1}$.

(v) Form Padé approximants to

$$F_5(z) = \frac{\frac{d}{dz}\left(\ln \frac{d}{dz} f(z)\right)}{\frac{d}{dz} \ln f(z)} \approx 1 + \frac{1}{\gamma}, \tag{2.11}$$

and evaluate the Padé approximants at the assumed value of μ^{-1}. This process yields a biased estimate of γ, but as a practical matter it is frequently relatively insensitive to the choice of μ^{-1} [Baker et al., 1967].

(vi) In cases where there are two or more series with the same branch point, and it is the branch point closest to the origin, one may obtain an unbiased estimate of the difference of the exponents by the method of 'critical-point renormalization' as follows: we have

$$f(z) = A(z)(1 - \mu z)^{-\gamma} + B(z) = \sum_{j=0}^{\infty} c_j z^j,$$

$$g(z) = C(z)(1 - \mu z)^{-\varepsilon} + D(z) = \sum_{j=0}^{\infty} c'_j z^j,$$
(2.12)

and $A(z)$, $B(z)$, $C(z)$, and $D(z)$ are analytic at $z = \mu^{-1}$. We define

$$h(z) = \sum_{j=0}^{\infty} \frac{c_j}{c'_j} z^j \propto (1 - z)^{-\gamma+\varepsilon-1} \quad \text{for } z \approx 1. \quad (2.13)$$

Form Padé approximants to

$$F_6(z) = (1 - z)\frac{d}{dz} \ln h(z) \approx \gamma - \varepsilon + 1 \quad (2.14)$$

to obtain an unbiased value of $\varepsilon - \gamma$ by evaluating the Padé approximants at $z = 1$. In (2.12) it is possible[†] to choose $g(z) = [f(z)]^p$, where p is an integer, $p \geq 2$. Then $\varepsilon = p\gamma$ and $F_6(z) \sim 1 + \gamma(1 - p)$. A similar scheme was proposed by Sheludyak and Rabinovich [1979].

(vii) The Baker–Hunter method [1973] sometimes allows one to detect subdominant confluent singularities. This procedure is a subtle one, and sometimes the method is interfered with by strong, nonconfluent singularities. Suppose, instead of (2.6), that

$$f(z) = \sum_{n=0}^{\infty} c_n z^n \approx \sum_{j=1}^{m} A_j (1 - \mu z)^{-\gamma_j} \quad (2.15)$$

near $z = \mu^{-1}$, and that $\gamma_1 > \gamma_2 > \cdots > \gamma_m$. Then $f(z)$ has m confluent singularities at $z = \mu^{-1}$. Make the change of variable

$$z = \mu^{-1}(1 - e^{-\xi}), \quad (2.16)$$

so that

$$g(\xi) = f(z(\xi)) = \sum_{n=0}^{\infty} g_n \xi^n \approx \sum_{j=1}^{m} A_j e^{\gamma_j \xi}. \quad (2.17)$$

[†] D. Bessis, private communication.

2.2 Decipherment of singularities from Padé approximants and apparent errors

A transform of $g(\xi)$ is defined by

$$G(s) = \int_0^\infty e^{-\xi} g(s\xi)\, d\xi. \tag{2.18}$$

From (2.17), we find

$$G(s) = \sum_{n=0}^\infty n!\, g_n s^n \approx \sum_{j=1}^m \frac{A_j}{1 - \gamma_j s}, \tag{2.19}$$

and hence we may compute biased estimates of γ_j and A_j to determine at least some of the stronger confluent, subdominant singularities.

By way of a caution, we add that when confluent singularities exist, they can, and often do, bias the results of the other methods (i)–(vi) listed above as well as appreciably slow down the rate of convergence. Furthermore, it has been frequently observed that other significant singularities interfere not only with the confluent-singularity analysis, but dramatically reduce the rate of convergence. Where feasible, steps to suppress their competing influence are rewarding.

With any approximation scheme, one must be able to estimate the approximation error. Using Padé approximants, there are three principal sources of error: (i) the given coefficients c_j are known only to limited accuracy, (ii) accuracy is lost in forming the coefficients of the Padé approximant and in forming its value, and (iii) the Padé approximant is not the function itself, leading to the fundamental approximation error. There is little to be done about (i) except to note that accuracy in the given coefficients is essential. The Padé approximant is necessarily misguided by errors in the given series, no matter what their source. The variation of the coefficients of the given series within their accuracy limits provides a very useful and instant error estimate of the accuracy of Padé approximation. For (ii), loss of accuracy in the formation of Padé approximants is, in practice, usually inexcusable. Multiple precision on a modern computer gives 20 or 30 decimal places, and this precision should be used if necessary. Of course, the approximation problem is often ill-conditioned, and the accuracy of the input coefficients is usually better than the accuracy of the output coefficients even with exact arithmetic. A rough and ready working guide is that one extra decimal place of working accuracy should be kept for every decimal place of accuracy required in the answer (see Section 2.1). As a purely empirical anthropological observation, most inexperienced Padé approximators use sufficiently high order approximants—often too high—and insufficient working accuracy to justify them.

We now turn to the problem of estimating the mathematical error of the approximation. In practice, this estimate must be based on likely hypotheses, and estimating these apparent errors is currently an art as well as a science.

In the previous section, we encountered defects in the approximants. They are nearby pole and zero combinations which are significant when they occur within the region of interest. The region is frequently most conveniently taken to be the largest circle containing the origin and whatever singularities are of interest, but excluding, with a margin to spare, all other nonpolar singularities. A defect would then be any pole in that region with a residue less than some preassigned value, say 0.003 times the expected residue. We are inclined to exclude these defective Padé approximants from the set of Padé approximants expected to be useful approximations. By so doing, we expect to obtain a set which is uniformly bounded on a bounded region which excludes singularities of the given function. Provided that there are sufficiently many Padé approximants in this set, the theorems of Section 6.4 assure convergence. In the assessment of apparent errors, it should be noted that, as a practical matter, the occurrence of defects seems often to cause successive Padé approximants ($\Delta L = 1$ or $\Delta M = 1$ or both) to be approximately equal, and one can be misled about the rate of convergence if the defects are not detected either directly or indirectly. The occurrence of defects can be thought of as a near miss at the existence of a block in the Padé table, where the determinant does not vanish but is anomalously small. In the case where there is a block, certain consecutive Padé approximants are, of course, exactly equal (Section 1.4). The existence of defects shows that it is important to analyze as much as possible of the Padé table to decide which poles and zeros are significant, and which are defects indicating that the Padé approximant in question is unreliable. Furthermore, it is plain that a blind calculation of Padé approximants at any one value (e.g., by using the ε-algorithm) ignores much of the information provided by the Padé approximants themselves about their convergence in the z-plane.

Let us now turn our attention to functions which have a dominant singularity of the form, see (2.6),

$$A(1 - \mu z)^{-\gamma}, \qquad (2.20)$$

and let us suppose that we have estimated this singularity, probably by using the methods of (2.7)–(2.19) to be

$$A'(1 - \mu' z)^{-\gamma'}. \qquad (2.21)$$

The three parameters A, μ, and γ are in principle determined by three equations, which originate from accuracy-through-order equations. Let

2.2 Decipherment of singularities from Padé approximants and apparent errors

these equations be accurate to orders z^J, z^{J+1}, and z^{J+2}, and then we know that

$$A\binom{-\gamma}{j}\mu^j = A'\binom{-\gamma'}{j}\mu'^j(1 + \eta_j), \quad j = J, J+1, J+2, \quad (2.22)$$

where η_j are small percentage errors. Obviously, if $\eta_j = 0$, the parameters are identical, and so the η_j are to be regarded as the source of the error in the approximation scheme. We consider percentage errors, because μ determines the scale of the z-plane; if μ is very different from unity, the magnitude of the terms in (2.22) can vary rapidly with j. If the approximation is a good one, we may use first-order expansions

$$A' = A + \delta A, \quad \gamma' = \gamma + \delta\gamma, \quad \mu' = \mu + \delta\mu. \quad (2.23)$$

A first-order analysis and logarithmic differentiation of (2.22) give

$$\frac{\delta A}{A} + j\frac{\delta\mu}{\mu} + \delta\gamma\left(\sum_{k=0}^{j-1}\frac{1}{\gamma + k}\right) = \eta_j, \quad j = J, J+1, J+2. \quad (2.24)$$

These three linear equations can easily be solved to give

$$-\frac{\delta\mu}{\mu} = (2\gamma + 2J + 1)\eta_{J+1} - (\gamma + J + 1)\eta_{J+2} - (\gamma + J)\eta_J, \quad (2.25a)$$

$$\delta\gamma = (\gamma + J)\left(\eta_{J+1} - \eta_J - \frac{\delta\mu}{\mu}\right), \quad (2.25b)$$

$$\frac{\delta A}{A} = \eta_J - J\frac{\delta\mu}{\mu} - \left(\sum_{k=0}^{J-1}\frac{1}{\gamma + k}\right)\delta\gamma. \quad (2.25c)$$

From (2.25a), we see that $\delta\mu/\mu$ is of order J compared to η_J, provided there is no unusual cancellation. From (2.25b) it follows that $\delta\gamma$ is of order $J\delta\mu/\mu$ or $J^2\eta_J$. Finally (2.25c) shows that $\delta A/A$ is of order $J^2 \ln J\eta_J$. We conclude that the errors in this determination of A, μ, and γ are in the ratio

$$\frac{\delta\mu}{\mu}:\delta\gamma:\frac{\delta A}{A} = 1:J:J\ln J, \quad (2.26)$$

where J is the order of the active terms in the series.

The determination of the apparent absolute size of the errors is more difficult. We will be explicit about our procedure when it is used in conjunction with a method such as (i) or (iv) in (2.7) and (2.10). In those cases the parameter μ is just a zero of the reciprocal of the function, and so $\delta\mu/\mu$ is proportional to the error of estimation of (e.g.) $1/F_1(x)$. Thus we can relate the error $\delta\mu/\mu$ to the errors in a table of values, which are in practice relatively easy to compute.

First let us look at the structure of the difference between two adjacent Padé approximants. By the (∗ ∗) identity (3.4.5),

$$[L/M] - [L-1/M] = \frac{C(L/M+1)}{C(L-1/M)} \frac{z^{L+M}}{B^{[L/M]}(z)B^{[L-1/M]}(z)}, \quad (2.27)$$

where the determinants are defined by (1.4.8). Since $[L/M]$ is exact through order z^{L+M}, the right-hand side of (2.27) gives the error in the coefficient of z^{L+M} in $[L-1/M]$. The next step is to relate this error to the magnitude of the η's. The hypothesis involved here is that there are no unusual cancellations, so that a reasonable estimate for the magnitude of η is obtained.

Since $B^{[L/M]}(0) = 1$, one can estimate the coefficient of z^{L+M}, and at the same time to some extent take into account higher-order terms, by forming a table of the left-hand side of (2.27) over the range $0 < z \leq \mu^{-1}$ and fitting it to the monomial z^{L+M} at the point in that range which gives the largest coefficient.

Now, in the treatment of method (i) of this section (F_1), a different reduction of (2.24) is convenient, namely,

$$\delta\gamma = (\gamma + J + 1)(\gamma + J)(2\eta_{J+1} - \eta_J - \eta_{J+2}),$$

$$\frac{\delta\mu}{\mu} = -\delta\gamma(\gamma + J)^{-1} + \eta_{J+1} - \eta_J,$$

$$\frac{\delta A}{A} = -\frac{J\delta\mu}{\mu} - \left(\sum_{k=0}^{J-1} \frac{1}{\gamma + k}\right)\delta\gamma + \eta_J.$$

Here $\delta\gamma = 0$ as $\gamma = \gamma' = 1$. Thus $\delta\mu/\mu$ is of order η, and $\delta A/A$, which by (2.7) plays the role of $\delta\gamma$, is of order $J\delta\mu/\mu$, in harmony with (2.26). Thus by this analysis we have, by (2.27), that η is given by

$$A\mu^{L+M}\eta \approx \frac{C(L/M+1)}{C(L-1/M)},$$

where we have used

$$\binom{-1}{n} = 1.$$

2.3 Some explicit forms for Padé denominators

In this section we consider the calculation of the denominators of Padé approximants directly from the explicit determinantal formula (1.1.8) for

2.3 Some explicit forms for Padé denominators

$Q^{[L/M]}(z)$, for $L \geq M - 1$. We are able to do this principally for the class of functions which may be represented as

$$f(z) = {}_2F_1(\alpha, 1, \gamma, z). \tag{3.1a}$$

Results for functions with the representations

$$f(z) = {}_1F_1(1, \gamma, z) \tag{3.1b}$$

and

$$f(z) = {}_2F_0(\alpha, 1; z) \tag{3.1c}$$

follow as corollaries. In Section 1.2 we showed that explicit calculation of the numerator and denominator for each Padé approximant of the exponential function is possible using the direct method, but existence of a simple explicit form for the numerator polynomial is special to this case. Once the denominator of an $[L/M]$ Padé approximant is constructed, the formula (1.1.9) leads directly to the numerator coefficients; equivalently, we may consider $A^{[L/M]}(z)$ to be defined by truncation of $B^{[L/M]}f(z)$ beyond terms of order z^L.

Of course, we do not suggest that numerical calculations are to be made using determinantal formulas, and in Section 2.1 we considered these problems. Nor do we suggest that the method of this section is always the best when the coefficients c_i of $f(x)$ are given algebraically. Use of the Q.D. algorithm (see Sections 3.8, 4.4) may well be simpler, but it leads to a continued-fraction representation of the Padé approximant. Explicit formulas for the numerator and denominator polynomials would have to be derived from the recursion formulas (4.2.7).

The coefficients of the Maclaurin expansion of (3.1a) are

$$c_i = \frac{(\alpha)_i}{(\gamma)_i} = \frac{(\alpha + i - 1)!}{(\alpha - 1)!} \frac{(\gamma - 1)!}{(\gamma + i - 1)!} = \frac{\Gamma(\alpha + i)}{\Gamma(\alpha)} \frac{\Gamma(\gamma)}{\Gamma(\gamma + i)}. \tag{3.2}$$

Substituting in (1.4.8) for $L \geq M - 1$ we find that

$$C(L/M) = \begin{vmatrix} \dfrac{(\alpha)_{L-M+1}}{(\gamma)_{L-M+1}} & \cdots & \dfrac{(\alpha)_L}{(\gamma)_L} \\ \vdots & & \vdots \\ \dfrac{(\alpha)_L}{(\gamma)_L} & \cdots & \dfrac{(\alpha)_{L+M-1}}{(\gamma)_{L+M-1}} \end{vmatrix}. \tag{3.3}$$

Reduction of (3.3) is simplified by defining p to be the product of the leading elements of the rows

58   Elementary developments

$$p = \prod_{i=1}^{M} \frac{(\alpha)_{L-M+i}}{(\gamma)_{L-M+i}},$$

so that

$$C(L/M) = p \cdot \begin{vmatrix} 1 & \dfrac{\alpha + L - M + 1}{\gamma + L - M + 1} & \cdots & \dfrac{(\alpha + L - M + 1)_{M-1}}{(\gamma + L - M + 1)_{M-1}} \\ 1 & \dfrac{\alpha + L - M + 2}{\gamma + L - M + 2} & \cdots & \dfrac{(\alpha + L - M + 2)_{M-1}}{(\gamma + L - M + 2)_{M-1}} \\ \vdots & \vdots & & \vdots \\ 1 & \dfrac{\alpha + L}{\gamma + L} & \cdots & \dfrac{(\alpha + L)_{M-1}}{(\gamma + L)_{M-1}} \end{vmatrix}.$$

(3.4)

Sequential row subtraction and expansion about the (1, 1) element lead to

$$C(L/M) = \frac{p(\alpha - \gamma)^{M-1}(M-1)!}{\prod_{i=1}^{M-1}(\gamma + L - M + i)_2}$$

$$\times \begin{vmatrix} 1 & \dfrac{\alpha + L - M + 2}{\gamma + L - M + 3} & \cdots & \dfrac{(\alpha + L - M + 2)_{M-2}}{(\gamma + L - M + 3)_{M-2}} \\ 1 & \dfrac{\alpha + L - M + 3}{\gamma + L - M + 4} & \cdots & \dfrac{(\alpha + L - M + 3)_{M-2}}{(\gamma + L - M + 4)_{M-2}} \\ \vdots & \vdots & & \vdots \\ 1 & \dfrac{\alpha + L}{\gamma + L + 1} & \cdots & \dfrac{(\alpha + L)_{M-2}}{(\gamma + L + 1)_{M-2}} \end{vmatrix}.$$

(3.5)

By using this technique recursively,

$$C(L/M) = \frac{p \prod_{i=1}^{M-1}\{(\alpha - \gamma - i + 1)^{M-i}(M-i)!\}}{\prod_{k=1}^{M-1}\prod_{i=k}^{M-1}(\gamma + L - i + 2k - 2)_2}. \qquad (3.6)$$

To calculate the coefficient of z^j in $Q^{[L/M]}(z)$, we need the following determinant in which the dotted lines enclose a deleted column:

2.3 Some explicit forms for Padé denominators

$$(-)^j q_j^{[L/M]} = \begin{vmatrix} \dfrac{(\alpha)_{L-M+1}}{(\gamma)_{L-M+1}} & \cdots & \vdots & \dfrac{(\alpha)_{L-j+1}}{(\gamma)_{L-j+1}} & \vdots & \cdots & \dfrac{(\alpha)_{L+1}}{(\gamma)_{L+1}} \\ \vdots & & \vdots & \vdots & \vdots & & \vdots \\ \dfrac{(\alpha)_L}{(\gamma)_L} & \cdots & \vdots & \dfrac{(\alpha)_{L-M-j}}{(\gamma)_{L-M-j}} & \vdots & \cdots & \dfrac{(\alpha)_{L-M}}{(\gamma)_{L-M}} \end{vmatrix} = p \times$$

$$\begin{vmatrix} 1 & \dfrac{\alpha+L-M+1}{\gamma+L-M+1} & \cdots & \vdots & \dfrac{(\alpha+L-M+1)_{M-j}}{(\gamma+L-M+1)_{M-j}} & \vdots & \cdots & \dfrac{(\alpha+L-M+1)_M}{(\gamma+L-M+1)_M} \\ 1 & \dfrac{\alpha+L-M+2}{\gamma+L-M+2} & \cdots & \vdots & \dfrac{(\alpha+L-M+2)_{M-j}}{(\gamma+L-M+2)_{M-j}} & \vdots & \cdots & \dfrac{(\alpha+L-M+2)_M}{(\gamma+L-M+2)_M} \\ \vdots & \vdots & & \vdots & \vdots & \vdots & & \vdots \\ 1 & \dfrac{\alpha+L}{\gamma+L} & \cdots & \vdots & \dfrac{(\alpha+L)_{M-j}}{(\gamma+L)_{M-j}} & \vdots & \cdots & \dfrac{(\alpha+L)_M}{(\gamma+L)_M} \end{vmatrix}.$$

Making $M - j$ simplifications of the type leading from (3.4) to (3.5), i.e., expansion by top left-hand element after row reduction, we are led to define the product of common factors

$$p_j = p \frac{\prod_{i=1}^{M-j} \{(\alpha - \gamma - i + 1)^{M-i}(M - i + 1)!\}}{(M - j)! \prod_{k=1}^{M-j} \prod_{i=k}^{M-1} (\gamma + L - i + 2k - 2)}, \tag{3.7}$$

and then

$$(-)^j q_j^{[L/M]} = p_j \cdot \begin{vmatrix} \dfrac{\alpha+L-j+1}{\gamma+L+M+1-2j} & \cdots & \dfrac{(\alpha-L-j+1)_j}{(\gamma+L+M+1-2j)_j} \\ \vdots & & \vdots \\ \dfrac{\alpha+L}{\gamma+L+M-j} & \cdots & \dfrac{(\alpha+L)_j}{(\gamma+L-M-j)_j} \end{vmatrix}.$$

Using the same technique of reduction, we obtain

$$(-)^j q_j^{[L/M]} =$$

$$p_j \frac{\prod_{i=1}^{j}\{(\alpha + L - i + 1)(j - i)!\} \prod_{i=M-j+1}^{M}(\alpha - \gamma - i + 1)^{M-i}}{\prod_{i=1}^{j}(\gamma + L + M - j - i + 1) \prod_{k=1}^{j-1}\prod_{i=k}^{j-1}(\gamma + L + M + 2k - i - j - 1)}.$$

$$\tag{3.8}$$

From (3.6), (3.7), and (3.8) there is substantial cancellation, and

$$(-)^j q_j^{[L/M]} = C(L/M) \frac{M!}{(M-j)!} \frac{(\alpha+L)\cdots(\alpha+L-j+1)}{(\gamma+L+M-1)\cdots(\gamma+L+M-j)}.$$

Hence, for $L \geq M - 1$,

$$Q^{[L/M]}(z) = C(L/M) \left\{ 1 + \frac{(-\alpha - L)(-M)z}{(1-\gamma-L-M)1!} \right.$$
$$\left. + \frac{(-\alpha-L)(-\alpha-L+1)(-M)(-M+1)z^2}{(1-\gamma-L-M)(2-\gamma-L-M)2!} + \cdots \right\}$$
$$= C(L/M)\,_2F_1(-M, -\alpha - L, 1 - \gamma - L - M; z), \qquad (3.9)$$

completing the derivation of the Padé denominator. We may derive a corollary from this result, by noting that

$$_1F_1(\beta, \gamma; z) = \lim_{\alpha \to \infty} {}_2F_1\left(\alpha, \beta, \gamma; \frac{z}{\alpha}\right).$$

Consequently the Padé denominators of

$$f(z) = {}_1F_1(1, \gamma; z)$$

are given by

$$B^{[L/M]}(z) = \lim_{\alpha \to \infty} {}_2F_1\left(-M, -\alpha - L, 1 - \gamma - L - M; \frac{z}{\alpha}\right)$$
$$= {}_1F_1(-M, 1 - \gamma - L - M; -z), \qquad L \geq M - 1. \quad (3.10)$$

Furthermore, by choosing $\gamma = 1$, we obtain the special case of the exponential function of Section 1.2.

Likewise, the Padé denominators of the asymptotic expansion of $_2F_0(\alpha, 1; z)$ are given by replacing z with γz and letting $z \to \infty$. We find

$$B^{[L/M]}(z) = {}_2F_0(-M, -\alpha - L; z), \qquad L \geq M - 1.$$

There is another class of formal series, derived from coefficients

$$c_i = \prod_{j=0}^{i-1} \frac{A - q^{j+\alpha}}{B - q^{j+\gamma}}, \qquad i = 0, 1, 2, \ldots, \qquad (3.11)$$

which generate the power series

$$f(z) = 1 + \frac{A - q^\alpha}{B - q^\gamma} z + \frac{(A - q^\alpha)(A - q^{\alpha+1})}{(B - q^\gamma)(B - q^{\gamma+1})} z^2 + \cdots, \qquad (3.12)$$

for which explicit expressions for the $[L/M]$ Padé approximants (with $L \geq M - 1$) can be given. As in (3.4), we construct the determinant $C(L/M)$ by defining

2.3 Some explicit forms for Padé denominators

$$C(L/M) = D(M, L - M + 1 + \alpha, L - M + 1 + \beta) \prod_{j=0}^{L-M} \left\{ \frac{A - q^{j+\alpha}}{B - q^{j+\gamma}} \right\}^M,$$

where

$$D(M, \mu, \nu) = \begin{vmatrix} 1 & \cdots & \prod_{r=0}^{M-1} \frac{A - q^{\mu+r}}{B - q^{\nu+r}} \\ \vdots & & \vdots \\ \prod_{r=0}^{M-1} \frac{A - q^{\mu+r}}{B - q^{\nu+r}} & \cdots & \prod_{r=0}^{2M-2} \frac{A - q^{\mu+r}}{B - q^{\nu+r}} \end{vmatrix}.$$

The determinants may be evaluated recursively from the recursion relation

$$D(M, \mu, \nu)$$
$$= \frac{(q-1)(q^2-1)\cdots(q^{M-1}-1)(Aq^\nu - Cq^\mu)^{M-1}(A-q^\mu)^{M-1}}{(C-q^\nu)(C-q^{\nu+1})\cdots(C-q^{\nu+M-1})(C-q^{\nu+1})^{M-1}}$$
$$\times D(M-1, \mu+1, \nu+2) q^{(M-1)(M-2)/2} / (C - q^\nu)^{M-1}.$$

Hence, using the same methods as for the hypergeometric function (3.2), we may obtain explicit expressions for the Padé approximants.

If the coefficients are given by

$$c'_i = \prod_{j=0}^{i-1} (A - q^{j+\alpha}) \tag{3.13}$$

instead of by (3.11) substitute $w = z/B$ and let $B \to \infty$.

If the coefficients are given by

$$c''_i = \prod_{j=0}^{i-1} (B - q^{j+\gamma})^{-1}, \tag{3.14}$$

substitute $w = Az$ and let $A \to \infty$. Thus results from series generated by (3.13) or (3.14) are special cases of results derived from (3.12). We pursue the direct calculation no further, because the approach based on the Q.D. algorithm (4.4.31, 32) is algebraically simpler.

The convergence of sequences of Padé approximants for $f(z)$ defined by (3.12) is discussed by Driver and Lubinsky [1990, 1991, 1993]. They show that the function $f(z)$ satisfies the identity

$$f(z)(C - zA) = C - q^{\gamma-1} + f(qz)(q^{\gamma-1} - zq^\alpha),$$

and they explain its connections to the generalized or q-hypergeometric function [Gasper and Rahman, 1990]. Determinantal formulas and continued-fraction representations of the q-extensions of (4.6.15) and (4.6.16) are given by Baker, Bessis, and Moussa [1992], who also show that the q-extension is the most general which preserves the contiguous relations (4.6.17), (4.6.21) of the hypergeometric functions. Explicit formulas for Padé approximants of the q-exponential and q-logarithmic functions are given by Borwein [1988]; continued-fraction representations of ratios of q-extensions of multiparameter hypergeometric functions have been found by numerous authors, e.g., Gupta, Ismail, and Masson [1992], Ismail and Rahman [1991], and Wimp [1987].

2.4 Bigradients and Hadamard's formula

Hitherto we have considered the problem of finding rational functions which satisfy

$$f(z) = \sum_{i=0}^{\infty} c_i z^i = \frac{P^{[L/M]}(z)}{Q^{[L/M]}(z)} + O(z^{L+M+1}), \tag{4.1}$$

where c_i are given coefficients and $c_0, c_1, \ldots, c_{L+M}$ are actually needed. A natural generalization of this problem is that of finding polynomials $P^{[L/M]}(z)$, $Q^{[L/M]}(z)$ of orders L, M, respectively, which satisfy

$$f(z) = \frac{g(z)}{d(z)} = \frac{\sum_{i=0}^{\infty} g_i z^i}{\sum_{i=0}^{\infty} d_i z^i} = \frac{P^{[L/M]}(z)}{Q^{[L/M]}(z)} + O(z^{L+M+1}). \tag{4.2}$$

Here g_i, d_i are the given coefficients, of which we need to know $g_0, g_1, \ldots, g_{L+M}$ and $d_0, d_1, \ldots, d_{L+M}$. If $d(z)$ is a polynomial of order m and $m < L + M$, we complete the definition by taking $d_{m+1} = d_{m+2} = \cdots = d_{L+M} = 0$. We assume that $d_0 \neq 0$, and leave the question of what modifications are needed in the presence of degeneracy as an exercise. With these conventions, and in the absence of degeneracy,

$$P^{[L/M]}(z) = - \begin{vmatrix} d_0 & 0 & \cdots & 0 & 0 & \cdots & 0 & g_0 \\ d_1 & d_0 & \cdots & 0 & 0 & \cdots & g_0 & g_1 \\ \vdots & \vdots & \ddots & \vdots & \vdots & \ddots & \vdots & \vdots \\ & & & & g_0 & \cdots & g_{M-1} & g_M \\ d_L & d_{L-1} & \cdots & d_0 & g_1 & \cdots & g_M & g_{M+1} \\ d_{L+1} & d_L & \cdots & d_1 & \vdots & & \vdots & \vdots \\ \vdots & \vdots & & \vdots & & & & \\ d_{L+M} & d_{L+M-1} & \cdots & d_M & g_L & \cdots & g_{L+M-1} & g_{L+M} \\ 1 & z & \cdots & z^L & 0 & \cdots & 0 & 0 \end{vmatrix}, \tag{4.3}$$

2.4 Bigradients and Hadamard's formula

$$Q^{[L/M]}(z) = \begin{vmatrix} d_0 & 0 & \cdots & 0 & 0 & \cdots & 0 & g_0 \\ d_1 & d_0 & & 0 & 0 & & g_0 & g_1 \\ \cdot & \cdot & & \cdot & \cdot & \cdot & \cdot & \cdot \\ \cdot & \cdot & \ddots & \cdot & \cdot & & \cdot & \cdot \\ \cdot & \cdot & & \cdot & g_0 & \cdots & g_{M-1} & g_M \\ d_L & d_{L-1} & \cdots & d_0 & g_1 & \cdots & g_M & g_{M+1} \\ d_{L+1} & d_L & \cdots & d_1 & \cdot & & \cdot & \cdot \\ \cdot & \cdot & & \cdot & \cdot & & \cdot & \cdot \\ \cdot & \cdot & & \cdot & \cdot & & \cdot & \cdot \\ d_{L+M} & d_{L+M-1} & \cdots & d_M & g_L & \cdots & g_{L+M-1} & g_{L+M} \\ 0 & 0 & \cdots & 0 & z^M & \cdots & z & 1 \end{vmatrix}, \quad (4.4)$$

We may verify that the proposed solution of (4.3) and (4.4) satisfies (4.2) by forming

$$g(z)Q^{[L/M]}(z) - d(z)P^{[L/M]}(z) \equiv E(z) =$$

$$\begin{vmatrix} d_0 & 0 & \cdots & 0 & 0 & \cdots & 0 & g_0 \\ d_1 & d_0 & & 0 & 0 & & g_0 & g_1 \\ \cdot & \cdot & & \cdot & \cdot & \cdot & \cdot & \cdot \\ \cdot & \cdot & \ddots & \cdot & \cdot & & \cdot & \cdot \\ \cdot & \cdot & & \cdot & g_0 & \cdots & g_{M-1} & g_M \\ d_L & d_{L-1} & \cdots & d_0 & \cdot & & \cdot & \cdot \\ \cdot & \cdot & & \cdot & \cdot & & \cdot & \cdot \\ \cdot & \cdot & & \cdot & \cdot & & \cdot & \cdot \\ d_{L+M} & d_{L+M-1} & \cdots & d_M & g_L & \cdots & g_{L+M-1} & g_{L+M} \\ d(z) & zd(z) & \cdots & z^L d(z) & z^M g(z) & \cdots & zg(z) & g(z) \end{vmatrix}, \quad (4.5)$$

By subtraction of the first row of (4.5) from its last row, followed by subtraction of z times the second row from the new last row, etc., we find that

$$E(z) = \sum_{k=L+M+1}^{\infty} z^k \times \begin{vmatrix} d_0 & 0 & \cdots & 0 & 0 & \cdots & g_0 \\ d_1 & \cdot & & 0 & 0 & & \cdot \\ \cdot & & \ddots & \cdot & \cdot & \cdot & \cdot \\ \cdot & & & \cdot & g_0 & \cdots & g_M \\ d_L & \cdot & \cdots & d_0 & \cdot & & \cdot \\ \cdot & \cdot & & \cdot & \cdot & & \cdot \\ \cdot & \cdot & & \cdot & \cdot & & \cdot \\ d_{L+M} & \cdot & \cdots & d_M & g_L & \cdots & g_{L+M} \\ d_k & \cdot & \cdots & d_{k-L} & g_{k-M} & \cdots & g_k \end{vmatrix},$$

which proves (4.2).

The determinants in (4.3) and (4.4) are called polynomial bigradients. An ordinary bigradient $\Delta_{L,M}(d_i, g_i)$ is an $(L+M) \times (L+M)$ determinant formed from the coefficients d_i arranged in its first L columns with a negative gradient and from the coefficients g_i arranged in the next M

columns with a positive gradient. Specifically, define

$$\Delta_{L,M}(d_i, g_i) = \begin{vmatrix} d_0 & 0 & \cdots & 0 & 0 & \cdots & 0 & g_0 \\ d_1 & d_0 & \cdots & 0 & 0 & \cdots & g_0 & g_1 \\ \vdots & \vdots & \ddots & \vdots & \vdots & \iddots & \vdots & \vdots \\ & & & & g_0 & \cdots & g_{M-2} & g_{M-1} \\ d_{L-1} & d_{L-2} & \cdots & d_0 & g_1 & \cdots & g_{M-1} & g_M \\ d_L & d_{L-1} & \cdots & d_1 & \cdot & & \cdot & \cdot \\ \vdots & \vdots & & \vdots & \vdots & & \vdots & \vdots \\ d_{L+M-1} & d_{L+M-2} & \cdots & d_M & g_L & \cdots & g_{L+M-2} & g_{L+M-1} \end{vmatrix}$$

(4.6)

As a preliminary to proving Hadamard's theorem, we define

$$e(z) = \frac{1}{d(z)} = \sum_{i=0}^{\infty} e_i z^i.$$

From (4.2), it follows that

$$e(z)g(z) = f(z).$$

Consequently,

$$\sum_{j=0}^{i} e_j d_{i-j} = \delta_{i0}, \quad i = 0, 1, 2, \ldots,$$

and

$$\sum_{j=0}^{i} e_j g_{i-j} = c_i, \quad i = 0, 1, 2, \ldots.$$

These equations justify the following identities:

$$\begin{bmatrix} e_0 & 0 & 0 \\ e_1 & e_0 & 0 \\ e_2 & e_1 & e_0 \end{bmatrix} \begin{bmatrix} d_0 & 0 & 0 \\ d_1 & d_0 & 0 \\ d_2 & d_1 & d_0 \end{bmatrix} = \begin{bmatrix} 1 & 0 & 0 \\ 0 & 1 & 0 \\ 0 & 0 & 1 \end{bmatrix} \quad (4.7)$$

and

$$\begin{bmatrix} e_0 & 0 & 0 \\ e_1 & e_0 & 0 \\ e_2 & e_1 & e_0 \end{bmatrix} \begin{bmatrix} 0 & 0 & g_0 \\ 0 & g_0 & g_1 \\ g_0 & g_1 & g_2 \end{bmatrix} = \begin{bmatrix} 0 & 0 & c_0 \\ 0 & c_0 & c_1 \\ c_0 & c_1 & c_2 \end{bmatrix}. \quad (4.8)$$

Now let us consider the $(L + M) \times (L + M)$ matrix

2.4 Bigradients and Hadamard's formula

$$E = \begin{bmatrix} e_0 & & & 0 \\ e_1 & e_0 & & \\ \vdots & & \ddots & \\ e_{L+M-1} & e_{L+M-2} & \cdots & e_0 \end{bmatrix},$$

for which $\det E = e_0^{L+M}$. Using the matrix E as a left multiplier for the bigradient matrix, and regarding (4.7) and (4.8) as truncations of large-matrix operations, we find that

$$e_0^{L+M}\Delta_{L,M}(d_i, g_i) = \begin{vmatrix} 1 & 0 & \cdots & 0 & 0 & \cdots & 0 & c_0 \\ 0 & 1 & \cdots & 0 & 0 & \cdots & c_0 & c_1 \\ \cdot & & \ddots & \cdot & \cdot & \ddots & \cdot & \cdot \\ \cdot & & & \cdot & \cdot & \ddots & \cdot & \cdot \\ \cdot & & & & c_0 & \cdots & c_{M-2} & c_{M-1} \\ 0 & 0 & \cdots & 1 & c_1 & \cdots & c_{M-1} & c_M \\ 0 & 0 & \cdots & 0 & \cdot & & \cdot & \cdot \\ \cdot & \cdot & & \cdot & \cdot & & \cdot & \cdot \\ \cdot & \cdot & & \cdot & \cdot & & \cdot & \cdot \\ 0 & 0 & \cdots & 0 & c_L & \cdots & c_{L+M-2} & c_{L+M-1} \end{vmatrix}.$$

Hence $\Delta_{L,M}(d_i, g_i) = d_0^{L+M} C(L/M)$, with the Hankel determinant defined by (1.4.8). This result enables us to prove Hadamard's theorem.

Theorem 2.4.1. *Let* $f(z) = \sum_{i=0}^{\infty} c_i z^i$, *and let* $C(L/M)$ *be its Hankel determinant defined in the usual way by* (1.4.8). *Let* $\{f(z)\}^{-1} = \sum_{i=0}^{\infty} c_i' z^i$, *and let* $C'(l/m)$ *be its Hankel determinant. Then*

$$C'(M/L) = (-1)^{(L+M)(L+M-1)/2} C(L/M) c_0^{-(L+M)}. \tag{4.9}$$

Proof. For the bigradient (4.6) relating to (4.1) and (4.2) we have established that

$$\Delta_{L,M}(d_i, g_i) = d_0^{L+M} C(L/M).$$

Had we started with the totally compatible *ansatz*

$$\{f(z)\}^{-1} = \frac{d(z)}{g(z)} = \sum_{i=0}^{\infty} c_i' z^i \tag{4.10}$$

we would have constructed the bigradient

$$\Delta'_{M,L}(g_i, d_i) = g_0^{L+M} C'(M/L).$$

The two bigradients are, in fact, the same except for the order of the columns, and careful inspection shows that $\frac{1}{2}(L + M)(L + M - 1)$ column interchanges are needed to identify the bigradients. Hence

$$C'(M/L) = (-1)^{(L+M)(L+M-1)/2} \left(\frac{d_0}{g_0}\right)^{L+M} C(L/M).$$

Since $c_0 = g_0/d_0$, (4.9) is proved. By virtue of the algebraic nature of the result, (4.9) holds good whenever it is well defined, which is whenever c_0, g_0, and d_0 are nonzero.

Trudi's theorem is probably the best-known result which uses bigradients explicitly. It also identifies the nature of degeneracies encountered with bigradients.

Theorem 2.4.2 [Trudi, 1862]. *Let $g(z)$ be a polynomial of degree l, and let $d(z)$ be a polynomial of degree m. Define the bigradient $\Delta_{l,m}(d_i, g_i)$ as in (4.6). If*

$$\Delta_{l,m}(d_i, g_i) = \Delta_{l-1,m-1}(d_i, g_i) = \cdots = \Delta_{l-j+1,m-j+1}(d_i, g_i) = 0$$

and

$$\Delta_{l-j,m-j}(d_i, g_i) \neq 0,$$

then $d(z)$ and $g(z)$ have a common polynomial divisor of order j.

Proof. If $\Delta_{l,m}(d_i, g_i) = 0$, then $C(l/m) = 0$ and the Padé equations for $f(z) = g(z)/d(z)$ are degenerate. However, $C(l - j/m - j) \neq 0$, and so a nondegenerate $[l - j/m - j]$ Padé approximant of $f(z)$ exists. By the theory of Section 1.4, this Padé approximant is also the $[l/m]$ Padé approximant of $f(z)$, and so the theorem follows.

For further properties of bigradients, we refer to Householder [1970, 1971], Householder and Stewart [1969], and Padé [1900].

3

Padé approximants and numerical methods

3.1 Aitken's Δ^2 method as [L/1] Padé approximants

One of the best-known and simplest techniques of accelerating the convergence of a sequence is Aitken's Δ^2 method. Given a sequence of real or complex numbers,

$$\mathscr{S} = \{S_n, n = 0, 1, 2, \ldots\}, \tag{1.1}$$

such that $S_n \to S$ as $n \to \infty$, the problem is to find a new sequence which converges faster to S.

Define

$$\Delta S_n = S_{n+1} - S_n,$$

$$\Delta^2 S_n = \Delta(\Delta S_n) = S_{n+2} - 2S_{n+1} + S_n,$$

which are the usual forward differences, and the new sequence

$$\mathscr{T} = \{T_n, n = 0, 1, 2, \ldots\},$$

where

$$T_n = S_n - \frac{(\Delta S_n)^2}{\Delta^2 S_n}. \tag{1.2}$$

It is clear from (1.2) why Aitken's method is called the Δ^2 method. There are many reasons for expecting in general that $\mathscr{T}$ converges to a limit, that this limit is S, and that convergence has been accelerated. But we must add an early word of caution: Aitken's method does not work for any arbitrary convergent sequence $\mathscr{S}$. Like all algorithms of numerical analysis, Aitken's method has its own domain of validity, and in certain circumstances it should not be used. An important example is where all the S_n are identical, so that the T_n are undefined. A more insidious example is the one in which the S_n are equal up to rounding errors, so that the T_n are meaningless noise. This is a notorious situation to beware of. But, in general, the method is safe if it is empirically convergent, and it has wide applicability.

Basically, Aitken's method [1926] is designed to treat sequences with geometric convergence. Suppose that

$$S_n = S - a\alpha^n \tag{1.3}$$

with $a \neq 0$ and $|\alpha| < 1$. Then

$$\Delta S_n = a\alpha^n(1 - \alpha),$$

$$\Delta^2 S_n = -a\alpha^n(1 - \alpha)^2,$$

and from (1.2)

$$T_n = (S - a\alpha^n) + (a\alpha^n) = S. \tag{1.4}$$

We see that in this case. Aitken's method yields the exact answer at every stage. More generally, for a sequence $\mathscr{S}$ which is dominated by one geometrically convergent component, we expect that Aitken's method accelerates convergence by 'taking out the geometrically convergent part'.

As a practical example, we consider the numerical evaluation of

$$S = \int_0^1 x^{-1/2}(1-x)^{-1/2} \exp x \, dx. \tag{1.5}$$

The integrand of (1.5) is infinite at the end points, but the integral is well defined. We define S_n as the value of the integral obtained by using 2^n equally spaced integration points. These Riemann sums, obtained by doubling the number of integration points at each successive evaluation, converge to S and are obtained with great ease. It turns out that Aitken's algorithm is a very effective technique of estimating S.

The connection between Padé approximants and Aitken's Δ^2 method is made, as in Section 1.3, by using the series derived from the sequence. Define

$$c_{n+1} = \Delta S_n = S_{n+1} - S_n, \quad n = 0, 1, 2, \ldots.$$
$$c_0 = S_0. \tag{1.6}$$

It follows that S_n are the partial sums of the series, and of course the series converges to S. We form the power series

$$f(z) = \sum_{i=0}^{\infty} c_i z^i. \tag{1.7}$$

Remember that formal power series have a radius of convergence which may be zero, finite, or infinite. We wish to evaluate $f(1) = S$. The

3.1 Aitken's Δ^2 method as [L/1] Padé approximants

method of finding $f(1)$ using the second row of the Padé table is to evaluate

$$[L/1]_f(1), \quad L = 0, 1, 2, \ldots, \tag{1.8}$$

and to determine the limit as $L \to \infty$.

From (1.1.8)–(1.1.10)

$$[L/1]_f(1) = \begin{vmatrix} c_L & c_{L+1} \\ \sum_{i=0}^{L-1} c_i & \sum_{i=0}^{L} c_i \end{vmatrix} \div \begin{vmatrix} c_L & c_{L+1} \\ 1 & 1 \end{vmatrix}$$

$$= \frac{(S_L - S_{L-1})S_L - (S_{L+1} - S_L)S_{L-1}}{(S_L - S_{L-1}) - (S_{L+1} - S_L)}$$

$$= \frac{S_{L-1}(S_{L+1} - 2S_L + S_{L-1}) - (S_L - S_{L-1})^2}{S_{L+1} - 2S_L + S_{L-1}}$$

$$= S_{L-1} - \frac{(\Delta S_{L-1})^2}{\Delta^2 S_{L-1}}.$$

which agrees with (1.2) and shows Aitken's method to be the equivalent of using [L/1] Padé approximants. An even more rapid proof of this result is given by taking $M = 1$ in (1.3.8).

A few sequences of numerical analysis are of the special type

$$S_{n+1} = f(S_n). \tag{1.9}$$

The function f is called a one-point iteration function in this context [Traub, 1964]. For example, the geometric sequence

$$S_n = S - a\alpha^n$$

in (1.3) is generated by

$$S_{n+1} = S - a\alpha^{n+1}$$
$$= S + \alpha(S_n - S)$$
$$= f(S_n)$$

with the identification $f(z) = S + \alpha(z - S)$. We see that in this case we have a convergent sequence when $\alpha = f'(S)$ satisfies $|\alpha| < 1$. Further, the geometric sequence corresponds to $f(z)$ being linear. If $\alpha = 1$, then $S_n = S - a$, and Aitken's method is inapplicable in this situation. We get further confirmation of the power of Aitken's method and so also of the [L/1] Padé method from the following theorem.

Theorem 3.1.1 [Henrici, 1964]. *Let $S_{n+1} = f(S_n)$ define a convergent real sequence with limit S, let $f(x)$ be twice differentiable at S, and let $f'(S) \neq 1$. Then, with the definition (1.2),*

$$T_n - S = O((S_n - S)^2).$$

Proof

$$\begin{aligned} S_{n+1} - S_n &= f(S_n) - S_n \\ &= f(S) + (S_n - S)f'(S) + \tfrac{1}{2}(S - S_n)^2 f''(\xi_n) - S_n \end{aligned} \quad (1.10)$$

for some ξ_n lying between S and S_n. Continuity of $f(x)$ and convergence of the sequence imply that $f(S) = S$. Hence (1.10) may be written as

$$\Delta S_n = A(S_n - S) + O((S - S_n)^2), \qquad (1.11)$$

where

$$A = f'(S) - 1 \neq 0.$$

Similarly, from (1.10) and (1.11),

$$\Delta S_{n+1} = A(S_{n+1} - S) + O((S - S_n)^2). \qquad (1.12)$$

Therefore from (1.11) and (1.12),

$$\Delta^2 S_n = A \Delta S_n (1 + O(S - S_n)^2).$$

Hence

$$\begin{aligned} T_n &= S_n - \frac{(\Delta S_n)\{A(S_n - S) + O((S - S_n)^2)\}}{A \Delta S_n (1 + O(S - S_n)^2)} \\ &= S_n - (S_n - S) + O((S - S_n)^2) \\ &= S + O((S - S_n)^2), \end{aligned}$$

so proving the theorem.

The previous theorem makes quite precise the statement that Aitken's method and the $[L/1]$ Padé method accelerate convergence of a sequence dominated by a geometrically convergent component, of the type given by (1.9). In the next section we turn our attention to generalizing these basic ideas. For further details of the general theory, we refer to Brezinski [1977].

If both sequences $\mathcal{S}$ and $\mathcal{T}$ given by (1.1) and (1.2) converge, then they converge to the same limit [Lubkin, 1952]. For an account of recent progress in convergence theory, we refer to Cordellier [1979a], Germain-Bonne [1979], Delahaye [1988], and Brezinski and Redivo-Zaglia [1991].

Example. Consider the series

$$\sum_{i=0}^{\infty} c_i = \frac{1}{2} + \frac{1}{3} - \frac{5}{6} + \frac{1}{4} + \frac{1}{5} - \frac{9}{20} + \cdots,$$

where

$$c_{3m-3} = \frac{1}{2m}, \quad c_{3m-2} = \frac{1}{2m+1}, \quad \text{and} \quad c_{3m-1} = -\frac{4m+1}{4m^2+2m}$$

for $m = 1, 2, 3, \ldots$. Define $S_n = \sum_{i=0}^{n} c_i$ and T_n by (1.2). Then we have the following convergence and divergence results [Marx, 1963]:

(i) $S_{3m-1} = 0$ for $m = 1, 2, 3, \ldots$,
(ii) $S_n \to 0$ as $n \to \infty$,
(iii) $T_{3m} \to 0$, $T_{3m-1} \to 1$, and $T_{3m-2} \to 0$ as $m \to \infty$.

Notice the implication that $\{S_n\}$ converges, yet $\{T_n\}$ diverges by oscillation.

3.2 Acceleration and overacceleration of convergence

It is natural to ask how the accelerated sequence derived from Aitken's Δ^2 method may be improved upon. The natural answer is to iterate Aitken's method. This answer is not entirely satisfactory, because of the lack of justification based on principle, as the following remarks will make clear.

Aitken's scheme works well if the original sequence converges geometrically; the accelerated sequence takes full account of the dominant terms in the original sequence, and one should wonder what is the reason for accelerating again. Let us suppose that the original sequence is a geometric sequence rounded to given accuracy. Then the accelerated sequence, according to (1.4), is the limit but contains small rounding errors. Further acceleration by Aitken's method (and the Padé method for that matter) requires differencing, and consequently the results depend entirely on rounding errors in the original sequence. Thus, some sort of theoretical basis or an empirical numerical criterion is an essential prerequisite before iterating acceleration schemes. It is all too tempting to try to extract too much information by acclerating a few terms of a sequence too fast.

Consider the partial sums

$$S_n = \sum_{r=0}^{n} (1 + \varepsilon_r)(0.5)^r, \quad n = 1, 2, 3, \ldots.$$

where the numbers $\{\varepsilon_r\}$ represent floating-point rounding errors with $|\varepsilon_i| \le \varepsilon$. We consider estimation of the quantity S_∞ using the Padé-approximant method by forming diagonal approximants to

$$f(z) = \sum_{r=0}^{\infty}(1 + \varepsilon_r)z^r$$

and evaluating the approximants at $z = 0.5$. (Note the trivial variation on the method described in Sections 3.1 and 1.1, where the approximants are evaluated at $z = 1$.) Working to first order in ε in (1.1.8) and (1.1.9), we find the error bounds

$$[1/1] = 2 \pm 7\varepsilon,$$

showing that this approximant is sensitive, but not unduly sensitive, to rounding errors in the data coefficients. However, when we come to consider the [2/2] Padé approximant, we find that

$$Q^{[2/2]}(z) = z^2(\varepsilon_2 + \varepsilon_4 - 2\varepsilon_3) - z(\varepsilon_1 + \varepsilon_4 - \varepsilon_2 - \varepsilon_3) + (\varepsilon_1 + \varepsilon_3 - 2\varepsilon_2)$$

to first order. We see that the value of $Q^{[2/2]}(z)$ is completely controlled by rounding error in this case; the zeros of $Q^{[2/2]}(z)$, which are the poles of the approximant, are distributed all over the complex plane. (We do not suggest that the distribution is random, and we would expect more zeros of $Q^{[2/2]}(z)$ near $|z| = 1$ than near $z = 0$, for example.) Since the value of the [2/2] Padé approximant depends primarily on rounding error, whereas the [1/1] approximant is accurate within errors, use of the [2/2] approximant is an example of overacceleration of convergence. In this case, the moral is that we should use the lower-order approximant.

The Padé method has the following interpretation (among others): the given sequence has a certain number of geometric components which dominate. Let us suppose that

$$S_n = a\sum_{r=0}^{n}\alpha^r + b\sum_{r=0}^{n}\beta^r + \text{much smaller terms} \tag{2.1}$$

with $a \ne 0$, $b \ne 0$, $\alpha \ne \beta$, $|\alpha| < 1$, $|\beta| < 1$, and $n = 0, 1, 2, \ldots$. This expression may be rewritten as

$$S_n = \frac{a}{1-\alpha}(1 - \alpha^{n+1}) + \frac{b}{1-\beta}(1 - \beta^{n+1}) + \text{much smaller terms.} \tag{2.2}$$

Note that S_n is derived from the series

$$c_0 \equiv S_0 = a + b + \text{a much smaller term},$$

$$c_n \equiv S_n - S_{n-1} = a\alpha^n + b\beta^n + \text{much smaller terms}, \quad n = 1, 2, 3, \ldots.$$

$$\tag{2.3}$$

The third row of the Padé table takes account of the explicit leading terms in (2.1), (2.2), and (2.3), whereas direct calculation shows that the once-iterated Aitken method does not. We assumed in (2.1) that $a \neq 0$ and $b \neq 0$, so that there are genuinely two geometric components which dominate the remainder, and this assumption is crucial. We are a bit vague about the size of the remainder terms, so as not to prejudice the development, and to admit possibilities such as $c_n = \alpha^n + (-\alpha)^n$ for which the odd terms vanish. We assume that $|\alpha| < 1$ and $|\beta| < 1$ which is conventional but not entirely necessary. The Padé method does make sense of well-posed problems with divergent sequences, such as (2.1) with $|\alpha| > 1$ or $|\beta| > 1$. If $\alpha = 1$ or $\beta = 1$, corresponding to a divergent sequence with one component derived from an arithmetic progression, the Padé method gives $S = \infty$ with an obvious interpretation.

As in Section 1.3, to justify the Padé method, we form the function

$$f(z) \equiv \sum_{r=0}^{\infty} c_r z^r = \frac{a}{1 - \alpha z} + \frac{b}{1 - \beta z} + \text{correction terms}, \quad (2.4)$$

which we wish to evaluate at $z = 1$. Formation of $[L/2]$ approximants is suggested by the explicit dominant terms of (2.4), and we apply de Montessus's theorem. Borrowing from Section 6.2, we quote the theorem in context here.

Let $R > |\alpha|^{-1}$ and $R > |\beta|^{-1}$. We assume that the remainder terms in (2.3) are small, and explicitly we require that $c_n = o(R^{-n})$. Hence, $f(z)$ is meromorphic with precisely two poles within $|z| < R$. Since $|\alpha| > 1$, then $R > 1$. Now de Montessus's theorem asserts that $[L/2]$ approximants converge to $f(z)$ at $z = 1$, which is not a pole of $f(z)$, by assumption. Hence, the $[L/2]$ approximants converge for sequences such as (2.1) or series such as (2.3), with the stated hypothesis about the residuals.

3.3 The ε-algorithm and the η-algorithm

In this section we describe the ε-algorithm for sequence transformations and show that one of its columns is the sequence of Aitken's Δ^2 method. Then we describe the η-algorithm, which is the corresponding algorithm for series transformations.

The ε-algorithm originates with Shanks [1955] and Wynn [1956]. It involves the two-dimensional array called the ε-table (Table 3.3.1). The subscript k of $\varepsilon_k^{(j)}$ denotes the column, and the superscript j measures the progression down the column. The table is constructed iteratively from its first two columns. Define $\varepsilon_{-1}^{(j)}$ to be zero and $\varepsilon_0^{(j)}$ to be the given sequence,

Table 3.3.1. *The ε-table*.

$\varepsilon_{-1}^{(0)}$				
	$\varepsilon_0^{(0)}$			
$\varepsilon_{-1}^{(1)}$				
	$\varepsilon_0^{(1)}$	$\varepsilon_1^{(0)}$		
$\varepsilon_{-1}^{(1)}$				$\varepsilon_2^{(0)}$
	$\varepsilon_0^{(1)}$	$\varepsilon_1^{(1)}$		
$\varepsilon_{-1}^{(3)}$			⋮	⋮
⋮	⋮			

for $j = 0, 1, 2, \ldots$. Then all the other elements may be calculated from the ε-algorithm, which is

$$\varepsilon_{k+1}^{(j)} = \varepsilon_{k-1}^{(j+1)} + [\varepsilon_k^{(j+1)} - \varepsilon_k^{(j)}]^{-1}. \tag{3.1}$$

To see more clearly how this rule should be applied, we note that it connects the elements in the rhombus pattern of Figure 3.3.1, which shows how the right-hand member $\varepsilon_{k+1}^{(j)}$ is derived from the other three members. It is now plain that the ε-algorithm allows the whole ε-table to be calculated. It is further plain that if $\varepsilon_k^{(j)} = \varepsilon_k^{(j+1)}$, i.e., two successive members of the same column are equal, $\varepsilon_{k+1}^{(j)}$ does not exist. We assume, unless explicitly stated otherwise, that all elements exist. Otherwise the table is said to be degenerate. We will show that the sequence of the fourth column, namely, $\{\varepsilon_2^{(j)}, j = 0, 1, 2, \ldots\}$, is the same as that obtained from Aitken's Δ^2 rule.

From (3.1),

$$\varepsilon_1^{(j)} = [\varepsilon_0^{(j+1)} - \varepsilon_0^{(j)}]^{-1}$$
$$= [S_{j+1} - S_j]^{-1} = [\Delta S_j]^{-1}.$$

Again from (3.1),

$$\varepsilon_2^{(j)} = \varepsilon_0^{(j+1)} + [\varepsilon_1^{(j+1)} - \varepsilon_1^{(j)}]^{-1}$$
$$= S_{j+1} + \frac{1}{[\Delta S_{j+1}]^{-1} - [\Delta S_j]^{-1}}$$
$$= S_j + \Delta S_j + \frac{\Delta S_j \Delta S_{j+1}}{\Delta S_j - \Delta S_{j+1}}$$
$$= S_j - \frac{(\Delta S_j)^2}{\Delta^2 S_j}, \quad \text{where } \Delta^2 S_j = \Delta S_{j+1} - \Delta S_j.$$

This formula is precisely Aitken's Δ^2 method (1.2) applied to the sequence $\{S_j, j = 0, 1, 2, \ldots\}$, and is also the result of using the second row of the Padé table as a Padé method for sequence acceleration.

3.3 The ε-algorithm and the η-algorithm

$$\varepsilon_{k-1}^{(j+1)} \quad \varepsilon_k^{(j)} \quad \varepsilon_{k+1}^{(j)}$$
$$\varepsilon_k^{(j+1)}$$

Figure 3.3.1. A rhombus pattern.

After we have established Wynn's identity in Section 3.4, we then show in Section 3.5 that the ε-table and the Padé table are identified by the formula

$$\varepsilon_{2k}^{(j)} = [k + j/k]_f(1).$$

What we have just achieved is the proof of this result for $k = 1$ and $j = 0, 1, 2, \ldots$.

Example 1. We consider Gregory's notoriously slowly convergent series for π.

$$\pi = 4 - \tfrac{4}{3} + \tfrac{4}{5} - \tfrac{4}{7} + \cdots.$$

In Table 3.3.2, we exhibit the even columns of the ε-table for this series. The first column seems scarcely convergent, whereas the correctness of the final extrapolations is instantly recognizable.

Example 2. We consider a familiar divergent series, namely, the one given by the Maclaurin series of $\ln(1 + z)$ with $z = 2$:

$$\ln 3 = 2 - 2 + \frac{8}{3} - 4 + \cdots.$$

This has a remarkable ε-table, with even columns given by Table 3.3.3. For comparison, $\ln 3 = 1.098612\ldots$. We see, by example, that the ε-algorithm may be used to sum divergent series.

Table 3.3.2. *Even columns of an ε-table for π.*

$n =$ 0	4.0000000				
1	2.6666667	3.1666667			
2	3.4666667	3.1333333	3.1423423		
3	2.8952381	3.1452381	3.1413919	3.1416149	
4	3.3396825	3.1396825	3.1416627	3.1415873	3.1415933
5	2.9760462	3.1427129	3.1415634	3.1415943	3.1415925
6	3.2837385	3.1408813	3.1416065	3.1415921	
7	3.0170718	3.1420718	3.1415854		
8	3.2523659	3.1412548			
9	3.0418396				

Table 3.3.3. *Even columns of an ε-table for ln 3.*

n = 0	.000000				
1	2.000000	1.000000			
2	.000000	1.142857	1.090909		
3	2.666667	1.066667	1.101449	1.098039	
4	−1.333333	1.128205	1.097046	1.098805	1.098570
5	5.066667	1.066667	1.099725	1.098521	1.098626
6	−5.600000	1.136842	1.097674	1.098667	
7	12.685714	1.049351	1.099507		
8	−19.314286	1.165714			
9	37.574603				

As an amusing paradox, we briefly mention Hardy's puzzle, which has an entirely straightforward solution. Let

$$f(z) = 1 + \frac{1}{2}\left(\frac{2z}{1+z^2}\right)^2 + \frac{1}{2}\cdot\frac{3}{4}\cdot\left(\frac{2z}{1+z^2}\right)^4 + \cdots. \quad (3.2)$$

Consider the domains $\mathcal{D}$ and $\mathcal{E}$ defined by

$$z \in \mathcal{D} \quad \text{provided } |z - i| < \sqrt{2} \text{ and } |z + i| < \sqrt{2},$$
$$z \in \mathcal{E} \quad \text{provided } |z - i| > \sqrt{2} \text{ and } |z + i| > \sqrt{2}.$$

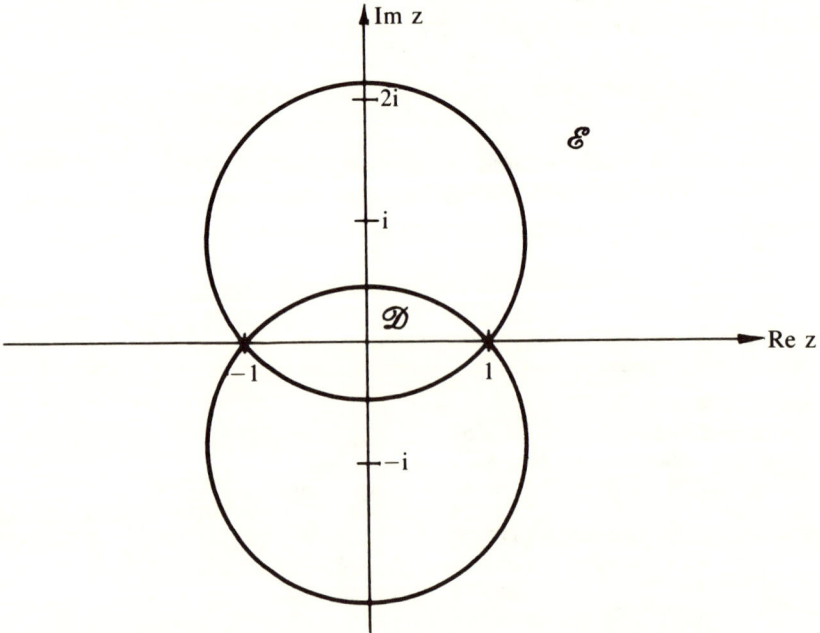

Figure 3.3.2. The z-plane, showing the domains $\mathcal{D}$ and $\mathcal{E}$.

3.3 The ε-algorithm and the η-algorithm

Define

$$g(z) = \frac{1+z^2}{1-z^2}.$$

Notice that (3.2) is a binomial series which converges if $|2z| < |1+z^2|$. The boundary of convergence is given by

$$(zz^* + iz^* - iz - 1)(zz^* + iz - iz^* - 1) = 0$$

showing that (3.2) converges for $z \in \mathscr{D}$ and $z \in \mathscr{E}$. Therefore

$$f(z) = \frac{1+z^2}{1-z^2} = g(z) \quad \text{if } z \in \mathscr{D}$$

and

$$f(z) = \frac{z^2+1}{z^2-1} = -g(z) \quad \text{if } z \in \mathscr{E}.$$

The new problem is to decide what result is given by the ε-algorithm for (3.2) with $z = 2i$. Does it converge to $g(2i) = -\frac{3}{5}$ or $-g(2i) = +\frac{3}{5}$?

At this stage, since the expansion parameter is invariant under $z \to 1/z$, the correct answer should seem clear, and it can be empirically verified by calculating $\varepsilon_2^{(0)}$, which is quite close. Proof of the veracity of this solution requires the technique of the theorem of Baker, Gammel, and Wills in Section 6.7.

As a final remark on the ε-algorithm, we note the possibility that the sequence $\mathscr{S} = \{\varepsilon_k^{(0)}, k = 0, 1, 2, \ldots\}$ does not converge, and yet an iteration of the algorithm using the derived sequence $\mathscr{S}$ as a new initializing sequence $\{\varepsilon_0'^{(J)}, J = 0, 1, 2, \ldots\}$ may lead to a convergent sequence $\{\varepsilon_0'^{(0)}\}$, e.g., as in Table 10.1.1. But no theoretical justification for iterating the ε-algorithm is known yet.

The ε-algorithm may be regarded as a sequence-to-sequence transformation: it is an algorithm for transforming the elements of the given series in the second column to the elements on the principal diagonal. Specifically, the sequence $\{\varepsilon_0^{(j)}, j = 0, 1, 2, \ldots\}$ is transformed to a new sequence $\{\varepsilon_k^{(0)}, k = 0, 1, 2, \ldots\}$. Bauer's η-algorithm is the equivalent series-to-series transformation.

The η-algorithm [Bauer, 1959]. The series c_i, $i = 0, 1, 2, \ldots$, is given. The η-algorithm is initialized by assigning these values to the second column of the η-table:

$$\eta_0^{(i)} = c_i, \quad i = 0, 1, 2, \ldots. \tag{3.3}$$

$$\eta_{-1}^{(0)}$$
$$\eta_0^{(0)}$$
$$\eta_{-1}^{(1)} \quad \eta_1^{(0)}$$
$$\eta_0^{(1)} \quad \eta_2^{(0)}$$
$$\eta_{-1}^{(2)} \quad \eta_1^{(1)} \quad \eta_3^{(0)}$$
$$\eta_0^{(2)} \quad \eta_2^{(1)}$$
$$\eta_{-1}^{(3)} \quad \eta_1^{(2)}$$
$$\eta_0^{(3)}$$
$$\eta_{-1}^{(4)} \quad \vdots$$
$$\vdots$$

Figure 3.3.3. The η-table.

The elements of the first column of the η-table are defined by the artificial values

$$\eta_{-1}^{(i)} = \infty, \quad i = 0, 1, 2, \ldots . \tag{3.4}$$

The recursion scheme is defined by

$$\frac{1}{\eta_{2k+1}^{(i)}} = \frac{1}{\eta_{2k-1}^{(i+1)}} + \frac{1}{\eta_{2k}^{(i+1)}} - \frac{1}{\eta_{2k}^{(i)}} \tag{3.5}$$

and

$$\eta_{2k}^{(i)} = \eta_{2k-2}^{(i+1)} + \eta_{2k-1}^{(i+1)} - \eta_{2k-1}^{(i)}. \tag{3.6}$$

Equations (3.5), (3.6) are rhombus rules connecting the entries shown in Figure 3.3.4. These equations enable the rightmost entries of the rhombus to be calculated; hence the entire η-table of Figure 3.3.3 can be constructed from its first two columns given by (3.3), (3.4). With these definitions (3.3)–(3.6), the η-algorithm defines a transformation of the series $c_i = \eta_0^{(i)}$, $i = 0, 1, 2, \ldots$, to a new series $c'_k = \eta_k^{(0)}$, $k = 0, 1, 2, \ldots$. One purpose of this algorithm is the construction from a convergent series $\sum_{i=0}^{\infty} c_i$ of a new series $\sum_{k=0}^{\infty} c'_k$ which converges faster to the same limit. As an empirical example of this we consider Gregory's series again:

$$\frac{\pi}{4} = 1 - \frac{1}{3} + \frac{1}{5} - \frac{1}{7} + \frac{1}{9} - \cdots . \tag{3.7}$$

Example 3. The second column of the η-table is constructed from (3.3) using the terms of the series (3.7). The first column is artificial, and the other columns are constructed from the rhombus rules (3.5) and (3.6). From Table 3.3.4 we find that the series (3.7) has been transformed to

$$\frac{\pi}{4} = 1 - \frac{1}{4} + \frac{1}{24} - \frac{1}{136} + \frac{4}{3145} - \cdots ; \tag{3.8}$$

3.3 The ε-algorithm and the η-algorithm

Table 3.3.4. *The η-table for Gregory's series.*

		1				
∞			$-\frac{1}{4}$			
	$-\frac{1}{3}$			$\frac{1}{24}$		
∞		$\frac{1}{8}$			$-\frac{1}{136}$	
	$\frac{1}{5}$		$-\frac{1}{120}$			$\frac{4}{3145}$
∞		$-\frac{1}{12}$		$\frac{1}{444}$		⋮
	$-\frac{1}{7}$		$\frac{1}{336}$		⋮	⋯
∞		$\frac{1}{16}$		⋮		
	$\frac{1}{9}$		⋮			
⋮	⋮					

$$
\begin{array}{ccc}
& \eta^{(i)}_{2k} & \\
\eta^{(i+1)}_{2k-1} & & \eta^{(i)}_{2k+1} \\
& \eta^{(i+1)}_{2k} &
\end{array}
\qquad
\begin{array}{ccc}
& \eta^{(i)}_{2k-1} & \\
\eta^{(i+1)}_{2k-2} & & \eta^{(i)}_{2k} \\
& \eta^{(i+1)}_{2k-1} &
\end{array}
$$

Elements of (3.3.5)   Elements of (3.3.6)

Figure 3.3.4. Rhombus rules for the η-table.

we have assumed that the transformed series converges to the same limit. From the figures quoted, the series (3.8) appears to converge faster than (3.7), as expected. To justify these manipulations, we will prove first that the ε-algorithm and the η-algorithm are equivalent in the sense that

$$\varepsilon^{(0)}_{2k} = \sum_{r=0}^{2k} \eta^{(0)}_r. \tag{3.9}$$

In the context of Example 3, this means that the odd partial sums of (3.8) yield the diagonal estimates of π given by the principal diagonal of Table 3.3.2, as is easily checked.

Theorem 3.3.1 *The identities*

$$\eta^{(i)}_{2k} = \varepsilon^{(i)}_{2k} - \varepsilon^{(i-1)}_{2k} = [\varepsilon^{(i-1)}_{2k+1} - \varepsilon^{(i)}_{2k-1}]^{-1}, \tag{3.10}$$

$$\eta^{(i)}_{2k+1} = \varepsilon^{(i-1)}_{2k+2} - \varepsilon^{(i)}_{2k} = [\varepsilon^{(i)}_{2k+1} - \varepsilon^{(i-1)}_{2k+1}]^{-1} \tag{3.11}$$

hold so long as the quantities involved are well defined.

Proof. As indicated in (3.10) and (3.11), we use the identity (3.1), which constitutes the ε-algorithm, whenever necessary. Since either (3.5) and (3.6) or (3.10) and (3.11) uniquely define the η-table, our method of proof consists of using (3.10) and (3.11) to establish (3.5) and (3.6). It is convenient to extend the domain of definition by assigning $\eta_0^{(-1)} = \varepsilon_0^{(-1)} = c_{-1} = 0$ in the second column. We show that the elements of the η-table defined by (3.10) and (3.11) are identical to those defined by (3.3)–(3.6). For the second column of the η-table, the definition (3.3) yields the same values as (3.10) with $k = 0$, because

$$\varepsilon_0^{(i)} = \sum_{j=0}^{i} c_i.$$

For the third column of the table, (3.11) becomes

$$[\eta_1^{(i)}]^{-1} = \varepsilon_1^{(i)} - \varepsilon_1^{(i-1)}$$
$$= [\varepsilon_0^{(i+1)} + \varepsilon_0^{(i)}]^{-1} - [\varepsilon_0^{(i)} - \varepsilon_0^{(i-1)}]^{-1}$$
$$= [\eta_0^{(i+1)}]^{-1} - [\eta_0^{(i)}]^{-1}$$

which yields the same values as are defined by (3.5) when $k = 0$. To justify (3.5) and (3.6) as defining equations for the fourth and subsequent columns, we consider an identity among elements in a rectangular array in the ε-table. Let A, B, C, D be the elements shown in Figure 3.3.5. The identity $(D - B) - (C - A) = (D - C) - (B - A)$ is interpreted by (3.10) and (3.11) as

$$[\eta_{2k+1}^{(i)}]^{-1} - [\eta_{2k-1}^{(i+1)}]^{-1} = [\eta_{2k}^{(i+1)}]^{-1} - [\eta_{2k}^{(i)}]^{-1},$$

proving (3.5). Let W, X, Y, Z be the elements shown in Figure 3.3.6. The identity $(Z - X) - (Y - W) = (Z - Y) - (X - W)$ is interpreted by (3.10) and (3.11) as

$$\eta_{2k}^{(i)} - \eta_{2k-2}^{(i)} = \eta_{2k-1}^{(i+1)} - \eta_{2k-1}^{(i)},$$

proving (3.6). Hence the theorem is proved.

$$A = \varepsilon_{2k-1}^{(i)} \qquad\qquad B = \varepsilon_{2k+1}^{(i-1)}$$
$$\varepsilon_{2k}^{(i)}$$
$$C = \varepsilon_{2k-1}^{(i+1)} \qquad\qquad D = \varepsilon_{2k-1}^{(i)}$$

Figure 3.3.5. Five elements of the ε-table.

$$W = \varepsilon_{2k-2}^{(i)} \qquad\qquad X = \varepsilon_{2k}^{(i-1)}$$
$$\varepsilon_{2k-1}^{(i)}$$
$$Y = \varepsilon_{2k-2}^{(i-1)} \qquad\qquad Z = \varepsilon_{2k}^{(i)}$$

Figure 3.3.6. Five elements of the ε-table.

It only remains to observe that an immediate consequence of (3.10) and (3.11) is that

$$\eta_{2k+2}^{(0)} = \varepsilon_{2k+2}^{(0)} - \varepsilon_{2k+2}^{(-1)},$$
$$\eta_{2k+1}^{(0)} = \varepsilon_{2k+2}^{(-1)} - \varepsilon_{2k}^{(0)}.$$

Hence $\eta_{2k+1}^{(0)} + \eta_{2k+2}^{(0)} = \varepsilon_{2k+2}^{(0)} - \varepsilon_{2k}^{(0)}$, and by summation $\varepsilon_{2k}^{(0)} = \sum_{r=0}^{2k} \eta_r^{(0)}$, proving (3.9). We have thereby established its equivalence to the ε-algorithm as a sequence-to-sequence transformation.

A number of numerical examples, similar to Examples 1 and 2, and a full explanation of Shanks's transformation are given by Bender and Orszag [1978].

For further details about the ε-algorithm and η-algorithm, we refer to Bauer et al. [1963], Gekeler [1972], Mills [1975], Wynn [1960, 1961a], and Brezinski [1977]. Applications of the ε-algorithm to vector and matrix-valued quantities are treated in Section 8.4.

3.4 Wynn's identity and the ε-algorithm

Wynn's identity is an identity connecting neighboring Padé approximants in the Padé table:

$$([L/M + 1] - [L/M])^{-1} + ([L/M - 1] - [L/M])^{-1} =$$
$$([L - 1/M] - [L/M])^{-1} + ([L + 1/M] - [L/M])^{-1}. \quad (4.1)$$

It is easy to remember this identity from the identification with compass points in the Padé table, shown in Figure 3.4.1. With this mnemonic, the identity is written as

$$(S - C)^{-1} + (N - C)^{-1} = (W - C)^{-1} + (E - C)^{-1}.$$

It is valid when all the indicated Padé approximants exist and are nondegenerate. This section is mostly devoted to a self-contained proof of Wynn's identity; in Section 3.5, there is a more complete derivation of the various other identities. At the end of this section, we use Wynn's identity to prove that entries in even columns of the ε-table are, in fact, Padé approximants.

$$\begin{array}{ccccccc}
& [L/M - 1] & & & & N & \\
[L - 1/M] & [L/M] & [L + 1/M] & \rightarrow & W & C & E \\
& [L/M + 1] & & & & S & \\
\end{array}$$

Figure 3.4.1. Compass points in the Padé table.

We will derive two Frobenius identities in the course of this proof. Consider the determinant

$$Q^{[L/M+1]}(z) = \begin{vmatrix} c_{L-M} & c_{L-M+1} & \cdots & c_L & c_{L+1} \\ c_{L-M+1} & c_{L-M+2} & \cdots & c_{L+1} & c_{L+2} \\ \vdots & \vdots & & \vdots & \vdots \\ c_L & c_{L+1} & \cdots & c_{L+M} & c_{L+M+1} \\ z^{M+1} & z^M & \cdots & z & 1 \end{vmatrix}.$$

We apply Sylvester's identity and consider the deletion of the first and last rows and the first and last columns. Each determinant defined by these deletions is of a standard type, and the identity is

$$Q^{[L/M+1]}(z)C(L+1/M) = Q^{[L+1/M]}(z)C(L/M+1) \\ - zQ^{[L/M]}(z)C(L+1/M+1).$$

We will use this identity in the form

$$\frac{Q^{[L+1/M]}(z)}{C(L+1/M)} - \frac{Q^{[L/M+1]}(z)}{C(L/M+1)} = \frac{zQ^{[L/M]}(z)C(L+1/M+1)}{C(L+1/M)C(L/M+1)}. \quad (4.2)$$

We also consider the following determinant for $Q^{[L-1/M]}(z)$, which is contrived to give the desired result:

$$Q^{[L-1/M]}(z) = (-1)\begin{vmatrix} c_{L-M} & c_{L-M+1} & \cdots & c_{L-1} & c_L & 0 \\ c_{L-M+1} & c_{L-M+2} & \cdots & c_L & c_{L+1} & 0 \\ \vdots & \vdots & & \vdots & \vdots & \vdots \\ c_{L-1} & c_L & \cdots & c_{L+M-2} & c_{L+M-1} & 0 \\ c_L & c_{L+1} & \cdots & c_{L+M-1} & c_{L+M} & 1 \\ z^M & z^{M-1} & \cdots & z & 1 & 0 \end{vmatrix}.$$

Application of Sylvester's identity with deletion of the first and last rows and the first and last columns yields another identity among the above quantities:

$$Q^{[L-1/M]}(z)C(L+1/M) = Q^{[L/M-1]}(z)C(L/M+1) \\ + Q^{[L/M]}(z)C(L/M),$$

which we use in the form

$$\frac{Q^{[L-1/M]}(z)}{C(L/M+1)} - \frac{Q^{[L/M-1]}(z)}{C(L+1/M)} = \frac{Q^{[L/M]}(z)C(L/M)}{C(L/M+1)C(L+1/M)}. \quad (4.3)$$

We are now in a position to prove Wynn's identity. We consider first

$$[L+1/M] - [L/M] = \frac{P^{[L+1/M]}(z)Q^{[L/M]}(z) - P^{[L/M]}(z)Q^{[L+1/M]}(z)}{Q^{[L+1/M]}(z)Q^{[L/M]}(z)}.$$

3.4 Wynn's identity and the ε-algorithm

The numerator has degree $L + M + 1$, but the approximants agree to order z^{L+M} by their definition. Hence

$$[L + 1/M] - [L/M] = \frac{z^{L+M+1}}{Q^{[L+1/M]}(z)Q^{[L/M]}(z)} \cdot \text{const.}$$

The origin of the constant is the coefficient of z^{L+1} in $P^{[L+1/M]}(z)$ and of z^M in $Q^{[L/M]}(z)$, because all other terms are of lower order. Hence, using (1.1.8) and (1.1.9), we have

$$[L + 1/M] - [L/M] = \frac{C(L + 1/M)C(L + 1/M + 1)z^{L+M+1}}{Q^{[L+1/M]}(z)Q^{[L/M]}(z)}, \quad (4.4)$$

and replacing L by $L - 1$ gives

$$[L/M] - [L - 1/M] = \frac{C(L/M)C(L/M + 1)z^{L+M}}{Q^{[L/M]}(z)Q^{[L-1/M]}(z)}. \quad (4.5)$$

We consider next

$$[L/M + 1] - [L/M] = \frac{P^{[L/M+1]}(z)Q^{[L/M]}(z) - Q^{[L/M+1]}(z)P^{[L/M]}(z)}{Q^{[L/M+1]}(z)Q^{[L/M]}(z)}.$$

The numerator has degree $L + M + 1$, and the approximants agree to order z^{L+M} by their definition. Hence

$$[L/M + 1] - [L/M] = \frac{\text{const} \cdot z^{L+M+1}}{Q^{[L/M+1]}(z)Q^{[L/M]}(z)}.$$

The origin of the constant is the coefficient of z^{M+1} in $Q^{[L/M+1]}(z)$ and z^L in $P^{[L/M]}(z)$. Hence

$$[L/M + 1] - [L/M] = \frac{C(L/M + 1)C(L + 1/M + 1)z^{L+M+1}}{Q^{[L/M]}(z)Q^{[L/M + 1]}(z)}, \quad (4.6)$$

and replacing M by $M - 1$ gives

$$[L/M] - [L/M - 1] = \frac{C(L/M)C(L + 1/M)z^{L+M}}{Q^{[L/M-1]}(z)Q^{[L/M]}(z)}. \quad (4.7)$$

Equations (4.4)–(4.7) are all that is required for Wynn's identity. We find

$$\{[L + 1/M] - [L/M]\}^{-1} - \{[L/M + 1] - [L/M]\}^{-1}$$

$$= \frac{Q^{[L/M]}(z)}{C(L + 1/M + 1)z^{L+M+1}} \left\{ \frac{Q^{[L+1/M]}(z)}{C(L + 1/M)} - \frac{Q^{[L/M+1]}(z)}{C(L/M + 1)} \right\}$$

$$= \frac{\{Q^{[L/M]}(z)\}^2}{z^{L+M}C(L + 1/M)C(L/M + 1)}, \quad (4.8)$$

using (4.4), (4.6), and then (4.2).

We also find that

$$\{[L - 1/M] - [L/M]\}^{-1} - \{[L/M - 1] - [L/M]\}^{-1}$$

$$= \frac{Q^{[L/M]}(z)}{C(L/M)z^{L+M}} \left\{ -\frac{Q^{[L-1/M]}(z)}{C(L/M + 1)} + \frac{Q^{[L/M-1]}(z)}{C(L + 1/M)} \right\}$$

$$= \frac{-\{Q^{[L/M]}(z)\}^2}{z^{L+M}C(L + 1/M)C(L/M + 1)}, \qquad (4.9)$$

using (4.5), (4.7), and (4.3). Since (4.8) = −(4.9), we deduce

$$\{[L + 1/M] - [L/M]\}^{-1} + \{[L - 1/M] - [L/M]\}^{-1}$$
$$= \{[L/M + 1] - [L/M]\}^{-1} + \{[L/M - 1] - [L/M]\}^{-1},$$

which is Wynn's identity.

Hitherto in this section, we have assumed the relevant entries of the Padé table to exist and be nondegenerate. A modification of (4.1) which takes explicit account of the presence of blocks is Cordellier's identity, which is treated comprehensively in Section 3.7.

We next show that the entries in the even columns of the ε-table are entries in the rows of the Padé table. Specifically we will prove that

$$\varepsilon_{2k}^{(j)} = [k + j/k]_f(1) \qquad (4.10)$$

provided the indicated quantities exist. In Section 3.2, we defined the second column of the ε-table, namely, $\{\varepsilon_0^{(j)}, j = 0, 1, 2, \ldots\}$, to be the sequence of $[j/0]$ approximants, which is the sequence of truncated Taylor series, both evaluated at $z = 1$. We proved that the fourth column, namely, $\{\varepsilon_2^{(j)}, j = 0, 1, 2, \ldots\}$, is the sequence of $[j + 1/1]$ Padé approximants evaluated at $z = 1$. We will prove (4.10) by induction, noting that we have already established the result for $k = 0$ and $k = 1$. We use the ε-algorithm repeatedly:

$$\varepsilon_{k+1}^{(j)} = \varepsilon_{k-1}^{(j+1)} + (\varepsilon_k^{(j+1)} - \varepsilon_k^{(j)})^{-1}, \qquad (4.11)$$

as indicated by the rhombus rule in Figure 3.4.2. We will prove the

	n					$\varepsilon_{2k}^{(j-1)}$	
	nw	ne		nw		ne	
w	c	e	≡	$\varepsilon_{2k-2}^{(j+1)}$	$\varepsilon_{2k}^{(j)}$		$\varepsilon_{2k+2}^{(j-1)}$
	sw	se		sw		se	
	s				$\varepsilon_{2k}^{(j+1)}$		

Figure 3.4.2. Part of the ε-table.

connection among the $[L/M]$, $[L/M\pm 1]$, and $[L\pm 1/M]$ Padé approximants, and we expect it from (4.10) to involve $\varepsilon_{2k}^{(j)}$, $\varepsilon_{2k-2}^{(j+1)}$, $\varepsilon_{2k+2}^{(j-1)}$, $\varepsilon_{2k}^{(j+1)}$, and $\varepsilon_{2k}^{(j-1)}$ which are the corresponding epsilons. These are indicated in the figure. Note that because the columns of the ε-table will be shown to correspond to rows of the Padé table, the compass points do not correspond, and to emphasize this we use lower case letters in the ε-table.

Application of the ε-algorithm (4.11) gives the formulas

$$ne - nw = (c - n)^{-1},$$

$$se - sw = (s - c)^{-1},$$

$$(sw - nw)^{-1} = c - w,$$

$$(se - ne)^{-1} = e - c.$$

Simple manipulation yields

$$(n - c)^{-1} + (s - c)^{-1} = (w - c)^{-1} + (e - c)^{-1},$$

which is Wynn's identity for Padé approximants with the identification (4.10). We have only to observe that the ε-algorithm is used to calculate columns of the ε-table working from left to right, and that Wynn's algorithm calculates Padé approximants working from the first and second rows down. Then we see by direct construction of individual elements by induction that formula (4.10) is valid whenever the indicated quantities exist.

Notice that odd columns of the ε-table are not Padé approximants and also that even columns of the ε-table are Padé approximants on and above the diagonal, evaluated at the particular value $z=1$. The connection between the ε-algorithm and the Padé table may be made directly using (1.1.8), (1.1.9), and (1.3.8) [Shanks, 1955; Wynn, 1961b], but the extension to Cordellier's identity is obscured. An interesting consequence of the symmetry with respect to inversion [Cordellier, 1987] is that, for a normal Padé table,

$$(N^{-1} - C^{-1})^{-1} + (S^{-1} - C^{-1})^{-1} = (W^{-1} - C^{-1})^{-1} + (E^{-1} - C^{-1})^{-1},$$

defined with the notation of Figure 3.4.1.

3.5 Common identities and recursion formulas

The identities we discuss in this section apply either to the Padé approximants themselves, or to the numerators and denominators; consequently there are two quite different types of relationships to be distinguished.

One of the most remarkable relationships which occurs in the theory is that the numerators and denominators of neighboring Padé approximants obey the same recursion relations. This fact is the key to the connection with continued-fraction theory. The other relations we will prove have diverse applications elsewhere in this book.

We start with the basic definition

$$Q^{[L/M]}(z) = \begin{vmatrix} c_{L-M+1} & c_{L-M+2} & \cdots & c_L & c_{L+1} \\ c_{L-M+2} & c_{L-M+3} & \cdots & c_{L+1} & c_{L+2} \\ \vdots & \vdots & & \vdots & \vdots \\ c_L & c_{L+1} & \cdots & c_{L+M-1} & c_{L+M} \\ z^M & z^{M-1} & \cdots & z & 1 \end{vmatrix}. \quad (5.1)$$

This is an $(M+1) \times (M+1)$ determinant; its *general structure* is preserved after the deletion of the first or last columns and the first row. With either of these pairs of deletions, we end up with another $Q^{[l/m]}(z)$. This discussion is the precursor to applying Sylvester's identity with deletion of the first and last rows and columns. This action gives [Frobenius, 1881]

$$Q^{[L/M]}(z)C(L + 1/M - 1) = Q^{[L+1/M-1]}(z)C(L/M)$$
$$- zQ^{[L/M-1]}C(L + 1/M). \quad (5.2)$$

Equation (5.2) is referred to as a

$$\begin{pmatrix} * & & * \\ & * & \end{pmatrix}$$

identity, because it connects denominators of Padé approximants with the configuration shown in Figure 3.5.1.

We obtain a similar result by allowing deletion of rows, M, $M + 1$ and the first and last columns of (5.1), which is

$$Q^{[L/M]}(z)C(L/M - 1) = C(L + 1/M)Q^{[L/M-1]}(z)$$
$$- C(L + 1/M)zQ^{[L-1/M-1]}(z). \quad (5.3)$$

This is a

$$\begin{pmatrix} * & & * \\ & & * \end{pmatrix}$$

identity.

By rewriting (5.2) as

$$C(L/M)zQ^{[L-1/M-1]}(z) - C(L - 1/M)Q^{[L/M-1]}(z)$$
$$+ C(L/M - 1)Q^{[L-1/M]}(z) = 0 \quad (5.4)$$

3.5 Common identities and recursion formulas

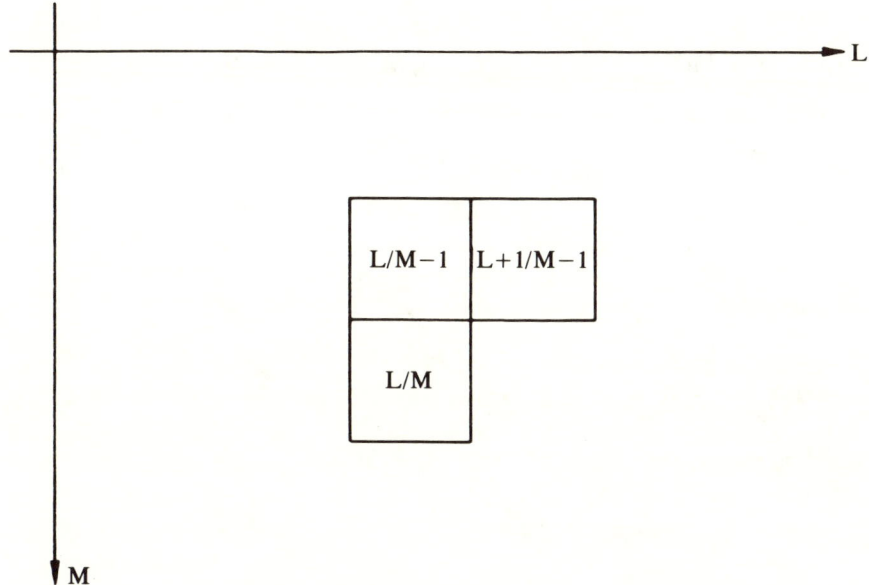

Figure 3.5.1. The locations in the Padé table of the denominators in (5.2).

and writing (5.3) as

$$C(L + 1/M)zQ^{[L-1/M-1]}(z) - C(L/M)Q^{[L/M-1]}(z) \\ + C(L/M - 1)Q^{[L/M]}(z) = 0, \quad (5.5)$$

we see that we may eliminate terms to obtain two new identities. These are

$$\{C(L/M)^2 - C(L + 1/M)C(L - 1/M)\}Q^{[L/M-1]}(z) \\ + C(L + 1/M)C(L/M - 1)Q^{[L-1/M]}(z) \\ - C(L/M - 1)C(L/M)Q^{[L/M]}(z) = 0$$

and

$$\{C(L/M)^2 - C(L + 1/M)C(L - 1/M)\}zQ^{[L-1/M-1]}(z) \\ + C(L/M)C(L/M - 1)Q^{[L-1/M]}(z) \\ - C(L - 1/M)C(L/M - 1)Q^{[L/M]}(z) = 0.$$

Using the simplest Sylvester identity, (1.4.11),

$$C(L/M + 1)C(L/M - 1) = C(L - 1/M)C(L + 1/M) - C(L/M)^2,$$

we find

$$-C(L/M + 1)Q^{[L/M-1]}(z) + C(L + 1/M)Q^{[L-1/M]}(z)$$
$$- C(L/M)Q^{[L/M]}(z) = 0, \qquad (5.6)$$

which is a

$$\begin{pmatrix} & * \\ * & * \end{pmatrix}$$

identity, and

$$-C(L/M + 1)zQ^{[L-1/M-1]}(z) + C(L/M)Q^{[L-1/M]}(z)$$
$$- C(L - 1/M)Q^{[L/M]}(z) = 0, \qquad (5.7)$$

which is a

$$\begin{pmatrix} * & \\ * & * \end{pmatrix}$$

identity.

Equations (5.4)–(5.7) are the Frobenius identities for the Padé denominators. For the numerators,

$$P^{[L/M]}(z) = \begin{vmatrix} c_{L-M+1} & c_{L-M+2} & \cdots & c_L & c_{L+1} \\ c_{L-M+2} & c_{L-M+3} & \cdots & c_{L+1} & c_{L+2} \\ \vdots & \vdots & & \vdots & \vdots \\ c_L & c_{L+1} & \cdots & c_{L+M-1} & c_{L+M} \\ \sum_{j=M}^{L} c_{j-M} z^j & \sum_{j=M-1}^{L} c_{j-M+1} z^j & \cdots & \sum_{j=1}^{L} c_{j+1} z^j & \sum_{j=0}^{L} c_j z^j \end{vmatrix}.$$

(5.8)

and Sylvester's identity with deletion of the first and last rows and columns leads to

$$P^{[L/M]}(z)C(L + 1/M - 1) =$$
$$P^{[L+1/M-1]}(z)C(L/M) - zP^{[L/M-1]}(z)C(L + 1/M). \quad (5.9)$$

Equation (5.9) has a form precisely similar to (5.2). Thus it is normal to write the Frobenius identities using $S^{[L/M]}(z)$, where we choose

either $\qquad S^{[L/M]}(z) = Q^{[L/M]}(z) \qquad$ (5.10a)

or $\qquad S^{[L/M]}(z) = P^{[L/M]}(z) \qquad$ (5.10b)

or $\qquad S^{[L/M]}(z) = G(z)P^{[L/M]}(z) + H(z)Q^{[L/M]}(z). \qquad$ (5.10c)

3.5 Common identities and recursion formulas

The generalization to (5.10c) is possible if $G(z)$ and $H(z)$ are functions of z only, and are independent of L and M; it is easily justified because the Frobenius identities are linear in the sense that (5.10c) is linear. Our conclusion is that with the definitions (5.10), we have identities among the elements $S^{[l/m]}$ of Figure 3.5.2 as follows:

Frobenius identities

$$C(L/M)zS^{[L-1/M-1]}(z) - C(L-1/M)S^{[L/M-1]}(z)$$
$$+ C(L/M-1)S^{[L-1/M]}(z) = 0 \quad \begin{pmatrix} * & * \\ & * \end{pmatrix},$$

$$C(L+1/M)zS^{[L-1/M-1]}(z) - C(L/M)S^{[L/M-1]}(z)$$
$$+ C(L/M-1)S^{[L/M]}(z) = 0 \quad \begin{pmatrix} * & * \\ & * \end{pmatrix},$$

$$C(L/M+1)S^{[L/M-1]}(z) - C(L+1/M)S^{[L-1/M]}(z)$$
$$+ C(L/M)S^{[L/M]}(z) = 0 \quad \begin{pmatrix} & * \\ * & * \end{pmatrix},$$

$$C(L/M+1)zS^{[L-1/M-1]}(z) - C(L/M)S^{[L-1/M]}(z)$$
$$+ C(L-1/M)S^{[L/M]}(z) = 0 \quad \begin{pmatrix} * & \\ * & * \end{pmatrix}, \quad (5.11)$$

From these important results, we may obtain further identities. Using the symbol $\cdot$ to denote entries to be eliminated, we examine the configuration

$$\begin{pmatrix} \cdot & * \\ \cdot & * \\ * & \end{pmatrix}$$

using

$$\begin{pmatrix} \cdot & * \\ * & \end{pmatrix}$$

twice and

$$\begin{pmatrix} & \cdot \\ \cdot & * \end{pmatrix}.$$

$$\begin{bmatrix} S^{[L-1/M-1]} & S^{[L/M-1]} \\ S^{[L-1/M]} & S^{[L/M]} \end{bmatrix} \leftrightarrow \begin{bmatrix} * & * \\ * & * \end{bmatrix}$$

Figure 3.5.2. Scheme for the Frobenius identities.

This leads to a single identity for the

$$\begin{pmatrix} & & * \\ & * & \\ * & & \end{pmatrix}$$

configuration which turns out to be

$$C(L+1/M)^2 S^{[L-1/M+1]}(z) + C(L/M+1)^2 S^{[L+1/M-1]}(z)$$
$$= \{C(L+1/M-1)C(L/M+2) - C(L-1/M+1)C(L+2/M)$$
$$+ zC(L+1/M)C(L/M+1)\} S^{[L/M]}(z), \qquad (5.12)$$

which is useful for Kronecker's algorithm (see Section 3.7). Similarly, there is a

$$\begin{pmatrix} * & & \\ & * & \\ & & * \end{pmatrix}$$

identity, which is

$$\frac{C(L/M)}{C(L+1/M+1)} S^{[L+1/M+1]}(z) + \frac{C(L+1/M+1)}{C(L/M)} z^2 S^{[L-1/M-1]}(z) =$$
$$\frac{\{C(L+1/M+2)C(L/M-1) - zC(L+2/M+1)C(L-1/M)\} S^{[L/M]}(z)}{C(L+1/M+1)C(L/M)}.$$

There are also

$$\begin{pmatrix} * \\ * \\ * \end{pmatrix}$$

and $(* \ * \ *)$ identities, given in Baker [1975a].

This identity concludes our survey of identities for $Q^{[L/M]}(z)$ and $P^{[L/M]}(z)$. Next we turn to identities for the Padé approximants themselves. First, there are the fundamental two-term identities between neighboring approximants. Consider

$$f(z) - \frac{P^{[L/M]}(z)}{Q^{[L/M]}(z)} = O(z^{L+M+1}) \qquad (5.13)$$

and

$$f(z) - \frac{P^{[L+1/M]}(z)}{Q^{[L+1/M]}(z)} = O(z^{L+M+2}). \qquad (5.14)$$

If the Padé approximants are degenerate, then (5.13), (5.14) are to be

3.5 Common identities and recursion formulas

understood as being multiplied through, in which case they become correct. The results, being algebraic, are essentially unchanged. Subtraction of (5.13) from (5.14) yields

$$\frac{P^{[L+1/M]}(z)}{Q^{[L+1/M]}(z)} - \frac{P^{[L/M]}(z)}{Q^{[L/M]}(z)} = O(z^{L+M+1}).$$

Therefore

$$Q^{[L/M]}(z)P^{[L+1/M]}(z) - Q^{[L+1/M]}(z)P^{[L/M]}(z) = O(z^{L+M+1}). \quad (5.15)$$

Since the left-hand side of (5.15) is a polynomial of order $L + M + 1$, (5.15) becomes

$$Q^{[L/M]}(z)P^{[L+1/M]}(z) - Q^{[L+1/M]}(z)P^{[L/M]}(z) = Kz^{L+M+1}. \quad (5.16)$$

By inspection of the determinantal forms (5.1), (5.8), we find the leading coefficients to be as follows:

The coefficient of z^M in $Q^{[L/M]}(z)$ is $(-1)^M C(L + 1/M)$.

The coefficient of z^L in $P^{[L/M]}(z)$ is $(-1)^M C(L/M + 1)$.

Substituting in (5.16) for the leading coefficients, we obtain

$$C(L + 1/M)C(L + 1/M + 1) = K,$$

and so (5.16) after division by $Q^{[L/M]}(z)Q^{[L+1/M]}(z)$ finally becomes

$$[L + 1/M] - [L/M] = \frac{C(L + 1/M)C(L + 1/M + 1)z^{L+M+1}}{Q^{[L/M]}(z)Q^{[L+1/M]}(z)}, \quad (5.17)$$

which is a $(*\ *)$ identity. By working in a precisely similar way, and using Sylvester's identity for (5.18), we find

$$[L + 1/M + 1] - [L/M] = \frac{C(L + 1/M + 1)^2 z^{L+M+1}}{Q^{[L+1/M+1]}(z)Q^{[L/M]}(z)} \quad \begin{pmatrix} * & \\ & * \end{pmatrix}, \quad (5.18)$$

$$[L/M + 1] - [L/M] = \frac{C(L/M + 1)C(L + 1/M + 1)z^{L+M+1}}{Q^{[L/M+1]}(z)Q^{[L/M]}(z)} \quad \begin{pmatrix} * \\ * \end{pmatrix},$$
$$(5.19)$$

$$[L/M + 1] - [L + 1/M] = \frac{C(L + 1/M + 1)^2 z^{L+M+2}}{Q^{[L/M+1]}(z)Q^{[L+1/M]}(z)} \quad \begin{pmatrix} & * \\ * & \end{pmatrix}. \quad (5.20)$$

This identity concludes the derivation of the two-term identities

(5.17)–(5.20). From these may be derived some invariants, called cross ratios, which are independent of z. We find

$$\frac{[L/M] - [L/M + 1]}{[L/M] - [L + 1/M]} \frac{[L + 1/M] - [L + 1/M + 1]}{[L/M + 1] - [L + 1/M + 1]} =$$
$$\frac{C(L/M + 1)}{C(L + 1/M)} \frac{C(L + 2/M + 1)}{C(L + 1/M + 2}$$

which is a

$$\begin{pmatrix} * & - & * \\ | & & | \\ * & - & * \end{pmatrix}$$

identity. It is not the only

$$\begin{pmatrix} * & * \\ * & * \end{pmatrix}$$

identity; there are others given by

$$\begin{pmatrix} * & - & * \\ & \times & \\ * & - & * \end{pmatrix}$$

interconnections and by

$$\begin{pmatrix} * & & * \\ | & \times & | \\ * & & * \end{pmatrix}$$

and by more complicated patterns. Again we refer to Baker [1975a] for details.

3.6 Recursive calculation of the coefficients of Padé approximants

This section is concerned with the numerical calculation of the coefficients of the numerator and denominator polynomials of Padé approximants of high order. The results (2.1.6), (2.1.7) may present an unduly pessimistic estimate of the extent of the inevitable loss of numerical precision in solving Hankel systems. These condition numbers are directly related to the likely loss of numerical precision using the direct methods described in Section 2.1 in which it is tacitly assumed that the errors in the numerical representation of the coefficient matrix are uncorrelated. Nevertheless, correlation of the errors in the numerical representation of the coefficient matrix can be taken into account by exploiting the Hankel structure of the system.

3.6 Recursive calculation of the coefficients of Padé approximants

The particular Padé approximation problem in which $L = M$ and M is large is of great interest. The denominator-coefficient problem takes the form (1.1.6)

$$\begin{bmatrix} c_1 & \cdots & c_M \\ \vdots & & \vdots \\ c_M & \cdots & c_{2M-1} \end{bmatrix} \begin{bmatrix} q_M \\ \vdots \\ q_1 \end{bmatrix} = -\begin{bmatrix} c_{M+1} \\ \vdots \\ c_{2M} \end{bmatrix} \quad (6.1)$$

and we define

$$H_{M;M} = \begin{bmatrix} c_1 & \cdots & c_M \\ \vdots & & \vdots \\ c_M & \cdots & c_{2M-1} \end{bmatrix}, \quad (6.2)$$

so that $H_{M;M}$ represents a general Hankel matrix. We visualize adding further rows and columns to this matrix which preserve its structure. In this context, row exchanges are unhelpful because the structure is destroyed and there is now a significant difference between Hankel problems and Toeplitz problems. For the reasons given above, there is the hope and belief that the information that the coefficient matrix has Hankel structure can be exploited to provide greater precision for the solution than is possible with direct methods. Because of the existence of formulas such as (6.15), there is much interest in the solution of (6.1) as a means to the accurate computation of $H_{M;M}^{-1}$.

Our next theorem shows how two consecutive Padé approximants on a diagonal sequence can be related to two other consecutive approximants further along the same diagonal sequence. The secret is to form the two remainder series associated with the first two Padé approximants, and then construct the appropriate approximants of the ratio of these two remainder series.

Theorem 3.6.1 (A progression theorem). *Let $m, k \geq 1$ be two integers that are given, together with the power series $f(z) = \sum_{i=0}^{\infty} c_i z^i$.*

Assume that $P^{[m+J-1/m-1]}(z)$, $Q^{[m+J-1/m-1]}(z)$ are numerator and denominator Padé polynomials of type $[m+J-1/m-1]$ for the power series $f(z)$, and similarly for $P^{[m+J/m]}(z)$ and $Q^{[m+J/m]}(z)$, but also with

$$C(m + J/m) \neq 0. \quad (6.3)$$

With these polynomials, two 'remainder series' are defined by

$$d(z) = \{f(z)Q^{[m+J-1/m-1]}(z) - P^{[m+J-1/m-1]}(z)\}/z^{2m+J-1}, \quad (6.4)$$

$$e(z) = \{f(z)Q^{[m+J/m]}(z) - P^{[m+J/m]}(z)\}/z^{2m+J+1}. \quad (6.5)$$

Then $d(0) \neq 0$. Assume that $A^{[k/k+1]}(z)$, $B^{[k/k+1]}(z)$ are nondegenerate numerator and denominator Padé polynomials of type $[k/k+1]$ for the rational power series $e(z)/d(z)$.

Then the Padé polynomials of types $[m+k+J/m+k]$ and $[m+k+J+1/m+k+1]$ for $f(z)$ are given by

$$\begin{bmatrix} P^{[m+k+J+1/m+k+1]}(z) & P^{[m+k+J/m+k]}(z) \\ Q^{[m+k+J+1/m+k+1]}(z) & Q^{[m+k+J/m+k]}(z) \end{bmatrix}$$
$$= \begin{bmatrix} P^{[m+J/m]}(z) & P^{[m+J-1/m-1]}(z) \\ Q^{[m+J/m]}(z) & Q^{[m+J-1/m-1]}(z) \end{bmatrix} \begin{bmatrix} B^{[k/k+1]}(z) & B^{[k-1/k]}(z) \\ -z^2 A^{[k/k+1]}(z) & -z^2 A^{[k-1/k]}(z) \end{bmatrix},$$
(6.6)

and the approximant of type $[m+k+J+1/m+k+1]$ is nondegenerate.

Proof. The condition $C(m+J/m) \neq 0$ ensures that the rows of the determinantal representation (1.1.8) of $Q^{[m+J-1/m-1]}(z)$ are linearly independent, and so the representation itself is nontrivial. In fact $C(m+J/m)$ is the leading coefficient of the error formula (1.1.11) for an approximant of type $[m+J-1/m-1]$, and so $d(0) = C(m+J/m) \neq 0$.

With the notation used, it is straightforward to check that the degrees of the polynomials in the matrix product on the right-hand side of (6.6) agree with those designated for the four polynomials on the left-hand side.

Using the second column of (6.6), we have

$P^{[m+k+J/m+k]}(z) - f(z)Q^{[m+k+J/m+k]}(z)$

$\quad = \{P^{[m+J/m]}(z) - f(z)Q^{[m+J/m]}(z)\} B^{[k-1/k]}(z)$

$\quad\quad - z^2 \{P^{[m+J-1/m-1]}(z) - f(z)Q^{[m+J-1/m-1]}(z)\} A^{[k-1/k]}(z)$

$\quad = z^{2m+J+1} \{d(z) A^{[k-1/k]}(z) - e(z) B^{[k-1/k]}(z)\}$

$\quad = O(z^{2m+2k+J+1})$ (6.7)

which has the accuracy-through-order required for the second column of the left-hand side. Similarly,

$$P^{[m+k+J+1/m+k+1]}(z) - f(z)Q^{[m+k+J+1/m+k+1]}(z) = O(z^{2m+2k+J+3}) \quad (6.8)$$

as is required for the first column of the left-hand side of (6.6). By hypothesis, $Q^{[m+J/m]}(0) \neq 0$, $B^{[k/k+1]}(0) \neq 0$, and so, from (6.6),

$$Q^{[m+k+J+1/m+k+1]}(0) = Q^{[m+J/m]}(0) B^{[k/k+1]}(0) \neq 0. \quad (6.9)$$

Therefore $P^{[m+k+J+1/m+k+1]}(z)$, $Q^{[m+k+J+1/m+k+1]}(z)$ are the polynomials associated with a nondegenerate Padé approximant of the indicated type.

3.6 Recursive calculation of the coefficients of Padé approximants

The previous theorem is remarkable because the same condition (6.3) that ensures that a particular approximant is nondegenerate is also the one that ensures that the leading term of the error expansion of its predecessor is nonzero. In fact, it is also the condition that the leading coefficient of $P^{[m+J+1/m-1]}(z)$ as given by (1.1.9) is nonzero.

We are now in a position to outline a reliable method for the recursive calculation of Padé approximants. Later, we will see that it can easily be modified for robustness too. The method is based on forming, at some stage, the remainder functions $d(z)$ and $e(z)$ associated with two successive Padé approximants of $f(z)$ in a paradiagonal sequence. A nondegenerate Padé approximant of type $[k/k+1]$ for $e(z)/d(z)$ is located and constructed, along with its polynomial predecessors. From the polynomials constructed, the next nondegenerate Padé approximant for $f(z)$ in the sequence is found, and it is located k places further down in the original paradiagonal sequence. In turn, the remainder functions associated with this approximant and its predecessor can be found, and the whole scheme becomes recursive. As such, this recursion scheme is the basis of a reliable procedure for the calculation of Padé approximants in a paradiagonal sequence.

We have glossed over two significant points. The first is the initialization of the algorithm, which we will deal with later in the section. The second is that of formation of the approximant of type $[k/k+1]$ to $e(z)/d(z)$. This can be done either by the inversion of the series for $d(z)$, or else by solving the linear (Sylvester) system of equations for the polynomial bigradient. Both these methods implicitly involve the solution of linear systems of equations which are likely to be ill conditioned, and a robust linear solver must be used. This brings us naturally to the most substantial theme of this section, namely, a robust method of calculation of high-order Padé approximants.

The outstanding problem is to decide which is the optimal value of k to choose for formation of the Padé approximant of type $[k/k+1]$ for $e(z)/d(z)$. To answer this question, we could use a 'black-box' routine which calculates or estimates the condition number κ of the Hankel matrix $H_{k;k+1}$ used in forming the $[k/k+1]$-type approximant for $e(z)/d(z)$, and use the least value of k for which $\kappa(H_{k;k+1}) \leq 10^d$, where d denotes the number of decimal places of floating-point precision which we are prepared to lose. Alternatively, we can estimate $\kappa(H_{k;k+1})$ using the following interesting representation of $H^{-1}_{k;k+1}$. We begin this development with a lemma.

Lemma 1. Let $A(z) = \sum_{i=0}^{l} a_i z^i$, $B(z) = \sum_{i=0}^{m} b_i z^i$ be Padé polynomials of type $[l/m]$ for $f(z)$, and let

$$H_{l;m} = \begin{bmatrix} c_{l-m+1} & \cdots & c_l \\ \vdots & \ddots & \vdots \\ c_l & \cdots & c_{l+m-1} \end{bmatrix}. \tag{6.10}$$

Then

$$H_{l;m} \begin{bmatrix} b_{m-1} & \cdots & b_0 \\ \vdots & \ddots & \\ b_0 & & 0 \end{bmatrix}$$

$$= \begin{bmatrix} a_l & \cdots & a_{l-m+1} \\ & \ddots & \vdots \\ 0 & & a_l \end{bmatrix} - H_{l-m;m} \begin{bmatrix} 0 & & b_m \\ & \ddots & \vdots \\ b_m & \cdots & b_1 \end{bmatrix}. \tag{6.11}$$

Proof. We use the convention that $c_j, a_j = 0$ if $j < 0$. The Padé equations are displayed as

$$\begin{bmatrix} 0 & & & c_0 \\ & & c_0 & c_1 \\ & \ddots & & \vdots \\ c_0 & \cdots & & c_m \\ \vdots & & & \vdots \\ c_{l-m+1} & \cdots & & c_{l+1} \\ \vdots & & & \vdots \\ c_l & \cdots & & c_{l+m} \end{bmatrix} \begin{bmatrix} b_m \\ \vdots \\ \\ b_0 \end{bmatrix} = \begin{bmatrix} a_0 \\ \vdots \\ \\ a_l \\ 0 \\ \vdots \\ 0 \end{bmatrix} \tag{6.12}$$

for $l \geq m$ and similarly for $l < m$. For any integer $k \in [0, m-1]$, we extract from (6.11) the $(l-k+1)$th through the $(l-k+m)$th equations:

$$\begin{bmatrix} c_{l-k-m} & \cdots & c_{l-k} \\ \vdots & \ddots & \vdots \\ c_{l-k-1} & \cdots & c_{l+m-k-1} \end{bmatrix} \begin{bmatrix} b_m \\ \vdots \\ b_0 \end{bmatrix} = \begin{bmatrix} a_{l-k} \\ \vdots \\ a_l \\ 0 \\ \vdots \\ 0 \end{bmatrix}$$

corresponding to accuracy through orders $z^{l-k}, \ldots, z^{l+m-k-1}$. These equations are split as

$$\begin{bmatrix} c_{l-m+1} & \cdots & c_{l-k} \\ \vdots & & \vdots \\ c_l & \cdots & c_{l+m-k-1} \end{bmatrix} \begin{bmatrix} b_{m-k-1} \\ \vdots \\ b_0 \end{bmatrix} =$$

3.6 Recursive calculation of the coefficients of Padé approximants

$$\begin{bmatrix} a_{l-k} \\ \vdots \\ a_l \\ 0 \\ \vdots \\ 0 \end{bmatrix} - \begin{bmatrix} c_{l-k-m} & \cdots & c_{l-m} \\ \vdots & & \vdots \\ c_{l-k-1} & \cdots & c_{l-1} \end{bmatrix} \begin{bmatrix} b_m \\ \vdots \\ b_{m-k} \end{bmatrix}$$

and they are expanded to become

$$H_{l;m} \begin{bmatrix} b_{m-k-1} \\ \vdots \\ b_0 \\ 0 \\ \vdots \\ 0 \end{bmatrix} = \begin{bmatrix} a_{l-k} \\ \vdots \\ a_l \\ 0 \\ \vdots \\ 0 \end{bmatrix} - H_{l-m;m} \begin{bmatrix} 0 \\ \vdots \\ 0 \\ b_m \\ \vdots \\ b_{m-k} \end{bmatrix}.$$

The cases $k = 0, 1, \ldots, m - 1$ constitute the identity (6.11) columnwise.

Corollary. *Let* $P(z) = \sum_{i=0}^{l-1} p_i z^i$, $Q(z) = \sum_{i=0}^{m-1} q_i z^i$ *be Padé approximants of type* $[l - 1/m - 1]$ *for* $f(z)$. *Then*

$$H_{l;m} \begin{bmatrix} q_{m-2} & \cdots & q_0 & 0 \\ \vdots & \cdots & & \\ q_0 & & & \\ 0 & & & 0 \end{bmatrix} =$$

$$\begin{bmatrix} p_{l-1} & \cdots & p_{l-m} \\ & \ddots & \vdots \\ 0 & & p_{l-1} \end{bmatrix} - H_{l-m;m} \begin{bmatrix} 0 & & q_{m-1} \\ & \cdots & \vdots \\ q_{m-1} & \cdots & q_0 \end{bmatrix}. \quad (6.13)$$

Proof. The method of proof is the same.

The following important theorem evolved sequentially from earlier theorems of Baxter and Hirschman [1964], Gohberg and Semencul [1972], and Heinig and Rost [1984, Theorem 1.2]. This analysis follows Labahn, Choi, and Cabay [1990].

Theorem 3.6.2 (Paradiagonal representation). *Let* $A(z)/B(z)$ *be a nondegenerate Padé approximant of type* $[l/m]$ *for* $f(z)$, *and let* $B(z) = \sum_{i=0}^{m} b_i z^i$ *with* $b_0 \neq 0$ *denotes the denominator polynomial. Let* $Q(z) = \sum_{i=0}^{m-1} q_i z^i$ *denote the Padé denominator polynomial of type* $[l - 1/m - 1]$ *for* $f(z)$. *Then*

$$f(z)Q(z) - P(z) = r_0 z^{l+m-1} + O(z^{l+m}), \quad (6.14)$$

where $r_0 \neq 0$, and

$$H_{l;m}^{-1} = \frac{1}{r_0 b_0} \left\{ \begin{bmatrix} b_{m-1} & \cdots & b_0 \\ \vdots & \reflectbox{$\ddots$} & \\ b_0 & & 0 \end{bmatrix} \begin{bmatrix} q_{m-1} & \cdots & q_0 \\ & \ddots & \vdots \\ 0 & & q_{m-1} \end{bmatrix} \right.$$

$$\left. - \begin{bmatrix} q_{m-2} & \cdots & q_0 & 0 \\ \vdots & \reflectbox{$\ddots$} & & \\ q_0 & & & \\ 0 & & & \end{bmatrix} \begin{bmatrix} b_m & \cdots & b_1 \\ & \ddots & \vdots \\ 0 & & b_m \end{bmatrix} \right\}. \qquad (6.15)$$

Remarks. In this section only, we waive our convention that $b_0 = 1$, and note that $B(z)$, $Q(z)$ are only unique up to constant multipliers. Notice that the result (6.15) does not depend on the normalization chosen for $B(z)$, $Q(z)$.

Proof. We begin with an identity which corresponds to (5.18), but express it in a way which is as independent of the normalization as possible. In the notation of the theorem, we can prove that

$$A(z)Q(z) - P(z)B(z) = r_0 b_0 z^{l+m-1}. \qquad (6.16)$$

First, notice that the left-hand side is a polynomial of degree $l + m - 1$ at most, and then that

$$A(z)Q(z) - P(z)B(z) = B(z)\{f(z)Q(z) - P(z)\} + O(z^{l+m+1})$$
$$= b_0 r_0 z^{l+m-1} + O(z^{l+m}).$$

The result (6.16) follows immediately with $r_0 \neq 0$, using the error formula (1.1.11), and the same argument as in Theorem 3.6.1. It is also an immediate consequence of (5.18) using the normalization of that section. The identity

$$\begin{bmatrix} a_l & \cdots & a_{l-m+1} \\ & \ddots & \vdots \\ 0 & & a_l \end{bmatrix} \begin{bmatrix} q_{m-1} & \cdots & q_0 \\ & \ddots & \vdots \\ 0 & & q_{m-1} \end{bmatrix} - \begin{bmatrix} p_{l-1} & \cdots & p_{l-m} \\ & \ddots & \vdots \\ 0 & & p_{l-1} \end{bmatrix} \begin{bmatrix} b_m & \cdots & b_1 \\ & \ddots & \vdots \\ 0 & & b_m \end{bmatrix} = b_0 r_0 I.$$

follows as a consequence of (6.16), because the (i, j) element of this identity represents equality of the coefficients of $z^{l+m-1+i-j}$ in (6.16).

If we write the result of the previous lemma as

$$H \triangleleft b = \triangleleft a - H_{l-m;m} \triangleleft b,$$

3.6 Recursive calculation of the coefficients of Padé approximants

of its corollary as

$$H\left\{\triangleright_q \triangleleft_p\right\} = \triangleright_p \triangleleft \;- H_{l-m;m} \triangleright \triangleleft_q,$$

and of (6.16) as

$$\triangleright_a \triangleleft_q - \triangleright_p \triangleleft_b = b_0 r_0 I,$$

we have

$$H\left\{\triangleright_b \triangleleft_q - \triangleright_q \triangleleft_b\right\} =$$

$$\triangleright_a \triangleleft_q - \triangleright_p \triangleleft_b - H_{l-m;m}\left\{\triangleright_b \triangleleft_q - \triangleright_q \triangleleft_b\right\}.$$

Because $b(z)q(z) = q(z)b(z)$, we find that

$$\triangleright_b \triangleleft_q - \triangleright_q \triangleleft_b = 0,$$

using the previous method. By row permutation, we find

$$\triangleright_b \triangleleft_q - \triangleright_q \triangleleft_b = 0,$$

and hence

$$H\left\{\triangleright_b \triangleleft_q - \triangleright_q \triangleleft_b\right\} = b_0 r_0 I,$$

as required to prove (6.15).

For comparison, we state the following theorem which is based on the coefficients of neighboring denominator polynomials in a column sequence.

Theorem 3.6.3 (Gohberg and Heinig, 1974]. *Let $A(z)/B(z)$ be a nondegenerate Padé approximant of type $[l/m-1]$ for $f(z)$, and let $B(z) = \sum_{i=0}^{m-1} b_i z^i$ denote the denominator polynomial. Likewise, let $P(z)/Q(z)$ be a nondegenerate Padé approximant of type $[l-1/m-1]$ for $f(z)$ and let $Q(z) = \sum_{i=0}^{m} q_i z^i$ denote its denominator polynomial. Then*

$$H_{l;m}^{-1} = \frac{1}{r_0 b_0}\left\{\begin{bmatrix} b_{m-1} & \cdots & b_0 \\ \vdots & \ddots & \\ b_0 & & 0 \end{bmatrix}\begin{bmatrix} q_{m-1} & \cdots & q_0 \\ & \ddots & \vdots \\ 0 & & q_{m-1} \end{bmatrix}\right.$$

$$\left. - \begin{bmatrix} q_{m-2} & \cdots & q_0 & 0 \\ \vdots & \ddots & & \\ q_0 & & & \\ 0 & & & \end{bmatrix}\begin{bmatrix} 0 & b_{m-1} & \cdots & b_1 \\ & \ddots & \ddots & \vdots \\ 0 & & & b_{m-1} \\ & & 0 & 0 \end{bmatrix}\right\},$$

where r_0 is the residual defined by

$$f(z)Q(z) - P(z) = r_0 z^{l+m-1} + O(z^{l+m}).$$

For the proof of this result (which is similar to the proof of Theorem 3.6.2), and for details of the antidiagonal and related identities of this kind, we refer to Heinig and Rost [1984] and Labahn, et al. [1990]

As we have previously noted, iterative use of formula (6.6) requires the coefficients of the Padé approximant of type $[k/k+1]$ for $e(z)/d(z)$, and so these coefficients are required accurately. Let $\|v\|$ denote the supremum norm $\|v\|_\infty$ for a generic vector v. The usage of matrix and vector norms is well explained by Golub and van Loan [1989]. If v is the vector of coefficients of a polynomial $v(z)$, then let $\|v\| = \|v\|$. Then we obtain from (6.15) the very rough and ready estimate that

$$\|H_{l;m}^{-1}\| \leq \frac{2m\|b\|\|q\|}{|r_0||b_0|}.$$

At this stage, it is convenient to normalize the Padé polynomials with $\|b\| = \|q\| = 1$, so that $|b_0|$ is uniquely determined, and $|r_0|$ is uniquely given by (6.14). We then have

$$\kappa(H_{l;m}) \leq \frac{2m}{|r_0 b_0|}\|H_{l;m}\| \tag{6.17}$$

as a very crude but easily obtainable estimate of the condition number of $H_{l;m}$. Suppose that we are prepared to lose about d decimal places of floating-point precision in the calculation of a Padé approximant, in which (6.6) is used as part of the iterative procedure. Then a test such as

$$\kappa(H_{k;k+1}) \lesseqgtr 10^d, \tag{6.18}$$

where $H_{k;k+1}$ denotes the Hankel matrix associated with Padé approximation of type $[k/k+1]$ for $e(z)/d(z)$, is essential. With the upper sign, the approximant is acceptable and iteration continues, whereas the lower sign indicates that the next approximant in the sequence should be tried. Meleshko and Cabay [1991] and Cabay and Meleshko [1993] have found criteria of this kind to be very effective. A more sophisticated error analysis has been given by Gutknecht [1993]. The point is that it is foolish to use a formula such as (6.6) recursively without some kind of cumulative running error estimate of the loss of precision of the computation.

We proceed to formulate the initialization of the algorithm for a reliable (and robust) method of calculation of high-order Padé approximants. Suppose that it is an approximant of type $[L/M]$ which is ultimately required. Then $J = L - M$ is the offset which determines the appropriate paradiagonal sequence. We introduce the artificial initializations given by Gragg [1972]. If $J \geq 0$, the procedure is regularly initialized with $l_1 = J$, $m_1 = 0$, and its predecessor is of type $[J - 1/-1]$ which is necessarily artificial. We take

$$P^{[J/0]}(z) = \sum_{i=0}^{J} c_i z^i, \qquad Q^{[J/0]}(z) = 1,$$
$$P^{[J-1/-1]}(z) = -z^{J-1}, \qquad Q^{[J-1/-1]}(z) = 0, \quad (6.19)$$

as a suitable initialization, in the sense that they formally satisfy the relevant identities (6.14, 6.16), despite the fact that, for example, $P^{[J-1/-1]}(z)$ is not even a polynomial if $J = 0$. Using these formulas, we find that

$$d(z) = 1, \quad e(z) = c_{J+1} + z c_{J+2} + \cdots.$$

If, for example, $c_{J+1} \neq 0$, $k = 1$, $m = 0$, then the last column of the identity (6.6) takes the form

$$\begin{bmatrix} P^{[J+1/1]}(z) \\ Q^{[J+1/1]}(z) \end{bmatrix} = \begin{bmatrix} \sum_{i=0}^{J} c_i z^i & -z^{J-1} \\ 1 & 0 \end{bmatrix} \begin{bmatrix} 1 - z c_{J+2}/c_{J+1} \\ -z^2 c_{J+1} \end{bmatrix},$$

which is easily verified as correct.

If $J < 0$, then it is essential that $c_0 \neq 0$, so that the reciprocal series can be formed, and we formally set

$$\sum_{i=0}^{\infty} c'_i z^i = \left[\sum_{i=0}^{\infty} c_i z^i \right]^{-1}.$$

We take $m = m_1 = -J$, and the procedure is regularly initialized with $l_1 = 0$, $m_1 = -J$, and its predecessor is of type $[-1/m_1 - 1]$ which is again artificial. We take

$$P^{[0/m]}(z) = 1, \qquad Q^{[0/m]}(z) = \sum_{i=0}^{m} c'_i z^i,$$
$$P^{[-1/m-1]}(z) = 0, \qquad Q^{[-1/m-1]}(z) = -z^{m-1}, \quad (6.20)$$

as a suitable initialization. Using these formulas, we find that

$$d(z) = -f(z), \quad e(z) = \left[f(z) \sum_{i=0}^{m} c'_i z^i - 1 \right] \bigg/ z^{m+1},$$

and

$$e(z)/d(z) = c'_{m+1} + z c'_{m+2} + \cdots.$$

If $c'_{m+1} \neq 0$, $k = 1$, $J = -m$, then the last column of (6.6) is

$$\begin{bmatrix} P^{[1/m+1]}(z) \\ Q^{[1/m+1]}(z) \end{bmatrix} = \begin{bmatrix} 1 & 0 \\ \sum_{i=0}^{m} c'_i z^i & -z^{m-1} \end{bmatrix} \begin{bmatrix} 1 - zc'_{m+2}/c'_{m+1} \\ -z^2 c'_{m+1} \end{bmatrix},$$

which can be verified as correct using the duality theorem 1.5.1.

All the basic ingredients for the robust calculation of a high-order Padé approximant of type $[M + J/M]$ have now been given [Cabay and Meleshko, 1993], and they are assembled into the following highly schematic calculational procedure. We refer to Meleshko and Cabay [1991] for their detailed implementation. Within the procedure, $H_{k;k+1}$ is the Hankel matrix defined by (6.12) and associated with the Maclaurin series of $d(z)/e(z)$.

Procedure 3.6.1.

1. Initialize using (6.19) or (6.20) to obtain an approximant of type $[m + J/m]$ and its predecessor form. Let $k = 1$.
2. Form $d(z)/e(z)$ using (6.4) and (6.5).
3. If $k = 1$ and $H_{0;1}$ is not small, construct the approximant of type $[m + J + 1/m + 1]$ using the right-hand column of (6.6). This approximant has been accurately constructed, and so has its predecessor of type $[m + J/m]$. Increment $m := m + 1$ and return to stage 2 to iterate.
4. If $H_{k;k+1}$ is well conditioned, construct the polynomials $B^{[k-1/k]}(z)$, $A^{[k-1/k]}(z)$, $B^{[k/k+1]}(z)$, and $A^{[k/k+1]}(z)$ directly, and then the approximant of type $[m + k + J + 1/m + k + 1]$ and its predecessor form using (6.6). Increase $m := m + k + 1$ and return to stage 2 to iterate.
5. We reach this stage if $k = 1$ and $H_{0;1}$ is small, or if $k \geq 2$ and $H_{k;k+1}$ is ill conditioned. In this case, the approximant of type $[m + k + J/m + k]$ cannot be constructed accurately, possibly because it does not exist. Increment $k := k + 1$ and return to stage 2 to iterate.

All operations in this procedure are to be understood as referring to the coefficients of the functions involved.

We conclude this section with a brief mention of superfast algorithms in the context of Padé approximation of Hamburger functions. They are introduced by another progression theorem.

Theorem 3.6.4. *Let m, $k \geq 1$ be two integers that are given, together with the two power series*

3.6 Recursive calculation of the coefficients of Padé approximants

$$g(z) = g_0 + g_1 z + \cdots + g_N z^N + \cdots, \quad (6.21)$$

$$h(z) = h_0 + h_1 z + \cdots + h_N z^N + \cdots \quad (6.22)$$

whose ratio $f(z) = g(z)/h(z)$ is a Hamburger function (see Section 5.6). Assume that $P^{[m-1/m]}(z)$, $Q^{[m-1/m]}(z)$ are numerator and denominator Padé polynomials of type $[m-1/m]$ for the rational power series $g(z)/h(z)$ and similarly for $P^{[m/m+1]}(z)$ and $Q^{[m/m+1]}(z)$. With these quantities, two 'remainder series' are defined by

$$d(z) = \{g(z)Q^{[m-1/m]}(z) - h(z)P^{[m-1/m]}(z)\}/z^{2m}, \quad (6.23)$$

$$e(z) = \{g(z)Q^{[m/m+1]}(z) - h(z)P^{[m/m+1]}(z)\}/z^{2m+2}. \quad (6.24)$$

Then $\tilde{e}(z) = e(z)/d(z)$ is also a Hamburger function. Let $A^{[k-1/k]}(z)$, $B^{[k-1/k]}(z)$ be Padé polynomials of type $[k-1/k]$ for the rational power series $\tilde{e}(z) = e(z)/d(z)$, and similarly for type $[k/k+1]$. Then the Padé polynomials of type $[m+k/m+k+1]$ and the Padé approximant of type $[m+k+1/m+k+2]$ for $f(z)$ are given by

$$\begin{bmatrix} P^{[m+k+1/m+k+2]}(z) & P^{[m+k/m+k+1]}(z) \\ Q^{[m+k+1/m+k+2]}(z) & Q^{[m+k/m+k+1]}(z) \end{bmatrix}$$

$$= \begin{bmatrix} P^{[m/m+1]}(z) & P^{[m-1/m]}(z) \\ Q^{[m/m+1]}(z) & Q^{[m-1/m]}(z) \end{bmatrix} \begin{bmatrix} B^{[k/k+1]}(z) & B^{[k-1/k]}(z) \\ -z^2 A^{[k/k+1]}(z) & -z^2 A^{[k-1/k]}(z) \end{bmatrix}.$$

$$(6.25)$$

Proof. The proof is essentially the same as that of Theorem 3.6.1, except that certain existence properties must be established. To this end, let $fc\{g(z)\}$ denote generically the first coefficient of a series $g(z)$, so $fc\{g(z)\} \neq 0$ except when $g(z) = 0$. Because $f(z)$ is a Hamburger function,

$$C(k-1/k) > 0, \quad k = 1, 2, \ldots. \quad (6.26)$$

It follows from (1.1.11) that

$$fc\{h(z)P^{[m+k/m+k+1]}(z) - g(z)Q^{[m+k/m+k+1]}(z)\} =$$

$$-h(0)C(m+k+1/m+k+2)/C(m+k/m+k+1)$$

in the notation of (1.4.8). Likewise, from (6.23), (6.24),

$$fc\{d(z)\} = d(0) = h(0)C(m/m+1)/C(m-1/m), \quad (6.27)$$

$$fc\{e(z)\} = e(0) = h(0)C(m+1/m+2)/C(m/m+1), \quad (6.28)$$

Substituting into (6.7), we obtain

$$\frac{C(m+k+1/m+k+2)}{C(m+k/m+k+1)} = \frac{C(m/m+1)}{C(m-1/m)} \frac{C_{\tilde{e}}(k/k+1)}{C_{\tilde{e}}(k-1/k)}, \quad (6.29)$$

where $C_{\tilde{e}}(l/m)$ denotes the equivalent of the determinant in (1.4.8) for the series $\tilde{e}(z)$ instead of $f(z)$.

It follows from (6.27), (6.28) that $\tilde{e}(0) = C_{\tilde{e}}(0/1) > 0$, and then from (6.29) that $C_{\tilde{e}}(k/k+1) > 0$ for all k.

Notice that recursive application of (6.29) leads to the general result

$$C_{\tilde{e}}(k/k+1) = \left[\frac{C(m-1/m)}{C(m/m+1)}\right]^k \frac{C(m+k+1/m+k+2)}{C(m+1/m+2)} \tilde{e}(0)$$

provided the denominators do not vanish.

Superfast algorithms. If high-order Padé approximants are required in applications, it is essential that rounding error in the calculation be controllable. There is no difficulty about this if the coefficients c_i belong to a finite field. Alternatively, it may be that positivity properties of, for example, Hamburger functions, can be used to establish numerical stability of well-designed numerical methods. With this strong caveat, we proceed to consider superfast algorithms.

In this section, the next, and Sections 4.2, 4.4, we see how Padé approximants of type $[m/m]$ in a normal Padé table can be calculated in $O(m^2)$ operations (where constants of proportionality in the order notation are being omitted). Such computations are referred to as fast algorithms. We now show how a divide and conquer strategy can be used to achieve the same results in $O(m \log_2^2 m)$ operations, which are therefore called superfast.

Suppose that the coefficients $\{c_i\}_{i=0}^{2n}$ are given, and that the Padé approximant of type $[n-1/n]$ is required for $f(z) = \sum c_i z^i$. Then a procedure based on the following plan can be used, where all computations are to be understood as implying that evaluation is required of the coefficients of the polynomials.

Procedure 3.6.2

1. $m := \left[\dfrac{n}{2}\right] - 1, \ k := \left[\dfrac{n+1}{2}\right] - 1.$

2. $g(z) := f(z), \ h(z) := 1.$

3. Compute $P^{[m-1/m]}(z)$, $Q^{[m-1/m]}(z)$, $P^{[m/m+1]}(z)$, $Q^{[m/m+1]}(z)$, for $g(z)/h(z)$.
4. Compute $d(z)$, $e(z)$ using (6.23), (6.24).
5. Compute $A^{[k-1/k]}$, $B^{[k-1/k]}(z)$, $A^{[k/k+1]}(z)$, $B^{[k/k+1]}$ for $e(z)/d(z)$.
6. Compute $P^{[m+k/m+k+1]}(z)$, $Q^{[m+k/m+k+1]}(z)$, $P^{[m+k+1/m+k+2]}(z)$, $Q^{[m+k+1/m+k+2]}(z)$ using (6.25).
7. End.

Notice that $m + k + 2 = n$, so step 6 yields an approximant of type $[n - 1/n]$. We see that steps 3, 5 involve the fundamental computation, and so the computation must be structured recursively, and a proper initialization must be provided. The algorithm is designed for the computation of two sequential Padé approximants for a rational power series, because the procedure is to be recursive and step 5 requires formation of two sequential Padé approximants for a rational function.

We notice that steps 3, 5 each involve a computation of approximately half the size of the given problem. Fast Fourier transform methods for polynomial multiplication in steps 4, 6 involve $O(n \log_2 n)$ operations, and hence the 'time count' $t(n)$ obeys a rule of the form

$$t(n) = 2t(n/2) + O(n \log_2 n).$$

Hence $t(n) = O(n \log_2^2 n)$.

Methods of this kind are remarkably elegant, but because they are so prone to failure we do not elaborate the details here. The fundamental papers on the superfast algorithm are by Brent, Gustavson, and Yun [1980] and Gragg et al. [1982]; strong caveats on their usage are expressed by Ammar and Gragg [1986] and Bunch [1985], and for this reason no attempt at generality has been made.

The identities (6.25) form a generalization of the recursion relations for a continued fraction of the form (4.5.4), and the reproducing property for Hamburger functions generalizes that of Example 1 in Section 5.6.

In conclusion, notice that the methods of this section not only address the problem of the recursive calculation of a paradiagonal sequence of Padé approximants, but also solve the problem of the recursive solution of a Hankel system of linear equations of the form

$$H_{M;M}\mathbf{x} = \begin{bmatrix} c_1 & \cdots & c_M \\ \vdots & & \vdots \\ c_M & \cdots & c_{2M-1} \end{bmatrix} \mathbf{x} = \mathbf{d}. \qquad (6.30)$$

Procedure 3.6.1 (with $L = M$, $J = 0$) is a computational procedure for calculating the elements of $H_{M;M}^{-1}$ in the form (6.15). This representation of $H_{M;M}^{-1}$ is precisely what is needed to solve (6.30).

3.7 Kronecker's algorithm and Cordellier's identity

The identities for calculating Padé approximants that have been given so far in this book all have the property of achieving accuracy-through-order progressively. In this respect, Kronecker's algorithm is quite different. It is used to construct approximants of types $[N - m/m]$ of the antidiagonal sequence with N fixed, $m = 0, 1, 2, \ldots$, and all these approximants have the same accuracy through order.

Beginning with $N + 1$ terms of the usual data series (1.1.1), Kronecker's algorithm is expressed by

Initialization

$$p_{-1}(z) = z^{N+1}, \qquad q_{-1}(z) = 0, \qquad (7.1)$$

$$p_0 = \sum_{i=0}^{N} c_i z^i, \qquad q_0(z) = 1. \qquad (7.2)$$

Recursion. For $j = 0, 1, 2, \ldots, \omega$,

$$\rho_j(z) = \underset{z \to \infty}{\text{principal part}} \{p_{j-1}(z)/p_j(z)\}, \qquad (7.3)$$

$$p_{j+1}(z) = p_j(z)\rho_j(z) - p_{j-1}(z), \qquad (7.4)$$

$$q_{j+1}(z) = q_j(z)\rho_j(z) - q_{j-1}(z). \qquad (7.5)$$

Termination. The algorithm is terminated at the index ω for which $p_{\omega+1}(z) = 0$.

Formula (7.1) is an artificial initialization which is used for its simplicity. Formulas (7.3), (7.4) together imply that $\rho_j(z)$ is the polynomial quotient of the polynomials $p_{j-1}(z)$ divided by $p_j(z)$ and that $-p_{j+1}(z)$ is the remainder. Thus recursive use of (7.3), (7.4) amounts to implementation of the Euclidean algorithm. Equations (7.1)–(7.5) express the algorithm in polynomial form, but the underlying algorithm is an algorithm for the coefficients of these polynomials.

The next theorem establishes the principal properties of the approximants constructed.

Theorem 3.7.1. *The polynomials* $(p_j(z), q_j(z))$ *constructed using Kronecker's algorithm have the accuracy-through-order property that*

$$f(z)q_j(z) - p_j(z) = O(z^{N+1}). \qquad (7.6)$$

For $k = 1, 2, \ldots, \omega$,

$$\deg\{p_k(z)\} = N + 1 - \sum_{j=0}^{k} \deg\{\rho_j(z)\}, \qquad (7.7)$$

$$\deg\{q_k(z)\} = \sum_{j=0}^{k-1} \deg\{\rho_j(z)\}, \qquad (7.8)$$

and
$$\deg\{\rho_k(z)\} \geq 1. \qquad (7.9)$$

If $q_j(0) \neq 0$, $p_j(z)/q_j(z)$ is the Padé approximant of type $[N - j/j]$.

Proof. From the initialization, it is clear that (7.6) holds for the cases $j = -1, 0$. Assume that (7.6) holds for the cases $j = k - 1, k$. From (7.4), (7.5),

$$\begin{aligned} f(z)q_{k+1}(z) - p_{k+1}(z) &= [f(z)q_k(z) - p_k(z)]\rho_k(z) \\ &\quad - [f(z)q_{k-1}(z) - p_{k-1}(z)] \\ &= O(z^{N+1}), \end{aligned}$$

and hence (7.6) is proved by induction.

By (7.3) of the construction,

$$\deg\{\rho_j(z)\} = \deg\{p_{j-1}(z)\} - \deg\{p_j(z)\}.$$

The rest of the results follow immediately.

The polynomials $(p_j(z), q_j(z))$ constructed using Kronecker's algorithm do not have either the determinantal or the Baker normalization. Their normalization is given by the next theorem, in which $\dot\pi$ denotes the leading coefficient of a polynomial $\pi(z)$.

Theorem 3.7.2. *The leading coefficients of $p_j(z)$, $q_j(z)$ are given by*

$$\dot p_j = [\dot\rho_j \dot\rho_{j-1} \cdots \dot\rho_0]^{-1}, \qquad (7.10)$$

$$\dot q_j = \dot\rho_0 \dot\rho_1 \cdots \dot\rho_{j-1}, \qquad (7.11)$$

and consecutive approximants are related by

$$p_{j-1}(z)q_j(z) - p_j(z)q_{j-1}(z) = z^{N+1}. \qquad (7.12)$$

Proof. Equations (7.10), (7.11) follow directly from (7.4), (7.5), as does also

$$\det\begin{vmatrix} p_{j+1}(z) & p_j(z) \\ q_{j+1}(z) & q_j(z) \end{vmatrix} = \det\begin{vmatrix} p_j(z) & p_{j-1}(z) \\ q_j(z) & q_{j-1}(z) \end{vmatrix}. \qquad (7.13)$$

From the initialization,

$$\det\begin{vmatrix} p_0(z) & p_{-1}(z) \\ q_0(z) & q_{-1}(z) \end{vmatrix} = -z^{N+1},$$

and (7.12) follows by using (7.13) iteratively.

Theorem 3.7.3. *Consider $(p_k(z), q_k(z))$ formed using Kronecker's algorithm. Let*

$$L = \deg\{p_k(z)\}, \quad M = \deg\{q_k(z)\}, \quad v = N - L - M + 1.$$

Then
$$v = \deg\{\rho_k(z)\}. \tag{7.14}$$

If $q_k(0) \neq 0$, $(p_k(z), q_k(z))$ constitute the minimal entry of a block in the Padé table. This block is nontrivial if $v > 1$.

Proof. Property (7.14) follows from (7.7) and (7.8).

Suppose that $p_k(z), q_k(z)$ share a nontrivial common factor $d(z)$. Since $q_k(0) \neq 0$, z is not a factor of $q_k(z)$, nor is it a factor of $d(z)$. By hypothesis, $d(z)$ factors the left-hand side of (7.12), but not the right-hand side. Therefore $d(z)$ can only be a constant.

If $v > 1$, it is readily seen that all approximants of $f(z)$ of types $[l/m]$ with $l \geq L$, $m \geq M$, $l + m \leq N$ lie in part of a nontrivial block whose elements all equal $p_k(z)/q_k(z)$. Since $p_k(z), q_k(z)$ have no nontrivial common factor, the uniqueness theorem shows that they form the minimal element of the block.

Theorem 3.7.3 elucidates one way in which Kronecker's algorithm traverses a block, and this situation is described pictorially in Figure 3.7.1. In the other case, it happens that $q_k(0) = 0$, and this case is treated next.

Theorem 3.7.4. *Consider $(p_k(z), q_k(z))$ formed using Kronecker's algorithm and let $L = \deg\{p_k(z)\}$, $M = \deg\{q_k(z)\}$. In this case, suppose that $q_k(0) = 0$, and let μ be the greatest power of z such that z^μ divides $q_k(z)$. Then $p_k(z), q_k(z)$ share no other polynomial factors. The fraction $p_k(z)/q_k(z)$ is the minimal element of precise type $[L - \mu/M - \mu]$ of a nontrivial block in the Padé table; it does not have accuracy through order z^N, and it is not a Padé approximant according to the Baker definition.*

Proof. Adapting the previous proof, it follows that z^μ is the greatest nontrivial common factor of $p_k(z)$ and $q_k(z)$. After canceling this factor in $p_k(z)/q_k(z)$, irreducibility follows as in the previous proof. The remainder of the proof follows from the block structure theorem, Theorem 1.4.3.

This theorem elucidates the other way in which Kronecker's algorithm traverses a block, and the situation is shown diagrammatically in Figure 3.7.2.

3.7 Kronecker's algorithm and Cordellier's identity 109

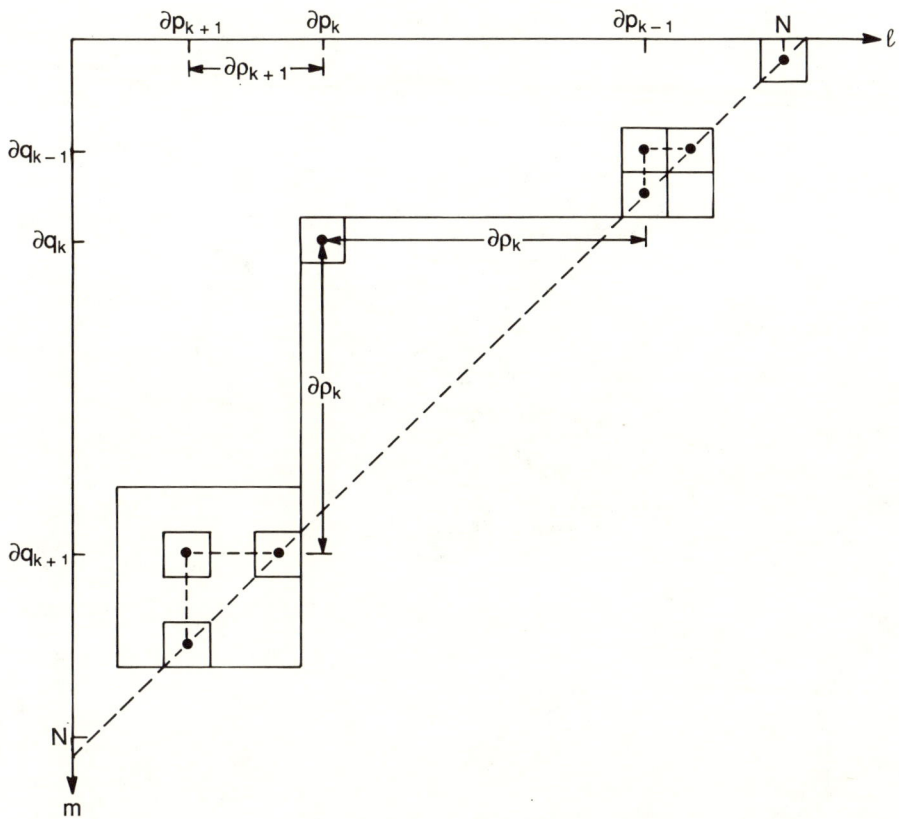

Figure 3.7.1. Display of the degrees of successive numerator and denominator polynomials generated using Kronecker's algorithm when $b_k(0) \neq 0$. The symbol ∂ denotes 'degree of'.

Example. Construct the approximants of types $[5 - m/m]$ for $m = 0, 1, 2, 3$ for

$$f(z) = 1 - z + 2z^2 + z^5. \tag{7.15}$$

Solution. Using (7.1)–(7.5), we obtain

$p_{-1}(z) = z^6, \quad q_{-1} = 0$ \hfill (artificial);

$p_0(z) = 1 - z + 2z^2 + z^5, \quad q_0(z) = 1$ \hfill (type 5/0);

$\rho_0(z) = \text{PP}\left\{\dfrac{z^6}{1 - z + 2z^2 + z^5}\right\} = z;$

$p_1 = z - z^2 + 2z^3, \quad q_1(z) = z$ \hfill (type 4/1);

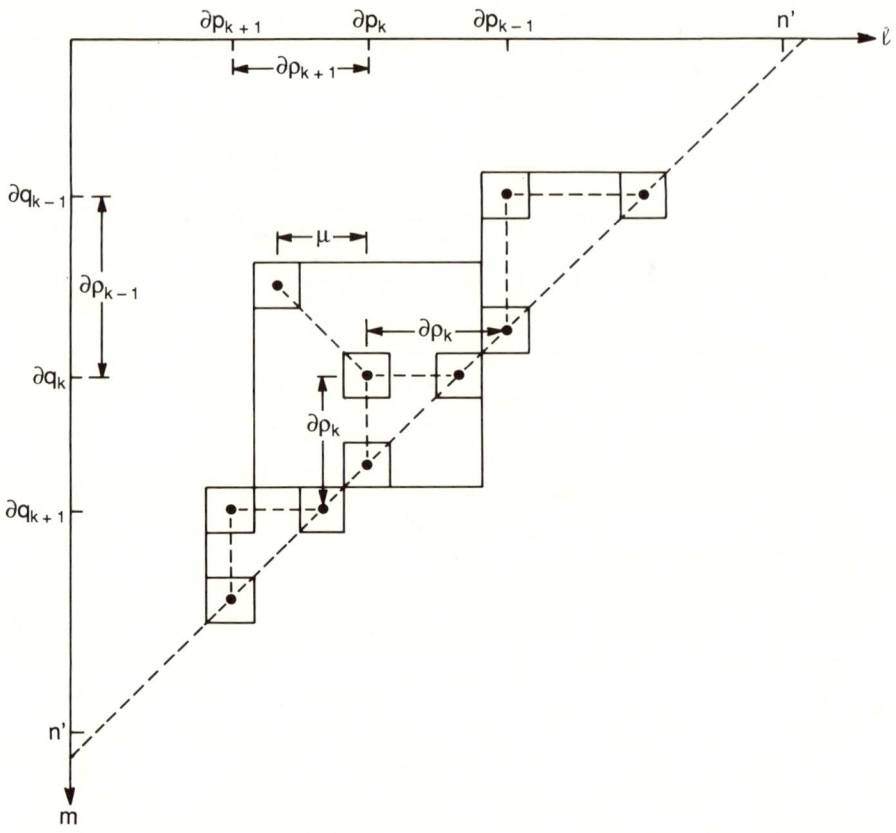

Figure 3.7.2. Display of the degrees of successive numerator and denominator polynomials generated using Kronecker's algorithm when $b_k(z) = O(z^\mu)$ precisely and $\mu > 0$.

$$p_1(z) = \mathrm{PP}\left\{\frac{1 - z + 2z^2 + z^5}{z - z^2 + 2z^3}\right\} = \frac{z^2}{2} + \frac{z}{4} - \frac{1}{8};$$

$$p_2(z) = -1 + \frac{7z}{8} - \frac{13z^2}{8}, \quad q_2(z) = -1 - \frac{z}{8} + \frac{z^2}{4} + \frac{z^3}{2} \quad \text{(type 2/3)};$$

We observe that $p_0(z)/q_0(z)$, $p_2(z)/q_2(z)$ are the Padé approximants of types [5/0], [2/3] for $f(z)$. The Padé approximants of types [4/1], [3/2] do not exist according to the Baker definition; see Theorem 3.7.4 and Figure 3.7.1. $p_1(z)/q_1(z) = 1 - z + 2z^2$ is the Padé approximant of type [2/0], and it does not have accuracy through order z^5.

Theorem 3.7.5. *Kronecker's algorithm is a reliable algorithm: each Padé approximant which exists in the antidiagonal sequence is constructed by the algorithm.*

3.7 Kronecker's algorithm and Cordellier's identity

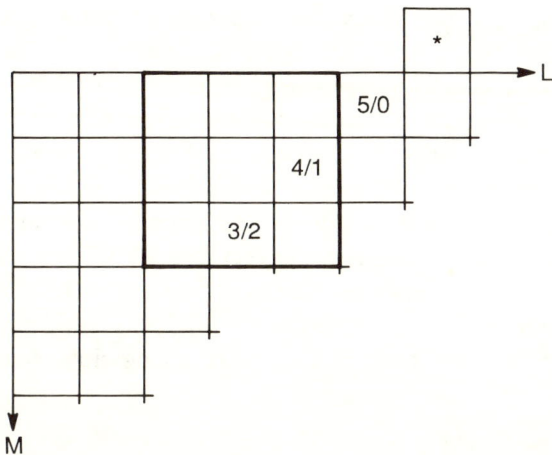

Figure 3.7.3. The block structure for the example of Kronecker's algorithm.

Proof. Suppose that $A(z)/B(z)$ is the Padé approximant of type $[N - \hat{M}/\hat{M}]$ whose existence is known a priori. Let $p_k(z)$, $q_k(z)$ be the polynomials constructed using Kronecker's algorithm such that $q_k(z)$ has the greatest degree $m \leq \hat{M}$. If $A(z)/B(z)$ is not found in the sequence, it must be that $\deg\{q_{k+1}(z)\} > \hat{M}$. Then

$$\deg\{\rho_k(z)\} = \deg\{q_{k+1}(z)\} - \deg\{q_k(z)\} > \hat{M} - m.$$

If $q_k(0) \neq 0$, the situation is governed by Theorem 7.3 and is depicted in Figure 3.7.1; otherwise the situation is governed by Theorem 7.4 and is depicted in Figure 3.7.2. In either case, because

$$\hat{M} < m + \deg\{\rho_k(z)\},$$

it must be that $A(z)/B(z)$ lies in the block containing $p_k(z)/q_k(z)$, and hence $A(z)/B(z) = p_k(z)/q_k(z)$.

Partly because Kronecker's algorithm is a reliable algorithm, it has a famous application in signal processing [McEliece and Shearer, 1978]. In this application, the polynomial coefficients are binary numbers, and so the decision about whether or not a particular coefficient is zero is easily made automatically. When the coefficients belong to the field $\mathbb{R}$ or $\mathbb{C}$, identification of a numerical zero is harder, and suitable precautions must be taken. The issues of overflow and underflow described in Section 4.1 are also relevant. Nevertheless, with these caveats, Kronecker's algorithm has many attractive features.

The formalism of Kronecker's algorithm is now going to be used to prove Cordellier's identity. This is an identity connecting four elements on the edges of a block in the Padé table with the minimal element of the block. Suppose that the block has dimension r (in the sense that the trivial block has size $r = 1$). Cordellier's identity is a generalization of Wynn's identity to blocks of size $r > 1$.

For a block whose central or common entry is $C(z)$, let $N(z)$ be any entry on the north side of the block. Let $W(z)$ denote the entry on the west side of the block on the same antidiagonal as $N(z)$, and let $E(z)$ denote the entry on the east side of the block on the same diagonal as $N(z)$. Finally, let $S(z)$ complete the rectangle WNES on the south side. Because $N(z)$ may be chosen in r different places on the north side of the block, there are in fact r different Cordellier identities. The general situation is shown in Figure 3.7.4, where all possible degeneracies are allowed for.

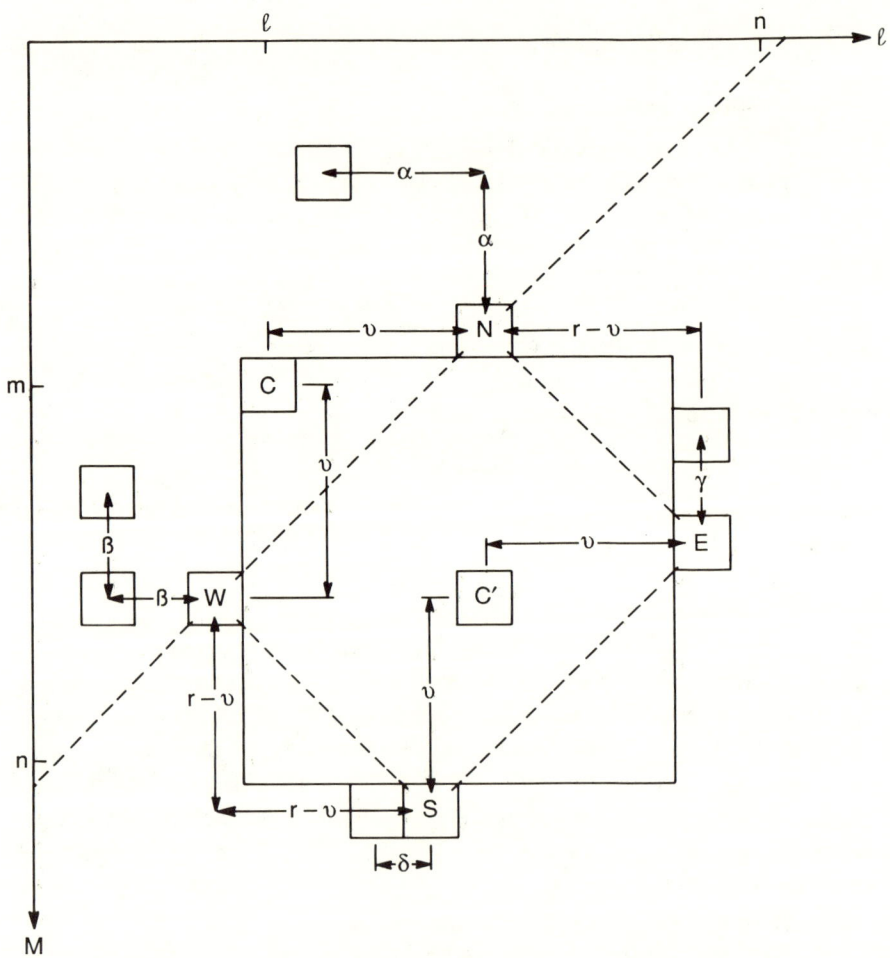

Figure 3.7.4. The general block structure for Cordellier's identity.

3.7 Kronecker's algorithm and Cordellier's identity

If $C(z)$ is of type $[l/m]$ and $N(z)$ is of type $[l + v/m - 1]$, then

$$\deg\{p_C(z)\} = l, \quad \deg\{p_N(z)\} = l + v, \quad \deg\{p_W(z)\} = l + 1 - \beta,$$
$$\deg\{q_C(z)\} = m, \quad \deg\{q_N(z)\} = m + 1 - \alpha, \quad \deg\{q_W(z)\} = m + v$$

(7.16)

for some $\alpha, \beta \geqslant 0$. From (7.5), it also follows that

$$v = \deg\{\rho_C(z)\}; \tag{7.17}$$

the entries $N(z)$, $W(z)$ are located on the antidiagonal

$$n = l + m + v - 1. \tag{7.18}$$

Likewise, the other entries associated with the Cordellier rectangle have degrees

$$\deg\{p_E(z)\} = l + r, \quad \deg\{p_{C'}(z)\} = l + r - v,$$
$$\deg\{p_S(z)\} = l + r - v - \delta - 1,$$
$$\deg\{q_E(z)\} = m + r - v - \gamma - 1, \quad \deg\{q_{C'}(z)\} = m + r - v,$$
$$\deg\{q_S(z)\} = m + r \tag{7.19}$$

for some $\gamma, \delta \geqslant 0$, where C' labels the polynomials $p_{C'}(z)$, $q_{C'}(z)$ corresponding to the element $C'(z)$ and is associated with the antidiagonal

$$n' = n + 2(r - v) \tag{7.20}$$

on which $E(z)$ and $S(z)$ lie.

With this specification we can derive the following theorem.

Theorem 3.7.6 [Cordellier, 1974, 1979b, 1989]

$$[N(z) - C(z)]^{-1} + [S(z) - C(z)]^{-1} = [E(z) - C(z)]^{-1}$$
$$+ [W(z) - C(z)]^{-1}. \tag{7.21}$$

Proof. The identities

$$\left.\begin{array}{l} p_N(z)q_C(z) - q_N(z)p_C(z) = z^{n+1} \\ p_C(z)q_W(z) - q_C(z)p_W(z) = z^{n+1} \\ p_E(z)q_{C'}(z) - q_E(z)p_{C'}(z) = z^{n'+1} \\ p_{C'}(z)q_{S'}(z) - q_{C'}(z)p_S(z) = z^{n'+1} \end{array}\right\} \tag{7.22}$$

all represent (7.12). By using (7.4), (7.5) as well, one obtains

$$\left.\begin{array}{l} p_N(z)q_W(z) - q_N(z)p_W(z) = z^{n+1}\rho_C(z) \\ p_E(z)q_S(z) - q_E(z)p_S(z) = z^{n'+1}\rho_{C'}(z) \end{array}\right\}. \tag{7.23}$$

From (7.22) one obtains

$$N(z) - C(z) = \frac{p_N(z)q_C(z) - q_N(z)p_C(z)}{q_N(z)q_C(z)} = \frac{z^{n+1}}{q_N(z)q_C(z)} \quad (7.24)$$

and similarly

$$W(z) - C(z) = -z^{n+1}/(q_W(z)q_C(z)). \quad (7.25)$$

From (7.5), (7.24), and (7.25) one obtains

$$[N(z) - C(z)]^{-1} - [W(z) - C(z)]^{-1} = \frac{q_C(z)^2 p_C(z)}{z^{n+1}} \quad (7.26)$$

and similarly

$$[E(z) - C(z)]^{-1} - [S(z) - C(z)]^{-1} = \frac{q_{C'}(z)^2 p_{C'}(z)}{z^{n'+1}}. \quad (7.27)$$

To show that (7.26) and (7.27) are equal, the quantities $p_C(z)$, $p_{C'}(z)$ must be related to each other. To do this, introduce the notation $\overset{v}{\simeq}$ to indicate that the $v+1$ terms of leading order on each side of the relationship are equal. From (7.22), (7.23), it follows that

$$p_N(z)q_C(z) \overset{v}{\simeq} z^{n+1}, \qquad p_E(z)q_{C'}(z) \overset{v}{\simeq} z^{n'+1},$$

$$p_C(z)q_W(z) \overset{v}{\simeq} z^{n+1}, \qquad p_{C'}(z)q_{C'}(z) \overset{v}{\simeq} z^{n'+1},$$

$$p_N(z)q_W(z) \overset{v}{\simeq} z^{n+1}p_C(z), \qquad p_E(z)q_S(z) \overset{v}{\simeq} z^{n'+1}p_{C'}(z).$$

Hence,

$$\left.\begin{array}{l}\dfrac{p_C(z)q_C(z)^2}{z^{n+1}} \overset{v}{\simeq} \dfrac{p_N(z)q_W(z)q_C(z)^2}{z^{2n+2}} \overset{v}{\simeq} [C(z)]^{-1}, \\[2mm] \dfrac{p_{C'}(z)q_{C'}(z)^2}{z^{n'+1}} \overset{v}{\simeq} \dfrac{p_E(z)q_S(z)q_{C'}(z)^2}{z^{2n'+2}} \overset{v}{\simeq} [C'(z)]^{-1} = [C(z)]^{-1}.\end{array}\right\} \quad (7.28)$$

According to Theorem 3.7.4, $q_{C'}(z)$ is to be divided by z^{r-v} to become the denominator of the minimal Padé approximant of the block, namely, $q_C(z)$, up to a constant factor:

$$\frac{q_C(z)}{\dot{q}_C} = \frac{q_{C'}(z)}{\dot{q}_{C'} z^{r-v}}. \quad (7.29)$$

By substituting (7.29) into (7.28), one obtains

$$p_C(z)\dot{q}_C^2 \overset{v}{\simeq} p_{C'}(z)\dot{q}_{C'}^2. \quad (7.30)$$

Because $p_C(z)$, $p_{C'}(z)$ are both polynomials of precise degree v, (7.30) becomes an equality connecting them:

$$p_C(z)\dot{q}_C^2 = p_{C'}(z)\dot{q}_{C'}^2. \quad (7.31)$$

From (7.20), (7.29), and (7.31),

$$\frac{q_C(z)^2 \rho_C(z)}{z^{n+1}} = \frac{q_{C'}(z)^2 \rho_{C'}(z)}{z^{n'+1}}$$

and equality of (7.26), (7.27) follows directly, as required.

The original proof of the remarkable identity (7.21) by Cordellier [1974] appeared in final form in his thesis [1989]. It is an elegant, geometric proof which has the advantage of generalizing directly to the vector case, but the disadvantage of involving limiting procedures whose independence in cases such as (7.16) with $\alpha, \beta \geq 0$ is hard to follow. Determinantal proofs were given by Cordellier [1979b] and Beckermann, Neuber, and Muhlbach [1992] which apply only to the scalar case. The vector case was first proved analytically by Graves-Morris and Jenkins [1986]; the method given here is that of Graves-Morris and Roberts [1994], which generalizes to the vector and matrix cases (see Section 8.4).

3.8 The Q.D. algorithm and the root problem

If we are given the formal expansion of $f(z)$,

$$f(z) = c_0 + c_1 z + c_2 z^2 + \cdots, \tag{8.1}$$

and we know that $f(z)$ is a meromorphic function which is analytic at the origin, it is natural to wonder if the Padé method is useful for locating the poles and zeros of $f(z)$. Indeed, a vast theory of root solving exists; we will give a simple method which works when the poles of $f(z)$ have distinct moduli, and so also do the zeros. In this case, we may order the poles as $u_1, u_2, \ldots$ with

$$|u_1| < |u_2| < \cdots < |u_M| < \cdots \tag{8.2}$$

and the zeros as $v_1, v_2, \ldots$ with

$$|v_1| < |v_2| < \cdots < |v_L| < \cdots. \tag{8.3}$$

Then $f(z)$ has the representation

$$f(z) = \frac{\sum_{i=1}^{L}\left(1 - \frac{z}{v_i}\right)}{\sum_{j=1}^{M}\left(1 - \frac{z}{u_j}\right)} h(z), \tag{8.4}$$

where $h(z)$ is analytic and zero-free in $|z| < \min(|v_{L+1}|, |u_{M+1}|)$. The special cases where $f(z)$ has either no poles or no zeros cause no problems in principle. However, the condition that the poles and zeros respectively have distinct moduli is important both in principle and in practice. To find the poles of $f(z)$, which means the numerical values of u_j, we form $[L/M]$ Padé approximants to $f(z)$. Keeping M fixed, and with increasing order of approximation, we expect that

$$Q^{[L/M]}(z) \to C(L/M) \sum_{j=1}^{M} \left(1 - \frac{z}{u_i}\right) \text{ as } L \to \infty. \tag{8.5}$$

In fact, this result is a consequence of de Montessus's theorem.

If we write

$$Q^{[L/M]}(z) = q_M^{[L/M]} z^M + q_{M-1}^{[L/M]} z^{M-1} + \cdots + q_0^{[L/M]}$$

and use the explicit determinantal formula (5.1) for $q_M^{[L/M]}$, $q_0^{[L/M]}$, it follows from the usual expression for the product of the roots that

$$\prod_{j=1}^{M} u_j = \lim_{L \to \infty} \frac{C(L - 1/M)}{C(L/M)}. \tag{8.6}$$

Similarly, by considering $M - 1$ roots,

$$\prod_{j=1}^{M-1} u_j = \lim_{L \to \infty} \frac{C(L/M - 1)}{C(L + 1/M - 1)}. \tag{8.7}$$

We define

$$u(L/M) = \frac{C(L + 1/M - 1)C(L - 1/M)}{C(L/M - 1)C(L/M)}. \tag{8.8}$$

For the particular case of $M = 1$, the appropriate definition is

$$u(L/1) = c_{L-1}/c_L. \tag{8.9}$$

Then, from (8.6), (8.7), and (8.8),

$$u_M = \lim_{L \to \infty} u(L/M). \tag{8.10}$$

Recalling the duality theorem (Theorem 1.5.1), we find that the reciprocal of $[L/M]$ is the $[M/L]$ Padé approximant of $\{f(z)\}^{-1}$. Let $C'(L/M)$ denote the Hankel determinant of the $[M/L]$ Padé approximant of $g(z) = \{f(z)\}^{-1}$. With this notation, we find, as in (8.6), that

$$\prod_{j=1}^{L} v_j = \lim_{M \to \infty} \frac{C'(M - 1/L)}{C'(M/L)}, \tag{8.11}$$

and, corresponding of (8.8), we define

3.8 The Q.D. algorithm and the root problem

$$v(L/M) = \frac{C'(M-1/L)}{C'(M/L)} \frac{C'(M+1/L-1)}{C'(M/L-1)} \qquad (8.12)$$

$$= \frac{C(L-1/M+1)C(L/M-1)}{C(L-1/M)C(L/M)}, \qquad (8.13)$$

where (8.13) follows from (8.12) by Hadamard's formula (2.4.9). For the particular case of $L = 1$, the appropriate definition is

$$v(1/M) = g_{M-1}/g_M, \qquad (8.14)$$

where $g(z) = \sum_{i=0}^{\infty} g_i z^i$. By the construction (8.12), the Lth zero of $g(z)$ is given by

$$\lim_{M \to \infty} v(L/M) = v_L. \qquad (8.15)$$

To determine the numerical values of u_M, v_l, we need some computational rules.

Product rule

$$u(L/M)v(L/M) = u(L/M+1)v(L+1/M). \qquad (8.16)$$

Proof. From the definitions (8.8), (8.13),

$$u(L/M)v(L/M) = \frac{C(L+1/M-1)C(L-1/M+1)}{C(L/M)^2}$$

$$= u(L/M+1)v(L+1/M).$$

Addition rule

$$u(L/M+1) + v(L+1/M) = u(L+1/M+1) + v(L+1/M+1). \qquad (8.17)$$

Proof. Using the definitions (8.8), (8.13), and Sylvester's identity,

$$u(L/M+1) - v(L+1/M+1)$$

$$= \frac{C(L+1/M)C(L-1/M+1)C(L+1/M+1)}{C(L/M)C(L/M+1)C(L+1/M+1)}$$

$$- \frac{C(L/M)C(L/M+2)C(L+1/M)}{C(L/M)C(L/M+1)C(L+1/M+1)}$$

$$= \frac{C(L+1/M)[C(L/M+1)]^2}{C(L/M)C(L/M+1)C(L+1/M+1)}$$

$$= \frac{C(L+1/M)C(L/M+1)}{C(L/M)C(L+1/M+1)}.$$

Similarly, we find $u(L+1/M+1) - v(L+1/M)$ yields the same answer.

Having established these identities, we display them pictorially in Figure 3.8.1. All the quantities $u(L/M)$, $v(L/M)$, for $L, M = 0, 1, 2, \ldots$, can be constructed with the aid of these identities (8.16), (8.17), the initializing values $u(L/1) = c_{L-1}/c_L$ given by (8.9), and the artificial initializing values

$$v(L/0) = 0, \quad L = 0, 1, 2, \ldots, \tag{8.18}$$

for the first column and

$$u(0/M) = 0, \quad M = 1, 2, 3, \ldots, \tag{8.19}$$

for the first row. We verify that (8.18) gives the correct initializing values by using (8.17) with $M = 0$ together with (8.8) and (8.13). Likewise, (8.19) is verified by using (8.17) with $L = 0$ together with (8.8) and (8.13). The quantities $u(L/M)$, $v(L/M)$ are usually displayed in the u–v table [Gragg, 1972], shown in Table 3.8.1.

Summary. The first two columns and the first row initialize the construction of the u–v table, according to (8.9), (8.18), and (8.19). The other entries are constructed by the Q.D. algorithm, expressed by (8.16) and (8.17). The poles $u_1, u_2, u_3, \ldots$ of $f(z)$ are shown as the limiting values of the 'u-columns', and the zeros $v_1, v_2, v_3, \ldots$ of $f(z)$ are shown as limiting values of the 'v-rows'.

In principle, the rate of convergence of the Q.D. algorithm for poles and zeros is geometric. To see this, we refer to the analysis of Section 6.2. In (6.2.11), we find that

$$|\det C| = |\det C(L/M)|, \tag{8.20}$$

and from (6.2.22) we note that the dominant part of (6.20) is given by

$$|\det D| = K \prod_{i=1}^{M} |u_i|^{-L}, \tag{8.21}$$

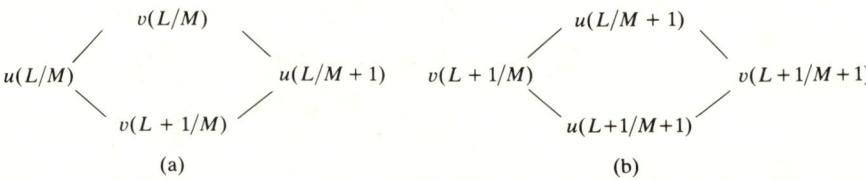

Figure 3.8.1. Pictorial representation of the rhombus rules: (a) product rule, (b) addition rule.

3.8 The Q.D. algorithm and the root problem

Table 3.8.1. *The u–v table.*

	0		0		0		0		
0		$v(1/1)$		$v(1/2)$		$v(1/3)$		$v(1/4)$	... v_1
	$u(1/1)$		$u(1/2)$		$u(1/3)$		$u(1/4)$		
0		$v(2/1)$		$v(2/2)$		$v(2/3)$		$v(2/4)$	... v_2
	$u(2/1)$		$u(2/2)$		$u(2/3)$		$u(2/4)$		
0		$v(3/1)$		$v(3/2)$		$v(3/3)$		$v(3/4)$	... v_3
	$u(3/1)$		$u(3/2)$		$u(3/3)$		$u(3/4)$		
	$\vdots$		$\vdots$		$\vdots$		$\vdots$		
	u_1		u_2		u_3		u_4		

where K is a nonzero constant (independent of L). An inspection of the equations (6.2.11), (6.2.14)–(6.2.22), and (6.2.33) shows that

$$|\det C(L/M)| = K \left|\prod_{i=1}^{M} u_i^{-L}\right| \left\{1 + O\left(\left(\frac{|u_M|}{R}\right)^L\right)\right\}. \tag{8.22}$$

Equation (8.22) is Hadamard's formula. It holds under the conditions that $f(z)$ is analytic in the disc $|z| \leq R$ except for precisely M poles, counting multiplicities, in the annulus $0 < |z| < R$.

Let us assume that the poles at u_{M-1}, u_M, and u_{M+1} have distinct moduli:

$$|u_{M-1}| < |u_M| < |u_{M+1}|. \tag{8.23}$$

Substitute (8.22) into (8.8) to show that

$$u(L/M) = u_M \{1 + O(\theta^L)\}, \tag{8.24}$$

and

$$\theta = \max\left\{\left|\frac{u_{M-1}}{u_M}\right|, \left|\frac{u_M}{u_{M+1}}\right|\right\}.$$

The hypothesis (8.23) ensures that $0 < \theta < 1$, and (8.24) shows that the u-columns of the u–v table converge geometrically, in principle.

In practice, the Q.D. algorithm, as expressed in the summary preceding, is unstable. Rounding error accumulates in the u-columns whose limits are theoretically $u_2, u_3, u_4, \ldots$, according to (8.10) and (8.24). To demonstrate this, we consider

$$\frac{v(L+1/M)}{v(L/M)} = \frac{C(L/M+1)}{C(L-1/M+1)} \frac{C(L+1/M-1)}{C(L/M-1)} \frac{C(L-1/M)}{C(L+1/M)},$$

which is obtained from (8.13). Hence, from (8.10),

$$\lim_{L \to \infty} \frac{v(L+1/M)}{v(L/M)} = \frac{u_M}{u_{M+1}},$$

Table 3.8.2. *Elements of the u–v table showing schematically the direction of calculation with the progressive form of the Q.D. algorithm.*

		0		0		0		0
0		v(1/1)		v(2/1)		v(3/1)		v(4/1)
	u(1/1)		(1)		(2)		(3)	
0		(11)		(12)		(13)		
	u(2/1)		(21)		(22)			
0		(31)		(32)				
	u(3/1)		(41)					
	⋮		⋮					
	u_1		u_2					

and we crudely estimate $v(L/M)$ by the formula

$$v(L/M) \approx (u_M/u_{M+1})^L.$$

Consequently, $v(L/M) \to 0$ as $L \to \infty$. We deduce that calculation of the second column $\{v(L/1), L = 0, 1, 2, \ldots\}$ necessarily involves substantial rounding error (low relative precision), introduced by cancellation of comparable quantities in the first u-column. This low relative precision is directly transmitted by the multiplication rule (8.16) to each element of the column $\{u(L/2), L = 0, 1, 2, \ldots\}$. Hence calculation of u_2 by the basic Q.D. algorithm is necessarily unstable, and a similar argument extends to the poles u_3, u_4, ... To remedy this instability, Henrici [1958] proposed the progressive form of the Q.D. algorithm. The order of calculation is changed so as to avoid the necessarily inaccurate arithmetic computations.

Progressive Q.D. algorithm. This algorithm may be used for the computation of the poles of $f(z)$, provided the moduli of these poles are distinct. It is initialized by construction of the coefficients g_i of the power series

$$g(z) = \sum_{i=0}^{\infty} g_i z^i = [f(z)]^{-1} = \left[\sum_{i=0}^{\infty} c_i z^i\right]^{-1}$$

from the coefficients $\{c_i\}$. The $\{g_i, i = 1, 2, \ldots\}$ are constructed iteratively using the identity

$$g_i = c_0^{-1} \sum_{k=0}^{i-1} g_k c_{i-k}.$$

Hence the quantities

$$v(1/M) = g_{M-1}/g_M, \quad M = 1, 2, \ldots,$$

3.8 The Q.D. algorithm and the root problem

are constructed, according to (8.14). The other initializing equations (8.9), (8.18), and (8.19) are retained. The progressive form of the Q.D. algorithm requires that the order of calculation using (8.16), (8.17) be as indicated in Table 3.8.2. The entries in this schematic section of the u–v table are calculated in numerical order 1, 2, 3, ..., 11, 12, 13, ..., 21, 22, ... as indicated there.

Notice that the progressive form of the Q.D. algorithm may be naturally regarded as a practical method for finding the zeros of a meromorphic function $g(z) = \sum_{i=0}^{\infty} g_i z^i$, defined by its power series coefficients g_i, provided that these zeros of $g(z)$ are known to have distinct moduli. For a treatment of the difficult case where the zeros of $g(z)$ have equal moduli, we refer to Henrici [1974, p. 642], or to Rutishauser [1954, p. 35].

4
Connection with continued fractions

4.1 Definitions, recursion relations, and computation

In this chapter we do not aspire to summarize the companion volume of Jones and Thron [1980] which is devoted to the general theory of continued fractions. Here, we set out to present a working knowledge of the basic concepts of continued fractions, so that we may give a self-contained account of how continued-fraction theory supplements our understanding of Padé approximation. The discovery of continued fractions in the West seems to have been made by Bombelli [1572]; Jones and Thron [1980] and Brezinski [1990] give historical surveys. In Section 4.7, we quote the basic convergence theorems for general continued fractions, and refer to the companion volume for the proofs. We are primarily concerned with continued fractions associated with power series, for which the continued fractions happen to be Padé approximants. Indeed, in the next chapter we will see that S-fractions are associated with Stieltjes series and that real J-fractions are associated with Hamburger series. The convergents of these fractions form simple sequences in the Padé table.

There is no doubt that part of Padé approximation theory grew out of continued-fraction theory. We choose to regard the Padé table as the fundamental set of rational approximants, and the convergents of various continued fractions derived from power series as particular subsequences of the Padé table. We suggest that which continued-fraction representation is the most useful is often seen most clearly by considering first which sequence from the Padé table has the desired asymptotic behavior or rate of convergence. This view does not mirror the historical development of the subject.

4.1 Definitions, recursion relations, and computation

A continued fraction has the general form

$$b_0 + \cfrac{a_1}{b_1 + \cfrac{a_2}{b_2 + \cfrac{a_3}{b_3 + \cdots}}} \qquad (1.1)$$

The entries in (1.1), a_i and b_i, are called the elements of the continued fraction. They are usually real or complex numbers. The fraction may be written more compactly as

$$b_0 + \frac{a_1}{b_1 +} \frac{a_2}{b_2 +} \frac{a_3}{b_3 +} \cdots \qquad (1.2)$$

with precisely the same meaning as (1.1). By truncating the fraction, we define its convergents, which we denote by ratios A_i/B_i for $i = 0, 1, 2, \ldots$. We find

$$\frac{A_0}{B_0} = b_0, \quad \frac{A_1}{B_1} = b_0 + \frac{a_1}{b_1} = \frac{b_0 b_1 + a_1}{b_1},$$

$$\frac{A_2}{B_2} = b_0 + \frac{a_1}{b_1 + a_2/b_2} = \frac{b_0 b_1 b_2 + a_2 b_0 + a_1 b_2}{b_1 b_2 + a_2}. \qquad (1.3)$$

If the fraction has only a finite number of elements, it takes the form

$$b_0 + \frac{a_1}{b_1 +} \frac{a_2}{b_2 +} \cdots \frac{a_n}{+ b_n}. \qquad (1.4)$$

This is equivalent to (1.2) with $a_{n+1} = 0$; (1.4) is called a terminating fraction. The value of a terminating fraction is defined by finite arithmetic. We also note that (1.4) is the $(n + 1)$st convergent of (1.2).

In general, a continued fraction is said to converge and have the value v if

$$\lim_{n \to \infty} \left(\frac{A_n}{B_n} \right) = v. \qquad (1.5)$$

In other words, provided the limit of the ratios A_n/B_n as $n \to \infty$ exists, it defines a value of the continued fraction. Otherwise, the continued fraction is said to diverge.

The name 'series' is used to describe $\sum_{i=0}^{\infty} c_i$, meaning a set of numbers $c_0, c_1, c_2, \ldots$ to be added. The word 'series' is also used as the value of the sum indicated, provided this value is finite. The same verbal ambiguity arises with continued fractions. Expressions such as (1.1), (1.2), or

$$K\left(\frac{a_n}{b_n}\right) \quad \text{or} \quad \underset{i=1}{\overset{\infty}{\Phi}} \frac{a_i}{b_i}$$

are to be found in the literature. They denote the fact that the pairs (a_1, b_1), (a_2, b_2), (a_3, b_3), ... define the continued fraction, or else the value of the fraction if it converges. Once noticed, the ambiguity causes no confusion, and is unimportant in practice.

Part of the definition (1.5) of convergence of a continued fraction refers to the ratios A_n/B_n which are the values of the convergents of the continued fraction. It is possible to construct different continued fractions which have all their convergents equal in value. Such fractions are called equivalent, and, by definition, equivalent fractions all have the same value.

As an example of an equivalence transformation, consider

$$\cfrac{a_1}{b_1 + \cfrac{a_2}{b_2 + \cfrac{a_3}{b_3 + \cdots}}} = \cfrac{(a_1/b_1)}{1 + \cfrac{(a_2/b_1 b_2)}{1 + \cfrac{(a_3/b_2 b_3)}{1 + \cdots}}}. \qquad (1.6)$$

By division of the 'first' numerator and denominator within (1.6) by b_1, and by division of the 'nth' numerator and denominator within (1.6) by b_n for $n = 2, 3, 4, \ldots$, the denominator elements have been reduced to unity. The values of the convergents are unaltered, but the elements of the derived fraction are

$$\left(\frac{a_1}{b_1}, 1\right), \quad \left(\frac{a_2}{b_1 b_2}, 1\right), \quad \left(\frac{a_3}{b_2 b_3}, 1\right), \quad \ldots.$$

Another simple example shows that the numerator elements may be reduced to unity. We find that

$$\frac{a_1}{b_1 +} \frac{a_2}{b_2 +} \frac{a_3}{b_3 +} \cdots = \frac{1}{b_1/a_1 +} \frac{1}{b_2 a_1/a_2 +} \frac{1}{b_3 a_2/(a_3 a_1) +} \frac{1}{+ b_4 a_1 a_3/(a_2 a_4) +} \cdots. \qquad (1.7)$$

This freedom of representation of the continued fractions using different elements constitutes a group of equivalence transformations. A general member of the group is represented by

$$\frac{e_1 a_1}{e_1 b_1 +} \quad \frac{e_1 e_2 a_2}{e_2 b_2 +} \quad \frac{e_2 e_3 a_3}{e_3 b_3 +} \quad \frac{e_3 e_4 a_4}{e_4 b_4 +} \cdots \quad (1.8)$$

in terms of the parameters $\{e_1, e_2, e_3 \ldots\}$ which are required to be invertible.

The convergents of the continued fractions are ratios A_n/B_n, as is emphasized by (1.3) and (1.5). However, it is useful to define the (partial) numerators A_n and denominators B_n separately, but consistently, so that $(n+1)$th convergent is given by (1.4). The definitions and consistency are expressed by

Theorem 4.1.1 [Euler, 1737]. *For the continued fraction*

$$b_0 + \frac{a_1}{b_1 +} \quad \frac{a_2}{b_2 +} \quad \frac{a_3}{b_3 +} \cdots \quad \frac{a_m}{+ b_m +} \quad \frac{a_{m+1}}{b_{m+1} +} \cdots, \quad (1.9)$$

we define numerators A_i by $A_0 = b_0$, $A_1 = b_1 b_0 + a_1$, and

$$A_i = b_i A_{i-1} + a_i A_{i-2} \quad \text{for } i = 2, 3, 4, \ldots. \quad (1.10)$$

The denominators are defined by $B_0 = 1$, $B_1 = b_1$, and

$$B_i = b_i B_{i-1} + a_i B_{i-2} \quad \text{for } i = 2, 3, 4, \ldots. \quad (1.11)$$

With this definition, the ratio A_n/B_n is the $(n+1)$th convergent of (1.9).

Proof. By inspection, A_0/B_0 and A_1/B_1 are the values given by (1.3). We prove (1.10) and (1.11) by induction. Suppose that they hold for $i = m$. Then the $(m+1)$th convergent of (1.9) is

$$\frac{A_m}{B_m} = \frac{b_m A_{m-1} + a_m A_{m-2}}{b_m B_{m-1} + a_m B_{m-2}}. \quad (1.12)$$

To obtain the $(m+2)$th convergent of (9) from the $(m+1)$th, we replace b_m by $b_m + a_{m+1}/b_{m+1}$ wherever it appears in the algebraic expression for A_m/B_m. Since b_m does not occur in the algebraic expressions for A_{m-1}, A_{m-2}, B_{m-1}, or B_{m-2} defined by (1.10) and (1.11), we find that

$$\frac{A_{m+1}}{B_{m+1}} = \frac{(b_m + a_{m+1}/b_{m+1})A_{m-1} + a_m A_{m-2}}{(b_m + a_{m+1}/b_{m+1})B_{m-1} + a_m B_{m-2}} = \frac{b_{m+1} A_m + a_{m+1} A_{m-1}}{b_{m+1} B_m + a_{m+1} B_{m-1}}, \quad (1.13)$$

where the induction hypothesis has been used to obtain (1.13). Clearly, the definitions (1.10) and (1.11) for $i = m+1$ are consistent with the values of the $(m+2)$th convergent derived in (1.13).

In this theorem we have derived the most important formula needed for continued-fraction theory: the recursion relation for the numerators A_i and the denominators B_i defined by (1.10) and (1.11). As an example, we may inspect A_2/B_2 in (1.3) and see that it is given correctly by the recursion formula. The trivial modification of taking $b_0 = 0$ allows the recursion to apply to the fraction (1.6). A consequence of Theorem 4.1.1 is that it shows that the following alternative definition of convergence of a continued fraction is entirely equivalent to the previous one.

Alternative definition. The continued fraction

$$b_0 + \frac{a_1}{b_1} + \frac{a_2}{b_2} + \frac{a_3}{b_3} + \cdots$$

is said to converge and have value v if the ratio A_n/B_n of the quantities A_n and B_n defined recursively by (1.10) and (1.11) tends to v as $n \to \infty$.

The other main question we consider in this section is that of how to evaluate a continued fraction numerically. There does not seem to be an agreed best method of computing continued fractions, and so we state three principal methods without making any clear recommendation about which is 'best'. The problem is, given the elements a_i, b_i, and a value of the variable z, to compute

$$f(z) = \frac{a_1 z}{b_1} + \frac{a_2 z}{b_2} + \frac{a_3 z}{b_3} + \cdots, \qquad (1.14)$$

which we assume to be a convergent fraction.

Forward recursion method. The numerators $A_i(z)$ and denominators $B_i(z)$ are calculated from the recursion relation (1.10), (1.11). Because of the ease of computing the values of successive convergents, the forward recursion method is the standard method for computing continued fractions.

It is sometimes the case in practice that the crude approximation

$$A_n \simeq \alpha r_1^n + \beta r_2^n,$$
$$B_n \simeq \gamma r_1^n + \delta r_2^n \qquad (1.15)$$

gives a rough and ready estimate of the rate of growth of A_n and B_n. Equation (1.15) is exact in the special case that the coefficients a_i, b_i in (1.9) are constant. If $|r_1| > |r_2|$, the α, γ terms are the dominant components and the β, δ terms become less significant as $n \to \infty$. Such considerations assist in the understanding of the following numerical hazards of the forward recursion method.

4.1 Definitions, recursion relations, and computation

(a) *Floating-point overflow of A_n, B_n.* If $|r_1| \gg 1$, floating-point overflow may occur at some point during the actual computation. Rescaling of A_n, A_{n-1}, B_n, and B_{n-1} at the critical point in the obvious way is recommended. Underflow must likewise be anticipated.

(b) *Suppression of required solution by a dominant solution.* If the true solutions A_n, B_n of the exact recursion relation (2.7) are the subdominant solutions, they may be suppressed numerically. Other solutions, $\tilde{A}_n$, $\tilde{B}_n$, generated by different initial conditions (e.g. $\tilde{A}_0 \neq A_0$) for which $A_n/\tilde{A}_n \to 0$, $B_n/\tilde{B}_n \to 0$, exist in this case. Rounding error introduced in the initial stages always becomes a dominant effect in such calculations. The connection with the concepts of Miller's backward recursion algorithm is obvious.

(c) *Accumulation of roundoff error in a dominant solution.* If $|A_n|$ and $|B_n|$ are decreasing sequences, rounding error may accumulate using the forward recursion method. This has been discovered empirically [Jones and Thron, 1974a] for the following continued fraction:

$$f(z) = \frac{z}{1} - \frac{z}{1} - \frac{z}{1} - \frac{z}{1} - \cdots \quad (1.16)$$

evaluated at $z = 1/4$, using fixed *decimal* precision in the floating-point operations. (Why would binary arithmetic be different?) The fraction is being evaluated on the edge of its domain of convergence, and and so it is a natural prototype for investigation of the buildup of roundoff error.

Backward recursion method. This method of evaluating the nth convergent of (1.14) starts at the 'tail' end of the convergent. We define

$$f_{n+1}^{(n)} = 0,$$

$$f_i^{(n)}(z) = \frac{a_i z}{b_i + f_{i+1}^{(n)}(z)}, \quad i = n, n-1, n-2, \ldots, 1, \quad (1.17)$$

and then $f_1^{(n)}(z)$ is the nth convergent of $f(z)$. The major drawback of this method is that of deciding in advance which is the appropriate value of n to choose so that the nth convergent is an adequate approximation to $f(z)$.

Euler's summation formula. *The nth convergent of (1.14) is given by the formula*

$$\frac{A_n(z)}{B_n(z)} = \sum_{i=1}^{n} \prod_{j=1}^{i} \rho_i, \quad (1.18)$$

where the quantities $\rho_1, \rho_2, \rho_3, \ldots, \rho_n$ are defined in terms of the elements of the continued fraction by

$$\rho_1 = \frac{a_1 z}{b_1}, \quad 1 + \rho_2 = \frac{1}{1 + a_2 z/(b_1 b_2)} \qquad (1.19)$$

and recursively by

$$1 + \rho_i = \left(1 + \frac{a_i z (1 + \rho_{i-1})}{b_i b_{i-1}}\right)^{-1}, \quad i = 3, 4, \ldots. \qquad (1.20)$$

Proof. We construct a summation formula by using the identity

$$\frac{A_n}{B_n} = \left(\frac{A_n}{B_n} - \frac{A_{n-1}}{B_{n-1}}\right) + \left(\frac{A_{n-1}}{B_{n-1}} - \frac{A_{n-2}}{B_{n-2}}\right) + \left(\frac{A_{n-2}}{B_{n-2}} - \frac{A_{n-3}}{B_{n-3}}\right) + \cdots$$

$$+ \frac{A_1}{B_1} = \frac{a_1 z}{b_1} + \sum_{i=2}^{n} \frac{(-1)^{i+1}}{B_i B_{i-1}} \prod_{j=1}^{i} (a_i z); \qquad (1.21)$$

a formula equivalent to (1.21) is derived as (7.6). A comparison of equations (1.18), (1.21) leads us to define

$$\rho_i = \frac{-a_i z B_{i-2}}{B_i}, \quad i = 3, 4, 5, \ldots. \qquad (1.22)$$

Equation (1.19) defines the first two convergents correctly. To establish (1.20), we use the recursion relation

$$B_i = b_i B_{i-1} + a_i z B_{i-2}$$

to prove that

$$\frac{B_{i-1}}{B_{i-2}} = \frac{-a_i z}{b_i} \left\{\frac{1}{\rho_i} + 1\right\}, \qquad (1.23)$$

and the recursion

$$B_{i-1} = b_{i-1} B_{i-2} + a_{i-1} z B_{i-3}$$

to prove that

$$\frac{B_{i-2}}{B_{i-1}} = \frac{1}{b_{i-1}} \{1 + \rho_{i-1}\}. \qquad (1.24)$$

Equation (1.20) is proved by multiplying (1.23) and (1.24).

A consequence of (1.19), (1.20), and the equivalence transformation (1.6) is the new representation of (1.14) by

$$\frac{A_n(z)}{B_n(z)} = \frac{\rho_1}{1} - \frac{\rho_2}{1 + \rho_2} - \frac{\rho_3}{1 + \rho_3} - \cdots - \frac{\rho_n}{1 + \rho_n}. \qquad (1.25)$$

The summation formula expressed by (1.18)–(1.20) has the advantage that it is suitable for iterative computation, which is important if the rate of convergence is not known a priori.

For further details of the numerical methods, and especially for error estimates, we refer to Jones and Thron's companion volume [1980], to Blanch [1964], and to Gautschi [1967].

4.2 Continued fractions derived from Maclaurin series

A formal power series may be manipulated into the form of a continued fraction very easily. In this section, we ignore all questions of convergence. To begin with, assume that all the inverses we need exist. Later on, reliable algorithms that can deal with degenerate cases are given.

The given power series is

$$f(z) = c_0 + c_1 z + c_2 z^2 + \cdots. \tag{2.1}$$

We calculate the reciprocal of the series

$$1 + \frac{c_2 z}{c_1} + \frac{c_3 z^2}{c_1} + \cdots = (1 + c_1^{(1)} z + c_2^{(1)} z^2 + \cdots)^{-1},$$

which allows the reexpansion

$$c_0 + c_1 z + c_2 z^2 + \cdots = c_0 + \frac{c_1 z}{1 + c_1^{(1)} z + c_2^{(1)} z^2 + \cdots}. \tag{2.2}$$

Next we calculate the reciprocal of the series

$$1 + \frac{c_2^{(1)} z}{c_1^{(1)}} + \frac{c_3^{(1)} z}{c_1^{(1)}} + \cdots = (1 + c_1^{(2)} z + c_2^{(2)} z^2 + \cdots)^{-1},$$

which allows another reexpansion

$$c_0 + c_1 z + c_2 z^2 + \cdots = c_0 + \cfrac{c_1 z}{1 + \cfrac{c_1^{(1)} z}{1 + c_1^{(2)} z + c_2^{(2)} z^2 + \cdots}}. \tag{2.3}$$

It is clear [Salzer, 1962] that by forming the reciprocal series, we have devised an iterative procedure which allows us to write formally

$$f(z) = c_0 + \frac{c_1 z}{1} + \frac{c_1^{(1)} z}{1} + \frac{c_1^{(2)} z}{1} + \frac{c_1^{(3)} z}{1} + \cdots, \tag{2.4}$$

which corresponds to the series (2.1). The convergents of (2.4) are rational fractions in the variable z. To be quite general, we assume only that the resultant fraction representing the power series takes the form

$$f(z) = b_0 + \frac{a_1 z}{b_1} + \frac{a_2 z}{b_2} + \frac{a_3 z}{b_3} + \cdots \qquad (2.5)$$

and (2.4) is just a special case of (2.5). The first few convergents of (2.5) may be easily calculated:

$$\frac{A_0(z)}{B_0(z)} = b_0, \quad \frac{A_1(z)}{B_1(z)} = \frac{b_0 b_1 + a_1 z}{b_1},$$
$$\frac{A_2(z)}{B_2(z)} = \frac{b_0 b_1 b_2 + (a_2 b_0 + a_1 b_2) z}{b_1 b_2 + a_2 z}. \qquad (2.6)$$

We see that (2.6) is equivalent to (1.3) with the replacement $a_i \to a_i z$ for all i. Following the analysis of Section 4.1, especially Theorem 4.1.1, we see that the numerators and denominators of (2.5) are generated by

$$A_0(z) = b_0, \quad A_1(z) = b_0 b_1 + a_1 z,$$
$$A_i(z) = b_i A_{i-1}(z) + a_i z A_{i-2}(z), \quad i = 2, 3, 4, \ldots, \qquad (2.7a)$$

and

$$B_0(z) = 1, \quad B_1(z) = b_1,$$
$$B_i(z) = b_i B_{i-1}(z) + a_i z B_{i-2}(z), \quad i = 2, 3, 4, \ldots. \qquad (2.7b)$$

The connection between the convergents of the fractions (2.4) and (2.5) and the entries in the Padé table is expressed by

Theorem 4.2.1. *Provided that c_1 and every coefficient $c_1^{(j)}$ are nonzero, the continued fraction (2.4) has the Maclaurin expansion (2.1). In this case, the Padé approximants of (2.1) are identified with the convergents of the continued fraction by*

$$[M/M]_f(z) = \frac{A_{2M}(z)}{B_{2M}(z)} \quad \text{and} \quad [M + 1/M]_f(z) = \frac{A_{2M+1}(z)}{B_{2M+1}(z)} \qquad (2.8)$$

for $M = 0, 1, 2, \ldots$.

Remarks. No statement is implied by this theorem about the domains of convergence in the z-plane, if any, of either the series (2.1) or the fraction (2.4). If nontrivial domains of convergence exist, they are likely to be different. Even if (2.1) and (2.4) are convergent, the theorem does not directly assert equality of these values.

Proof. Since each expression (2.1), (2.2), (2.3) has the same formal Maclaurin expansion, the first part of the theorem is true by induction.

4.2 Continued fractions derived from Maclaurin series

To establish (2.8), we note that

$$\deg\{A_0(z)\} = 0, \quad \deg\{B_0(z)\} = 0,$$
$$\deg\{A_1(z)\} \leq 1, \quad \deg\{B_1(z)\} = 0. \tag{2.9}$$

We prove (2.8) by induction. Suppose that

$$\deg\{A_{2m}(z)\} \leq m, \quad \deg\{B_{2m}(z)\} \leq m,$$
$$\deg\{A_{2m+1}(z)\} \leq m+1, \quad \deg\{B_{2m+1}(z)\} \leq m, \tag{2.10}$$

for $m = 0, 1, 2, \ldots, M$. Then

$$\deg\{A_{2m+2}(z)\} = \deg\{b_{2m+2}A_{2m+1}(z) + a_{2m+2}zA_{2m}(z)\} \leq m+1,$$
$$\deg\{A_{2m+3}(z)\} = \deg\{b_{2m+3}A_{2m+2}(z) + a_{2m+3}zA_{2m+1}(z)\} \leq m+2,$$
$$\deg\{B_{2m+2}(z)\} = \deg\{b_{2m+2}B_{2m+1}(z) + a_{2m+2}zB_{2m}(z)\} \leq m+1,$$
$$\deg\{B_{2m+3}(z)\} = \deg\{b_{2m+3}B_{2m+2}(z) + a_{2m+3}zB_{2m+1}(z)\} \leq m+1.$$

Two of these relationships are, in fact, equalities: see Theorem 4.2.4. Hence the fractions $A_m(z)/B_m(z)$ have numerators and denominators of the requisite orders for all m, and power series which agree with (2.1) to order z^m inclusive. Consequently the fractions $A_m(z)/B_m(z)$ are the Padé approximants of (2.1) of the orders indicated by (2.8).

Notice that the fractions $\{A_m(z)/B_m(z)\}$ defined in this section occupy a descending staircase sequence in the Padé table, which starts with a horizontal tread, as shown in Figure 4.2.1.

As an example of a continued fraction of this type, we may use five terms of the Maclaurin expansion of $\exp(z)$ to show that

$$\exp(z) = 1 + \frac{z}{1-} \frac{z}{2+} \frac{z}{3-} \frac{z}{2+} \cdots. \tag{2.11}$$

This example shows an advantage of using (2.5) rather than (2.4), because the elements of the fraction may be taken to be integers with this representation. We will derive the general term of (2.11) in Section 4.6. If we consider the reciprocal of (2.11) and replace z by $-z$, we get a different representation:

$$\exp(z) = \frac{1}{1-} \frac{z}{1+} \frac{z}{2-} \frac{z}{3+} \cdots. \tag{2.12}$$

This is of the general type

$$f(z) = \frac{a_1}{b_1+} \frac{a_2 z}{b_2+} \frac{a_3 z}{b_3+} \cdots. \tag{2.13}$$

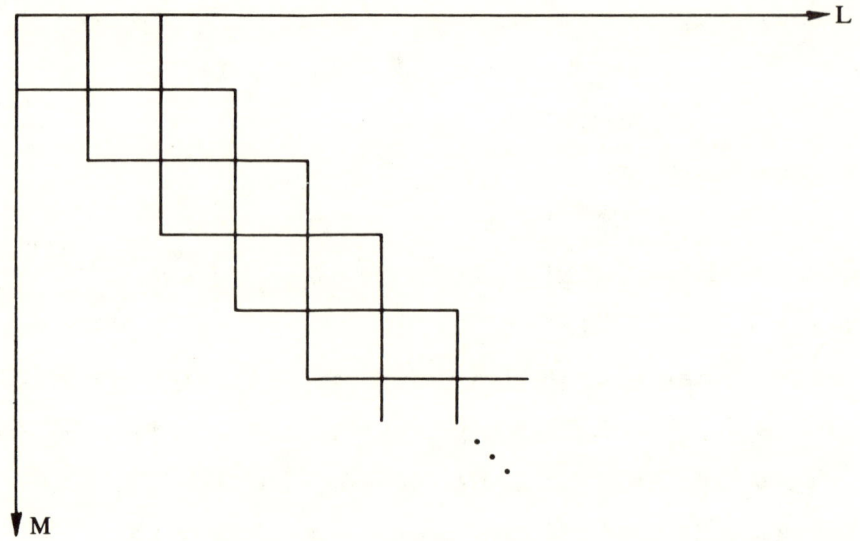

Figure 4.2.1. A descending staircase sequence in the Padé table corresponding to convergents of (2.4).

The numerators and denominators of (2.13) are derived from

$$A_1(z) = a_1, \quad A_2(z) = b_2 a_1,$$
$$A_{i+1}(z) = b_{i+1} A_i(z) + a_{i+1} z A_{i-1}(z), \quad i = 2, 3, 4, \ldots, \quad (2.14a)$$

and

$$B_1(z) = b_1, \quad B_2(z) = b_1 b_2 + a_2 z,$$
$$B_{i+1}(z) = b_{i+1} B_i(z) + a_{i+1} z B_{i-1}(z), \quad i = 2, 3, 4, \ldots, \quad (2.14b)$$

The Padé approximants which are the convergents of (2.13) are given by

$$[M/M]_f(z) = \frac{A_{2m}(z)}{B_{2m}(z)} \quad \text{and} \quad [M/M+1]_f(z) = \frac{A_{2m+1}(z)}{B_{2m+1}(z)}. \quad (2.15)$$

These occupy a descending sequence in the Padé table which begins with a stair, as shown in Figure 4.2.2.

The comparison of (2.5) or (2.13) with the sequence of Padé approximants indicates that (2.5) is to be preferred in particular asymptotic regions of the z-plane where $|f(z)|$ is increasing, and (2.13) is to be preferred where $|f(z)|$ is decreasing as $|z|$ increases.

If functions are even or odd, they are degenerate in a rather trivial way,

4.2 Continued fractions derived from Maclaurin series

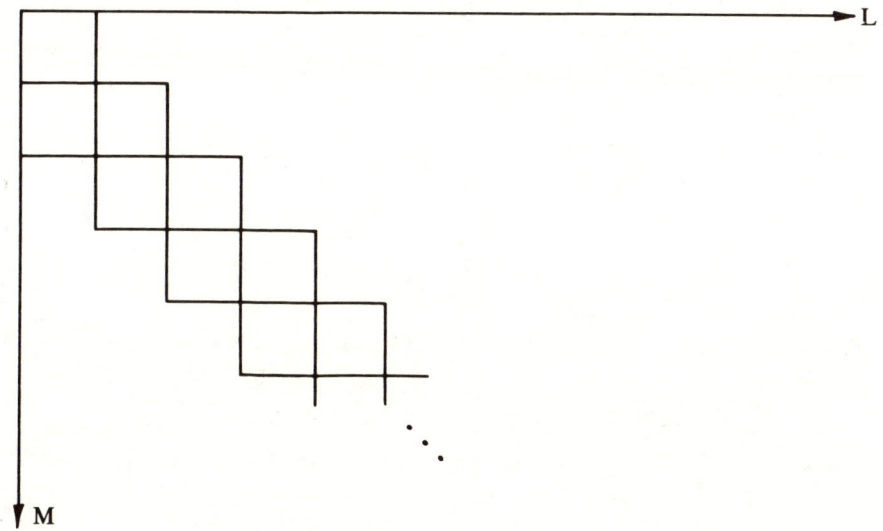

Figure 4.2.2. A descending staircase sequence in the Padé table corresponding to convergents of (2.13).

and there is no purpose in making a great issue of this. If the function is even, either one uses

$$f^{(\text{even})}(z) = \frac{a_1}{b_1} + \frac{a_2 z^2}{b_2} + \frac{a_3 z^2}{b_3} + \cdots$$

or

$$f^{(\text{even})}(z) = b_0 + \frac{a_1 z^2}{b_1} + \frac{a_2 z^2}{b_2} + \cdots.$$

If the given function is odd, normally one chooses

$$f^{(\text{odd})}(z) = \frac{a_1 z}{b_1} + \frac{a_2 z^2}{b_2} + \frac{a_3 z^2}{b_3} + \cdots. \qquad (2.16)$$

It is customary to adopt the most convenient form of continued fraction without discussing alternative possible representations. Consideration of the sequence of convergents as they appear in the Padé table with the desired asymptotic properties and consideration of any known degeneracies give a guide to the best continued-fraction representation to use.

Viskovatov's method [1803] is a handy method for constructing continued-fraction expansions, or equivalently staircase sequences of Padé approximants, from the given power series. It avoids having to reciprocate any series numerically, as was part of the construction of (2.2) and (2.3). It is a labor saving device which is as useful for people as for computers. The simplest form of the method is based on the formal identity

$$\frac{\sum_{r=0}^{\infty} a_r z^r}{\sum_{r=0}^{\infty} b_r z^r} = \frac{a_0}{b_0} + \frac{z}{\dfrac{\sum_{r=0}^{\infty} b_r z^r}{\sum_{r=0}^{\infty} (a_{r+1} - a_0 b_{r+1}/b_0) z^r}} \qquad (2.17)$$

The use of the method is best described by an illustrative example.

$$\exp(z) = 1 + z + \frac{z^2}{2} + \frac{z^3}{6} + \frac{z^4}{24} + \cdots$$

$$= 1 + \frac{z}{\left(\dfrac{1}{1 + \dfrac{z}{2} + \dfrac{z^2}{6} + \dfrac{z^3}{24} + \cdots}\right)}$$

$$= 1 + \frac{z}{1+} \frac{z}{\left(\dfrac{1 + \dfrac{z}{2} + \dfrac{z^2}{6} + \cdots}{-\dfrac{1}{2} - \dfrac{z}{6} - \dfrac{z^2}{24} + \cdots}\right)}$$

$$= 1 + \frac{z}{1+} \frac{z}{-2+} \frac{z}{\dfrac{-\dfrac{1}{2} - \dfrac{z}{6} - \dfrac{z^2}{24} + \cdots}{\dfrac{1}{6} + \dfrac{z}{12} + \cdots}}$$

$$= 1 + \frac{z}{1+} \frac{z}{-2+} \frac{z}{-3+} \cdots . \qquad (2.18)$$

Viskovatov's method breaks down if any of the coefficients $c_1^{(n)}$ in (2.3) vanish. In turn, if any $c_1^{(n)} = 0$, then $f(z)$ cannot be represented by a continued fraction of the form (2.4), which is called a regular C-fraction. In cases of this kind, Viskovatov's method generates continued fractions of the form

$$C(z) = b_0 + \frac{a_1 z^{\alpha_1}}{1+} \frac{a_2 z^{\alpha_2}}{1+} \frac{a_3 z^{\alpha_3}}{1+} \cdots \qquad (2.19)$$

4.2 Continued fractions derived from Maclaurin series

with $a_i \neq 0$ and $\alpha_i \geq 1$ for all i. Such fractions are called general C-fractions or general corresponding fractions because they maintain correspondence (accuracy through order) with $f(z)$ [Leighton and Scott, 1939; Scott and Wall, 1940a, b]. The convergents of fractions such as (2.19) do not necessarily lie in a simple staircase of the Padé table. Consider the series

$$f(z) = 1 + z^2 - z^5 + \lambda z^6. \tag{2.20}$$

The third convergent of the general C-fraction corresponding to (2.20) is

$$\frac{A_3(z)}{B_3(z)} = 1 + \frac{z^2}{1 + z^3} = \frac{1 + z^2 + z^3}{1 + z^3} \tag{2.21}$$

and its fourth convergent is

$$\frac{A_4(z)}{B_4(z)} = 1 + \frac{z^2}{1} + \frac{z^3}{1 + \lambda z};$$

the [3/3] Padé approximant for $f(z)$ is

$$[3/3](z) = \frac{1 + \lambda z + z^2 + (1 + \lambda)z^3}{1 + \lambda z + z^3}.$$

We see that $A_3(z)/B_3(z)$ is not the Padé approximant of type [3/3] (unless $\lambda = 0$), but at higher order they coincide:

$$A_4(z)/B_4(z) = [3/3](z).$$

It turns out that continued fractions of the form

$$W(z) = \pi_0(z) + \frac{z^{\mu_1}}{\pi_1(z)} + \frac{z^{\mu_2}}{\pi_2(z)} + \frac{z^{\mu_3}}{\pi_3(z)} + \cdots \tag{2.22}$$

are the most satisfactory representations of Padé approximants in the general case. This conclusion has been reached independently by several authors, and notably by Magnus [1962b] and Werner [1979, 1980]. Continued fractions of the form (2.22) are equivalent to Magnus' P-fractions. The fraction (2.22) can be constructed using the generalized Viskovatov method. In the form given below, it is a method for constructing Padé approximants of types $[m + J/m]$, with $J \geq -1$, in a paradiagonal sequence, for a Maclaurin series (2.1). For an approximant of type $[m + J/m]$, the actual degree of the numerator may be less than $m + J$ and the degree of the denominator may be less than m, but the approximant has full accuracy through order, with error $O(z^{2m+J+1})$.

The following specification of the generalized Viskovatov algorithm is similar to that given by Bultheel [1980b].

Specification. Two formal power series $S_0(z)$, $S_1(z)$ are given, and $S_1(0) \neq 0$. The algorithm provides all the approximants of type $[m + J/m]$ for $S_0(z)/S_1(z)$ for $J \geq -1$.

Initialization. A polynomial $\pi_0(z)$ and integers v_0, v_1, and μ_1 are defined by

$$\pi_0(z) = [S_0(z)/S_1(z)]_0^J, \tag{2.23}$$

$$\mu_1 = O\{S_0(z) - \pi_0(z)S_1(z)\}, \tag{2.24}$$

$$v_0 = J, \quad v_1 = \mu_1 - J. \tag{2.25}$$

Recursion. For $i = 1, 2, 3, \ldots$, a formal power series $S_{i+1}(z)$, a polynomial $\pi_i(z)$ of degree v_i at most, and positive integers μ_{i+1}, v_{i+1} are defined by

$$S_{i+1}(z) = z^{-\mu_i}\{S_{i-1}(z) - \pi_{i-1}(z)S_i(z)\}, \tag{2.26}$$

$$\pi_i(z) = [S_i(z)/S_{i+1}(z)]_0^{v_i}, \tag{2.27}$$

$$\mu_{i+1} = O\{S_i(z) - \pi_i(z)S_{i+1}(z)\}, \tag{2.28}$$

$$v_{i+1} = \mu_{i+1} - v_i. \tag{2.29}$$

Termination. If $\mu_{N+1} = \infty$, the algorithm terminates and

$$f(z) = \pi_0(z) + \frac{z^{\mu_1}}{\pi_1(z)\ +\ } \frac{z^{\mu_2}}{\pi_2(z)\ +\ } \cdots \frac{z^{\mu_N}}{+\ \pi_N(z)}. \tag{2.30}$$

Otherwise, the algorithm may be initialized using just the coefficients c_0, $c_1, \ldots, c_{2M+J}$ of $f(z)$, and terminating in accordance with Theorem 4.2.

The numerators and denominators of the convergents are constructed using the recursions

$$p_j(z) = \pi_j(z)p_{j-1}(z) + z^{\mu_j}p_{j-2}(z), \tag{2.31}$$

$$q_j(z) = \pi_j(z)q_{j-1}(z) + z^{\mu_j}q_{j-2}(z) \tag{2.32}$$

for $j = 1, 2, \ldots$, with the initialization

$$p_0(z) = \pi_0(z), \quad p_{-1}(z) = 1, \tag{2.33}$$

$$q_0(z) = 1, \quad q_{-1}(z) = 0. \tag{2.34}$$

The rational fraction

$$r_k(z) = p_k(z)/q_k(z) \tag{2.35}$$

4.2 Continued fractions derived from Maclaurin series

is the Padé approximant of type $[\tau_k + J/\tau_k]$ for $f(z)$, where $\tau_0 = 0$ and

$$\tau_k = \sum_{j=1}^{k} v_j, \quad k = 1, 2, 3, \ldots. \tag{2.36}$$

Remarks. The definitions (2.23), (2.27) involve Nuttall's notation:

$$[\varphi(z)]_\alpha^\beta = \sum_{i=\alpha}^{\beta} \varphi_i z^i \tag{2.37}$$

denotes the section of the Maclaurin series between powers z^α, z^β inclusive and $[\varphi(z)]_\alpha^\beta = 0$ if $\beta < \alpha$. These definitions require construction of the first few terms of the quotient of the two power series. Synthetic (long) division is a much better method than reciprocating $S_1(z)$ and $S_{i+1}(z)$ in these equations. For example, if

$$\frac{S_0(z)}{S_1(z)} = c_0 + c_1 z + \cdots + c_{j-1} z^{j-1} + \frac{T(z) z^j}{S_1(z)}$$

has been obtained, the next step is to take $c_j = T(0)/S_1(0)$ and construct

$$U(z) = \{T(z) - c_j S_1(z)\}/z$$

and

$$\frac{S_0(z)}{S_1(z)} = c_0 + c_1 z + \cdots + c_j z^j = \frac{U(z) z^{j+1}}{S_1(z)}.$$

If a subdiagonal sequence, or approximant of type $[L/M]$ with $L < M - 1$ is required, then the procedure above should be used for forming the approximant of type $[M/L]$ for $f(z)^{-1}$, according to the duality principle of Theorem 1.5.1.

Theorem 4.2.2. *The generalized Viskovatov algorithm as formulated above constructs Padé approximants $r_k(z)$ of type $[\tau_k + J/\tau_k]$ as detailed in (2.35), (2.36).*

Proof. First, notice that a consequence of (2.24), (2.26)–(2.28) is that

$$S_k(z) S_{k+1}(z)^{-1} = \pi_k(z) + O(z^{\mu_{k+1}}), \tag{2.38}$$

for each $k \geq 0$, where the order is precise. Consider the convergent of index k of (2.22). Using the recursions (2.31), (2.32), it is

$$r_k(z) = \pi_0(z) + \frac{z^{\mu_1}}{\pi_1(z) +} \frac{z^{\mu_2}}{\pi_2(z) +} \cdots \frac{z^{\mu_k}}{+ \pi_k(z)}. \tag{2.39}$$

For this value of k, and some $i \in \{0, 1, \ldots, k\}$, make the inductive hypothesis that

$$S_i(z)S_{i+1}(z)^{-1} = \pi_i(z) + \frac{z^{\mu_{i+1}}}{\pi_{i+1}(z)\ +} \ \frac{z^{\mu_{i+2}}}{\pi_{i+2}(z)\ +} \cdots \frac{z^{\mu_k}}{+\ \pi_k(z)} + O(z^{\theta_{ik}}), \qquad (2.40)$$

where the order is precise and

$$\theta_{ik} = v_{k+1} + 2v_k + 2v_{k-1} + \cdots + 2v_{i+1} + v_i. \qquad (2.41)$$

Equation (2.38) justifies the hypothesis (2.40) for the case $i = k$. Combining (2.26) and (2.40), one obtains

$$S_{i-1}(z)S_i(z)^{-1} = \pi_{i-1}(z) + \frac{z^{\mu_i}}{\pi_i(z)\ +} \ \frac{z^{\mu_{i+1}}}{\pi_{i+1}(z)\ +} \cdots \frac{z^{\mu_k}}{+\ \pi_k(z)} + O(z^{\theta_{ik} + \mu_i}),$$

and (2.40) is established with $i \to i - 1$. Therefore (2.40) holds for all $i \in \{0, 1, \ldots, k\}$. Because $v_0 = J$, we have the principal result that

$$S_0(z)S_1(z)^{-1} = \pi_0(z) + \frac{z^{\mu_1}}{\pi_1(z)\ +} \ \frac{z^{\mu_2}}{\pi_2(z)\ +} \cdots \frac{z^{\mu_k}}{+\ \pi_k(z)} + O(z^{2\tau_k + J + v_{k+1}}), \qquad (2.42)$$

where the order of the remainder is precise, and $v_{k+1} \geq 1$.

From (2.24), (2.28), and the recursion (2.32), one obtains

$$q_k(0) = \pi_k(0)\pi_{k-1}(0) \cdots \pi_1(0) = S_1(0)/S_{k+1}(0) \neq 0$$

and hence $r_k(z)$ is the Padé approximant of type $[\tau_k + J/\tau_k]$ for $f(z)$.

Our next task is an investigation of the reliability of the generalized Viskovatov algorithm, and to discover how the sequence traverses blocks in the Padé table. Notice that (2.42) established accuracy through order by a factor of $z^{v_{k+1}-1}$ more than is necessary for the proof. This fact signals the existence of a block in the Padé table all of whose entries equal $r_k(z)$. As such, the block includes the type $[\tau_k + J/\tau_k]$, and Theorem 1.4.2 states that the block is square. Because, by construction, the accuracy-through-order result (2.42) cannot be strengthened, the block cannot include both the types $[\tau_k + J + v_{k+1}/\tau_k]$ and $[\tau_k + J/\tau_k + v_{k+1}]$. Thus all approximants of types $[L/M]$ lying in the triangle

$$T_g: L \geq \tau_k + J, \quad M \geq \tau_k, \quad L + M < 2\tau_k + J + v_{k+1}$$

exist and equal $r_k(z)$. If $v_{k+1} > 1$, the block is nontrivial, and approximants which would lie in the triangle

$$T_b: L < \tau_k + J + v_{k+1}, \quad M < \tau_k + v_{k+1}, \quad L + M \geq 2\tau_k + J + v_{k+1}$$

cannot exist under the modern definition, because they cannot attain the desired accuracy through order for the given degree constraints.

If it is the approximant of type $[\tau_k + J + v_{k+1}/\tau_k]$ which does not belong to the block, the situation is shown in Figure 4.2.3.

If it is the approximant of type $[\tau_k + J/\tau_k + J]$ which does not belong to the block, the situation is shown in Figure 4.2.4.

In both cases, the approximants of type

$$[\tau_k + J/\tau_k], [\tau_k + J + 1/\tau_k + 1], \ldots, [\tau_k + v_{k+1} + J - 1/\tau_k + v_{k+1} - 1]$$

lie in the block, but the next approximant is of type

$$[\tau_k + v_{k+1} + J/\tau_k + v_{k+1}] = [\tau_{k+1} + J/\tau_{k+1}]$$

and it lies in the next block. Recalling the previous theorem, we have proved the following result.

Theorem 4.2.3. *The generalized Viskovatov algorithm as formulated in (2.23)–(2.36) is a reliable algorithm for calculation of a paradiagonal*

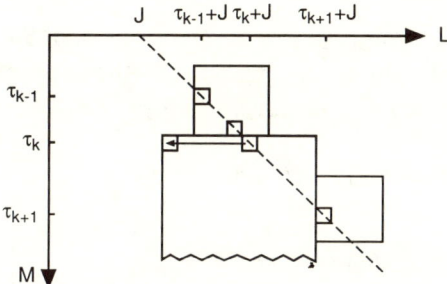

Figure 4.2.3. The case in which $\deg\{p_k(z)\} < \tau_k + J$ and $[\tau_k + J + v_{k+1}/\tau_k]$ lies outside the block of index k.

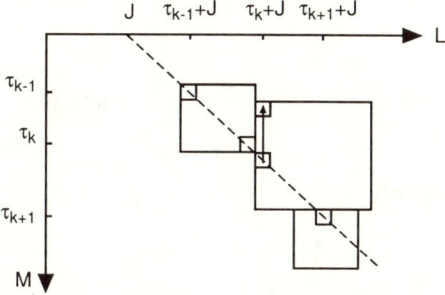

Figure 4.2.4. The case in which $\deg\{q_k(z)\} < \tau_k$ and $[\tau_k + J/\tau_k + J]$ lies outside the block of index k.

sequence: every Padé approximant of type $[m + J/m]$, $m = 0, 1, 2, \ldots$, $J \geq -1$, which exists according to the Baker definition, is constructed by the algorithm.

There are several alternative methods of deriving the result of Theorem 4.2.3 which are not based so directly on the block structure theorem. For example, by using the method of proof of Theorem 3.7.3, it is readily found that $p_k(z)$ and $q_k(z)$ can have no nontrivial common factors. From the foregoing, it is *impossible* that

$$\deg\{p_k(z)\} < \tau_k + J \quad \text{and} \deg\{q_k(z)\} < \tau_k,$$

because $r_k(z)$ would lie in the block of $r_{k-1}(z)$. This circumstance is reflected by Figures 4.2.3 and 4.2.4. In Figure 4.2.3, $\deg\{p_k(z)\} < \tau_k + J$ is allowed, whereas in Figure 4.2.4, $\deg\{q_k(z)\} < \tau_k$ is allowed.

The other form of the generalized Viskovatov algorithm is used to construct the appropriate generalization of a staircase sequence. The setting is in many ways similar to that of the previous one specified in (2.23)–(2.36), the principal exception being that the types of the approximants generated are different.

Specification. Two formal power series $S_0(z)$ and $S_1(z)$ are given, and $S_1(0) \neq 0$. The algorithm constructs all the approximants of types $[m + J/m]$ and $[m + J + 1/m]$ for $S_0(z)/S_1(z)$ for $J \geq -1$.

Initialization. A polynomial $\pi_0(z)$ and integers ν_0, ν_1, and μ_1 are defined by

$$\left. \begin{aligned} \pi_0(z) &= [S_0(z)/S_1(z)]_0^J, \\ \mu_1 &= O\{S_0(z) - \pi_0(z)S_1(z)\}, \\ \nu_0 &= J, \\ \nu_1 &= \mu_1 - J - 1. \end{aligned} \right\} \quad (2.43)$$

Recursion. For $i = 1, 2, 3, \ldots$, a formal power series $S_{i+1}(z)$, a polynomial $\pi_i(z)$ of degree ν_i at most, and nonnegative integers μ_{i+1}, ν_{i+1} are defined by

$$\left. \begin{aligned} S_{i+1}(z) &= z^{-\mu_i}\{S_{i-1}(z) - \pi_{i-1}(z)S_i(z)\}, \\ \pi_i(z) &= [S_i(z)/S_{i+1}(z)]_0^{\nu_i}, \\ \mu_{i+1}(z) &= O\{S_i(z) - \pi_i(z)S_{i+1}(z)\}, \\ \nu_{i+1} &= \mu_{i+1} - \nu_i - 1. \end{aligned} \right\} \quad (2.44)$$

Termination. The termination is identical to (2.31)–(2.35).

The character of the polynomials constructed using this form of the generalized Viskovatov algorithm is clear from the following theorem.

Theorem 4.2.4. *The polynomials constructed using* (2.43), (2.44), (2.32)–(2.34) *possess the following properties*

$$\deg \{p_{2j+1}(z)\} = j + 1 + \sum_{i=0}^{2j+1} v_i, \quad \deg \{p_{2j}(z)\} \leq j + \sum_{i=0}^{2j} v_i, \quad (2.45)$$

$$\deg \{q_{2j+1}(z)\} \leq j + \sum_{i=1}^{2j+1} v_i, \quad \deg \{q_{2j}(z)\} = j + \sum_{i=0}^{2j} v_i, \quad (2.46)$$

$$q_k(0) = \pi_k(0)\pi_{k-1}(0) \cdots \pi_1(0) = S_1(0)/S_{k+1}(0) \neq 0,$$

and the leading coefficients of $p_{2j+1}(z)$ *and* $q_{2j}(z)$ *are both unity.*

Proof. The proof is by induction based on (2.31)–(2.32) and the initializations (2.33)–(2.34).

This form of Viskovatov's algorithm is a reliable algorithm for computing the approximants in the sequence $[m + J/m]$ and $[m + J + 1/m]$ for $m = 0, 1, 2, \ldots$. The proof of this statement is similar to that of Theorem 4.2.3 and is omitted for brevity. The proof depends on comparing the degree results (2.45), (2.46) with the constraints of block structure; see Figures 4.2.3, 4.2.4. The detail has been elucidated in a similar context in Section 3.6

We refer to Arndt [1980] and Section 7.1 for further details of how these methods are incorporated into the general scheme of Hermite-Padé approximants, and to Busonnais [1978] and Bultheel [1987] for details of other implementations.

4.3 Various representations of continued fractions

We are primarily concerned with continued fractions whose convergents form a sequence of Padé approximants. We will consider continued fractions containing the complex variable z explicitly, so as to maintain the connection with Padé approximation. The formulas remain valid with $z = 1$; this may be more useful for other purposes connected with the general theory of continued fractions. We will first establish the existence of tridiagonal determinantal formulas for the numerators and denominators of a continued fraction.

Theorem 4.3.1. *The convergents of the continued fraction*

$$f(z) = b_0 + \frac{a_1 z}{b_1} + \frac{a_2 z}{b_2} + \frac{a_3 z}{b_3} + \cdots \quad (3.1)$$

have numerators given, for $n = 0, 1, 2, \ldots$, by

$$A_n(z) = \begin{vmatrix} b_0 & -a_1 z & & & & & 0 \\ 1 & b_1 & -a_2 z & & & & \\ & 1 & b_2 & -a_3 z & & & \\ & & \ddots & \ddots & \ddots & & \\ & & & 1 & b_{n-2} & -a_{n-1} z & \\ & & & & 1 & b_{n-1} & -a_n z \\ 0 & & & & & 1 & b_n \end{vmatrix}, \quad (3.2)$$

and denominators given, for $n = 1, 2, 3, \ldots$, by

$$B_n(z) = \begin{vmatrix} b_1 & -a_2 z & & & & & 0 \\ 1 & b_2 & -a_3 z & & & & \\ & 1 & b_3 & -a_4 z & & & \\ & & \ddots & \ddots & \ddots & & \\ & & & 1 & b_{n-2} & -a_{n-1} z & \\ & & & & 1 & b_{n-1} & -a_n z \\ 0 & & & & & 1 & b_n \end{vmatrix}. \quad (3.3)$$

Proof. We expand the determinant (3.2) by its last column. This leads to the equation

$$A_n(z) = b_n A_{n-1}(z) + a_n z A_{n-2}(z), \quad (3.4)$$

which is the standard recursion relation (2.7a). For $n = 0$ we have $A_0(z) = b_0$, and for $n = 1$ we have $A_1(z) = b_0 b_1 + a_1 z$. From (2.14a) we deduce that the tridiagonal representation (3.2) is valid for the numerators of the convergents of the continued fraction (3.1).

We deduce from (2.14b) that the tridiagonal representation (3.3) is valid for the denominators of the convergents of the continued fraction (3.2).

Theorem 4.3.2. *The convergents of the continued fraction*

$$f(z) = \frac{a_1}{b_1} + \frac{a_2 z}{b_2} + \frac{a_3 z}{b_3} + \cdots \quad (3.5)$$

have numerators, given for $n = 2, 3, \ldots,$ *by*

$$A_n = a_1 \begin{vmatrix} b_2 & -a_3 z & & & & 0 \\ 1 & b_3 & -a_4 z & & & \\ & \ddots & \ddots & \ddots & & \\ & & 1 & & b_{n-1} & -a_n z \\ 0 & & & & 1 & b_n \end{vmatrix} \quad (3.6)$$

and denominators given by (3.3).

Proof. The method of proof is identical to that of Theorem 4.3.1.

The determinants (3.2) and (3.3) express the numerator and denominators of Padé approximants of types $[M/M]$ or $[M + 1/M]$, according to the discussion of Section 4.2. Equations (3.2) and (3.3) involve the elements a_i and b_i of the continued fraction (3.1). Next, we reexpress (3.1) in terms of Hankel determinants, so as to relate (3.1), (3.2), and (3.3) more directly to the coefficients c_i of the Maclaurin expansion of $f(z)$.

Theorem 4.3.3. *The continued fraction* (4.3.1) *may be expressed as*

$$f(z) = c_0 + \frac{c_1 z}{1} - \frac{c_2 z}{c_1} - \frac{zC(2/2)/C(1/1)}{C(2/1)/C(1/1)} - \frac{zC(3/2)/C(2/1)}{C(2/2)/C(2/1)} - \cdots . \quad (3.7)$$

The initial elements of (3.1) *are*

$$b_0 = c_0, \quad a_1 = c_1, \quad b_1 = 1, \quad a_2 = -c_2, \quad b_2 = c_1. \quad (3.8)$$

The other elements are given, for $M = 1, 2, 3, \ldots,$ *by*

$$a_{2m+1} = -\frac{C(M + 1/M + 1)}{C(M/M)}, \quad (3.9)$$

$$b_{2m+1} = \frac{C(M + 1/M)}{C(M/M)}, \quad (3.10)$$

and, for $M = 2, 3, 4, \ldots,$ *by*

$$a_{2m} = -\frac{C(M + 1/M)}{C(M/M - 1)}, \quad (3.11)$$

$$b_{2m} = \frac{C(M/M)}{C(M/M - 1)}. \quad (3.12)$$

With these values (3.8)–(3.12) *for the elements of the continued fraction* $f(z)$, *the determinantal formulas* (3.2) *and* (3.3) *satisfy for all M*

$$A_{2M}(z) = P^{[M/M]}(z), \qquad B_{2M}(z) = Q^{[M/M]}(z),$$
$$A_{2M+1}(z) = P^{[M+1/M]}(z), \quad B_{2M+1}(z) = Q^{[M+1/M]}(z), \qquad (3.13)$$

where we assume $B_0(z) \equiv 1$.

Remark. The elements a_i, b_i given in (3.8)–(3.12) are only determined up to equivalence transforms if one only requires that (3.1) should have a given Maclaurin expansion; they are determined uniquely by the conditions (3.13).

Proof. In (3.3), we set $z = 0$ and obtain

$$B_n(0) = \prod_{i=1}^{n} b_i. \qquad (3.14)$$

Thus we can arrange for the conventional normalization to hold, namely,

$$\prod_{i=1}^{2M} b_i = Q^{[M/M]}(0) = C(M/M) \qquad (3.15a)$$

and

$$\prod_{i=1}^{2M+1} b_i = Q^{[M+1/M]}(0) = C(M+1/M), \qquad (3.15b)$$

by choosing the coefficients b_i according to (3.10) and (3.12). By inspection, we verify that the initial values (3.8) are chosen so as to satisfy

$$A_0(z) = b_0 = c_0,$$
$$B_0(z) = 1,$$
$$A_1(z) = \begin{vmatrix} b_0 & -a_1 z \\ 1 & b_1 \end{vmatrix} = c_0 + c_1 z, \qquad (3.16)$$
$$B_1(z) = b_1 = 1.$$

To establish (3.9) and (3.11) for the elements a_i, we use Frobenius identities (3.5.3) and (3.5.7).

The

M/M	$M+1/M$
	$M+1/M+1$

identity is

$$C(M+1/M)Q^{[M+1/M+1]} = C(M+1/M+1)Q^{[M+1/M]}(z)$$
$$- zC(M+2/M+1)Q^{[M/M]}(z). \qquad (3.17)$$

The recursion relation (2.7b) for the even-order denominator $B_{2M+2}(z)$ is

$$B_{2M+2}(z) = b_{2M+2} B_{2M+1}(z) + a_{2M+2} z B_{2M}(z).$$

By comparing this equation with (3.17), we see that the definition (3.11) of a_{2M+2} ensures that

$$B_{2M+2}(z) = Q^{[M+1/M+1]}(z) \qquad (3.18)$$

provided that all the lower-order denominators are identical.

The

M/M − 1	
M/M	M + 1/M

identity is

$$C(M/M)Q^{[M+1/M]}(z) = C(M + 1/M)Q^{[M/M]}(z)$$
$$- zC(M + 1/M + 1)Q^{[M/M-1]}(z). \qquad (3.19)$$

The odd-order denominator $B_{2M+1}(z)$ is also generated by (2.7b), which we write as

$$B_{2M+1}(z) = b_{2M+1}B_{2M}(z) + a_{2M+1}zB_{2M-1}(z).$$

By comparing this equation with (3.19), we see that the definition (3.9) of a_{2M+1} ensures that

$$B_{2M+1}(z) = Q^{[M+1/M]}(z) \qquad (3.20)$$

provided that all the lower-order denominators are identical. By combining (3.18), (3.20), and (3.8), the representation (3.7) is proved by induction. The Padé numerators $P^{[M/M]}$ and $P^{[M+1/M]}$ satisfy the same recursions (3.17), (3.19) as the denominators $Q^{[M/M]}$ and $Q^{[M+1/M]}$, as also do the numerators A_n, (3.5).

Hence (3.13) is established by induction, using the initial conditions expressed by (3.16).

Theorem 4.3.4. *A power series and a continued fraction may be formally identified by*

$$\sum_{i=0}^{\infty} c_i z^i = \frac{a_1}{b_1 +} \frac{a_2 z}{b_2 +} \frac{a_3 z}{b_3 +} \cdots \qquad (3.21)$$

if $a_1 = c_0$, $a_2 = -c_1$, $b_1 = 1$, $b_2 = c_0$, *and, for* $M = 1, 2, 3, \ldots$,

$$\begin{aligned}
b_{2M+1} &= \frac{C(M/M)}{C(M-1/M)}, \\
a_{2M+1} &= -\frac{C(M/M+1)}{C(M-1/M)}, \\
b_{2M+2} &= \frac{C(M/M+1)}{C(M/M)}, \\
a_{2M+2} &= -\frac{C(M+1/M+1)}{C(M/M)}.
\end{aligned} \qquad (3.22)$$

The numerators and denominators generated by recursion are identical to those of Theorem 4.3.2.

Proof. The method of proof is identical to that of Theorem 4.3.3.

Corollary. *Using an equivalence transformation, we may reexpress* (3.21) *as*

$$\sum_{i=0}^{\infty} c_i z^i = \frac{a_1'}{1} + \frac{a_2' z}{1} + \frac{a_3' z}{1} + \cdots, \qquad (3.23)$$

where

$$\begin{aligned} a_1' &= c_0, \\ a_2' &= -c_1/c_0, \\ a_3' &= -\frac{C(1/2)}{C(1/1)c_0}, \end{aligned} \qquad (3.24)$$

and for $M = 2, 3, 4, \ldots,$

$$\begin{aligned} a_{2M}' &= -\frac{C(M/M)C(M-2/M-1)}{C(M-1/M)C(M-1/M-1)}, \\ a_{2M+1}' &= -\frac{C(M/M+1)C(M-1/M-1)}{C(M/M)C(M-1/M)}. \end{aligned} \qquad (3.25)$$

As we see in (5.5.25) and (5.6.33), such a representation and its contraction are of especial importance in Stieltjes series.

In fact, (3.23) is a formal identity, and it represents the connection between a formal power series and a sequence of $[M/M]$ and $[M-1/M]$ Padé approximants. The elements a_i' of (3.23) are relatively easily calculated directly from the coefficients c_i using the Q.D. algorithm; no one would contemplate using (3.25) for iterative numerical computations in view of the likely ill conditioning of the Hankel determinants. In the context of numerical computation, it is more usual to write (3.23) as

$$\sum_{i=0}^{\infty} c_i z^i = \frac{c_0}{1} - \frac{q_1^0 z}{1} - \frac{e_1^0 z}{1} - \frac{q_2^0 z}{1} - \frac{e_2^0 z}{1} - \cdots, \qquad (3.26)$$

where

$$\begin{aligned} q_1^0 &= c_1/c_0, \\ e_1^0 &= \frac{C(1/2)}{c_1 c_0}, \end{aligned} \qquad (3.27)$$

and for $M = 2, 3, 4, \ldots$,

$$q_M^0 = \frac{C(M/M)C(M - 2/M - 1)}{C(M - 1/M - 1)C(M - 1/M)}, \quad (3.28a)$$

$$e_M^0 = \frac{C(M/M + 1)C(M - 1/M - 1)}{C(M - 1/M)C(M/M)}. \quad (3.28b)$$

More generally, for $J \geq 0$, we use the expansion

$$\sum_{i=0}^{\infty} c_i z^i = \sum_{i=0}^{J-1} c_i z^i + \frac{c_J z^J}{1} - \frac{q_1^J z}{1} - \frac{e_1^J z}{1} - \frac{q_2^J z}{1} - \frac{e_2^J z}{1} - \cdots. \quad (3.29)$$

Equation (3.26) is a special case of (3.29) with $J = 0$. With our usual convention that $L = M + J$, it follows from (3.28) that, for $M = 2, 3, 4, \ldots$ and $J \geq 0$,

$$q_M^J = \frac{C(L/M)C(L - 2/M - 1)}{C(L - 1/M - 1)C(L - 1/M)} \quad (3.30a)$$

and

$$e_M^J = \frac{C(L/M + 1)C(L - 1/M - 1)}{C(L - 1/M)C(L/M)}. \quad (3.30b)$$

Theorem 4.3.5. *The elements of* (3.29) *satisfy, for* $J \geq 0$ *and* $M = 1, 2, 3, \ldots$,

$$e_M^J q_{M+1}^J = e_M^{J+1} q_M^{J+1} \quad (3.31)$$

and

$$q_M^J + e_M^J = e_{M-1}^{J+1} + q_M^{J+1}. \quad (3.32)$$

First proof. Substituting from (3.30),

$$e_M^J q_{M+1}^J = \frac{C(L - 1/M - 1)C(L + 1/M + 1)}{C(L/M)^2} = e_M^{J+1} q_M^{J+1}$$

for $M = 2, 3, 4, \ldots$ and $J \geq 0$. The case for $M = 1$ is easily verified explicitly, and one may also treat it using the convention that $C(L/0) \equiv 1$. Likewise one verifies (3.32):

$$e_M^J - q_M^{J+1}$$

$$= \frac{C(L - 1/M - 1)\{C(L/M + 1)C(L/M - 1) - C(L + 1/M)C(L - 1/M)\}}{C(L - 1/M)C(L/M)C(L/M - 1)}$$

$$= \frac{C(L/M)C(L - 1/M - 1)}{C(L - 1/M)C(L/M - 1)},$$

$$e_{M-1}^{J+1} - q_M^J$$
$$= \frac{C(L/M)\{C(L-2/M-1)C(L/M-1) - C(L-1/M-2)C(L-1/M)\}}{C(L-1/M-1)C(L/M)C(L/M-1)}$$
$$= \frac{C(L/M)C(L-1/M-1)}{C(L-1/M)C(L/M-1)} = e_M^J - q_M^{J+1}.$$

Second proof. We consider the algebraic identity from two sequential continued-fraction expansions, namely,

$$\sum_{i=J}^{\infty} c_i z^i = \frac{c_J z^J}{1} - \frac{q_1^J z}{1} - \frac{e_1^J z}{1} - \frac{q_2^J z}{1} - \frac{e_2^J z}{1} - \cdots$$
$$= c_J z^J + \frac{c_{J+1} z^{J+1}}{1} - \frac{q_1^{J+1} z}{1} - \frac{e_1^{J+1} z}{1} - \frac{q_2^{J+1} z}{1} - \cdots . \quad (3.33)$$

We make a contraction, given by (5.3) and explained in Section 4.5, on these fractions, and it follows that

$$c_J z^J + \frac{c_J q_1^J z^{J+1}}{1 - (q_1^J + e_1^J)z} - \frac{e_1^J q_2^J z^2}{1 - (q_2^J + e_2^J)z} - \frac{e_2^J q_3^J z^2}{1 - (q_3^J + e_3^J)z} - \cdots$$
$$= c_J z^J + \frac{c_{J+1} z^{J+1}}{1 - q_1^{J+1} z} - \frac{q_1^{J+1} e_1^{J+1} z^2}{1 - (e_1^{J+1} + q_2^{J+1})z} - \frac{q_2^{J+1} e_2^{J+1} z^2}{1 - (e_2^{J+1} + q_3^{J+1})z} - \cdots .$$
(3.34)

By identifying the coefficients in these expansions, (3.31) and (3.32) are proved.

The Q.D. algorithm. This algorithm, due to Rutishauser [1954, 1957], may be used to construct the coefficients q_i^0, e_i^0 in the continued-fraction representation (3.26) of the Padé approximants of a given power series. In this context, the initializing values are taken to be

$$\left.\begin{array}{l} e_0^J = 0, \qquad J = 1, 2, 3, \ldots, \\ q_1^J = c_{J+1}/c_J, \qquad J = 0, 1, 2, \ldots, \end{array}\right\} \quad (3.35)$$

as is required by (3.34). Equations (3.31) and (3.32) constitute the body of the algorithm. The elements are normally exhibited in the Q.D. table as shown in Table 4.3.1.

The Q.D. algorithm of (3.31), (3.32) is often the most powerful algorithmic method for obtaining closed-form representations of continued fractions corresponding to power series whose coefficients are given analytically. This representation is, in fact, of wider importance. The coefficients of the continued fraction are also the coefficients of the recursion relation of the denominator polynomials of the convergents of

4.3 Various representations of continued fractions

Table 4.3.1. *The Q.D. table. Its two left columns are specified by (3.35), and the remaining elements are determined by (3.31) and (3.32). The elements along any diagonal are the elements of a continued fraction (3.29).*

e_0^1	q_1^0						
		e_1^0					
	q_1^1		q_2^0				
e_0^2		e_1^1		e_2^0			
	q_1^2		q_2^1		q_3^0		
e_0^3		e_1^2		e_2^1			$\cdots$
	q_1^3		q_2^2		$\vdots$	$\vdots$	
e_0^4		e_1^3		$\vdots$			
	q_1^4		$\vdots$				
e_0^5		$\vdots$					
$\vdots$	$\vdots$						

the fraction. These denominator polynomials of (say) even order may happen to be associated with a set of orthogonal polynomials as defined by (5.3.18) and (5.3.20) for Stieltjes functions. In some cases, such as case 3 below with $0 < q < 1$, the polynomials are orthogonal on the unit circle with respect to a nonnegative weight function. In other cases, the orthogonality is purely formal in the sense of the umbral calculus [Roman and Rota, 1978; Van Rossum, 1953; Hendriksen and Van Rossum, 1982]. In cases where the weight function is nonnegative, the recursion relation for the orthogonal polynomials contains a great deal of information about the character of the weight function.

The Q.D. table of $\exp(z)$. The left-hand column of Table 4.3.1 consists of zeros and the next column contains

$$q_1^j = \frac{c_{j+1}}{c_j} = \frac{1}{j+1},$$

according to (3.35). The rules (3.31) and (3.32) connect entries at the vertices of rhombuses in the Q.D. table:

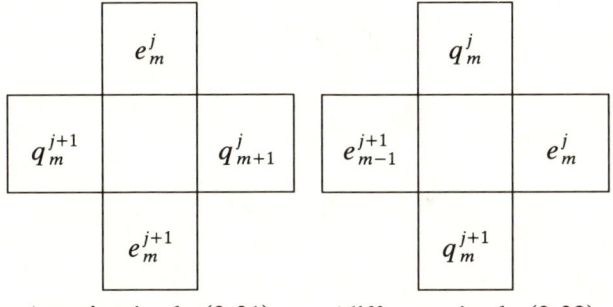

'quotient' rule (3.31)   'difference' rule (3.32)

For $j = 0, 1, 2, \ldots$ and $m = 1, 2, 3, \ldots$, one finds that

$$q_m^{(j)} = \frac{j + m - 1}{(j + 2m - 2)(j + 2m - 1)}, \quad e_m^{(j)} = \frac{-m}{(j + 2m - 1)(j + 2m)}.$$

From these formulas Table 4.3.2 is constructed and the parentheses emphasize that j is an index on the left-hand side. By putting $j = 0$, we obtain the representation (3.26) as

$$\exp(z) = \frac{1}{1-} \frac{z}{1+} \frac{\frac{1}{2}z}{1-} \frac{\frac{1}{6}z}{1+} \frac{\frac{1}{6}z}{1-} \frac{\frac{1}{10}z}{1+} \cdots,$$

which agrees with (6.1a).

The example of Section 2.3 which admits a large number of interesting special cases is $f(z) = \sum_{i=0}^{\infty} c_i z^i$ with

$$c_i = \prod_{j=0}^{i-1} \frac{A - q^{\alpha+j}}{C - q^{\gamma+j}}, \quad i = 1, 2, 3, \ldots, \tag{3.36}$$

and $c_0 = 1$ [Wynn, 1967]. Using (3.31), (3.22), and (3.35), it is a matter of algebra to check that

$$q_m^{(j)} = q^{m-1} \frac{[A - q^{\alpha+j+m-1}][C - q^{\gamma+j+m-2}]}{[C - q^{\gamma+j+2m-3}][C - q^{\gamma+j-2m-2}]}, \tag{3.37}$$

$$e_m^{(j)} = q^{m+j-1} \frac{[1 - q^m][Cq^\alpha - Aq^{\gamma+m-1}]}{[C - q^{\gamma+j+2m-2}][C - q^{\gamma+j+2m-1}]} \tag{3.38}$$

for $j = 0, 1, 2, \ldots$ and $m = 1, 2, 3, \ldots$. The elements $q_m^{(0)}$, $e_m^{(0)}$ are required for (3.26), giving

$$f(z) = \frac{1}{1-} \frac{[A - q^\alpha][C - q^\gamma]^{-1}z}{1-} \frac{(1 - q)[Cq^\alpha - Aq^\gamma][C - q^\gamma]^{-1}[C - q^{\gamma+1}]^{-1}z}{1} - \cdots$$

formally. The special cases are found as follows.

Case 1. Replace z by Cz and let $C \to \infty$ in (3.36) so that the Maclaurin coefficients become

$$c_i = \prod_{j=0}^{i-1} [A - q^{\alpha+j}], \quad i = 0, 1, 2, \ldots, \tag{3.39}$$

and $f(z) = 1 + [A - q^\alpha]z + [A - q^\alpha][A - q^{\alpha+1}]z^2 + \cdots$.

Table 4.3.2. *Part of the Q.D. table of* $\exp(z)$.

0		1						
	$\frac{1}{2}$		$-\frac{1}{2}$					
0				$\frac{1}{6}$				
	$\frac{1}{3}$		$-\frac{1}{6}$		$-\frac{1}{6}$			
0				$\frac{1}{6}$			$\frac{1}{10}$	
	$\frac{1}{4}$		$-\frac{1}{12}$		$\frac{1}{10}$			
0				$\frac{3}{20}$				
	$\frac{1}{5}$		$-\frac{1}{20}$					
0								

From (3.37), (3.38),

$$q_m^{(j)} = q^{m-1}[A - q^{\alpha+j+m-1}], \qquad (3.40)$$

$$e_m^{(j)} = q^{\alpha+m+j-1}[1 - q^m] \qquad (3.41)$$

for $j = 0, 1, 2, \ldots$ and $m = 1, 2, 3, \ldots,$
and formally

$$f(z) = \frac{1}{1-} \frac{[A-q^\alpha]z}{1} - \frac{q^\alpha(1-q)z}{1} - \frac{q[A-q^{\alpha+1}]z}{1} - \frac{q^{\alpha+1}[1-q^2]z}{1} - \cdots. \qquad (3.42)$$

Case 2. Take $A = 0$ and replace z by $-z$ so that (3.39) becomes

$$c_i = q^{i(i-1)/2}, \quad i = 0, 1, 2, \ldots, \qquad (3.43)$$

and

$$f(z) = 1 + z + qz^2 + q^3z^3 + q^6z^4 + \cdots \qquad (3.44)$$

which is the partial theta function of Rogers and Szegö. From (3.40) and (3.41), the Q.D. coefficients for (3.44) are

$$q_m^{(j)} = q^{j+2m-2}, \; e_m^{(j)} = -q^{m+j-1}(1-q^m). \qquad (3.45)$$

By taking $j = 1$ in (3.29), (3.45), we find

$$f(z) = 1 + \frac{z}{1-} \frac{qz}{1+} \frac{q(1-q)z}{1} - \frac{q^3z}{1+} \frac{q^2(1-q^2)z}{1} - \cdots. \qquad (3.46)$$

This form is of particular interest [Lubinsky and Saff, 1987; Balk, 1960; Heine, 1878] and we return to it in Sections 6.1, 6.7. The corresponding denominator polynomials are related to Stieltjes–Wigert polynomials [Wigert, 1923].

Case 3. Replace z by z/A in (3.36) and let $A \to \infty$ so that the Maclaurin coefficients become

$$c_i = \prod_{j=0}^{i}[C - q^{\gamma+j}]^{-1}$$

and the continued-fraction coefficients are given by

$$q_m^{(j)} = \frac{q^{m-1}[C - q^{\gamma+j+m-2}]}{[C - q^{\gamma+j+2m-3}][C - q^{\gamma+j+2m-2}]},$$

$$e_m^{(j)} = -\frac{q^{\gamma+j+2m-2}[1 - q^m]}{[C - q^{\gamma+j+2m-2}][C - q^{\gamma+j+2m-1}]},$$

similarly to case 1.

Case 4. Take $A = C = 1$ in (3.36) and let $q \to 1$ so that the limiting case has Maclaurin coefficients

$$c_i = \prod_{j=0}^{i-1} \frac{\alpha + j}{\gamma + j} \tag{3.47}$$

and continued-fraction coefficients

$$q_m^{(j)} = \frac{(\alpha + j + m - 1)(\gamma + j + m - 2)}{(\gamma + j + 2m - 3)(\gamma + j + 2m - 2)}, \tag{3.48}$$

$$e_m^{(j)} = \frac{m(\gamma - \alpha + m - 1)}{(\gamma + j + 2m - 2)(\gamma + j + 2m - 1)}. \tag{3.49}$$

The coefficients (3.47) are those of a hypergeometric function, and it follows from (3.26) and (3.47)–(3.49) that

$$_2F_1(\alpha, 1, \gamma; z) = \frac{1}{1-} \frac{\alpha z}{\gamma -} \frac{1(\gamma - \alpha)z}{\gamma + 1 -} \cdots$$
$$- \frac{(\alpha + m - 1)(\gamma + m - 2)z}{\gamma + 2m - 2}$$
$$- \frac{m(\gamma - \alpha + m - 1)z}{\gamma + 2m - 1 -} \cdots$$

as in (6.16) with $b = 0$, $a = \alpha$, $c = \gamma - 1$.

Other special cases follow from case 4 using the limiting procedures of cases 1, 3 and yield the fractions (6.14), (6.15) similarly.

A comprehensive compilation of all the continued fractions that can be constructed in this way, together with the coefficients of the associated (orthogonal) polynomials, as defined in (5.3.20) for Stieltjes functions, is given by Wynn [1967]. Notice that fractions of the form

$$f(z) = c_0 + \frac{c_1 z}{1} - \frac{q_1^{(1)} z}{1} - \frac{e_1^{(1)} z}{1} - \frac{q_2^{(1)} z}{1} - \frac{e_2^{(1)} z}{1} - \cdots$$

also follow directly from the formulas such as (3.37), (3.38) on taking $j = 1$, as exemplified in case 3.

One may extend the Q.D. table above its diagonal by defining $q_{i+1}^{(-i)} = 0$ for $i = 1, 2, 3, \ldots$. In this way all the elements of the top row except $q_1^{(0)}$ are defined to be zero, and the rhombus rules allow completion of the table. If the coefficients of the reciprocal series are used by setting $e_{i+1}^{(i)} = c'_{i+1}/c'_i$ for $i = 0, 1, 2, \ldots$, in the notation of (2.4.10), and the appropriate order of calculation is followed, then the whole table may be calculated more stably by the progressive form of the Q.D. algorithm described in Section 3.8; otherwise the Q.D. algorithm is notoriously unstable. If $f(z)$ is meromorphic, the 'q' columns converge to the reciprocals of the poles of $f(z)$, provided the moduli of the poles are distinct. Similarly the 'e' rows converge to the reciprocals of the zeros of $f(z)$, provided they have distinct moduli. This property follows from (3.8.4) and (3.8.6) et seq. It is instructive to compare this form of the Q.D. algorithm with that described in Section 3.8: they are not identical.

Throughout this section we have assumed that all the inverses we need do in fact exist. It is quite possible that $b_k = 0$ for some value of k in (4.1) and so only the first k convergents are defined; convergence of the fraction is meaningless. Likewise the Q.D. algorithm may break down if a zero divisor is encountered. Worse, from a numerical point of view, is the possibility that the computed results are generated by rounding error in a zero or near-zero entry. We refer to Claessens and Wuytack [1979] for an extension of (3.31) and (3.32) to the case of a nonnormal Q.D. table.

4.4 The Berlekamp–Massey algorithm and an application of it

The Berlekamp–Massey algorithm is used to calculate the coefficients of the denominator polynomials in a staircase sequence of Padé approximants. We begin with a description of the algorithm in a straightforward way for a nondegenerate case, and then consider how the algorithm is modified to treat other cases.

As usual, the data consist of the coefficients of

$$f(z) = c_0 + c_1 z + c_2 z^2 + \cdots \tag{4.1}$$

and the construction is initialized with

$$A_0(z) = c_0, \quad B_0(z) = 1; \quad A_{-1}(z) = 0, \quad B_{-1}(z) = 1. \quad (4.2)$$

At each stage of the construction, a remainder function $r_i(z)$ is defined by

$$f(z)B_i(z) - A_i(z) = z^{i+1}r_i(z), \quad i = -1, 0, 1, \ldots, \quad (4.3)$$

where $r_i(0)$ is well defined. For the moment, we consider only the cases in which

$$r_i(0) \neq 0, \quad i = -1, 0, 1, \ldots. \quad (4.4)$$

At stage i of the algorithm, both (4.3) and

$$f(z)B_{i-1}(z) - A_{i-1}(z) = z^i r_{i-1}(z), \quad i = 0, 1, 2, \ldots, \quad (4.5)$$

hold. The Berlekamp–Massey algorithm is based on the principle of eliminating the leading residuals on the right-hand sides of (4.3), (4.5) recursively. This is done by forming

$$B_{i+1}(z) = B_i(z) - zB_{i-1}(z)r_i(0)/r_{i-1}(0), \quad (4.6)$$

$$A_{i+1}(z) = A_i(z) - zA_{i-1}(z)r_i(0)/r_{i-1}(0), \quad (4.7)$$

so that

$$f(z)B_{i+1}(z) - A_{i+1}(z) = z^{i+2}r_{i+1}(z),$$

where $r_{i+1}(0)$ is well defined. It is straightforward to prove by induction that the algorithm expressed by (4.6), (4.7) with the initialization (4.2) leads to

$$\deg\{B_i(z)\} \leq \left[\frac{i+1}{2}\right], \quad \deg\{A_i(z)\} \leq \left[\frac{i}{2}\right] \quad (4.8)$$

and $B_i(0) = 1$ for all i. The notation $[R]$ used in (4.8) denotes the greatest integer not exceeding R. By storing the coefficients of the denominator polynomials, and calculating the values of $r_i(0)$ sequentially from

$$r_i(0) = \sum_{j=0}^{\alpha_i} (B_i)_j c_{i+1-j}, \quad (4.9)$$

where $\alpha_i = \deg\{B_i(z)\}$, we have an efficient algorithm for constructing the denominator polynomials of the staircase sequence of types $[0/0]$, $[0/1]$, $[1/1]$, $[1/2]$, ...

Other staircase sequences can be generated by modifying the initialization (4.2).

This algorithm was originally designed for use in circuit synthesis for BCH decoding and in this context the coefficients c_k lie in a finite field. Thus the hypothesis that all $r_i(0) \neq 0$ is not sufficiently general to be

4.4 The Berlekamp–Massey algorithm and an application of it

realistic. We proceed to a reliable version of the algorithm which calculates all the entries of a staircase sequence, whether the table is normal or not. The entries obtained in locations within blocks of the Padé table corresponding to nonexistent Padé approximants are obviously not Padé approximants. In fact, the polynomials $(A_i(z), B_i(z))$ forming these entries have higher order than is allowed by (4.8), but possess the correct accuracy-through-order property of (4.3). Thus all members of the sequence can be used for circuit design.

Specification. The algorithm constructs the coefficients of the denominator polynomials $\{B_i(z)\}_{i=0}^{\infty}$ of Padé approximants of type $[m/m]$ if $i = 2m$ and of type $[m/m+1]$ if $i = 2m+1$, if such approximants exist. Otherwise, a rational fraction is constructed for which

$$f(z) - A_i(z)/B_i(z) = O(z^{i+1}) \tag{4.10}$$

and $A_i(z)$, $B_i(z)$ are of minimal degree.

Construction. On entry to the ith stage of the construction, the polynomials $A_i(z)$, $B_i(z)$ have been (at least implicitly) constructed, and a predecessor index $\sigma(i)$ has been set. The remainder function $r_i(z)$ is then defined by

$$f(z)B_i(z) - A_i(z) = z^{i+1} r_i(z) \tag{4.11}$$

at every stage. The next branch of the construction depends on whether $r_i(0) = 0$ or not; the predecessor index $\sigma(i)$ has been set with the vital property that $r_{\sigma(i)}(0) \neq 0$. In fact, the coefficients of $A_i(z)$ are not required for the sequential implementation of the algorithm, and so their construction can be omitted if they are not required. Likewise, the first term $r_i(0)$ of $r_i(z)$ is required but not its series development. It is obtained as

$$r_i(0) = \sum_{j=0}^{\alpha_i} c_{i+1-j}(B_i)_j \tag{4.12}$$

where

$$\alpha_i = \deg\{B_i(z)\} \quad \text{and} \quad B_i(z) = \sum_{j=0}^{\alpha_i} (B_i)_j z^j. \tag{4.13}$$

If, at stage i, $r_i(0) = 0$ but $r_{i-1}(0) \neq 0$, this signals the entry of the staircase sequence into a block of the Padé table of which $(A_i(z), B_i(z))$ is the minimal entry. Figures 4.2.3 and 4.2.4 display the possible degeneracies, but in the context of a diagonal sequence of approximants. Let $\delta(i)$ be the least positive integer for which $r_{i+\delta(i)}(0) \neq 0$, so $\delta(i) + 1$ is the dimension of the block; $\delta(i)$ is also the degree of oversatisfaction r

used in the proof of Theorem 1.4.3. The situation is shown in Figure 4.4.1 for the case $\delta(i) = 2$. Entries in the block which are (trivially) Padé approximants are given by

$$B_{i+j}(z) = B_i(z), \quad A_{i+j}(z) = A_i(z), \quad j = 1, 2, \ldots, \delta(i). \quad (4.14)$$

Following the principle of eliminating the residual, the next entry is defined by

$$B_{i+\delta(i)+1}(z) = B_i(z) - z^{\delta(i)+1} B_{i-1}(z) r_{i+\delta(i)}(0)/r_{i-1}(0), \quad (4.15)$$

$$A_{i+\delta(i)+1}(z) = A_i(z) - z^{\delta(i)+1} A_{i-1}(z) r_{i+\delta(i)}(0)/r_{i-1}(0). \quad (4.16)$$

Closer inspection reveals that these steps lead inevitably to a pronounced jump in the degrees of the polynomials, and the exact amounts depend on the way in which the staircase sequence is located relative to the block. Because of this, the stage (4.15), (4.16) is referred to as a major change. The other entries in the staircase sequence whose indices lie within the block and their successors are defined by

$$B_{i+\delta(i)+j+1}(z) = B_{i+\delta(i)+j}(z) - z^j B_{i+\delta(i)}(z) r_{i+\delta(i)+j}(0)/r_{i+\delta(i)}(0), \quad (4.17)$$

$$A_{i+\delta(i)+j+1}(z) = A_{i+\delta(i)+j}(z) - z^j A_{i+\delta(i)}(z) r_{i+\delta(i)+j}(0)/r_{i+\delta(i)}(0). \quad (4.18)$$

for $j = 1, 2, \ldots, \delta(i)$. These equations do not lead to degree increase, and so are referred to as minor changes. The entry $(A_{i+2\delta(i)+1}(z), B_{i+2\delta(i)+1}(z))$ is the first Padé approximant outside the block, and its successor is calculated using $\sigma(i + 2\delta(i) + 1) = i + \delta(i)$.

In Table 4.4.1, we give an example of how the algorithm passes

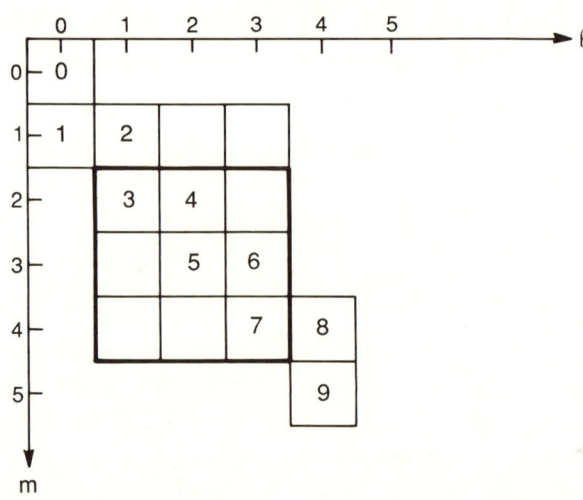

Figure 4.4.1. The staircase sequence corresponding to Table 4.4.1.

4.4 The Berlekamp–Massey algorithm and an application of it

Table 4.4.1. *An example of some properties of the polynomials generated by the Berlekamp–Massey algorithm.*

$B_i(z)$	deg$\{B_i(z)\}$	deg$\{A_i(z)\}$	Residual property
$B_1(z)$	1	0	$r_1(0)z^2$
$B_2(z)$	1	1	$r_2(0)z^2$
$B_3(z)$	2	1	$r_3(0) = 0$
$B_4(z) = B_3(z)$	2	1	$r_4(0) = 0$
$B_5(z) = B_3(z)$	2	1	$r_5(0)z^6$
$B_6(z) = B_5(z) - z^3 B_2(z) r_5(0)/r_2(0)$	4	4	$r_6(0)z^7$
$B_7(z) = B_6(z) - zB_5(z)r_6(0)/r_5(0)$	4	4	$r_7(0)z^8$
$B_8(z) = B_7(z) - z^2 B_5(z) r_7(0)/r_5(0)$	4	4	$r_8(0)z^9$
$B_9(z) = B_8(z) - z^3 B_5(z) r_8(0)/r_5(0)$	5	4	$r_9(0)z^{10}$

through a block. We suppose that $r_1(0) \neq 0$, $r_2(0) \neq 0$ and that $r_3(0) = 0$ signals that $(A_3(z), B_3(z))$ is the minimal entry of a nontrivial block. Then $r_4(0) = 0$ but $r_5(0) \neq 0$ indicates that $\delta(3) = 2$ and the block has size 3. Because $r_5(0) \neq 0$, $B_6(z)$ and $A_6(z)$ are constructed with a major change, as is evident from the degree columns. Whether $r_6(0)$, $r_7(0)$, and $r_8(0)$ are zero or not is immaterial. The predecessor labels are $\sigma(3) = \sigma(4) = \sigma(5) = 2$, $\sigma(6) = \sigma(7) = \sigma(8) = 5$.

If $c_0 = c_1 = \cdots = c_{k-1} = 0$ and $c_k \neq 0$ in (4.1), a nontrivial initialization stage is required. It involves steps corresponding to minor changes, and as such it amounts to reciprocation of the first $k + 1$ nontrivial terms of the series (4.1).

Example. Consider the example $k = 3$, $f(z) = c_3 z^3 + c_4 z^4 + \cdots$, for which we have the following table.

Table 4.4.2. *An example of properties of some polynomials which initiate the Berlekamp–Massey algorithm.*

$B_i(z)$	$A_i(z)$	Residual
$B_{-1} = B_0 = B_1 = B_2 = 1$	$A_{-1} = A_0 = A_1 = A_2 = 0$	$O(z^3)$
$B_3 = 1$	$A_3 = c_3 z^3$	$O(z^4)$
$B_4 = 1 - zc_4/c_3$	$A_4 = A_3$	$O(z^5)$
$B_5 = B_4 - z^2 r_4(0)/c_3$	$A_5 = A_3$	$O(z^6)$
$B_6 = B_5 - z^3 r_5(0)/c_3$	$A_6 = A_3$	$r_6(z)z^7$
$B_7 = B_6 - z^4 r_6(0)/c_3$	$A_7 = A_3$	$r_7(z)z^8$

The entries $(A_6z), B_6(z))$ and $(A_7(z), B_7(z))$ normally suffice to start the iteration stage of the algorithm. If we specialize the example of the initialization even more, the method described leads to a representation of $f(z) = c_3z^3 + c_8z^8$ by

$$f(z) = \frac{c_3z^3}{1} - \frac{(c_8/c_3)z^5}{1 + (c_8/c_3)z^5}.$$

It is instructive to locate the convergents of $f(z)$ in Figure 4.4.2.

The following algorithm specifies the construction. It is to be understood as being numerically implemented on the coefficients of the polynomials.

Initialization. Let k be the least nonnegative integer for which $c_k \neq 0$. Set

$$B_{-1}(z) = B_0(z) = \cdots = B_k(z) = 1, \quad A_k(z) = \cdots = A_{2k+1}(z) = c_kz^k,$$
$$A_{-1}(z) = A_0(z) = \cdots = A_{k-1}(z) = 0,$$
$$\sigma(k) = \sigma(k+1) = \cdots = \sigma(2k+1) = k-1,$$

and, for $j = k, k+1, \ldots, 2k$,

$$B_{j+1}(z) = B_j(z) - z^{j-k+1}r_j(0)/c_k. \tag{4.19}$$

Iteration. For $j = 2k+1, 2k+2, \ldots$, construct $r_j(0)$ using (4.12).

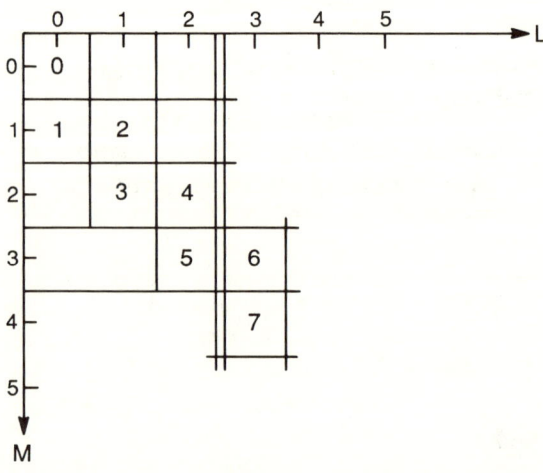

Figure 4.4.2. The staircase sequence corresponding to Table 4.4.2.

4.4 The Berlekamp–Massey algorithm and an application of it

If $r_j(0) = 0$, then

$$B_{j+1}(z) = B_j(z), \quad A_{j+1}(z) = A_j(z), \qquad (4.20)$$
$$\sigma(j+1) = \sigma(j);$$

else we have $r_j(0) \neq 0$.
  If $j > \sigma(j) + 1$ (major change), then

$$B_{j+1}(z) = B_j(z) - z^{j-\sigma(j)} B_{\sigma(j)}(z) r_j(0)/r_{\sigma(j)}(0), \qquad (4.21)$$
$$A_{j+1}(z) = A_j(z) - z^{j-\sigma(j)} A_{\sigma(j)}(z) r_j(0)/r_{\sigma(j)}(0), \qquad (4.22)$$
$$\sigma(j+1) = j;$$

else we have $j = \sigma(j) + 1$ (minor change). Then

$$B_{j+1}(z) = B_j(z) - z^{j-\sigma(j)} B_{\sigma(j)}(z) r_j(0)/r_{\sigma(j)}(0), \qquad (4.23)$$
$$A_{j+1}(z) = A_j(z) - z^{j-\sigma(j)} A_{\sigma(j)}(z) r_j(0)/r_{\sigma(j)}(0), \qquad (4.24)$$
$$\sigma(j+1) = \sigma(j).$$

Next j.

Notice that it is only the predecessor statement that differs between the two branches that are called major and minor changes, and this subtle change controls the degree growth of the polynomials.

Proof of the construction. By inspection, we find that the initialization stage of the algorithm terminates with $\deg\{B_{2k}(z)\} \leq k$, $\deg\{B_{2k+1}(z)\} \leq k+1$, and

$$B_j(0) = 1, \quad f(z)B_j(z) - A_j(z) = O(z^{j+1}), \quad j = 2k, 2k+1. \qquad (4.25)$$

Thus these two entries are Padé approximants of $f(z)$ of types $[k/k]$ and $[k/k+1]$, respectively. Moreover, $r_{\sigma(2k+1)}(0) = c_k \neq 0$.

The degree bounds

$$\left.\begin{array}{l} \deg\{B_j(z)\} \leq \left[\dfrac{i+1}{2}\right], \\[2mm] \deg\{A_j(z)\} \leq \left[\dfrac{i}{2}\right], \end{array}\right\} j = i, i+1, \ldots, i+\delta(i); \qquad (4.26)$$

$$\left.\begin{array}{l} \deg\{B_j(z)\} \leq \left[\dfrac{i+1}{2}\right] + \delta(i), \\[2mm] \deg\{A_j(z)\} \leq \left[\dfrac{i}{2}\right] + \delta(i), \end{array}\right\} j = i+\delta(i)+1, \ldots, i+2\delta(i); \qquad (4.27)$$

apply to the polynomials constituting the entries of indices $i, i+1, \ldots, i+2\delta(i)$ within the block specified by (4.15)–(4.18). The inequalities

$$\deg\{B_j(z)\} \le \left[\frac{j+1}{2}\right], \quad \deg\{A_j(z)\} \le \left[\frac{j}{2}\right] \tag{4.28}$$

apply to the next two entries $j = i + 2\delta(i) + 1$, $i + 2\delta(i) + 2$. These results are easily derived by induction from (4.15)–(4.18). In fact, the bounds (4.28) apply to all polynomials except those generated by (4.19) for $j \le 2k - 2$ and (4.21)–(4.24) for $j \le i + 2\delta(i) - 2$. The fact that all these polynomials are of minimal degree for the purpose of satisfying the accuracy-through-order property (4.10) is more easily seen using the methods of Section 4.2.

It is worth noting that the core of this algorithm can be compacted in the following way. A nontrivial block is entered with

$$f(z)B_{i-1}(z) - A_{i-1}(z) = r_{i-1}(z)z^i, \quad r_{i-1}(0) \ne 0, \tag{4.29}$$

and its dimension is determined by

$$f(z)B_i(z) - A_i(z) = O(z^{i+1}), \quad \text{with } r_i(0) = 0,$$
$$= r_{i+\delta(i)}(z)z^{i+\delta(i)+1}, \tag{4.30}$$

with $r_{i+\delta(i)}(0) \ne 0$ and $\sigma(i) = i - 1$. For values of i satisfying (4.29), we define

$$\hat{r}_i(z) = [r_{i-1}(z)/r_{i+\delta(i)}(z)]_0^{\delta(i)}, \tag{4.31}$$

where Nuttall's truncation notation $[\varphi(z)]_0^{\delta(i)}$ in (4.31) denotes the first $\delta(i) + 1$ terms of the Maclaurin series of $\varphi(z)$. The recursions defined by

$$\left.\begin{array}{l} B_{i+2\delta(i)+1}(z) = \hat{r}_i(z)B_i(z) - z^{\delta(i)+1}B_{\sigma(i)}(z), \\ A_{i+2\delta(i)+1}(z) = \hat{r}_i(z)A_i(z) - z^{\delta(i)+1}A_{\sigma(i)}(z) \end{array}\right\} \tag{4.32}$$

are motivated by the successive elimination of residuals from (4.29) and (4.30) and justify the result that

$$f(z)B_{i+2\delta(i)+1}(z) - A_{i+2\delta(i)+1}(z) = r_{i+2\delta(i)+1}(z)z^{i+2\delta(i)+2}, \tag{4.33}$$

where the remainder function $r_{i+2\delta(i)+1}(z)$ is regular at $z = 0$. Thus, with the initialization of (4.19), $A_{i+2\delta(i)+1}(z)/B_{i+2\delta(i)+1}(z)$ is a Padé approximant of type $[[(i+1)/2] + \delta(i)/[i/2] + 1 + \delta(i)]$, using an obviously mixed notation. As such, it is the next genuine Padé approximant in the staircase sequence containing $A_{i-1}(z)/B_{i-1}(z)$ and $A_i(z)/B_i(z)$. Equations of the form (4.32), supplemented by (4.29)–(4.31), specify an algorithm which both uses and generates only the genuine Padé approximants in the

4.4 The Berlekamp–Massey algorithm and an application of it

staircase sequence. It follows from (4.33) that the power series (4.1) can be formally reexpressed as

$$f(z) = \frac{c_k z^k}{\hat{r}_k(z)} - \frac{z^{\delta(i_2)+1}}{\hat{r}_{i_2}(z)} - \frac{z^{\delta(i_3)+1}}{\hat{r}_{i_3}(z)} - \cdots, \tag{4.34}$$

where the sequence $\{i_j\}_{j=1}^{\infty}$ is defined initially by $i_1 = k$ and recursively by

$$i_{j+1} = i_j + 2\delta(i_j) + 1, \quad j = 1, 2, \ldots.$$

Notice that each $\hat{r}_{i_j}(0) \neq 0$. This form compares directly with (4.2.30), the calculation of which was a main aim of that section, but for which a more direct approach (based on ratios of series) to the calculation of the $\hat{r}_i(z)$ was adopted. Because the Berlekamp–Massey algorithm and the generalized Viskovatov algorithm have the same conceptual basis, they share with Kronecker's algorithm the capacity to 'hop across' a block. However, it is not suitable for incorporation into rectangular-matrix Padé approximation because the accuracy through order is achieved with jumps.

Applications of this work are mostly found in signal processing. In this area of application, it is usual to use the variable $s = z^{-1}$, and there are several changes of presentation of the results, but not of their mathematical content. We begin with the power series expressed as

$$\tilde{f}(s) = c_0 + c_1 s^{-1} + c_2 s^{-2} + \cdots. \tag{4.35}$$

By comparison with the previous developments, we use the polynomials $A_i(z)$, $B_i(z)$ to form a reversed numerator and denominator polynomials defined by

$$q_i(s) = s^{\alpha_i} B_i(s^{-1}), \quad p_i(s) = s^{\alpha_i} A_i(s^{-1}), \tag{4.36}$$

where $\alpha_i = \deg\{B_i(s)\}$, so

$$\tilde{f}(s)q_i(s) - p_i(s) = O(s^{\alpha_i - i - 1}), \quad s \to \infty. \tag{4.37}$$

Since $B_i(0) = 1$, $q_i(s)$ is a monic polynomial of degree α_i. Remainder functions $\tilde{r}_i(s)$ are defined analogously to (4.11) with

$$\tilde{f}(s)q_i(s) - p_i(s) = s^{\alpha_i - i - 1}\tilde{r}_i(s) \tag{4.38}$$

and $\tilde{r}_i(s) = r_i(z^{-1})$ is regular at $s = \infty$. For example, we have

$$i = 0: \quad \tilde{f}(s) \cdot 1 - c_0 = c_1 s^{-1} + O(s^{-2}),$$

$$i = 1: \quad \tilde{f}(s)(s - c_1/c_0) - c_0 s = s^{-1}(c_2 - c_1^2/c_0) + O(s^{-2}),$$

$$i = 2: \quad \tilde{f}(s)(s - c_2/c_1) - \left[s - \frac{c_1^2 - c_2 c_0}{c_1 c_0}\right]c_0 = s^{-2}\tilde{r}_2(s),$$

when $c_0 \neq 0$, $c_1 \neq 0$ in (4.35). Again, the Berlekamp–Massey algorithm is based on successive elimination of the leading residuals in these equations and their successors. In nondegenerate cases, the iteration takes the form:

If $\alpha_i = \alpha_{i-1} + 1$, then $q_{i+1}(s) = q_i(s) - q_{i-1}(s)\tilde{r}_i(\infty)/\tilde{r}_{i-1}(\infty)$,

else we have $\alpha_i = \alpha_{i-1}$. Then $q_{i+1}(s) = sq_i(s) - q_{i-1}(s)\tilde{r}_i(\infty)/\tilde{r}_{i-1}(\infty)$.

From these, it follows that $q_i(s)$ is monic, $p_i(0) = 0$ if i is odd, and

$$\deg\{q_i(s)\} = \deg\{p_i(s)\} = \left[\frac{i+1}{2}\right],$$

so

$$\tilde{f}(s)q_i(s) = p_i(s) + \tilde{r}_i(s)s^{-1-[i/2]}.$$

In cases where degeneracy of the form $c_0 = c_1 = \cdots = c_k = 0$, the initialization of the s-form of the Berlekamp–Massey algorithm differs in detail from the z-form previously given, but the structure of the recursion is identical. This difference occurs because, in the example of Figure 4.4.2, $f(z) = z^3 = s^{-3}$, and so the approximant of type [3/0] is exact for the z-form, whereas the approximant of type [0/3] is exact for the s-form.

The accuracy-through-order condition seems more natural in the z-form, whereas the initialization seems more natural in the s-form.

Initialization. Let k be the least nonnegative integer for which $c_k \neq 0$. Define

$$q_{-1}(s) = q_0(s) = \cdots = q_{k-1}(s) = 1,$$

$$p_{-1}(s) = p_0(s) = \cdots = p_{k-1}(s) = 0, \quad \tilde{r}_{k-1}(\infty) = c_k,$$

$$q_k(s) = s^k, \quad p_k(s) = c_k, \quad \sigma(k) = k - 1.$$

Iteration. [Remark: on entry to stage j, the predecessor index $\sigma(j)$ has been set and $q_j(s) = \sum_{i=0}^{\alpha_j}(q_j)_i s^i$, $q_{\sigma(j)}(s)$ are available.] As in (4.11), form

$$\tilde{r}_j(\infty) = \sum_{i=0}^{\alpha_j} c_{i-\alpha_j+j+1}(q_j)_i \tag{4.39}$$

and also

$$\pi_j = j + \alpha_{\sigma(j)} - \sigma(j) \tag{4.40}$$

which is called the potential degree at stage j.

For $j = k, k+1, \ldots,$
If $\tilde{r}_j(\infty) = 0$, then

$$q_{j+1}(s) = q_j(s), \quad p_{j+1}(s) = p_j(s), \quad \text{and } \sigma(j+1) = \sigma(j);$$

else we have $\tilde{r}_j(\infty) \neq 0$.

If $\pi_j > \alpha_j$, then make the major change

$$q_{j+1}(s) = s^{\pi_j - \alpha_j} q_j(s) - q_{\sigma(j)} \tilde{r}_j(\infty)/r_{\sigma(j)}(\infty), \quad (4.41)$$

$$p_{j+1}(s) = s^{\pi_j - \alpha_j} p_j(s) - p_{\sigma(j)} \tilde{r}_j(\infty)/r_{\sigma(j)}(\infty), \quad (4.42)$$

$$\sigma(j+1) = j;$$

else we have $\pi_j \leq \alpha_j$. Then make the minor change

$$q_{j+1}(s) = q_j(s) - s^{\alpha_j - \pi_j} q_{\sigma(j)} \tilde{r}_j(\infty)/\tilde{r}_{\sigma(j)}(\infty), \quad (4.43)$$

$$p_{j+1}(s) = p_j(s) - s^{\alpha_j - \pi_j} p_{\sigma(j)} \tilde{r}_j(\infty)/\tilde{r}_{\sigma(j)}(\infty), \quad (4.44)$$

$$\sigma(j+1) = \sigma(j).$$

Next k.

This s-form of the Berlekamp–Massey algorithm is justified on the grounds that it performs in exactly the same way as the z-form previously given. The main difference is that part of the initialization of the z-form could be subsumed as minor changes in the s-form of the algorithm, and the logic is controlled by degree tests.

As an example of how the theory of this section has been applied, we consider the synthesis of a linear system which delivers six consecutive outputs $c_1, c_2, \ldots, c_6$ subsequent to a unit input at the time origin. The system is composed of shift elements (delays) denoted by blocks and multiplier elements denoted by triangles, as shown in Figure 4.4.3. At a

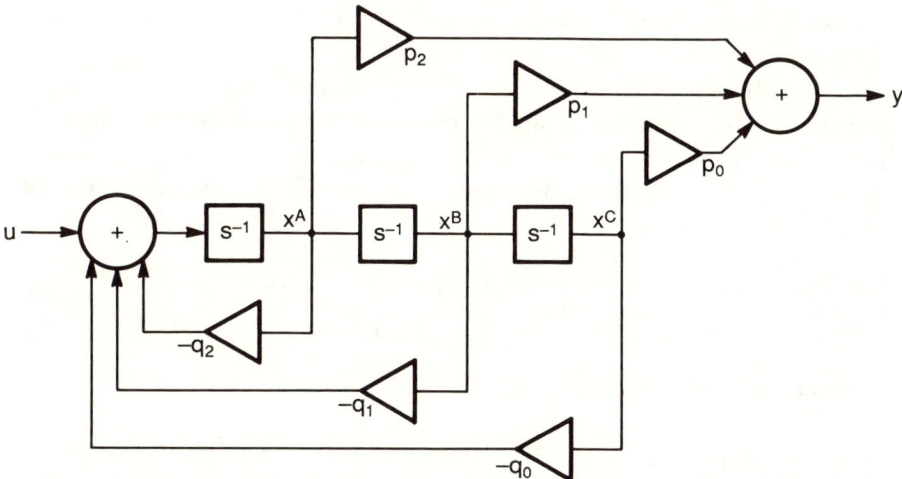

Figure 4.4.3. Controller canonical form of a single-input, single-output digital system.

generic node J, the power series $x^J(s) = \sum_{i=0}^{\infty} x_i^J s^{-1}$ represents signal values $\{x_i^J\}_{i=0}^{\infty}$ at times 0, 1, 2, ... Thus the input is $u(s) = 1$ and the desired output is given in terms of Markov parameters by

$$y(s) = c_1 s^{-1} + c_2 s^{-2} + \cdots + c_6 s^{-6} + \cdots \quad (4.45)$$

and in this context it is called the transfer function of the system. By inspection of the system, we find that

$$x^A(s) = [u(s) - q_2 x^A(s) - q_1 x^B(s) - q_0 x^C(s)]s^{-1},$$
$$x^B(s) = s^{-1} x^A(s), \quad x^C(s) = s^{-1} x^B(s),$$
$$y(s) = p_0 x^C(s) + p_1 x^B(s) + p_2 x^A(s)$$

and hence

$$y(s) = (p_0 + p_1 s + p_2 s^2) x^C(s),$$
$$s x^A(s) = 1 - (q_2 s^2 + q_1 s + q_0) x^C(s),$$

and

$$y(s) = \frac{p_0 + p_1 s + p_2 s^2}{s^3 + q_2 s^2 + q_1 s + q_0}. \quad (4.46)$$

Except in some cases of degeneracy, the values of the parameters p_0, p_1, p_2, q_0, q_1, q_2 in this expression are given directly by applying the previous version of the Berlekamp–Massey algorithm to the series (4.45). It is instructive to check quickly that the circuit of Figure 4.4.3 has the outputs desired in the $s \to \infty$ limit. As a practical matter, it is probably necessary to check the stability of any system proposed before implementing it.

A further development of the theory of this section is based on V-matrices. A V-matrix is an operator which advances the matrix of polynomials (and their predecessors) from stage j to stage $j + 1$, as shown in (4.49). For the moment, note that (4.41) and (4.42) may be expressed as

$$\begin{bmatrix} p_{\sigma(j)}(s) & p_j(s) \\ q_{\sigma(j)}(s) & q_j(s) \end{bmatrix} \begin{bmatrix} 0 & -\tilde{r}_j(\infty)/r_j(\infty) \\ 1 & s^{\pi_j - \alpha_j} \end{bmatrix} = \begin{bmatrix} p_{\sigma(j+1)}(s) & p_{j+1}(s) \\ q_{\sigma(j+1)}(s) & q_{j+1}(s) \end{bmatrix} \quad (4.47)$$

and that (4.43), (4.44) may be expressed as

$$\begin{bmatrix} p_{\sigma(j)}(s) & p_j(s) \\ q_{\sigma(j)}(s) & q_j(s) \end{bmatrix} \begin{bmatrix} 1 & -s^{\alpha_j - \pi_j} \tilde{r}_j(\infty)/r_j(\infty) \\ 0 & 1 \end{bmatrix} = \begin{bmatrix} p_{\sigma(j+1)}(s) & p_{j+1}(s) \\ q_{\sigma(j+1)}(s) & q_{j+1}(s) \end{bmatrix}$$

$$(4.48)$$

These results may be combined into the form

$$\begin{bmatrix} p_{\sigma(j)}(s) & p_j(s) \\ q_{\sigma(j)}(s) & q_j(s) \end{bmatrix} V_j = \begin{bmatrix} p_{\sigma(j+1)}(s) & p_{j+1}(s) \\ q_{\sigma(j+1)}(s) & q_{j+1}(s) \end{bmatrix}, \quad (4.49)$$

where V_j is a unimodular matrix, in the sense that its determinant is a nonzero constant.

The form of this algorithm in nondegenerate cases was first given by Berlekamp [1968], and extended to degenerate cases by Massey [1969]. Its close connection with the generalized Viskovatov algorithm and Magnus's P-fractions in the previous section was pointed out by Gragg and Lindquist [1983], who also explore the connections with the factorization of Hankel matrices. The details of the connection with the various staircase sequences in the Padé table were fully updated by Bultheel [1980a, b] and Bultheel and Van Barel [1986]. There are numerous different forms of circuits which synthesize a given transfer function, and these are amply described by Kailath [1980]. The use of V-matrices as a basis for the theoretical development was made by Antoulas [1986] and subsequently amplified by Bultheel and Van Barel [1990b].

4.5 Different types of continued fractions

The types of continued fractions which are fundamental to the representation of power series are the *regular C-fractions*. These have the form

$$C(z) = b_0 + \frac{a_1 z}{1} + \frac{a_2 z}{1} + \frac{a_3 z}{1} + \cdots, \quad (5.1)$$

with $a_i \neq 0$ for $i = 1, 2, 3, \ldots$. They may be constructed from a given power series by Viskovatov's method, the Q.D. algorithm, or any other convenient method. They are called C-fractions because they correspond to the given power series, and the regularity condition is that $a_i \neq 0$ for all i. An iterative reexpansion of the convergents of (5.1) shows that the successive convergents correspond to the [0/0], [1/0], [1/1], [2/1], [2/2], ... sequence of Padé approximants to the given power series, as shown in Section 4.2.

If, during the construction of (5.1) from the power series as in (2.1)–(2.4), an a_i is found to be zero, a different representation, such as the general C-fraction (2.19), must be used.

An alternative form of the regular C-fraction has the representation

$$C(z) = \frac{a_0}{1} + \frac{a_1 z}{1} + \frac{a_2 z}{1} + \frac{a_3 z}{1} + \cdots \quad (5.2)$$

with different elements a_i from those in (1); we still require that $a_i \neq 0$ for all i, for regularity. Equation (5.2) corresponds to the [0/0], [0/1], [1/1], [1/2], [2/2], ... sequence of Padé approximants.

The simple algebraic identity

$$1 + \frac{pz}{1 + \frac{qz}{D}} = 1 + pz - \frac{pqz^2}{qz + D} \tag{5.3}$$

leads to a *contraction* of the continued fraction. By taking $p_i = a_{2i}$, $q_i = a_{2i+1}$, we may contract (5.1) and generate its *associated fraction*:

$$A(z) = b_0 + \frac{a_1 z}{1 + a_2 z} - \frac{a_2 a_3 z^2}{1 + (a_3 + a_4)z} - \frac{a_4 a_5 z^2}{1 + (a_5 + a_6)z} - \cdots . \tag{5.4}$$

The convergents of $A(z)$ are alternate convergents of $C(z)$, and occupy the diagonal of the Padé table in this case. A particular case of the regular C-fraction is the Stieltjes or *S-fraction*, which is

$$s(z) = \frac{a_1}{1\ +} \ \frac{a_2 z}{1\ +} \ \frac{a_3 z}{1\ +} \cdots \tag{5.5}$$

with $a_i > 0$, $i = 1, 2, 3, \ldots$. The properties of the convergents and convergence of S-fractions are discussed extensively in the next chapter in the context of Padé approximation of Stieltjes functions. We will see that if the S-fraction converges (e.g., if the divergence condition of Section 4.7 is satisfied), then

$$s(z) = \int_0^\infty \frac{d\phi(u)}{1 + zu}, \quad |\arg(z)| < \pi, \tag{5.6}$$

where $\phi(t)$ is a bounded and nondecreasing function defined on $0 \leq u < \infty$. Using the variable $\omega = z^{-1}$, (5.6) is frequently expressed in the form

$$S(\omega) = zs(z) = \frac{a_1}{\omega\ +} \ \frac{a_2}{1\ +} \ \frac{a_3}{\omega\ +} \ \frac{a_4}{1\ +} \cdots \tag{5.7}$$

which is generated by a simple equivalence transformation, and the theory of Sections 5.5 and 5.6 shows that

$$S(\omega) = \int_0^\infty \frac{d\phi(u)}{\omega + u}, \quad |\arg(\omega)| < \pi. \tag{5.8}$$

Continued fractions of the type

$$J(\omega) = \frac{k_1}{l_1 + \omega} - \frac{k_2}{l_2 + \omega} - \frac{k_3}{l_3 + \omega} - \cdots \tag{5.9}$$

4.5 Different types of continued fractions

in which $k_i \neq 0$, $i = 1, 2, 3, \ldots$, are called *J-fractions*. If $k_i > 0$ and l_i are real for $i = 1, 2, 3, \ldots$, then (5.9) is called a *real J-fraction* [Wall, 1931, 1932a, b, 1948]. Such a fraction may be derived from (5.7) by the contraction formula (5.3), with the identifications

$$k_1 = a_1, \quad l_1 = a_2,$$

$$\left.\begin{array}{l} k_i = a_{2i-2}a_{2i-1}, \\ l_i = a_{2i-1}a_{2i} \end{array}\right\} \quad \text{for } i = 2, 3, 4, \ldots.$$

Thus we see that the convergents of (5.9) correspond to alternate convergents of (5.7). We will see in Section 5.6 that any convergent real J-fraction has a representation

$$J(\omega) = \int_{-\infty}^{\infty} \frac{d\psi(u)}{\omega + u}, \quad \text{Im } \omega \neq 0, \tag{5.10}$$

where $\psi(u)$ is a bounded and nondecreasing function defined on $-\infty < u < \infty$. Equation (5.8) is a special case of (5.10) where $\psi(u)$ is constant on $-\infty < u \leq 0$.

Euler used a method of writing an equivalent continued fraction for Maclaurin series. This fraction has convergents which reduce to truncated Maclaurin series.

A short calculation with the recursion relations (4.4) reveals that

$$\sum_{i=0}^{n} c_i z^i = c_0 + \frac{c_1 z}{1} - \frac{c_2 z}{c_1 + c_2 z} - \frac{c_1 c_3 z}{c_2 + c_3 z} - \cdots - \frac{c_n c_{n-2} z}{c_{n-1} + c_n z}. \tag{5.11}$$

Needless to say, the convergence of the continued fraction can be no different from that of the original series, and these expansions do not seem to be of much use.

Two-point Padé approximants are designed for the (possibly formal) approximation of a function which is given by Taylor expansions about two different points. It is usually convenient for these points to be taken as $z = 0$ and $z = \infty$. In Section 7.1, we consider the general problem of N-point Padé approximation, and here we restrict our attention to an overview of the continued-fraction representations of T-fraction (Thronfraction) approximants generated by expansions of the form

$$f(z) = c_0 + c_1 z + c_2 z^2 + \cdots \tag{5.12}$$

$$= -c_{-1} z^{-1} - c_{-2} z^{-2} - c_{-3} z^{-3} + \cdots, \tag{5.13}$$

where (5.12) is the usual Maclaurin expansion L_0 of $f(z)$ and (5.13) is a similar expansion L_∞ about $z = \infty$. The reason for adopting an unusual sign convention in (5.13) is to preserve the standard format of the determinantal representation of the denominator of the T-fraction; see Section 7.1. The procedure to be described for construction of the approximants can be regarded as a set of formal operations on power series. Alternatively, it may be assumed that a function $f(z)$ exists and that it is faithfully represented by expansions of the form (5.12) and (5.13), in which case the procedures generate approximations to $f(z)$ which should have some reasonable domain of applicability. We consider two examples to illustrate these points.

Example 1

$$\frac{1}{\sqrt{1+z^2}} = \frac{1}{1} + \frac{z^2}{2} + \frac{z^2}{2} + \frac{z^2}{2} + \cdots \qquad (5.14)$$

$$= \frac{1}{z} + \frac{1}{2z} + \frac{1}{2z} + \frac{1}{2z} + \cdots \qquad (5.15)$$

$$= \frac{1}{1+z} - \frac{z}{1+z} - \frac{1}{2(1+z)} - \frac{z}{1+z} - \frac{z}{2(1+z)} - \cdots, \qquad (5.16)$$

where

(i) expansion (5.14) converges for all z except on cuts from $\pm i$ to $\pm\infty$,
(ii) expansion (5.15) converges for all z except on a cut from i to $-i$, and
(iii) expansion (5.16) converges for all z except on the semicircle $|z| = 1$, $\mathrm{Re}\, z < 0$.

Remarks. The expansion (5.14) is designed to be accurate for small $|z|$, expansion (5.15) is designed to be accurate for large $|z|$, and expansion (5.16) is designed to be accurate for all $|z|$. We note that the former two objectives are overachieved, whereas the latter is underachieved.

Notice that the expansion (5.14) can be constructed by iterative substitution for r in

$$r = z^2/(2 + r).$$

This equation has the solution

$$r = -1 + \sqrt{1 + z^2},$$

and (5.14) follows formally because

$$(1 + z^2)^{-\frac{1}{2}} = 1/(1 + r).$$

The expansion (5.16) arises from making an equivalence transformation on

$$(1+z^2)^{-\frac{1}{2}} = \frac{1}{1+z-} \frac{2z}{2(1+z)-} \frac{2z}{2(1+z)-} \cdots \qquad (5.17)$$

and the convergents of (5.17) are the same as those of (5.16). The expansion (5.17) can be produced by iterative substitution for r in

$$r = \frac{2z}{2(1+z)-r}.$$

The equation has the solution

$$r = 1 + z - \sqrt{1+z^2},$$

and (5.17) follows formally because

$$(1+z^2)^{-\frac{1}{2}} = 1/(1+z-r).$$

Unfortunately, none of these remarks helps directly to discover or prove the domain of convergence of the continued fraction.

Proof. For the expansion (5.14), the recursion relations for the partial numerators and denominators are

$$\left.\begin{array}{l} A_i = 2A_{i-1} + z^2 A_{i-2}, \\ B_i = 2B_{i-1} + z^2 B_{i-2} \end{array}\right\} \quad i = 2, 3, \ldots.$$

The solutions of these recursions are

$$A_i = \alpha_+(\rho_+)^i + \alpha_-(\rho_-)^i, \qquad (5.18)$$

$$B_i = \beta_+(\rho_+)^i + \beta_-(\rho_-)^i, \qquad (5.19)$$

where ρ_+, ρ_- are the roots of

$$\rho^2 - 2\rho - z^2 = 0 \qquad (5.20)$$

and $\alpha_+, \alpha_-, \beta_+, \beta_-$ follow by requiring that

$$A_0 = 0, \quad B_0 = 1, \quad A_1 = 1, \quad B_1 = 1.$$

From (5.20), $\rho_\pm = 1 \pm \sqrt{1+z^2}$. The ratio A_i/B_i has a well-determined limiting value if $|\rho_+| \neq |\rho_-|$. The exceptional set on which $|\rho_+| = |\rho_-|$ is given by $\operatorname{Re} z = 0$ and $|\operatorname{Im} z| > 1$.

The expression (5.15) is analyzed similarly. The expansion (5.16) is equivalent to (5.17), for which the recursions are

$$A_i = 2(1+z)A_{i-1} - 2zA_{i-2},$$

$$B_i = 2(1+z)B_{i-1} - 2zB_{i-2}.$$

The geometric ratios ρ_+, ρ_- are the roots of
$$\rho^2 - 2(1 + z)\rho + 2z = 0$$
as in (5.18)–(5.20). Then
$$\rho_\pm = 1 + z \pm \sqrt{1 + z^2}.$$
These ratios have equal moduli if
$$\left|1 + \frac{\sqrt{1 + z^2}}{1 + z}\right| = \left|1 - \frac{\sqrt{1 + z^2}}{1 + z}\right|$$
which implies that
$$\sqrt{1 + z^2} = it(1 + z) \tag{5.21}$$
for $t \in \mathbb{R}$. With $x = \mathrm{Re}\, z$, $y = \mathrm{Im}\, z$, (5.21) implies that
$$x = \frac{-t^2}{1 + t^2}, \quad y = \frac{\sqrt{2t^2 + 1}}{1 + t^2}$$
which is a parametrized form of the semicircle
$$x^2 + y^2 = 1, \quad x \leq 0.$$

Example 2. [Thron, 1948].
$$T(z) = \frac{1}{1 - z} + \frac{z}{1 - z} + \frac{z}{1 - z} + \cdots. \tag{5.22}$$
This continued fraction takes the values
$$T(z) = 1, \qquad \text{for } |z| < 1, \tag{5.23}$$
$$T(z) = -z^{-1}, \quad \text{for } |z| > 1; \tag{5.24}$$
all the poles and zeros of the convergence of $T(z)$ lie on the unit circle.

Proof. The partial numerators and denominators of $T(z)$ satisfy the recursions
$$A_i = (1 - z)A_{i-1} + zA_i$$
$$B_i = (1 - z)B_{i-1} + zB_i$$
with
$$A_0 = 0, \quad B_0 = 1, \quad A_1 = 1, \quad B_1 = 1 - z.$$
It is readily verified that the solution of these recursions are polynomials $A_i(z)$ and $B_i(z)$ given by
$$B_i(z) = A_{i+1}(z) = (1 - (-z)^{i+1})/(1 + z), \quad i = 0, 1, 2, \ldots. \tag{5.25}$$

The zeros of $A_i(z)$, $B_i(z)$ are (respectively) located uniformly on the unit circle.

It follows from (5.25) that $A_i(z)/B_i(z) \to 1$ for $|z| < 1$, whereas $A_i(z)/B_i(z) \to -z$ for $|z| > 1$, as $i \to \infty$.

Example 2 shows what may happen (at best) when the two series specifying the approximant, namely,

$$f(z) = 1, \quad z \to 0, \qquad (L_0)$$

and

$$f(z) = -z^{-1}, \quad z \to \infty, \qquad (L_\infty)$$

represent, in a natural sense, different functions. The convergents of the continued fraction converge to one function in one region of the z-plane containing the origin and to another function in another region containing the point at infinity and they diverge all along the common boundary (except at $z = -1$). In this sense, Examples 1 and 2 demonstrate the ideal behavior of T-fractions in making sense of given asymptotic expansions of the forms (5.12) and (5.13). The connected domain of convergence for the T-fraction (5.15) of Example 1, in which the expansions L_0 and L_∞ are naturally compatible, should be contrasted with that for Example 2, in which the expressions L_0 and L_∞ are naturally incompatible.

The continued fractions (5.16), (5.22) are examples of T-fractions represented by

$$T(z) = \frac{F_1}{1 + G_1 z} \genfrac{}{}{0pt}{}{}{+} \frac{F_2 z}{1 + G_2 z} \genfrac{}{}{0pt}{}{}{+} \frac{F_3 z}{1 + G_3 z} \genfrac{}{}{0pt}{}{}{+} \cdots. \qquad (5.26)$$

It is readily found that $T(z)$ possesses expansions

$$T(z) = F_1 - F_1(G_1 + F_2)z + \cdots, \quad z \to 0, \qquad (5.27)$$

$$T(z) = \frac{F_1}{G_1 z} - \frac{(G_2 + F_2)F_1}{G_1^2 G_2 z^2} + \cdots, \quad z \to \infty, \qquad (5.28)$$

which correspond to expansion L_0 in (5.12) and L_∞ in (5.13). The method of calculating F_i, G_i for given expansions L_0, L_∞ is described in Section 7.1, along with a more general analysis.

T-fraction expansions are much the same as the M-fractions of Murphy and McCabe [McCabe, 1974, 1975; McCabe and Murphy, 1976; Cooper, Magnus, and McCabe, 1986]. T-fractions are often expressed with the roles of 0, ∞ in (5.12)–(5.13) or (5.27)–(5.28) interchanged, so as to match

$$f(z) = -c_{-1}z - c_{-2}z^2 - c_{-3}z^3 - \cdots$$

$$= c_0 + c_1 z^{-1} + c_2 z^{-2} + \cdots.$$

A constant and other terms can easily be included by adding them in explicitly, but this is somewhat unnatural in the context of nonlinear methods of approximation. The fact that Stieltjes series such as

$$f(z) = \int_a^b \frac{w(t)}{1+zt}\, dt, \quad 0 < a < b < \infty, \tag{5.29}$$

possess expansions of the form (5.12)–(5.13) justifies and motivates the representation (5.26) as a standard form for T-fractions [Gutknecht, 1993]. Conditions on the c_j which allow a representation of the (5.29) when $a = 0$, $b = \infty$ constitute the strong Stieltjes moment problem, which has been fully solved [Jones, Thron, and Waadeland, 1980].

T-fractions can be introduced as contractions of PC-fractions. Perron–Carathéodory (PC) fractions are defined as an expansion of the form

$$P(w) = \beta_0 + \frac{\alpha_1}{\beta_1} + \frac{1}{\beta_2 w} + \frac{\alpha_3 w}{\beta_3} + \frac{1}{\beta_4 w} + \frac{\alpha_5 w}{\beta_5} + \frac{1}{\beta_6 w} + \cdots, \tag{5.30}$$

where α_{2i+1}, β_i are complex constants satisfying

$$\alpha_{2i+1} = 1 - \beta_{2i}\beta_{2i+1} \neq 0, \quad i = 1, 2, 3, \ldots. \tag{5.31}$$

Contraction of (5.30) can be made by the procedure based on (5.3). Alternatively, following Jones, Njåstad, and Thron [1986], note that the partial denominators of (5.30) satisfy

$$Q_{2n}(w) = \beta_{2n} w Q_{2n-1}(w) + Q_{2n-2}(w), \tag{5.32}$$

$$Q_{2n+1}(w) = \beta_{2n+1} Q_{2n}(w) + \alpha_{2n+1} w Q_{2n-1}(w). \tag{5.33}$$

Substitute for $Q_{2n-1}(w)$ and $Q_{2n+1}(w)$ from (5.32) into (5.33) and use the equality (5.31) to obtain

$$Q_{2n+2}(w) = \left(1 + \frac{\beta_{2n+2} w}{\beta_{2n}}\right) Q_{2n}(w) - \frac{\beta_{2n+2} \alpha_{2n+1}}{\beta_{2n}} Q_{2n-2}(w) \tag{5.34}$$

as the recursion for the partial denominators of even order. The partial numerators obey the same recursion. They are initialized with

$$Q_0(w) = 1,\ Q_2(w) = \beta_2 w + 1;\ P_0(w) = \beta_0,\ P_2(w) = (\alpha_1 + \beta_0)\beta_2 w + \beta_0.$$

From (5.30) and (5.34) it follows that $P_{2n}(w)/Q_{2n}(w)$ are the convergents of

$$t(w) = \beta_0 + \frac{\alpha_1 \beta_2 w}{1 + \beta_2 w} - \frac{(\alpha_3 \beta_4/\beta_2) w}{1 + (\beta_4/\beta_2)w} - \frac{(\alpha_5 \beta_6/\beta_4) w}{1 + (\beta_6/\beta_4)w} - \cdots.$$

With the substitution $w = z^{-1}$, $T(z) = t(w) - \beta_0$, we obtain

$$T(z) = \frac{\alpha_1 \beta_1}{z + \beta_2} - \frac{(\alpha_3 \beta_4/\beta_2)z}{z + \beta_4/\beta_2} - \frac{(\alpha_5 \beta_6/\beta_4)z}{z + \beta_6/\beta_4} - \cdots$$

in the standard form of (5.26).

T-fractions are introduced here primarily as a continued-fraction form of an important class of two-point Padé approximants. However, as the contractions of positive PC-fractions, in which

$$\alpha_1 = -2\beta_0 \leq 0, \quad \beta_{2n} = \beta_{2n+1}^*, \quad |\beta_{2n}| < 1,$$

they play an important role in the trigonometric moment problem; we refer to the review by Jones and Thron [1988] and the books by Lorentzen and Waadeland [1992] and Jones and Thron [1980] for references and further details of this and all topics in this section; other key references are given at the end of Section 7.1.

4.6 Examples of continued fractions which are Padé approximants

We present here some examples of continued fractions which are also Padé approximants. The examples are either staircase or diagonal sequences in the Padé table, obtained from J-fractions or S-fractions. We do not give continued fractions which are merely Euler's corresponding fractions whose convergents are identical to truncated Maclaurin series. We quote a quite comprehensive set of formulas for the functions which are known to have useful continued-fraction expansions. We conclude the section with their formal algebraic derivation, which consists of showing that the Maclaurin series of each continued fraction is the same as that of the given function. The question of convergence is left to Section 4.7.

Exponential function

$$\exp(z) = \frac{1}{1} - \frac{z}{1+} \frac{z}{2-} \frac{z}{3+} \frac{z}{2-} \cdots \frac{z}{+2} - \frac{z}{2n+1+} \cdots \quad (6.1a)$$

$$= 1 + \frac{z}{1-} \frac{z}{2+} \frac{z}{3-} \frac{z}{2+} \frac{z}{5-} \frac{z}{2+} \cdots \frac{z}{-2} + \frac{z}{2n+1-} \cdots \quad (6.1b)$$

$$= 1 + \frac{z}{1 - z/2 +} \frac{z^2/(4 \times 3)}{1 +} \frac{z^2/(4 \times 15)}{1 +} + \frac{z^2/(4 \times 35)}{1 +} \cdots \frac{z^2/(4(4n^2 \times 1))}{1 +} \cdots. \quad (6.1c)$$

These expansions converge for all z.

Tangent function

$$\tan z = \frac{z}{1} - \frac{z^2}{3} - \frac{z^2}{5} - \frac{z^2}{7} - \cdots - \frac{z^2}{2n+1} - \cdots. \quad (6.2)$$

This converges for all z except $z = (2n+1)\pi/2$, n integral

Hyperbolic tangent

$$\tanh z = \frac{z}{1} + \frac{z^2}{3} + \frac{z^2}{5} + \frac{z^2}{7} + \cdots + \frac{z^2}{2n+1} + \cdots. \quad (6.3)$$

This converges for all z except $z = (2n+1)i\pi/2$, n integral.

Binomial function

$$(1+z)^\nu = 1 + \frac{\nu z}{1} + \frac{(1-\nu)z}{2} + \frac{(1+\nu)z}{3} + \frac{(2-\nu)z}{2} + \cdots$$

$$+ \frac{(n-\nu)z}{2} + \frac{(n+\nu)z}{2n+1} + \cdots. \quad (6.4)$$

This converges for all z except $-\infty < z \leq 1$, unless ν is integral, when the continued fraction terminates and the result is exact.

Inverse tangent

$$\tan^{-1} z = \frac{z}{1} + \frac{1 \times z^2}{3} + \frac{4 \times z^2}{5} + \cdots + \frac{n^2 z^2}{2n+1} + \cdots. \quad (6.5)$$

This converges for all z in the z-plane cut from i to $i\infty$ and from $-i$ to $-i\infty$.

Inverse hyperbolic tangent

$$\tanh^{-1}(z) = \tfrac{1}{2}\ln\frac{1+z}{1-z} = \frac{z}{1} - \frac{1 \times z^2}{3} - \frac{4 \times z^2}{5} - \cdots - \frac{n^2 z^2}{2n+1} - \cdots. \quad (6.6)$$

This fraction converges in the whole z-plane cut by $(-\infty, -1]$ and $[1, +\infty)$.

Natural logarithm

$$\ln(1+z) = \frac{z}{1} + \frac{1^2 z}{2} + \frac{1^2 z}{3} + \frac{2^2 z}{4} + \frac{2^2 z}{5} + \cdots + \frac{n^2 z}{2n} + \frac{n^2 z}{2n+1} + \cdots. \quad (6.7)$$

This fraction converges for all z except $-\infty < z \leq -1$.

Exponential integral

$$E_n(z) = \int_1^\infty \frac{e^{-zt}}{t^n} dt = e^{-z}\left(\frac{1}{z+n} - \frac{n}{z+n+2} - \frac{2(n+1)}{z+n+4} - \cdots\right.$$
$$\left. - \frac{((r+1)(n+r))}{z+n+2r+2} - \cdots\right). \quad (6.8)$$

This is valid in the entire z-plane cut along $-\infty < z \leq 0$; the integral representation is only valid for Re $z > 0$; see also (6.23).

Complementary error function

$$\text{erfc}(z) = 1 - \text{erf}(z) = \frac{2}{\sqrt{\pi}} \int_z^\infty e^{-t^2} dt$$
$$= \frac{e^{-z^2}}{\sqrt{\pi}}\left(\frac{1}{z} + \frac{1/2}{z} + \frac{1}{z} + \frac{3/2}{z} + \frac{2}{z} + \cdots + \frac{n/2}{z} + \cdots\right). \quad (6.9)$$

This converges for Re $z > 0$.

Prym's incomplete Gamma function

$$\Gamma(a, z) = \int_z^\infty t^{a-1} e^{-t} dt = e^{-z} z^a \left(\frac{1}{z+1-a} - \frac{1-a}{z+3-a} - \frac{2(2-a)}{z+5-a}\right.$$
$$\left. - \cdots - \frac{n(n-a)}{z+2n+1-a} - \cdots\right). \quad (6.10)$$

This is valid for all z except in $-\infty < z \leq 0$. If a is a positive integer, the fraction terminates and so the representation is valid for all z. The connection with (6.8) is that $E_n = z^{n-1}\Gamma(1-n, z)$; see also (6.24).

Error function

$$\text{erf}(z) = \frac{2}{\sqrt{\pi}} \int_0^z e^{-t^2} dt =$$
$$\frac{2ze^{-z^2}}{\sqrt{\pi}}\left(\frac{1}{1} - \frac{2z^2}{3} + \frac{4z^2}{5} - \frac{6z^2}{7} + \cdots + \frac{4nz^2}{4n+1} - \frac{(4n+2)z^2}{4n+3} + \cdots\right).$$
$$(6.11)$$

This converges for all z. However, convergence is not fast for Re $z \gg 2$, and erfc(z) and its continued fraction (6.10) are more useful in such applications. The relation with Dawson's integral, $e^{-z^2}\int_0^z e^{t^2} dt$, is given in (7.1.80), and its T-fraction comes from (7.1.79) and (7.1.83).

Incomplete Gamma function

$$\gamma(a, z) = \int_0^z t^{a-1} e^{-t}\, dt$$

$$= z^a e^{-z} \left(\frac{1}{a} - \frac{az}{a+1} + \frac{z}{a+2} - \frac{(a+1)z}{a+3} + \frac{2z}{a+4} - \cdots \right.$$

$$\left. + \frac{nz}{a+2n} - \frac{(a+n)z}{a+2n+1} + \cdots \right), \qquad (6.12)$$

where a is not a negative integer or zero. If a is a strictly positive integer, this converges for all z. If a is not integral, the continued fraction in (6.12) converges, but $\gamma(a, z)$ is only defined in the z-plane cut by $-\infty < z \leq 0$. A T-fraction representation is given by (7.1.81) and (7.1.79), and an error formula is given by Luke [1975].

Definition of hypergeometric functions. We use the hypergeometric function $_pF_q(a_1, a_2, \ldots, a_p, b_1, b_2, \ldots, b_q; z)$ with p numerator parameters and q denominator parameters. The examples

$$_0F_1(a; z) = 1 + \frac{z}{a} + \frac{z^2}{a(a+1)2!} + \cdots$$

$$+ \frac{z^n}{a(a+1)\cdots(a+n-1)n!} + \cdots,$$

$$_2F_1(a, b, c; z) = 1 + \frac{abz}{c} + \frac{a(a+1)b(b+1)z^2}{c(c+1)2!} + \cdots$$

$$+ z^n \frac{a(a+1)\cdots(a+n-1)b(b+1)\cdots(b+n-1)}{c(c+1)\cdots(c+n-1)n!}$$

$$+ \cdots,$$

$$_2F_0(a, b; z) = 1 + abz + \frac{a(a+1)b(b+1)z^2}{2!} + \cdots$$

$$+ z^n \frac{a(a+1)\cdots(a+n-1)b(b+1)\cdots(b+n-1)}{n!}$$

$$+ \cdots$$

make the definition clear. The definitions are valid formally provided the denominator parameters are not negative integers or zero. Notice that $_0F_1(a; z)$ is an entire function, $_2F_1(a, b, c; z)$ is analytic in the z-plane cut by $1 \leq z < \infty$, and $_2F_0(a, b; z)$ has a purely formal definition, since the radius of convergence of the series is zero. The given expansion is a formal expansion of

$$_2F_0(a, b; z) = \frac{1}{\Gamma(a)} \int_0^\infty \frac{e^{-t}t^{a-1}}{(1-zt)^b} dt,$$

which is the proper definition, valid for Re $a > 0$ and z not on the positive real axis.

$_0F_1$-hypergeometric-function relation

$$\frac{_0F_1(a+1; z)}{_0F_1(a; z)} = \frac{a}{a} + \frac{z}{a+1} + \frac{z}{a+2} + \frac{z}{a+3} + \cdots + \frac{z}{a+n} + \cdots .$$

(6.13a)

This converges for all z not a zero of $_0F_1(a; z)$. A relation for Bessel functions follows from the formula relating $_0F_1$ hypergeometric functions to Bessel functions,

$$J_\nu(z) = \frac{(\tfrac{1}{2}z)^\nu {}_0F_1(1+\nu, -\tfrac{1}{4}z^2)}{\Gamma(\nu+1)}.$$

We deduce that

$$\frac{J_\nu(z)}{J_{\nu-1}(z)} = \frac{z}{2\nu} - \frac{z^2}{2(\nu+1)} - \frac{z^2}{2(\nu+2)} - \cdots - \frac{z^2}{2(\nu+n)} - \cdots .$$

(6.13b)

Confluent hypergeometric-function relation

$$\frac{_1F_1(a+1, b+1; z)}{_1F_1(a, b; z)} = \frac{b}{b} - \frac{(b-a)z}{b+1} + \frac{(a+1)z}{b+2} - \frac{(b-a+1)z}{b+3}$$

$$+ \cdots \frac{(a+n)z}{+ b+2n} - \frac{(b-a+n)z}{b+2n+1} + \cdots . \quad (6.14)$$

This converges for all z except for the zeros of $_1F_1(a, b; z)$.

$_2F_0$-hypergeometric-function relation

$$\frac{_2F_0(a, b+1; z)}{_2F_0(a, b; z)} = \frac{1}{1} - \frac{az}{1} - \frac{(b+1)z}{1} - \frac{(a+1)z}{1} - \frac{(b+2)z}{1} - \cdots$$

$$- \frac{(b+n)z}{1} - \frac{(a+n)z}{1} - \cdots . \quad (6.15)$$

This converges in the cut z-plane except in the cut $0 \leq z < \infty$.

Hypergeometric-function relation [Gauss, 1813].

$$\frac{{}_2F_1(a, b+1, c+1; z)}{{}_2F_1(a, b, c; z)} = \frac{c}{c} - \frac{a(c-b)z}{c+1} - \frac{(b+1)(c-a+1)z}{c+2} - \cdots$$
$$- \frac{(a+n)(c-b+n)z}{c+2n+1} - \frac{(b+n+1)(c-a+n+1)z}{c+2n+2} - \cdots . \quad (6.16)$$

This converges in the cut z-plane except on the cut $1 \leq z < \infty$, and except for the zeros of ${}_2F_1(a, b, c; z)$.

Other continued-fraction developments which are Padé approximants for special functions are known [Wall, 1948, p. 369; Rogers, 1907]; they mostly involve integrals of hyperbolic and elliptic functions.

The derivation of each of the preceding formulas (6.1)–(6.16) consists of both an algebraic and an analytical part. First we show that the Maclaurin series of the two sides of the equations agree term by term. Since the results (6.1)–(6.12) are corollaries of (6.13), (6.14), and (6.15), we discuss these cases first.

Proof of (6.1). Take $a = 0$ in (6.14). ${}_1F_1(0, b; z) = 1$ by definition. This step is used in most of the corollaries. Accordingly, we find that

$${}_1F_1(1, b+1; z) = \frac{1}{1} - \frac{z}{b+1} + \frac{1 \times z}{b+2} - \frac{(b+1)z}{b+3} + \cdots + \frac{nz}{b+2n}$$
$$- \frac{(b+n)z}{b+2n+1} + \cdots .$$

Taking $b = 0$, the left-hand side is $\exp(z)$, and (6.1a) follows. To prove (6.1b), write $\exp(z) = \{\exp(-z)\}^{-1}$ and use the representation (6.1a) for $\exp(-z)$. To prove (6.1c), use

$$\frac{e^y - e^{-y}}{e^y + e^{-y}} = \tanh y.$$

Hence, (6.1c) follows from (6.3) by taking $2y = z$ and using the formula

$$e^z = 1 - \frac{2}{1 - [\tanh(z/2)]^{-1}}.$$

Proof of (6.2).

$$\sin z = z\,{}_0F_1(\tfrac{3}{2}; -z^2/4),$$
$$\cos z = {}_0F_1(\tfrac{1}{2}; -z^2/4),$$

and the result follows from (6.13).

4.6 Examples of continued fractions which are Padé approximants

Proof of (6.3).
$$\sinh z = z\,_0F_1(\tfrac{3}{2}; z^2/4),$$
$$\cosh z = \,_0F_1(\tfrac{1}{2}; z^2/4),$$
and the result follows from (6.13).

Proof of (6.4)–(6.7).
$$(1 + z)^{-\nu} = \,_2F_1(\nu, 1, 1; -z),$$
$$\tan^{-1} z = z\,_2F_1(\tfrac{1}{2}, 1, \tfrac{3}{2}; -z^2),$$
$$\tanh^{-1} z = z\,_2F_1(\tfrac{1}{2}, 1, \tfrac{3}{2}; z^2),$$
$$\ln(1 + z) = z\,_2F_1(1, 1, 2; -z),$$
and (6.4)–(6.7) follow from (6.16) with $b = 0$.

Proof of (6.8). Use the representation
$$_2F_0(a, b; z') = \frac{1}{\Gamma(a)} \int_0^\infty \frac{e^{-t} t^{a-1}\, dt}{(1 - z't)^b}$$
taking $a = 1$, $b = n$, $z' = -1/z$. For $\operatorname{Re} z > 0$,
$E_n(z) = e^{-z} z^{-1}\,_2F_0(1, n; -1/z)$, and (6.8) follows from (6.15).

Proof of (6.9).
$$\operatorname{erfc}(z) = \frac{1}{\sqrt{\pi}} \Gamma(\tfrac{1}{2}, z^2) = \frac{1}{\sqrt{\pi}} \int_{z^2}^\infty t^{-1/2} e^{-t}\, dt,$$
and so (6.9) follows from (6.10).

Proof of (6.10). This is the same as for (6.8).

Proof of (6.11).
$$\operatorname{erf}(z) = \frac{2}{\sqrt{\pi}} \int_0^z e^{-t^2}\, dt = \frac{2ze^{-z^2}}{\sqrt{\pi}} \int_0^1 e^{z^2(1-u^2)}\, du$$
$$= \frac{2ze^{-z^2}}{\sqrt{\pi}}\,_1F_1(1, \tfrac{3}{2}; z^2),$$
and so (6.11) follows from (6.14).

Proof of (6.12).
$$\gamma(a, z) = \int_0^z t^{a-1} e^{-t}\, dt$$
$$= z^a a^{-1}\,_1F_1(a, 1 + a; -z) \quad \text{(by expansion)}$$

$$= e^{-z} \int_0^z (z-u)^{a-1} e^u \, du \quad \text{(by setting } t = z - u\text{)}$$
$$= z^a e^{-z} a^{-1} {}_1F_1(1, 1+a; z) \quad \text{(by expansion).}$$

The continued-fraction expansion then follows from (6.14).

Proof of (6.13). Series expansion of the hypergeometric function shows that

$$_0F_1(a+1; z) = {}_0F_1(a; z) - \frac{z}{a(a+1)} {}_0F_1(a+2; z).$$

Therefore

$$\frac{_0F_1(a+1; z)}{_0F_1(a; z)} = \frac{1}{1 + \dfrac{z}{a(a+1)} \dfrac{_0F_1(a+2; z)}{_0F_1(a+1; z)}}.$$

This formula is simple to iterate and is used to generate the continued-fraction expansion (6.13a).

Proof of (6.14). Series expansion of the confluent hypergeometric function shows that

$$_1F_1(a+1, b+1; z) = {}_1F_1(a, b; z) + \frac{z(b-a)}{b(b+1)} {}_1F_1(a+2, b+2; z).$$

Therefore

$$\frac{_1F_1(a+1, b+1; z)}{_1F_1(a, b; z)} = \frac{1}{1 - \dfrac{z(b-a)}{b(b+1)} \dfrac{_1F_1(a+2, b+2; z)}{_1F_1(a+1, b+1; z)}}.$$

Again, this formula is simple to iterate and is used to generate the continued-fraction expansion (6.14).

Proof of (6.15). Formal operations with the power series similar to the previous operations lead to the formula

$$_2F_0(a, b+1; z) = {}_2F_0(a, b; z) + az \, {}_2F_0(a+1, b+1; z). \quad (6.17)$$

However, we must use the representation

$$_2F_0(a, b; z) = \frac{1}{\Gamma(a)} \int_0^\infty \frac{e^{-t} t^{a-1}}{(1-zt)^b} \, dt, \quad (6.18)$$

which is valid for Re $a > 0$ and z not on the positive real axis, to establish the result (6.17). The identity

4.6 Examples of continued fractions which are Padé approximants

$$\frac{t^{a-1}}{(1+zt)^b} = \frac{t^{a-1}}{(1+zt)^{b+1}} + \frac{zt^a}{(1+zt)^{b+1}}$$

leads to equality of the integrands and provides the proof of (6.15) for Re $a > 0$. Extension to complex values of a is by analytic continuation. Hence

$$\frac{_2F_0(a, b+1; z)}{_2F_0(a, b; z)} = \frac{1}{1 - az\dfrac{_2F_0(a+1, b+1; z)}{_2F_0(a, b+1; z)}}. \qquad (6.19)$$

Since $_2F_0(a, b; z) = {_2F_0}(b, a; z)$, it follows from (6.19) that

$$\frac{_2F_0(a+1, b+1; z)}{_2F_0(a, b+1; z)} = \frac{1}{1 - (b+1)z\dfrac{_2F_0(a+1, b+2; z)}{_2F_0(a+1, b+1; z)}}. \qquad (6.20)$$

Equations (6.19) and (6.20) together provide a formula which may be iterated to yield (6.15).

Proof of (6.16). The expansion of the hypergeometric function leads to the identity

$$_2F_1(a, b, c; z) = {_2F_1}(a, b+1, c+1; z)$$
$$- \frac{a(c-b)}{c(c+1)} z \, _2F_1(a+1, b+1, c+2; z).$$

This may be rewritten as

$$\frac{_2F_1(a, b+1, c+1; z)}{_2F_1(a, b, c; z)} = \frac{1}{1 - \dfrac{a(c-b)}{c(c+1)} z \dfrac{_2F_1(a+1, b+1, c+2; z)}{_2F_1(a, b+1, c+1; z)}}.$$

$$(6.21)$$

Since $_2F_1(a, b, c; z) = {_2F_1}(b, a, c; z)$, we may rewrite (6.21), replacing c by $c+1$, etc., as

$$\frac{_2F_1(a+1, b+1, c+2; z)}{_2F_1(a, b+1, c+1; z)} =$$

$$\frac{1}{1 - z\dfrac{(b+1)(c-a+1)\,_2F_1(a+1, b+2, c+3; z)}{(c+1)(c+2)\,_2F_1(a+1, b+1, c+2; z)}}. \qquad (6.22)$$

Together (6.21) and (6.22) yield a formula which connects ratios of hypergeometric functions in which the numerator parameters a and b are increased by 1 and the denominator parameter c is increased by 2. The result (6.16) follows by iteration.

To summarize this section, we observe that a variety of familiar functions have continued-fraction expansions given by (6.1)–(6.16). Using the algebraic results (6.17), (6.19)–(6.22), we have proved that the Maclaurin expansion of each function is the same as that of the corresponding continued fraction. In this sense, the results (6.1)–(6.16) are formal equalities. In the next section, we find the domain of values of z for which (6.1)–(6.16) are true equalities.

Example 1. The contracted form of

$$e^z E_n(z) = \frac{1}{z} + \frac{n}{1} + \frac{1}{z} + \frac{n+1}{1} + \frac{2}{z} + \cdots + \frac{n+r}{1} + \frac{r+1}{z} + \cdots \quad (6.23)$$

is given by (6.8).

Example 2. The contracted form of

$$e^z z^{-a} \Gamma(a, z) = \frac{1}{z} + \frac{1-a}{1} + \frac{1}{z} + \frac{2-a}{1} + \cdots + \frac{n}{z} + \frac{n+1-a}{1} + \cdots \quad (6.24)$$

is given by (6.10).

4.7 Convergence of continued fractions

The derivation given in Section 4.6 of the continued-fraction expansions of the familiar functions of mathematics is purely algebraic. We can justify the usual meaning of the equality signs in Section 4.6 if we can show that

(i) each continued fraction converges to a limit function, and
(ii) this limit function is the same function as the one from which the fraction originated.

We simply state the relevant theorems here, and refer to the companion volume of Jones and Thron and to Lorentzen and Waadeland [1992] for the proofs.

Since each convergent of a continued fraction of the types discussed in Section 4.6 is a rational function of the variable z, one expects to be able to specify conditions which are sufficient to ensure that the limit function is meromorphic in z. Normally, most authors prove theorems by showing that the convergents of a continued fraction [such as (6.14)] converge on bounded domains of the z-plane which do not contain the poles or other singularities, if any, of the limit function. However, some authors prefer to discuss convergence of continued fractions in terms of the chordal metric, which means the same as convergence on the Riemann sphere. This ambiguity rarely leads to confusion, provided the trap is anticipated. We use convergence in the ordinary sense, unless otherwise stated.

We consider a continued fraction in the reduced standard form

$$b_0 + \frac{a_1 z}{1} + \frac{a_2 z}{1} + \frac{a_3 z}{1} + \cdots. \tag{7.1}$$

Naturally enough, convergence criteria for continued fractions are always essentially properties of the 'tail' of the fraction; the value of b_0 is totally immaterial. The *divergence condition* is an important criterion for the convergence of (7.1).

Theorem 4.7.1. *If the continued fraction (7.1) converges for any nonzero value of z, then either*

$$\sum_{n=1}^{\infty} \left| \frac{a_2 a_4 \cdots a_{2n}}{a_3 a_5 \cdots a_{2n+1}} \right| \quad \text{or} \quad \sum_{n=1}^{\infty} \left| \frac{a_3 a_5 \cdots a_{2n-1}}{a_4 a_6 \cdots a_{2n}} \right| \tag{7.2}$$

must diverge.

Remarks. Note that divergence of one of the series (7.2) is a necessary but not a sufficient condition for convergence of (7.1). It must be supplemented by further conditions on the elements a_i if convergence is to be proved, as is done in Theorem 4.7.3.

We do not prove Theorem 4.7.1. Instead we show the scope of such proofs by proving a similar result due to Seidel which is self-contained and very relevant to Stieltjes series.

Theorem 4.7.2. *Consider the continued fraction*

$$\frac{1}{b_1} + \frac{1}{b_2} + \frac{1}{b_3} + \cdots, \tag{7.3}$$

where $b_i > 0$ for all i. The fraction (7.3) converges if and only if $\sum_{i=1}^{\infty} b_i$ diverges.

Proof. We first prove that convergence of (7.3) implies divergence of $\sum b_i$. First we prove an identity which shows that the sequence of convergents of (7.3) may be written as an alternating series. The nth convergent of (7.3) is

$$\frac{A_n}{B_n} = \left(\frac{A_n}{B_n} - \frac{A_{n-1}}{B_{n-1}}\right) + \left(\frac{A_{n-1}}{B_{n-1}} - \frac{A_{n-2}}{B_{n-2}}\right) + \left(\frac{A_{n-2}}{B_{n-2}} - \frac{A_{n-3}}{B_{n-3}}\right) + \cdots$$
$$+ \frac{A_1}{B_1}. \tag{7.4}$$

By using the recurrences

$$A_i = b_i A_{i-1} + A_{i-2}, \tag{7.5a}$$
$$B_i = b_i B_{i-1} + B_{i-2}, \tag{7.5b}$$

we find by iteration that

$$A_n B_{n-1} - B_n A_{n-1} = (-1)^n (A_2 B_1 - B_2 A_1) = (-1)^{n+1}.$$

Substituting this result in (7.4), we find that

$$\frac{A_n}{B_n} = \frac{1}{b_1} + \sum_{i=2}^{n} \frac{(-1)^{i+1}}{B_i B_{i-1}}, \quad n = 2, 3, 4, \ldots. \tag{7.6}$$

Since $B_1 = b_1$, $B_2 = 1 + b_1 b_2$, it follows from (7.5b) that

$$B_i \geq m = \min(1, b_1) > 0 \quad \text{for all } i \geq 1.$$

Therefore, each convergent of (7.3) may be expressed by (7.6) as an alternating series.

The alternating-series test states [Ferrar, 1938, p. 47] that if (i) u_n is a decreasing sequence, (ii) $u_n \to 0$ as $n \to \infty$, and (iii) $u_n > 0$ for all n, then the series $\sum_{n=0}^{\infty}(-1)^n u_n$ is called an alternating series and it converges.

We show that the condition that $\sum b_n$ diverges implies that $A_n/B_n \to f$ as $n \to \infty$. Note that

$$B_i B_{i-1} = (b_i B_{i-1} + B_{i-2}) B_{i-1} \geq m^2 b_i + B_{i-1} B_{i-2} \quad \text{for all } i,$$

and this shows that

$$B_n B_{n-1} \geq m^2 \sum_{i=2}^{n} b_i.$$

Hence condition (ii) of the alternating-series test is valid, and we deduce that $A_n/B_n \to f$ as $n \to \infty$.

If $\sum b_n$ is not divergent, let

$$\sum_{n=1}^{\infty} b_n = B.$$

We may easily prove by induction that
$$B_n < (1 + b_1)(1 + b_2) \cdots (1 + b_n)$$
using (7.5b), and hence
$$\ln B_n < B \quad \text{and} \quad B_n < e^B \quad \text{for all } n.$$

Consequently, the terms of (7.6) do not decrease in modulus, and so the ratios A_n/B_n do not converge to any limiting value.

An interesting application of Theorem 4.7.2 is that we may show directly that the Stieltjes fraction (5.5.24) converges on the positive real z-axis. The connection between Theorems 4.7.1 and 4.7.2 becomes evident from an equivalence transformation, as we state in Example 1.

Our next theorem provides a sufficient condition for the convergence of a continued fraction.

Theorem 4.7.3 (Parabola theorem) [Scott and Wall, 1940b; Thron, 1974]. *The continued fraction*

$$b_0 + \frac{a_1 z}{1} + \frac{a_2 z}{1} + \frac{a_3 z}{1} + \cdots \tag{7.7}$$

converges provided that

(i) α *may be found in the range* $-\pi/2 < \alpha < \pi/2$, *and* n_0 *exists such that the elements of (7.7) satisfy*

$$|a_n z| - \operatorname{Re}(a_n z e^{-2i\alpha}) \leq \tfrac{1}{2} \cos^2 \alpha \tag{7.8}$$

for all $n > n_0$; *and*
(ii) *the divergence condition (7.2) is satisfied.*

If it so happens that the sequence $\{a_n\}$ has a nonzero limit, let $a_n \to a$. We consider the case $\alpha = 0$. The key condition (7.8) is then satisfied if we impose a constraint on z, namely,

$$|z| - \operatorname{Re}\{z e^{i \arg(a)}\} < \frac{1}{2|a|}.$$

This defines a domain whose boundary is a parabola with its focus at $z = 0$ and its axis running through $z = -1/(4a)$. The geometric interpretation of the constraint is that z must be nearer the origin than the directrix of the parabola. The hypothesis that $a_n \to a$, $a \neq 0$, is also sufficient to satisfy the divergence condition, and so this extra hypothesis allows a simple corollary of the parabola theorem.

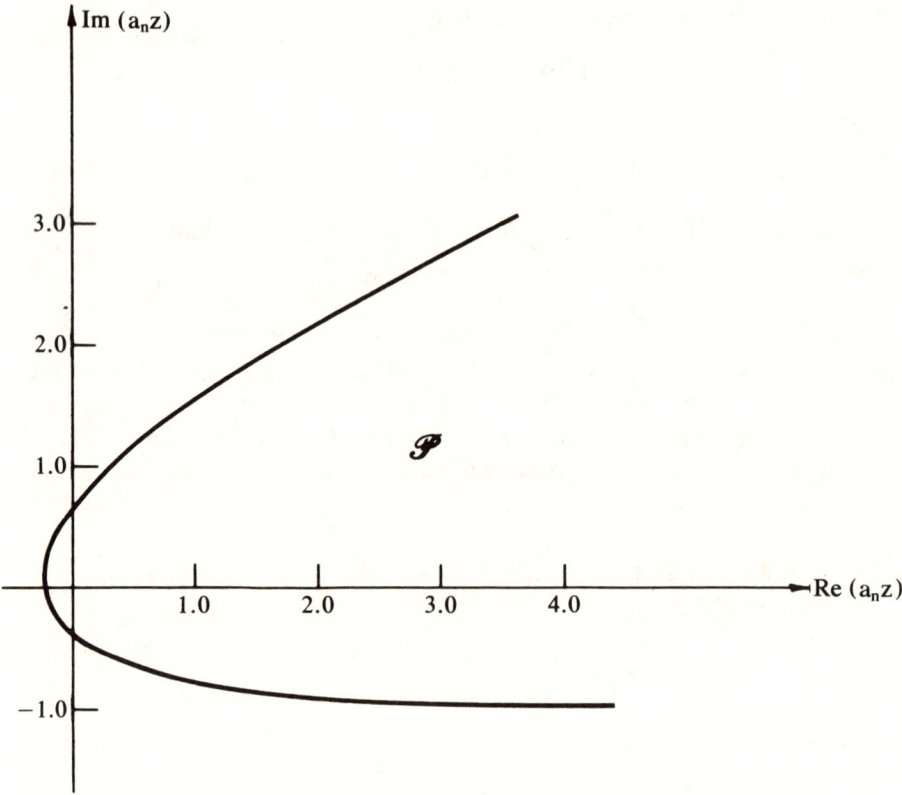

Figure 4.7.1. The parabolic domain $\mathcal{P}$ of Theorem 4.7.3 corresponding to $a = e^{-i\pi/12}$.

Theorem **4.7.4 (Cardioid theorem)** [Paydon and Wall, 1942; Dennis and Wall, 1945; Thron, 1974]. *Provided that n_0 and k may be found such that*

(a) $k > 0$,
(b) $|a_n| - \operatorname{Re} a_n \leq 1/(2k)$ *for all $n > n_0$, and*
(c) *the divergence condition (7.2) is satisfied,*

then the fraction (7.7) converges for all z in the region enclosed by the cardioid

$$|z| = k[1 + \cos(\arg(z))]. \tag{7.9}$$

Interpretation. Condition (b) of the theorem requires that all the partial numerator coefficients a_n of the continued fraction lie in a parabolic domain $\mathcal{P}_1$, shown in Figure 4.7.2. If the conditions (a), (b), (c) of the

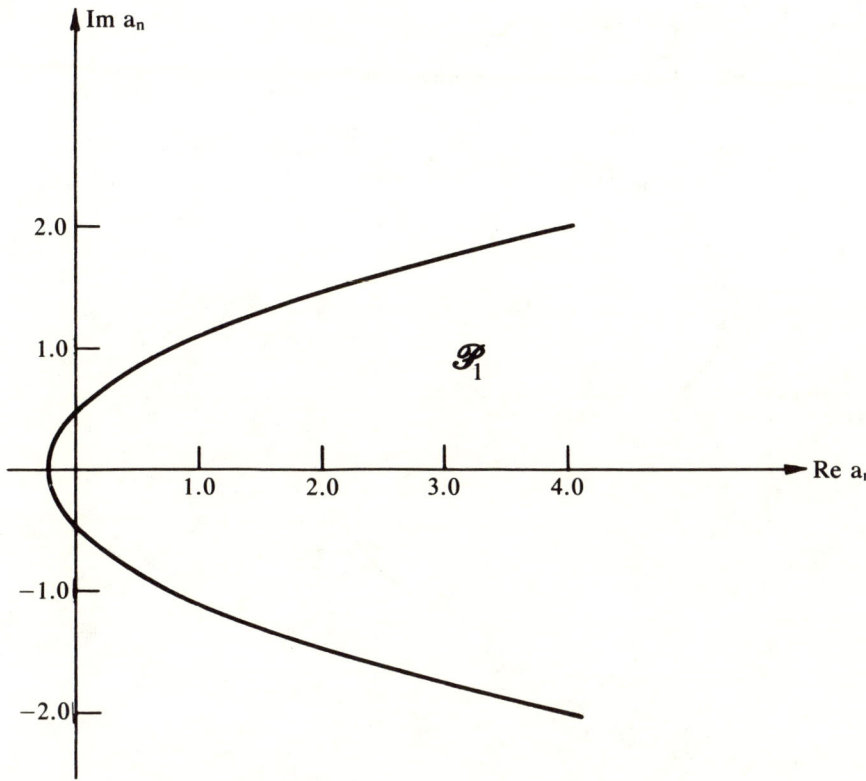

Figure 4.7.2. The parabolic domain $\mathcal{P}_1$ of Theorem 4.7.4 corresponding to $k = 1.07$.

theorem are satisfied, convergence of the fraction is assured in the cardioid shown in Figure 4.7.3. Note that a larger value of k gives a more restrictive parabolic constraint and a larger cardioid domain of convergence for the fraction.

Theorem 4.7.5. *If the sequence of convergents of* (7.7) *converges uniformly in* $|z| < R$, *with* $R > 0$, *to a limit function* $f(z)$, *then* $f(z)$ *is analytic in* $|z| < R$ *and its power series generates the fraction* (7.7).

Remark. This theorem is a consequence of Theorem 4.7.3, and is proved using Weierstrass's theorem [Titchmarsh, 1939, p. 95; Copson, 1948, p. 97].

Theorem 4.7.6 [Van Vleck, 1904]. *If* $a_n \to 0$ *as* $n \to \infty$, *the fraction* (7.7) *converges to a meromorphic function of* z. *If* $a_n \to a$ *and* $n \to \infty$ *with* $a \neq 0$, *then* (7.7) *converges to a function* $f(z)$ *which is meromorphic in the*

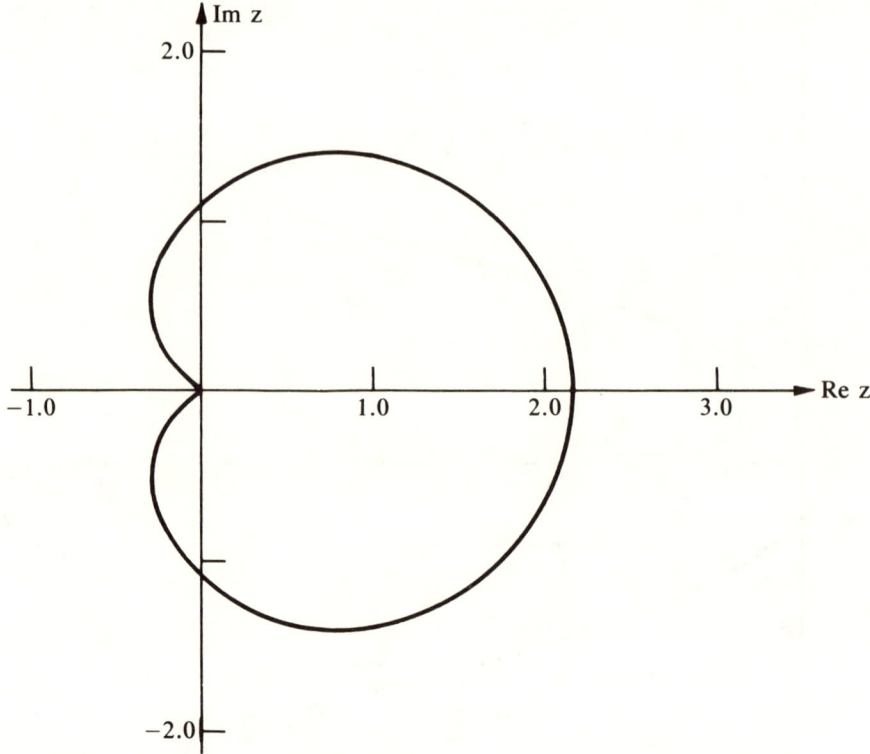

Figure 4.7.3. The cardioid domain of Theorem 4.7.4 corresponding to $k = 1.07$.

cut z-plane. The cut is placed in the shadow of $-(4a)^{-1}$ from the origin, as shown in Figure 4.7.4. In each case, convergence is uniform on any compact set containing no poles of the limit function, and the limit function has the continued-fraction expansion (7.7).

Next, we will use some of these theorems to prove the quoted results of the previous section. Van Vleck's theorem is used to prove (6.13), (6.14), and (6.16); (6.15) is proved by using the cardioid theorem.

Proof of (6.13). We have shown that both the left- and right-hand sides of the equation

$$\frac{{}_0F_1(a+1;z)}{{}_0F_1(a;z)} = \frac{1}{1+} \; \frac{a_1 z}{1+} \; \frac{a_2 z}{1+} \cdots \frac{a_n z}{1+} \cdots \quad (7.10)$$

with $a_n = \{(a+n-1)(a+n)\}^{-1}$ have the same formal power-series expansion. Since $a_n \to 0$, Theorem 4.7.6 states that (7.10) is an identity for all z not a zero of ${}_0F_1(a;z)$.

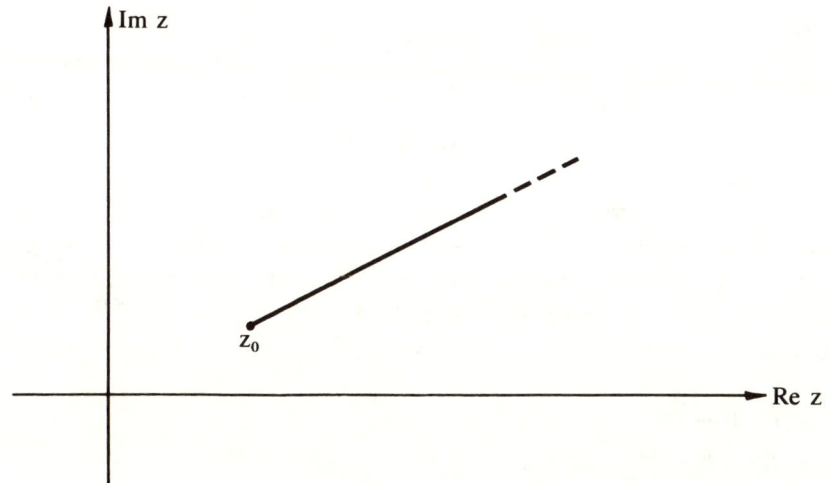

Figure 4.7.4. The cut from $z_0 = -(4a)^{-1}$ to ∞ in the complex z-plane for Theorem 4.7.6.

Proof of (6.14). We have shown that both the left- and right-hand sides of the equation

$$\frac{{}_1F_1(a+1, b+1; z)}{{}_1F_1(a, b; z)} = \frac{1}{1+} \frac{a_1 z}{1+} \frac{a_2 z}{1+} \cdots \frac{a_k z}{1+} \cdots \quad (7.11)$$

with

$$a_{2n} = (a+n)\{(b+2n-1)(b+2n)\}^{-1}, \quad n = 1, 2, 3, \ldots,$$

and

$$a_{2n+1} = (a-b-n)\{(b+2n)(b+2n+1)\}^{-1}, \quad n = 1, 2, 3, \ldots,$$

have the same formal power-series expansion. Since $a_n \to 0$, Theorem 4.7.6 asserts that the right-hand side of (7.11) is convergent and that (7.11) is an identity for all z not a zero of ${}_1F_1(a, b; z)$.

Partial proof of (6.15). We have shown that both the left- and right-hand sides of the identity

$$\frac{{}_2F_0(a, b+1; z)}{{}_2F_0(a, b; z)} = \frac{1}{1+} \frac{a(-z)}{1+} \frac{(b+1)(-z)}{1+} \frac{(a+1)(-z)}{1+} \cdots$$
$$\frac{-a_n z}{+ \ 1 \ +} \cdots$$

with

$$a_{2n} = b + n, \quad n = 1, 2, 3, \ldots,$$
$$a_{2n+1} = a + n, \quad n = 0, 1, 2, \ldots, \qquad (7.12)$$

have the same formal expansion. By noting that the ratio of successive numerator coefficients $a_k/a_{k+1} \to 1$, we may show that the divergence condition is satisfied, which is one necessary condition for the convergence of (7.12). It is more convenient to use the variable $z' = -z$, so that (7.12) becomes formally

$$\frac{{}_2F_0(a, b+1; -z')}{{}_2F_0(a, b; -z')} = \frac{1}{1+} \; \frac{a_1 z'}{1+} \; \frac{a_2 z'^2}{1+} \; \cdots \; \frac{a_k z'^k}{1+} \; \cdots. \qquad (7.13)$$

This equation has the status of a formal algebraic identity, and we seek to show that it represents an identity between function values in the cut z'-plane.

The sequence a_n is shown in Figure 4.7.5. We see that for any $k \geq 0$, $n_0 = n_0(k)$ exists such that

$$|a_n| - \operatorname{Re} a_n \leq \frac{1}{2k} \quad \text{for all } n > n_0.$$

Hence the continued fraction (7.13) converges for all z' interior to the cardioid

$$|z'| = k[1 + \cos(\arg z')].$$

Since this is true for any $k > 0$, the continued fraction converges for all z' except on $-\infty < z' < 0$, which corresponds to the positive real z-axis.

To establish equality between the left- and right-hand sides of (7.13), we note that for the special case of $b = 0$, $a > 0$, we have a strict Stieltjes series in the variable z', and the theory of Section 5.5 is applicable. Using Carleman's theorem, (7.13) is established directly as a true equality for $|\arg(z')| < \pi$. To extend this argument to the cases in question, we refer to Wall [1945].

Proof of (6.16). We have shown that both the left- and right-hand sides of the equation

$$\frac{{}_2F_1(a, b+1; c+1; z)}{{}_2F_1(a, b, c; z')} = \frac{1}{1+} \; \frac{a_1 z}{1+} \; \frac{a_2 z}{1+} \; \cdots \; \frac{a_n z}{1+} \; \cdots. \qquad (7.14)$$

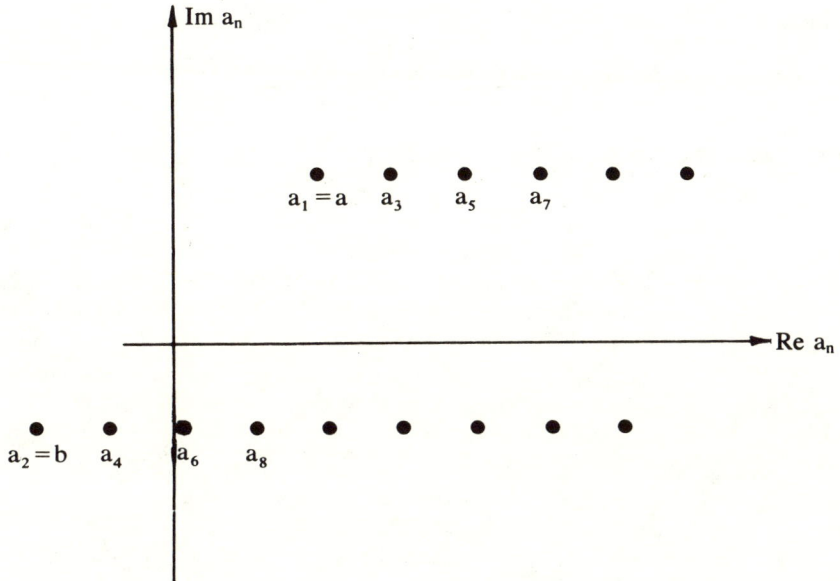

Figure 4.7.5. The numerator elements a_n in the complex plane defined by (7.12).

with

$$a_{2n} = \frac{(b+n)(a-c-n)}{(c+2n-1)(c+2n)}$$

and

$$a_{2n+1} = \frac{(a+n)(b-c-n)}{(c+2n)(c+2n+1)}$$

have the same formal expansion. Since $a_n \to -\frac{1}{4}$, Theorem 4.7.6 asserts that (7.14) is an identity valid in the z-plane cut along $1 \leq z < \infty$. This is, of course, the usual domain of definition of a $_2F_1$ hypergeometric function. Hence the fraction (7.14) converges for all z not on the cut and not a zero of $_2F_1(a, b; c; z)$.

Example 1. Consider the fraction

$$b_0 + \frac{z}{b_1 +} \frac{z}{b_2 +} \frac{z}{b_3 +} \cdots$$

with $b_i > 0$ for all i. The divergence condition (7.2) is precisely the condition that $\sum_{i=1}^{\infty} b_i$ diverges.

Example 2. The condition that $a_n/a_{n+1} \to c$ as $n \to \infty$ is sufficient to satisfy the divergence condition (7.2).

Example 3. The fraction

$$\frac{(1\times 2)^v}{1} + \frac{(2\times 3)^v}{1} + \cdots + \frac{j(j+1)^v}{1} + \cdots$$

converges for $0 \leqslant v \leqslant 1$ and diverges for $v > 1$.

5
Stieltjes series and Pólya series

5.1 Introduction to Stieltjes series

Definition. A Stieltjes function is defined by the Stieltjes-integral[†] representation

$$f(z) = \int_0^\infty \frac{d\phi(u)}{1 + zu}, \quad (1.1)$$

where $\phi(u)$ is a bounded, nondecreasing function (taking infinitely many different values) on $0 \leq u < \infty$ and with finite real-valued moments given by

$$f_j = \int_0^\infty u^j \, d\phi(u), \quad j = 0, 1, 2, \ldots. \quad (1.2)$$

From (1.1), it follows immediately that $f(z)$ is a real symmetric function, defined in the cut z-plane with the cut along the negative real axis as shown in Figure 5.1.1; real symmetric functions are defined by (1.6).

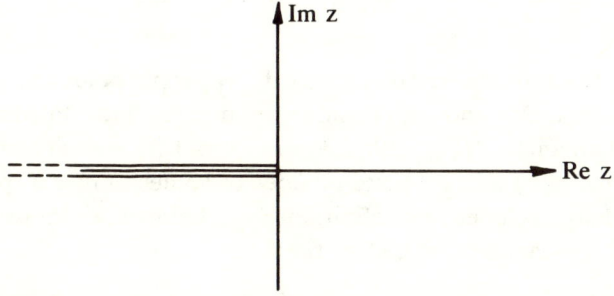

Figure 5.1.1. The cut z-plane in which $f(z)$ is defined by (1.1).

[†] A good explanation of Stieltjes integrals is given in Perron [1957, Vol. 2, p. 180] or Rudin [1976, Chapter 6].

A formal expansion of (1.1) always provides a series expansion of $f(z)$, called a Stieltjes series and given by

$$f(z) = \sum_{j=0}^{\infty} f_j(-z)^j. \tag{1.3}$$

The series is called formal because it may not converge for any z (except $z = 0$); nevertheless it is a useful representation of the function $f(z)$, if properly reinterpreted, as we will show in this chapter. It is easier to use the positive-definite coefficients $\{f_j\}$ in the expansion (1.3) rather than our standard notation, $f(z) = \sum_{j=0}^{\infty} c_j z^j$, because the determinantal inequalities of Theorem 5.1.2 take on a simpler form.

The phase 'taking infinitely many different values' in the definition of a Stieltjes series is made part of the definition so as to exclude the following special case. If $\phi(u)$ takes on a finite number of values, say $m + 1$ distinct values, then $\phi(u)$ is piecewise constant on $m + 1$ intervals covering the range $0 \le u < \infty$. Suppose that

$\phi(u) = 0 \quad$ on $0 \le u < u_1$,

$\phi(u) = \phi_i \quad$ on $u_i < u < u_{i+1} \quad$ for $i = 1, 2, \ldots, m - 1$,

$\phi(u) = \phi_m \quad$ on $u > u_m$.

Then $d\phi(u) = 0$ except in neighborhoods of $u = u_i$, $i = 1, 2, \ldots, m$, and so

$$f(z) = \int_0^\infty \frac{d\phi(u)}{1 + zu_i} = \sum_{i=1}^m \frac{1}{1 + zu_i} \int_{\text{nhd. of } u_i} d\phi(u)$$

$$= \sum_{i=1}^m \frac{\phi(u_i +) - \phi(u_i -)}{1 + zu_i}$$

$$= \sum_{i=1}^m \frac{\lambda_i}{1 + zu_i} \quad \text{with } \lambda_i > 0 \quad \text{for } i = 1, 2, \ldots, m. \tag{1.4}$$

Hence $f(z)$ is a rational function of z with m simple poles at $z = -u_i^{-1}$ on the negative real axis and with positive-definite residues. Furthermore, all Padé approximants of $f(z)$ with $L \ge m - 1$ and $M \ge m$ are exact. Thus, the case when $\phi(u)$ takes a finite number of values only is a special case, and it is usually excluded, by definition, from being a Stieltjes series.

If $\phi(u)$ is constant on $\lambda \le u < \infty$, then

$$f(z) = \int_0^\lambda \frac{d\phi(u)}{1 + zu}. \tag{1.5}$$

In this case, $f(z)$ is defined in the cut z-plane, cut along $-\infty < z < -\lambda^{-1}$.

5.1 Introduction to Stieltjes series

The power-series expansion of $f(z)$ given by (1.3) is then convergent in the disk

$$|z| < \lambda^{-1}$$

shown in Figure 5.1.2.

Whether or not the formal series $\sum_{j=0}^{\infty} f_j(-z)^j$ of $f(z)$ has a zero radius of convergence, the Padé approximants of the series are vital for its analysis and are useful for its numerical evaluation, as we will see in this chapter. We can prove convergence of the Padé approximants largely because we can prove that the poles of the Padé approximants lie on the cuts of the Stieltjes function. Stieltjes functions are real symmetric functions defined in the cut plane, with the negative real axis as the cut.

A function is defined to be real symmetric if it takes complex conjugate values when the variable is complex conjugated [Titchmarsh, 1939, p. 155]. This condition is that

$$f(z^*) = [f(z)]^*. \tag{1.6}$$

An important and immediate consequence of this applies to a function $f(z)$ which is analytic at a point $z = x_0$ on the real axis, so the expansion

$$f(z) = \sum_{i=0}^{\infty} d_i (z - x_0)^i$$

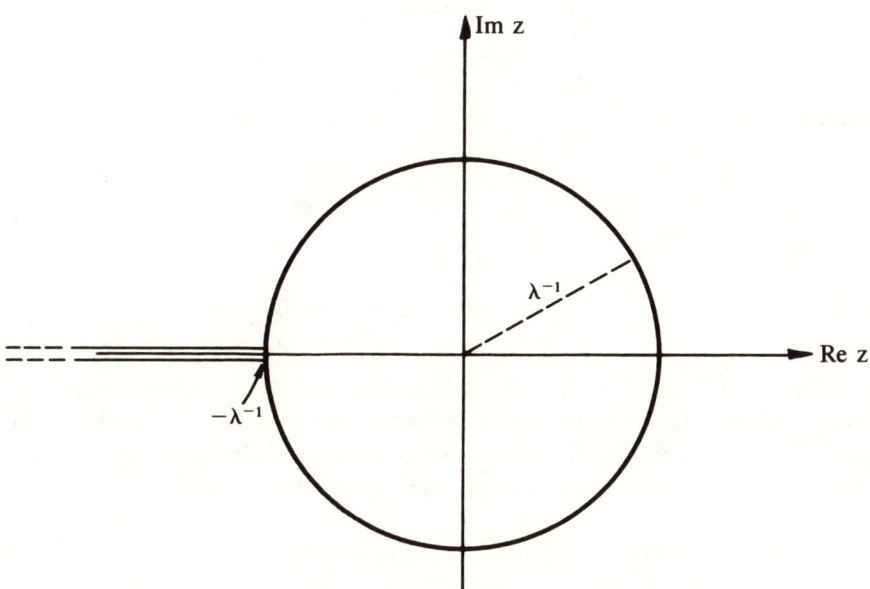

Figure 5.1.2. The cut z-plane in which $f(z)$ is defined by (1.5).

is convergent in some small disk enclosing $z = x_0$. The coefficients d_i, $i = 0, 1, 2, \ldots$, are real if and only if $f(z)$ is real symmetric, which justifies the name 'real symmetric.'

Stieltjes functions are real symmetric and, as can be shown from (1.1), have a negative imaginary part on the cut in the sense that

$$\operatorname{Im} f(x + i\varepsilon) = -\operatorname{Im} f(x - i\varepsilon) = \frac{\pi}{x} \phi'\left(-\frac{1}{x}\right) \quad \text{if } x = \operatorname{Re} z < 0, \quad (1.7)$$

provided the implied limit ($\varepsilon \to 0$) exists (cf. Lemma 3 in Section 5.6). We distinguish the three cases (i) $\phi(u)$ is differentiable, (ii) $\phi(u)$ is continuous but not differentiable, and (iii) $\phi(u)$ is discontinuous at a point u [Riesz and Szökefalvi-Nagy, 1955].

Before embarking on the proofs of the properties of Stieltjes series, let us consider an illustrative example of a Stieltjes series. The function is

$$f(z) = \frac{1}{z} \ln(1 + z)$$
$$= 1 - \tfrac{1}{2}z + \tfrac{1}{3}z^2 + \cdots \quad \text{for } |z| < 1.$$

Its coefficients are given by

$$f_j = \frac{1}{j+1}.$$

Hence the density function defined by

$$\phi(u) = u, \quad \phi'(u) = 1 \quad \text{on } 0 \leq u \leq 1,$$
$$\phi(u) = 1, \quad \phi'(u) = 0 \quad \text{on } 1 < u < \infty \quad (1.8)$$

ensures that

$$f_j = \int_0^\infty u^j \, d\phi(u) = \frac{1}{j+1}. \quad (1.9)$$

Further,

$$f(z) = \int_0^1 \frac{du}{1 + zu} = \int_0^\infty \frac{d\phi(u)}{1 + zu},$$

so $f(z)$ is a Stieltjes series according to (1.1), (1.2), where $\phi(u)$ is a nondecreasing function taking on infinitely many values [given by (1.8)] and all the moments are well defined by (1.9). The real symmetry is evident from the real coefficients in the expansion, and also

$$\operatorname{Im} f(z + i\varepsilon) = \frac{\pi}{x} \quad \text{for all } x = \operatorname{Re} z < -1,$$

illustrating (1.7).

The following theorem shows the effect on a Stieltjes series of deletion of the first $J+1$ terms of the series.

Theorem 5.1.1. *Let $f(z)$ be a Stieltjes series with a formal expansion and representation*

$$f(z) = \sum_{j=0}^{\infty} f_j(-z)^j, \quad f(z) = \int_0^{\infty} \frac{d\phi(u)}{1+zu}.$$

Then $g(z)$ given by the formal expansion

$$g(z) = (-z)^{-J-1} \sum_{j=J+1}^{\infty} f_j(-z)^j \tag{1.10}$$

is also a Stieltjes series represented by

$$g(z) = \int_0^{\infty} \frac{u^{J+1} d\phi(u)}{1+zu}. \tag{1.11}$$

Proof. The results follow immediately from the definitions (1.1)–(1.3).

Stieltjes series may be recognized by virtue of the determinantal conditions satisfied by the power-series coefficients f_j. We use the following definition, which is more convenient than using the determinants $C(L/M)$ defined in (1.4.8) in this context:

$$D(m, n) = \begin{vmatrix} f_m & f_{m+1} & \cdots & f_{m+n} \\ f_{m+1} & f_{m+2} & \cdots & f_{m+n+1} \\ \vdots & \vdots & & \vdots \\ f_{m+n} & f_{m+n+1} & \cdots & f_{m+2n} \end{vmatrix}. \tag{1.12}$$

These definitions are related by the identities (see Example 1)

$$D(L - M + 1, M - 1) = C(L/M) \quad \text{if } L - M \text{ is odd},$$

$$D(L - M + 1, M - 1) = (-)^M C(L/M) \quad \text{if } L - M \text{ is even}.$$

Theorem 5.1.2. *A necessary condition for $f(z)$ to be a Stieltjes series is that all the determinants $D(m, n)$ with $m \geq 0$, $n \geq 0$ are positive.*

Remark. The condition is also sufficient, provided that $f(z)$ is uniquely determined by its power series in principle. The proof of this result follows later.

Proof. We use the properties of Stieltjes series

$$f(z) = \int_0^{\infty} \frac{d\phi(u)}{1+uz}, \quad f_j = \int_0^{\infty} u^j d\phi(u),$$

and we must prove that the coefficients f_j satisfy $D(m, n) > 0$. Notice that each coefficient f_m is defined to be positive, and so $D(m, 0) > 0$. Let us define

$$G(x_0, x_1, \ldots, x_n) = \int_0^\infty u^m (x_0 + x_1 u + \cdots + x_n u^n)^2 \, d\phi(u)$$

$$= \int_0^\infty \left(\sum_{i=0}^n \sum_{j=0}^n u^{m+i+j} x_i x_j \right) d\phi(u)$$

$$= \sum_{i=0}^n \sum_{j=0}^n f_{i+j+m} x_i x_j, \tag{1.13}$$

which is seen to be a real positive quadratic form in the $n + 1$ real variables $x_0, x_1, \ldots, x_n$. Therefore $G(x_0, \ldots, x_n)$ has a minimum value on the hypersphere

$$S(x_0, \ldots, x_n) \equiv \sum_{i=0}^n x_i^2 \equiv x_0^2 + x_1^2 + \cdots + x_n^2 = 1. \tag{1.14}$$

The values of the $\{x_i\}$ at this minimum are given by Lagrange's undetermined-multiplier method, following variation of the n independent coordinates. The equations are

$$\frac{\partial G}{\partial x_i} - \lambda \frac{\partial S}{\partial x_i} = 0, \quad i = 0, 1, \ldots, n, \tag{1.15}$$

and $S(\mathbf{x}) = 1$.

Hence from (1.15),

$$\sum_{j=0}^n f_{i+j+m} x_j - \lambda x_i = 0 \quad \text{for } i = 0, 1, \ldots, n, \tag{1.16}$$

which is a set of $n + 1$ homogeneous equations for $x_0, \ldots, x_n$ with a consistency condition that λ must be an eigenvalue of the real and symmetric matrix

$$H = \begin{bmatrix} f_m & f_{m+1} & \cdots & f_{m+n} \\ f_{m+1} & f_{m+2} & \cdots & f_{m+n+1} \\ \vdots & & & \vdots \\ f_{m+n} & f_{m+n+1} & \cdots & f_{m+2n} \end{bmatrix}. \tag{1.17}$$

Let $\mathbf{x}^{(k)}$ be the eigenvector associated with the eigenvalue $\lambda^{(k)}$. Then we find from (1.13) that

$$G(\mathbf{x}) = \mathbf{x}^T H \mathbf{x},$$

and from (1.14) that

$$G(\mathbf{x}^{(k)}) = \mathbf{x}^{(k)T}H\mathbf{x}^{(k)} = \lambda^{(k)}.$$

If any eigenvector $\mathbf{x}^{(k)}$ of H had a strictly negative eigenvalue $\lambda^{(k)}$, then the quadratic form $G(\mathbf{x}^{(k)})$ of (1.13) would be negative, which is impossible. Hence all the eigenvalues of H are nonnegative, det H is the product of these eigenvalues, and so $D(m, n) \geq 0$ for all $m \geq 0$, $n > 0$.

This argument completes the proof except for the special case in which it might happen that $D(m, n) = 0$. This is associated with the existence of a nontrivial solution of the homogeneous linear equations

$$H\mathbf{x} = 0,$$

and therefore

$$G(\mathbf{x}) = \mathbf{x}^T H \mathbf{x} = 0 \quad \text{for } \mathbf{x}^T \cdot \mathbf{x} = 1.$$

From (1.13), we discover a nontrivial polynomial $p_n(u)$ for which

$$\int_0^\infty u^m [p_n(u)]^2 \, d\phi(u) = 0.$$

This can only occur if $d\phi(u) = 0$ except at the zeros of $p_n(u)$, contradicting the hypothesis that $\phi(u)$ takes on infinitely many values, which is a requirement for $f(z)$ to be a Stieltjes series.

Corollary 1. *All $[L/M]$ Padé approximants to Stieltjes series exist and are nondegenerate, if $L \geq M - 1$.*

The proof follows immediately from the basic representation (1.1.8). The sequences of Padé approximants to Stieltjes series of the next section are characterized by $J = L - M = $ const., with the natural condition that $J \geq -1$.

Corollary 2. *For any given Stieltjes series, there exists a regular C-fraction with the same formal expansion denoted by*

$$\sum_{i=0}^\infty c_i z^i = \frac{a_1'}{1} + \frac{a_2' z}{1} + \frac{a_3' z}{1} + \cdots \qquad (1.18)$$

with $a_i' > 0$ for all $i \geq 1$. This justifies the nomenclature (4.5.5) for the S-fraction, and demonstrates the identification with Stieltjes series.

Proof. In Chapter 4, we saw from (4.2.13) and (4.3.5) the possibility of expressing a formal power series as a regular C-fraction. Theorem 5.1.2

enables the signs of the elements of the fraction to be determined. We find that $a_{2M+2} > 0$, $a_{2M+1} < 0$ and

$$\text{sign}\,(b_{2M+2}) = \text{sign}\,(b_{2M+1}) = (-1)^M.$$

Elementary equivalence transformations of the form (4.1.6) may be used to show that the fraction has the representation (1.18) with positive-definite elements $\{a'_i, i = 1, 2, \ldots\}$. For example, note that $a'_1 = c_0 > 0$, $a'_2 = -c_1/c_0 > 0$, etc.

As general historical references for Stieltjes series (and not just for this section) we cite Stieltjes [1889, 1894], Tchebycheff [1858], Markov [1884], and Van Vleck [1903]. An excellent review is given by Perron [1957], and material related to continued fractions is treated by Wall [1948].

Example 1. Using the definitions (1.3), (1.12), (1.1.1), and (1.4.8), it follows that

$$D(m, n) = (-1)^{m(n+1)} C(m + n/n + 1). \tag{1.19}$$

Example 2. Using the definitions (1.2), (1.12), it follows that

$$D(m, n) = \int_0^\infty d\phi(u_0) \int_0^\infty d\phi(u_1) \cdots \int_0^\infty d\phi(u_n)$$

$$\times u_0^m u_1^{m+1} \cdots u_n^{m+n} \begin{vmatrix} 1 & u_0 & \cdots & u_0^n \\ 1 & u_1 & \cdots & u_1^n \\ \vdots & \vdots & & \vdots \\ 1 & u_n & \cdots & u_n^n \end{vmatrix}. \tag{1.20}$$

By permuting the variables and summing, we find that

$$(n + 1)! D(m, n) = \int_0^\infty d\phi(u_0) \int_0^\infty d\phi(u_1) \cdots \int_0^\infty d\phi(u_n)$$

$$\times u_0^m u_1^m \cdots u_n^m \begin{vmatrix} 1 & u_0 & \cdots & u_0^n \\ 1 & u_1 & \cdots & u_1^n \\ \vdots & \vdots & & \vdots \\ 1 & u_n & \cdots & u_n^n \end{vmatrix}^2.$$

Hence,

$$D(m, n) = \frac{1}{(n+1)!} \int_0^\infty u_0^m \, d\phi(u_0) \int_0^\infty u_1^m \, d\phi(u_1) \cdots \int_0^\infty u_n^m \, d\phi(u_n)$$

$$\times \prod_{i=0}^{n} \prod_{j=0}^{i-1} (u_i - u_j)^2 \tag{1.21}$$

and

$$D(m, n) > 0 \text{ [Bessis, 1979]}.$$

5.2 Convergence of Stieltjes series

The convergence properties of Padé approximants of Stieltjes series hinge on the fact that all the poles of the $[M + J/M]$ Padé approximants (with $J \geq -1$) to Stieltjes series lie on the negative real axis and have positive residues. As a special exception (1.4) shows that if $f(z)$ is a Stieltjes series, *except* that $\phi(u)$ has precisely m points of increase, then $f(z)$ is a rational function with poles on the negative real axis with positive residues. In this case, all the $[M + J/M]$ Padé approximants to $f(z)$ with $J \geq -1$ and $M \geq m$ are identical to $f(z)$. For the case of genuine Stieltjes series, we have the following important theorem.

Theorem 5.2.1. *If $f(z)$ is a Stieltjes series, then the poles of the $[M + J/M]$ Padé approximants (with $J \geq -1$) to $f(z)$ are simple poles which lie on the negative real axis and have positive residues.*

Remark. The proof uses the determinantal inequalities of Theorem 5.1.2 for the coefficients f_j, and does not require the fundamental representation of $\{f_j\}$ given by (1.2) and based on the existence of $\phi(u)$.

Proof. For $J \geq -1$, we define $\Delta_0^{(J)}(x) = 1$, and then for $M = 1, 2, 3, \ldots,$

$$\Delta_M^{(J)}(x) = \begin{vmatrix} f_{1+J} + xf_{2+J} & f_{2+J} + xf_{3+J} & \cdots & f_{M+J} + xf_{M+J+1} \\ f_{2+J} + xf_{3+J} & f_{3+J} + xf_{4+J} & \cdots & f_{M+J+1} + xf_{M+J+2} \\ \vdots & \vdots & & \vdots \\ f_{M+J} + xf_{M+J+1} & f_{M+J+1} + xf_{M+J+2} & \cdots & f_{2M+J-1} + xf_{2M+J} \end{vmatrix}.$$

(2.1)

Apart from a sign, these quantities are compact expressions for the denominators of a diagonal sequence of Padé approximants (see Section 1.3). They are related by

$$Q^{[M+J/M]}(x) = (-1)^{M(J+1)} \Delta_M^{(J)}(x).$$

(2.2)

Notice that $\Delta_M^{(J)}(x)$ is a real-valued function of the real variable x. When we consider sequences $\{\Delta_M^{(J)}(x), M = 0, 1, 2, \ldots\}$ with J fixed, we often omit the superscript (J). We first show that the functions

$$\Delta_0(x), \Delta_1(x), \Delta_2(x), \ldots$$

(2.3)

form a Sturm sequence. This means that if, for some x, $\Delta_j(x) = 0$, then $\Delta_{j-1}(x)$ and $\Delta_{j+1}(x)$ have opposite signs. We apply Sylvester's determinantal identity to $A = \Delta_{j+1}(x)$ given by (2.1), (2.2). Using subscripts to denote the deleted rows and columns, it follows from Sylvester's identity that

$$A \times A_{M-1,M;M-1,M} = A_{M;M} A_{M-1;M-1} - A_{M;M-1} A_{M-1;M}.$$

By symmetry,

$$A_{M;M-1} A_{M-1;M} = (A_{M;M-1})^2 \geq 0.$$

We identify

$$A_{M-1,M;M-1,M} = \Delta_{j-1}(x) \quad \text{and} \quad A_{M,M} = \Delta_j(x),$$

and therefore

$$\Delta_{j-1}(x) \Delta_{j+1}(x) \leq 0 \quad \text{if} \quad \Delta_j(x) = 0. \tag{2.4}$$

Also, (2.4) is valid if we identify $\Delta_0(x) = 1$ and take $j = 1$. We exclude the possibility that

$$\Delta_{j-1}(x) \Delta_{j+1}(x) = 0 \quad \text{and} \quad \Delta_j(x) = 0$$

by using the Frobenius

$$\begin{bmatrix} * & & \\ & * & \\ & & * \end{bmatrix}$$

identity, which would imply that $\Delta_j(x) = 0$ for all j if any two successive Δ's vanish at the same value of x. Therefore (2.3) is a Sturm sequence.

To locate the roots of $\Delta_M(x)$, which are the poles of the $[M + J/M]$ Padé approximant, we consider the Sturm sequence (2.3) and recall that $f_i > 0$. The first few members of the sequence are

$$\Delta_0(x) = 1,$$
$$\Delta_1(x) = f_{1+J} + x f_{2+J},$$
$$\Delta_2(x) = \Delta_1(x)(f_{3+J} + x f_{4+J}) - (f_{2+J} + x f_{3+J})^2.$$

From the representation (2.1), we have the general properties that

$$\left. \begin{array}{l} \Delta_j(0) > 0, \\ \Delta_j(-\infty) = (-)^j \cdot \infty \end{array} \right\} \quad \text{for all } j. \tag{2.5}$$

We define $x_{j,k}$ to be the kth zero of $\Delta_j(x)$, as is shown in Figure 5.2.1 and according to the following interlacing scheme:

$$\Delta_1(x) = 0 \quad \text{at } x = x_{1,1} = -\frac{f_{1+J}}{f_{2+J}}.$$

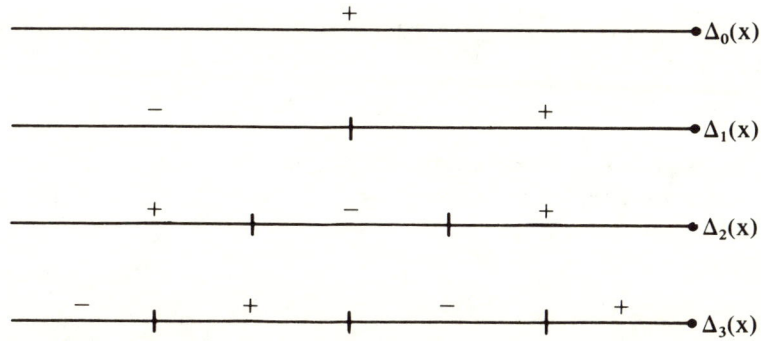

Figure 5.2.1. The sign of the first four Padé denominators of Stieltjes series for $x \leq 0$.

From (2.4)

$$\Delta_2(x_{1,1}) < 0$$

and from (2.5)

$$\Delta_2(0) > 0;$$

therefore

$$\Delta_2(x) = 0 \quad \text{at } x = x_{2,1} \in (x_{1,1}, 0).$$

From (2.4) and (2.5)

$$\Delta_2(x_{1,1}) < 0 \quad \text{and} \quad \Delta_2(-\infty) = +\infty;$$

therefore $\Delta_2(x) = 0$ at $x = x_{2,2} \in (-\infty, x_{1,1})$.

The complete argument for the interlacing of the zeros of $\Delta_j(x)$ follows by induction in a similar way. In short, $\Delta_j(x) = 0$ has roots $x = x_{j,k}$, for $k = 1, 2, \ldots, j$, which lie on the negative real axis and are distinct, because each zero of $\Delta_{j-1}(x)$ lies in an interval $(x_{j,k+1}, x_{j,k})$. Hence

$$\text{sign}[\Delta_{j-1}(x_{j,k})] = (-)^{k+1}$$

and the poles of the Padé approximants are distinct and lie on the negative real axis. To prove that the residues are positive, we use the identity (3.5.18),

$$[M + J + 1/M + 1] - [M + J/M] = \frac{(-x)^{2M+J+1}[D(1 + J, M)]^2}{\Delta_{M+1}(x)\Delta_M(x)}.$$

(2.6)

Writing $P^{[M+J/M]}(x) = \Gamma_M(x)$ for short, it follows from (2.6) that

$$\Gamma_{M+1}(x)\Delta_M(x) - \Gamma_M(x)\Delta_{M+1}(x) > 0 \quad \text{for } x < 0.$$

If $\Delta_{M+1}(x) = 0$, $\text{sign}[\Gamma_{M+1}(x)] = \text{sign}[\Delta_M(x)]$ and at the particular root $x = x_{M+1,k}$,

$$\text{sign}[\Gamma_{M+1}(x_{M+1,k})] = (-)^{k+1}.$$

Because $\Delta_{M+1}(x)$ is a polynomial of degree $M + 1$, $\Delta'_{M+1}(x)$ has precisely one zero in each interval $(x_{M+1,k}, x_{M+1,k+1})$. Therefore, since $\Delta_{M+1}(0) > 0$,

$$\left. \frac{d}{dx} \Delta_{M+1}(x) \right|_{x = x_{M+1,k}} = (-)^{k+1},$$

and so

$$\left. \frac{\Gamma_{M+1}(x)}{\Delta'_{M+1}(x)} \right|_{\Delta_{M+1}(x)=0} > 0.$$

Thus the residues of the poles of the Padé approximants are positive, and the theorem is proved.

We now proceed to consider properties of the diagonal and first subdiagonal sequence of Padé approximants to Stieltjes series for $x > 0$.

Theorem 5.2.2. Let $\sum_{j=0}^{\infty} f_j(-z)^j$ be a Stieltjes series. For z real and positive, its Padé approximants obey the following inequalities:

$$[M/M + 1] > [M - 1/M] \qquad \begin{bmatrix} * & \\ & * \end{bmatrix}, \qquad (2.7)$$

$$[M + 1/M + 1] < [M/M] \qquad \begin{bmatrix} * & \\ & * \end{bmatrix}, \qquad (2.8)$$

$$[M - 1/M] > [M/M - 1] \qquad \begin{bmatrix} & * \\ * & \end{bmatrix}, \qquad (2.9)$$

$$[M/M] < [M + 1/M - 1] \qquad \begin{bmatrix} & * \\ * & \end{bmatrix}, \qquad (2.10)$$

$$[M/M] > [M - 1/M] \qquad [* \ *], \qquad (2.11)$$

$$[M/M]' > [M - 1/M]' \qquad [* \ *]. \qquad (2.12)$$

Remark. The proof uses properties based on the determinantal inequalities for the coefficients f_j which are used to construct the Padé approximants. Like the other theorems of this section, it does not assume the representation (1.2) which characterizes the basic function $f(z)$; corresponding inequalities involving $f(z)$ are given by (4.3).

Proof. From (3.5.18), the

$$\begin{bmatrix} * & \\ & * \end{bmatrix}$$

identity is

$$[M+J+1/M+1] - [M+J/M] = \frac{x^{2M+J+1}[D(1+J,M)]^2}{Q^{[M+J+1/M+1]}(x)Q^{[M+J/M]}(x)}. \quad (2.13)$$

From the proof of Theorem 5.2.1, the denominator functions $\Delta_M^{(J)}(x)$ (and $\Delta_{M+1}^{(J)}(x)$) are positive for $x \geq 0$. By virtue of (2.2), (2.7) and (2.8) follow from (2.13) with $J=-1$ and $J=0$, respectively.

From (3.5.20), the

$$\begin{bmatrix} & * \\ * & \end{bmatrix}$$

identity is

$$[M+J/M] - [M+J+1/M-1] = \frac{x^{2M+J+1}[D(2+J,M-1)]^2}{Q^{[M+J/M]}(x)Q^{[M+J+1/M-1]}(x)}. \quad (2.14)$$

Hence (2.9) and (2.10) follow from (2.14) with $J=-1$ and $J=0$, respectively.

From (3.5.17), the $(* \quad *)$ identity becomes

$$[M/M] - [M-1/M] = \frac{x^{2M}D(0,M)D(1,M-1)}{\Delta_M^{(0)}(x)\Delta_M^{(-1)}(x)}. \quad (2.15)$$

Inequality (2.11) follows from (2.15). The coefficients of every power of x in $\Delta_M^{(0)}(x)$ and $\Delta_M^{(-1)}(x)$ are positive; see Example 1. We define

$$R(x) \equiv x^{-2M}\Delta_M^{(0)}(x)\Delta_M^{(-1)}(x)$$

$$= \sum_{i=1}^{2M} r_i x^{-i}$$

and note that each $r_i > 0$, which implies that

$$R'(x) < 0 \quad \text{for} \quad x > 0.$$

Hence $[M/M]' - [M-1/M]' > 0$ for $x > 0$.

Corollary. *With the hypotheses of the theorem,*

$$(-)^{J+1}\{[M+J+1/M+1] - [M+J/M]\} \geq 0,$$
$$(-)^{J+1}\{[M+J/M] - [M+J+1/M-1]\} \geq 0,$$

and these also hold when differentiated once.

Proof. The proof follows that of the theorem in every respect.

The essential result of Theorem 5.2.2 which is carried further is that the $[M-1/M]$ Padé approximants, for fixed $x > 0$, form a strictly increasing sequence. This leads to a convergence theorem after some preliminary theorems.

These inequalities [(2.11), for example] have natural generalizations in the complex plane. There they become nested inclusion regions for the higher-order approximants. They can be derived by exploiting Theorem 5.5.9 and allied results. The inclusion regions are convex, and lens shaped. The two vertices of the boundaries are the $[M/M]$ and $[M-1/M]$ approximants. The new lens-shaped region touches the boundary of the old in only two places, as shown in Figure 5.2.2. A more thorough discussion is given in Baker [1975a] and by Henrici and Pfluger [1966], Gargantini and Henrici [1967], and Baker [1969].

Theorem 5.2.3. *The sequence of $[M-1/M]$ Padé approximants to a Stieltjes series is uniformly bounded as $M \to \infty$ in the domain $\mathcal{D}(\Delta)$. $\mathcal{D}(\Delta)$ is a bounded region of the complex z-plane which is at least at a distance Δ from the cut $-\infty < z \leq 0$ along the negative real axis.*

Proof. From Theorem 5.2.1, we may write

$$[M - 1/M] = \sum_{i=1}^{M} \frac{\beta_i}{1 + \gamma_i z}, \qquad (2.16)$$

where $\beta_i > 0$, $\gamma_i > 0$ for $i = 1, 2, \ldots, M$. Each of the Padé approximants may be bounded by

$$|[M - 1/M]| \leq \sum_{i=1}^{M} \frac{\beta_i}{|1 + \gamma_i z|}, \qquad (2.17)$$

and therefore

$$|[M - 1/M]| \leq \sum_{i=1}^{M} \beta_i \quad \text{if Re } z \geq 0. \qquad (2.18)$$

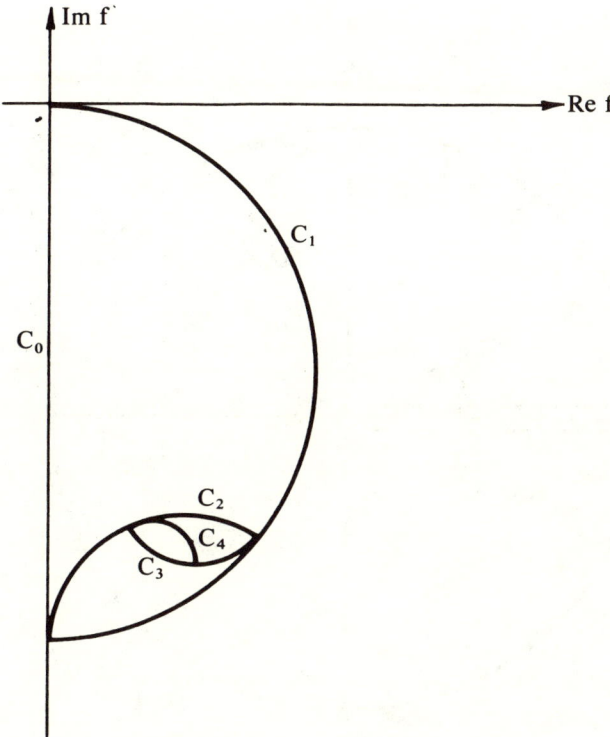

Figure 5.2.2. Nested inclusion regions for the values of a staircase sequence of Padé approximants of a Stieltjes series.

The right-hand side of (2.18) is interpreted by taking $z = 0$ in (2.16), so that

$$|[M - 1/M]|_{z=0} = f_0 = \sum_{i=1}^{M} \beta_i, \qquad (2.19)$$

and thus $[M - 1/M](z)$ is uniformly bounded for $\operatorname{Re} z \geq 0$. If $\operatorname{Re} z < 0$, because $\mathcal{D}(\Delta)$ is bounded we may assume that $\mathcal{D}(\Delta) \subset \{z, |z| < R_{\max}\}$ and then

$$|[M - 1/M]| \leq \sum_{i=1}^{M} \left| \frac{\beta_i}{1 + \gamma_i z} \right|$$

$$\leq \left(\sum_{i=1}^{M} \beta_i \right) \sup_{\gamma > 0} \left| \frac{1}{1 + \gamma z} \right|$$

$$= f_0 [\inf_{\gamma > 0} \sqrt{1 + 2\gamma \operatorname{Re}(z) + \gamma^2 |z|^2}]^{-1}.$$

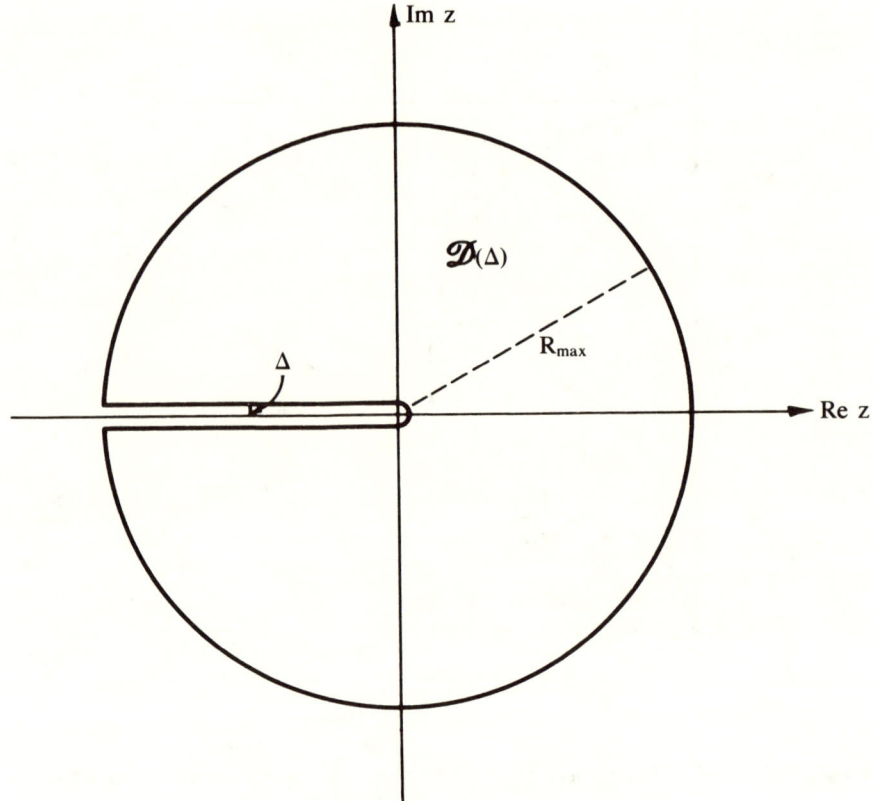

Figure 5.2.3. The bounded domain $\mathcal{D}(\Delta)$.

An elementary calculation shows that the minimum is achieved for $\gamma = -(\operatorname{Re} z)/|z|^2$, and also that

$$|[M - 1/M]| \leq \frac{f_0|z|}{|\operatorname{Im} z|} \leq f_0 \frac{R_{\max}}{\Delta}. \qquad (2.20)$$

Hence the theorem is proved.

Corollary. *All paradiagonal sequences $[M + J/M]$ of Padé approximants (with $J \geq -1$) to Stieltjes series are uniformly bounded in the domain $\mathcal{D}(\Delta)$.*

Proof. For $J \geq 0$, the polynomial $\tilde{f}(z) = \sum_{i=0}^{J} f_i(-z)^i$ is uniformly bounded on $\mathcal{D}(\Delta)$ and may be treated separately. Then $(-z)^{-J-1}[f(z) - \tilde{f}(z)]$ may be expressed in the form (2.16). The rest of the proof is straightforward.

5.2 Convergence of Stieltjes series

We have now established that the sequence of $[M - 1/M]$ Padé approximants to a Stieltjes series is strictly increasing and bounded at any given point on the positive real axis, and so it is convergent. In fact, we are building up to a much stronger result than pointwise convergence, and this requires the concept of equicontinuity of a sequence [Courant and Hilbert, 1953, Chapter 2].

Definition. A sequence of functions $f_m(z)$, $m = 0, 1, \ldots$, defined on a domain $\mathcal{D}$ is equicontinuous if, given any $\varepsilon > 0$, there exists $\delta > 0$, depending only on ε, such that

$$|f_m(z_1) - f_m(z_2)| < \varepsilon \quad \text{for all} \quad m = 0, 1, \ldots, \tag{2.21}$$

for any pair of points $z_1, z_2 \in \mathcal{D}$ satisfying $|z_1 - z_2| < \delta$.

The significant part of the latter definition is that $\delta = \delta(\varepsilon)$ does not depend on m, z_1, or z_2. Thus equicontinuity unites the property of uniform continuity of each member of the sequence with that of independence of which member of the sequence is selected.

Theorem 5.2.4. *The sequence of $[M - 1/M]$ Padé approximants to a Stieltjes series is equicontinuous on $\mathcal{D}(\Delta)$.*

Proof. From Theorem 5.2.1, we may write

$$[M - 1/M] = \sum_{i=1}^{M} \frac{\beta_i}{1 + \gamma_i z}, \quad \beta_i > 0, \gamma_i > 0, \quad \text{for} \quad i = 1, 2, \ldots, M.$$

Therefore

$$|[M - 1/M](z_1) - [M - 1/M](z_2)| \leq |z_1 - z_2| \sum_{i=1}^{M} \frac{\beta_i \gamma_i}{|1 + \gamma_i z_1||1 + \gamma_i z_2|}.$$

From (2.16), we find that $\sum_{i=1}^{M} \beta_i \gamma_i = f_1$. By inspection of (2.20), we deduce that

$$|[M - 1/M](z_1) - [M - 1/M](z_2)| \leq |z_1 - z_2| f_1 \frac{R_{\max}^2}{\Delta^2}$$

provided $z_1, z_2 \in \mathcal{D}(\Delta)$. Hence the sequence of $[M - 1/M]$ Padé approximants is equicontinuous on $\mathcal{D}(\Delta)$.

Corollary. *All paradiagonal sequences $[M + J/M]$ of Padé approximants (with $J \geq -1$) to Stieltjes series are equicontinuous on $\mathcal{D}(\Delta)$.*

Proof. The proof is completely parallel to that of the theorem.

The property of equicontinuity established is exploited by Arzela's theorem:

Theorem 5.2.5. *For any set of functions which are uniformly bounded and equicontinuous on $\mathcal{D}$, there exists a subsequence which converges uniformly to a continuous function defined on $\mathcal{D}$.*

Proof. Take a countable, dense set of points in $\mathcal{D}$. This is easily done by using a countable set of rationals $\{r_j\}$ which is dense on $(-\infty, \infty)$, and forming the countable set $\{z_{jk} = r_j + ir_k\}$. The subset of this contained in $\mathcal{D}$ is the point set required.

Let P_J be the set of the first J points, so $P_J = \{z_i, i = 1, 2, \ldots, J\}$ is a subset of $\mathcal{D}$.

Let the given equicontinuous functions form a sequence S,

$$S = \{f_n(z), n = 0, 1, 2, \ldots\}.$$

Because S is a sequence of functions which is uniformly bounded on $\mathcal{D}$, we may define S_1 to be a subsequence of S convergent at z_1, S_2 to be a subsequence of S_1 convergent at z_2, etc. Then, by construction, S_i is an infinite subsequence of S which converges at $z_1, z_2, \ldots, z_i$.

Given any $\varepsilon > 0$, define $\delta = \delta(\varepsilon)$, so that

$$|f_n(z) - f_n(z')| < \varepsilon \quad \text{for} \quad n = 0, 1, 2, \ldots$$

whenever $|z - z'| < \delta$ and $z, z' \in \mathcal{D}$. δ is independent of n by the equicontinuity hypothesis.

Choose J sufficiently large so that the δ-neighborhoods of all the points P_J cover $\mathcal{D}$. This condition means that

$$\mathcal{D} \subset \bigcup_{j=1}^{J} \{z, |z_j - z| < \delta\},$$

and this choice is possible because the z_j are dense in $\mathcal{D}$. Then, for the chosen ε and any given $z \in \mathcal{D}$, z_j exists with $j \le J$ and $|z_j - z| \le \delta$. Because the sequence S_J converges at z_j, N exists such that

$$|f_m(z_j) - f_n(z_j)| < \varepsilon$$

for all $f_m, f_n \in S_J$, $m, n > N$, and $j = 1, 2, \ldots, J$. By equicontinuity

$$|f_m(z_j) - f_m(z)| < \varepsilon,$$
$$|f_n(z_j) - f_n(z)| < \varepsilon,$$

5.2 Convergence of Stieltjes series

and hence

$$|f_m(z) - f_n(z)| < 3\varepsilon \quad \text{for all} \quad f_m, f_n \in S_J, \quad m, n > N.$$

By choosing the subsequence of functions to be the Jth element of S_J, $J = 1, 2, \ldots$, we find a subsequence of the given sequence which satisfies Cauchy's condition for uniform convergence, and prove Arzela's theorem.

We have proved in Theorem 5.2.3 that the sequence of $[M + J/M]$ Padé approximants to a Stieltjes are uniformly bounded on $\mathcal{D}(\Delta)$, and Theorem 5.2.4 establishes that the sequence is equicontinuous on $\mathcal{D}(\Delta)$. Thus Arzela's theorem asserts that a subsequence of $[M + J/M]$ Padé approximants to a Stieltjes series converges uniformly to a continuous limit function $f^{(J)}(z)$ on $\mathcal{D}(\Delta)$.

Next we need a familiar theorem on uniformly convergent sequences of analytic functions:

Weierstrass's theorem [Titchmarsh, 1939, p. 95]. *Let a sequence of functions $g_1(z), g_2(z), g_3(z), \ldots$ be such that each member is analytic in a domain $\mathcal{D}_1$ and it converges uniformly to a limit function $g(z)$ in any domain $\mathcal{D}_2$ in the interior of $\mathcal{D}_1$. Then $g(z)$ is analytic in $\mathcal{D}_2$.*

We apply Weierstrass's theorem directly to assert that $f^{(J)}(z)$ is analytic in $\mathcal{D}(2\Delta)$. But Δ was chosen (see Theorem 5.2.3) as an arbitrary small positive number, and can be replaced by $\frac{1}{2}\Delta$ without further implications. Thus we may deduce that for arbitrary positive Δ, $f^{(J)}(z)$ is analytic in $\mathcal{D}(\Delta)$.

Following Theorem 5.2.3, we noted that the entire sequence converges pointwise on the positive real axis, and so this pointwise limit is the real function $f^{(J)}(x)$. Since $f^{(J)}(x)$ is analytic on an interval of the real axis, the analytic continuation of $f^{(J)}(x)$ to $\mathcal{D}(\Delta)$, the domain of analyticity, is unique. Thus we have proved

Theorem 5.2.6. *The sequence of $[M + J/M]$ Padé approximants of a Stieltjes series (with $J \geq -1$) converges uniformly on $\mathcal{D}(\Delta)$, as shown in Figure 5.2.3, to a real symmetric function $f^{(J)}(z)$, analytic on $\mathcal{D}(\Delta)$.*

Having established convergence of the paradiagonal sequences (with $J \geq -1$), the obvious questions are what the limit function $f^{(J)}(z)$ is and how convergence is achieved. We answer the second question first by showing that the power series coefficients of $(-z)^j$ of the expansions of the $[M + J/M]$ Padé approximants approach the coefficients f_j from below.

Theorem 5.2.7. Let $f(z)$ be a Stieltjes series given by the formal power series

$$f(z) = \sum_{i=0}^{\infty} f_i(-z)^i.$$

Its $[L/M]$ Padé approximant has the power series expansion

$$[L/M] = \sum_{i=0}^{\infty} f_i^{[L/M]}(-z)^i. \tag{2.22}$$

Then, for all i and $L \geq M - 1$,

$$0 \leq f_i^{[L/M]} \leq f_i. \tag{2.23}$$

Proof. Each $[L/M]$ Padé approximant of $f(z)$ with $L \geq M - 1$ exists, so (2.22) is a well-defined series with a nontrivial circle of convergence for fixed L and M. If $i \leq L + M$, the Padé equations and (1.2) require that

$$0 \leq f_i^{[L/M]} = f_i.$$

Otherwise, for $i > L + M$, consider

$$\begin{aligned} f_i - f_i^{[L/M]} &= (f_i^{[L+1/M+1]} - f_i^{[L/M]}) \\ &\quad + (f_i^{[L+2/M+2]} - f_i^{[L+1/M+1]}) \\ &\quad + \cdots \\ &\quad + (f_i - f_i^{[L+K/M+K]}). \end{aligned} \tag{2.24}$$

Provided $2K \geq i - L - M$, the last term vanishes by virtue of the Padé equations. Now consider the expansion of (2.6),

$$[M + J + 1/M + 1] - [M + J/M] = \frac{(-z)^{2M+J+1}[D(1 + J, M)]^2}{\Delta_{M+1}(z)\Delta_M(z)}. \tag{2.25}$$

At $z = 0$, $\Delta_{M+1}(z)$ and $\Delta_M(z)$ are positive. The zeros of $\Delta_{M+1}(z)$ and $\Delta_M(z)$ occur at negative values. Thus we may write

$$\begin{aligned} [\Delta_M(z)]^{-1} &= \alpha_0^{-1} \prod_{i=1}^{M} (1 + \alpha_i z)^{-1} \quad \text{with } \alpha_0, \alpha_1, \ldots, \alpha_M > 0 \\ &= \alpha_0^{-1} \prod_{i=1}^{M} [1 - \alpha_i(-z)]^{-1} \\ &= \alpha_0^{-1} \prod_{j=1}^{M} [1 + \alpha_j(-z) + \alpha_j^2(-z)^2 + \cdots] \\ &= \alpha_0' + \alpha_1'(-z) + \alpha_2'(-z)^2 + \cdots \quad \text{with } \alpha_0', \alpha_1', \alpha_2', \ldots > 0, \end{aligned}$$

and similarly

$$[\Delta_M(z)\Delta_{M+1}(z)]^{-1} =$$

$$\alpha_0'' + \alpha_1''(-z) + \alpha_2''(-z)^2 + \cdots \text{ with } \alpha_0'', \alpha_1'', \alpha_2'', \cdots > 0.$$

This expansion, and therefore the expansion of (2.25), is a power series in $(-z)$ with positive coefficients. Hence every bracket of (2.24) is positive, and

$$f_i - f_i^{[L/M]} \geq 0,$$

proving the theorem.

As a general reference, we cite Baker [1975a, Chapter 15], and we also refer to Baker [1969], Common [1968], Wynn [1968b], and Brezinski [1977, p. 82].

Example 1. The proof of Theorem 5.2.1 shows that the coefficients of each power of x in the polynomials $\Delta_M^{(0)}(x)$ and $\Delta_M^{(-1)}(x)$ are positive.

5.3 Moment problems and orthogonal polynomials

The principal question left open in the previous section is to what the $[M + J/M]$ Padé approximants of Stieltjes series converge. This question is part of a wider question: to what extent do the coefficients f_j determine the measure $d\phi(u)$? The latter is a moment problem. If it is true that for some positive measure $d\phi(u)$ defined on $-\infty < u < \infty$ the coefficients

$$f_j = \int_{-\infty}^{\infty} u^j\, d\phi(u), \quad j = 0, 1, 2, \ldots, \tag{3.1}$$

are finite and well defined, then it is natural to call f_j the moments associated with the measure. The phraseology has a historical setting in which $\phi'(u)$ is a density per unit length, necessarily positive, of a linear mass distribution, such as a beam. If the beam has variable density, $\phi'(u)$ is not constant. If the beam has a weight attached at u_1, $\phi(u)$ has a positive jump discontinuity at u_1. The integration limits in (3.1) are set by the length of the beam, outside of which $d\phi(u) = 0$. The mathematical questions which emerge from this physical setting have the name of moment problems. Given the values of all the moments f_j, $j = 0, 1, 2, \ldots$, the problems are:

(i) *Existence.* Does a positive measure $d\phi(u)$ exist to allow the representation of $\{f_j\}$ by (3.1)?
(ii) *Determinacy.* Is $d\phi(u)$ uniquely determined?

(iii) *Nature.* Are the $\{f_j\}$ Stieltjes moments? Are they Hamburger moments or Hausdorff moments, etc?

The answers to these problems depend on various conditions on the moments. We will explain the solutions as well as the problems in their various settings. Stieltjes gave a clear answer to some of the outstanding problems in the form of the following theorem.

Theorem 5.3.1. *If the coefficients* $\{f_j, j = 0, 1, 2, \ldots\}$ *satisfy the determinantal conditions* $D(0, n) > 0$, $D(1, n) > 0$ *for all* $n = 0, 1, 2, \ldots$, *then a Stieltjes measure* $d\phi(u)$ *exists for which*

$$f_j = \int_0^\infty u^j \, d\phi(u). \tag{3.2}$$

Further,

$$f(z) = \int_0^\infty \frac{d\phi(u)}{1 + zu} \tag{3.3}$$

is a Stieltjes series with the given coefficients $\{f_j\}$ *in its formal expansion.*

Proof. An inductive proof based on Sylvester's identity establishes that $D(m, n) > 0$ for all $m, n > 0$ given that $D(0, n) > 0$ and $D(1, n) > 0$ for all n.

Consider the sequence of $[M - 1/M]$ Padé approximants to the formal power series

$$\sum_{j=0}^\infty f_j(-z)^j. \tag{3.4}$$

Theorems 5.2.1–5.2.7 all concern Padé approximants and are based entirely on properties of the coefficients f_j—in fact, the properties that the determinants $D(m, n)$ are positive. Consequently, the hypotheses of Theorem 5.2.6 are valid and the $[M - 1/M]$ Padé approximants of (3.4) converge uniformly to an analytic function $f^{(-1)}(z)$ in $\mathcal{D}(\Delta)$ shown in Figure 5.2.3. We deduce from Theorem 5.2.1 that

$$[M - 1/M] = \sum_{i=1}^M \frac{\beta_i}{1 + \gamma_i z}$$

$$= \sum_{i=1}^M \frac{\beta_i(1 + \gamma_i z^*)}{[1 + \gamma_i \operatorname{Re} z]^2 + [\operatorname{Im} z]^2}$$

with $\beta_i, \gamma_i > 0$ for $i = 1, 2, \ldots, M$. (3.5)

From (3.5),

$$\operatorname{Im}[M - 1/M](z) < 0 \quad \text{if} \quad \operatorname{Im} z > 0, \tag{3.6}$$

5.3 Moment problems and orthogonal polynomials

$$\text{Im}\,[M - 1/M](z) > 0 \quad \text{if} \quad \text{Im}\, z < 0, \tag{3.7}$$

$$\text{Re}\,[M - 1/M](z) = \text{Re}\,\{[M - 1/M](z)\}^*, \tag{3.8}$$

$$\text{Im}\,[M - 1/M](z) = -\text{Im}\,\{[M - 1/M](z)\}^*. \tag{3.9}$$

Let us apply Cauchy's theorem to $[M - 1/M]$ using a contour C which is the boundary of $\mathcal{D}(\Delta)$, and inside which $[M - 1/M](z)$ is analytic. This contour is shown in Figure 5.3.1, and Cauchy's theorem states that

$$[M - 1/M](z) = \frac{1}{2\pi i} \int_C \frac{[M - 1/M](\omega)\, d\omega}{\omega - z}.$$

We may now take the limit as $R' \to \infty$, noting that the contribution from the large circle tends to zero, and obtain, using (3.8) and (3.9),

$$[M - 1/M](z) = \frac{1}{\pi} \int_{-\infty}^{0} \frac{\text{Im}\,[M - 1/M](\omega + i\Delta)}{\omega - z}\, d\omega. \tag{3.10}$$

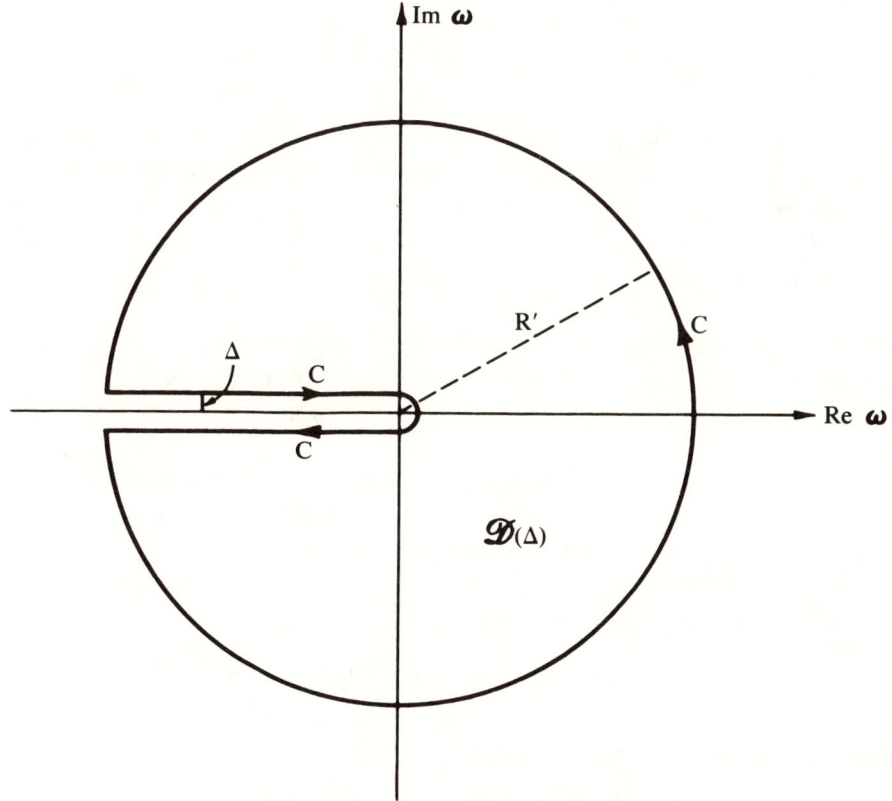

Figure 5.3.1. The contour C is the boundary of $\mathcal{D}(\Delta)$.

For given $\Delta > 0$, we may let $M \to \infty$, giving

$$\lim_{M \to \infty} [M - 1/M](z) = \frac{1}{\pi} \int_{-\infty}^{0} \frac{\operatorname{Im} f^{(-1)}(\omega + i\Delta)}{\omega - z} d\omega$$

$$= -\frac{1}{\pi} \int_{0}^{\infty} \operatorname{Im} f^{(-1)}\left(\frac{-1}{t + i\varepsilon}\right) \frac{dt}{t(1 + zt)}. \quad (3.11)$$

For any open interval (u, v) of the negative real axis, we may choose $\Delta \ll v - u$, $\varepsilon \ll v - u$ and consider

$$\phi(v-) - \phi(u+) = \int_{u}^{v} d\phi(u)$$

$$= \lim_{\varepsilon \downarrow 0} \int_{u}^{v} \left[-\frac{1}{\pi t} \operatorname{Im} f^{(-1)}\left(\frac{-1}{t + i\varepsilon}\right) \right] dt \quad (3.12)$$

provided the limit is well defined. The details are given in Section 5.6. Equation (3.12) provides a construction of a Stieltjes measure so that (3.11) may be written as

$$\lim_{M \to \infty} [M - 1/M](z) = f^{(-1)}(z) = \int_{0}^{\infty} \frac{d\phi(u)}{1 + zu} \quad (3.13)$$

for z not on the negative real axis, $-\infty < z \le 0$. We have avoided any discussion of the value of $\phi(u)$ at a jump discontinuity, because this value is usually of no importance, and we say that $\phi(u)$ is substantially determined by (3.12). Equally, we may take $\phi(0) = 0$ without loss of generality. The theorem is now proved.

It is important to realize that this theorem answers the questions of existence and nature, but not the question of uniqueness. Proofs of uniqueness are presently based on further hypotheses, as we discuss in Sections 5.4, 5.5.

Let us now consider a change of variable $w = -z^{-1}$ which reveals a remarkable connection between Padé approximants and orthogonal polynomials. It is, in fact, the historical approach to orthogonal polynomials, and rightly so.

We consider the basic initial representation

$$f(z) = \int_{0}^{\infty} \frac{d\phi(u)}{1 + zu} \quad (3.14)$$

and rearrange it and define $F(w)$ by

$$F(w) = \int_{0}^{\infty} \frac{d\phi(u)}{w - u} = \frac{1}{w} f\left(-\frac{1}{w}\right), \quad (3.15)$$

5.3 Moment problems and orthogonal polynomials

which has the formal expansion about $w = \infty$,

$$F(w) = \sum_{i=0}^{\infty} \frac{f_i}{w^{i+1}} = \frac{f_0}{w} + \frac{f_1}{w^2} + \cdots. \tag{3.16}$$

If we construct a set of polynomials $\{\pi_m(u), m = 0, 1, 2, \ldots\}$ orthogonal over $d\phi(u)$, we fundamentally require that

$$\int_0^{\infty} \pi_m(u) u^k \, d\phi(u) = 0 \quad \text{for} \quad k = 0, 1, 2, \ldots, m-1. \tag{3.17}$$

This equation is tantamount to the orthogonality condition

$$\int_0^{\infty} \pi_m(u) \pi_k(u) \, d\phi(u) = 0 \quad \text{for} \quad k = 0, 1, 2, \ldots, m-1, \tag{3.18}$$

and we have taken for granted the usual assumption that each $\pi_k(u)$ is a polynomial of degree precisely equal to k.

If we write

$$\pi_m(u) = u^m \sum_{i=0}^{m} \beta_i^{(m)} u^{-i}, \tag{3.19}$$

the equation (3.17) becomes

$$\sum_{i=0}^{m} \beta_i^{(m)} f_{m+k-i} = 0, \quad k = 0, 1, \ldots, m-1.$$

With the identification $f_j = (-1)^j c_j$ and by taking $\beta_0^{(m)} = 1$, this set of linear equations is seen to be the Padé equations (1.1.6), with the solution $b_i = (-1)^i \beta_i^{(m)}$ for $i = 0, 1, \ldots, m$. Hence (3.19) becomes

$$\pi_m(u) = u^m B^{[m-1/m]}(-1/u).$$

Since $C(m-1/m) \neq 0$ for Stieltjes series, we prefer to use the equivalent but conventionally normalized orthogonal polynomials given by $\pi_0(u) = 1$ and

$$\pi_m(u) = u^m Q^{[m-1/m]}(-1/u), \quad m = 1, 2, 3, \ldots. \tag{3.20}$$

These polynomials $\pi_m(u)$, defined by (3.20), satisfy the orthogonality condition (3.18). A natural observation at this point is that the

$$\begin{bmatrix} * & & \\ & * & \\ & & * \end{bmatrix}$$

identity (Section 3.5) for three consecutive $Q^{[m-1/m]}(z)$ denominators immediately is interpreted as the recursion relation for the orthogonal polynomials.

As a corollary to this development, we observe that

$$\pi_m^{(J)}(u) = u^m Q^{[m+J/m]}(-u^{-1}), \quad m = 0, 1, 2, \ldots, \tag{3.21}$$

are orthogonal polynomials over $u^{J+1} \, d\phi(u)$ for any $J \geq -1$.

Next, we proceed with the converse development. We assume the usual properties of the orthogonal polynomials $\{\pi_m(u), n = 0, 1, 2, \ldots\}$ and define polynomials (3.30) whose ratio is the Padé approximant of $f(z)$ defined by (3.14).

To determine the numerator of the Padé approximant to $f(z)$, or of $F(w)$ expanded about $w = \infty$, recall (3.15),

$$F(w) = \frac{1}{w} f\left(\frac{-1}{w}\right) = \int_0^\infty \frac{d\phi(u)}{w - u}, \quad f(z) = wF(w) = -\frac{1}{z} F\left(-\frac{1}{z}\right),$$

which leads us to expect

$$[m - 1/m]_f(z) = \frac{P^{[m-1/m]}(z)}{Q^{[m-1/m]}(z)} = \frac{w^m P^{[m-1/m]}(-1/w)}{w^m Q^{[m-1/m]}(-1/w)} = \frac{w\rho_m(w)}{\pi_m(w)}, \tag{3.22}$$

where $\rho_m(w)$ is a polynomial of degree $m - 1$. Thus we are led to consider

$$\pi_m(w) F(w) = \int_0^\infty \frac{d\phi(u)}{w - u} \pi_m(w)$$

$$= \int_0^\infty \frac{\pi_m(w) - \pi_m(u)}{w - u} \, d\phi(u) + \int_0^\infty \frac{\pi_m(u) \, d\phi(u)}{w - u}. \tag{3.23}$$

Equation (3.23) splits into two parts, and we find

$$\pi_m(w) F(w) = \rho_m(w) + \varepsilon(w), \tag{3.24}$$

where we will find that $\varepsilon(w)$ plays the role of an error. The first part is

$$\rho_m(w) \equiv \int_0^\infty \frac{\pi_m(w) - \pi_m(u)}{w - u} \, d\phi(u) \tag{3.25}$$

$$= \int_0^\infty \{\text{polynomial in } w, u \text{ of degree } m - 1\} \, d\phi(u)$$

$$= \text{polynomial in } w \text{ of degree } m - 1.$$

A glimpse forward to (3.30) explains why it is convenient to use a subscript m for a polynomial of degree $m - 1$ in this instance. The second part of (3.24) is

$$\varepsilon(w) \equiv \int_0^\infty \frac{\pi_m(u)}{w - u} \, d\phi(u)$$

5.3 Moment problems and orthogonal polynomials

$$= \frac{1}{w}\int_0^\infty \pi_m(u)(1-u/w)^{-1}\,d\phi(u)$$

$$= \frac{1}{w}\int_0^\infty \left\{\left[1 + \frac{u}{w} + \cdots + \left(\frac{u}{w}\right)^{m-1}\right]\right.$$
$$\left. + \left(\frac{u}{w}\right)^m\left(1-\frac{u}{w}\right)^{-1}\right\}\pi_m(u)\,d\phi(u)$$

$$= \frac{1}{w^{m+1}}\int_0^\infty \frac{u^m\,d\phi(u)}{1-u/w}\pi_m(u), \tag{3.26}$$

where the orthogonality property (3.17) has been used. From (3.23),

$$F(w) = \frac{\rho_m(w)}{\pi_m(w)} + \frac{\varepsilon(w)}{\pi_m(w)}. \tag{3.27}$$

Using the O notation to indicate the leading behavior for large $|w|$, we find from (3.20) and (3.26) that

$$\varepsilon(w) = O(w^{-m-1}), \quad \pi_m(w) = O(w^m),$$

and hence from (3.27)

$$F(w) = \frac{\rho_m(w)}{\pi_m(w)} + O(w^{-2m-1}). \tag{3.28}$$

Recalling from (3.25) and (3.20) that $\rho_m(w)$ and $\pi_m(w)$ are polynomials of degrees $m-1$ and m, respectively, we have proved that

$$f(z) = wF(w) = \frac{w\rho_m(w)}{\pi_m(w)} + O(w^{-2m}), \tag{3.29}$$

where

$$\frac{w\rho_m(w)}{\pi_m(w)} = \frac{w^{-m+1}\rho_m(w)}{w^{-m}\pi_m(w)} = \frac{(-z)^{m-1}\rho_m(-1/z)}{(-z)^m\pi_m(-1/z)}.$$

Hence, following (3.20), we define

$$\widetilde{P}^{[m-1/m]}(z) = (-z)^{m-1}\rho_m(-z^{-1}),$$
$$\widetilde{Q}^{[m-1/m]}(z) = (-z)^m\pi_m(-z^{-1}), \tag{3.30}$$

with ρ_m, π_m defined by (3.19) and (3.25). This proves that

$$[m-1/m]_f(z) = \frac{\widetilde{P}^{[m-1/m]}(z)}{\widetilde{Q}^{[m-1/m]}(z)}. \tag{3.31}$$

Hence we see that, except for normalization, $\widetilde{P}^{[m-1/m]}(z)$ and $\widetilde{Q}^{[m-1/m]}(z)$ are the numerator and denominator polynomials of Padé approximants.

Furthermore, from (3.15), (3.24), (3.26), (3.30), and (3.31) we have the explicit error formula

$$f(z) - [m - 1/m] = \frac{(-z)^m}{\pi_m(-z^{-1})} \int_0^\infty \frac{u^m \pi_m(u)\, d\phi(u)}{1 + zu}. \qquad (3.32)$$

Other formulas of this kind are given in Section 5.4.

Equations (3.14)–(3.31) show a different approach to the construction of Padé approximants. As a bonus, (3.25) indicates that the polynomials $\rho_m(w)$ satisfy exactly the same recursion relation as $\pi_m(w)$, the proof needing nothing more than the orthogonality property (3.17).

For general reviews of the scope of this section, we refer to Allen et al. [1975], and Karlsson and von Sydow [1976], who show that much of theory of Sections 5.1–5.5 can be derived using orthogonality methods.

5.4 Stieltjes series convergent in $|z| < R$

If the Stieltjes series discussed in the previous sections have a nonzero radius of convergence, which is to say that they are analytic in a neighborhood of the origin, then convergence theorems are easily proved and the moment problem is determinate. This section is devoted to the results ensuing from the hypothesis that $f(z)$ is a Stieltjes series with a nonzero radius of convergence, and they may be contrasted with the results of Section 5.5.

The property that the poles of the Padé approximants of $f(z)$ are on the cut of $f(z)$ is retained. In fact, we have the stronger result:

Theorem 5.4.1. *Let $f(z)$ be a Stieltjes series convergent in $|z| < R$. Then the poles of the $[M + J/M]$ Padé approximant, with $J \geq -1$, lie on the real axis in the interval $-\infty < z < -R$.*

We give two different proofs of this theorem.

Method 1. Suppose the contrary. From Theorem 5.2.1, there is a pole at $z = z_0$ with $-R \leq z_0 < 0$ of the $[M_1 + J_1/M_1]$ Padé approximant, with $J_1 \geq -1$. The interlacing property implies that every $[M + J_1/M]$ Padé approximant with $M \geq M_1$ has a pole in the interval $(z_0, 0)$, and let the limit of the poles nearest to $z = 0$ be at $z = z_1$. Then from Theorem 5.2.7,

$$R^{-1} \limsup_{m \to \infty} (f_m)^{1/m} \geq R^{-1} \limsup_{m \to \infty} (f_m^{[L/M]})^{1/m} = |z_1|^{-1}.$$

Hence, $|z_1| \geq R$, contradicting the hypothesis. Therefore, the poles of the Padé approximant lie on the open interval $(-\infty, -R)$.

5.4 Stieltjes series convergent in $|z| < R$

Method 2. If $f(z)$ is analytic in $|z| < R$, its Stieltjes-integral representation can have no singularities in this circle, and so becomes

$$f(z) = \int_0^{R^{-1}} \frac{d\phi(u)}{1 + zu}. \tag{4.1}$$

We may also take $d\phi(u) = 0$ on $R^{-1} < u < \infty$.

The poles of the $[m + J/m](z)$ Padé approximant occur at zeros of $\pi_m(u)$, where $z = -u^{-1}$ and $\pi_m(u)$ is a polynomial satisfying the orthogonality conditions (3.17):

$$\int_0^\infty u^j \pi_m(u) u^{J+1} \, d\phi(u) = 0 \quad \text{for} \quad j = 0, 1, 2, \ldots, n-1.$$

Suppose that m_1 zeros of $\pi_m(u)$ do not lie in $0 < u < R^{-1}$, but elsewhere in the complex u-plane at $u = u_1, u_2, \ldots, u_{m_1}$. Then $\pi_m(u)$ has the representation

$$\pi_m(u) = \kappa \prod_{i=1}^{m_1} (u - u_i) \prod_{j=m_1+1}^{m} (u - u_j),$$

where κ is a normalization constant. Consider

$$I = \int_0^{R^{-1}} c \left\{ \prod_{j=m_1+1}^{m} (u - u_j) \right\} \pi_m(u) u^{J+1} \, d\phi(u).$$

The integral of I is strictly positive, having no sign changes at the points $u_{m_1+1}, \ldots, u_m$, where it vanishes. But, by orthogonality, $I = 0$. Thus $m_1 = 0$, all zeros of $\pi_m(u)$ lie in $(0, R^{-1})$, and all poles of the $[M + J/M]$ Padé approximant lie on the cut of $f(z)$.

The next theorem concerns the limit functions $f^{(J)}(z)$ of the paradiagonal sequence $[M + J/M]$ of Padé approximants of a Stieltjes series $f(z)$, which are shown to be identical to $f(z)$. Again, we give two methods of proof, based on the integral representation and on orthogonal polynomials.

Theorem 5.4.2. *Let $f^{(J)}(z)$ be the limit functions of $[M + J/M]$ Padé approximants, with $J \geq -1$, to a Stieltjes series $f(z)$, which are analytic in $\mathcal{D}(\Delta)$. If $f(z)$ is analytic in $|z| < R$, then $f^{(J)}(z) = f(z)$ for all $J \geq -1$.*

Method 1. The hypothesis of the theorem implies that

$$f(z) = \int_0^{R^{-1}} \frac{d\phi(u)}{1 + zu}$$

as in (4.1). The analytic structure is shown in Figure 5.4.1.

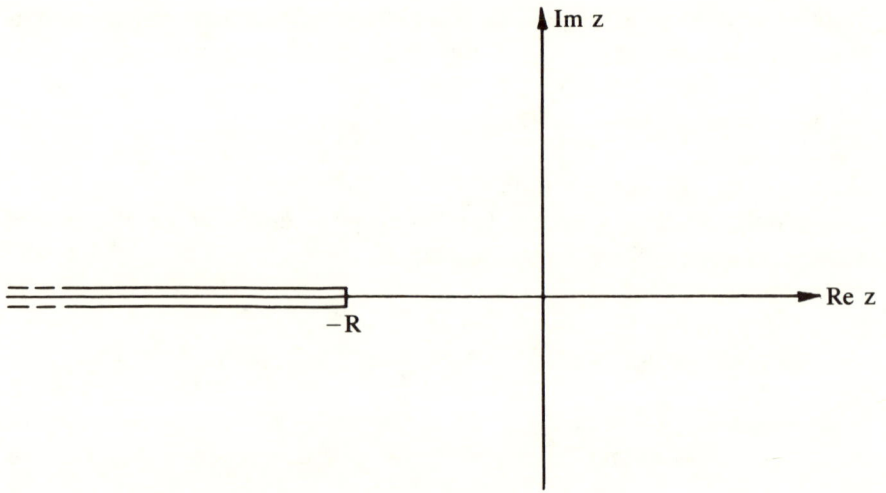

Figure 5.4.1. The domain of analyticity of $f(z)$.

From Theorem 5.2.6, we see that $f^{(J)}(z)$, $f(z)$ have an identical Maclaurin expansion. Since $f(z)$ is analytic in $|z| < R$, $f^{(J)}(z)$ is identical to $f(z)$ for any $J \geq -1$, and the theorem is proved.

Method 2. The explicit error formula of (3.27) is used. First we consider the case of $J = -1$ and make the usual changes of variable

$$w = -z^{-1}, \quad F(w) = -zf(z).$$

Equations (3.26) and (3.27) give

$$\left| F(w) - \frac{\rho_m(w)}{\pi_m(w)} \right| = \left| \frac{\varepsilon(w)}{\pi_m(w)} \right|$$

$$\leq \frac{1}{|w|^{m+1}} \int_0^{R^{-1}} \frac{u^m \, d\phi(u)}{|1 - u/w|} \left| \frac{\pi_m(u)}{\pi_m(w)} \right|.$$

From Theorem 5.4.1, the zeros of $\pi_m(u)$ occur at $u = u_i$, $i = 1, 2, \ldots, m$, and lie in $(0, R^{-1})$. Therefore

$$\left| F(w) - \frac{\rho_m(w)}{\pi_m(w)} \right| \leq \frac{1}{|wR|^m} \int_0^{R^{-1}} \frac{d\phi(u)}{|w - u|} \prod_{i=1}^m \left| \frac{u - u_i}{w - u_i} \right|.$$

Let k be an arbitrary constant greater than 1; then provided $|w| > kR^{-1}$,

$$\left| F(w) - \frac{\rho_m(w)}{\pi_m(w)} \right| \leq (wR)^{-m}(k-1)^{-m-1} R \int_0^{R^{-1}} d\phi(u). \quad (4.2)$$

This proves convergence of the sequence of $[m - 1/m]$ Padé approximants in $|w| > 2R^{-1}$ to $f(z)$. If $J > -1$, the first $J + 1$ terms of the series must be treated explicitly, so the problem of $[M + J/M]$ Padé approximants of $f(z)$ reduces to that of $[M - 1/M]$ Padé approximants to $[f(z) - \sum_{j=0}^{J}(-z)^j f_j]/z^{J+1}$.

The second method shows that the rate of convergence in $|z| < R/2$ is geometrical, as one would expect from the order notation. The sharper result of Theorem 5.4.4 is obtainable by using Schwarz's lemma in conjunction with the first method, but the sharpest results following it depend on a refinement of method 2.

An important consequence of Theorem 5.4.2 is that the inequalities (2.7) and (2.8) can be interpreted as bounds for the function $f(x)$ for x real and positive. For each m,

$$[m - 1/m](x) \leq f(x) \leq [m/m](x) \tag{4.3}$$

so the sequences $\{[m - 1/m]\}_{m=1}^{\infty}$ and $\{[m/m]\}_{m=1}^{\infty}$ form converging lower and upper bounds for $f(x)$. These bounds are attainable in the sense of (1.4). Other bounds are given by Baker [1975a, Chapter 16].

More recently, Gilewicz and Magnus [1991] have found inequalities such as

$$0 < f(x) - [n + k/n + 1](x) < \frac{|x|}{R} \{f(x) - [n + k/n](x)\}$$

for $-R < x < 0$ and $n, k \geq 0$.

Theorem 5.4.3 (Schwarz's lemma). *If $f(z)$ is analytic in $|z| < R$ and continuous on $|z| \leq R$, and further*

$$f(0) = f'(0) = \cdots = f^{(n)}(0) = 0$$

and

$$\max_{|z|=R} |f(z)| = M,$$

then

$$|f(z)| \leq M \left|\frac{z}{R}\right|^{n+1} \quad \text{if } |z| \leq R.$$

Proof. Apply the maximum-modulus theorem to

$$g(z) = z^{-(n+1)} f(z).$$

$g(z)$ is analytic in $|z| \leq R$, and so $|g(z)| < R^{-n-1} M$ on $|z| = R$. Therefore the maximum-modulus theorem asserts that

$$|g(z)| \leq R^{-(n+1)} M \quad \text{for } |z| \leq R$$

and

$$|f(z)| \le \left|\frac{z}{R}\right|^{n+1} M \quad \text{for } |z| \le R,$$

proving the theorem.

The convergence of Padé approximant to Stieltjes series is much faster than that suggested by the formulas of Theorem 5.4.2 for $\text{Re } z > 0$. The following theorem establishes an interesting result about the rate of convergence, and gives a clear idea of what can be expected in general, at best, from the Padé method. This is the first theorem of the chapter to extend beyond paradiagonal sequences, and is possible because convergence to a common limit is established. We are concerned with convergence in the bounded domain shown in Figure 5.4.2 and called $\mathscr{D}^+(\Delta)$ because convergence is proved in a domain larger than $\mathscr{D}(\Delta)$. $\mathscr{D}^+(\Delta)$ is defined by $|z| < R_{\max}$, but such that all points of $\mathscr{D}^+(\Delta)$ are at least a distance Δ from $-\infty < z < -R$.

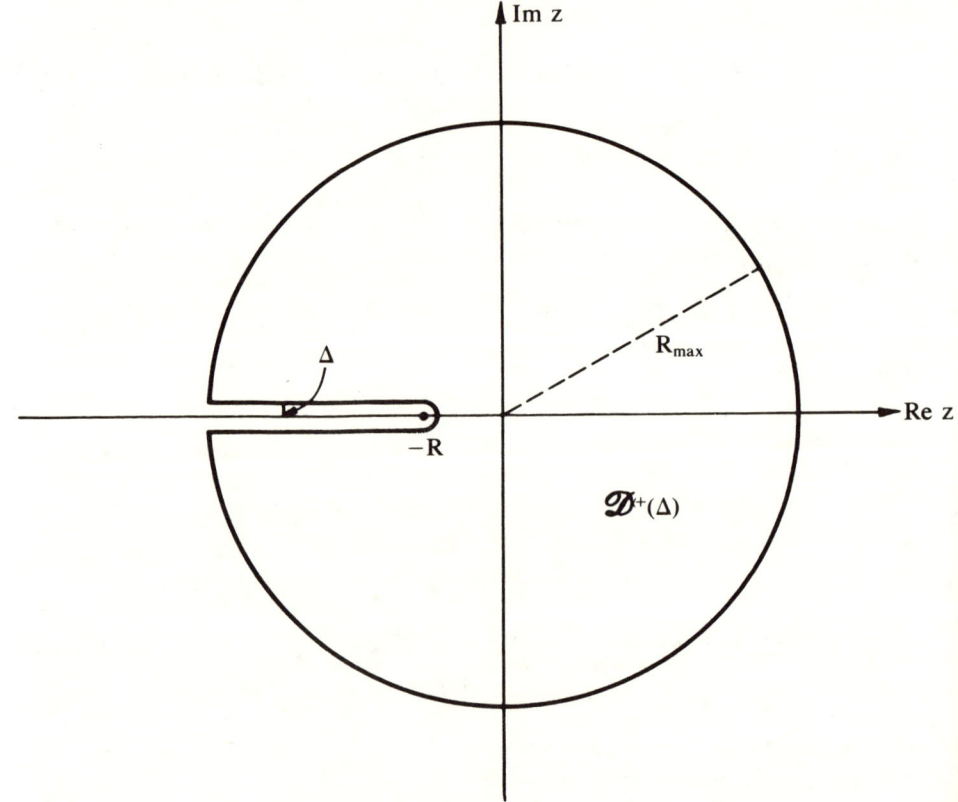

Figure 5.4.2. The domain $\mathscr{D}^+(\Delta)$.

5.4 Stieltjes series convergent in $|z| < R$

Theorem 5.4.4 **[Baker, 1975a, p. 220].** *Let $f(z)$ be a Stieltjes series with radius of convergence $R > 0$. Let $\{P_k(z)\}$ be any sequence of $[M_k + J_k/M_k]$ Padé approximants of $f(z)$ with $J_k \geq -1$. Let $\rho = R - \Delta > 0$. Then convergence in $\mathscr{D}^+(\Delta)$ is given by*

$$|P_k(z) - f(z)| \leq \left|\frac{z}{\rho}\right|^{J_k} \left|\frac{\sqrt{\rho + z} - \sqrt{\rho}}{\sqrt{\rho + z} + \sqrt{\rho}}\right|^{2M_k} \cdot \text{constant}.$$

Comment 1. For paradiagonal sequences, $[M + J/M]$, with $J_k = J$ fixed, convergence follows at a rate

$$|P_k(z) - f(z)| \leq \left|\frac{\sqrt{\rho + z} - \sqrt{\rho}}{\sqrt{\rho + z} + \sqrt{\rho}}\right|^{2M} \cdot \text{constant}. \tag{4.4}$$

Comment 2. For any ray sequence of $[L/M]$ Padé approximants, with $L = \lambda M$ and $\lambda \geq 1$, convergence follows at a rate

$$|P_k(z) - f(z)| \leq \left|\frac{\sqrt{\rho + z} - \sqrt{\rho}}{\sqrt{\rho + z} + \sqrt{\rho}}\right|^{2M} \left|\frac{z}{\rho}\right|^{(\lambda-1)M} \cdot \text{constant}. \tag{4.5}$$

Proof. Let us consider $J \geq 0$, and $[M + J/M]$ Padé approximants to $f(z)$. Then, from Section 5.2, the results

$$[M + J/M] - \sum_{i=0}^{J} f_i(-z)^i = \sum_{i=1}^{M} \frac{\beta_i}{1 + \gamma_i z} \tag{4.6}$$

and

$$\left|[M + J/M] - \sum_{i=0}^{J} f_i(-z)^i\right| \leq \text{constant} \tag{4.7}$$

follow for $z \in \mathscr{D}(\Delta)$ and all $M, J \geq 0$. Clearly (4.7) extends to $\mathscr{D}^+(\Delta)$ with a minor modification of the proof. Using Schwarz's lemma, (4.7) may be altered to

$$\left|[M + J/M] - \sum_{i=0}^{J} f_i(-z)^i\right| < \left|\frac{z}{\rho}\right|^{J+1} \cdot \text{constant}$$

for $|z| < \rho$ and so also for $z \in \mathscr{D}^+(\Delta)$ and all $M, J \geq 0$. Further, for $z \in \mathscr{D}^+(\Delta)$,

$$\left|\sum_{i=0}^{J} f_i(-z)^i - f(z)\right| < \left|\frac{z}{\rho}\right|^{J+1} \cdot \text{constant},$$

and hence

$$|[M + J/M] - f(z)| < \left|\frac{z}{\rho}\right|^{J+1} \cdot \text{constant}. \tag{4.8}$$

Consider the special mapping

$$z = \frac{4\rho w}{(1-w)^2}, \quad w = \frac{\sqrt{\rho+z} - \sqrt{\rho}}{\sqrt{\rho+z} + \sqrt{\rho}}. \tag{4.9}$$

The unit circle in the w-plane is given parametrically by $w = e^{i\theta}$, $0 \leq \theta < 2\pi$. This becomes

$$\sqrt{\rho+z} - \sqrt{\rho} = e^{i\theta}(\sqrt{\rho+z} + \sqrt{\rho}),$$

or

$$\sqrt{\rho+z} = (e^{i\theta} + 1)(1 - e^{i\theta})^{-1}\sqrt{\rho},$$

and by squaring,

$$z = -2\rho(1 - \cos\theta)^{-1},$$

which is the parametric equation of $-\infty < z \leq \rho$. The mapping is origin preserving and conformal except at $w = 1$. Let us reconsider (4.8) in the form

$$\left| \frac{[M + J/M](z(w)) - f(z(w))}{z(w)^{J+1}} \right| < \rho^{-J-1} c,$$

which is valid for all $z \in \mathcal{D}^+(\Delta)$, where c is a positive constant independent of $z(w)$, J, and M. For $z \to 0$, the left-hand side is of order z^{2M}, i.e., of order w^{2M}, and so the Schwarz lemma sharpens this result to

$$|[M + J/M](z(w)) - f(z(w))| < \left|\frac{z}{\rho}\right|^{J+1} |w|^{2M} c,$$

which is finally written as

$$|[M + J/M] - f(z)| < \left|\frac{z}{\rho}\right|^{J+1} \left|\frac{\sqrt{\rho+z} - \sqrt{\rho}}{\sqrt{\rho+z} + \sqrt{\rho}}\right|^{2M} c, \tag{4.10}$$

which is also valid for $z \in \mathcal{D}^+(\Delta)$. The case of $J = -1$ is simpler than the general case of $J \geq 0$, because subtraction of the first $J+1$ terms is unnecessary. The proof is otherwise unaltered, and the theorem is proved.

The next theorem, originally due to Markov [1884] is important because it establishes convergence of the paradiagonal sequences of types $[m + J/m]$, $J \geq -1$, $m = 1, 2, 3, \ldots$, to $f(z)$ for a rather more general class of functions than the sub-section title includes, namely,

$$f(z) = \int_a^b \frac{d\phi(u)}{1 + zu}, \tag{4.11}$$

where $-\infty < a \leq 0 \leq b < \infty$. For the special case of $a = 0$, $b = R^{-1}$, $f(z)$ is a Stieltjes series convergent in $|z| < R$. In this case, the content of Theorems 5.4.4 and 5.4.5 is much the same, but their proofs are quite different.

First, we consolidate the statements associated with equation (3.21) of the previous section: consider the polynomials $\{\pi_m^{(J)}(u)\}_{m=0}^{\infty}$ which are orthogonal over a fixed measure in the sense that

$$\int_a^b \pi_m^{(J)}(u) \pi_k^{(J)}(u) u^{J+1} \, d\phi(u) = \delta_{km}. \tag{4.12}$$

As in (3.30), a polynomial of degree m is defined by

$$Q^{[m+J/m]}(z) = (-z)^m \pi_m^{(J)}(-z^{-1}). \tag{4.13}$$

Note that $Q^{[m+J/m]}(0) \neq 0$. [In fact, $Q^{[m+J/J]}(0) > 0$ by convention.]

Shortly it will be proved that $Q^{[m+J/m]}(z)$ is a Padé denominator polynomial of the indicated type. For $J \geq -1$,

$$Q^{[m+J/m]}(z)\left\{f(z) - \sum_{j=0}^{J} f_j(-z)^j\right\} = (-z)^m \pi_m^{(J)}(-z^{-1}) \int_a^b \frac{(-zu)^{J+1}}{1+zu} \, d\phi(u), \tag{4.14}$$

where

$$f_j = \int_a^b u^j \, d\phi(u), \tag{4.15}$$

from (4.11). This representation suggests the definition

$$P^{[m+J/m]}(z) = Q^{[m+J/m]}(z) \sum_{j=0}^{J} f_j(-z)^j$$

$$+ \int_a^b \frac{(-z)^m \{\pi_m^{(J)}(-z^{-1}) - \pi_m^{(J)}(u)\}}{1+zu}(-zu)^{J+1} \, d\phi(u) \tag{4.16}$$

which is a polynomial of degree $m + J$ at most; from (4.10) it follows that $P^{[m+J/m]}(z)$ is the associated Padé numerator polynomial of type $[m + J/m]$.

The error function is defined as

$$E^{[m+J/m]}(z) = f(z) - P^{[m+J/m]}(z)/Q^{[m+J/m]}(z), \tag{4.17}$$

and then from (4.14), (4.16),

$$E^{[m+J/m]}(z) = \frac{(-z)^m}{Q^{[m+J/m]}(z)} \int_a^b \frac{\pi_m^{(J)}(u)}{1+zu}(-zu)^{J+1} \, d\phi(u)$$

$$= \frac{(-z)^m}{Q^{[m+J/m]}(z)} \int_a^b \frac{\pi_m^{(J)}(u)}{1+zu}(-zu)^{m+J+1} \, d\phi(u), \tag{4.18}$$

where the latter equality of (4.16) was derived using the fact that $\pi_m^{(J)}(u)$ is orthogonal to $1, u, \ldots, u^{m-1}$ in the sense of (4.12). Consequently

$$E^{[m+J/m]}(z) = O[z^{2m+J+1}] \tag{4.19}$$

and hence

$$[L/M](z) = P^{[m+J/m]}(z)/Q^{[m+J/m]}(z)$$

is the Padé approximant for $f(z)$ of the indicated type.

The error formula can be developed to give convergence results by further exploiting orthogonality. From (4.13), (4.18),

$$E^{[m+J/m]}(z) = \frac{(-z)^{2m+J-1}}{\{Q^{[m+J/m]}(z)\}^2} \int_a^b \frac{(-zu)^m \pi_m^{(J)}(-z^{-1})}{1+zu} \pi_m^{(J)}(u) u^{J+1} \, d\phi(u)$$

$$= \frac{(-z)^{2m+J+1}}{\{Q^{[m+J/m]}(z)\}^2} \int_a^b \frac{\{\pi_m^{(J)}(u)\}^2}{1+zu} u^{J+1} \, d\phi(u) \tag{4.20}$$

because $\{\pi_m^{(J)}(u) - (-zu)^m \pi_m^{(J)}(-z^{-1})\}/(1+zu)$ is a polynomial in u of degree $m-1$ at most, and so orthogonality justifies the second equality; see Allen et al. [1974]. This result is most of what is needed to prove

Theorem 5.4.5 [Markov, 1884]. *Each paradiagonal sequence of Padé approximants of types $[m + J/m]$, $m = 1, 2, 3, \ldots$, for any $J \geq -1$, for a Hamburger function of the type (4.11) converges in the open set formed from the complex plane cut along $(-\infty, -b^{-1}]$ and $[-a^{-1}, \infty)$. The rate of convergence is governed by*

$$\limsup_{m \to \infty} |E^{[m+J/m]}(z)|^{1/m} \leq \left| \frac{\sqrt{z^{-1}+b} - \sqrt{z^{-1}+a}}{\sqrt{z^{-1}+b} + \sqrt{z^{-1}+a}} \right| \tag{4.21}$$

with the phase convention that $\sqrt{z^{-1}+b}$, $\sqrt{z^{-1}+a}$ are positive for $z^{-1} > -a$.

Likewise, each paradiagonal sequence with $J \geq -1$ for a Stieltjes function of type (4.1) converges in the open cut plane $\mathbb{C}\setminus(-\infty, -R^{-1}]$ at a rate governed by

$$\limsup_{m \to \infty} |E^{[m+J/m]}(z)|^{1/m} \leq \left| \frac{\sqrt{1+z/R} - 1}{\sqrt{1+z/R} + 1} \right| \tag{4.22}$$

with the convention that $\sqrt{1+z/R} > 0$ for $z > -R^{-1}$.

Proof. The proof is based on the known asymptotic expansions of polynomials $\{p_n(z)\}_{n=0}^\infty$ which are orthogonal over a measure $d\nu(x)$ whose support lies in $[-1, 1]$. From Freud [1976, p. 117],

$$\limsup_{n \to \infty} |p_n(w)|^{1/n} \geq |w + \sqrt{w^2 - 1}| \tag{4.23}$$

5.4 Stieltjes series convergent in $|z| < R$

for $w \in \mathbb{C} \setminus [-1, 1]$, using the branch of $\sqrt{w^2 - 1}$ which is positive for $w > 1$. By changing the variable according to

$$w = (2t - a - b)/(b - a),$$

it follows that orthogonal polynomials $\pi_m^{(J)}(u)$, normalized by (4.13), satisfy

$$\liminf_{m \to \infty} |\pi_m^{(J)}(u)|^{1/m} = \left| \frac{\sqrt{t-b} + \sqrt{t-a}}{b-a} \right|^2, \quad (4.24)$$

where $\sqrt{t-b}$, $\sqrt{t-a}$ are positive for $t > b$, $t > a$, respectively.

From (4.20), we find that

$$|E^{[m+J/m]}(z)| < \frac{|z|^J}{|\pi_m^{(J)}(-z^{-1})|^2} \cdot \frac{1}{\delta} \cdot \int_a^b \{\pi_m^{(J)}(u)\}^2 u^{J+1} \, d\phi(u), \quad (4.25)$$

where δ is the least distance of z^{-1} from $[a, b]$. From (4.12), (4.24) and (4.25), we obtain

$$\limsup_{m \to \infty} |E^{[m+J/m]}(z)|^{1/m} \leq \frac{(b-a)}{|\sqrt{z^{-1}+b} + \sqrt{z^{-1}+a}|^2}$$

$$= \left| \frac{\sqrt{z^{-1}+b} - \sqrt{z^{-1}+a}}{\sqrt{z^{-1}+b} + \sqrt{z^{-1}+a}} \right| = \left| \frac{1-\theta}{1+\theta} \right|, \quad (4.26)$$

where

$$\theta = \sqrt{\frac{z^{-1}+a}{z^{-1}+b}}.$$

With the phase conventions stated, and for $z \in \mathbb{C} \setminus [-b^{-1}, -a^{-1}]$,

$$|\arg(z^{-1}+b)| < |\arg(z^{-1}+a)| \leq \pi$$

and hence $\operatorname{Re} \theta > 0$. Because $|1 - \theta| < |1 + \theta|$ in the open cut plane, (4.26) implies that

$$\lim_{m \to \infty} E^{[m+J/m]}(z) = 0,$$

so (4.21), (4.22) are convergence results.

Motivated by the results expressed by the previous theorems for Stieltjes functions, Baker [1975a] discovered empirically a remarkable heart-shaped region of divergence of ray sequences of approximants by plotting their zeros. For example, the symbols $\odot$, $+$, $\times$ in Figure 5.4.3 denote the zeros of the [4/1]-, [12/3]-, and [20/5]-type Padé approximants for $z^{-1} \ln(1+z)$, and these zeros necessarily indicate divergence. The

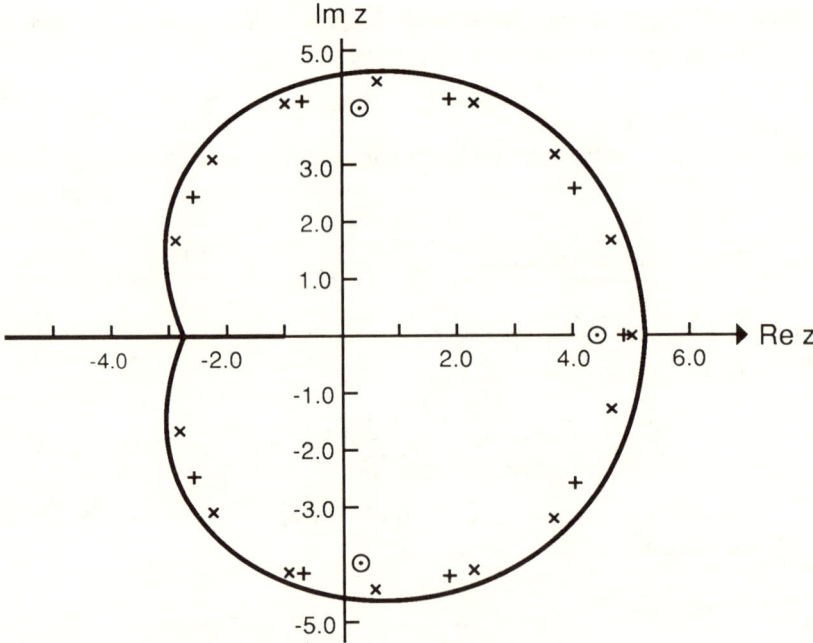

Figure 5.4.3. The proven boundary of convergence of the ray sequence S_3 of Padé approximants for Stieltjes functions. $\odot$, $+$, $\times$ denote zeros of the [4/1], [12/3], and [20/5] Padé approximants for $z^{-1}\ln(1+z)$.

correct boundary of convergence was proved using potential theory by Stahl [1976, 1980], and independently by Graves-Morris [1981a] as follows.

Consider the superdiagonal ray sequence S_γ of Padé approximants of types $[L_i/M_i]$ with $L_i/M_i \to \gamma + 1$ for the Stieltjes function

$$f(z) = \int_0^1 \frac{d\phi(u)}{1+zu}. \tag{4.27}$$

As usual, let $J_i = L_i - M_i$, and we see that we are concerned with the sequence for which

$$\lim_{i\to\infty} J_i/M_i = \gamma, \quad \gamma \geq 0. \tag{4.28}$$

Thus we are dealing with a larger class of polynomials $\{\{\pi_m^{(J_i)}\}_{m=0}^\infty\}_{i=0}^\infty$ than was used in (4.12), and (4.12) becomes

$$\int_0^1 \pi_m^{(J_i)}(u)\pi_k^{(J_i)}(u)u^{J_i+1}\,d\phi(u) = \delta_{km}, \quad i = 1, 2, \ldots. \tag{4.29}$$

The case of $d\phi(u) = du$ is an important instance of (4.29): if we substitute $x = 1 - 2u$ and

$$\pi_m^{(J_i)}(u) = (-1)^m\sqrt{2m + J_i + 2}\,P_m^{(J_i+1,0)}(x), \tag{4.30}$$

then (4.29) becomes

$$\int_{-1}^{1} P_m^{(J_i+1,0)}(x) P_k^{(J_i+1,0)}(x)(1-x)^{J_i+1} \, dx = \frac{2^{J_i+2}\delta_{km}}{2m+J_i+2} \quad (4.31)$$

which is the familiar orthonormality property of Jacobi polynomials [Szegö, 1967, p. 68]. For fixed α, β, the asymptotic properties of the Jacobi polynomials $P_m^{(\alpha,\beta)}(x)$ are well known [Szegö, 1967, p. 196]; for the case (4.28), the saddle-point method was used to prove that

$$\lim_{i\to\infty} |P_{M_i}^{(J_i,0)}(x)|^{1/M_i} = \frac{(t_+ - 1)^{\gamma+1}(t_+ + 1)}{2(t_+ - x)(x-1)^{\gamma}}, \quad (4.32)$$

where both implicit and explicit roots are positive for $x > 1$, and

$$t_+ = \frac{\gamma x - \gamma + 2x + \sqrt{\gamma^2(x+1)^2 + 4(x^2-1)(1+\gamma)}}{2(\gamma+1)}$$

with phase convention that $\mathrm{Im}(t_+) \mathrm{Im}(x) \geq 0$.

Using the interpolative methods described by Freud [1976, p. 117], it turns out that the result (4.32) is typical in the sense that

$$\liminf_{i\to\infty} |p_{m_i}^{(J_i,0)}(x)|^{1/M_i} \geq \frac{(t_+ - 1)^{\gamma+1}(t_+ + 1)}{2(t_+ - x)(x-1)^{\gamma}} \quad (4.33)$$

for general orthonormal polynomials satisfying

$$\int_{-1}^{1} p_m^{(J_i,0)}(x) p_k^{(J_i,0)}(x)(x+1)^{J_i} \, dv(x) = \frac{2^{J_i+1}\delta_{km}}{2m+J_i+1}$$

which is (4.31) with dx replaced by $dv(x)$.

With the substitution $x = 1 + 2/z$, (4.33) becomes

$$\liminf_{i\to\infty} \left| p_{m_i}^{(J_i,0)}\left[1 + \frac{2}{z}\right] \right|^{1/M_i} \geq \frac{|t_+ - 1|^{\gamma+1}|t_+ + 1||z|^{\gamma+1}}{2^{\gamma+1}|zt_+ - z - 2|}, \quad (4.34)$$

where

$$t_+(x(z)) = \frac{\gamma + z + 2 + \sqrt{\gamma^2(z+1)^2 + 4(z+1)(1+\gamma)}}{z(\gamma+1)}.$$

Finally, by defining

$$Q^{[m+J/m]}(z) = z^{m_i} p_{m_i}^{(J_i,0)}(1 + 2/z), \quad (4.35)$$

substituting (4.34), (4.35), into (4.20), and following the proof of Theorem 5.4.5, we obtain convergence in the heart-shaped region

$$2^{\gamma+1}|t_+ z - z - 2| < |t_+ - 1|^{\gamma+1}|t_+ + 1||z|^{1+\gamma/2}$$

shown in Figure 5.4.3. The connection between the heart-shaped region of Figure 5.4.3 and the domain on which normalized incomplete polynomials achieve their maximum modulus is explained by Saff [1983].

Our next theorem about Stieltjes series with a nonzero radius of convergence R is that the moment problem is determinate. This means that the coefficients f_j have the unique representation

$$f_j = \int_0^{R^{-1}} u^j \, d\phi(u). \tag{4.36}$$

This is a corollary of the following simple theorem.

Theorem 5.4.6. *The moment problem for a finite interval is determinate.*

Proof. Suppose $d\phi_1(u)$ and $d\phi_2(u)$ are two different measures for which

$$f_j = \int_a^b u^j \, d\phi_1(u) = \int_a^b u^j \, d\phi_2(u), \quad j = 0, 1, 2, \ldots.$$

Then $\int_a^b p(u) \, d(\phi_1 - \phi_2) = 0$ for every polynomial $p(u)$. Using Weierstrass's approximation theorem, $\int_a^b \psi(u) \, d(\phi_1 - \phi_2) = 0$ for every continuous function $\psi(u)$. Since $\phi_1(u) - \phi_2(u)$ is of bounded variation, $\phi_1(u) - \phi_2(u) = \text{constant}$, except possibly at the common points of discontinuity of ϕ_1 and ϕ_2.

In summary, if $f(z)$ is a Stieltjes series with a nonzero radius of convergence, we have found that $[M + J/M]$ paradiagonal sequences with $J \geq -1$ converge uniquely and determine a unique solution of the moment problem.

The contour integral methods of this section provide an "instant" proof [Zinn-Justin, 1970] of the orthogonality results for the polynomials (4.13) associated with Stieltjes and Hamburger series having nonzero radii of convergence, without the paraphernalia of the complete theory. We start with the Hamburger function

$$f(z) = \int_a^b \frac{d\phi(u)}{1 + zu},$$

with $-\infty \leq a < 0 \leq b < \infty$ and let $Q^{[m-1/m]}(z)$ be the denominator of its $[m - 1/m]$ Padé approximant. The mth-order polynomial, $\pi_m(u)$, is defined by

$$\pi_m(u) = u^m Q^{[m-1/m]}(-1/u), \quad m = 1, 2, 3, \ldots,$$

and $\pi_0(u) = 1$. Then the set $\{\pi_m(u), m = 0, 1, 2, \ldots\}$ is orthogonal in the sense that

$$I_{mk} = \int_a^b \pi_m(u) \pi_k(u) \, d\phi(u) = 0 \quad \text{for } m \neq k. \tag{4.37}$$

To prove this, let $w = -1/z$, as before, and define

$$F(w) = \int_a^b \frac{d\phi(u)}{w-u} = -z\int_a^b \frac{d\phi(u)}{1+uz} = -zf(z).$$

Using Cauchy's theorem,

$$\frac{1}{2\pi i}\int_{|w|=2R} \pi_m(w)\pi_k(w)F(w)\,dw = \int_a^b \pi_m(u)\pi_k(u)\,d\phi(u) = I_{mk},$$

where $R = \max\{|a|, |b|\}$. Substituting $z = -1/w$, we find

$$I_{mk} = \frac{1}{2\pi i}\int_{|z|=1/(2R)} (-z)^{-m-k-1} Q^{[m-1/m]}(z) Q^{[k-1/k]}(z) f(z)\,dz. \quad (4.38)$$

Without loss of generality, we may choose $m > k$. Since

$$Q^{[m-1/m]}(z)f(z) = P^{[m-1/m]}(z) + O[z^{2m}],$$

we may collapse the contour of (4.38) to the origin and find that $I_{mk} = 0$, establishing (4.37).

5.4.1 Hausdorff moment problem

If the Stieltjes series defined in (4.1) has radius of convergence $R = 1$, the moment problem is called the Hausdorff moment problem. In fact, a simple change of scale of the variable z in (4.1) allows us to assume that $R = 1$ without loss of generality and to consider Stieltjes series defined by

$$f(z) = \int_0^1 \frac{d\phi(u)}{1+zu} = \sum_{j=0}^\infty f_j(-z)^j, \quad (4.39)$$

where

$$f_j = \int_0^1 u^j\,d\phi(u), \quad j = 0, 1, 2, \ldots, \quad (4.39)$$

and $\phi(u)$ is a bounded nondecreasing function defined on $0 \leq u \leq 1$. The moments defined by (4.39) are called Hausdorff moments, and the Hausdorff moment problem consists of constructing $\phi(u)$ from the given sequence $f_0, f_1, f_2, \ldots$. This we have done in (3.12) using Padé approximants. To make the connection between the Hausdorff moment problem and totally monotone sequences, we define Δ, the forward difference operator, by

$$\Delta f_j = f_{j+1} - f_j, \quad j = 0, 1, 2, \ldots,$$

and higher differences are defined similarly (see Section 3.1).

Definition. A sequence $\{f_j\}$ is *totally monotone* if

$$(-)^k \Delta^k f_j \geq 0 \quad \text{for all } j, k \geq 0. \tag{4.40}$$

This definition immediately implies that a totally monotone sequence $\{f_j\}$ is a positive non-increasing sequence.

Theorem 5.4.7. *The sequence defined by (4.39) is totally monotone.*

Proof. By inspection of (4.39), we see that

$$(-)^k \Delta^k f_j = \int_0^1 (1-u)^k u^j \, d\phi(u) \geq 0$$

for all $j, k \geq 0$, proving the result.

The converse of this result—that if a sequence $\{f_j\}$ is totally monotone, then it has a Stieltjes integral representation (4.39)—is also true. The best proof does not use Padé approximants; we refer the reader to the proof in the books by Wall [1948, p. 267] and Widder [1972, p. 109].

For further details, we refer to Bernstein [1928], Brezinski [1978a], Gragg [1968], and Wynn [1966b].

5.4.2 Integer moment problem

If the Stieltjes moments defined by

$$f_j = \int_0^\Lambda u^j \, d\phi(u), \quad j = 0, 1, 2, \ldots, \tag{4.41}$$

are known to be integers, construction of the Stieltjes function

$$f(z) = \int_0^\Lambda \frac{d\phi(u)}{1 + zu}$$

is an integer moment problem [Barnsley, Bessis, and Moussa, 1979]. We consider the simplest cases, $\Lambda = 1$ and $\Lambda = 2$, as examples.

Example 1 ($\Lambda = 1$). In this case, it follows from (4.41) that the moments $\{f_j, j = 0, 1, 2, \ldots\}$ form a positive decreasing sequence. Therefore f_∞ exists such that $f_j \to f_\infty$, and f_∞ is an integer. Hence n exists such that $f_j = f_\infty$ for all $j \geq n$, and we find that

$$f_n - f_{n+1} = \int_0^1 (1-u) u^n \, d\phi(u) = 0,$$

$$d\phi(u) = [A\delta(u) + B\delta(u-1)] \, du,$$

and
$$f_j = f_1 \quad \text{for all } j \geq 1.$$

The function $f(z)$ is thus given exactly by its [1/1] Padé approximant as

$$f(z) = A + \frac{B}{1+z}.$$

Example 2 ($\Lambda = 2$). In this case, we use the moments (4.41) to construct the sequence of integers

$$m_k = \int_0^2 u^k (2-u)^k \, d\phi(u), \quad k = 0, 1, 2, \ldots,$$

which is a positive decreasing sequence. Therefore m_∞ exists such that $m_k \to m_\infty$, and so n exists such that $m_j = m_\infty$ for all $j \geq n$. We find that

$$m_n - m_{n+1} = \int_0^2 u^k (2-u)^k (1-u)^2 \, d\phi(u) = 0,$$

$$d\phi(u) = [A\delta(u) + B\delta(u-1) + C\delta(u-2)] \, du,$$

and
$$f_j = f_1(2 - 2^{j-1}) + f_2(2^{j-1} - 1) \quad \text{for all } j \geq 1.$$

Thus it happens again in this example that $f(z)$ is given exactly by one of its Padé approximants. Specifically, we find that

$$f(z) = \frac{A + B + C + (3A + 2B + C)z + 2Az^2}{1 + 3z + 2z^2}.$$

Notice that the possible positions of the poles of $f(z)$ are determined uniquely by the specification $\Lambda = 2$ for an integer moment problem.

Solutions of the integer moment problem expressed by (4.41) are known for $\Lambda \leq 4$. For any $\Lambda < 4$, the solution is a finite-order Padé approximant of $f(z)$, implying that $f(z)$ is rational. When $\Lambda = 4$, a new type of solution

$$f(z) = m(1 + 4z)^{-1/2}, \quad m \text{ integral},$$

becomes possible. It is also known that the general problem expressed by (4.41) can be reduced to the case of one with $\Lambda \leq 6$ [Barnsley et al., 1979].

We conclude this section by noting that the general question raised in Section 5.3 of the nature of a given sequence of moments is not always answered fully by categorizing them as Stieltjes moments or Hamburger moments. A more complete answer would include further specification of the support of the measure $d\phi(u)$.

5.5 Stieltjes series with zero radius of convergence

Let us start with an example of a function $f(z)$ which has a Maclaurin series with zero radius of convergence. Consider, for $a > 0$,

$$f(z) = {}_2F_0(a, 1, -z) = \frac{1}{\Gamma(a)} \int_0^\infty \frac{e^{-u} u^{a-1}}{1 + zu} du. \tag{5.1}$$

$f(z)$ is a Stieltjes function expressed in the standard form

$$f(z) = \int_0^\infty \frac{d\phi(u)}{1 + zu}, \tag{5.2}$$

where

$$\phi(u) = \frac{1}{\Gamma(a)} \int_0^u e^{-t} t^{a-1} dt. \tag{5.3}$$

We see that

$$f(z) = \sum_{j=0}^\infty f_j(-z)^j = 1 - az + a(a+1)z^2 + \cdots, \tag{5.4}$$

where

$$f_j = a(a+1) \cdots (a+j-1) = \int_0^\infty u^j \, d\phi(u). \tag{5.5}$$

The difficulty with the series (5.4) is that it converges for no values of z except $z = 0$, because the individual terms do not tend to zero. However, we know from Section 5.3 that the paradiagonal sequence of $[M + J/M]$ Padé approximants to the formal expansion (5.4) is convergent in the bounded domain $\mathcal{D}(\Delta)$ which does not include the origin. In this section, we describe how Padé approximants are useful for reconstructing functions from power series with zero radius of convergence. We restrict our attention to Stieltjes series, for which the convergence theorems can be established.

A basic precept of Padé approximation is that there is a function $f(z)$ which is determined by its Maclaurin expansion. In general, it is not true that an arbitrary function is determined by its Maclaurin expansion. This is demonstrated by the following example.

Example 1.

$$g(x) = \exp(-1/x), \quad 0 \leq x \leq \infty.$$

The function $g(x)$ of the real variable x is well defined, and so are all its (right-handed) derivatives for $x \geq 0$, and in this sense

$$\left. \frac{d^j g(x)}{dx^j} \right|_{x=0} = 0 \quad \text{for } j = 0, 1, 2, \ldots.$$

5.5 Stieltjes series with zero radius of convergence

Thus we see that any given function $f(x)$ with a Maclaurin expansion about $x = 0$ has the same expansion as $f(x) + g(x)$. Clearly, conditions must be imposed to define the classes of functions which are uniquely determined by their power-series expansions. If we assume that $f(z)$ is analytic at $z = 0$, which implies that the Maclaurin expansion of $f(z)$ has nonzero radius of convergence, then $f(z)$ is determined by analytic continuation, and may be uniquely defined in complex plane cut by the Mittag-Leffler star. Analyticity at the origin of $f(z)$ was a foundation of the development of the previous section, but it is an unnecessarily strong hypothesis for our present purposes. We must consider a weaker condition which will enable $f(z)$ to be determined by its Maclaurin expansion.

The second difficulty which we encounter is that the representation (5.2) is not necessarily unique even for Stieltjes series in the case where they have zero radius of convergence. This raises the question of determinacy mentioned in Section 5.3. The following examples establish this nonuniqueness by showing that a nontrivial measure $d\phi_0(u)$ exists, corresponding to a function $\phi_0(u)$ of bounded variation, for which

$$\int_0^\infty u^k \, d\phi_0(u) = 0, \quad k = 0, 1, 2, \ldots.$$

Example 2 (Rennison)

$$d\phi_0(u) = \sum_{n=0}^\infty \frac{(-)^n \pi^{2n+1}}{(2n+1)!} \delta(u - 2^{2n+1}) \, du.$$

This distribution corresponds to a piecewise continuous $\phi_0(u)$ with jumps of oscillating sign. $\phi_0(u)$ is of bounded variation, and

$$\int_0^\infty u^k \, d\phi_0(u) = \sin(2^k \pi) = 0 \quad \text{for } k = 0, 1, 2, \ldots.$$

Example 3 (Stieltjes).

$$d\phi_0(u) = u^{-\ln u} \sin(2\pi \ln u) \, du.$$

This distribution corresponds to a continuous $\phi_0(u)$ of bounded variation. The substitution

$$\ln u = t + \frac{n+1}{2}$$

may be used to show that

$$\int_0^\infty u^n \, d\phi_0(u) = \pm \exp\left[\left(\frac{n+1}{2}\right)^2\right] \int_{-\infty}^\infty e^{-t^2} \sin(2\pi t) \, dt = 0.$$

Thus we are led to impose certain extra conditions on the moments $\{f_j\}$ to ensure that the constructed function $f(x)$ and its generating measure $d\phi(u)$ are unique. These conditions are obviously weaker than the conditions which render $f(z)$ analytic at $z = 0$.

Our starting point is a series expansion

$$\sum_{j=0}^{\infty} c_j z^j = \sum_{j=0}^{\infty} f_j(-z)^j. \tag{5.6}$$

The usual definition of convergence of the power series (5.6) is that $\sum_{j=0}^{\infty} c_j z^j$ converges to $f(z_0)$ at $z = z_0$ if

$$\lim_{n \to \infty} \sum_{j=0}^{n} c_j z_0^j = f(z_0). \tag{5.7}$$

The existence of the limit implied by (5.7) in turn implies that the power series (5.6) is convergent for all z such that $|z| < |z_0|$, and a circle of convergence is established. This familiar definition is to be contrasted with the definition of asymptotic convergence needed in this section.

Definition. A power series $\sum_{j=0}^{\infty} c_j z^j$ is *asymptotically convergent* to $f(z)$ if

$$\lim_{z \to 0} \left| \left[f(z) - \sum_{j=0}^{n} c_j z^j \right] z^{-n} \right| = 0 \tag{5.8}$$

for $\arg(z) \in \mathcal{A}$ and for each $n = 0, 1, 2, \ldots$. $\mathcal{A}$ is an angular interval, such as $-\alpha < \arg(z) < \beta$, specifying a wedge domain of asymptotic convergence at $z = 0$, as shown in Figure 5.5.1 [Erdélyi, 1956, p. 22].

Asymptotic convergence is denoted by the special symbol $\simeq$. If (5.8) is satisfied, we write

$$f(z) \simeq \sum_{j=0}^{\infty} c_j z^j, \quad \alpha < \arg(z) < \beta.$$

Example 4. We quote the result that Euler's series

$$E(z) = {}_2F_0(1, 1, -z) \simeq 1 - (1!)z + (2!)z^2 - (3!)z^3 + \cdots \tag{5.9}$$

is an asymptotic expansion of the function given below in (5.11) and defined by this integral representation in $-\pi < \arg(z) < \pi$.

We next quote a powerful theorem of Carleman [1926], which enables us not only to prove the result of Example 4 but to establish the existence of a unique function $f(z)$ to be associated with certain power series with zero radius of convergence.

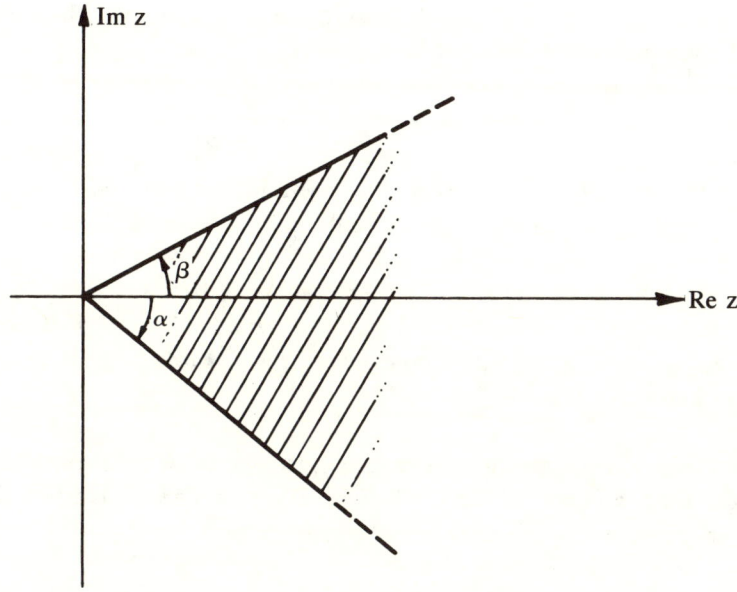

Figure 5.5.1. A wedge domain of asymptotic convergence.

Carleman's criterion. *Let $\{f_j\}$ satisfy the determinantal conditions (of Theorem 5.1.2) for Stieltjes series, and also the condition that*

$$\sum_{j=1}^{\infty} (f_j)^{-1/(2j)} \text{ diverges}. \tag{5.10}$$

Then there exists a unique Stieltjes function $f(z)$, analytic in $\operatorname{Re} z > 0$, such that the asymptotic equality

$$\sum_{j=0}^{\infty} f_j(-z)^j \simeq f(z)$$

holds in $|\arg(z)| < \pi$. (*In fact, this equality also holds in any disk $\mathscr{C}$ such as is shown in Figure* 5.6.2).

Returning to Example 4, which is (5.4) with $a = 1$, we see that Euler's series (5.9) is a Stieltjes series necessarily satisfying the determinantal conditions of Theorem 5.1.2. From Stirling's formula,

$$f_j = j! \simeq \left(\frac{j}{e}\right)^j \sqrt{2\pi j} \quad \text{as } j \to +\infty,$$

and

$$(f_j)^{-1/(2j)} \simeq \left(\frac{e}{j}\right)^{1/2} (2\pi j)^{-1/(4j)} \quad \text{as } j \to +\infty.$$

Since $(2\pi j)^{-1/(4j)} \to 1$, $\sum^\infty f_j^{-1/(2j)}$ diverges, and Carleman's theorem asserts the uniqueness of the Stieltjes function

$$E(z) = {}_2F_0(1, 1; -z) = \int_0^\infty \frac{e^{-u}}{1 + zu} du \qquad (5.11)$$

defined in $-\pi < \arg(z) < \pi$, with the asymptotic expansion (5.9). $E(z)$ has its branch cut along the negative real z-axis.

Theorem 5.5.1. *Let $f(z) = \sum_{j=0}^\infty f_j(-z)^j$ be a series satisfying Stieltjes determinantal conditions and Carleman's criterion. Then all paradiagonal sequences of $[M + J/M]$ Padé approximants with $J \geq -1$ converge to $f(z)$ in the domain $\mathcal{D}(\Delta)$.*

Proof. Carleman's criterion asserts the existence of at most one Stieltjes function $f(z)$, analytic in $\operatorname{Re} z > 0$, with right-handed derivatives at the origin specified by the asymptotic equality

$$f(z) \simeq \sum_{j=0}^\infty f_j(-z)^j.$$

If $J \geq 0$, we use the device of considering the first $J + 1$ terms of $f(z)$ explicitly (as in the corollary to Theorem 5.2.3), and so reduce the problem to that with $J = -1$. Theorem 5.3.1 asserts the existence of a limit function $f^{(-1)}(z)$ of the sequence of $[M - 1/M]$ Padé approximants, that $f^{(-1)}(z)$ is analytic in $\mathcal{D}(\Delta)$, and, from (3.13), that $f^{(-1)}(z) \simeq \sum_{j=0}^\infty f_j(-z)^j$. Thus $f(z) = f^{(-1)}(z)$ in $\operatorname{Re} z > 0$ and so throughout $\mathcal{D}(\Delta)$.

An interesting application of the techniques developed so far in this section occurs in the theory of Stirling's formula. In this context, we need Binet's second formula, which is

$$\ln \Gamma(z) = (z - \tfrac{1}{2}) \ln z - z + \tfrac{1}{2} \ln 2\pi + J(z), \qquad (5.12)$$

valid for $\operatorname{Re} z > 0$, where

$$J(z) = 2 \int_0^\infty \frac{\tan^{-1}(t/z) \, dt}{\exp(2\pi t) - 1} \qquad (5.13a)$$

$$= -\frac{1}{\pi} \int_0^\infty \ln(1 - e^{-2\pi t}) \frac{z \, dt}{z^2 + t^2}. \qquad (5.13b)$$

Proof [Ford, 1960, Chapter 1; Hardy, 1956, p. 339]. Define

$$R(n) = \sum_{j=1}^n \ln j - \tfrac{1}{2} \ln n - \int_1^n \ln x \, dx. \qquad (5.14)$$

5.5 Stieltjes series with zero radius of convergence

We derive a formula for $R(n)$ by considering the principal-value contour integral

$$\oint_{C_1+C_2} \frac{\cot \pi z}{2i} \ln z \, dz = \sum_{j=1}^{n} \ln j - \tfrac{1}{2}\ln n \tag{5.15}$$

over the contour of Figure 5.5.2, which has been evaluated using the residue theorem. Because $\ln z$ is a real symmetric function,

$$\int_1^n \ln x \, dx = \tfrac{1}{2}\int_{C_1} \ln z \, dz - \tfrac{1}{2}\int_{C_2} \ln z \, dz. \tag{5.16}$$

Substituting (5.15) and (5.16) into (5.14), we find that

$$R(n) = \int_{C_1} \psi_1 \ln z \, dz + \int_{C_2} \psi_2 \ln z \, dz,$$

where

$$\psi_1(z) = \frac{\cot \pi z}{2i} - \frac{1}{2} = \frac{1}{\exp(2\pi i z) - 1}$$

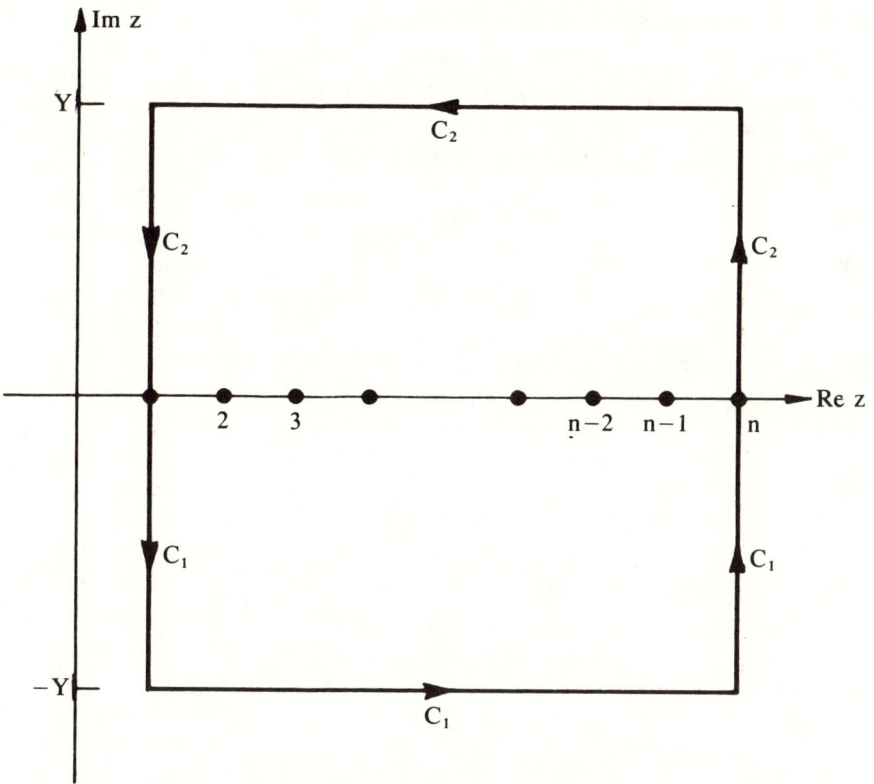

Figure 5.5.2. The complex z-plane showing the contours C_1 and C_2.

and
$$\psi_2(z) = \frac{\cot \pi z}{2i} + \frac{1}{2} = \frac{1}{1 - \exp(-2\pi i z)}.$$

By extending the contours to $Y = \infty$, we deduce that

$$R(n) = 2\int_0^\infty \frac{\tan^{-1}(t/n)}{\exp(2\pi t) - 1} dt - 2\int_0^\infty \frac{\tan^{-1} t}{\exp(2\pi t) - 1} dt. \qquad (5.17)$$

By direct integration of (5.14), we know that

$$R(n) = \ln(n!) - \tfrac{1}{2}\ln n - [x \ln x - x]_1^n. \qquad (5.18)$$

Using (5.17), (5.18), and the definition (5.13a), we deduce that

$$\ln \Gamma(n) = (n - \tfrac{1}{2})\ln n - n + J(n) + C,$$

where C is a constant, independent of n. The value of this constant C is given by Stirling's formula (see Titchmarsh [1939, p. 150]), since $J(+\infty) = 0$, and so (5.12) is established for all positive integers $z = 1, 2, 3, \ldots$. Carlson's uniqueness theorem (see Titchmarsh [1939, p. 186]) states that any function $f(z)$ which is analytic for Re $z > 0$, is bounded by $\exp(k|z|)$ with $k < \pi$ for Re $z \geq 0$, and which satisfies $f(z) = 0$ for $z = 0, 1, 2, \ldots$ is zero identically: $f(z) = 0$ for Re $z > 0$. Hence (5.12) is established.

We exhibit the connection with Stirling's formula by writing (5.12) as

$$\Gamma(z) = z^z e^{-z} \sqrt{2\pi z} \exp\{J(z)\} \quad \text{for Re } z > 0,$$

whereas Stirling's formula [Titchmarsh, 1939] is

$$\Gamma(z) = z^z e^{-z} \sqrt{2\pi z} \{1 + O(z^{-1})\} \quad \text{for } |\arg(z)| < \pi, |z| \to \infty.$$

Using (5.13a) and the expansion

$$\tan^{-1}\left(\frac{t}{z}\right) = \frac{t}{z} - \frac{1}{3}\left(\frac{t}{z}\right)^3 + \frac{1}{5}\left(\frac{t}{z}\right)^5 - \cdots$$
$$+ \frac{(-)^{n+1} t^{2n-1}}{(2n-1)z^{2n-1}} + \frac{(-)^n}{z^{n-1}} \int_0^t \frac{u^{2n} \, du}{u^2 + z^2},$$

we see that $J(z)$ has the asymptotic expansion

$$J(z) \simeq \sum_{j=0}^\infty (-)^j f_j z^{-(2j+1)}, \qquad (5.19)$$

where

$$f_j = \frac{-1}{\pi} \int_0^\infty u^{2j} \ln(1 - e^{-2\pi u}) \, du, \quad j = 0, 1, 2, \ldots. \qquad (5.20)$$

Notice also that $J(z)$ is analytic in $\operatorname{Re} z > 0$. This suggests a change of variable in (5.13) and (5.19). Define $y = z^{-2}$ and $K(y) = zJ(z)$. Then

$$K(y) = \sum_{j=0}^{\infty} f_j(-y)^j$$

$$= -\frac{1}{2\pi} \int_0^\infty \frac{\ln(1 - e^{-2\pi\sqrt{u}})}{\sqrt{u}} \frac{du}{1 + yu}. \tag{5.21}$$

Equation (5.21) shows that $K(y)$ is a Stieltjes series with zero radius of convergence. From (5.20),

$$f_j = \frac{(2j)!}{2^{2j+1}\pi^{2j+2}} \sum_{r=1}^{\infty} \frac{1}{r^{2j+2}}, \quad j = 0, 1, 2, \ldots. \tag{5.22}$$

It is easy to deduce that $\sum_{j=1}^{\infty} f_j^{-1/(2j)}$ diverges and that Carleman's criterion is satisfied. Hence $K(y)$ is uniquely determined by a convergent sequence of $[M - 1/M]$ Padé approximants and by its continued-fraction expansion in $|\arg(y)| < \pi$, corresponding to $\operatorname{Re} z > 0$. From Abramowitz and Stegun [1964, Chapter 23] and Wall [1948, p. 364], we see that the actual asymptotic expansion of $J(z)$ is given in terms of Bernoulli numbers by

$$J(z) \simeq \sum_{j=0}^{\infty} \frac{B_{2j+2}}{(2j+1)(2j+2)} z^{-(2j+1)}$$

$$= \frac{z^{-1}}{12} - \frac{z^{-3}}{360} + \frac{z^{-5}}{1260} - \frac{z^{-7}}{1680} + \frac{z^{-9}}{1188} - \cdots.$$

Our conclusion is that the continued-fraction expansion [Char, 1980]

$$J(z) = \frac{\frac{1}{12}}{z} + \frac{\frac{1}{30}}{z} + \frac{\frac{53}{210}}{z} + \frac{\frac{195}{371}}{z} + \cdots \tag{5.23}$$

converges in $\operatorname{Re} z > 0$. Convergence of this continued fraction is rapid until the eighth-order convergent is reached, when the rate of convergence becomes poor; this is understood in the context of the theory of inclusion regions (see Henrici and Pfluger [1966] or Baker [1975a] for further details). Numerical values for the diagonal sequence of Padé approximants for $\exp[J(z)]$ are given by Bender and Orszag [1978].

This example has led us to consider the connection between the S-fraction (4.5.7) and Stieltjes series defined in (1.1).

We have stated, in Theorem 5.5.1, sufficient conditions under which the sequences of $[M - 1/M]$ and $[M/M]$ Padé approximants of a Stieltjes series

$$f(z) = \int_0^\infty \frac{d\phi(u)}{1 + zu} = c_0 + c_1 z + c_2 z^2 + \cdots$$

converge to $f(z)$. These two sequences of approximations can also be expressed as the sequence of convergents of the continued fraction

$$f_c(z) = \frac{a_1}{1} + \frac{a_2 z}{1} + \frac{a_3 z}{1} + \cdots, \qquad (5.24)$$

where

$$a_1 = c_0 = f_0, \qquad (5.25)$$

$$a_2 = \frac{-c_1}{c_0} = \frac{f_1}{f_0},$$

$$a_3 = -\frac{C(1/2)}{c_0 c_1} = \frac{D(0,1)}{D(0,0)D(1,0)},$$

$$a_{2M} = -\frac{C(M/M)C(M-2/M-1)}{C(M-1/M-1)C(M-1/M)} = \frac{D(1, M-1)D(0, M-2)}{D(1, M-2)D(0, M-1)},$$

$$a_{2M+1} = -\frac{C(M/M+1)C(M-1/M-1)}{C(M-1/M)C(M/M)} = \frac{D(0, M)D(1, M-2)}{D(0, M-1)D(1, M-1)},$$

using the results of (4.3.25). Notice that $a_i > 0$ for $i = 1, 2, 3, \ldots$, so we have proved the following result.

Theorem 5.5.2. *If $f(z)$ is a Stieltjes function, the continued fraction (5.24) derived from the corresponding Stieltjes series is an S-fraction.*

The converse result follows from a theorem derived in Perron's book [1957, p. 208]. We quote it in the form that if $zf_c(z)$, derived from (5.24), is a convergent S-fraction, then its convergents form a sequence which converges to a function $f(z)$ which is a Stieltjes series, defined by (1.1), in which $\phi(u)$ is essentially unique. We do not prove this result, but note that (5.25) shows that the property that each $a_i > 0$ is sufficient to ensure that the determinants $D(m, n)$ are positive for all $m, n \geq 0$. Also we note that the hypothesis that the S-fraction converges is essential to establish the uniqueness property of the corresponding $\phi(u)$.

To conclude this section, we mention that there are various kinds of conditions on a real function $S(\omega)$, defined on $0 \leq \omega \leq \infty$, which are necessary and sufficient conditions to ensure that $S(\omega)$ has a Stieltjes representation of the form (4.5.8). As an example, we quote one result:

Theorem 5.5.3. *Necessary and sufficient conditions for $S(\omega)$ to have the integral representation*

$$S(\omega) = \int_0^\infty \frac{d\phi(t)}{\omega + t} \qquad (5.26)$$

with $\phi(t)$ nondecreasing and bounded are that $S(\omega) \geq 0$ and

$$(-1)^{k-1}\left(\frac{d}{d\omega}\right)^{2k-1}\{\omega^k S(\omega)\} \geq 0, \quad k = 1, 2, \ldots, \quad \text{on } 0 < \omega < \infty,$$

and that a finite limit of $\omega S(\omega)$ exists as $\omega \to +\infty$.

The proof of this theorem is given in Widder [1972, p. 364], and we stress that this is just one of several similar results which characterize Stieltjes functions [Widder, 1972, Chapter 8]. We omit any details because neither the statements of the theorems nor the proofs involve Padé approximation.

From the basic representation (1.1) of a Stieltjes function $f(z)$, it follows that Stieltjes functions form a subclass of the class of completely monotonic functions defined on $[0, \infty)$. As such, they are directly related to a subclass of the class of absolutely continuous functions [Widder, 1972, Chapter 4].

Example 1. $f(z)$ is a Stieltjes series, and $g(z)$ is defined by

$$f(z) = \frac{f(0)}{1 + zg(z)}.$$

Hadamard's formula, given in Theorem 2.4.1, establishes that the coefficients g_j defined by

$$g(z) = \sum_{j=0}^{\infty} g_j(-z)^j$$

satisfy the inequalities

$$D_g(m, n) \equiv \begin{vmatrix} g_m & g_{m+1} & \cdots & g_{m+n} \\ g_{m+1} & g_{m+2} & \cdots & g_{m+n+1} \\ \vdots & \vdots & & \vdots \\ g_{m+n} & g_{m+n+1} & \cdots & g_{m+2n} \end{vmatrix} > 0$$

for all $m, n \geq 0$.

5.6 Hamburger series and the Hamburger moment problem

A Hamburger function is defined to be a function with an integral representation

$$f(z) = \int_{-\infty}^{\infty} \frac{d\phi(u)}{1 + uz} \tag{6.1}$$

where the moments

$$f_j = \int_{-\infty}^{\infty} u^j \, d\phi(u), \quad j = 0, 1, 2, \ldots, \tag{6.2}$$

are finite and $\phi(u)$ is non-decreasing [Hamburger, 1920, 1921].

A Hamburger series is defined to be a series

$$\sum_{j=0}^{\infty} c_j z^j = \sum_{j=0}^{\infty} f_j(-z)^j$$

with moments f_j defined by (6.2). This is the series derived by a formal expansion of (6.1). Just as before, with Stieltjes series, we exclude the case where $\phi(u)$ is piecewise constant with a finite number of jump discontinuities and consequently $f(z)$ is a rational function. The characteristic feature of Hamburger series is the full range $(-\infty, \infty)$ of integration. The inverse problem, which is the determination of $f(z)$ from the moments, is called the Hamburger moment problem. Hamburger series, functions, moments, etc. are sometimes called extended Stieltjes series, functions, moments, etc.

The conditions for Hamburger series are weaker than those for Stieltjes series, and so the Hamburger moments satisfy fewer conditions than the Stieltjes moments. As before in (1.12), we define

$$D(m, n) = \begin{vmatrix} f_m & f_{m+1} & \cdots & f_{m+n} \\ f_{m+1} & f_{m+2} & \cdots & f_{m+n+1} \\ \vdots & \vdots & & \vdots \\ f_{m+n} & f_{m+n+1} & \cdots & f_{m+2n} \end{vmatrix}. \tag{6.3}$$

Theorem 5.6.1. *If f_j, $j = 0, 1, \ldots,$ are moments of a Hamburger series satisfying (6.2), the determinants $D(2m, n) > 0$ for all $m, n > 0$.*

Proof. The inequality

$$\int_{-\infty}^{\infty} u^{2m}(x_0 + x_1 u + \cdots + x_n u^n)^2 \, d\phi(u) > 0$$

is used, following precisely the method of Theorem 5.1.2.

Notice that unless $D(2m + 1, n) > 0$ for all $m, n \geq 0$, the series is not a Stieltjes series. The next theorem shows that Hamburger series have Padé approximants with poles on the real axis (not at the origin), positive residues on the negative real axis, and negative residues on the positive real axis.

Theorem 5.6.2. *If the coefficients $\{f_j\}$ satisfy the inequalities $D(0, n) > 0$ for all $n > 0$, then $[M - 1/M]$ Padé approximants of the formal power*

5.6 Hamburger series and the Hamburger moment problem

series $\sum_{j=0}^{\infty} f_j(-z)^j$ may be written as

$$[M - 1/M] = \sum_{i=1}^{M} \frac{\beta_i}{1 + \gamma_i z} \qquad (6.4)$$

with $\beta_i > 0$, γ_i real, for $i = 1, 2, \ldots, M$.

Proof. An inductive proof based on $D(0, n) > 0$ and Sylvester's identity establishes that $D(2m, n) > 0$ for all m, $n > 0$. Then the method of Theorem 5.2.1 is used, but with the change of variable $w = -z^{-1}$. The Padé denominator is

$$Q^{[M-1/M]}(z) = \begin{vmatrix} f_0 + zf_1 & f_1 + zf_2 & \cdots & f_{M-1} + zf_M \\ f_1 + zf_2 & f_2 + zf_3 & \cdots & f_M + zf_{M+1} \\ \vdots & \vdots & & \vdots \\ f_{M-1} + zf_M & f_M + zf_{M+1} & \cdots & f_{2M-2} + zf_{2M-1} \end{vmatrix}$$

$$= w^{-M} \begin{vmatrix} wf_0 - f_1 & wf_1 - f_2 & \cdots & wf_{M-1} - f_M \\ wf_1 - f_2 & wf_2 - f_3 & \cdots & wf_M - f_{M+1} \\ \vdots & \vdots & & \vdots \\ wf_{M-1} - f_M & wf_M - f_{M+1} & \cdots & wf_{2M-2} - f_{2M-1} \end{vmatrix}.$$

$$(6.5)$$

Following the ideas of the second method of proof of Theorem 5.4.1, we consider the polynomial

$$\pi_M(w) = w^M Q^{[M-1/M]}(-1/w).$$

Its leading coefficient is $D(0, M - 1)$, which is positive, and so each $\pi_M(w)$ is positive for w sufficiently large and positive. Sylvester's identity implies that

$$\pi_{M-1}(w)\pi_{M+1}(w) \leq 0 \quad \text{if } \pi_M(w) = 0.$$

Since $\pi_0(w) = 1$ and $\pi_1(w) = wf_0 - f_1$, the interlacing property of the zeros of $\pi_M(w)$ follows by induction. Note the possibility of a zero at $w = 0$. This situation corresponds to $Q^{[M-1/M]}(z)$ having true degree $M - 1$ instead of M, a situation not prohibited by the Hamburger conditions. The signs of the residues following using the method of Theorem 5.2.1, and the theorem is proved.

The motive for using $w = (-z)^{-1}$ in this proof is to borrow from the theory of orthogonal polynomials. Using the methods of Section 5.3, an identical proof shows that the polynomials $\pi_M(w)$ are orthogonal over a positive measure. The methods show that the zeros of $\pi_M(w)$ occur in a real interval including the origin, and that zeros of successive polynomials interlace.

Following the development of Section 5.2, we omit any analysis designed to prove convergence of a sequence of Padé approximants for $x > 0$, which is irrelevant, and are led to Theorem 5.2.3 concerning uniform boundedness in $\mathscr{D}(\Delta)$ of paradiagonal sequences. The analogue is

Theorem 5.6.3. *The sequence of $[M - 1/M]$ Padé approximants to a Hamburger series is uniformly bounded as $M \to \infty$ in the domain $\mathscr{D}^I(\Delta)$, a bounded, disconnected, two-component domain of the z-plane which is at least a distance Δ from the real axis.*

Remark. The two components of $\mathscr{D}^I(\Delta)$ are shown in Figure 5.6.1.

Proof. The method is identical to that of Theorem 5.2.3.

The analogue of Theorem 5.2.4 is

Theorem 5.6.4. *The sequence of $[M - 1/M]$ Padé approximants to a Hamburger series is equicontinuous in $\mathscr{D}^I(\Delta)$.*

Proof. The method is identical to that of Theorem 5.2.4.

Arzela's theorem is now applicable to the $[M - 1/M]$ sequence, and we prove, using Theorems 5.6.3 and 5.6.4, the following result:

Theorem 5.6.5. *At least an infinite subsequence of $[M - 1/M]$ Padé approximants to a Hamburger series is uniformly convergent in $\mathscr{D}^I(\Delta)$ to a function $\tilde{f}(z)$ analytic in that region.*

Theorem 5.6.5 leaves several questions unanswered. However, if the coefficients $\{f_j\}$ are such that $f(z)$ is analytic in $|z| < R$, the analysis is quite straightforward.

Theorem 5.6.6. *If the moments $\{f_j\}$ are such that $D(0, n) > 0$ for all $n > 0$ and $\sum_{j=0}^{\infty} f_j(-z)^j$ is convergent in $|z| < R$, then*

(i) *a positive measure $d\phi(u)$ exists such that*

$$f_j = \int_{-R^{-1}}^{R^{-1}} u^j \, d\phi(u), \quad j = 0, 1, 2, \ldots,$$

(ii) *$d\phi(u)$ is unique,*

(iii)
$$\tilde{f}(z) = f(z) = \int_{-R^{-1}}^{R^{-1}} \frac{d\phi(u)}{1 + zu}.$$

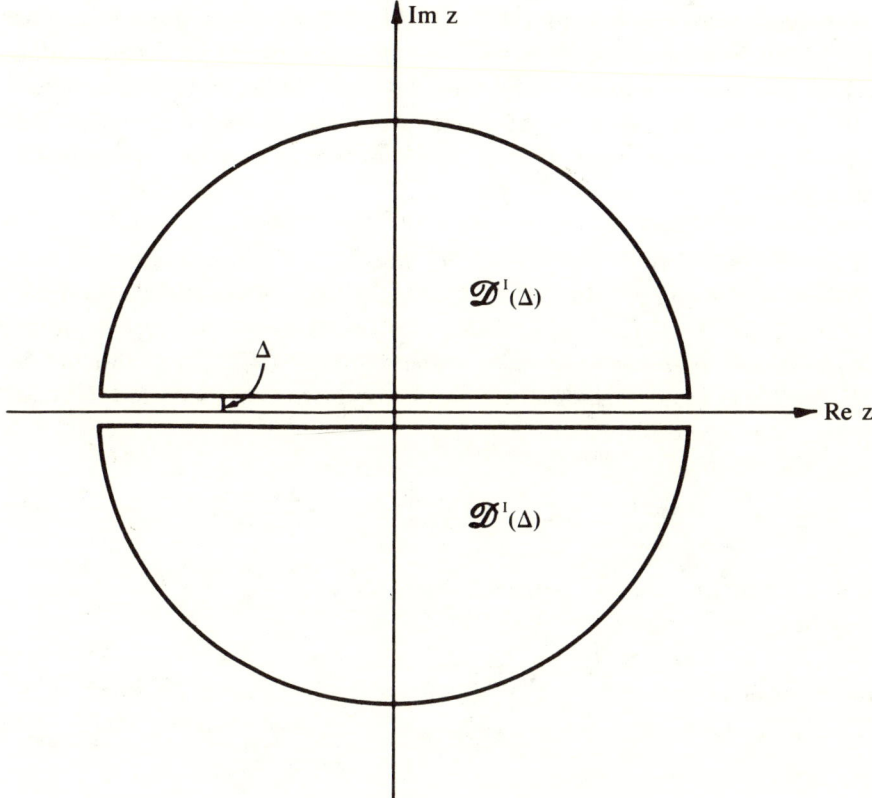

Figure 5.6.1. The domain $\mathscr{D}^I(\Delta)$.

Proof. The construction of Theorem 5.3.1 allows the representation

$$\tilde{f}(z) = \int_{-\infty}^{\infty} \frac{d\phi(u)}{1 + zu} \qquad (6.6)$$

since the determinantal conditions $D(0, n) > 0$ are satisfied. The methods of Theorem 5.4.1 imply that the poles of $[M - 1/M]$ Padé approximants of $\sum_{j=0}^{\infty} f_j(-z)^j$ are located on the cut, and method 2 of Theorem 5.4.2 implies that $f(z)$ and $\tilde{f}(z)$ have an identical Maclaurin expansion. By hypothesis, this is convergent in $|z| < R$, and so $f(z) = \tilde{f}(z)$.

By a construction similar to (3.12), we establish that $\phi(u)$ is essentially unique, and the results (i), (ii), and (iii) of the theorem are proved.

Next we consider the case where $\sum_{j=0}^{\infty} f_j(-z)^j$ is not convergent in $|z| < R$ for any strictly positive value of R, and we seek conditions under which the result of Theorem 5.6.5 may be strengthened. This theorem only guarantees the existence of a function $\tilde{f}(z)$ which is the limit of a convergent subsequence of the $[M-1/M]$ Padé approximants of the given Hamburger series. Theorem 5.6.5 does not imply that there are no other convergent subsequences with different limits, nor does it assert that $\tilde{f}(z)$ has any expansion at all, let alone the given power series. We fill this gap by showing in the next theorem that the given power series is an asymptotic expansion of $\tilde{f}(z)$.

Theorem 5.6.7. *Let the coefficients $\{f_j\}$ satisfy the determinantal conditions $D(0, n) > 0$ for all $n = 0, 1, 2, \ldots$. Let $\tilde{f}(z)$ be the limit of a convergent subsequence of $[M-1/M]$ Padé approximants of the formal series $\sum_{j=0}^{\infty} f_j(-z)^j$. Then $\tilde{f}(z)$ is a real symmetric function, $\tilde{f}(z)$ is analytic in the two half planes defined by $\operatorname{Im} z > 0$ and $\operatorname{Im} z < 0$, and $\tilde{f}(z)$ has the asymptotic expansion*

$$\tilde{f}(z) \simeq \sum_{j=0}^{\infty} f_j(-z)^j \quad \text{as } z \to 0, \, z \in \mathscr{C}. \tag{6.7}$$

$\mathscr{C}$ *is any disk of radius r, center $z = ir$, as shown in Figure 5.6.2, or else $\mathscr{C}$ may be centered at $z = -ir$.*

Remark. The existence of at least one convergent subsequence of $[M-1/M]$ Padé approximants is guaranteed by Theorem 5.6.5.

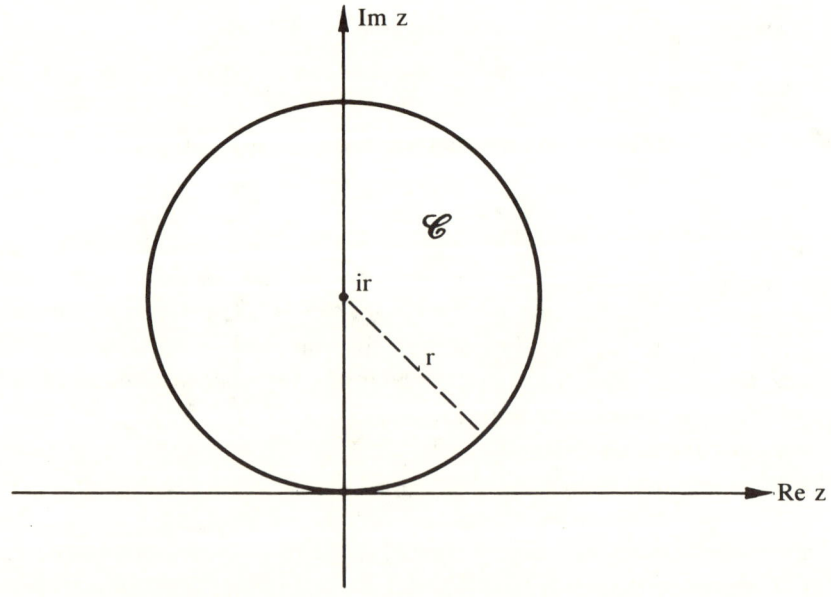

Figure 5.6.2. The disc $\mathscr{C}$ of radius r and centered at $z = ir$.

5.6 Hamburger series and the Hamburger moment problem

Proof. We first prove that, for any positive integer k,

$$z^{-2k}\left\{[M - 1/M] - \sum_{j=0}^{2k} f_j(-z)^j\right\} \to 0 \quad \text{as } z \to 0 \tag{6.8}$$

uniformly for all $M > k$ and $z \in \mathscr{C}$.

Each Padé approximant has the representation (6.4),

$$[M - 1/M] = \sum_{i=1}^{M} \frac{\beta_i}{1 + \gamma_i z} = \sum_{i=1}^{M} \sum_{j=0}^{\infty} \beta_i (-\gamma_i z)^j, \tag{6.9}$$

where the expansion is convergent in $|z| < R_M$, with

$$R_M = \min_{1 \leq i \leq M} |\gamma_i|^{-1}.$$

This expansion (6.9) agrees with the given series up to order z^{2M-1} inclusive, so

$$[M - 1/M] - \sum_{j=0}^{2k} f_j(-z)^j = \sum_{i=1}^{M} \beta_i \sum_{j=2k+1}^{\infty} (-z\gamma_i)^j$$

for $k < M$ and $|z| < R_M$. Therefore (6.8) holds for the particular case of $k = M - 1$, for $k < M - 1$, $|z| < R_M$, and $z \in \mathscr{C}$,

$$\left|[M - 1/M] - \sum_{j=0}^{2k} f_j(-z)^j\right|$$

$$= \left|\sum_{i=1}^{M} \beta_i \sum_{j=2k+1}^{\infty} (-z\gamma_i)^j\right|$$

$$\leq \left|\sum_{i=1}^{M} \beta_i(-z\gamma_i)^{2k+1}\right| + \left|\sum_{i=1}^{M} \sum_{j=2k+2}^{\infty} \beta_i(-z\gamma_i)^j\right|$$

$$\leq |z|^{2k+1}|f_{2k+1}| + \sum_{i=1}^{M} \beta_i \left|\sum_{j=2k+2}^{\infty} (-z\gamma_i)^j\right|$$

$$< |z|^{2k+1}|f_{2k+1}| + |z|^{2k+2}\left(\sum_{i=1}^{M} \beta_i \gamma_i^{2k+2}\right) \max_{1 \leq i \leq m} \left|\frac{1}{1 + z\gamma_i}\right|. \tag{6.10}$$

If $z \in \mathscr{C}$, then $\operatorname{Im} z \geq |\operatorname{Re} z|^2/(2r)$. Using an analysis similar to that of Theorem 5.2.3, we find that

$$\sup_{\gamma} \left|\frac{z}{1 + \gamma z}\right| = \frac{|z|^2}{|\operatorname{Im} z|} < 4r \quad \text{for } z \in \mathscr{C}.$$

Hence we have shown that $|z(1 + \gamma_i z)^{-1}|$ is uniformly bounded for all $z \in \mathscr{C}$ and independently of M.

By hypothesis, $k < M - 1$, so

$$\sum_{i=1}^{M} \beta_i \gamma_i^{2k+2} = f_{2k+2}.$$

Thus we deduce that, for $z \in \mathcal{C}$ and all $M > k$,

$$z^{-2k}\left|[M - 1/M] - \sum_{j=0}^{2k} f_j(-z)^j\right| \to 0 \quad \text{as } z \to 0,$$

uniformly in M. Taking the limit as $M \to \infty$, though values for which $[M - 1/M]$ converges to $\tilde{f}(z)$ in the domain $\mathcal{D}'(\Delta)$, we discover that $\tilde{f}(z)$ has the property that

$$z^{-2k}\left|\tilde{f}(z) - \sum_{j=0}^{2k} f_j(-z)^j\right| \to 0 \quad \text{as } z \to 0 \qquad (6.11)$$

for $z \in \mathcal{C}$. Hence

$$z^{-k}\left|\tilde{f}(z) - \sum_{j=0}^{k} f_j(-z)^j\right| \to 0 \quad \text{as } z \to 0$$

for $z \in \mathcal{C}$ and each positive integer k. This establishes the asymptotic expansion

$$\tilde{f}(z) \simeq \sum_{j=0}^{\infty} f_j(-z)^j, \quad z \to 0, \quad z \in \mathcal{C}.$$

$\tilde{f}(z)$ is, by Theorem 5.6.5, the uniform limit of a sequence of real symmetric Padé approximants each analytic in $\operatorname{Im} z \neq 0$. Therefore $\tilde{f}(z)$ is real symmetric, and by Weierstrass's theorem (quoted in Section 5.2), $\tilde{f}(z)$ is analytic in $\operatorname{Im} z \neq 0$.

In order to obtain an integral representation of $\tilde{f}(z)$, it is more convenient to use the variable $w = -z^{-1}$. We define

$$\tilde{F}(w) = \frac{1}{w}\tilde{f}\left(\frac{-1}{w}\right). \qquad (6.12)$$

The sequence of approximants (6.9),

$$[M - 1/M](z) = \sum_{i=1}^{M} \frac{\beta_i}{1 + \gamma_i z}, \quad M = 1, 2, 3, \ldots,$$

defined by the formal series $\sum_{j=0}^{\infty} f_j(-z)^j$, now becomes

$$J_M(w) = \frac{1}{w}[M - 1/M]\left(\frac{-1}{w}\right) = \sum_{i=1}^{M} \frac{\beta_i}{w - \gamma_i}, \quad M = 1, 2, 3, \ldots. \qquad (6.13)$$

Theorem 5.6.5 now implies that a subsequence of (6.13) converges in a domain in the w-plane corresponding to $z \in \mathcal{D}^I(\Delta)$. We can prove

Theorem 5.6.8. *Let the coefficients $\{f_j\}$ satisfy the determinantal conditions $D(0, n) > 0$ for all $n = 0, 1, 2, \ldots$. Let $\mathcal{D}^w(\delta)$ be a two-component domain in the w-plane,*

$$\mathcal{D}^w(\delta) = \{w: |\operatorname{Im} w| > \delta\}.$$

Then, for $w \in \mathcal{D}^w(\delta)$, there exists a function $\tilde{F}(w)$ which is the limit of a convergent subsequence of $\{\tilde{J}_M(w), M = 0, 1, 2, \ldots\}$, where $\tilde{J}_M(w)$ are Padé approximants associated with the formal series $\sum_{j=0}^{\infty} f_j w^{-j-1}$. $\tilde{F}(w)$ is a real symmetric function, $\tilde{F}(w)$ is analytic in $\operatorname{Im} w \neq 0$, and it has the asymptotic expansion

$$\tilde{F}(w) \simeq \sum_{j=0}^{\infty} f_j w^{-j-1}, \quad w \to \infty, \quad |\operatorname{Im} w| > \delta, \tag{6.14}$$

where ε is arbitrarily small but positive. Also,

$$\{\operatorname{Im} \tilde{F}(w)\} \operatorname{Im} w < 0 \quad \text{for } \operatorname{Im} w \neq 0. \tag{6.15}$$

Proof. The condition $\operatorname{Im} w > \delta$ corresponds to the disk $\mathscr{C}$ of Figure 5.6.2 of radius $(2\delta)^{-1}$. Thus we see that Theorem 5.6.8 is the expression of Theorem 5.6.7 in terms of the variable $z = -w^{-1}$.

We use $\mathcal{M}$ to denote the values of M for which the subsequence of $J_M(w)$ converges in some specified domain $\mathcal{D}^w(\delta)$.

Theorem 5.6.9. *The approximants $\tilde{J}_M(w)$, defined by (6.13), are uniformly bounded by the inequality*

$$\left| \tilde{J}_M(w) - \frac{f_0}{w} \right| < \frac{\sqrt{f_0 f_2}}{w|\operatorname{Im} w|}. \tag{6.16}$$

The limit function $\tilde{F}(w)$, defined by Theorem 5.6.8, is uniformly bounded by

$$\left| \tilde{F}(w) - \frac{f_0}{w} \right| \leq \frac{\sqrt{f_0 f_2}}{w|\operatorname{Im} w|}. \tag{6.17}$$

Proof. Using (6.13), we find that

$$\left| \tilde{J}_M(w) - \frac{f_0}{w} \right| = \left| \sum_{i=1}^{M} \frac{\beta_i \gamma_i}{w(w - \gamma_i)} \right|$$

$$< \frac{1}{w|\operatorname{Im} w|} \sum_{i=1}^{M} \beta_i |\gamma_i|.$$

Equation (6.16) follows from the Cauchy-Schwarz inequality

$$\left(\sum_{i=1}^{M} a_i b_i\right)^2 \leq \left(\sum_{i=1}^{M} a_i^2\right)\left(\sum_{i=1}^{M} b_i^2\right),$$

with $a_i = \sqrt{\beta_i}$, $b_i = \sqrt{\beta_i}|\gamma_i|$. Since (6.16) is a uniform bound, (6.17) follows by letting $M \to \infty$, $M \in \mathcal{M}$.

With Theorems 5.6.8 and 5.6.9 established, we proceed to our goal of constructing a representation of $\tilde{F}(w)$ as a Stieltjes integral. We need some preliminary results first.

Lemma 1. *For any $\delta > 0$,*

$$\int_{-\infty}^{\infty} \operatorname{Im} \tilde{F}(u + i\delta)\, du = -\pi f_0. \tag{6.18}$$

Proof. Because $\tilde{F}(w)$ is analytic in the upper half plane, we are led to consider the contour integral

$$\int_C \tilde{F}(w)\, dw = 0$$

where C is the contour $ABCD$ shown in Figure 5.6.3. Point A is the point $w = d^2 + i\delta$, and we define

$$w_A = d^2 + i\delta, \qquad w_B = d^2 + id,$$
$$w_C = -d^2 + id, \qquad w_D = -d^2 + i\delta.$$

The arc BC is the arc of a circle, centered at $w = 0$ and of radius $d\sqrt{d^2 + 1}$. Using (6.17), we estimate the integral over AB:

$$\left|\int_A^B \tilde{F}(w)\, dw\right| \leq f_0 \int_\delta^d \frac{dy}{|d^2 + iy|} + \sqrt{f_0 f_2} \int_\delta^d \frac{dy}{y|d^2 + iy|}$$

$$\leq \frac{f_0}{d} + \frac{\sqrt{f_0 f_2}}{d^2} \ln\left(\frac{\delta}{d}\right).$$

Likewise, we estimate the integral over the arc BC:

$$\left|\int_B^C \left\{\tilde{F}(w) - \frac{f_0}{w}\right\} dw\right| \leq \sqrt{f_0 f_2} \int_B^C \frac{|dw|}{|w|d} \leq \frac{\pi \sqrt{f_0 f_2}}{d}.$$

We evaluate the integral

$$\int_B^C \frac{f_0}{w}\, dw = if_0\left\{\pi - 2\tan^{-1}\left(\frac{1}{d}\right)\right\}.$$

5.6 Hamburger series and the Hamburger moment problem

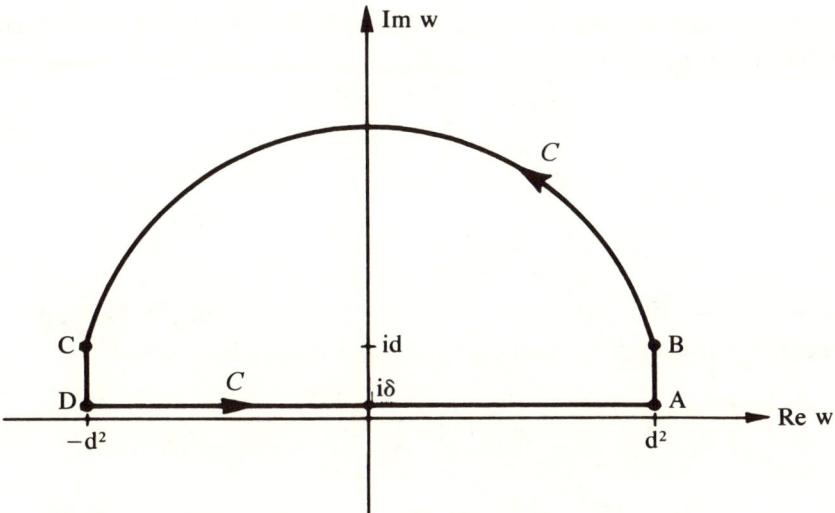

Figure 5.6.3. The contour C in the w-plane.

Using these results, it is straightforward to deduce that

$$\lim_{d \to \infty} \int_D^A \tilde{F}(w)\, dw = -i\pi f_0.$$

Equation (6.18) is the imaginary part of this formula.

As a consequence of Lemma 1, we are led to define

$$\phi(\delta, u) = -\frac{1}{\pi}\int_{-\infty}^u \tilde{F}(t + i\delta)\, dt. \qquad (6.19)$$

Theorem 5.6.8 and Lemma 1 imply that

$$\frac{d\phi(\delta, u)}{du} = -\frac{1}{\pi}\tilde{F}(u + i\delta) > 0 \quad \text{for } \delta > 0,$$

and that

$$\int_{-\infty}^{\infty} \phi(\delta, u) = f_0.$$

Consequently, $d\phi(\delta, u)$ is a bounded Hamburger measure.

Lemma 2. *For* $\operatorname{Im} w > \delta > 0$,

$$\tilde{F}(w) = \frac{1}{\pi}\int_{-\infty}^{\infty} \frac{\operatorname{Im}\tilde{F}(u + i\delta)\, du}{u + i\delta - w} = \int_{-\infty}^{\infty} \frac{d\phi(\delta, u)}{w - u - i\delta}. \qquad (6.20)$$

Proof. We use the contour C of Figure 5.6.3, and let w be an interior point of C. Then

$$\tilde{F}(w) = \frac{1}{2\pi i} \int_C \frac{\tilde{F}(w')\, dw'}{w' - w}.$$

Using the previous analysis, we let $d \to \infty$ and deduce that

$$\tilde{F}(w) = \frac{1}{2\pi i} \int_{-\infty}^{\infty} \frac{\tilde{F}(u + i\delta)\, du}{u + i\delta - w}. \tag{6.21}$$

If w lies within C, the point $w' = w^* + 2i\delta$ lies outside C; Cauchy's theorem shows that

$$0 = \frac{1}{2\pi i} \int_{-\infty}^{\infty} \frac{\tilde{F}(u + i\delta)\, du}{u + i\delta - (w^* + 2i\delta)}.$$

Taking the complex conjugate,

$$0 = \frac{1}{2\pi i} \int_{-\infty}^{\infty} \frac{\{\tilde{F}(u + i\delta)\}^*\, du}{u + i\delta - w}. \tag{6.22}$$

Subtracting (6.22) from (6.21) yields

$$\tilde{F}(w) = \frac{1}{\pi} \int_{-\infty}^{\infty} \frac{\operatorname{Im} \tilde{F}(u + i\delta)\, du}{u + i\delta - w}.$$

Equation (6.20) is established by the definition (6.19).

Lemma 3. *Let $\phi(\delta, u)$ be a family of bounded nondecreasing functions of u defined on $-\infty < u < \infty$ for $\delta > 0$. Then there exists a sequence $\{\delta_i, i = 1, 2, 3, \ldots\}$ such that $\delta_i \to 0$ and $\phi(\delta_i, u) \to \phi(u)$ as $i \to \infty$, where $\phi(u)$ is bounded and nondecreasing on $-\infty < u < \infty$.*

Remark. $\phi(\delta, u)$, defined by (6.19), satisfies the conditions of Lemma 3.

Proof. Take a set of points $U = \{u_k\}$ which is dense on $(-\infty, \infty)$. Let D_0 be the sequence $\{1/i,\ i = 1, 2, 3, \ldots\}$. The sequence of values $\{\phi(\delta, u_1), \delta \in D_0\}$ is a bounded sequence, and we may choose D_1 to be a subsequence of D_0 for which

$$\{\phi(\delta, u_1),\ \delta \in D_1\} \text{ is convergent.}$$

Let $\lim_{\delta \to 0,\, \delta \in D_1} \phi(\delta, u_1) = \phi(u_1)$.

Likewise, we choose a subsequence D_2 of D_1 for which

$$\{\phi(\delta, u_2),\ \delta \in D_2\} \to \phi(u_2).$$

Similarly, we choose D_k so that $D_k \subset D_{k-1} \subset \cdots \subset D_1$ and for which

$$\{\phi(\delta, u_k),\ \delta \in D_k\} \to \phi(u_k).$$

Thus we define iteratively a bounded nondecreasing function $\phi(u)$ on a dense set of points in $-\infty < u < \infty$.

Although $\phi(\delta, u)$ is continuous, the limit function $\phi(u)$ need not be continuous. If $\phi(u)$ is discontinuous at $u = v$, v is called a point of discontinuity of ϕ. Because $\phi(u)$ is increasing, we may define

$$\phi(v_-) = \lim_{u \to v} \phi(u) \quad \text{for } u \in U, u < v,$$

$$\phi(v_+) = \lim_{u \to v} \phi(u) \quad \text{for } u \in U, u > v,$$

and

$$\Delta\phi(v) = \phi(v_+) - \phi(v_-).$$

There are only, at most, a countable set of points of discontinuity of ϕ: not more than two points with $\Delta\phi \geq f_0/2$, not more than four points with $f_0/4 \leq \Delta\phi < f_0/2$, not more than eight points with $f_0/8 \leq \Delta\phi < f_0/4$, etc. The points of discontinuity may be dense on the entire interval; nevertheless, Lemma 3 still defines $\phi(u)$ at these points. At all points in $-\infty < u < \infty$ other than its points of discontinuity, $\phi(u)$ is continuous; such points are naturally called points of continuity of $\phi(u)$. We have now set up the framework for a substantial theorem:

Theorem 5.6.10. *Let the coefficients $\{f_j\}$ satisfy the determinantal conditions $D(0, n) > 0$ for all $n = 0, 1, 2, \ldots$. Let $\widetilde{F}(w)$ be the limit of a convergent subsequence of $\{\widetilde{J}_M(w), M = 0, 1, 2, \ldots\}$ of Padé approximants defined for the formal series*

$$\sum_{j=0}^{\infty} f_j w^{-j-1}.$$

Then $\widetilde{F}(w)$ has the Hamburger integral representation

$$\widetilde{F}(w) = \int_{-\infty}^{\infty} \frac{d\phi(u)}{w - u}, \tag{6.23}$$

where $\phi(u)$ is bounded and nondecreasing on $-\infty < u < \infty$.

Proof. The fact that the definition (6.19) fits the conditions of Lemma 3 allows $\phi(u)$ to be constructed. We obtain (6.23) from (6.20) in the limit $\delta \to 0$. Because $\widetilde{F}(w)$ is real symmetric, it is valid for all Im $w \neq 0$.

We take the opportunity to quote the Stieltjes inversion formula, a simple consequence of Lemma 2 and Lemma 3, which expresses $\phi(u)$ in terms of $\tilde{F}(w)$ for Hamburger functions. The formula is

$$\frac{\phi(t_+) + \phi(t_-)}{2} = \frac{1}{\pi} \lim_{\delta \to 0} \int_{-\infty}^{t} \operatorname{Im} \tilde{F}(u + i\delta) \, du.$$

With the integral representation (6.23) established, it is convenient to revert to our standard variable $z = -1/w$. Under the conditions of Theorem 5.6.7, we have proved that

$$\lim_{\substack{M \to \infty \\ M \in \mathcal{M}}} [M - 1/M] = \tilde{f}(z) = \int_{-\infty}^{\infty} \frac{d\phi(u)}{1 + zu} \simeq \sum_{j=0}^{\infty} f_j(-z)^j. \quad (6.24)$$

Next, we prove the standard integral formula for each coefficient f_j, which follows from Hamburger's theorem:

Theorem 5.6.11. *Let $f(z)$ be defined by the Hamburger integral representation*

$$f(z) = \int_{-\infty}^{\infty} \frac{d\phi(u)}{1 + zu}. \quad (6.25)$$

Then a necessary and sufficient condition for the moments

$$f_j = \int_{-\infty}^{\infty} u^j \, d\phi(u) \quad (6.26)$$

to be finite is that $f(z)$ has the asymptotic expansion

$$f(z) \simeq \sum_{j=0}^{\infty} f_j(-z)^j \quad (6.27)$$

defined in the sectors $\varepsilon \leq |\arg(z)| \leq \pi - \varepsilon$ for arbitrarily small $\varepsilon > 0$.

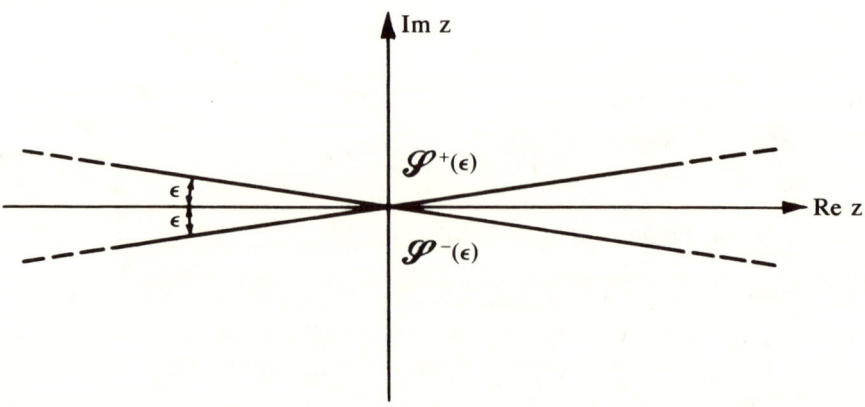

Figure 5.6.4. The sectors $\mathscr{S}^+(\varepsilon)$ and $\mathscr{S}^-(\varepsilon)$. $\mathscr{S}^+(\varepsilon)$ is the sector $\varepsilon \leq \arg(z) \leq \pi - \varepsilon$, and $\mathscr{S}^-(\varepsilon)$ is the sector $-\pi + \varepsilon \leq \arg(z) \leq \varepsilon$.

5.6 Hamburger series and the Hamburger moment problem

Remarks. In the context of this theorem, we understand the purpose in making the distinction between a Stieltjes function and a Stieltjes series, and between a Hamburger function and a Hamburger series. A Hamburger function is defined by the representation (6.25), provided the moments (6.26) are finite. A Hamburger series is defined by the right-hand side of (6.27), provided the moments (6.26) are finite. Hamburger's theorem asserts that a Hamburger function $f(z)$ defined by (6.25) has a Hamburger series defined uniquely by the right-hand side of (6.27) using the definition (6.26). The theorem also asserts that at least one Hamburger function, defined by (6.25), may be associated with a given Hamburger series defined by (6.26) and (6.27).

Proof. Assuming the representation (6.25) and the definition (6.26), it follows that

$$R_k(z) \equiv z^{-k}\left[f(z) - \sum_{j=0}^{k} f_j(-z)^j\right]$$

$$= (-)^{k+1} z \int_{-\infty}^{\infty} \frac{u^{k+1}\, d\phi(u)}{1 + zu}$$

$$= (-)^{k+1}\left[zf_{k+1} - z^2 \int_{-\infty}^{\infty} \frac{u^{k+2}\, d\phi(u)}{1 + zu}\right]. \quad (6.28)$$

Hence, if k is odd,

$$|R_k(z)| \leq |z|f_{k+1} + |z|^2 \int_{-\infty}^{\infty} \frac{|u|^{k+2}\, d\phi(u)}{|u|\,|\mathrm{Im}\, z|}$$

$$\leq |z|f_{k+1} + \frac{|z|^2 f_{k+1}}{|\mathrm{Im}\, z|}$$

$$\leq |z|f_{k+1}(1 + |\csc \varepsilon|).$$

We deduce that for k odd, $R_k(z) \to 0$ uniformly as $|z| \to 0$ in the sectors $\mathcal{S}^+(\varepsilon)$ or $\mathcal{S}^-(\varepsilon)$. If k is even, an extra term must be included in (6.28), with the consequence that

$$|R_k(z)| \leq |z|\,|f_{k+1}| + |z|f_{k+2}(1 + |\csc \varepsilon|),$$

and (6.27) is proved.

For the converse, we assume that an asymptotic Hamburger series is given, with moments defined by (6.26). The representation (6.26) allows the construction of $f(z)$ for $\mathrm{Im}\, z \neq 0$ according to (6.25). The estimates of $|R_k(z)|$ establish that (6.27) is the asymptotic series of $f(z)$.

We now state our principal result:

Theorem 5.6.12. *Let the coefficients $\{f_j\}$ satisfy the determinantal conditions $D(0, n) > 0$ for all $n = 0, 1, 2, \ldots$. Let $\mathcal{D}^I(\Delta)$ be a bounded domain, at a distance of at least Δ from the real axis, as shown in Figure 5.6.1. Then the sequence of $[M - 1/M]$ Padé approximants, $M = 1, 2, 3, \ldots$, of the formal series $\sum_{j=0}^{\infty} f_j(-z)^j$ contains a subsequence, denoted by $M \in \mathcal{M}$, which converges to $\tilde{f}(z)$ in $\mathcal{D}^I(\Delta)$. The function $\tilde{f}(z)$ is real symmetric, is analytic in $\mathcal{D}^I(\Delta)$, and has the representations*

$$\lim_{\substack{M \to \infty \\ M \in \mathcal{M}}} [M - 1/M] = \tilde{f}(z) = \int_{-\infty}^{\infty} \frac{d\phi(u)}{1 + zu} \simeq \sum_{j=0}^{\infty} f_j(-z)^j, \quad (6.29)$$

where Figure 5.6.4 shows the sectors for the asymptotic expansion and

$$f_j = \int_{-\infty}^{\infty} u^j \, d\phi(u), \quad j = 0, 1, 2, \ldots, \quad (6.30)$$

and $\phi(u)$ is bounded and nondecreasing on $-\infty < u < \infty$.

Proof. Given $\Delta > 0$, notice that a positive δ may always be found such that $\mathcal{D}^I(\Delta) \subset \{z : -z^{-1} \equiv w \in \mathcal{D}^w(\delta)\}$; see also Example 2. Theorem 5.6.12 then follows from Theorems 5.6.5, 5.6.7, and 5.6.11 and (6.24).

What has not been established in Theorem 5.6.12 is that $\tilde{f}(z)$ is uniquely represented by (6.29) and (6.30). In the previous section we gave examples to show that, even in the case of Stieltjes series, the conditions of the theorem are insufficient to define $\phi(u)$ and $\tilde{f}(z)$ uniquely. A sufficient condition for uniqueness is Carleman's criterion for Hamburger series, which we quote:

Carleman's criterion. *If the coefficients $\{f_j\}$ are real and satisfy the determinantal inequalities $D(0, n) > 0$ for all $n \geq 0$ and also the condition that*

$$\sum_{j=1}^{\infty} f_{2j}^{-1/(2j)} \text{ diverges}, \quad (6.31)$$

then there exists a unique Hamburger function $f(z)$ with the asymptotic expansion

$$f(z) \simeq \sum_{j=0}^{\infty} f_j(-z)^j, \quad \text{Im } z \neq 0.$$

5.6 Hamburger series and the Hamburger moment problem

If Carleman's criterion (6.31) is satisfied, the representations (6.29) and (6.30) are uniquely determined and the moment problem is said to be determinate. If Carleman's criterion is not satisfied, we refer the reader to the books of Akhiezer [1965] and Wall [1948] for further details.

The connection between the sequence of $[M - 1/M]$ Padé approximants which converge to $\tilde{f}(z)$ and the J-fractions of (4.4.9) is established using the variable $\omega = z^{-1}$. Following (6.29), we define

$$\bar{J}(\omega) = z^{-1}\tilde{f}(z) = \int_{-\infty}^{\infty} \frac{d\phi(u)}{\omega + u}. \tag{6.32}$$

Consequently $\bar{J}(\omega)$ is the limit of a sequence of convergents given by

$$J_M(\omega) = z^{-1}[M - 1/M]_{\tilde{f}}(z), \quad M \in \mathcal{M}, \quad \text{Im } \omega \neq 0.$$

The algebraic derivation of (4.5.9) and (4.5.10) is valid (provided the fractions are well defined) for Hamburger series as well as for Stieltjes series. Using (4.4.23)–(4.4.25), (4.5.1)–(4.5.4), and (1.12), we deduce that

$$J_M(\omega) = \frac{k_1}{l_1 + \omega} - \frac{k_2}{l_2 + \omega} - \frac{k_3}{l_3 + \omega} - \cdots - \frac{k_M}{l_M + \omega},$$

where

$$k_1 = a_1 = f_0,$$
$$l_1 = a_2 = f_1/f_0, \tag{6.33a}$$
$$k_2 = a_2 a_3 = \frac{D(0, 1)}{D(0, 0)^2},$$
$$l_2 = a_3 + a_4 = \frac{D(0, 1)}{D(0, 0)D(1, 0)} + \frac{D(1, 1)D(0, 0)}{D(1, 0)D(0, 1)}, \tag{6.33b}$$

and for $j = 3, 4, 5, \ldots$,

$$k_j = \frac{D(0, j - 3)D(0, j - 1)}{D(0, j - 2)^2},$$
$$l_j = \frac{D(0, j - 1)D(1, j - 3)}{D(0, j - 2)D(1, j - 2)} + \frac{D(1, j - 1)D(0, j - 2)}{D(1, j - 2)D(0, j - 1)}. \tag{6.33c}$$

We use (6.33) to prove

Theorem 5.6.13. *Given the real J-fraction*

$$J(\omega) = \frac{k_1}{l_1 + \omega} - \frac{k_2}{l_2 + \omega} - \frac{k_3}{l_3 + \omega} - \cdots \tag{6.34}$$

with $k_i > 0$ and l_i real for all $i \geq 1$, the sequence of convergents of $J(\omega)$ contains a convergent subsequence, with the property

$$\lim_{\substack{M \to \infty \\ M \in \mathcal{M}}} J_M(\omega) = \bar{J}(\omega) = \int_{-\infty}^{\infty} \frac{d\phi(u)}{\omega + u}, \qquad (6.35)$$

where $\phi(u)$ is bounded and nondecreasing on $-\infty < u < \infty$.

If $J(\omega)$ given by (6.34) is a convergent J-fraction, then it has the Hamburger representation

$$J(\omega) = \int_{-\infty}^{\infty} \frac{d\phi(u)}{\omega + u}. \qquad (6.36)$$

Proof. By expansion of (6.34), we find that $J(\omega)$ has the formal expansion

$$J(\omega) = \frac{1}{\omega} \sum_{j=0}^{\infty} f_j(-\omega)^{-j}.$$

An inductive proof based on (6.33) shows that the coefficients f_j satisfy the determinantal conditions that $D(0, n) > 0$ and $D(1, n)$ is real. Therefore the coefficients f_j are real and satisfy the Hamburger series conditions. By Theorem 5.6.12, a subsequence of the convergents of $J(\omega)$ is convergent. If the fraction $J(\omega)$ is convergent, then all subsequences converge to a common limit.

It is interesting to notice in the latter proof that the hypothesis of convergence of the fraction (6.34) replaces Carleman's criterion as the condition for uniqueness.

Herglotz functions are a class of functions which are intimately connected with Hamburger series. We define these functions, and conclude this section by quoting the main theorem.

Definition. $h(z)$ is defined to be a Herglotz function if

(i) $h(z)$ is analytic for $\text{Im}(z) \neq 0$,
(ii) $\text{Im } h(z)$ and $\text{Im } z$ have the same sign for all z.

Provided that $h(z)$ has the representation (6.27) (which obviously includes the case that $h(z)$ is actually analytic at the origin) then $h(z)$ is necessarily real symmetric. We quote Herglotz's theorem in the form [Stone, 1932, p. 573]:

Theorem 5.6.14. *A Herglotz function has the representation*

$$h(z) = \alpha z + \beta + \int_{-\infty}^{\infty} \frac{zu + 1}{u - z} d\sigma(u), \qquad (6.37)$$

where $\sigma(u)$ is bounded and nondecreasing on $-\infty < u < \infty$, β is real, and $\alpha \geq 0$. When this representation (6.37) exists, it is unique.

Corollary. *Provided that the first moment*

$$\mu_1 = \int_{-\infty}^{\infty} u \, d\sigma(u)$$

is finite, $h(z)$ has the representation

$$h(z) = \alpha z + \beta' \int_{-\infty}^{\infty} \frac{d\phi(u)}{u - z},$$

where $\beta' = \beta - \mu_1$ and $d\phi(u) = (1 + u^2) \, d\sigma(u)$.

The connection between Herglotz and Stieltjes functions is relatively straightforward. If $f(z)$ is such that

(i) $f(z)$ is analytic in the cut plane $|\arg(z)| < \pi$,
(ii) $f(z) \to f_\infty$ as $z \to \infty$, with $f_\infty \geq 0$,
(iii) $f(z)$ has the representation (6.27),
(iv) $-f(z)$ is a Herglotz function,

then $f(z)$ is a Stieltjes function. A neat proof of this based on continued-fraction methods is given by Bender and Orszag [1978].

For further details of the applications of Herglotz functions, we refer to Narcowich and Allen [1975] and Allen and Narcowich [1975].

For further details of the construction of the density function $\phi(u)$ in the context of this section, we refer to the books of Perron [1957] and Riesz and Szökefalvi-Nagy [1955].

Techniques and results are known for treating a series which is the difference of two Stieltjes series [Baker and Gammel, 1971; Barnsley, 1976].

Example 1. Let $f(z)$ have the formal power series

$$f(z) = \sum_{j=0}^{\infty} f_j(-z)^j,$$

in which the coefficients f_j satisfy the conditions $D(0, n) > 0$ for all $n \geq 0$. Define $g(z)$ by the identity

$$f(z) = \frac{f_0}{1 + f_1 f_0^{-1} z - z^2 g(z)}$$

so that

$$g(z) = \sum_{j=0}^{\infty} g_j(-z)^j, \quad \text{formally.}$$

Hadamard's formula (1.6.9) shows that the coefficients g_j satisfy the conditions

$$D_g(0, n) = \begin{vmatrix} g_0 & g_1 & \cdots & g_n \\ g_1 & g_2 & \cdots & g_{n+1} \\ \vdots & \vdots & & \vdots \\ g_n & g_{n+1} & \cdots & g_{2n} \end{vmatrix} > 0.$$

Example 2. Hamburger's theorem is valid if the condition that z lies in the sector $\mathscr{S}^+(\varepsilon)$ or $\mathscr{S}^-(\varepsilon)$ is replaced by the condition that z lies in the disk $\mathscr{C}$ of Figure 5.6.2.

Example 3. Let $\sum_{j=0}^{\infty} f_j(-z)^j$ be a Hamburger series, let $\omega = z^{-1}$, and define, according to (6.4),

$$J_M(\omega) = z^{-1}[M - 1/M](z) = \sum_{i=1}^{M} \frac{\beta_i}{\omega + \gamma_i}.$$

Then, for $M = 1, 2, 3, \ldots$,

(i) the poles of $J_M(\omega)$ have positive residues,
(ii) the zeros of $J_M(\omega)$ are real,
(iii) the poles and zeros of $J_M(\omega)$ interlace.

5.7 Pólya frequency series

We have already seen that explicit forms for the numerator $A^{[L/M]}(z)$ and the denominator $B^{[L/M]}(z)$ of the Padé approximant of $\exp(z)$ are known. The results may be obtained from the explicit forms (1.2.12) or from collocation (10.3.21); continued fractions (4.6.1) provide yet another equivalent representation. The formula (1.2.12) for the Padé approximant of $\exp(z)$ may be derived from the determinantal representations (1.1.8) and (1.1.9), using (1.2.6):

$$C(L/M) = (-1)^{M(M-1)/2} \prod_{k=1}^{M} \frac{1}{k!(k+1)! \cdots (k+L-1)!}$$

with $L, M \geq 1$, using the methods of Section 5.2. Notice that the sign of $C(L/M)$ is $(-1)^{M(M-1)/2}$, which does not depend on L; for $M = 1, 2, 3, \ldots$, the signs are $+, -, -, +, +, -, -, \ldots$, and this pattern of signs of $C(L/M)$ characterizes a class of functions known as Pólya frequency series [Schoenberg, 1951]. (They are also known as totally positive series: see Example 1.) Every function of the class has the representation

$$f(z) = a_0 e^{\gamma z} \prod_{j=1}^{\infty} (1 + \alpha_j z)(1 - \beta_j z)^{-1} \qquad (7.1)$$

with $a_0 > 0$, $\gamma \geq 0$, $\alpha_j \geq 0$, $\beta_j \geq 0$, and

$$\sum_{j=1}^{\infty} (\alpha_j + \beta_j) \text{ convergent.}$$

The present point of interest is that these functions have not only convergent Padé approximants, but also the numerators and denominators of the Padé approximants separately converge along any ray of the Padé table.

Theorem 5.7.1 [Arms and Edrei, 1970]. *If $f(z)$ is a Pólya frequency series defined by (7.1), and $L_k \to \infty$, $M_k \to \infty$ with $(M_k/L_k) \to \omega$ as $k \to \infty$, then*

$$A^{[L_k/M_k]}(z) \to a_0 \exp\left(\frac{\gamma}{1+\omega} z\right) \prod_{j=1}^{\infty} (1 + \alpha_j z), \tag{7.2}$$

$$B^{[L_k/M_k]}(z) \to \exp\left(\frac{-\gamma\omega}{1+\omega} z\right) \prod_{j=1}^{\infty} (1 - \beta_j z) \tag{7.3}$$

uniformly on any compact region of the z-plane.

Corollary 1. *If $f(z) = e^{\gamma z}$,*

$$A^{[L_k/M_k]} \to \exp\left(\frac{\gamma z}{1+\omega}\right) \quad \text{and} \quad B^{[L_k/M_k]} \to \exp\left(\frac{-\gamma\omega z}{1+\omega}\right). \tag{7.4}$$

Corollary 2. *Provided $\alpha_i \neq -\beta_j$ for all i, j, a finite number of α_i and β_j may be arbitrary and complex.*

Corollary 1 is a special case of the theorem, and Corollary 2 is a generalization. The proofs are in Baker [1975a] and in Arms and Edrei [1970]. Excellent accounts of the basic theory are given by Edrei [1953] and Karlin [1968].

Partly because the exponential function has known explicit forms for its Padé approximants and partly because of its role as a solution of the most elementary ordinary differential equation $dy/dx = y$, considerable attention has been paid to the location of poles and zeros of the Padé approximants of $\exp(z)$. If the $[L/M]$ Padé approximant of $\exp(z)$ has a pole in the left half plane, then this Padé approximant does not define an A-stable method for the solution of ordinary differential equations (see Sections 10.3, 10.4). Simply because $\exp(-z) = [\exp(z)]^{-1}$, the duality theorem (Theorem 1.5.1) shows that a zero of the $[L/M]$ Padé approximant of $\exp(z)$ is a pole of the $[M/L]$ Padé approximant of $\exp(-z)$, and so results about poles of Padé approximants translate directly to results about zeros.

A number of theorems can be derived from the geometry of order stars introduced in Section 1.2. The key to this analysis is the argument principle [Titchmarsh, 1939]. Because e^{-z} is an analytic function,

$$g(z) = e^{-z}[L/M](z) \qquad (7.5)$$

is meromorphic in $\mathbb{C}$. Further because $e^{-z} \neq 0$ in $\mathbb{C}$, the poles and zeros of $g(z)$ are the same as those of $[L/M](z)$. The region A_0 was defined in Section 1.2 as that for which $|g(z)| = 1$; let $\phi(z)$ be defined (up to a multiple of 2π) by

$$g(z) = e^{i\phi(z)}. \qquad (7.6)$$

For $z \in A_0$, $\phi(z)$ is real and is called the phase or argument of $g(z)$. As z encircles a finite simply connected A_- region along a path from A_0 (once, and counterclockwise) from any starting point P, its phase is required to change continuously. On returning to P, the phase must have changed by $2n\pi$ for some integer n.

From (7.6), it follows that $\phi(z)$ is analytic in any (sufficiently small) neighborhood of A_0 (except at P). Use local coordinates s, v where s is the arc length along A_0, and v is the distance along the normal to A_0, taken outward from A_-. Define a local complex variable $w = v + is$, and then use it to express the second Cauchy–Riemann equation as

$$\frac{\partial(\text{Re }\phi)}{\partial s} = -\frac{\partial(\text{Im }\phi)}{\partial v}.$$

Because $\|g(z)\|_{z\in A_+} > \|g(z)\|_{z\in A_-}$, it follows from (7.6) that $\partial(\text{Im }\phi)/\partial v < 0$ and hence $\partial(\text{Re }\phi)/\partial s > 0$. Therefore the phase $\phi(z)$ is an increasing function as the above A_- region is encircled counterclockwise, except possibly at confluent interpolation points such as the origin where local coordinates are ill defined. This case was treated in Section 1.2; from (1.2.16) we see that the phase changes smoothly at the origin, which is taken as a single interpolation point for the A_- region under discussion. Interpolation points on A_0 are uniquely identified by the values

$$\arg g(z) = 2k\pi, \quad k = 0, \pm 1, \pm 2, \ldots.$$

Hence, as this finite A_- region is encircled, its phase increases by $2n\pi$, where n is the number of interpolation points on the boundary. Because A_- regions contain no poles (see (1.2.14)), n is also the number of zeros of $[L/M](z)$ in A_-. We conclude that, as the simply connected finite A_- region is encircled once and counterclockwise, the phase $\phi(z)$, as defined by (7.6) plus continuity, increases by $2n\pi$, where n equals both the number of zeros in A_- and the number of interpolating points on the boundary, counting multiplicities for both except for the origin, which counts singly. Similarly, it follows that the phase $\phi(z)$ decreases by $2v\pi$

as, a similar simply connected A_+ region abutting the origin is encircled, where v equals the number of poles in A_+ and v also equals the number of interpolating points on the boundary, counting multiplicities except at the origin.

Let us return to consider the behavior of $g(z)$ near the origin. From (1.1.11) and (1.2.6), it follows that

$$f(z) - [L/M](z) = O[z^{L+M+1}],$$

where the order $N+1 = L+M+1$ of the approximation error is precise. Hence there are $L+M+1$ segments of type A_- abutting the origin. The numerator polynomial $P^{[L/M]}(z)$ has at most L zeros, and hence at least $M+1$ of the A_- segments abutting the origin ultimately merge into the infinite A_- region. In so doing, they confine at least M of the A_+ regions. Because there are at most M zeros of $Q^{[L/M]}(z)$, and each of these occurs in an A_+ region, it must be that each finite A_+ region contains precisely one pole of $[L/M](z)$, and each finite A_- region contains precisely one zero. Moreover, the only point of interpolation is the origin. This argument was given by Iserles and Powell [1981] as a simpler proof of Iserles maximal interpolation theorem [Iserles, 1979b].

For reasons given in Sections 10.3, 10.4, rational approximations to $\exp(z)$ which are A-acceptable are of great importance in the theory of ordinary and partial differential equations: a rational function $R(z)$ is A-acceptable if $|R(z)| < 1$ for $\text{Re } z < 0$. The functions are characterized by a lemma:

Lemma [Wanner, Hairer, and Nørsett, 1978]. *$R(z)$ is A-acceptable if and only if the whole A_+ region of its order star has no intersection with the imaginary axis and $R(z)$ has no poles with $\text{Re } z < 0$.*

Proof. If $R(z)$ is A-acceptable, $|R(z)| \leq 1$ for $\text{Re } z \leq 0$. Therefore $R(z)$ has no poles in $\text{Re } z \leq 0$, and the infinite A_+ region does not extend to $\text{Re } z > 0$. Any other A_+ region which might occur in the left half plane would necessarily be finite and hence contain a pole, which has been excluded. Therefore the finite A_+ regions which contain poles are all in the right half plane.

Conversely, because $|R(z)| \leq 1$ on $\text{Im } z = 0$, $R(\infty) \leq 1$. By assumption, $R(z)$ is analytic in $\text{Re } z \leq 0$, and so, by the maximum-modulus theorem $|R(z)| \leq 1$ in $\text{Re } z \leq 0$, and $|R(z)| < 1$ in $\text{Re } z < 0$.

The following elegant proof of Ehle's [1973] conjecture, proved in parts by Birkhoff and Varga [1965], Ehle [1968, 1973], and Nørsett [1974], is due to Iserles and Powell [1981] and Iserles and Nørsett [1991].

Theorem. *The A-acceptable Padé approximants for the exponential function are precisely those of types $[L/M]$ with $M = L$, $L + 1$, or $L + 2$.*

Proof. Assume that $[L/M](z) = R(z)$ is A-acceptable. In terms of the previous Lemma, all segments of its A_+ region in the right half plane are bounded and contain poles. Suppose that there are n_+ of these segments in the right half plane and n_- in the left half plane. Then

$$n_+ \leq M. \tag{7.7}$$

We have previously shown that there are precisely $L + M + 1$ segments abutting the origin, equidistributed in angle. Hence

$$n_+ + n_- = L + M + 1, \quad n_- \leq n_+ + 1. \tag{7.8}$$

Equations (7.7) and (7.8) imply that

$$L \leq M. \tag{7.9}$$

In the left half plane, all the n_- segments of A_+ at the origin ultimately merge into the infinite A_+ region, and thereby confine $n_- - 1$ segments of A_-. These contain zeros of $R(z)$ and hence

$$n_- - 1 \leq L. \tag{7.10}$$

Again, equidistribution in angle implies that

$$n_+ + n_- = L + M + 1, \quad n_+ \leq n_- + 1, \tag{7.11}$$

and (7.10), (7.11) imply that

$$M \leq L + 2.$$

This establishes that A-acceptable Padé approximants are necessarily of types $M = L$, $L + 1$, or $L + 2$; the proof of the converse is straightforward because the poles and zeros have been shown to be located in the finite A_+ and A_- regions, respectively.

Because $\exp(+\infty) = \infty$ and $\exp(-\infty) = 0$, we expect the poles of the Padé approximants to accumulate near $x = +\infty$ and the zeros near $x = -\infty$ with increasing order of approximation. If $L_1 > L_2$, there are more zeros of $[L_1/M]$ than $[L_2/M]$ to be accommodated, and so we expect a smaller zero-free region for $[L_1/M]$. These simple ideas are borne out by the previous geometrical results. They are made much more precise by the following theorem.

Theorem 5.7.2 (Sectorial theorem) [Saff and Varga, 1975]. *Let $z = re^{i\theta}$ denote a generic point in the complex plane. The $[L/M]$ Padé approxi-*

mant to $\exp(z)$ is zero-free in the sector given by

$$\cos \theta \geq \frac{L - M - 2}{L + M}, \quad \text{provided } M \geq 2.$$

Example 1. If $L = 3M + 2$, the $[3M + 2/M]$ Padé approximants are zero-free in the region $\cos \theta \geq \frac{1}{2}$, i.e. $|\theta| \leq 60°$. Clearly, all Padé approximants with $L \leq 3M + 2$ are also zero-free in this region, as shown in Figure 5.7.1.

The range of $|z|$ in which the zeros of the approximants can occur is governed by

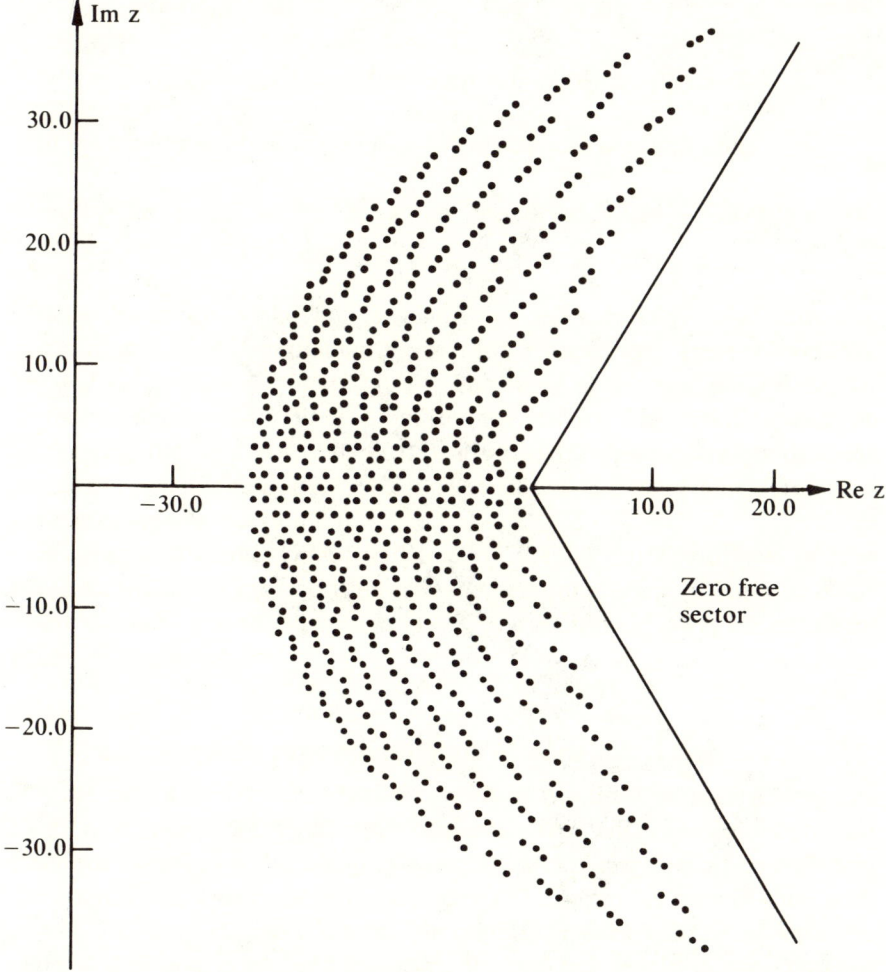

Figure 5.7.1. Zeros of $[L/M]$ Padé approximants of $\exp(z)$ with $M \leq [\frac{1}{3}L]$, for $L = 1, 2, 3, \ldots, 36$.

***Theorem* 5.7.3 (Annulus theorem) [Saff and Varga, 1977b].** *For any $L \geq 1$ and any $M \geq 0$, all the zeros of the $[L/M]$ Padé approximants of $\exp(z)$ lie in the annulus $(L + M)\mu < |z| < L + M + \frac{4}{3}$, where μ is the (unique) positive root of $te^{1+t} = 1$, which occurs at $t = \mu = 0.278465$.*

While these theorems were being proved, a number of numerical experiments were done by Ruttan, Varga, and Saff, motivated by the following result.

***Theorem* 5.7.4 (Parabola theorem) [Saff and Varga, 1976b].** *Let $z = x + iy$ denote a generic point in the complex plane. The $[L/M]$ Padé approximant to $\exp(z)$ is zero-free in the parabolic region*

$$y^2 \leq 4(M + 1)(x + M + 1). \tag{7.12}$$

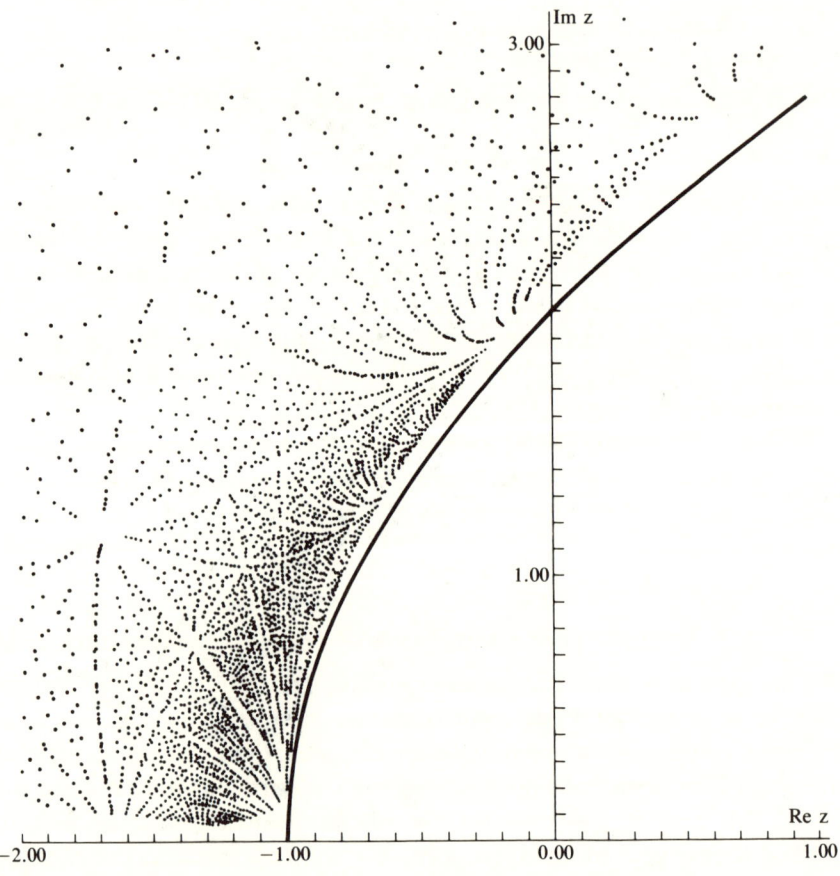

Figure 5.7.2. The zero-free region of $[L/M]$ Padé approximants of $\exp\{(M + 1)z\}$.

5.7 Pólya frequency series

Theorem 5.7.4 can be reviewed by rescaling the axes by the factor $M + 1$, leading to the following restatement:

Corollary. *The $[L/M]$ Padé approximant of $\exp\{(M + 1)z\}$ is zero-free in the region*

$$\left(\frac{y}{2}\right)^2 \leq x + 1.$$

The locations of the zeros of $[L/M]$ Padé approximants with $L \leq 25$, $M \leq 25$ are shown in Figure 5.7.2, demonstrating visually the result of Theorem 5.7.4. The remarkable trajectories of the zeros are not, as yet, fully explained. However, it turns out that scaling by the factor $L + M$ as suggested by the annulus theorem leads to sharper results than the parabolic boundary of Figure 5.7.2 for the zeros. With this scaling, all the limit points are accounted for by the following theorem.

Theorem 5.7.5 (Eye theorem) [Saff and Varga, 1978a]. *Consider a ray sequence $[L_j/M_j](z)$, $j = 1, 2, 3, \ldots$, of Padé approximants for $\exp(z)$, for which*

$$\lim_{j \to \infty} (M_j/L_j) = \sigma, \quad L_j \to \infty,$$

for any $\sigma \geq 0$. Let $N_j = L_j + M_j$, and define

$$g_\sigma(z) = \sqrt{1 + z^2 - 2z\left(\frac{1 - \sigma}{1 + \sigma}\right)}$$

so that $g_\sigma(0) = 1$ in the cut plane shown in Figure 5.7.3. The branch points z_σ^+ and z_σ^- of $g_\sigma(z)$ are marked on the figure.

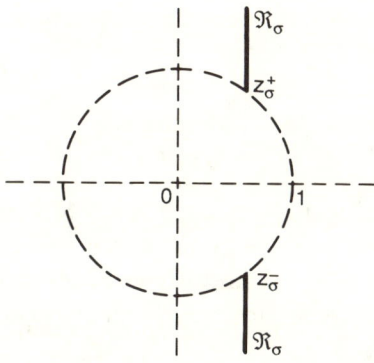

Figure 5.7.3. The complex plane cut for $w_\sigma(z)$ and $g_\sigma(z)$.

In turn, define

$$w_\sigma(z) = \frac{4\sigma^{(\sigma/1+\sigma)}ze^{g_\sigma(z)}}{(1+\sigma)(1+z+g_\sigma(z))^{(2/1+\sigma)}(1-z+g_\sigma(z))^{(2\sigma/1+\sigma)}}$$

in the same cut plane, taking each term in the denominator to be real and positive at $z = 0$. Then

(i) $\hat{z}$ is a limit point of the zeros of the Padé approximants $\{[L_j/M_j](N_jz)\}_{j=1}^\infty$ if and only if

$$\hat{z} \in D_\sigma = \{z: |\arg z| \geq \cos^{-1}\frac{1-\sigma}{1+\sigma}, |w_\sigma(z)| = 1, |z| \leq 1\};$$

(ii) when $\sigma > 0$, $\hat{z}$ is a limit point of poles of the approximants $\{[L_j/M_j](N_jz)\}_{j=1}^\infty$ if and only if

$$\hat{z} \in E_\sigma = \{z: |\arg z| \leq \cos^{-1}\frac{1-\sigma}{1+\sigma}, |w_\sigma(z)| = 1, |z| \leq 1\};$$

(iii) $\hat{z}$ is a limit point of nontrivial zeros of the remainders $\{e^{N_jz} - [L_j/M_j](N_jz)\}_{j=1}^\infty$ if and only if

$$\hat{z} \in F_\sigma = \{z: |w(z)| = 1 \text{ and } |z| \geq 1\}.$$

The curves D_σ, E_σ, F_σ form the eye shape shown in Figure 5.7.4 for the case of $L = 24$, $M = 8$, $N = 32$, $\sigma = 1/3$.

The proofs of these theorems are too long to give here, and we refer to the original papers for details. Suffice it to say that the formulas

$$A^{[L/M]}(z) = {}_1F_1(-L, -L-M; z) = \frac{1}{(L+M)!}\int_0^\infty e^{-t}(t+z)^L t^M \, dt$$

$$B^{[L/M]}(z) = {}_1F_1(-M, -L-M; -z) = \frac{1}{(L+M)!}\int_0^\infty e^{-t}(t-z)^M t^L \, dt$$

which follow from (1.2.11), and

$$e^z - [L/M](z) = \frac{e^z z^{L+M+1}\int_0^1 e^{-zt}t^L(t-1)^M \, dt}{(L+M)! \, {}_1F_1(-M, -L-M; -z)},$$

which is proved in Section 10.3, figure prominently in the proofs.

Our last theorem about the exponential function concerns the behavior of Padé approximants lying on rays in the Padé table with z confined to the negative real axis. We define an approximation error on this axis by

$$\eta^{[L/M]} = \sup_{-\infty \leq x \leq 0} |\exp(x) - [L/M](x)|.$$

5.7 Pólya frequency series

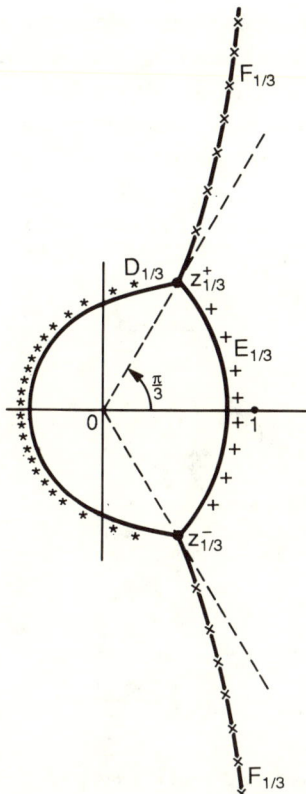

Figure 5.7.4. Zeros of $[24/8](32z)$ and $e^{32z} - [24/8](32z)$, and poles of $[24/8](32z)$. $*$ = zeros of $[24/8](32z)$; $+$ = poles of $[24/8](32z)$; $\times$ = zeros of $e^{32z} - [24/8](32z)$.

Theorem 5.7.6 [Saff, Varga, and Ni, 1976]. *If a ray sequence is defined by $(L_k/M_k) \to \omega$ as $k \to \infty$, with $0 \le \omega \le 1$, then*

$$\{\eta^{[L_k/M_k]}\}^{1/M_k} \to \omega^\omega \left(\frac{1-\omega}{2}\right)^{1-\omega} \equiv g(\omega).$$

Corollary. *Observe that* $\min\{g(\omega)\} = \frac{1}{3}$ *for* $0 \le \omega \le 1$, *with minimization achieved at* $\omega = \frac{1}{3}$, $g(\omega) = \frac{1}{3}$. *The ray sequence of $[L/3L]$ Padé approximants to $\exp(z)$ has the best rate of convergence on the negative real axis, with convergence characterized by*

$$\eta^{[L/3L]} \simeq (\tfrac{1}{3})^L.$$

The reader is referred to the original papers for complete statements and proofs of the foregoing theorems. The proofs are based on the explicit forms and the interrelations which exist for the Padé numerator and denominator. This concludes our discussion on the location of poles and zeros of Padé approximants to $\exp(z)$.

We turn our attention to trigonometric polynomials having real roots. The discussion is based on the formulas

$$\sin z = z \prod_{n=1}^{\infty} \left(1 - \frac{z^2}{n^2\pi^2}\right), \tag{7.13}$$

$$\cos z - 1 = -2\sin^2\left(\frac{z}{2}\right) = \frac{-z^2}{2} \prod_{n=1}^{\infty} \left(1 - \frac{z^2}{4n^2\pi^2}\right)^2, \tag{7.14}$$

$$\cos z - \cos \alpha = (1 - \cos \alpha) \prod_{n=-\infty}^{\infty} \left(1 - \frac{z^2}{(2n\pi + \alpha)^2}\right), \quad 0 < \alpha \leq \pi. \tag{7.15}$$

Any real cosine polynomial

$$C(z) = c_0 + c_1 \cos z + c_2 \cos 2z + \cdots + c_N \cos Nz \tag{7.16}$$

or any real sine polynomial

$$S(z) = s_1 \sin z + s_2 \sin 2z + \cdots + s_N \sin Nz, \tag{7.17}$$

if it has only real zeros, may be re-expressed using (7.15). We may write

$$C(z) = K(\cos z - 1)^p \prod_{k=1}^{k_{\max}} (\cos z - \cos \alpha_k)^{p_k}, \tag{7.18}$$

where p, p_k are nonnegative integer powers. $S(z)$ may be expressed quite generally as

$$S(z) = C(z) \sin z, \tag{7.19}$$

and so we consider (7.15), (7.16), and (7.18). We see that

$$C(z) = K\left(\frac{-z^2}{2}\right)^p \left\{\prod_{n=1}^{\infty} \left(1 - \frac{z^2}{4n^2\pi^2}\right)^{2p}\right\}$$

$$\times \prod_{k=1}^{k_{\max}} \left\{(1 - \cos \alpha_k)^{p_k} \prod_{n=-\infty}^{\infty} \left(1 - \frac{z^2}{(2n\pi + \alpha_k)^2}\right)^{p_k}\right\}, \tag{7.20}$$

and (7.20) fits the bill for the Arms–Edrei theorem, leading to the following result:

***Theorem 5.7.7** [Edrei, 1975a]. The denominators of the $[L_k/M_k]$ Padé approximants of the normalized real cosine polynomial having real zeros, $C(z)z^{-2p}$, defined by (7.16), (7.18), and (7.20) tend to unity uniformly on any compact domain of the z-plane provided $L_k \to \infty$ and $M_k \to \infty$ as*

$k \to \infty$. *An identical result under identical conditions holds for $S(z)z^{-2p-1}$ defined by (7.17), (7.19), and (7.20).*

For the function $T(z) = z^{-2}\tan^2 z$,

$$A^{[L_k/M_k]} \to z^{-2}\sin^2 z \quad \text{and} \quad B^{[L_k/M_k]} \to \cos^2 z$$

under the same conditions.

In this section, we have considered several theorems about what appear to be special cases. Primarily for this reason, no proofs have been given. However, it is tempting to speculate that a theory as complete as that for Stieltjes series will eventually be forthcoming, and that methods of general applicability will emerge in the process.

Example 1. The Toeplitz determinants of the coefficients of Pólya frequency series, defined by

$$T(L/M) \equiv \begin{vmatrix} c_L & c_{L+1} & \cdots & c_{L+M-1} \\ c_{L-1} & c_L & \cdots & c_{L+M-2} \\ \vdots & \vdots & & \vdots \\ c_{L-M+1} & c_{L-M+2} & \cdots & c_L \end{vmatrix}$$

satisfy $T(L/M) \geq 0$ for all $L, M \geq 0$.

6
Convergence theory

6.1 Introduction to convergence theory: rows

This chapter is concerned with what is known about convergence of sequences of Padé approximants to complex functions.

For row sequences, de Montessus's theorem proves convergence for functions meromorphic in a disk, as explained in Section 6.2.

Diagonal sequences are the natural choice for meromorphic functions in the absence of further information. Simply because meromorphy of a function implies meromorphy of its reciprocal, the symmetric choice of diagonal approximants is natural, especially in view of the duality theorem. Paradiagonal sequences, $[M + J/M]$ with J fixed and $M \to \infty$, are an obvious generalization, usually motivated by the requirement of an asymptotic approximation of z^J as $z \to \infty$.

Ray sequences of $[L/M]$ Padé approximants with $L = \lambda M$, λ fixed, are useful in special circumstances, and parabolic sequences, such as $[M^2/M]$, are worth considering sometimes. Hence interest settles on general sequences $[L_k/M_k]$ which may be particularized to suit special needs. The most natural convergence theorems for general sequences of Padé approximants involve convergence in capacity, a difficult concept. Instead of attempting to prove pointwise convergence for a class of functions, the theorems prove that the region of bad approximation becomes arbitrarily small. In no way do these theorems imply pointwise convergence, but they do prove that the Padé method converges in a real sense in very general circumstances.

Recognizing that Padé approximants to meromorphic functions are simply rational approximations, another natural development is the notion of convergence of the function values on the Riemann sphere, which is treated in Section 6.4. This allows the value ∞ to be treated on an equal footing with any other function value.

6.1 Introduction to convergence theory: rows

The counterexamples of Perron and others are clear warnings about what may not be proved concerning convergence of Padé approximants to entire functions, and an understanding of these surprising results saves us from ill-founded and wasted optimism about mythical theorems. In fact, the Padé conjecture, due to Baker, Gammel, and Wills, has been widely accepted, and is the foundation of many calculations and applications. It asserts the convergence of a subsequence of diagonal approximants, and is more fully discussed in Section 6.7.

It is interesting to consider a few theorems about rows of the Padé table; with the advantage of having explicit expressions for the Padé approximants, one or two optimal results and several significant theorems have been proved which seem to act as signposts to the complete theory. A discussion of the convergence of row sequences indicates clearly the type of results expected in general.

The first row of the Padé table consists of $[L/0]$ Padé approximants, which are truncated Maclaurin series. They converge within (but not necessarily on) a circle of convergence of which the radius may be zero, finite, or infinite.

The second row of $[L/1]$ approximants is governed by a theorem of Beardon [1968b].

Theorem 6.1.1. Let $f(z)$ be analytic in $|z| \leq R$. Then an infinite subsequence of $[L/1]$ Padé approximants converges to $f(z)$ uniformly in $|z| \leq R$.

Proof. By hypothesis, $f(z)$ is analytic in $|z| \leq R$, and consequently within a larger circle, $|z| < \rho$, for some ρ' with $\rho > \rho' > R$. Let

$$f(z) = \sum_{i=0}^{\infty} c_i z^i \quad \text{with } c_i = O((\rho')^{-i}). \tag{1.1}$$

The second row of Padé approximants is given by

$$[L/1] = c_0 + c_1 z + \cdots + c_{L-1} z^{L-1} + \frac{c_L z^L}{1 - c_{L+1} z / c_L}. \tag{1.2}$$

If a subsequence of coefficients $\{c_{L_j}, j = 1, 2, \ldots\}$ is zero, then $[L_j - 1/1]$ are truncated Maclaurin expansions which converge to $f(z)$ uniformly in $|z| \leq R < \rho'$. So we assume that no infinite subsequence of $\{c_L\}$ vanishes, and consider $r_L = c_{L+1}/c_L$ which is well defined for all sufficiently large L. Because

$$c_L z^L = O\left(\left(\frac{R}{\rho'}\right)^L\right),$$

the sequence of $[L/1]$ Padé approximants given by (1.2) converges uniformly in $|z| \leq R$ *unless* there exists a sequence of values of L for which $1 - c_{L+1}zc_L^{-1} = 0$ within $|z| < \rho'$. Thus either a subsequence of the second row converges uniformly, or else for some L_0 and all $L > L_0$, $|c_L/c_{L+1}| > \rho'$. In the latter case

$$\left|\frac{c_{L_0}}{c_L}\right| = \prod_{i=L_0}^{L-1}\left|\frac{c_i}{c_{i+1}}\right| > (\rho')^{L-L_0},$$

contradicting (1.1), so the theorem is proved.

Perron's counterexample [Perron, 1957, Chapter 4]. A function $f(z) = \sum_{i=0}^{\infty} c_i z^i$ is defined by selecting a sequence of points $\{z_n, n = 1, 2, \ldots\}$ in the complex plane, and the coefficients c_i are defined in triples by

$$\left.\begin{array}{l} c_{3n} = z_n/(3n + 2)!, \\ c_{3n+1} = 1/(3n + 2)!, \\ c_{3n+2} = 1/(3n + 2)!, \end{array}\right\} \text{ if } |z_n| \leq 1,$$

or

$$\left.\begin{array}{l} c_{3n} = 1/(3n + 2)!, \\ c_{3n+1} = 1/(3n + 2)!, \\ c_{3n+2} = z_n^{-1}/(3n + 2)!, \end{array}\right\} \text{ if } |z_n| > 1.$$

Since $|c_i| \leq (i!)^{-1}$, the comparison test shows that $f(z)$ is entire. Since

$$\frac{c_{3n}}{c_{3n+1}} = z_n \text{ if } |z_n| < 1 \quad \text{or} \quad \frac{c_{3n+1}}{c_{3n+2}} = z_n \text{ if } |z_n| > 1,$$

(1.2) shows that either $[3n/1]$ or $[3n + 1/1]$ has a pole at the selected point z_n. By choosing $\{z_n\}$ dense in the plane, repeating its values if necessary, we construct a holomorphic function for which the $[L/1]$ Padé approximants converge in no open set, however small, in the z-plane.

Perron's counterexample shows that Beardon's theorem is an optimal result. Unfortunately, results for the sequences of $[L/1]$ approximants to functions which are analytic in a disk do not point helpfully to the true mathematical characteristics of other row sequences. For $f(z)$ analytic in $z \in \mathbb{C}$, Baker and Graves-Morris [1977] proved that a subsequence of the row sequence of type $[L/2]$ converges. However, the hypothesis that $f(z)$ is analytic in the disk $|z| < R$ but not in any larger disk is too weak to secure convergence of even a subsequence of a row sequence of Padé approximants in $|z| < R$. This was first proved by Buslaev, Gončar, and Suetin [1983] who constructed

$$f(z) = \frac{1 + \sqrt[3]{2}z}{1 - z^3} = 1 + \sqrt[3]{2}z + z^3 + \sqrt[3]{2}z^4 + \cdots.$$

This function is analytic in $|z| < 1$, yet every approximant of type $[L/2]$ contains a pole in $|z| < 1$. This result was a precursor of a most instructive counterexample.

Counterexample of Lubinsky and Saff. These authors investigated the convergence of general sequences of Padé approximants for the partial theta function $h_q(z)$ with $q = e^{i\theta}$ and $\theta/(2\pi)$ real and irrational. For such q, $h_q(z)$ is defined by (4.3.44), it is analytic in $|z| < 1$, and it has a natural boundary on $|z| = 1$. For a row sequence of Padé approximants of types $[L/M]$, M fixed, Lubinsky and Saff [1987] identify a disk $|z| < \Delta_{M,q}$ of convergence of the sequence, but with the property that its circumference

$$|z| = \Delta_{M,q} < 1$$

contains limit points of the poles of the row sequence. For example, they find that

$$\Delta_{2,q_2} = \inf_{|q|=1} \Delta_{2,q} = 0.58\ldots,$$

$$\Delta_{17,q_{17}} = \inf_{|q|=1} \Delta_{17,q} = 0.24\ldots.$$

Lubinsky and Saff show that no subsequence of a row sequence of type $[L/M]$ with $M \geq 2$ for $h_q(z)$ as specified above converges in all of $|z| < 1$, by locating a pole of each approximant within $|z| < 1$. For example, for $M = 2$, it follows from the determinantal formula (1.1.8) and the normalization (1.4.7) that

$$B^{[L/2]}(z) = z^2 q^{2L} - zq^L(q+1) + 1. \tag{1.3}$$

Substitute

$$w = -zq^L. \tag{1.4}$$

Then the poles of the approximants correspond to the roots w_+, w_- of

$$w^2 + w(q+1) + 1 = 0.$$

By trying $w_+ = e^{i\varphi}$, it follows that $|w_+| = 1$ is impossible and because $w_+ w_- = 1$, we may take $|w_+| < 1$. Hence a root of $B^{[L/2]}(z)$ occurs at

$$z_+ = -w_+ q^{-L}.$$

Because $\theta/2\pi$ is irrational, we conclude from (1.4) that poles of the row sequence of Padé approximants of type $[L/2]$ lie densely on the circle $|z| = |w_+|$, with $|w_+| < 1$.

From the constructive point of view, the largest domain in which Lubinsky and Saff establish convergence of general sequences for $h_q(z)$ with $|q| = 1$ is

$$|z| < 3 - 2\sqrt{2} = 0.1715\ldots.$$

The case (1.3) above is an instance of their more general result that

$$B^{[L/M]}(z) = G_M(-zq^L), \tag{1.5}$$

where

$$G_n(z) = \sum_{j=0}^{n} \begin{bmatrix} n \\ j \end{bmatrix} z^j \tag{1.6}$$

and the q-binomial coefficient is defined by

$$\begin{bmatrix} n \\ j \end{bmatrix} = \frac{1-q^n}{1-q} \cdot \frac{1-q^{n-1}}{1-q^2} \cdots \frac{1-q^{n-j+1}}{1-q^j}. \tag{1.7}$$

From (1.5), we see how zeros of the Rogers–Szegö polynomial $G_M(w)$ determine the poles of the row sequence of type $[L/M]$ for $L = 1, 2, 3, \ldots$.

More recently, interest has settled on some inverse problems for row sequences. Suppose that the poles $\{\zeta_i^{[L/M]}\}_{i=1}^{M}$ of the Padé approximants of types $[L/M]$, M fixed, for a formal power series $\sum_{i=0}^{\infty} c_i z^i$ converge to not necessarily distinct points $\zeta_1, \zeta_2, \ldots, \zeta_M$. Let $Z = \max_{1 \leq i \leq M} |\zeta_i|$. Following previous fundamental work by Fabry [1896] and Gončar [1981], Suetin [1983] has shown that the power series is the Maclaurin series of a function $f(z)$ which is meromorphic in $|z| < Z$. Let R_M denote the radius of the actual disk D_M of meromorphy of $f(z)$ containing $\zeta_1, \zeta_2, \ldots, \zeta_M$ but no other singularities. Suetin [1984] then proved that

(i) if ζ_i is an interior point of D_M, i.e., $|\zeta_i| < R_M$, then ζ_i is a pole of $f(z)$ and the convergence $\zeta_i^{[L/M]} \to \zeta_i$ is geometric;
(ii) if ζ_i is a boundary point of D_M, i.e., $|\zeta_i| = Z = R_M$, then ζ_i is necessarily a singularity of $f(z)$.

A sophisticated criterion for determining R_M based on differencing successive approximants has been given by Vavilov, Prokhorov, and Suetin [1983], following those of Wilson [1927]; see also Baker [1975a, Chapter 11].

6.2 de Montessus's theorem

Before embarking on de Montessus's theorem, we will state the Cauchy–Binet formula. This is a formula for calculating the determinant of the product of two matrices. It combines a number of familiar results, which we give as examples, in a unified scheme [Gragg, 1972].

We define a multiindex $\boldsymbol{\alpha} = (\alpha_1 \alpha_2 \ldots \alpha_k)$ which belongs to the class

$$\mathcal{I}\binom{m}{k}$$

of multiindices with k elements chosen from $(1, 2, \ldots, m)$, and which obeys the extra condition

$$1 \leq \alpha_1 < \alpha_2 < \cdots < \alpha_k \leq m.$$

We define $A(\boldsymbol{\alpha}, \boldsymbol{\beta})$ to be the submatrix of A formed from the rows $\{\alpha_i\}$ and columns $\{\beta_j\}$ of the original matrix A.

We define $\mathcal{M}(m \times n)$ to be the class of matrices with m rows and n columns. If

$$\boldsymbol{\alpha} \in \mathcal{I}\binom{m}{l}, \quad \boldsymbol{\beta} \in \mathcal{I}\binom{n}{l}, \text{ and } A \in \mathcal{M}(m \times n),$$

then $A(\boldsymbol{\alpha}, \boldsymbol{\beta})$ is an $l \times l$ matrix belonging to $\mathcal{M}(l \times l)$.

The Cauchy–Binet theorem. Let $A \in \mathcal{M}(m \times n)$, $B \in \mathcal{M}(m \times k)$, $C \in \mathcal{M}(k \times n)$, $A = BC$,

$$\boldsymbol{\alpha} \in \mathcal{I}\binom{m}{l}, \boldsymbol{\beta} \in \mathcal{I}\binom{n}{l}, \text{ and } \boldsymbol{\gamma} \in \mathcal{I}\binom{k}{l}.$$

These conditions ensure that $A(\boldsymbol{\alpha}, \boldsymbol{\beta})$, $B(\boldsymbol{\alpha}, \boldsymbol{\gamma})$, and $C(\boldsymbol{\gamma}, \boldsymbol{\beta})$ are $l \times l$ matrices, and

$$\det A(\boldsymbol{\alpha}, \boldsymbol{\beta}) = \sum_{\boldsymbol{\gamma} \in \mathcal{I}\binom{k}{l}} \det B(\boldsymbol{\alpha}, \boldsymbol{\gamma}) \det C(\boldsymbol{\gamma}, \boldsymbol{\beta}).$$

Example 1. Let A, B, C be $m \times m$ matrices, and let

$$\boldsymbol{\alpha}, \boldsymbol{\beta}, \boldsymbol{\gamma} \in \mathcal{I}\binom{m}{m}$$

be the full m-dimensional multiindex. Then the Cauchy–Binet formula reduces to

$$\det A = \det B \det C.$$

Example 2. Let A, B, C be square $m \times m$ matrices, and let $l = 1$, so that $\boldsymbol{\alpha}, \boldsymbol{\beta}, \boldsymbol{\gamma}$ reduce to ordinary indices. Then the formula becomes

$$A_{\alpha\beta} = \sum_{\gamma} B_{\alpha\gamma} C_{\gamma\alpha}.$$

which is the matrix multiplication rule.

Example 3. If $l > k$, the set

$$\mathcal{I}\binom{k}{l}$$

is empty and

$$\det A(\boldsymbol{\alpha}, \boldsymbol{\beta}) = 0.$$

This corresponds to the property that if the rows of a matrix are linearly dependent, then its determinant is zero.

Example 4. Let $A, B, C \in \mathcal{M}(3, 3)$, $A = BC$, $\boldsymbol{\alpha} = (1, 2)$, and $\boldsymbol{\beta} = (1, 3)$. Then

$$\det \begin{bmatrix} a_{11} & a_{12} & a_{13} \\ a_{21} & a_{22} & a_{23} \\ a_{31} & a_{32} & a_{33} \end{bmatrix} = \det\left(\begin{bmatrix} b_{11} & b_{12} & b_{13} \\ b_{21} & b_{22} & b_{23} \\ b_{31} & b_{32} & b_{33} \end{bmatrix} \begin{bmatrix} c_{11} & c_{12} & c_{13} \\ c_{21} & c_{22} & c_{23} \\ c_{31} & c_{32} & c_{33} \end{bmatrix}\right)$$

$$= \sum_{j=1}^{3} \det \begin{bmatrix} b_{11} & b_{12} & b_{13} \\ b_{21} & b_{22} & b_{23} \\ b_{31} & b_{32} & b_{33} \end{bmatrix} \det \begin{bmatrix} c_{11} & c_{12} & c_{13} \\ c_{21} & c_{22} & c_{23} \\ c_{31} & c_{32} & c_{33} \end{bmatrix} \longrightarrow [\text{row } j]$$

$$\downarrow$$
$$[\text{col } j]$$

The Cauchy–Binet theorem is useful in the proof we give of de Montessus's theorem.

de Montessus's theorem applies to a function $f(z)$ which is meromorphic with M poles within a circle. In practical terms, provided M is known, de Montessus's theorem asserts convergence of the row of $[L/M]$ approximants within the circle, which is all that can possibly be established. If the number of poles of $f(z)$ within the circle is not known a priori, the theorem is usually unhelpful. The complete statement of de Montessus's theorem [1902] for simple poles is:

Theorem 6.2.1. *Let $f(z)$ be a function which is meromorphic in the disk $|z| \leq R$, with precisely M simple poles at distinct points $z_1, z_2, \ldots z_M$, where*

$$0 < |z_1| \leq |z_2| \leq \cdots \leq |z_M| < R.$$

Then

$$\lim_{L \to \infty} [L/M] = f(z) \qquad (2.1)$$

uniformly on any compact subset of

$$\mathcal{D}_M = \{z, |z| \leq R, z \neq z_i, i = 1, 2, \ldots, M\}. \qquad (2.2)$$

6.2 de Montessus's theorem

Proof. Since $|z_1| > 0$ and $f(z)$ is analytic in $|z| < |z_1|$, $f(z)$ has a power-series expansion

$$f(z) = \sum_{i=0}^{\infty} c_i z^i. \tag{2.3}$$

Define the polynomial

$$B(z) = \sum_{i=0}^{M} B_i z^i = \prod_{i=1}^{M} \left(1 - \frac{z}{z_i}\right). \tag{2.4}$$

Then the series for $A(z)$ defined by

$$\left(\sum_{i=0}^{M} B_i z^i\right)\left(\sum_{i=0}^{\infty} c_i z^i\right) = \sum_{i=0}^{\infty} A_i z^i \equiv A(z) \tag{2.5}$$

is convergent in $|z| \leq R$, and we have thereby re-expressed the function $f(z)$ as

$$f(z) = A(z)/B(z) \quad \text{with } B_0 = 1. \tag{2.6}$$

Equating coefficients of z^i in (2.5) for $i = L+1, L+2, \ldots, L+M$ with $L \geq M$ gives

$$\sum_{j=0}^{M} B_j c_{i-j} = A_i, \quad i = L+1, L+2, \ldots, L+M. \tag{2.7}$$

Now the $[L/M]$ Padé approximant is

$$[L/M] = \frac{\sum_{i=0}^{L} a_i^{[L/M]} z^i}{\sum_{j=0}^{M} b_j^{[L/M]} z^j} = \frac{A^{[L/M]}(z)}{B^{[L/M]}(z)}$$

and is defined by the Padé equations

$$\sum_{j=0}^{M} b_j^{[L/M]} c_{i-j} = 0, \quad i = L+1, L+2, \ldots, L+M, \tag{2.8}$$

with $b_0^{[L/M]} = 1$. A major part of the proof of de Montessus's theorem consists of proving that $b_j^{[L/M]} \to B_j$ as $L \to \infty$. Define

$$\Delta_j = B_j - b_j^{[L/M]}, \quad j = 0, 1, \ldots, M, \tag{2.9}$$

and note that $\Delta_0 = 0$. From (2.7) and (2.8),

$$\sum_{j=1}^{M} \Delta_j c_{i-j} = A_i, \, i = L+1, L+2, \ldots, L+M, \tag{2.10}$$

which are M equations for $\Delta_1, \Delta_2, \ldots, \Delta_M$. We define

$$C = \begin{bmatrix} c_L & \cdots & c_{L-M+1} \\ \vdots & & \vdots \\ c_{L+M-1} & \cdots & c_L \end{bmatrix}, \quad \Delta = \begin{bmatrix} \Delta_1 \\ \vdots \\ \Delta_M \end{bmatrix}, \quad \mathbf{A} = \begin{bmatrix} A_{L+1} \\ \vdots \\ A_{L+M} \end{bmatrix}, \qquad (2.11)$$

so that (2.10) becomes $C\Delta = \mathbf{A}$, and $\Delta = C^{-1}\mathbf{A}$ if C is nonsingular. Then

$$\Delta_i = \sum_{j=1}^{M}(C^{-1})_{ij}A_j = \frac{\sum_{j=1}^{M} C(j,i)A_j}{\det C} = \frac{\det C(i \to \mathbf{A})}{\det C}, \qquad (2.12)$$

where $C(j, i)$ is the cofactor of the element of C in row j and column i, and $C(i \to \mathbf{A})$ is the matrix with column i replaced by $\mathbf{A}$. To establish that $\Delta_i \to 0$, we must consider the behavior of $C(j, i)$ and $\det C$ as $L \to \infty$ [Hadamard, 1892].

Because $f(z)$ has precisely M simple poles in $|z| < R$, we consider explicitly the contribution of these poles to the Maclaurin coefficients by writing

$$\sum_{i=0}^{\infty} c_i z^i = \sum_{k=1}^{M} r_k \left(1 - \frac{z}{z_k}\right)^{-1} + \sum_{i=0}^{\infty} c_i' z^i, \qquad (2.13)$$

where $\sum_{i=0}^{\infty} c_i' z^i$ is convergent in $|z| \leq R$, and $c_i' = O(R^{-i})$. By expanding $(1 - z/z_k)^{-1}$, it follows from (2.13) that

$$c_i = d_i + O(R^{-i}), \qquad (2.14)$$

where

$$d_i = \sum_{k=1}^{M} r_k z_k^{-i} \qquad (2.15)$$

is the dominant part of the coefficients c_i. We define a matrix D by

$$D_{ij} = d_{L+i-j} = \sum_{k=1}^{M} r_k z_k^{-L-i+j}, \quad i, j = 1, 2, \ldots, M, \qquad (2.16)$$

which is the dominant part of C.

Example. For $M = 2$

$$D = \begin{bmatrix} r_1 z_1^{-L} + r_2 z_2^{-L} & r_1 z_1^{-L+1} + r_2 z_2^{-L+1} \\ r_1 z_1^{-L-1} + r_2 z_2^{-L-1} & r_1 z_1^{-L} + r_2 z_2^{-L} \end{bmatrix}$$

$$= \begin{bmatrix} 1 & 1 \\ z_1^{-1} & z_2^{-1} \end{bmatrix} \begin{bmatrix} r_1 z_1^{-L} & r_1 z_1^{-L+1} \\ r_2 z_2^{-L} & r_2 z_2^{-L+1} \end{bmatrix}$$

$$= \begin{bmatrix} 1 & 1 \\ z_1^{-1} & z_2^{-1} \end{bmatrix} \begin{bmatrix} r_1 z_1^{-L} & 0 \\ 0 & r_2 z_2^{-L} \end{bmatrix} \begin{bmatrix} 1 & z_1 \\ 1 & z_2 \end{bmatrix}.$$

We are led to define Vandermonde matrices V, V' and a diagonal matrix D' by

$$V_{ij} = (z_j)^{1-i}, \quad V'_{ij} = (z_i)^{j-1}, \quad D'_{ij} = r_i z_i^{-L} \delta_{ij} \quad (i, j = 1, 2, \ldots, M), \quad (2.17)$$

so that D may be written as

$$D = VD'V' \tag{2.18}$$

and

$$\det D = \det V \det D' \det V'. \tag{2.19}$$

We need the formulas

$$\det V = \prod_{i=2}^{M} \prod_{j=1}^{i} (z_i^{-1} - z_j^{-1}) \tag{2.20}$$

and

$$\det V' = \prod_{i=2}^{M} \prod_{j=1}^{i} (z_i - z_j), \tag{2.21}$$

which are the well-known expansions of Vandermonde determinants. Their important characteristic is that they are nonzero constants, independent of L. Hence

$$\det D = K \prod_{i=1}^{M} (z_i)^{-L}, \tag{2.22}$$

where K is a nonzero constant.

Now consider the evaluation of $\det C$. Its dominant contribution is given by (2.23) below, but (2.14) shows that there are also a large but finite number of terms given by replacing a d_i by an $O(R^{-i})$ term. Hence we define $p = |z_N|$ to be the modulus of the pole furthest from the origin in $|z| < R$, and

$$\det C = K \prod_{i=1}^{M} (z_i)^{-L} \left(1 + O\left(\frac{p}{R}\right)^m\right). \tag{2.23}$$

In a similar way, using the factorization (2.18) and the Cauchy–Binet theorem, we may show that

$$C(j, i) = O\left(\prod_{i=1}^{M-1} |z_i|^{-L}\right) = O\left(p^L \prod_{i=1}^{M} |z_i|^{-L}\right). \tag{2.24}$$

Since (2.5) converges in $|z| \leq R$, $A_i = O(R^{-i})$. Hence, from (2.12), (2.23), and (2.24),

$$\Delta_i = O\left(\frac{p}{R}\right)^L, \quad i = 1, 2, \ldots, M, \tag{2.25}$$

which establishes that $\Delta_i \to 0$ uniformly as $L \to \infty$.

It now remains to consider

$$f(z) - [L/M] = \frac{A(z)B^{[L/M]}(z) - B(z)A^{[L/M]}(z)}{B(z)B^{[L/M]}(z)}. \tag{2.26}$$

Now

$$A(z)B^{[L/M]}(z) = \left(\sum_{i=0}^{\infty} A_i z^i\right)\left(\sum_{j=0}^{M} b_j^{[L/M]} z^j\right)$$

$$= \sum_{j=0}^{M} b_j^{[L/M]} \sum_{i=j}^{\infty} A_{i-j} z^i.$$

By using the formal identity defining the approximants and noting that $B(z)A^{[L/M]}$ is a polynomial of degree $L + M$, it follows that

$$A(z)B^{[L/M]}(z) - A^{[L/M]}(z)B(z) = \sum_{j=0}^{M} b_j^{[L/M]} \sum_{i=L+M+1}^{\infty} A_{i-j} z^i.$$

Hence

$$|A(z)B^{[L/M]}(z) - A^{[L/M]}(z)B(z)| \leq \sum_{j=0}^{M} |b_j^{[L/M]}| \cdot \left|\sum_{i=L+M+1}^{\infty} A_{i-j} z^i\right|.$$

Since $\Delta_i \to 0$, $b_j^{[L/M]}$ are bounded as $L \to \infty$. Since $A(z)$ is analytic in $|z| \leq R$, for some $R' > R$

$$|A(z)B^{[L/M]}(z) - B(z)A^{[L/M]}(z)| = O((z/R')^L), \tag{2.27}$$

and the numerator of (2.26) tends to zero as $L \to \infty$ for $|z| \leq R$. To treat the denominator, we consider

$$|B(z) - B^{[L/M]}(z)| \leq \sum_{j=1}^{M} |B_j - b_j^{[L/M]}| |z^j|. \tag{2.28}$$

The right-hand side of (2.28) tends to zero as $L \to \infty$ for $|z| \leq R$. Let z belong to a compact subset of this disk for which $|B(z)| > 2\varepsilon$ for any given positive ε. Hence $L(\varepsilon)$ exists such that $|B^{[L/M]}(z)| > \varepsilon$ for all $L \to L(\varepsilon)$. Then the modulus of the denominator of (2.26) is bounded below by $2\varepsilon^2$ independently of L for $L > L(\varepsilon)$. Together with (2.27), it follows from (2.26) that $f(z) - [L/M] \to 0$ as $L \to \infty$ uniformly on any compact subset of $\mathcal{D}_M$, and the theorem is proved.

Now we proceed with the full theorem of R. de Montessus de Ballore. This theorem allows for the possibility of multiple poles instead of simple poles, and this inclusion makes the proof technically complicated.

Theorem 6.2.2 [de Montessus, 1902]. *Let $f(z)$ be a function which is meromorphic in the disk $|z| \leq R$, with m poles at distinct points z_1, z_2, ..., z_m with*

6.2 de Montessus's theorem

$$0 < |z_1| \leq |z_2| \leq \cdots \leq |z_m| < R.$$

Let the pole at z_k have multiplicity μ_k, and let the total multiplicity $\sum_{k=1}^{m} \mu_k = M$ precisely. Then

$$f(z) = \lim_{L \to \infty} [L/M]$$

uniformly on any compact subset of

$$\mathcal{D}_m = \{z, |z| \leq R, z \neq z_k, k = 1, 2, \ldots, m\}.$$

Proof. The beginning and the end of the proof are substantially the same as for Theorem 6.2.1.

We define

$$B(z) = \sum_{i=0}^{M} B_i z^i = \prod_{k=1}^{m} \left(1 - \frac{z}{z_k}\right)^{\mu_k} \quad (2.29)$$

so that $f(z) = A(z)/B(z)$ with $A(z)$ analytic in $|z| \leq R$. We replace (2.13) by

$$\sum_{i=0}^{\infty} c_i z^i = \sum_{k=1}^{m} \sum_{\tau=1}^{\mu_k} r_\tau^{(k)} \left(1 - \frac{z}{z_k}\right)^{-\tau} + \sum_{i=0}^{\infty} c'_i z^i, \quad (2.30)$$

so that $c_i = d_i + O(R^{-i})$, with

$$d_i = \sum_{k=1}^{m} \sum_{\tau=1}^{\mu_k} r_\tau^{(k)} \binom{-\tau}{i} (-z_k)^{-i}.$$

The binomial coefficient is

$$\binom{-\tau}{i} = \frac{(-\tau)(-\tau-1)\cdots(-\tau-i+1)}{1 \times 2 \times \cdots \times i}$$

$$= \frac{(i+1)(i+2)\cdots(i+\tau-1)}{1 \times 2 \times \cdots \times (\tau-1)} (-1)^i$$

and it obeys the Vandermonde identity [Riordan, 1966, p. 9], which follows by equating coefficients of $x^{\tau-1}$ in $(1+x)^{L+i-j+\tau-1} = (1+x)^{L+i+\tau-1}(1+x)^{-j}$:

$$(-1)^{L+i-j} \binom{-\tau}{L+i-j} = \sum_{\sigma=0}^{\tau-1} \binom{-j}{\sigma} \binom{L+i+\tau-1}{\tau-\sigma-1}. \quad (2.31)$$

Hence the elements $d_{L+i-j}(i, j = 1, 2, \ldots, M)$ of the matrix D are

$$d_{L+i-j} = \sum_{k=1}^{m} \sum_{\tau=1}^{\mu_k} r_\tau^{(k)} (z_k)^{-L-i+j} \sum_{\sigma=0}^{\tau-1} \binom{-j}{\sigma} \binom{L+i+\tau-1}{\tau-\sigma-1},$$

which may be cast as the elements of a matrix product

$$d_{L+i-j} = \sum_{k=1}^{m}\sum_{\sigma=0}^{\mu_k-1}\left[\underbrace{\sum_{\tau=\sigma+1}^{\mu_k} r_\tau^{(k)}(z_k)^{-L-i}\binom{L+i+\tau-1}{\tau-\sigma-1}}_{\text{row suffix } i, \text{ column label } (k,\sigma)}\right]\left[\underbrace{(z_k)^j\binom{-j}{\sigma}}_{\text{row label } (k,\sigma), \text{ column suffix } j}\right].$$

Furthermore,

$$\sum_{\tau=\sigma+1}^{\mu_k} r_\tau^{(k)}(z_k)^{-L-i}\binom{L+i+\tau-1}{\tau-\sigma-1}$$

$$= \sum_{\tau=\sigma+1}^{\mu_k} r_\tau^{(k)}(z_k)^{-L-i} \sum_{\sigma'=0}^{\tau-\sigma-1}\binom{L+i}{\sigma'}\binom{\tau-1}{\tau-1-\sigma-\sigma'}$$

$$= \sum_{\sigma'=0}^{\mu_k-\sigma-1}(z_k)^{-i}\binom{L+i}{\sigma'}\sum_{\tau=\sigma'+\sigma+1}^{\mu_k} r_\tau^{(k)}(z_k)^{-L}\binom{\tau-1}{\sigma+\sigma'}$$

$$= \underbrace{\sum_{\sigma'=0}^{\mu_k-1}(z_k)^{-i}\binom{L+i}{\sigma'}}_{\text{row label } i, \text{ column label } \sigma'}$$

$$\times \underbrace{\sum_{\tau=\sigma+\sigma'+1}^{\mu_k}\theta(\mu_k-\sigma-\sigma'-1)r_\tau^{(k)}(z_k)^{-L}\binom{\tau-1}{\sigma+\sigma'}}_{\text{row label } \sigma', \text{ column label } (k,\sigma)}.$$

This provides a complete breakdown of D as

$$D_{ij} = \sum_{(k'\sigma')}\sum_{(k\sigma)} V_{i,(k'\sigma')} D'_{(k'\sigma')(k\sigma)} V'_{(k\sigma)j},$$

where

$$D'_{(k'\sigma')(k\sigma)} = \delta_{kk'}\sum_{\tau=\sigma'+\sigma+1}^{\mu_k}\theta(\mu_k-\sigma-\sigma'-1)r_\tau^{(k)}(z_k)^{-L}\binom{\tau-1}{\sigma+\sigma'},$$

$$V_{i,(k'\sigma')} = (z_{k'})^{-i}\binom{L+i}{\sigma'}, \tag{2.32a}$$

and

$$V'_{(k\sigma)j} = (z_k)^j\binom{-j}{\sigma}. \tag{2.32b}$$

V and V' are basically Vandermonde matrices, which have nonzero determinants. The matrix D' takes block diagonal form induced by the Kronecker factor of $\delta_{kk'}$. Each block of D' has triangular form, expressed

6.2 de Montessus's theorem

by the Heaviside factor $\theta(\mu_k - \sigma - \sigma' - 1)$. Hence it follows that

$$\det D = K \prod_{k=1}^{m} [(z_k)^{\mu_k}]^{-L}, \qquad (2.33)$$

where K is a nonzero constant. Notice that only the residues $r_{\mu_k}^{(k)}$ are required to be nonzero. Equation (2.31) is the equivalent of (2.22) in the theorem for simple poles, and the remainder of the proof follows that of the main theorem very closely from that point onward.

Example. Let $f(z)$ have a triple pole at $z = z_1$ and a double pole at $z = z_2$. Then $\mu_1 = 3$, $\mu_2 = 2$, and the $(k'\sigma')$ indices are $(1, 0)$, $(1, 1)$, $(1, 2)$, $(2, 0)$, $(2, 1)$. We have

$V_{i,(k'\sigma')}$

$$= \begin{bmatrix} z_1^{-1} & (L+1)z_1^{-1} & \frac{1}{2}(L+1)(L+2)z_1^{-1} & z_2^{-1} & (L+1)z_2^{-1} \\ z_1^{-2} & (L+2)z_1^{-2} & \frac{1}{2}(L+2)(L+3)z_1^{-2} & z_2^{-2} & (L+2)z_2^{-2} \\ z_1^{-3} & (L+3)z_1^{-3} & \frac{1}{2}(L+3)(L+4)z_1^{-3} & z_2^{-3} & (L+3)z_2^{-3} \\ z_1^{-4} & (L+4)z_1^{-4} & \frac{1}{2}(L+4)(L+5)z_1^{-4} & z_2^{-4} & (L+4)z_2^{-4} \\ z_1^{-5} & (L+5)z_1^{-5} & \frac{1}{2}(L+5)(L+6)z_1^{-5} & z_2^{-5} & (L+5)z_2^{-5} \end{bmatrix}.$$

(2.34)

Determinants of these matrices are sometimes called confluent alternants [Aitken, 1964, p. 120]; for confluence, see Graves-Morris and Johnson [1990, 1995]. The $(1, \sigma') \times (1, \sigma)$ block of $D_{(k'\sigma')(k\sigma)}$ is

$$\begin{array}{c} \\ \sigma' = 0 \\ \sigma' = 1 \\ \sigma' = 2 \end{array} \begin{bmatrix} \overset{\sigma=0}{[r_1^{(1)} + r_2^{(1)} + r_3^{(1)}]z_1^{-L}} & \overset{\sigma=1}{-[r_2^{(1)} + 2r_3^{(1)}]z_1^{-L}} & \overset{\sigma=2}{r_3^{(1)} z_1^{-1}} \\ -[r_2^{(1)} + 2r_3^{(1)}]z_1^{-L} & r_3^{(1)} z_1^{-L} & 0 \\ r_3^{(1)} z_1^{-L} & 0 & 0 \end{bmatrix},$$

and the triangular structure is self-evident.

There are several features of de Montessus's theorem which are worth emphasizing. Firstly, we assumed meromorphy in $|z| \leq R$, and that all the poles lie in $|z| < R$. Although it is unusual to assume meromorphy on closed sets, and indeed our felony is compounded by using the implication of analyticity in a slightly larger annulus, it is quite convenient, and we prove the usual result about uniform convergence in $|z| \leq R$ except in neighborhoods of the poles. The alternative assumption (analyticity on an open set) leads to uniform convergence on a compact subset.

Secondly, there are many generalizations of the theorem. Much work has been done on the problem of one mild singularity on the boundary $|z| = R$, for which the dominant part of the coefficients c_i is a 'smooth' function of i, and controllable in the spirit of the proof given [Wilson, 1928a, b, 1930].

Thirdly, we repeat that the number of poles M in the circle $|z| < R$ is fixed precisely. The necessity of this condition is seen by the counter-example $f(z) = (1 - z^2)^{-1}$. Inside $|z| < 1$, $f(z)$ is analytic and the $[L/0]$ Padé approximants converge. For all z, the $[L/2]$ Padé approximants are exact. But the $[L/1]$ Padé approximants are Maclaurin polynomials if L is odd and nonexistent if L is even. In fact, the $[L/1]$ are less useful than any other row, and in general an ill-chosen row sequence for a function having poles equidistant from the origin is best avoided.

Lastly, we observe that we have given a kind of constructive proof of de Montessus's theorem. Indeed, an error formula follows by this method [Gragg, 1972]. There is also an intimate connection with symmetric polynomials [Baker, 1975a, p. 135]. This proof generalizes, in a way, to more than one dimension [Chisholm and Graves-Morris, 1975; Graves-Morris, 1977]. For these reasons, the preceding proof is important, but in the next section we give a simpler, implicit proof.

6.3 Hermite's formula and de Montessus's theorem

In this section we prove Hermite's formula for Maclaurin's expansion and show that it leads to a useful formula for the error incurred in Padé approximation. Then these results are used in several elegant proofs of de Montessus's theorem.

Hermite's formula. *If $f(z)$ is analytic inside a contour Γ enclosing the origin and $f(z)$ is continuous on Γ, then its $[L/0]$ Padé approximant is given by*

$$[L/0] = \frac{1}{2\pi i} \int_\Gamma \frac{v^{L+1} - z^{L+1}}{v - z} \frac{f(v)}{v^{L+1}} \, dv. \qquad (3.1)$$

Comment. This formula is the special case of Hermite's more general formula, Theorem 7.1.1, which is needed in this section.

Proof.

$$\frac{v^{L+1} - z^{L+1}}{v - z} = \sum_{j=0}^{L} v^{L-j} z^j,$$

and so $[L/0]$ defined by (3.1) is a polynomial in z of degree L. In fact,

$$[L/0] = \sum_{j=0}^{L} z^j \frac{1}{2\pi i} \int_\Gamma v^{-j-1} f(v) \, dv = \sum_{j=0}^{L} z^j \frac{f^{(j)}(0)}{j!},$$

which proves the formula for the truncated Maclaurin expansion of $f(z)$.

6.3 Hermite's formula and de Montessus's theorem

Corollary. *An error formula for Padé approximation is given by*

$$f(z) - [L/M] = \frac{z^{L+M+1}}{2\pi i Q^{[L/M]}(z) R_M(z)} \int_\Gamma \frac{f(v) Q^{[L/M]}(v) R_M(v)}{v^{L+M+1}(v-z)} dv \quad (3.2)$$

with the hypothesis that $f(z)$ is analytic in and continuous on a contour Γ which encloses both z and the origin. $R_M(z)$ is an arbitrary polynomial of degree at most M, but not identically zero. [$R_M(z)$ may be taken to be $R_M(z) = 1$, or whatever is most convenient in context.]

Proof by method 1. Interpolate $f(z)Q^{[L/M]}(z)R_M(z)$ to order $L+M$ using Hermite's formula

$$\pi_{L+M}(z) = \frac{1}{2\pi i} \int_\Gamma \frac{v^{L+M+1} - z^{L+M+1}}{v-z} \frac{f(v) Q^{[L/M]}(v) R_M(v)}{v^{L+M+1}} dv.$$

But, by Cauchy's theorem,

$$f(z) Q^{[L/M]}(z) R_M(z) = \frac{1}{2\pi i} \int_\Gamma \frac{1}{v-z} f(v) Q^{[L/M]}(v) R_M(v) \, dv,$$

and subtraction yields

$$f(z) Q^{[L/M]}(z) R_M(z) - \pi_{L+M}(z) = \frac{1}{2\pi i} \int \left(\frac{z}{v}\right)^{L+M+1} \frac{f(v) Q^{[L/M]}(v) R_M(v)}{v-z} dv.$$

(3.3)

The Padé equations require

$$f(z) Q^{[L/M]}(z) R_M(z) = \{P^{[L/M]}(z) + O(z^{L+M+1})\} R_M(z)$$
$$= P^{[L/M]}(z) R_M(z) + O(z^{L+M+1}),$$

and it follows that

$$\pi_{L+M}(z) = P^{[L/M]}(z) R_M(z) + O(z^{L+M+1})$$
$$= P^{[L/M]}(z) R_M(z),$$

since $R_M(z)$ is a polynomial of order not greater than M. The result (3.2) now follows from (3.3) by division.

Proof by method 2. We present a second method of proof to explain a paradox occurring in the use of this equation. Apply Cauchy's theorem to the function:

$$\phi(z) = R_M(z)\{f(z) Q^{[L/M]}(z) - P^{[L/M]}(z)\} z^{-L-M-1}.$$

The Padé equations ensure that $\phi(z)$ is analytic in a neighborhood of the origin, so

$$\phi(z) = \frac{1}{2\pi i} \int_\Gamma \frac{\phi(v)}{v - z} dv,$$

where Γ is any simple closed contour enclosing z and the origin. Hence

$$f(z) - [L/M] = \frac{z^{L+M+1}}{2\pi i Q^{[L/M]}(z) R_M(z)}$$

$$\times \left\{ \int_\Gamma \frac{R_M(v) f(v) Q^{[L/M]}(v) \, dv}{(v-z) v^{L+M+1}} + E(z) \right\}, \quad (3.4)$$

where

$$E(z) = \int_\Gamma \frac{R_M(v) P^{[L/M]}(v) \, dv}{v^{L+M+1}(v - z)}.$$

The integrand of $E(z)$ is analytic in $|v| > |z|$, and so the contour Γ may be expanded to infinity. The integrand is $O(v^{-2})$, since $R_M(v)$ has order no greater than M, and therefore $E(z) = 0$, explaining the paradox of (3.4) and proving the error formula (3.2).

We now apply these results to two other proofs of de Montessus's theorem.

Proof by method 3. [Saff, 1972]. The hypothesis of de Montessus's theorem is that $f(z)$ is meromorphic in $|z| \leq R$ with poles of total multiplicity M, and it is convenient to choose the polynomial $R_M(z)$ to annihilate these poles.

We consider first the case when the poles of $f(z)$ are simple and are located at $z_1, z_2, \ldots, z_M$. Form the polynomials

$$R_0(z) = 1,$$

$$R_k(z) = (z - z_1)(z - z_2) \cdots (z - z_k) \quad \text{for } k = 1, 2, \ldots, M.$$

Then it follows that $R_M(z) f(z)$ is analytic in $|z| < R$. We will also need the monic polynomial $t^{(L,M)}(z) = R_M(z) + \sum_{k=1}^{M} a_k^{(L,M)} R_{k-1}(z)$ of degree M defined in terms of the arbitrary coefficients $a_k^{(L,M)}$, which will be fixed presently.

Hermite's formula is used to construct the Maclaurin polynomial of degree $L + M$ which interpolates $t^{(L,M)}(z) R_M(z) f(z)$ at the origin. Since this function is analytic in $|z| \leq R$,

$$\pi_{L+M}(z) = \frac{1}{2\pi i} \int_{|v|=R} \frac{v^{L+M+1} - z^{L+M+1}}{v - z} \frac{t^{(L,M)}(v) R_M(v) f(v)}{v^{L+M+1}} dv.$$

6.3 Hermite's formula and de Montessus's theorem

$$= \frac{1}{2\pi i} \int_{|v|=R} \frac{1-(z/v)^{L+M+1}}{v-z} \left\{ R_M(v) + \sum_{k=1}^{M} a_k^{(L,M)} R_{k-1}(v) \right\}$$

$$\times R_M(v) f(v) \, dv. \tag{3.5}$$

The idea is to choose the coefficients $a_k^{(L,M)}$ so that $R_M(z)$ is a factor of $\pi_{L+M}(z)$. This factorization is accomplished by setting $\pi_{L+M}(z_j) = 0$ for $j = 1, 2, \ldots, M$, and leads to the set of linear equations for $a_k^{(L,M)}$

$$\sum_{k=1}^{M} c_{jk}^{(L,M)} a_k^{(L,M)} = d_j^{(L,M)}, \tag{3.6}$$

where

$$c_{jk}^{(L,M)} = \frac{1}{2\pi i} \int_{|v|=R} \frac{1-(z_j/v)^{L+M+1}}{v-z_j} R_{k-1}(v) R_M(v) f(v) \, dv \tag{3.7}$$

and

$$d_j^{(L,M)} = \frac{-1}{2\pi i} \int_{|v|=R} \frac{1-(z_j/v)^{L+M+1}}{v-z_j} [R_M(v)]^2 f(v) \, dv. \tag{3.8}$$

The principal feature of (3.7) and (3.8) is their well-defined limit as $L \to \infty$. From (3.8),

$$d_j^{(L,M)} \to d_j = \frac{-1}{2\pi i} \int_{|v|=R} \frac{[R_M(v)]^2 f(v) \, dv}{v-z_j}$$

$$= 0, \quad \text{using Cauchy's theorem,}$$

and from (3.7),

$$c_{jk}^{(L,M)} \to c_{jk} = \frac{1}{2\pi i} \int_{|v|=R} \frac{R_k(v) R_M(v) f(v) \, dv}{v-z_j}. \tag{3.9}$$

If $j < k$, $c_{jk} = 0$. If $j = k$, $c_{jj} \neq 0$, and this value depends only on the residues and positions of the M poles of $f(z)$. We discover, first, that for L sufficiently large, the system of equations (3.6) for $a_k^{(L,M)}$ is nonsingular, and second, that $a_k^{(L,M)} \to 0$ as $L \to \infty$. Consequently, $\{a_k^{(L,M)}\}$ exists such that $R_M(z)$ is a factor of $\pi_{L+M}(z)$. From the specification of (3.5), we can identify the Padé polynomials separately in

$$\frac{\pi_{L+M}(z)/R_M(z)}{t^{(L,M)}(z)} = \frac{P^{[L/M]}(z)}{Q^{[L/M]}(z)} = f(z) + O(z^{L+M+1}),$$

and furthermore $t^{(L,M)}(z) \to R_M(z)$ as $L \to \infty$. To complete the proof, the error formula gives

$$f(z) - [L/M] = \frac{z^{L+M+1}}{2\pi i Q^{[L/M]}(z) R_M(z)} \int_{|v|=R} \frac{f(v) Q^{[L/M]}(v) R_M(v)}{v^{L+M+1}(v-z)} \, dv$$

$$= O(z^{L+M+1}) \tag{3.10}$$

and

$$\limsup_{L \to \infty} |f(z) - [L/M]|^{1/L} \leq |z|/R, \tag{3.11}$$

showing convergence for $|z| < R$.

Finally we consider the case in which at least one of the poles is a multipole. Suppose that z_1 is a pole of order p. Then $(z - z_1)^p$ is a factor of $\pi_{L+M}(z)$ provided

$$\left(\frac{d}{dz_1}\right)^n \pi_{L+M}(z_1) = 0, \quad n = 0, 1, \ldots, p-1,$$

in (3.5). The polynomials $\{R_k(z), k = 0, 1, \ldots, M\}$ are defined by confluence; for example, $R_k(z) = (z - z_1)^k$, $k = 0, 1, \ldots, p$. The equivalent of (3.9) is

$$c_{jk}^{(L,M)} \to c_{jk} = \frac{(j-1)!}{2\pi} \int_{|v|=R} \frac{R_k(v) R_M(v) f(v) \, dv}{(v - z_1)^j}, \ j, k = 1, 2, \ldots, p.$$

If $j < k$, $c_{jk} = 0$ and if $j = k$, $c_{jj} \neq 0$. Any other multipoles are treated similarly, and the proof for the case of simple poles is otherwise unchanged.

This method enables de Montessus's theorem to be generalized to Saff's theorem about convergence of row sequences of multipoint Padé approximants (see Section 7.1) on a simply connected point set $\mathscr{E}$, except at the M poles of the given function and subject to an appropriate selection of the interpolation points within the region of analyticity of the given function. Saff's method also generalizes to deal with the case in which $f(z)$ is a vector- or matrix-valued function [Roberts, 1995].

The fourth, and most powerful, method of proof of de Montessus's theorem characterizes the convergence of the denominator polynomial, and has considerable capacity for generalization and extension.

Proof by method 4. Again, let $f(z)$ have m poles at points z_k, $k = 1, 2, \ldots, m$, each of multiplicity m_k in $|z| < R$ so that $M = \sum_{k=1}^{m} m_k$ is their total multiplicity. Define

$$R_M(z) = \prod_{k=1}^{m} (z - z_k)^{m_k} \tag{3.12}$$

so that

$$g(z) = f(z) R_M(z) \tag{3.13}$$

is analytic in $|z| \leq R$, and

$$g(z_k) \neq 0, \quad k = 1, 2, \ldots, m. \tag{3.14}$$

6.3 Hermite's formula and de Montessus's theorem

Suppose that $p_N(z)$, $q_N(z)$ are classical Padé polynomials of type $[L/M]$ with $L + M = N$ obtained (say) by solving the homogeneous linear system (1.1.5), (1.1.7).

Next, consider a Hermite basis of polynomials

$$B = \{B_{k,s}(z),\ k = 1, 2, \ldots, m,\ s = 0, 1, \ldots, m_k - 1\}$$

such that both

$$\deg\{B_{k,s}(z)\} \leq M - 1 \quad \text{for all } k, s$$

and the polynomials interpolate at the points z_i in the Hermite sense:

$$\left[\left(\frac{d}{dz}\right)^j B_{k,s}(z)\right]_{z=z_i} = \delta_{ik} \cdot \delta_{js}, \quad 1 \leq i \leq m, 0 \leq j \leq m_i - 1. \quad (3.15)$$

Then we can express

$$q_N(z) = \sum_{k=1}^{m} \sum_{s=0}^{m_k-1} q_N^{(s)}(z_k) B_{k,s}(z) + q_{N,M} R_M(z). \quad (3.16)$$

We suppose that the solution of (1.1.5) for the coefficients of $q_N(z)$ has been normalized so that both

$$\sum_{k=1}^{m} \sum_{s=0}^{m_k-1} |q_N^{(s)}(z_k)| + q_{N,M} = 1 \quad (3.17)$$

and $q_{N,M} \geq 0$, for all N.

Then, from (3.13) and Padé approximation,

$$q_N(z)g(z) = q_N(z)f(z)R_M(z)$$
$$= p_N(z)R_M(z) + O(z^{N+1}). \quad (3.18)$$

Because $p_N(z)R_M(z)$ is a polynomial of degree N at most, Hermite's error formula can be used:

$$q_N(z)g(z) = p_N(z)R_M(z) + \frac{z^{N+1}}{2\pi i}\int_\Gamma \frac{q_N(t)g(t)\,dt}{t^{N+1}(t-z)}, \quad (3.19)$$

where Γ is the circle $|t| = R$. On taking $z = z_k$ in (3.19),

$$q_N(z_k)g(z_k) = \frac{z_k^{N+1}}{2\pi i}\int_\Gamma \frac{q_N(t)g(t)\,dt}{t^{N+1}(t-z_k)}. \quad (3.20)$$

Because (3.17) leads to an upper bound on $q_N(t)$ on Γ, it follows from (3.14) and (3.20) that

$$\limsup_{N\to\infty} |q_N(z_k)|^{1/N} \leq \frac{|z_k|}{R}. \quad (3.21)$$

Now make the inductive hypothesis that the jth derivative of $q_N(z)$ evaluated at the pole z_k satisfies

$$\limsup_{N\to\infty} |q_N^{(j)}(z_k)|^{1/N} \le \frac{|z_k|}{R}, \quad j = 0, 1, \ldots, l-1, \qquad (3.22)$$

for some $l \ge 1$ and all k for which $l < m_k$. This result has just been proved in (3.21) for the case $l = 1$. Differentiate (3.19) l times and substitute $z = z_k$:

$$q_N^{(l)}(z_k)g(z_k) + \sum_{i=0}^{l-1}\binom{l}{i}q_N^{(i)}(z_k)g^{(l-i)}(z_k)$$

$$= [p_N(z)R_M(z)]^{(l)}(z_k) + \left[\left(\frac{d}{dz}\right)^l \frac{z^{N+1}}{2\pi i}\int_\Gamma \frac{q_N(t)g(t)\,dt}{t^{N+1}(t-z)}\right]_{z=z_k}. \qquad (3.23)$$

Provided $l \le m_k - 1$, the first term on the right-hand side vanishes. With (3.21), the second term on the left-hand side obeys

$$\limsup_{N\to\infty} \left|\sum_{i=0}^{l-1}\binom{l}{i}q_N^{(i)}(z_k)g^{(l-i)}(z_k)\right|^{1/N} \le \frac{|z_k|}{R}.$$

Using Leibnitz formula and then by direct estimation,

$$\limsup_{N\to\infty} \left|\left[\left(\frac{d}{dz}\right)^l \frac{z^{N+1}}{2\pi i}\int_\Gamma \frac{q_N(t)g(t)}{t^{N+1}(t-z)}\,dt\right]_{z=z_k}\right|^{1/N} \le \frac{|z_k|}{R},$$

because (3.16), (3.17) imply that $q_N(z)$ is uniformly bounded. Hence, from (3.23) and these estimates,

$$\limsup_{N\to\infty} |q_N^{(l)}(z_k)|^{1/N} \le \frac{|z_k|}{R},$$

and therefore

$$\limsup_{N\to\infty} |q_N^{(j)}(z_k)|^{1/N} \le \frac{|z_k|}{R}, \quad \begin{cases} k = 1, 2, \ldots, m, \\ j = 0, 1, \ldots, m_k - 1. \end{cases} \qquad (3.24)$$

Consequently, (3.17) implies that $\lim_{N\to\infty} q_{N,M} = 1$, and (3.16) implies that

$$\lim_{N\to\infty} q_N(z) = R_M(z). \qquad (3.25)$$

The remainder of the proof follows directly from Hermite's error formula, as in method 2.

It is formula (3.24) which specifies the way in which $q_N(z) \to R_M(z)$ as $N \to \infty$ which is the advantage of this proof. The basic method is due to

Wallin [1984] and Graves-Morris and Saff [1984]. Wallin used it to establish a divergence result for the approximants on the boundary of the region of meromorphy enclosing the M poles, and his proof covers the case of regular multipoint Padé approximation too (see Section 7.1). Graves-Morris and Saff used the method to derive an analogue of de Montessus's theorem for simultaneous Padé approximation (see Section 8.1).

6.4 Uniqueness of convergence

A striking feature of any Padé approximant is that it is a rational function. Consequently, if the limit of a sequence of Padé approximants is to be at all useful, this limit should be meromorphic (or even holomorphic) in some substantial region of the complex plane. The expectation is that the pole positions and the residues of the Padé approximants converge to those of the limit function. But because the values of the Padé approximants near the poles of the limit function are arbitrarily large, ordinary convergence at the pole is not possible. The easiest technical device which circumvents this difficulty is the use of the chordal metric. The notion of convergence is replaced by that of convergence on the sphere. Many theorems about convergence of Padé approximants to meromorphic functions are more elegantly re-expressed in terms of convergence on the sphere, and then the poles are not classed as belonging to an exceptional set [Baker, 1974]. For example, de Montessus's theorem takes the form

If $f(z)$ is meromorphic in $|z| \leq R$, has poles of total multiplicity M in $|z| < R$, and is analytic at the origin, then the row sequence $[L/M] \to f(z)$ uniformly on the sphere in $|z| \leq R$.

To define the chordal metric [Ostrowki, 1925; Hille, 1959, p. 42], let v_1 and v_2 be two complex numbers. These might be the values $v_1 = f(z_1)$, $v_2 = f(z_2)$ of a meromorphic function. Then $\chi(v_1, v_2)$, which defines the distance between v_1 and v_2 in the chordal metric, is the chord length P_1P_2 on the Riemann sphere in Figure 6.4.1. P_1, P_2 are the points where the unit sphere centered at the origin intersects the line from its north pole N to v_1, v_2 in the equatorial plane. The origin, $z = 0$, is mapped to the south pole S, and $z = \infty$ is mapped to the unique point N. A geometrical calculation shows that

$$\chi(v_1, v_2) = P_1P_2 = \frac{2|v_1 - v_2|}{\sqrt{|1 + v_1^* v_2|^2 + |v_1 - v_2|^2}}$$
$$= \frac{2|v_1 - v_2|}{\sqrt{1 + |v_1|^2}\sqrt{1 + |v_2|^2}}. \qquad (4.1)$$

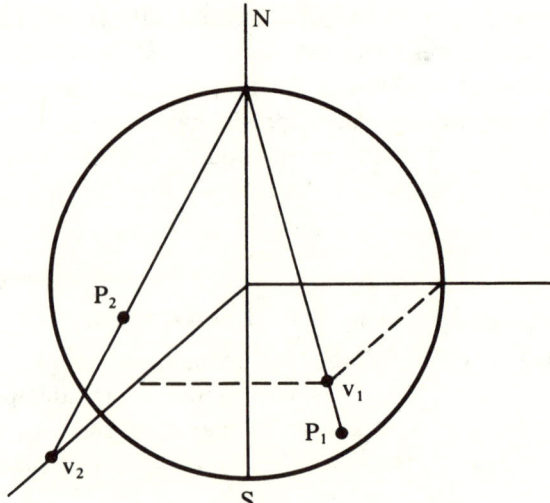

Figure 6.4.1. The Riemann sphere.

Proof. Figure 6.4.2 shows a vertical plane containing the north and south poles, and the complex point v_1. By simple trigonometry $NV_1 = \sec \alpha_1$ and $NP_1 = 2\cos \alpha_1$, and therefore $NV_1 \cdot NP_1 = 2$.

Figure 6.4.3 shows the plane of the lines $NV_1 P_1$ and $NV_2 P_2$ as well as the chord $P_1 P_2$ required.

Because $NV_1 \cdot NP_1 = NV_2 \cdot NP_2 = 2$, the triangles $NV_1 V_2$ and $NP_2 P_1$ are similar and so

$$\frac{NV_1}{NP_2} = \frac{NV_2}{NP_1} = \frac{V_1 V_2}{P_1 P_2}.$$

Thus $P_1 P_2 = V_1 V_2 \cos \alpha_1 (2\cos \alpha_2)$, and Equation (4.1) follows immediately.

The mapping of the Riemann sphere onto the complex plane is especially useful when the limit $z \to \infty$ is uniquely defined, because this limit corresponds to the unique limit $\mathbf{r} \to \mathbf{ON}$.

The chordal metric obeys an inequality which follows directly from (4.1):

$$\chi(v_1, v_2) < 2|v_1 - v_2|. \tag{4.2}$$

This proves mathematically that nearby points in the complex plane are mapped to nearby points on the sphere. Elementary geometry can also be used to prove that

$$\chi(\infty, v) = \frac{2}{\sqrt{1 + |v|^2}} \tag{4.3}$$

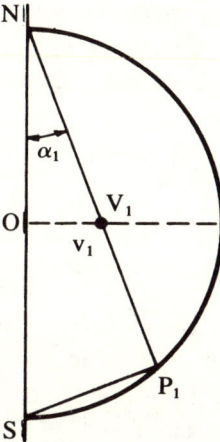

Figure 6.4.2. A section of the Riemann sphere. The point V_1 corresponds to the value v_1 in the complex plane.

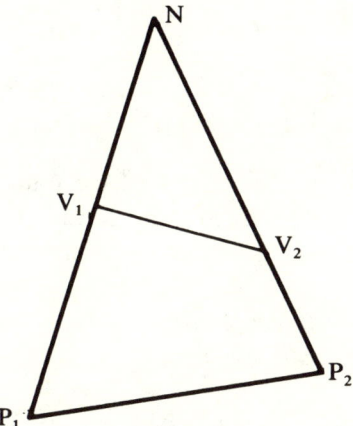

Figure 6.4.3. The triangle NP_1P_2. The points V_1, V_2 correspond to values v_1, v_2 in the complex plane.

and that

$$\chi(v_1, v_2) \leq 2 \tag{4.4}$$

for all v_1 and v_2. The geometrical proof of (4.4) is quite obvious from Figure 1.

From the definition of the chordal metric, we will prove

Theorem 6.4.1. *Any meromorphic function is continuous on the sphere.*

Proof. Let $f(z)$ be the meromorphic function. First, let z_0 be any regular point of $f(z)$, so that $f(z)$ is analytic in $|z - z_0| < \rho$ for some $\rho > 0$. Then $f(z)$ is continuous at $z = z_0$, and for any $\varepsilon > 0$, a δ may be found such that

$$|f(z) - f(z_0)| < \tfrac{1}{2}\varepsilon \quad \text{for } |z - z_0| < \delta,$$

and from (4.2),

$$\chi(f(z), f(z_0)) < \varepsilon \quad \text{for } |z - z_0| < \delta,$$

proving continuity in the chordal metric at regular points.

If z_0 is not a regular point, it is a pole of $f(z)$. In this case, m exists such that $\phi(z) = (z - z_0)^m f(z)$ is regular at $z = z_0$, $\phi(z_0) \neq 0$, and m is an integer with $m \geq 1$. Given any $\varepsilon > 0$,

$$|f(z)| > 2/\varepsilon \quad \text{and} \quad \chi(f(z), \infty) < \varepsilon$$

for all z such that both

$$|z - z_0| < \left| \frac{\varepsilon}{4} \phi(z_0) \right|^{1/m}$$

and

$$|\phi(z) - \phi(z_0)| < \tfrac{1}{2} |\phi(z_0)|.$$

Since $\phi(z)$ is continuous at $z = z_0$, we have found implicitly a neighborhood in which $\chi(f(z), \infty) < \varepsilon$, proving continuity at any pole. In other words, the mapping $z \to f(z)$ is continuous in the chordal metric.

Corollary. *Any meromorphic function defined on a compact region is uniformly continuous on the sphere.*

Proof. The mapping $z \to P(f(z))$, where $P(f(z))$ is the point P on the Riemann sphere representing the complex number $f(z)$, is continuous on a compact region and so is uniformly continuous.

We now turn to three theorems which apply when z is confined to a compact region and uniform continuity may be exploited.

Theorem 6.4.2. *If $\{P_n(z)\}$ is any sequence of meromorphic functions which converge uniformly on the sphere for all z in a compact region $\mathfrak{R}$, then*

(i) *the limit function is uniformly continuous on the sphere, in the region $\mathfrak{R}$,*

6.4 Uniqueness of convergence

(ii) $P_n(z)$ are equicontinuous on the sphere and in $\mathfrak{R}$,

(iii) *the limit function is meromorphic in the interior of* $\mathfrak{R}$.

Proof. The chordal metric satisfies the triangle inequality. Let the limit of $\{P_n(z)\}$ be $f(z)$. To prove continuity on the sphere of $f(z)$ at $z = z_0$, consider

$$\chi(f(z), f(z_0)) \leq \chi(f(z), P_N(z)) + \chi(P_N(z), P_N(z_0)) + \chi(P_N(z_0), f(z_0)).$$

Given any $\varepsilon > 0$, uniform convergence on the sphere of $P_n(z)$ to $f(z)$ gives an N for which

$$\left.\begin{array}{l} \chi(f(z), P_n(z)) < \varepsilon/3, \\ \chi(f(z_0), P_n(z_0)) < \varepsilon/3 \end{array}\right\} \quad \text{for all } n \geq N \text{ and for all } z, z_0 \in \mathfrak{R}.$$

Uniform continuity on the sphere of $P_N(z)$ gives a δ for which $\chi(P_N(z), P_N(z_0)) < \varepsilon/3$ for all $z, z_0 \in \mathfrak{R}$ such that $|z - z_0| < \delta$. Hence $\chi(f(z), f(z_0)) < \varepsilon$, proving (i), which is that $f(z)$ is uniformly continuous on the sphere for $z \in \mathfrak{R}$.

To prove (ii), the equicontinuity property, consider the inequality

$$\chi(P_n(z), P_n(z_0)) < \varepsilon. \tag{4.5}$$

The triangle inequality shows that (4.5) is true, provided

$$\chi(P_n(z), f(z)) < \varepsilon/3, \tag{4.6}$$

$$\chi(f(z), f(z_0)) < \varepsilon/3, \tag{4.7}$$

and

$$\chi(f(z_0), P_n(z_0)) < \varepsilon/3. \tag{4.8}$$

Now (4.6) and (4.8) are true for all $n > N(\varepsilon)$ independently of z, z_0, and (4.7) is true for all $z, z_0 \in \mathfrak{R}$ such that $|z - z_0| < \delta(\varepsilon)$. Hence (4.5) is true for all $n > N(\varepsilon)$, provided $z, z_0 \in \mathfrak{R}$ and $|z - z_0| < \delta(\varepsilon)$.

Since $P_k(z)$ is uniformly continuous on the sphere for $z \in \mathfrak{R}$, k fixed,

$$\chi(P_k(z), P_k(z_0)) < \varepsilon \text{ for } |z - z_0| < \delta_k(\varepsilon).$$

By choosing

$$\delta_{\min} = \min\{\delta_1, \delta_2, \ldots \delta_{N(\varepsilon)}, \delta(\varepsilon)\},$$

it follows that $\chi(P_n(z), P_n(z_0)) < \varepsilon$ for all $z, z_0 \in \mathfrak{R}$ with $|z - z_0| < \delta_{\min}$, and (ii) is proved.

To show that $f(z)$ is meromorphic in the interior of $\mathfrak{R}$, take any point z_0 in the interior. If $f(z_0)$ is finite, there is a closed neighborhood, $|z - z_0| \leq \delta$, in which $f(z)$ is the uniform limit of a sequence of analytic functions, and so $f(z)$ is analytic in the interior of this region by

Weierstrass's theorem. If $f(z_0)$ is infinite, we consider the reciprocal functions, $(P_n(z))^{-1}$, which are analytic at $z = z_0$ for sufficiently large n; hence $(f(z))^{-1}$ is analytic at $z = z_0$ for an identical reason based on Weierstrass's theorem, and hence $f(z)$ is meromorphic in the interior of $\mathcal{R}$.

Next we use Arzela's theorem (applied to compact regions rather than bounded domains) to establish a sequence to which Theorem 6.4.2 is applied.

Theorem 6.4.3 [Hille, 1962, p. 241]. *If $\{P_n(z)\}$ is an infinite sequence of meromorphic functions which is equicontinuous on the sphere for z in a compact region $\mathcal{R}$, then at least a subsequence of $\{P_n(z)\}$ converges uniformly to a limit $f(z)$ which is uniformly continuous on the sphere for $z \in \mathcal{R}$ and is meromorphic in the interior of $\mathcal{R}$.*

Proof. Using the construction of the proof of Arzela's theorem (Theorem 5.2.5), we obtain a subsequence of $\{P_n(z)\}$ which converges uniformly in $\mathcal{R}$ and satisfies the conditions of Theorem 6.4.2, which in turn proves the result.

This theorem completes the buildup necessary for the convergence uniqueness theorem. It follows the style of the theorems for Stieltjes series to the extent that Arzela's theorem leads to a convergent subsequence. This result is combined with the proof that the whole sequence converges at the origin to yield convergence in a larger domain of the whole sequence.

Theorem 6.4.4. *Let $P_k(z) = [L_k/M_k]$ be a sequence of Padé approximants of a function $f(z)$ which is regular at the origin, such that $L_k + M_k \to \infty$ as $k \to \infty$. If $\{P_k(z)\}$ is equicontinuous on the sphere in a simply connected and compact region $\mathcal{R}$, of which the origin is an interior point, then the domain of definition of $f(z)$ may be extended so that $f(z)$ is meromorphic in the interior of $\mathcal{R}$ and $[L_k/M_k]$ converges to $f(z)$ on the sphere for all $z \in \mathcal{R}$.*

Proof. Theorem 6.4.3 asserts the existence of at least one limit function $\tilde{f}(z)$, which is meromorphic in the interior of $\mathcal{R}$ and with $f(0) = \tilde{f}(0)$. Hence ρ exists such that $f(z)$ and $\tilde{f}(z)$ are analytic in $|z| < \rho$; $\tilde{f}(z)$ is the uniform limit of an equicontinuous subsequence and so is meromorphic in the interior of $\mathcal{R}$. Consider now the entire sequence $\{P_k(z),$

$k = 1, 2, 3, \ldots\}$. Given any $\varepsilon > 0$, equicontinuity of the sequence on the sphere requires that $\delta = \delta(\varepsilon)$ exists such that

$$|P_k(z) - f(0)| < \varepsilon \quad \text{for all } |z| \leq \delta < \rho.$$

Thus $|P_k(y)|$ is uniformly bounded on $|y| = \delta$. Using Cauchy's theorem,

$$\frac{P_k(z) - f(z)}{z^{L_k + M_k}} = \frac{1}{2\pi i} \int_C \frac{P_k(y) - f(y)}{y^{L_k + M_k}} \frac{dy}{y - z},$$

where C is the circle $|y| = \delta$ and z is restricted to $|z| \leq \frac{1}{2}\delta$. Hence

$$|P_k(z) - f(z)| < (\tfrac{1}{2})^{L_k + M_k} \cdot \text{constant}$$

for $|z| \leq \frac{1}{2}\delta$ and all k. By the uniqueness of analytic continuation, this proves that the entire sequence $\{P_k(z)\}$ converges on the sphere to $f(z) = \tilde{f}(z)$ as $k \to \infty$ and that $f(z)$ is meromorphic in the interior of $\mathcal{R}$.

Omitting the detailed conditions, we see that establishing convergence of a sequence of Padé approximants to a meromorphic function depends mainly on proving that the approximants form an equicontinuous sequence. Such results are most naturally expressed in terms of convergence on the Riemann sphere. Our final theorem is a similar result, but involving ordinary convergence.

Theorem 6.4.5 [Baker, 1965]. *Let $P_k(z) = [L_k/M_k]$ be a member of a sequence of Padé approximants of a function $f(z)$ which is analytic at the origin, and such that $|P_k(z)|$ is uniformly bounded on a simply connected bounded domain $\mathcal{D}$ containing the origin. Provided $L_k + M_k \to \infty$ as $k \to \infty$, then $P_k(z) \to f(z)$ uniformly on any compact region $\mathcal{R} \subset \mathcal{D}$, thereby extending the domain of definition of $f(z)$, and $f(z)$ is analytic in $\mathcal{R}$.*

Proof. We first prove that the sequence $\{P_k(z)\}$ is equicontinuous on $\mathcal{D}$. If $|P_k(z)|$ is bounded on $\mathcal{D}$, then $P_k(z)$ is analytic on $\mathcal{D}$ because it is rational. We use the formula

$$\frac{dP_k(z)}{dz} = \frac{1}{2\pi i} \int_C \frac{P_k(w)\, dw}{(z - w)^2},$$

where C is a circle with center z, radius δ. We have arranged it so that we may choose $\delta > 0$ independently of z for $z \in \mathcal{R}$. Hence $|P_k'(z)|$ is uniformly bounded in $\mathcal{R}$. Consequently $\{P_k(z)\}$ is an equicontinuous sequence. Theorem 6.4.4 shows that $\{P_k(z)\} \to f(z)$ uniformly on the sphere. As the limit of a uniformly bounded sequence, $f(z)$ is not merely meromorphic but analytic in $\mathcal{R}$, and the theorem is proved. See also Baker and Graves-Morris [1982].

Theorem 6.4.4 asserts that the limit function is meromorphic in $\mathfrak{R}$. If the given function, which is proved to be the limit function, is not meromorphic in $\mathfrak{R}$, but the other requirements are met, it follows from the theorem that the Padé approximants are not equicontinuous on the sphere. Likewise, if $f(z)$ is not analytic, but the other conditions of the theorem are met, the Padé approximants of $f(z)$ cannot be uniformly bounded. A similar result is given by Jones and Thron [1975], and the generalization to a class of finite Laurent series is given in Jones and Thron [1979]. Other important theorems governing the convergence of Padé approximants of functions analytic at the origin and with various regions of meromorphy have been given by Chisholm [1966], Beardon [1968a], and Baker [1970].

An interesting problem is the 'disk problem'. The problem is simply stated. Suppose that the sequence of diagonal $[M/M]$ Padé approximants to $f(z)$ has no poles in a disk centered about the origin of radius R. Prove that this sequence converges uniformly on compact subsets of the disk. In spite of the apparent simplicity of this problem, it is not yet fully resolved. Baker [1965] proved convergence for $|z| < 0.3R$; this was improved by Chisholm [1966] to $|z| < (\sqrt{2} - 1)R$ and Zinn-Justin [1971] further improved the result to $|z| < 1/\sqrt{3}R$. Baker [1976] proved, with the further assumption that the number of zeros of the approximants in the disk divided by M tends to zero, that there is convergence of general sequences in compact subsets of $|z| < R$. In 1982, Gončar [1982] was able to remove this restriction on the number of zeros for diagonal sequences, but at the expense of the existence of a possible exceptional set of capacity zero (see Section 6.6). For a given formal power series with $f(0) = 1$, he assumes: (i) All the $[M/M]$ Padé approximants exist for all $M > M_0$ for some M_0. (ii) There exists a domain Ω of the form $\Omega = D \backslash E$ with $\partial D \subset \partial \{[\text{Convex Hull}(\partial D)]'\}$, where ∂ denotes the boundary and ' denotes the complement in the extended complex plane and E is a set of capacity zero. An open disk satisfies the conditions on D, for example. (iii) That there exists an M_1 such that for all $M > M_1$ the $[M/M]$ have no poles in Ω. Under these hypotheses, he shows that the $[M/M]$ converge uniformly on compact subsets of Ω to a function holomorphic in Ω. The method of proof uses the identity (3.5.18). With it we may write

$$[M + 1/M + 1] = 1 + \sum_{m=0}^{M} \frac{C(m + 1/m + 1)^2 z^{2m+1}}{Q^{[m+1/m+1]}(z) Q^{[m/m]}(z)}. \tag{4.9}$$

The first observation is that the constant in the numerator becomes

$$A_m = \frac{C(m + 1/m + 1)}{C(m/m)}$$

when we change the normalization of the Q's to $Q(0) = 1$. He then shows that

$$\limsup_{m \to \infty} |A_m|^{1/m} = A < \infty, \quad (4.10)$$

which in turn implies convergence when $|z| < 1/(1 + \sqrt{A})$. He proves, by a long potential-theoretic argument, that, possibly excepting a set E of capacity zero, the series (4.9) continues to converge in the whole disk, which leads to the conclusion of his theorem.

6.5 Convergence in measure

We will make repeated use of the formula (3.2):

$$f(z) - [L/M] = \frac{z^{L+M+1}}{2\pi i Q^{[L/M]}(z) R_M(z)} \int_\Gamma \frac{f(t) Q^{[L/M]}(t) R_M(t)\, dt}{t^{L+M+1}(t-z)} \quad (5.1)$$

for the error of Padé approximation. The idea is that we should bound $|Q^{[L/M]}(z)|^{-1}$ in (5.1) using a known bound for an arbitrary polynomial of degree M. This bound is given by

$$|q_m(z)| \geq \eta^m, \quad (5.2)$$

where $q_m(z)$ is a polynomial with leading coefficient unity, and is true *except* for z belonging to a set $\mathscr{E}$ of measure at most $\pi\eta^2$. Usually we know nothing about the location of this exceptional set, but we have the given bound on its area.

The theorems we are about to prove next are valid except on sets of arbitrarily small measure. This statement means that this section contains no assertion about convergence at any particular point in the z-plane. The results are all to the effect that the 'area of disruption' caused by unwanted pole-and-zero combinations of the approximants is arbitrarily small.

As an example of (5.2), we consider $q_m(z) = (z - a)^m$, which is a polynomial of degree m with leading coefficient unity. Then $|q_m(z)| \geq \eta^m$ except for $|z - a| < \eta$, which is a disk of area $\pi\eta^2$. This is an extreme example, in which all the roots of $q_m(z)$ are at $z = a$, and also shows that the theorem is the best result of its kind. If the roots of $q_m(z)$ are well separated, the inequality measure $(\mathscr{E}) < \pi\eta^2$ gives a substantial overestimate of the area of the multicomponent region in which $|q_m(z)| < \eta^m$. Equation (5.1) leads directly to a bound on $|f(z) - [L/M]|$ if the inequality (5.2) is used. This approach [Zinn-Justin, 1971] leads to a variety of interesting theorems about convergence of various sequences of Padé approximants. They seem, at the time of writing, to be the most natural general results for convergence of Padé approximants to analytic functions.

Throughout this section, we assume that the needed Padé approximants exist, and the theorems are to be understood in this sense. There is nothing in the right-hand side of (5.1) which implies the existence of $[L/M]$. Because, for example, a subsequence of Padé approximants of any row exists, the theorem about convergence in measure of a row may be rephrased as a theorem about convergence in measure of an infinite subsequence of extant Padé approximants in the row. To avoid this cumbersome phrasing, we state clearly that the results of this section apply to Padé approximants which exist, and no theorem is to be understood as asserting the existence of a particular Padé approximant.

The first theorem is an analogue of de Montessus's theorem in the context of convergence in measure, and then we turn to analogues of diagonal sequences. An interesting feature of the proofs is the possibility of bounding the area of disruption, δ_k, of the $[L_k/M_k]$ Padé approximant of a sequence so that $\sum_{k=1}^{\infty} \delta_k < \infty$ and $\sum_{k=K}^{\infty} \delta_k < \delta$, where δ is a preassigned arbitrarily small positive number. Then one has the convergence in measure of the sequence $[L_k/M_k]$, $k = 1, 2, \ldots$, provided the Padé approximants exist. It is important to notice which theorems assert convergence in measure of an entire sequence of extant Padé approximants and which theorems refer to subsequences only.

The first theorem asserts convergence in measure of a row sequence of Padé approximants of a meromorphic function. It is a weaker form of de Montessus's theorem applicable when the degree of the denominator is known to be greater than or equal to (instead of precisely equal to) the number of poles within the circle of convergence of the row of Padé approximants.

Theorem 6.5.1. *Let $f(z)$ be analytic at the origin and also in a given disk $|z| \leq R$, except for m poles, counting multiplicity. Consider a row of $[L/M]$ Padé approximants of $f(z)$ with M fixed, $M \geq m$, and $L \to \infty$. Suppose that arbitrarily small, positive ε and δ are given. Then L_0 exists such that $|f(z) - [L/M]| < \varepsilon$ for any $L > L_0$ and for all $|z| < R$ except for $z \in \mathscr{E}_L$, where $\mathscr{E}_L$ is a set of points in the z-plane of measure less than δ.*

Proof. Let $f(z)$ have poles at $z = \alpha_1, z = \alpha_2, \ldots, z = \alpha_m$ within $|z| < R$, and define
$$R_m(z) = (z - \alpha_1)(z - \alpha_2) \cdots (z - \alpha_m),$$
so that $R_m(z)f(z)$ is analytic in $|z| \leq R$. Then Hermite's formula is directly applicable as

$$f(z) - [L/M] = \frac{z^{L+M+1}}{2\pi i Q_M(z) R_m(z)} \int_{|t|=R} \frac{f(t) Q_M(t) R_m(t)}{t^{L+M+1}(t-z)} dt, \quad (5.3)$$

where $Q_M(z) \equiv Q^{[L/M]}(z)$. Since $|z| < R$, define

$$K = \frac{1}{1 - |z/R|} \sup_{|t|=R} |f(t)| \quad (5.4)$$

and then (5.3) and (5.4) yield

$$|f(z) - [L/M]| \leq \frac{K}{|Q_M(z) R_m(z)|} \left|\frac{z}{R}\right|^{L+M+1} \sup_{|t|=R} |Q_M(t) R_m(t)|. \quad (5.5)$$

The next step is to bound some factors on the right-hand side of (5.5). Separate the roots of $Q_M(z)$ into those with

$$|z_i| < 2R, \quad i = 1, 2, \ldots, M',$$

and those with

$$|z_i| \geq 2R, \quad i = M'+1, M'+2, \ldots, M.$$

Then

$$\sup_{|t|=R} \prod_{i=M'+1}^{M} \left|\frac{t - z_i}{z - z_i}\right| \leq \prod_{i=M'+1}^{M} \frac{1 + R/|z_i|}{1 - R/|z_i|} \leq 3^{M-M'}$$

and

$$\sup_{|t|=R} \left|\frac{Q_M(t)}{Q_M(z)}\right| = \sup_{|t|=R} \prod_{i=1}^{M'} \frac{|t - z_i|}{|z - z_i|} \prod_{i=M'+1}^{M} \frac{|t - z_i|}{|z - z_i|}$$

$$\leq \left\{\prod_{i=1}^{M'} \frac{3R}{|z - z_i|}\right\} 3^{M-M'} = \frac{3^M R^{M'}}{\prod_{i=1}^{M'} |z - z_i|}.$$

Because the zeros of $R_m(t)$ lie in $|t| < R$,

$$\frac{\sup_{|t|=R} |Q_M(t) R_m(t)|}{|Q_M(z) R_m(z)|} < \frac{R^{M'} 3^M \times (2R)^m}{|R_m(z)| \prod_{i=1}^{M'} |z - z_i|} \quad (5.6)$$

with $M' \leq M$. The denominator of the right-hand side of (5.6) contains a polynomial with leading coefficient unity and is bounded by

$$|R_m(z)| \prod_{i=1}^{M'} |z - z_i| > \eta^{M'+m} \quad (5.7)$$

except for $z \in \mathscr{E}_L$ where measure $(\mathscr{E}_L) \leq \pi\eta^2 = \delta$. Clearly, this set $\mathscr{E}_L$ depends on the values of z_i, which in turn depend on L. Assembling (5.5), (5.6), and (5.7),

$$|f(z) - [L/M]| \leq \left|\frac{z}{R}\right|^{L+M+1} \left[\frac{KR^{M'}3^M \times (2R)^m}{\eta^{M'+m}}\right], \quad (5.8)$$

except for $z \in \mathscr{E}_L$. The factor in square brackets on the right-hand side of (5.8) is bounded independently of L, and as $L \to \infty$ for $|z| < R$,

$$|f(z) - [L/M]| \to 0$$

except for $z \in \mathscr{E}_L$ where measure $(\mathscr{E}_L) \leq \delta$. The theorem is now proved.

We will now extend the theorem with two corollaries.

Corollary 1. *With the hypotheses of the theorem, the more general sequence $[L_k/M_k]$ satisfies*

$$|f(z) - [L_k/M_k]| < \varepsilon$$

for any $k > k_0$ and for all $|z| < R$ excepting $z \in \mathscr{E}_k$, where $\mathscr{E}_k$ is a set of measure less than δ, provided

(i) $L_k/M_k \to \infty$ as $k \to \infty$ $(M_k \neq 0)$, and
(ii) $M_k \geq M$ for all $k > k_0$.

Proof. We assume, without loss of generality, that $R/\eta > 1$, because a smaller value of η gives a stronger result. Since $M_k \geq M$, (5.8) is valid, and since $M' < M$,

$$|f(z) - [L_k/M_k]| \leq \left|\frac{z}{R}\right|^{L_k+M_k+1} \left(\frac{3R}{\eta}\right)^{M_k} K \left(\frac{2R}{\eta}\right)^m.$$

Because

(i) $L_k/M_k \to \infty$,
(ii) $|z/R| < 1$,

given any $\varepsilon' > 0$, we may choose k_0 such that

$$\frac{L_k}{M_k} \ln\left|\frac{z}{R}\right| + \ln\left(\frac{3R}{\eta}\right) < \ln \varepsilon'$$

for all $k > k_0$. Hence

$$\left|\frac{z}{R}\right|^{L_k} \left(\frac{3R}{\eta}\right)^{M_k} < (\varepsilon')^{M_k}.$$

By choosing $\varepsilon' = \min\{1, (\varepsilon/K)(\eta/2R)^m\}$, which is positive, it follows that

$$|f(z) - [L_k/M_k]| < \varepsilon$$

for any $k > k_0$.

Corollary 2. *With the hypotheses of the theorem for $f(z)$, ε, δ, and M, an L_0 exists such that*

$$|f(z) - [L/M]| < \varepsilon$$

for all $L > L_0$ and for all $|z| < R$ except for $z \in \mathscr{E}$, where $\mathscr{E}$ is a set of measure less than δ.

Remark. This result is substantially stronger than that of the main theorem, being a result about convergence of all the extant approximants of the $(M+1)$th row beyond $L = L_0$.

Proof. From (5.8), the $[l + L_1/M]$ Padé approximant satisfies

$$|f(z) - [l + L_1/M]| < \left|\frac{z}{R}\right|^{l+L_1+M+1} \left[\frac{KR^{M'}3^M(2R)^m}{\eta_l^{M'+m}}\right] \quad (5.9)$$

except on a set $\mathscr{E}_l$ of measure at most $\pi\eta_l^2$. We assume, without loss of generality, that $R/\eta_l > 1$. Recalling that $M' \leq M$, we choose L_1 such that

$$\left|\frac{z}{R}\right|^{L_1} (3R)^M K(2R)^m < \varepsilon,$$

where ε is a preassigned small positive number.

Then, the requirement on η_l that

$$\left|\frac{z}{R}\right|^l \left(\frac{1}{\eta_l}\right)^{M+m} = 1 \quad (5.10)$$

is sufficient for (5.9) to reduce to

$$|f(z) - [l + L_1/M]| < \varepsilon$$

except on a set $\mathscr{E}_l$ of measure at most $\pi\eta_l^2$. Equation (5.10) may be written as

$$2\pi\eta_l^2 = 2\pi \left|\frac{z}{R}\right|^{2l/(M+m)}$$

and

$$\sum_{l=l'}^{\infty} 2\pi\eta_l^2 = 2\pi \frac{|z/R|^{2l'/(M+m)}}{1 - |z/R|^{2/(M+m)}}. \quad (5.11)$$

We may certainly choose l' so that

$$\sum_{l=l'}^{\infty} 2\pi\eta_l^2 < \delta,$$

and this asserts that the total measure of all the individual exceptional sets of all the $[l + L/M]$ Padé approximants with $l > l'$ is arbitrarily small. Thus convergence in measure of a row is proved.

Theorem 6.5.2. *Let $f(z)$ be analytic at the origin, and also in the circle $|z| \leq R$, except for m poles, counting multiplicity. Consider a sequence $[L_k/M_k]$ of Padé approximants of $f(z)$ with $M_k \geq m$ and $L_k/M_k \to \infty$ as $k \to \infty$ ($M_k \neq 0$). Let ε, δ be arbitrarily small positive given numbers. Then k_0 exists such that*

$$|f(z) - [L_k/M_k]| < \varepsilon$$

for all $k > k_0$ and all $|z| < R$ except for $z \in \mathscr{E}$, where $\mathscr{E}$ is a set of points of the z-plane of measure less than δ.

Theorem 6.5.2 embodies the previous theorem and its two corollaries. The method of proof is a conflation of the proofs given.

Corollary 2 of Theorem 6.5.1 about convergence of rows is a powerful result. Recall that the exceptional sets have the poles of the approximants in $|z| < R$ as interior points. Convergence of a row is expected in $|z| < R$ except in neighborhoods of limit points of poles of the approximants and in neighborhoods of the poles of $f(z)$. In this light, the theorem may be regarded as a statement that the exceptional set of an entire row is a set of arbitrarily small measure which is in turn the union of sets enclosing poles of $f(z)$ and other limit points of the poles of the approximants. In this language, it is plain that Corollary 2 embodies many earlier results expressed in terms of ordinary convergence.

Having treated row sequences and their relatives, we turn to diagonal sequences, ray sequences, and their relatives. As an introduction, we prove Theorem 6.5.3, which is a simple theorem, and subsequently we generalize it substantially. This theorem was originally proved by Nuttall [1970b], using Szegö's theorem, but the proof given by Zinn-Justin [1971], based on Hermite's formula, is somewhat simpler.

Theorem 6.5.3. *Let $f(z)$ be a meromorphic function. Suppose that ε, δ are given positive numbers. Then M_0 exists such that any $[M/M]$ Padé approximant satisfies*

$$|f(z) - [M/M]| < \varepsilon$$

6.5 Convergence in measure

for all $M > M_0$ on any compact set of the z-plane except for a set $\mathscr{E}_M$ of measure less than δ.

Proof. We will set the scale of the z-plane by proving convergence for $|z| < 1$ except on a set $\mathscr{E}_M$ of measure less than δ. Define $\eta = \tfrac{1}{2}\sqrt{\delta/\pi}$. We may assume $0 < \eta < 1$ without loss of generality. Define

$$R_{\min} \equiv (3/\eta)^3.$$

For some Δ, no matter how small, in the range $0 < \Delta < 1$, R exists satisfying

(i) $f(z)$ is analytic in $R - \Delta < |z| < R + \Delta$,
(ii) $R > R_{\min}$, and
(iii) $f(z)$ has $m = m(R)$ poles located at $z = u_i$, $i = 1, 2, \ldots, m$, in $|z| < R$.

Next we need Hermite's interpolation formula using the polynomial

$$R_m(z) = \prod_{i=1}^{m}(z - u_i),$$

and consequently for $M \geq m$, $Q_M(z) \equiv Q^{[M/M]}(z)$,

$$f(z) - [M/M] = \frac{z^{2M+1}}{2\pi i Q_M(z) R_m(z)} \oint_{|t|=R} \frac{Q_M(t) R_m(t) f(t)}{(t-z) t^{2M+1}} dt.$$

We have chosen to consider $|z| < 1$, $R > R_{\min} > 2$, and so

$$|f(z) - [M/M]| \leq K R^{-(2M+1)} \frac{\sup\limits_{|t|=R} |Q_M(t) R_m(t)|}{|Q_M(z) R_m(z)|}, \tag{5.12}$$

where $K = \sup_{|t|=R} |f(t)|$. From (5.6), with some $M' \leq M$, we proved that

$$\frac{\sup\limits_{|t|=R} |Q_M(t) R_m(t)|}{|Q_M(z) R_m(z)|} < \frac{R^{M'} 3^M \times (2R)^m}{|R_m(z)| \prod\limits_{i=1}^{M'} |z - u_i|}. \tag{5.13}$$

The denominator of the right-hand side of (5.13) is a polynomial with leading coefficient unity, and is bounded by

$$|R_m(z)| \prod_{i=1}^{M'} |z - z_i| > \eta^{M'+m} \tag{5.14}$$

except for z in a set $\mathscr{E}_M$ of measure $\pi\eta^2 = \delta$. Assembling (5.12), (5.13), and (5.14),

$$|f(z) - [M/M]| < \left(\frac{1}{R}\right)^{2M+1} \frac{(3R)^M (2R)^m}{\eta^{M+n}}$$

$$< \left(\frac{3}{\eta}\right)^{M+m} \frac{1}{R^{M-m}}.$$

Provided $M > 2m$, and recalling that $R > (3/\eta)^3$,

$$|f(z) - [M/M]| < \left(\frac{3}{\eta}\right)^{3M/2} \frac{1}{R^{M/2}} < \varepsilon$$

for any $M > $ (some M_0), except on the set $\mathscr{E}_M$ of measure less than δ.

As already stated, Theorem 6.5.3 is a weak form of both what is known to be true and what is expected to be true about convergence in measure of ray sequences. Nonetheless, it provides a basis for further development.

First, the diagonal sequence may be replaced by the sequence $[L_k/M_k]$, $k = 1, 2, \ldots$, provided that, for any λ in the range $0 < \lambda < 1$, however small,

$$\lambda < \frac{L_k}{M_k} < \lambda^{-1}. \tag{5.15}$$

Equation (5.15) confines the Padé approximants to a fan-shaped region of the Padé table as shown in Figure 6.5.1. Provided (5.15) holds and $L_k + M_k \to \infty$, this weaker constraint is sufficient to allow convergence in measure.

Second, $f(z)$ need not be meromorphic, but may also have a countable number of isolated essential singularities. This means that $\exp[-(1-z)^{-1}]$ and $\exp[z\Gamma(z)]$ are allowable functions, but not functions whose singularities have a limit point in the finite z-plane.

Theorem 6.5.4 [Pommerenke, 1973]. *Let $f(z)$ be a function which is analytic at the origin and analytic in the entire z-plane except for a countable number of isolated poles and essential singularities. Suppose $\varepsilon > 0$ and $\delta > 0$ are given. Then M_0 exists such that any $[L/M]$ Padé approximant of the ray sequence with $L/M = \lambda$ ($\lambda \neq 0$, $\lambda \neq \infty$) satisfies*

$$|f(z) - [L/M]| < \varepsilon$$

for any $M \geq M_0$, on any compact set of the z-plane except for a set $\mathscr{E}_M$ of measure less than δ.

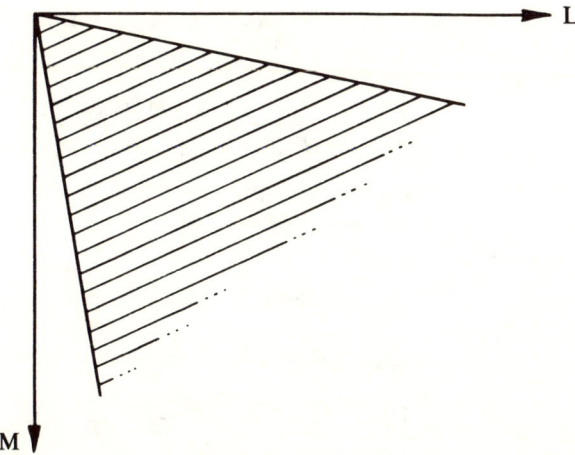

Figure 6.5.1. The allowed domain of the Padé table for the approximants in Theorem 6.5.4.

Proof. As in Theorem 6.5.3, we take $|z| < 1$, define $\eta = \frac{1}{2}\sqrt{\delta/\pi}$, and assume $0 < \eta < 1$. We choose

$$R_{\min} = \left(\frac{3}{\eta}\right)^{1+2/\lambda} \tag{5.16}$$

and some $\Delta > 0$, $R > R_{\min}$, such that

(i) $f(z)$ is analytic in $R - \Delta < |z| < R + \Delta$,
(ii) $f(z)$ has $m = m(R)$ poles located at $z = u_i$, $i = 1, 2, \ldots, m$, in $|z| < R$, and
(iii) $f(z)$ has $\mu = \mu(R)$ essential singularities located at $z = w_i$, $i = 1, 2, \ldots, \mu$, in $|z| < R$.

Define the polynomial

$$R_M(z) = \prod_{i=1}^{m}(z - u_i)\prod_{i=1}^{\mu}(z - w_i)^p, \tag{5.17}$$

where $p = p(M)$ is defined so that the ratio $\rho = p/M$ satisfies

$$0 < \frac{p}{M} < \min\left\{-\frac{\ln 2}{\ln \Delta}, \frac{\frac{1}{2}\lambda}{\mu}\right\}. \tag{5.18}$$

The purpose of this is to be able subsequently to let $M \to \infty$ and $p \to \infty$ simultaneously but keep $R_M(z)$ of sufficiently low degree. At any rate,

$R_M(z)$ has leading coefficient unity and degree less then M. We use Hermite's formula, with $Q_M(z) \equiv Q^{[L/M]}(z)$,

$$f(z) - [L/M] = \frac{z^{L+M+1}}{2\pi i Q_M(z) R_M(z)} \oint_C \frac{Q_M(t) R_M(t) f(t)}{(t-z) t^{L+M+1}} dt,$$

where C is a closed contour containing z and the origin and no singularities of $Q_M(z) f(z)$. By enlarging the contour so as to enclose the essential singularities, we find

$$f(z) - [L/M] = \frac{z^{L+M+1}}{2\pi i Q_M(z) R_M(z)}$$

$$\times \left\{ \int_{|t|=R} \frac{Q_m(t) R_M(t) f(t) \, dt}{(t-z) t^{L+M+1}} - \sum_{k=1}^{\mu} I_k(z) \right\}, \quad (5.19)$$

where

$$I_k(z) = \int_{|t-w_k|=\delta_k} \frac{R_M(t) Q_M(t) f(t)}{(t-z) t^{L+M+1}} dt. \quad (5.20)$$

Equation (5.20) is a contour integral round a small circle of radius δ_k enclosing the essential singularity at $z = w_k$. To bound $I_k(z)$, we require that $|z - w_k| > 2\delta_k$. Using the maximum-modulus theorem for the polynomial $Q_M(t) R_M(t) (t - w_k)^{-p}$,

$$\left| \frac{Q_M(t) R_M(t)}{(t - w_k)^p} \right| < \sup_{|t|=R} \left\{ \frac{|Q_M(t) R_M(t)|}{|t - w_k|^p} \right\} < \frac{\sup_{|t|=R} |Q_M(t) R_M(t)|}{\Delta^p}.$$

Hence, for $|z - w_k| > 2\delta_k$,

$$|I_k(z)| \leq 2\pi \frac{\sup_{|t|=R} |R_M(t) Q_M(t)|}{\Delta^p} \frac{\sup_{|t-w_k|=\delta_k} |f(t)| \delta_k^p}{(|w_k| - \delta_k)^{L+M+1}}. \quad (5.21)$$

We now specify the radii δ_k of the small circles by defining

$$\delta_k = \min \left\{ \left| \frac{w_k}{2R} \right|^{(\lambda+1)/p}, \frac{1}{2} |w_k| \right\}. \quad (5.22)$$

Equation (5.16) ensures that the regions $|z - w_k| < 2\delta_k$ surrounding the essential singularities in $|z| < 1$ have arbitrarily small total measure.

From (5.22),

$$\delta_k^p < \left(\frac{|w_k| - \delta_k}{R} \right)^{L+M}.$$

We define

$$\sup_{|t-w_k|=\delta_k} |f(t)| = K_k,$$

where K_k is independent of M, and then

$$|I_k(z)| < K' \frac{\sup_{|t|=R} |R_M(t)Q_M(t)|}{\Delta^p R^{L+M+1}}, \qquad (5.23)$$

where K' is independent of M and k. Assembling (5.19) and (5.23), we find

$$|f(z) - [L/M]| < \left|\frac{z}{R}\right|^{L+M+1} K'' \frac{\sup_{|t|=R} |Q_M(t)R_M(t)|}{|Q_M(z)R_M(z)|} \left\{1 + \sum_{k=1}^{\mu} \frac{1}{\Delta^p}\right\}, \qquad (5.24)$$

where K'' is also independent of M.

Again, when $Q_M(t)$ has M' zeros within $|t| \leq 2R$, we note that $M' \leq M$ and $R_M(t)$ is a polynomial of degree $m + p\mu$, so (5.6) and (5.7) yield

$$\frac{\sup_{|t|=R} |Q_M(t)R_M(t)|}{|Q_M(z)R_M(z)|} < \frac{(3R)^M (2R)^{m+p\mu}}{\eta^{m+p\mu+M}} \qquad (5.25)$$

provided $z \notin \mathscr{E}_M$, a set of measure less than $\pi\eta^2$. Since $L = \lambda M$ and $|z| < 1$, (5.18), (5.24), and (5.25) yield

$$|f(z) - [L/M]| \leq \left(\frac{1}{R}\right)^{(\lambda+1)M} K''' \left(\frac{3R}{\eta}\right)^{M+m+p\mu}, \qquad (5.26)$$

where K''' is also independent of M. From (5.18), $p\mu < \frac{1}{2}\lambda M$, and then (5.26) gives

$$|f(z) - [L/M]| \leq K''' R^{-\frac{1}{2}\lambda M + m} \left(\frac{3}{\eta}\right)^{M(1+\frac{1}{2}\lambda)+m}.$$

Hence, provided $R > (3/\eta)^{1+2/\lambda}$ and M is sufficiently large, $|f(z) - [L/M]| < \varepsilon$ except on the set $\mathscr{E}_M$ and the small circles enclosing the essential singularities of $f(z)$ in $|z| < 1$.

Corollary 1. *This theorem can also be generalized to treat arbitrary sequences in the region of the Padé table shown in Figure 6.5.1.*

Corollary 2 [Zinn-Justin, 1971]. *Let $f(z)$ be meromorphic in $|z| \leq R$. Let the number of zeros of $Q_M(z)$ in $|z| < R$ be q_M. Then if*

$$\frac{q_M \ln M}{M} \to 0 \quad \text{as } M \to \infty,$$

the $[M/M]$ Padé approximants of $f(z)$ converge in measure in $|z| < R/\sqrt{3}$.

Corollary 3 [Zinn-Justin, 1971]. *If $f(z)$ is an analytic function of exponential order less than $2/\lambda$, then the sequence of $[\lambda M/M]$ approximants converges on any compact set of the z-plane except on a set of arbitrarily small measure.*

We present these corollaries without proof. The second and third are interesting because they show that further restrictions on the class of functions considered lead to stronger convergence results. However, no theorem yet proved gives convergence in measure of diagonal Padé approximants in $|z| < R$ for functions known only to be meromorphic in $|z| < R$. We refer to Edrei [1975b] for a result which allows the essential singularities to be limit points of pole sequences, rather than isolated essential singularities as in Pommerenke's theorem.

6.6 Lemniscates, capacity, and measure

The purpose of this section is not to give a detailed and rigorous account of the foundations of capacity and measure, but rather to indicate why the results of the previous section are much stronger if rephrased in terms of capacity or Hausdorff measure.

The basis of Section 6.5 is the result related to Cartan's lemma [Cartan, 1928; Nuttall, 1970b] that, for a polynomial $q_m(z)$ with leading coefficient unity, $|q_m(z)| > \eta^m$ except on a set $\mathscr{E}$ of measure at most $\pi\eta^2$. The boundary on which $|q_m(z)| = \eta^m$ is usually called a lemniscate, and so all the results of Section 6.5 apply except on lemniscatic regions which are arbitrarily small. We will see that the natural measure of the size of a lemniscatic region is its capacity.

To begin with, let us recall Tchebycheff's result for the minimax polynomial on an interval $-1 \leq x \leq 1$. The problem is to find the polynomial $T_n(x)$ in the class P_n of all polynomials $p_n(x)$ of degree n and leading coefficient unity for which the limit

$$\inf_{p_n(x) \in P_n} \sup_{a \leq x \leq b} |p_n(x)|$$

is attained. The solution is well known. For $n \geq 1$,

$$T_n(x) = \frac{1}{2^{n-1}} \cos(n \cos^{-1} x) \qquad (6.1a)$$

is a polynomial of degree n of leading coefficient unity, and

$$\inf_{p_n(x) \in P_n} \sup_{-1 \leq x \leq 1} |p_n(x)| = \frac{1}{2^{n-1}}; \qquad (6.1b)$$

The normalisation of $T_n(x)$ used in (6.1a) is unusual, and it is used purely for consistency with (6.5). A linear change of variable

$$x' = -\frac{a(x-1)}{2} + \frac{b(x+1)}{2} \tag{6.2}$$

leads to the result that

$$\inf_{p_n(x) \in P_n} \sup_{a \leqslant x \leqslant b} |p_n(x)| = \left(\frac{b-a}{4}\right)^{n-1}\left(\frac{b-a}{2}\right). \tag{6.3}$$

To generalize these ideas to an arbitrary compact set in the z-plane, we have the following theorem.

Theorem 6.6.1. *Let $\mathscr{E}$ be a compact set in the complex plane (containing infinitely many points). Then there is a unique Tchebycheff polynomial for which*

$$M_n \equiv \inf_{p_n(z) \in P_n} \sup_{z \in \mathscr{E}} |p_n(z)| \tag{6.4}$$

is attained. The minimax polynomial, $p_n(z) = T_n(z)$, may be written as

$$T_n(z) = \prod_{i=1}^{n}(z - z_i). \tag{6.5}$$

The zeros z_i of $T_n(z)$ lie in the convex hull of $\mathscr{E}$, and the maximum value M_n of $|T_n(z)|$ is achieved at least n times on the boundary of $\mathscr{E}$.

Discussion. We shall not prove this important theorem [Hille, 1962, p. 265], but elaborate it with a few remarks.

It is easy to see that all the z_i lie within the convex hull of $\mathscr{E}$. For suppose not, and let z_1 lie outside the convex hull $\mathscr{H}$ of $\mathscr{E}$ and the points $z_2, z_3, \ldots, z_n$. By considering a point z_1' nearer to $\mathscr{H}$, and the point $z = z'$ for which the

$$\sup_{z \in \mathscr{E}} \left\{ |z - z_1'| \prod_{i=2}^{n} |z - z_i| \right\} = m'$$

is attained, we find that

$$M_n = \sup_{z \in \mathscr{E}} \prod_{i=1}^{n} |z - z_i| > |z' - z_1| \prod_{i=2}^{n} |z' - z_i|$$

$$> |z' - z_1'| \prod_{i=2}^{n} |z' - z_i| = m',$$

which contradicts the minimum property of M_n with respect to variation of the z_i. Hence the zeros of $T_n(z)$ lie in the convex hull of $\mathscr{E}$.

The proof of the existence of a minimax polynomial is based on 'tracing' the roots to extremal positions in the closed convex hull. We omit the proofs of existence and uniqueness of $T_n(z)$ in the general case. Since $\mathscr{E}$ is compact, it is obvious that the maximum of $|T_n(z)|$ is attained in $\mathscr{E}$. Further, the maximum-modulus theorem shows that the maximum is attained on the boundary of $\mathscr{E}$. That $|T_n(z)|$ should equal M_n at n distinct points on the boundary is another significant result we state without proof.

If $\mathscr{E}$ is a finite point set, a possibility excluded by the hypothesis of the theorem, then the Tchebycheff polynomial of order n is zero on $\mathscr{E}$ for $n > N$. Furthermore, it is not unique for $n > N$, and so this degenerate case is naturally excluded by the hypothesis.

Corollary. *The Tchebycheff polynomial $T_n(z)$ for a set $\mathscr{E}$ defines a lemniscate by $|T_n(z)| = M_n$ and a lemniscatic region $\mathscr{L}_n$ by*

$$|T_n(z)| \leq M_n \quad \text{for all } z \in \mathscr{L}_n.$$

Then $\mathscr{E} \subset \mathscr{L}_n$, and the boundary of $\mathscr{L}_n$ has at least n points in common with $\mathscr{E}$.

Discussion. The important idea is that $\mathscr{E}$ becomes a subset of the lemniscatic region $\mathscr{L}_n$ defined by $|T_n(z)| \leq M_n$, and the proof follows immediately from the maximum-modulus theorem and the main theorem.

We now proceed to three theorems which we prove, because the proofs illustrate the structure of lemniscates and the idea of capacity. Repeated use is made of the *maximum-modulus theorem*, which states that the maximum modulus of an analytic function in a compact region is achieved on the boundary.

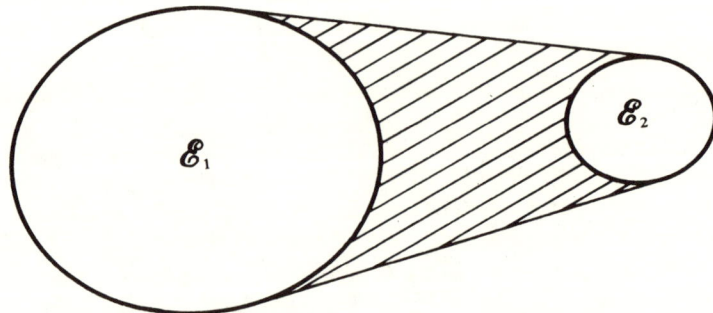

Figure 6.6.1. A set $\mathscr{E} = \mathscr{E}_1 \cup \mathscr{E}_2$. The union of the shaded region and the set $\mathscr{E}$ comprises the convex hull of $\mathscr{E}$.

6.6 Lemniscates, capacity, and measure

Theorem 6.6.2. *Let $\mathscr{E}$ be a compact set in the complex z-plane (containing infinitely many points). Let $T_n(z)$ be the Tchebycheff polynomials defined on $\mathscr{E}$, and let*

$$M_n(\mathscr{E}) = \sup_{z \in \mathscr{E}} |T_n(z)|. \tag{6.6}$$

Then the capacity of $\mathscr{E}$ is uniquely defined by

$$\lim_{n \to \infty} [M_n(\mathscr{E})]^{1/n} = \text{cap}(\mathscr{E}). \tag{6.7}$$

Proof. The only proof required is that the limit (6.7) be well defined. To do this, let

$$\alpha = \liminf [M_n(\mathscr{E})]^{1/n},$$

$$\beta = \limsup [M_n(\mathscr{E})]^{1/n}.$$

If δ is the maximum diameter of the compact set $\mathscr{E}$ in the usual sense, then from (6.4),

$$M_n(\mathscr{E}) = \sup_{z \in \mathscr{E}} |T_n(z)| = \inf_{z_i} \sup_{z \in \mathscr{E}} \prod_{i=1}^{n} |z - z_i| \leq \delta^n.$$

Hence $0 \leq \alpha \leq \beta \leq \delta$.

Given $\varepsilon > 0$, we may find N such that

$$\alpha + \varepsilon > M_N^{1/N}$$

and

$$|T_N(z)| < (\alpha + \varepsilon)^N \quad \text{for all } z \in \mathscr{E}.$$

For any positive integers m, k but with $m < N$,

$$|z^m [T_n(z)]^k| < K(\alpha + \varepsilon)^{Nk+m} \quad \text{for all } z \in \mathscr{E},$$

where

$$K = K(N, \mathscr{E})$$

is a constant independent of k. Recognizing $z^m[T_n(z)]^k$ as a polynomial of order $m + nk$, we see that

$$(M_{m+nk})^{1/(m+NK)} \leq K^{1/(m+Nk)}(\alpha + \varepsilon),$$

and by taking the limit as $k \to \infty$,

$$\limsup_{k \to \infty} [M_{m+Nk}]^{1/(m+Nk)} \leq \alpha + \varepsilon. \tag{6.8}$$

Equation (6.8) holds for any positive $m < N$, and any $\varepsilon > 0$. Hence $\beta = \alpha$, and cap $(\mathcal{E})$ is well defined by (6.7).

The quantity cap $(\mathcal{E})$ is a measure of the magnitude of the set $\mathcal{E}$. It is called the logarithmic capacity, colloquially abbreviated to capacity or transfinite diameter in different contexts.

The next theorem shows the key role of lemniscates in the theory of capacity.

Theorem 6.6.3 *If $\mathcal{E}$ is a lemniscatic region given by*

$$|p_n(z)| \leq \eta^n \quad \text{for } z \in \mathcal{E},$$

then cap $(\mathcal{E}) = \eta$.

Proof. The maximum-modulus theorem show that $\partial \mathcal{E}$ (the boundary of $\mathcal{E}$) is given by $|p_n(z)| = \eta^n$. Let $T_n(z)$ be the nth-order Tchebycheff polynomial defined on $\mathcal{E}$, so

$$|T_n(z)| \leq (\eta')^n \quad \text{for all } z \in \mathcal{E}$$

and consequently cap $(\mathcal{E}) = \eta'$. By definition of $T_n(z)$, $\eta' \leq \eta$. If $\eta' < \eta$, then

$$|T_n(z)| \leq |p_n(z)| \quad \text{for all } z \in \partial \mathcal{E}.$$

Recall

Rouché's theorem. *If $f(z)$ and $g(z)$ are analytic inside and on a closed contour C and $|g(z)| < |f(z)|$ on C, then $f(z)$ and $f(z) + g(z)$ have the same number of zeros inside C.*

In this case $f(z) = T_n(z)$, $g(z) = -p_n(z)$, and the theorem implies that a polynomial of degree $n - 1$ has n zeros in $\mathcal{E}$. This is impossible, hence $\eta' = \eta$ and cap $(\mathcal{E}) = \eta$, proving the theorem.

Before giving examples of the capacity of a set, we prove the major result which improves the theorems of the previous section.

Theorem 6.6.4. *Let $\mathcal{E}$ be a compact set. Then*

$$\text{meas}(\mathcal{E}) \leq \pi [\text{cap}(\mathcal{E})]^2. \tag{6.9}$$

Proof. Let $T_n(z)$ be the nth Tchebycheff polynomial defined on $\mathcal{E}$. For any $\varepsilon > 0$, (6.7) implies that a sufficiently large n exists such that

$$|T_n(z)| \leq [\text{cap}(\mathcal{E}) + \varepsilon]^n \quad \text{for all } z \in \mathcal{E}.$$

This inequality defines a lemniscatic region $\mathscr{L}_n$, of capacity $\eta = \mathrm{cap}\,(\mathscr{E}) + \varepsilon$, which has $\mathscr{E}$ as a subset. If we prove that

$$\mathrm{meas}\,(\mathscr{L}_n) \leq \pi[\mathrm{cap}\,(\mathscr{L}_n)]^2, \tag{6.10}$$

it follows that

$$\mathrm{meas}\,(\mathscr{E}) \leq \mathrm{meas}\,(\mathscr{L}_n) \leq \pi[\mathrm{cap}\,(\mathscr{L}_n)]^2 \leq \pi[\mathrm{cap}\,(\mathscr{E}) + \varepsilon]^2.$$

Hence the theorem is true provided we prove (6.10). For brevity, let the lemniscate be defined by

$$|p(z)| = \eta^n, \tag{6.11}$$

where $p(z) = T_n(z)$ and $\eta^n = M_n$. We will consider the family of lemniscates given by $|p(z)| = \rho$ for all positive ρ. This consists of at most

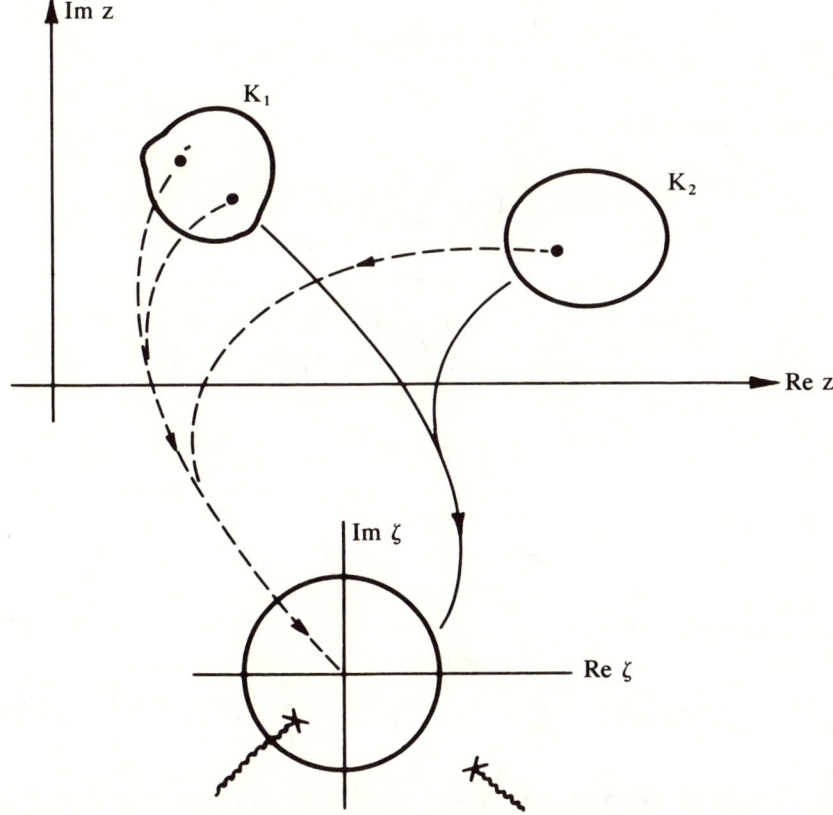

Figure 6.6.2. The mapping of K_1 and K_2 onto the circle $|\zeta| = \rho$, showing two branch points in the ζ-plane.

n disjoint closed curves $K_1, K_2, \ldots, K_m$. The maximum-modulus theorem shows that the curves lie outside each other. There is at least one root of $p(z)$ in each curve K_k: let there be precisely μ_k roots in K_k. As the point z moves around K_k, $p(z) = \prod_{i=1}^{n}(z - z_i) = \zeta$ moves μ_k times around the circle $|\zeta| = \rho$ and $\zeta' = \zeta^{1/\mu_k}$ moves once around the circle $|\zeta'| = \rho^{1/\mu_k}$.

Now consider the mappings $z \to p(z) = \zeta \to \zeta^{1/\mu_k} = \zeta'$ and the inverse mapping $\zeta' \to z = f(\zeta')$. Here $p(z)$ is clearly a single-valued function of z, but the inverse map $z = z(\zeta)$ is not, and $z = z(\zeta)$ has $n-1$ branch points occurring at $p'(z) = 0$. However, assuming that $|\zeta| = \rho$ avoids these branch points, one of the branches of $z = f(\zeta')$ is regular on $|\zeta'| = \rho^{1/\mu_k}$, and has the Laurent expansion

$$f(\zeta') = \sum_{j=-\infty}^{\infty} a_j^{(k)} \zeta'^j,$$

i.e.,

$$z = \sum_{j=-\infty}^{\infty} a_j^{(k)} \zeta^{j/\mu_k}. \tag{6.12}$$

To find the area of K_k,

$$\text{meas}(K_k) = \tfrac{1}{2}\oint_{K_k} (x\, dy - y\, dx)$$

$$= \tfrac{1}{2}\oint_{K_k} \text{Re}\, \frac{z^*\, dz}{i}$$

$$= \tfrac{1}{2} \text{Re} \oint_{K_k} \left(\sum_{j=-\infty}^{\infty} a_j^{(k)*} \rho^{j/\mu_k} e^{-ij\phi/\mu_k} \right)$$

$$\times \left(\sum_{j=-\infty}^{\infty} a_j^{(k)} \rho^{j/\mu_k} e^{ij\phi/\mu_k} \right) \frac{j\, d\phi}{\mu_k}$$

$$= \pi \sum_{j=-\infty}^{\infty} j |a_j^{(k)}|^2 \rho^{2j/\mu_k}.$$

Therefore

$$\sum_{k=1}^{m} \text{meas}(K_k) = \sum_{k=1}^{m} \pi \left[\sum_{j=1}^{\infty} j|a_j^{(k)}|^2 \rho^{2j/\mu_k} - \sum_{j=1}^{\infty} j|a_{-j}^{(k)}|^2 \rho^{-2j/\mu_k} \right]. \tag{6.13}$$

Equation (6.13) is an increasing and continuous function of ρ, and so is bounded by its value at $\rho = \infty$. For ρ sufficiently large, K_1 contains all n roots of $p(z)$, and therefore $\mu_1 = n$ and $m = 1$.

6.6 Lemniscates, capacity, and measure

Since $\zeta = p(z) = z^n + c_1 z^{n-1} + \cdots + c_n$,
$$z = f(\zeta) = \zeta^{1/n} + a_0 + a_1 \zeta^{-1/n} + \cdots, \tag{6.14}$$

which makes (6.12) more explicit, and (6.13) becomes

$$\sum_{k=1}^{m} \text{meas}(K_k) = \pi \rho^{2/n} - \sum_{j=1}^{\infty} j|a_{-j}|^2 \rho^{-2j/n} \quad \text{for large } \rho. \tag{6.15}$$

Hence $\text{meas}(\mathscr{L}_n) \leq \pi \eta^2$ if $\eta = \rho^{1/n}$.

The result is proved, but it is interesting to note from (6.15) that equality is attained when $a_0 = a_1 = \cdots = 0$ and (6.14) shows that this locus is the circle $z^n = \rho$.

We now turn to a few examples.

Example 1. The capacity of the interval $a \leq x \leq b$ is $(b - a)/4$.

Discussion. We assume the result (6.1) [Cheney, 1966, p. 61; Rivlin, 1969, Chapter 1] for the proof. Then result (6.3) follows from the substitution (6.2), and from (6.16) it follows that

$$\text{cap}(-1 \leq x \leq 1) = \tfrac{1}{2}$$

and the theorem is proved.

Example 2. The capacities of the disk $|z| \leq R$ and the circle $|z| = R$ are each equal to R.

Proof. The circle $|z| = R$ is a lemniscate, given by $|z^m| = R^m$, and so the results follow from Theorem 6.6.3 (and also from an inspection of the proof of Theorem 6.6.4).

Example 3. If $\mathscr{E}$ is a countable compact set, $\text{cap}(\mathscr{E}) = 0$. We leave the proof as an exercise.

We next state two very important theorems about capacity, without proof. Each gives considerable insight into the magnitude of the capacity of an arbitrary point set.

Theorem 6.6.5 [Hille, 1962, p. 268–73]. *Let $\mathscr{E}$ be a compact set. Then*

$$\text{cap}(\mathscr{E}) = \lim_{n \to \infty} \left[\max_{z_i \in \mathscr{E}} \prod_{1 \leq j < k \leq n} |z_j - z_k| \right]^{2/(n(n-1))}.$$

Discussion. The points $z_1, z_2, \ldots, z_n$ in $\mathscr{E}$ are chosen so as to maximize

$$\prod_{1 \leq j < k \leq n} |z_j - z_k|,$$

which contains $n(n-1)/2$ terms in its expansion. Hence it is clear that cap $(\mathscr{E})$ is bounded by the maximum diameter of $\mathscr{E}$. Furthermore, cap $(\mathscr{E})$ is some sort of geometric mean of the distance between the points of $\mathscr{E}$, and so it is called the transfinite diameter.

Two corollaries of the theorem are self-evident:

Corollary 1 (Monotonicity). *If $\mathscr{D}$ and $\mathscr{E}$ are compact sets with $\mathscr{D} \subset \mathscr{E}$, then* cap $(\mathscr{D}) \leq$ cap $(\mathscr{E})$.

Corollary 2 (Homogeneity). *If $z' = az + b$ maps $\mathscr{E}$ onto $\mathscr{E}'$, then* cap $(\mathscr{E}') = |a|$ cap $(\mathscr{E})$.

Theorem 6.6.6 [Hille, 1962, pp. 280–9]. *Let $\mu(z)$ be a normalized measure defined on $\mathscr{E}$, and define*

$$I[\mu] = \int_{\mathscr{E}} \int_{\mathscr{E}} \ln(|z_1 - z_2|^{-1}) \, d\mu(z_1) \, d\mu(z_2).$$

Let

$$V(\mathscr{E}) = \inf_{\mu} I[\mu].$$

Then

$$\text{cap}(\mathscr{E}) = \exp[-V(\mathscr{E})].$$

Discussion. We have in mind that $\mu(z)$ is a charge distribution on a two-dimensional surface $\mathscr{E}$, so $I[\mu]$ is the self-energy associated with the distribution μ. In physical equilibrium, this functional is a minimum, and so the potential due to the physical distribution of unit charge on $\mathscr{E}$ is

$$V(\mathscr{E}) = \inf_{\mu} I[\mu].$$

The further definition that $\ln(\text{cap}(\mathscr{E})) = -V(\mathscr{E})$ explains the name of logarithmic capacity for cap $(\mathscr{E})$.

Finally, we remark that we have emphasized the role of capacity as a point-set measure of the region of inaccurate approximation of Padé approximants to meromorphic functions. Another popular point-set measure is Hausdorff measure, and the results of the previous section generalize to convergence in α-dimensional Hausdorff measure [Wallin,

1974; Lubinsky, 1980a,b]. In this case, the measure is defined by

$$\Lambda_\alpha(\mathscr{E}) = \inf\left\{\sum_{i=1}^{\infty}[\delta(D_i)]^\alpha\right\},$$

where the infimum is taken over all possible denumerable families of circular disks D_i which cover the set $\mathscr{E}$, and $\delta(D_i)$ is the diameter of the disk D_i. Two-dimensional Hausdorff measure is similar to area in its properties, and one-dimensional Hausdorff measure is more like length.

The following Boutroux–Cartan lemma is most useful in the context of one-dimensional Hausdorff measure.

Theorem 6.6.7. *For any $\eta > 0$, the polynomial $p_n(z) = \prod_{i=1}^{n}(z - z_i)$ satisfies the inequality*

$$|p_n(z)| > (\eta/e)^n, \tag{6.16}$$

which is valid in the z-plane outside no more than n circles of radii r_i which obey the inequality

$$\sum_{i=1}^{n} r_i \leq 2\eta.$$

We do not prove this theorem, and we refer to Baker [1975a, Chapter 14] for a more complete account of the role of Hausdorff measure in convergence of Padé approximants. We merely note that (6.16) is the basic type of inequality required in this context. We also note that many of these results, also extend, mutatis mutandis, to the rational interpolation problem [Walsh, 1969, 1970; Karlsson, 1976].

We repeat that an inspection of the proofs of the theorems of Section 6.5 shows that they prove convergence in capacity, and we have shown that this is a substantially stronger result than convergence in measure. In particular, an interval has zero measure, but nonzero capacity (and nonzero one-dimensional Hausdorff measure), and so a finite interval, or any set containing a finite interval, is not a permissible exceptional set for the theorems of Section 6.5. Further, we draw attention to a result of Nuttall's, generalized below, that for certain functions with branch points connected by branch cuts, the Padé approximants converge in capacity in the z-plane except in a region which minimizes the capacity of the cuts in the z^{-1}-plane [Nuttall, 1977; Nuttall and Singh, 1977].

Stahl [1987] has obtained some important results on convergence in capacity of Padé approximants. His proof will not be given here, as it depends heavily on modern potential theory [Landkof, 1972], as did the results of Nuttall just described. Stahl has obtained two theorems. The first shows that there is a 'natural' place for the branch cuts of the 'near-diagonal' Padé approximants to be placed, and the second theorem describes the convergence of the Padé approximants away from the cuts. By near-diagonal we mean any sequence $[L/M]$ for which,

$$\lim_{M \to \infty} \frac{L}{M} = 1.$$

We use $\mathbb{C}$ to denote the complex plane and as a further bit of notation $\hat{\mathbb{C}}$ to denote the extended complex plane, i.e., the complex plane plus the point at infinity, and let $\omega = z^{-1}$.

***Theorem 6.6.8* (Stahl's extremal domain theorem).** *Let $f(z)$ be a given function, analytic at the origin. There exists a unique compact set $\mathcal{K}_0 \subset \mathbb{C}$, in the ω-plane such that*

(i) $\mathcal{D}_0 = \hat{\mathbb{C}} \backslash \mathcal{K}_0$ *is the domain in which $f(\omega^{-1})$ has a single-valued analytic continuation from $\omega = \infty$,*
(ii) $\text{cap}(\mathcal{K}_0) = \inf_{\mathcal{K}} \text{cap}(\mathcal{K})$, *where the infimum extends over all compact sets $\mathcal{K} \subseteq \mathbb{C}$ satisfying (i), and*
(iii) $\mathcal{K}_0 \subseteq \mathcal{K}$ *for all compact sets $\mathcal{K} \subseteq \mathbb{C}$ satisfying (i) and (ii).*

The set $\mathcal{K}_0$ is called the minimal set of $f(\omega^{-1})$ and the domain $\mathcal{D}_0$ is called the extremal domain. The 'natural' place for the cuts to go is on the minimal set $\mathcal{K}_0$ and we shall see from the next theorem that they do go there. The next theorem also has the important feature that it proves that the Padé approximants converge geometrically with order in the resulting extremal domain. In order to make this remark quantitative, we need to define the Green's function (Section 7.1) $g_{\mathcal{D}_0}(z, 0)$ with a logarithmic singularity at $z = 0$ for the extremal domain, and to define the derived function

$$F_f(z) = \exp\{-g_{\mathcal{D}_0}(z, 0)\}, \; z^{-1} \in \mathcal{D}_0. \tag{6.17}$$

The function $F_f(z)$ controls the rate of geometric convergence.

***Theorem 6.6.9* (Stahl's Padé convergence theorem).** *Let $f(z)$ be a given function, analytic at $z = 0$, and let the set $\mathcal{E} \subset \hat{\mathbb{C}}$, in the ω plane, of all the singularities of $f(\omega^{-1})$ be of capacity $\text{cap}(\mathcal{E}) = 0$. Then any near-diagonal sequence of Padé approximants to the function $f(z)$ converges in capacity to $f(z)$ in the extremal domain $\mathcal{D}_0$. More precisely: for any compact set $\mathcal{W} \subseteq \mathcal{D}_0$ in the ω-plane and $\varepsilon > 0$, with $z = \omega^{-1}$ we have*

$$\lim_{M \to \infty} \text{cap}\{\omega \in \mathcal{W}, |f(z) - [L/M](z)| > [F_f(z) + \varepsilon]^{L+M}\} = 0$$

and
$$\lim_{M\to\infty} \text{cap } \{\omega \in \mathcal{W}, |f(z) - [L/M](z)| < [F_f(z) - \varepsilon]^{L+M}\} = 0,$$

where $F_f(z)$ is given by (6.17).

A special case of this result is given by Nuttall [1990] using a more transparent derivation. In order to give a feeling for the above theorems of Stahl, we follow Nuttall in what follows.

First we remember the defining equations (1.1.1) and (1.1.10) for the Padé approximants

$$Q^{[L/M]}(z)f(z) - P^{[L/M]}(z) = O(z^{L+M+1}).$$

From this equation we can define a remainder function

$$R(z) = Q^{[L/M]}(z)f(z) - P^{[L/M]}(z). \tag{6.18}$$

Now the special case that we wish to consider is for the given function to be representable in the form

$$f(z) = f_0 + \frac{z}{2\pi i} \int_{\mathcal{L}} \frac{\sigma(\zeta)(\zeta^2 - 1)_+^{1/2}\, d\zeta}{z\zeta - 1}, \tag{6.19}$$

where $\mathcal{L} = \{\zeta, -1 \leq \zeta \leq 1\}$, f_0 is a constant, and $(\zeta^2 - 1)_+^{1/2}$ is the function, continuous on $\mathcal{L}$, which equals $-i$ at $\zeta = 0$. That is to say, $f(z)$ is analytic in the whole complex z-plane except for the cuts $-\infty \leq z \leq -1$ and $1 \leq z \leq \infty$. We are going to further restrict the nature of $\sigma(\zeta)$ by the following considerations. Let us define

$$y^2 = z^2 + y^2 z^2.$$

A useful way to parametrize the two solutions of this equation is by means of the uniformizing transformation,

$$z = \frac{2t}{1 + t^2}, \quad \text{so } y = \frac{2t}{1 - t^2}.$$

The two Riemann sheets are now simply described by

Sheet 1: $|t| < 1$,
  so that $\quad z = 0^{(1)}: \quad t = 0,$
Sheet 2: $|t| > 1$, $\quad z = 0^{(2)}: \quad t = \infty.$

This solution embodies the Riemann-sheet structure that we wish to impose on the function $f(z)$. We will do this in the following way. Let us define the auxiliary functions

$$G(t) = y(t)f(z(t)), \quad \text{for } t \in \text{Sheet 1,}$$

and

$$\Sigma(t) = G(t) + G(t^{-1}).$$

We then assume that $G(t)$ has an analytic continuation into $1 \le |t| \le 1 + a$, $a > 0$. This assumption implies that $\Sigma(t)$ is defined analytically in the set $1/(1 + a) \le |t| \le 1 + a$. We further assume that $\Sigma(t) \ne 0$ in that set.

To make further progress, we consider diagonal Padé approximants $[n/n]$ and define the quantities

$$\rho(t) = y(t)t^{-n}R(z(t)), \quad |t| < 1, \tag{6.20}$$

and

$$\Pi_1(t) = -t^{-n}P^{[n/n]}(z(t)), \quad \Pi_2(t) = t^{-n}Q^{[n/n]}(z(t)), \quad |t| > 1.$$

Note that ρ is bounded at $t = 0$ and the Π_j are bounded at $t = \infty$.

Using the properties of $y(t)$ we can write by the Padé equations for the case $|t| = 1$,

$$y(t)\Pi_1(t) + G(t)\Pi_2(t) = \rho(t), \quad -y(t)\Pi_1(t^{-1}) + G(t^{-1})\Pi_2(t^{-1}) = \rho(t^{-1}). \tag{6.21}$$

We may solve (6.21) by noting that $\Pi_j(t^{-1}) = t^{2n}\Pi_j(t)$. We obtain

$$\Sigma(t)\Pi_2(t) = \rho(t) + t^{-2n}\rho(t^{-1}), \quad |t| = 1. \tag{6.22}$$

The values of ρ in (6.22) are obtained by taking the limits $|t| \to 1$ in (6.20). The structure of (6.22) is the important thing to notice. We may solve it for $\Pi_2(t)$ and $\rho(t)$ if we first solve the inhomogeneous Hilbert equation for $h(t)$, $\chi_2(t)$,

$$\Sigma(t)\chi_2(t) = h(t), \quad |t| = 1, \tag{6.23}$$

where $h(t)$ and $\chi_2(t)$ are nonvanishing and analytic inside and outside the unit circle, respectively. The problem described by (6.23) is the classical Riemann boundary-value problem where (6.23) is the so-called connection formula, as it 'connects' the behavior between an analytic function inside $|t| = 1$ with that of one outside. See, for example, the treatise of Muskhelishvili [1977]. Nuttall gives the solution of (6.23) as

$$\left.\begin{matrix} |t| < 1, & h(t) \\ |t| > 1, & \chi_2(t) \end{matrix}\right\} = \exp\left[\frac{1}{2\pi i}\int_{|u|=1} \frac{\log \Sigma(u)\, du}{u - t}\right]. \tag{6.24}$$

The next step is to divide (6.22) by (6.23) which yields

$$\frac{\Pi_2(t)}{\chi_2(t)} = \frac{\rho(t)}{h(t)} + \frac{t^{-2n}\rho(t^{-1})}{h(t)}, \quad |t| = 1.$$

This formula shows that $\Pi_2(t)/\chi_2(t)$ and $\rho(t)/h(t)$, which are analytic outside and inside the unit circle, respectively, have boundary values on the unit circle that differ by $t^{-2n}\rho(t^{-1})/h(t)$. From this Nuttall is able to show that

$$\left.\begin{array}{l}|t|<1, \quad \rho(t)/h(t)\\ |t|>1, \quad \Pi_2(t)/\chi_2(t)\end{array}\right\} = \frac{-1}{2\pi i}\int_{|u|=1}\frac{u^{-2n}\rho(u^{-1})\,du}{h(u)(u-t)} + C_n \quad (6.25)$$

for some constant C_n. If $C_n \neq 0$ an appropriate normalization of the Padé polynomials can be found to make $C_n = 1$. If we now take the limit of the $|t|<1$ case as t approaches the unit circle, (6.25) implies the following singular integral equation for $\rho(t)$:

$$\frac{\rho(t)}{h(t)} = C_n - \frac{1}{2\pi i}\lim_{\varepsilon\to 0^+}\int_{|u|=1}\frac{u^{-2n}\rho(u^{-1})\,du}{h(u)(u-t(1-\varepsilon))}, \quad |t|=1. \quad (6.26)$$

By the assumptions we have made above, $\log\Sigma(u)$ is analytic in the region $1 \leq |u| \leq 1+a$, and so we can extend the contour of integration in (6.24) to give the result

$$h(t) = \exp\left[\frac{1}{2\pi i}\int_{|u|=1+a}\frac{\log\Sigma(u)\,du}{u-t}\right], \quad |t|<1.$$

This shows that $h(t)$ has a *nonzero* analytic continuation to the region $|t| \leq 1+a$. Consequently we can also extend (6.26) to

$$\frac{\rho(t)}{h(t)} = C_n - \frac{1}{2\pi i}\int_{|u|=1+a}\frac{u^{-2n}\rho(u^{-1})\,du}{h(u)(u-t)}, \quad |t|<1+a. \quad (6.27)$$

If we define

$$S_n \equiv \sup_{|t|=1}|\rho(t)/h(t)|, \quad (6.28)$$

then by the maximum-modulus principle of complex variable theory (see, e.g., Copson [1948]),

$$\sup_{|z|<r}|f(z)| \leq \sup_{|z|=r}|f(z)| \quad \text{when } f(z) \text{ is analytic in } |z| \leq r,$$

we have the results that

$$\left|\frac{\rho(u^{-1})}{h(u^{-1})}\right| \leq S_n \quad \text{for } |u|=1+a$$

and

$$\left|\frac{\rho(u^{-1})}{h(u)}\right| = \left|\frac{\rho(u^{-1})}{h(u^{-1})}\right| \cdot \left|\frac{h(u^{-1})}{h(u)}\right| \leq S_n O(1).$$

We can now deduce from (6.27) asymptotically as $n \to \infty$ that

$$\frac{\rho(t)}{h(t)} \sim C_n + (1+a)^{-2n}S_n O(1), \quad |t| \leq 1+a-\varepsilon, \quad \varepsilon > 0. \quad (6.29)$$

Now if we take the maximum modulus of (6.29) for $|t| = 1$, we have by (6.28)

$$S_n[1 + (1 + a)^{-2n}O(1)] = |C_n|, \quad n \to \infty.$$

By Theorems 1.4.3–1.4.5 either $f(z)$ is rational and so there is no convergence question, or there must exist an infinite subsequence of uniquely defined Padé approximants. For this subsequence, we may always select a Padé-approximant normalization so that $C_n = 1$. Hence it follows that asymptotically we have $S_n \sim 1$. Thus if we define

$$\chi_1(t) = -f(z)\chi_2(t), \quad |t| > 1,$$

we can conclude from (6.24)–(6.25) for the aforementioned subsequence that if $\varepsilon > 0$, and $0 < a' < a$ are fixed, then the asymptotic behavior of the polynomials is uniformly bounded by

$$\left.\begin{aligned}P^{[n/n]}(z) &= t^n\chi_1(t)[1 + (1 + a')^{-2n}O(1)],\\ Q^{[n/n]}(z) &= t^n\chi_2(t)[1 + (1 + a')^{-2n}O(1)],\end{aligned}\right\} \quad |t| > 1 + a' + \varepsilon,$$

$$\left.\begin{aligned}P^{[n/n]}(z) &= t^n\chi_1(t) + t^{-n}\chi_1(t^{-1}) + t^n(1 + a)^{-2n}O(1)],\\ Q^{[n/n]}(z) &= t^n\chi_2(t) + t^{-n}\chi_2(t^{-1}) + t^n(1 + a)^{-2n}O(1)],\end{aligned}\right\}$$

$$1 \leq |t| \leq 1 + a - \varepsilon.$$

It is to be noted that the positive nature of the χ_j forces the zeros to move asymptotically toward the set $\mathcal{L}$ defined for (6.19). This behavior is as would be expected from the discussion in Sections 6.7 and 2.2 and the ideas discussed above about the minimization of the capacity of the cut-set in the $1/z$-plane. It is to be noted that, for this special case, these results (due to Nuttall) are stronger than those in Stahl's theorem as no nth root is involved in the asymptotic results. Consequently, all the zeros of $Q^{[n/n]}(z)$ are eventually (as $n \to \infty$) forced out of any compact subset of $\mathbb{C}\setminus\mathcal{L}$.

6.7 The Padé conjecture

A conjecture of Baker, Gammel, and Wills, slightly rephrased, is popularly known as the Padé conjecture. The conjecture, which concerns convergence of diagonal Padé approximants to functions analytic in a disk, gave great impetus to the search for convergence theorems for the diagonal sequence, and also led to confidence in the usefulness of the diagonal sequence.

Before commencing the statement of the conjecture, we present Gammel's counterexample [Baker, 1973b] which shows why the conjecture takes its form and is not stronger.

6.7 The Padé conjecture

Gammel's counterexample. Let

$$f(z) = \sum_{n=0}^{\infty} c_n z^n = 1 + \sum_{k=1}^{\infty} \alpha_k \left\{ \sum_{n=n_k}^{2n_k} \left(\frac{z}{z_k}\right)^n \right\}$$

$$= 1 + \underbrace{\alpha_1\left(\frac{z}{z_1}\right) + \alpha_1\left(\frac{z}{z_1}\right)^2}_{k=1} + \underbrace{\alpha_2\left(\frac{z}{z_2}\right)^3 + \alpha_2\left(\frac{z}{z_2}\right)^4 + \cdots}_{k=2} + \cdots,$$

where the indices n_k are defined by $n_1 = 1$, $n_{k+1} = 2n_k + 1$. Let $c_n = \alpha_k z_k^{-n}$ if n is such that $n_k \leq n < n_{k+1}$. The coefficients $\{\alpha_k\}$ are specified by

$$\alpha_k = \frac{1}{(2k)!} \min\left(|z_k|^{n_k}, |z_k|^{2n_k}\right).$$

This choice of α_k ensures that

$$|c_n| \leq \frac{1}{n!},$$

so $f(z)$ is holomorphic (by the comparison test). The function may also be expressed by

$$f(z) = 1 + \sum_{k=1}^{\infty} \alpha_k \frac{\left(\frac{z}{z_k}\right)^{n_k} - \left(\frac{z}{z_k}\right)^{2n_k+1}}{1 - \frac{z}{z_k}}. \tag{7.1}$$

Inspection of (7.1) and the $[L/1]$ sequence shows that

$$[n_k/1] = [n_k + 1/1] = [n_k + 2/1] = \cdots = [2n_k - 1/1],$$

revealing a block of length n_k. This result implies that

$$[n_k/1] = [n_k/n_k]$$

(as is obvious by the accuracy-through-order criterion anyway), and so $[n_k/n_k]$ has a pole at $z = z_k$. By selecting the sequence $\{z_k, k = 1, 2, \ldots\}$ to be a set of points dense in the plane, and allowing such repetition as is necessary, we construct an analytic function with a subsequence of diagonal Padé approximants (in fact $[2^k - 1/2^k - 1]$ Padé approximants) which diverges on a set dense in the plane. In fact, a host of Gammel counterexamples can be constructed with Padé approximants which diverge in any desired region of the z-plane.

The only redeeming features of Gammel's counterexample are that the area of bad approximation at the poles of the $[n_k/n_k]$ Padé approximants tends to zero rapidly with increasing n, and that Baker has shown that another subsequence of diagonal approximants converges pointwise [Baker, 1973b].

Consideration of the implications of the counterexample shows why the Padé conjecture which follows is widely believed to be both true and as strong a result as possible.

Conjecture **[Baker, Gammel, and Wills, 1961].** *Let $f(z)$ be analytic in $|z| < R$ except for m poles at $z_1, z_2, \ldots, z_m$ with $0 < |z_1| \leq |z_2| \leq \cdots \leq |z_m| < R$, and except for one point z_0 on the boundary $|z| = R$. Further, given any $\varepsilon > 0$, there must exist a neighborhood $|z - z_0| < \delta$ in which $|f(z) - f(z_0)| < \varepsilon$, provided $|z| \leq R$, which means that $f(z)$ is continuous at z_0 within the circle. Then at least a subsequence of $[M/M]$ Padé approximants converges uniformly to $f(z)$ on any compact subset of*

$$\mathcal{D} = \{z, |z| \leq R, z \neq z_i, i = 0, 1, 2, \ldots, m\}.$$

It is regrettable that no proof yet exists. The importance of this conjecture depends on the homographic-invariance property of diagonal Padé approximants. The domain of pointwise convergence of a subsequence may be substantially extended by a development of the Padé conjecture. The strong form of this conjecture asserts that there exists a subsequence that converges in all those circles of meromorphy of $f(z)$ containing a finite number of poles of $f(z)$ and the origin, and it implies the following quasitheorem.

Quasitheorem **[Baker et al., 1961].** *Let $f(z)$ be analytic at $z = 0$, and let $\mathcal{D}$ be the union of all circles containing the origin in which $f(z)$ is meromorphic. Then at least a subsequence of $[M/M]$ Padé approximants converges to $f(z)$ pointwise on any compact subset of $\mathcal{D}$ which does not contain the poles of $f(z)$.*

'*Proof.*' Any point $z \in \mathcal{D}$ lies in the interior of a circle Γ with center c and radius R containing the origin O, as shown in Figure 6.7.1. Consider the mapping

$$w = \frac{|R/c|z}{z + c\{|R/c|^2 - 1\}}, \tag{7.2}$$

and let $f(z) = g(w)$. The circle $|w| = 1$ is the image of

$$|z| = |c/R| |z + c\{|R/c|^2 - 1\}|. \tag{7.3}$$

This is a circle for which the origin O is an interior point (if $|c| < R$), and O and P are inverse points provided $z_P = -c\{|R/c|^2 - 1\}$. Γ is given

6.7 The Padé conjecture

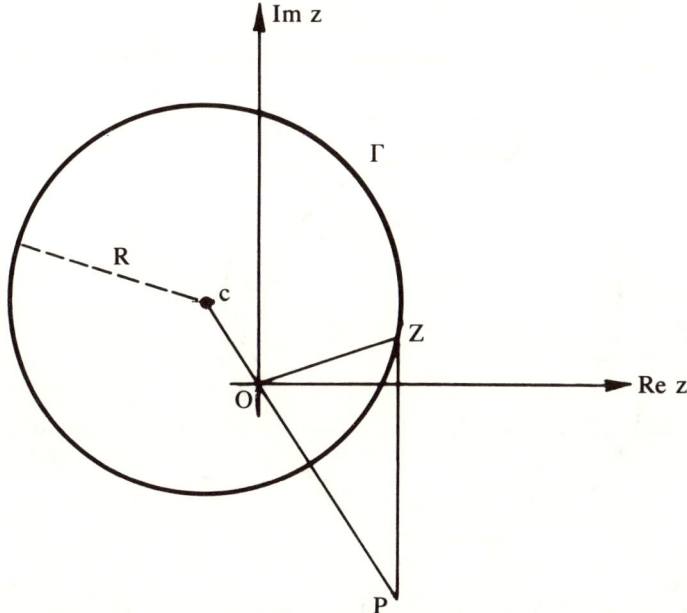

Figure 6.7.1. The circle Γ in the z-plane, center c and radius R.

parametrically by $OZ = |c/R| \cdot PZ$. The result 'follows' from Theorem 1.5.2.

This result justifies (7.3) and explains the choice (7.2). If $f(z)$ is meromorphic in Γ, then $g(w)$ is meromorphic in $|w| \leq 1$ and the Padé conjecture asserts that a subsequence of the diagonal sequence of Padé approximants converges to $g(w)$ in $|w| \leq 1$ (except at the poles). This quasitheorem is reinterpreted as asserting that the same subsequence of diagonal approximants to $f(z)$ converges at $z \in \Gamma$.

Application. Consider the function $f(z) = (z-a)^{3/2}(z-b)^{-3/2}$, where a, b are arbitrary (nonzero) points in the complex plane. The Padé conjecture asserts the convergence, in Γ_1 of Figure 6.7.2, of a subsequence. In this example, $z = \infty$ is a regular point, and so convergence at any point outside Γ_2 of a subsequence is asserted.

This example strongly suggests that the poles of the diagonal Padé approximants to $f(z)$ lie on the arc of a circle through O, a, and b. In this case, the mapping $t = z(a-b)/[(z-a)b]$ maps $z = a$ to $t = \infty$ and $z = b$ to $t = -1$. In fact $f(z) = h(t)$ is a Stieltjes series in t, proving that the poles of the diagonal approximants lie on the arc.

An alternative view is that the mapping $u = z^{-1}$ maps the points $z = a$

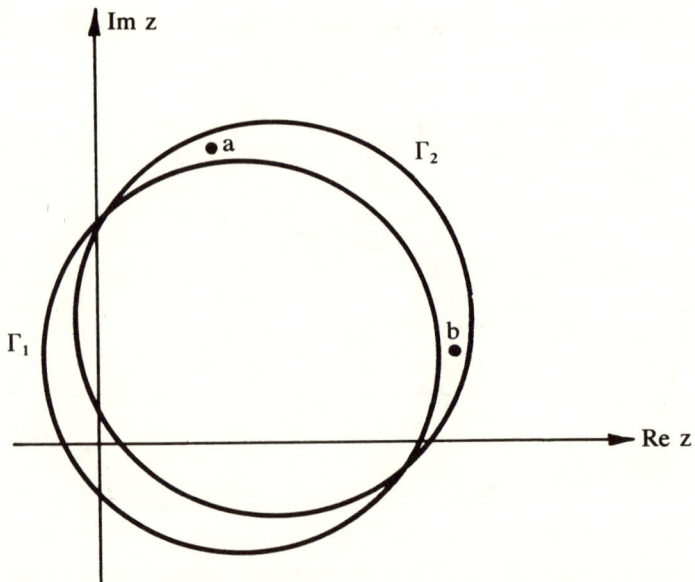

Figure 6.7.2. The complex z-plane, showing the circles Γ_1 and Γ_2.

and $z = b$ to $u = a^{-1}$ and $u = b^{-1}$. The exceptional set for diagonal Padé approximants to $f(z) = F(u)$ in the u-variable minimizes the capacity of the cut, and therefore is the straight line from a^{-1} to b^{-1}. This line is the image of the arc of the circle through $z = 0$, $z = a$, and $z = b$ from a to b.

The status of the Padé conjecture and its relation to open problems in rational approximation is surveyed by Walsh [1970] and Lubinsky [1992]. Progress toward a proof of the conjecture of Baker, Gammel, and Wills has been made by Lubinsky, who shows that, in the following context, most poles of the diagonal Padé approximants of $f(z)$ migrate out of the expected region of convergence. Let $f(z)$ be a function which is analytic and single-valued (uniform) in $\mathbb{C}\backslash\mathscr{E}$, where $\mathscr{E}$ is a set of capacity zero, and let $\mathscr{K}$ be a compact subset of $\mathbb{C}$ in which $f(z)$ is meromorphic. Then there exists an infinite subsequence S of positive integers such that the total multiplicity of the poles of the $[M/M]$ Padé approximants for $f(z)$ that lie in $\mathscr{K}$ is $o(M)$ for $M \in S$. We refer to Lubinsky [1991] for his more general result, which includes the case of multipoint approximants, and also to Lubinsky [1985a,b] for a proof of the conjecture of Baker, Gammel, and Wills for functions whose Maclaurin-series coefficients decrease very rapidly to zero.

7
Extensions of Padé approximants

7.1 Multipoint Padé approximants

A rational function which fits given function values at various points, not necessarily distinct, is called a multipoint Padé approximant. The associated problem of interpolation by rational functions is called the Cauchy–Jacobi problem. Multipoint Padé approximants are also called rational interpolants, N-point Padé approximants, or Newton Padé approximants, depending on the context. Interpolation at confluent points is sometimes called osculatory interpolation. For example, an appealing idea is that the $[N/N]$ rational form should satisfy N derivative conditions at $z = 0$ and $N - 1$ derivative conditions at $z = \infty$. Specifically, consider

$$f(z) = \left(\frac{1 + \frac{1}{2}z}{1 + 2z}\right)^{1/2}$$

$$= 1 - \tfrac{3}{4}z + O(z^2) \quad \text{as } z \to 0, \qquad (1.1)$$

$$= \tfrac{1}{2} + O(z^{-1}) \quad \text{as } z \to \infty. \qquad (1.2)$$

Linear algebra shows that the rational approximant so defined is

$$R^{[1/1]}(z) = \frac{1 + \tfrac{3}{4}z}{1 + \tfrac{3}{2}z},$$

which satisfies the accuracy-through-order conditions indicated by (1.1) and (1.2). At $z = 1$, the accuracy is 1%, which is the same absolute accuracy as is achieved by the [1/1] Padé approximant (cf. Section 1.1).

This analysis of a particular case indicates the approach to the general problem of rational interpolation at $z_0, z_1, z_2, \ldots$. The previous example is about 3-point interpolation with $z_0 = z_1 = 0$ and $z_2 = \infty$. The analysis of the general case is easier using finite points, and we are led to consider rational interpolation allowing confluence, and associated approximation problems. Space does not permit a full account of all these problems; we begin with an introduction to polynomial and rational interpolation, with particular emphasis on methods which allow confluence. First, we need the basic framework of Newtonian polynomial interpolation.

Divided Differences. For a function $f(z)$ satisfying such continuity properties as are necessary, we define

$$f[z_0] = f(z_0), \tag{1.3a}$$

$$f[z_0, z_1] = \frac{f(z_0) - f(z_1)}{z_0 - z_1}, \tag{1.3b}$$

and other divided differences are defined recursively by

$$f[z_0, z_1, \ldots, z_{r+1}] = \frac{f[z_0, z_1, \ldots, z_{r-1}, z_r] - f[z_0, z_1, \ldots, z_{r-1}, z_{r+1}]}{z_r - z_{r+1}},$$

$$r = 1, 2, \ldots. \tag{1.4}$$

Hermite's formula. *If $f(z)$ is analytic inside and continuous on a contour Γ enclosing $z_0, z_1, \ldots, z_k$, then*

$$f[z_0, z_1, \ldots, z_r] = \frac{1}{2\pi i} \int_\Gamma \frac{f(\zeta)}{\prod\limits_{k=0}^{r}(\zeta - z_k)} d\zeta. \tag{1.5}$$

Proof. The proof is by induction using (1.3) and (1.4).

For confluent points $z_0 = z_1 = \cdots = z_r$, it is natural to define

$$f[z_0, z_0, \ldots, z_0] = \frac{f^{(r)}(z_0)}{r!}. \tag{1.6}$$

Hermite's formula easily extends to cases of partial confluence.

Corollary. $f[z_0, z_1, \ldots, z_r]$ *is a totally symmetric function of all its arguments* $z_0, z_1, \ldots, z_r$.

Newton's formulas.

$$f(z) = \sum_{i=0}^{n} f[z_0, z_1, \ldots, z_i] \prod_{k=0}^{i-1}(z - z_k)$$

$$+ f[z_0, z_1, \ldots, z_n, z] \prod_{k=0}^{n}(z - z_k). \tag{1.7}$$

For $n > 0$, (1.7) is an identity expressing $f(z)$ as a Newton polynomial and a remainder term. One may 'deduce' the formal identity

$$f(z) = f[z_0] + (z - z_0)f[z_0, z_1] + (z - z_0)(z - z_1)f[z_0, z_1, z_2] + \cdots. \tag{1.8}$$

Whenever the remainder in (1.7) tends to zero, (1.8) becomes a true equality. The proof of (1.7) by induction is straightforward. It is the interpretation of (1.7) and (1.8) that is most significant. If $z_0 = z_1 = \cdots = z_i = \cdots$, (1.7) and (1.8) become

$$f(z) = f(z_0) + (z - z_0)f'(z_0) + \frac{(z - z_0)^2}{2!}f''(z_0) + \cdots \tag{1.9}$$

$$= \sum_{i=0}^{n} \frac{(z - z_0)^i}{i!} f^{(i)}(z_0) + \frac{(z - z_0)^{n+1}}{2\pi i} \int_{\Gamma} \frac{f(\zeta) \, d\zeta}{(\zeta - z_0)^{n+1}(\zeta - z)}. \tag{1.10}$$

Equation (1.10) holds provided Γ is a contour enclosing z, z_0 and $f(z)$ is analytic within Γ and continuous on Γ. In fact, (1.10) gives the Taylor series for $f(z)$ and its remainder. For conciseness, we make a further definition:

Definition

$$f_{i,j} = f[z_i, z_{i+1}, \ldots, z_j], \quad \text{for } j \geq i. \tag{1.11}$$

Then Newton's formula (1.8) becomes the formal identity

$$f(z) = f(z_0) + f_{0,1}(z - z_0) + f_{0,2}(z - z_0)(z - z_1) + \cdots.$$

We now proceed to consider interpolation of a given function $f(z)$ using rational fractions which are sometimes called interpolants. The basic problem is to find a rational fraction

$$r^{[L/M]}(z) = u^{[L/M]}(z)/v^{[L/M]}(z) \tag{1.12}$$

where $u^{[L/M]}(z)$ has maximum order L, $v^{[L/M]}(z)$ has maximum order M, and

$$r^{[L/M]}(z_i) = f(z_i), \quad i = 0, 1, 2, \ldots, L + M. \tag{1.13}$$

If a solution to this basic problem exists, it can be obtained by defining

$$u^{[L/M]}(z) = \sum_{j=0}^{L} u_j z^j, \quad v^{[L/M]}(z) = \sum_{k=0}^{M} v_k z^k \tag{1.14}$$

for specific values of L, M. Let us assume that $v_0 = 1$ is a permissible normalization for the moment. Substitution of (1.12) and (1.14) into (1.13) yields $L + M + 1$ linear equations for $L + M + 1$ unknown coefficients $u_0, u_1, \ldots, u_L, v_1, v_2, \ldots, v_M$. Normally there is a unique solution leading to a rational interpolant which is uniquely defined up to a constant common factor in the numerator and denominator of (1.12). Otherwise, the equations are said to be degenerate. If the equations are degenerate but consistent, and $v^{[L/M]}(z) \not\equiv 0$, then $u^{[L/M]}(z)$ and $v^{[L/M]}(z)$ have a common factor. Using (1.19) with $f(z) = r^{[L/M]}(z)$, it follows that the factors $(z - z_i)$, $i = 0, 1, \ldots, L + M$, are the only possible elementary common factors of $u^{[L/M]}(z)$ and $v^{[L/M]}(z)$. For each such factor $(z - z_k)$, (1.13) must be tested with $i = k$ for the proposed solution. If the linear equations are inconsistent, no rational function of type $[L/M]$ fits the data. As an example, we next show that no rational function of type $[1/1]$ fits the data

$$f(-1) = 1, \quad f(0) = 1, \quad f(1) = 3 \tag{1.15}$$

at the indicated points. The equations (1.12), (1.13), and (1.14) become

$$u_0 - u_1 = v_0 - v_1, \tag{1.16a}$$

$$u_0 = v_0, \tag{1.16b}$$

$$u_0 + u_1 = 3(v_0 + v_1). \tag{1.16c}$$

Equations (1.16a), (1.16b) imply that $u_0 = v_0$, $u_1 = v_1$, and so (1.16c) implies that $u_0 = u_1 = v_0 = v_1 = 0$. The equations (1.16) are degenerate. In fact, only the new value $f(1) = 1$ would render (1.16) consistent and allow rational interpolation to be effected by a (degenerate) interpolant of type $[1/1]$. A full analysis of possible degeneracies is given by Maehly and Witzgall [1960].

Since Padé approximation is rational approximation with complete confluence of the interpolation points, it is interesting to note the similarity between the previous analysis and that of the existence or nonexistence of Padé approximants in Section 1.4. Having briefly considered some of the hazards of using rational interpolation, we now give the standard solution in the nondegenerate case.

Theorem 7.1.1. *The N-point Padé approximant of type $[L/M]$ defined by interpolation at the points $z_0, z_1, \ldots, z_{L+M}$, allowing confluence, is normally given by*

$$r^{[L/M]}(z) = u^{[L/M]}(z)/v^{[L/M]}(z),$$

where $u^{[L/M]}(z)$ and $v^{[L/M]}(z)$ are defined by

7.1 Multipoint Padé approximants

$$u^{[L/M]}(z) = \begin{vmatrix} f_{M,L+1} & f_{M,L+2} & \cdots & f_{M,L+M} & \sum_{j=M}^{L} f_{M,j} \prod_{k=0}^{j-1}(z-z_k) \\ f_{M-1,L+1} & f_{M-1,L+2} & \cdots & f_{M-1,L+M} & \sum_{j=M-1}^{L} f_{M-1,j} \prod_{k=0}^{j-1}(z-z_k) \\ \vdots & \vdots & & \vdots & \vdots \\ f_{0,L+1} & f_{0,L+2} & \cdots & f_{0,L+M} & \sum_{j=0}^{L} f_{0,j} \prod_{k=0}^{j-1}(z-z_k) \end{vmatrix},$$

(1.17)

$$v^{[L/M]}(z) = \begin{vmatrix} f_{M,L+1} & f_{M,L+2} & \cdots & f_{M,L+M} & \prod_{k=0}^{M-1}(z-z_k) \\ f_{M-1,L+1} & f_{M-1,L+2} & \cdots & f_{M-1,L+M} & \prod_{k=0}^{M-2}(z-z_k) \\ \vdots & \vdots & & \vdots & \vdots \\ f_{0,L+1} & f_{0,L+2} & \cdots & f_{0,L+M} & 1 \end{vmatrix},$$

(1.18)

and the definition (1.11) has been used.
The remainder is given by

$$v^{[L/M]}(z)f(z) - u^{[L/M]}(z) = \prod_{k=0}^{L+M}(z-z_k)$$

$$\times \begin{vmatrix} f_{M,L+1} & f_{M,L+2} & \cdots & f_{M,L+M} & f[z_M, \ldots, z_{L+M}, z] \\ f_{M-1,L+1} & f_{M-1,L+2} & \cdots & f_{M-1,L+M} & f[z_{M-1}, \ldots, z_{L+M}, z] \\ \vdots & \vdots & & \vdots & \vdots \\ f_{0,L+1} & f_{0,L+2} & \cdots & f_{0,L+M} & f[z_0, \ldots, z_{L+M}, z], \end{vmatrix}$$

(1.19)

If 'impossible' entries in (1.17)–(1.19) occur, the following interpretation is intended: If $j < i$, then

$$f_{i,j} = 0, \quad \sum_{k=i}^{j} (term)_k = 0, \quad \text{and} \prod_{k=i}^{j}(factor)_k = 1.$$

A sufficient condition for the result that $u^{[L/M]}(z_i)/v^{[L/M]}(z_i) = f(z_i)$ is that $v^{[L/M]}(z_i) \neq 0$, $i = 0, 1, \ldots, L+M$.

Proof. The formulas (1.17) and (1.18) are polynomials of the appropriate orders. Using Newton's formula (1.7), it follows that

$$v^{[L/M]}(z)f(z) - u^{[L/M]}(z) = \prod_{k=0}^{L}(z - z_k) \times$$

$$\begin{vmatrix} f_{M,L+1} & f_{M,L+2} & \cdots & f_{M,L+M} & f[z_M, \ldots, z_L, z] \\ f_{M-1,L+1} & f_{M-1,L+2} & \cdots & f_{M-1,L+M} & f[z_{M-1}, \ldots, z_L, z] \\ \vdots & \vdots & & \vdots & \vdots \\ f_{0,L+1} & f_{0,L+2} & \cdots & f_{0,L+M} & f[z_0, \ldots, z_L, z] \end{vmatrix}. \quad (1.20)$$

Recalling the definitions (1.4) and (1.11), repeated subtraction of the jth column of (1.20) from the last for $j = 1, 2, \ldots, M$ yields (1.19). The right-hand side of (1.19) is manifestly zero at $z_0, \ldots, z_{L+M}$. Provided $v^{[L/M]}(z_i) \neq 0$, $i = 0, 1, \ldots, L + M$, the result follows.

Corollary. *If $z_0 = z_1 = \cdots = z_{L+M}$, (1.6), shows that $r^{[L/M]}(z)$ defined by (1.12), (1.17), and (1.18) reduces to a Padé approximant as in (1.1.8), (1.1.9), (1.1.11).*

Determinantal representations similar to (1.17), (1.18) but apparently unrelated to them are given by Graves-Morris [1984b]. These representations also resemble the barycentric formulas of Antoulas and Anderson [1986] who also discuss stability [1989]. Van Barel and Bultheel give a sophisticated analysis [1992b].

Note. Equations (1.17), (1.18), (1.19) can be adapted to solve the bigradient problem (Section 2.4) with confluent interpolation at each of two distinct points [Warner, 1974; Householder and Stewart, 1969].

Construction of the N-point Padé approximants using (1.17), (1.18) is cumbersome and quite unsuitable for numerical work. We now turn to consider some of the requirements of a satisfactory numerical algorithm.

Algorithms for rational interpolation may be classified into those designed to solve the coefficient problem and those designed to solve the value problem. The coefficient problem is the problem of evaluating the coefficients $\{u_0, u_1, \ldots, u_L; v_0, v_1, \ldots, v_M\}$ in (1.14) and thereby defining the rational form $r^{[L/M]}(z)$ in (1.12). The value problem entails evaluation of $r^{[L/M]}(z)$ at some prespecified value of z without necessarily finding the coefficients in (1.14) explicitly. For example, the ε-algorithm is a method of solving a value problem for Padé approximants without construction of the coefficients. The Q.D. algorithm for constructing continued fractions (see Section 4.3) is the solution of a coefficient problem.

If a table of values of an interpolant is required, it is usually more economical to solve the coefficient problem first and then to evaluate the interpolant at the various values of z. If only a single value of an interpolant is required, it is sometimes more efficient not to evaluate the coefficients explicitly. In fact, evaluation of continued fractions and polynomials is a comparatively rapid process. Consequently, the coefficient problem is especially important. Notice that computational economy is gained by using continued-fraction representations of interpolants of the type $r^{[M/M]}(z)$ rather than using polynomial ratios (1.12).

An important and desirable property of rational-interpolation methods is the property of allowing confluence of some or all of the interpolation points. Methods based directly on divided differences (1.3), (1.4) are intrinsically incapable of generalization to osculatory interpolation.

Reliability of rational-interpolation methods is another quality we seek. As stated earlier, no [1/1]-type interpolant fits the data of (1.15), and consequently any reliable interpolation method should recognize this problem as insoluble. Rational-interpolation algorithms must discriminate between soluble and insoluble problems, making due allowances for rounding and representation errors. This consideration leads us to the property of stability, which is closely related to reliability. If an algorithm is stable, small changes in the data lead to small changes in the results. A good rational-interpolation algorithm should be able to detect data which give unstable results.

We observe that sequential methods of constructing a particular rational interpolant depend on the existence of all the intermediate interpolants of the sequence. A reliable algorithm will work when the desired interpolant exists, even if the intermediate interpolants are degenerate or do not exist.

The qualities characterizing a good algorithm are notorious for not being entirely compatible, and the selection of a 'best algorithm' involves a compromise. For each of the various rational-interpolation problems, we seek algorithms which (i) are efficient, (ii) allow confluence, and (iii) are stable and reliable. We next turn to consider some of the best of the algorithms available.

Kronecker's algorithm **[Kronecker, 1881]**. To start the algorithm, the Newton interpolating polynomial (1.8) of order $L + M = m$ is used, with Newton coefficients derived using (1.4). Thus the initializing values are the coefficients in

$$p^{[m/0]}(z) = \sum_{i=0}^{m} f[z_0, z_1, \ldots, z_i] \prod_{k=0}^{i-1} (z - z_k), \qquad (1.21a)$$

$$q^{[m/0]}(z) = 1, \qquad (1.21b)$$

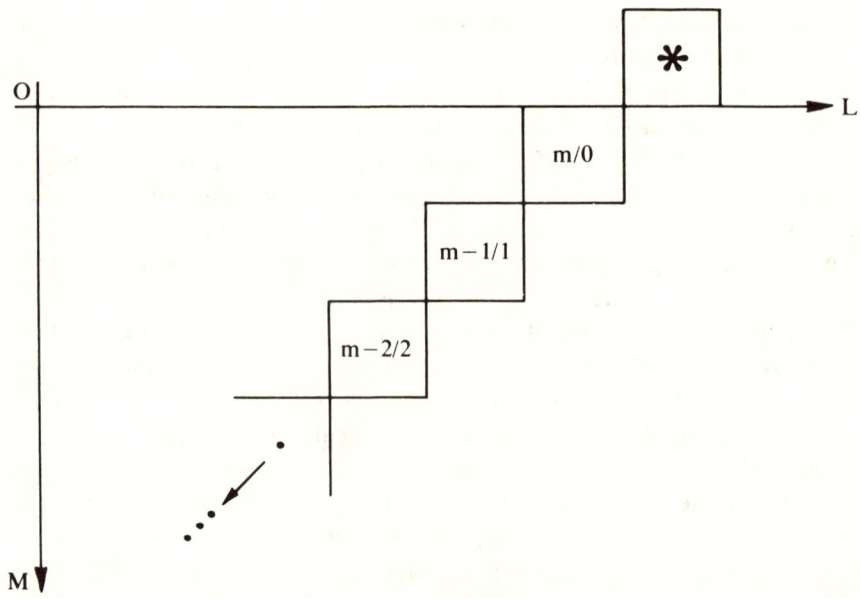

Figure 7.1.1. Elements of the N-point Padé table used in Kronecker's algorithm.

and the artificial initializing values are given by

$$p^{[m+1/-1]}(z) = \prod_{k=0}^{m}(z - z_k), \quad (1.22a)$$

$$q^{[m+1/-1]}(z) = 0. \quad (1.22b)$$

The recursion relations (which we will prove) are analogues of (3.5.12) and are

$$p^{[m-j/j]}(z) = (\alpha_j z + \beta_j)p^{[m-j+1/j-1]}(z) - p^{[m-j+2/j-2]}(z), \quad (1.23a)$$

$$q^{[m-j/j]}(z) = (\alpha_j z + \beta_j)q^{[m-j+1/j-1]}(z) - q^{[m-j+2/j-2]}(z). \quad (1.23b)$$

At stage (j), Equation (1.23a) is used to determine both α_j and β_j, so the apparent degree of the right-hand side of (1.23a) is reduced by two. Then (1.23a) gives the next numerator and (1.23b) the next denominator.

Verification. We assume that the algorithm is nonsingular, which is tantamount to assuming that the numerators $p^{[m-k/k]}(z)$ have the full indicated degree $(m - k)$ for all values of k needed. The coefficients α_j and β_j may be found at each stage j such that (1.23a) has order $m - j$, and we only have to prove that (1.23) defines approximants which interpolate correctly at $z_0, z_1, \ldots, z_m$ to the values given by (1.21). From (1.23), it follows that

$$D = \begin{vmatrix} p^{[m-j/j]}(z) & p^{[m-j+1/j-1]}(z) \\ q^{[m-j/j]}(z) & q^{[m-j+1/j-1]}(z) \end{vmatrix}$$

$$= \begin{vmatrix} p^{[m-j+1/j-1]}(z) & p^{[m-j+2/j-2]}(z) \\ q^{[m-j+1/j-1]}(z) & q^{[m-j+2/j-2]}(z) \end{vmatrix}.$$

D is therefore independent of j, and (1.21), (1.22) imply that

$$D = -\prod_{k=0}^{m}(z - z_k).$$

Thus, provided the algorithm is nonsingular,

$$\frac{p^{[m-j/j]}(z)}{q^{[m-j/j]}(z)} = \frac{p^{[m-j+1/j-1]}(z)}{q^{[m-j+1/j-1]}(z)} - \frac{\prod_{k=0}^{m}(z - z_k)}{q^{[m-j/j]}(z)q^{[m-j+1/j-1]}(z)},$$

showing that each entry interpolates as required. Notice that this algorithm is unaffected by confluence once it has been initialized.

An attractive feature of the Kronecker algorithm is that it is easily modified to become a reliable algorithm. If, for some value of j, $p^{[m-j/j]}(z)$ defined by (1.23a) does not have its full indicated order $(m - j)$, the $[m - j/j]$ interpolant is degenerate. The next interpolant in the sequence is defined by

$$p^{[m-j-k/j+k]}(z) = \pi_k(z)p^{[m-j/j]}(z) - p^{[m-j+1/j-1]}(z), \qquad (1.23c)$$

$$q^{[m-j-k/j+k]}(z) = \pi_k(z)q^{[m-j/j]}(z) - q^{[m-j+1/j-1]}(z), \qquad (1.23d)$$

where $\pi_k(z)$ is the polynomial of degree k such that the left-hand side of (1.23c) has its indicated degree $m - j - k$, in just the same way as in Section 3.7. It is easy to show [Graves-Morris, 1979] that (1.23c) and (1.23d) define the next nondegenerate interpolant in the antidiagonal sequence. Kronecker's algorithm would appear to be especially useful for the case where exact arithmetic is available and any degeneracy test must be decisive [Antoulas, 1988].

Thiele's reciprocal difference method [Thiele, 1909, Chapter 3; Hildebrand, 1956]. This method yields a continued-fraction representation of the N-point Padé approximant. Normally, the interpolation points should be distinct, although the elements of the fraction may be defined by continuity in the presence of confluence. Define the reciprocal differences

$$\rho_0 \equiv \rho[z_0] = f(z_0), \qquad (1.24a)$$

$$\rho_1 \equiv \rho[z_0, z_1] = (z_0 - z_1)\{f(z_0) - f(z_1)\}^{-1}, \qquad (1.24b)$$

$$\rho[z, z_0] = (z - z_0)\{f(z) - f(z_0)\}^{-1}, \qquad (1.24c)$$

and in general for $n > 1$

$$\rho_n \equiv \rho[z_0, z_1, \ldots, z_{n-1}, z_n], \tag{1.24d}$$

$$\rho[z_0, z_1, \ldots, z_{n-1}, z_n] = \frac{z_0 - z_n}{\rho[z_0, \ldots, z_{n-1}] - \rho[z_1, \ldots, z_{n-1}, z_n]} + \rho[z_1, \ldots, z_{n-1}]. \tag{1.24e}$$

Then the interpolant defined on $z_0, z_1, \ldots, z_n$ is given by

$$r_n(z) = \rho_0 + \frac{z - z_0}{\rho_1} + \frac{z - z_1}{\rho_2 - \rho_0} + \frac{z - z_2}{\rho_3 - \rho_1} + \cdots + \frac{z - z_{n-1}}{\rho_n - \rho_{n-2}}. \tag{1.25}$$

Notice that if $n = 2M$, which means that an odd number of interpolation points are used, then (1.25) defines an $[M/M]$-type approximant; if $n = 2M + 1$, an even number of points are used and an $[M + 1/M]$-type approximant is defined.

Verification. We first prove by induction the identity

$$f(z) = \rho_0 + \frac{z - z_0}{\rho_1} + \frac{z - z_1}{\rho_2 - \rho_0} + \frac{z - z_2}{\rho_3 - \rho_1} + \cdots$$
$$+ \frac{z - z_n}{\rho[z, z_0, \ldots, z_n] - \rho_{n-1}}. \tag{1.26}$$

For $n = 0$, (1.26) is interpreted as

$$f(z) = \rho_0 + \frac{z - z_0}{\rho[z, z_0]},$$

which is verified by (1.24c). For $n > 0$, we expand the final denominator of (1.26) using the identity

$$\rho[z, z_0, \ldots, z_n] - \rho_{n-1} = \rho_{n+1} - \rho_{n-1} + \frac{z - z_{n+1}}{\rho[z, z_0, \ldots, z_{n+1}] - \rho_n},$$

which is derived from (1.24e) in the form

$$\rho[z, z_0, \ldots, z_{n+1}] = \frac{z - z_{n+1}}{\rho[z, z_0, \ldots, z_n] - \rho_{n+1}} + \rho_n.$$

Hence (1.26) is proved by induction. If we let $z = z_0, z_1, \ldots, z_n$ sequentially, and provided no unusual cancellation occurs, (1.26) shows that (1.25) interpolates correctly at the $n + 1$ interpolation points, and therefore (1.25) is the $(n + 1)$-point Padé approximant. Its extension to the case of simultaneous rational interpolation, using a generalization of the Jacobi–Perron algorithm, has been given by Levrie and Bultheel [1993].

Thiele's method of rational interpolation is more interesting from an analytical point of view; for numerical values the following computational scheme is as efficient as any.

Modified Thacher–Tukey algorithm [Thacher and Tukey, 1960; Graves-Morris and Hopkins, 1981; Graves-Morris, 1981b]. We re-express (1.25) as

$$r_n(z) = f(z_0) + \frac{a_1(z - z_0)}{1} + \frac{a_2(z - z_1')}{1} + \frac{a_3(z - z_2')}{1} + \cdots$$
$$+ \frac{a_n(z - z_{n-1}')}{1}. \qquad (1.27)$$

The interpolation set is the set of distinct points

$$S_0 = \{z_0, z_1, \ldots, z_n\},$$

which are used in the order $z_0, z_1', z_2', \ldots, z_n'$.

To motivate the algorithm, we consider a function $f(z)$ defined on the interpolation set S_0 and interpolated by (1.27). We define functions $g_0(z)$, $g_1(z), \ldots, g_n(z)$ by

$$f(z) = f(z_0) + g_0(z), \qquad (1.28a)$$

and iteratively by

$$g_{i-1}(z) = \frac{a_i(z - z_{i-1}')}{1 + g_i(z)} \quad \text{for } i = 0, 1, 2, \ldots, n, \qquad (1.28b)$$

The special case of $f(z) = r_n(z)$ corresponds to $g_n(z) \equiv 0$. We expect, from (1.28b), that $g_{i-1}(z_{i-1}') = 0$, and hence also from this equation we expect that

$$g_{i-1}(z_i') = a_i(z_i' - z_{i-1}').$$

This equation is reexpressed in (1.29) to evaluate the coefficients $a_1, a_2, \ldots, a_n$, which must be finite and nonzero. It is part of the normal iterative step of the algorithm, called stage (a). If it turns out at some juncture (with $j = t + 1$) that $g_t(z) = 0$ for all $z \in S_{t+1}$, where S_{t+1} is a residual interpolation set, then this corresponds to stage (b) of the algorithm in which $r_t(z)$ is apparently the correct but degenerate interpolant. Certainly it is true that in all other cases, corresponding to stage (c) of the algorithm, there is no interpolating rational function. Having constructed (1.27) as the proposed interpolant, one must check that its denominator $q_t(z)$ given by (1.30) does not vanish at any of the interpolation points. One may show that this circumstance corresponds uniquely to the nonexistence of an interpolant. Notice that this algorithm is reliable in the sense that if an interpolant to the data exists, the algorithm finds it; if no such interpolant exists, the algorithm detects this situation via an error exit.

Initialization. We define the set

$$S_1 = \{z_1, z_2, \ldots, z_n\}.$$

We define values of a function $g_0(z)$ for $z \in S_1$ by

$$g_0(z) = f(z) - f(z_0) \quad \text{for } z \in S_1.$$

Iteration. The normal iterative step, used for $j = 1, 2, \ldots$ until termination, begins with stage (a). Otherwise, the iteration is concluded by stage (b) or stage (c).

Stage (a). If possible, choose z_j' from S_j such that

$$g_{j-1}(z_j') \neq 0, \infty.$$

Then we define

$$\left.\begin{aligned} a_j &= \frac{g_{j-1}(z_j')}{z_j' - z_{j-1}'}, \\ S_{j+1} &= S_j \backslash z_j', \\ g_j(z) &= \frac{a_j(z - z_{j-1}')}{g_{j-1}(z)} - 1 \quad \text{for all } z \in S_{j+1}. \end{aligned}\right\} \quad (1.29)$$

If $j = n$, we set $t = n$ and proceed to the termination stage; otherwise we proceed with the iteration with $j := j + 1$.

If it is not possible to choose z_j' from S_j such that

$$g_{j-1}(z_j') \neq 0, \infty,$$

it may be that we are at stage (b).

Stage (b). This stage is reached if, and only if,

$$g_{j-1}(z) = 0 \quad \text{for all } z \in S_j.$$

In this case, set $t = j - 1$ in (1.29). Exit from the iteration and proceed to termination for the denominator check.

Stage (c). This stage is reached if and only if $g_{j-1}(z) = 0, \infty$ for all $z \in S_j$, but $g_{j-1}(z) \neq 0$ for some $z \in S_j$. The algorithm terminates prematurely with an error exit, signifying that the desired interpolant does not exist.

Termination. If termination is reached with $t = 0$, then $r(z) = f_0$ is the correct rational interpolant. If termination is reached with $t = 1, 2, \ldots,$

n, we define

$$q_1(z) = 1, \quad q_2(z) = 1 + a_2(z - z_1') \tag{1.30a}$$

and for $i = 2, 3, \ldots, t - 1$,

$$q_{i+1}(z) = q_i(z) + a_{i+1}(z - z_i')q_{i-1}(z). \tag{1.30b}$$

Provided $q_t(z) \neq 0$ for all $z \in S_0$, the algorithm has a successful exit. Otherwise $q_t(z_j) = 0$ for some j, $0 \leq j \leq n$, and the algorithm has a failure exit in the termination stage, signifying that the required interpolant does not exist.

Gutknecht [1989] proposes a similar algorithm, except that infinite values of the intermediate interpolants are allowed. The practicalities of the barycentric method are discussed by Schneider and Werner [1991].

Generalized Q.D. algorithm [Wuytack, 1973; Graves-Morris, 1980b]. The Q.D. algorithm may be used to derive a continued-fraction representation of a Padé approximant from the coefficients of a Maclaurin series (see Section 4.3). We consider a generalization of this algorithm to the case in which the interpolation points may be distinct. We consider the following Thiele-type interpolant:

$$g_0(z) = \frac{c_0}{1-} \; \frac{q_1^0(z - z_0)}{1-} \; \frac{e_1^0(z - z_1)}{1-} \; \frac{q_2^0(z - z_2)}{1-}$$

$$\frac{e_2^0(z - z_3)}{1-} \cdots. \tag{1.31}$$

The coefficients of the nth convergent of $g_0(z)$ are derived from the coefficients of the Newton interpolating polynomial

$$\pi_n(z) = c_0 + c_1(z - z_0) + c_2(z - z_0)(z - z_1) + \cdots + c_n \prod_{i=0}^{n-1}(z - z_i)$$

using the following algorithm:

Initialization. For $J = 0, 1, 2, \ldots, n - 1$, define

$$Z_1^J = z_{J+1} - z_J, \tag{1.32a}$$

$$e_0^{J+1} = 0, \tag{1.32b}$$

$$q_1^J = \left[Z_1^J + \frac{c_J}{c_{J+1}}\right]^{-1}, \tag{1.32c}$$

$$e_1^J = -q_1^J - q_1^{J+1}(q_1^J Z_1^J - 1). \tag{1.32d}$$

Recursion. For $J = 0, 1, 2, \ldots$ and $i = 2, 3, \ldots$, we construct all well-defined quantities q_i^J, e_i^J recursively from the formulas

$$Z_i^J = z_{J+2i-1} - z_{J+2i-2}, \tag{1.33a}$$

$$q_i^J = \left[Z_i^J - \frac{e_{i-1}^J}{e_{i-1}^J + q_{i-1}^J} \frac{q_{i-1}^{J+1} + e_{i-2}^{J+1}}{q_{i-1}^{J+1}} \frac{Z_i^J e_{i-1}^{J+1} - 1}{e_{i-1}^{J+1}}\right]^{-1}, \tag{1.33b}$$

$$e_i^J = -q_i^J + (Z_i^J q_i^J - 1)(e_{i-1}^{J+1} + q_i^{J+1})(Z_i^J e_{i-1}^{J+1} - 1)^{-1}. \tag{1.33c}$$

Notice that this algorithm is well defined in the presence of confluence, in which case it reduces to a minor variant of the Q.D. algorithm (4.3.31), (4.3.32). Beckermann and Carstensen [1993] have fully analyzed the Q.D. algorithm for the non-normal Newton–Padé table.

Generalized ε-algorithm [Claessens, 1978c]. This algorithm is suitable for calculating the value of a rational interpolant. It is based on Claessens's identity, which, in the notation of (1.12), (1.13), and (1.14), is

$$[\{r^{[L+1/M]}(z) - r^{[L/M]}(z)\}^{-1} - \{r^{[L/M+1]}(z) - r^{[L/M]}(z)\}^{-1}](z - z_{L+M})$$
$$= [\{r^{[L/M-1]}(z) - r^{[L/M]}(z)\}^{-1}$$
$$- \{r^{[L-1/M]}(z) - r^{[L/M]}(z)\}^{-1}](z - z_{L+M+1}) \tag{1.34}$$

whenever the indicated quantities exist and are nondegenerate. Claessens's identity reduces to Wynn's identity in the confluent limit.

Outline of proof of (1.34). We follow the method of Section 3.4. We define

$$F_{1,L+M}^{[L/M]} = \begin{vmatrix} f_{M,L+1} & f_{M,L+2} & \cdots & f_{M,L+M} \\ f_{M-1,L+1} & f_{M-1,L+2} & \cdots & f_{M-1,L+M} \\ \vdots & \vdots & & \vdots \\ f_{1,L+1} & f_{1,L+2} & \cdots & f_{1,L+M} \end{vmatrix} \tag{1.35}$$

and note that $F_{1,L+M}^{[L/M]} = C(L/M)$ in the confluent limit. The subscripts of $F_{1,L+M}^{[L/M]}$ denote the indices $1, 2, \ldots, L + M$ of the interpolation points used in its construction. Using the methods of Section 3.4, we find that (3.4.4) generalizes directly to

$$r^{[L+1/M]}(z) - r^{[L/M]}(z) = \frac{(z - z_0) \cdots (z - z_{L+M+1}) F_{0,L+M+1}^{[L+1/M+1]} F_{0,L+M}^{[L+1/M]}}{v^{[L+1/M]}(z) v^{[L/M]}(z)},$$
$$\tag{1.36}$$

and (3.4.6) generalizes directly to

$$r^{[L/M+1]}(z) - r^{[L/M]}(z) = \frac{(z - z_0) \cdots (z - z_{L+M+1}) F_{0,L+M}^{[L/M+1]} F_{0,L+M+1}^{[L+1/M+1]}}{v^{[L/M+1]}(z) v^{[L/M]}(z)}.$$
$$\tag{1.37}$$

7.1 Multipoint Padé approximants

By reordering the points of (1.18), we find that

$$v^{[L/M+1]}(z)$$

$$= \begin{vmatrix} f_{M,L} & f_{M-1,L} & \cdots & f_{0,L} & \prod_{k=L+1}^{L+M+1}(z-z_k) \\ f_{M,L+1} & f_{M-1,L+1} & \cdots & f_{0,L+1} & \prod_{k=L+2}^{L+M+1}(z-z_k) \\ \vdots & \vdots & & \vdots & \vdots \\ f_{M,L+M+1} & f_{M-1,L+M+1} & \cdots & f_{0,L+M+1} & 1 \end{vmatrix}.$$

By applying Sylvester's identity to this, we find that

$$v^{[L/M+1]}(z) F_{0,L+M}^{[L+1/M]} = v^{[L+1/M]}(z) F_{0,L+M}^{[L/M+1]}$$
$$- (z - z_{L+M+1}) F_{0,L+M+1}^{[L+1/M+1]} v^{[L/M]}(z), \quad (1.38)$$

which is a generalization of (3.4.2).

Hence we deduce from (1.36), (1.37), and (1.38) that

$$\{r^{[L+1/M]}(z) - r^{[L/M]}(z)\}^{-1} - \{r^{[L/M+1]}(z) - r^{[L/M]}(z)\}^{-1}$$
$$= \frac{\{v^{[L/M]}(z)\}^2 (z-z_0)^{-1} \cdots (z-z_{L+M})^{-1}}{F_{0,L+M}^{[L+1/M]} F_{0,L+M}^{[L/M+1]}},$$

which is a generalization of (3.4.8). Equation (1.34) follows from a similar treatment of the right-hand side.

The generalized ε-algorithm is the formal identity

$$(z - z_{k+j+1})[\varepsilon_{k+1}^{(j)} - \varepsilon_{k-1}^{(j+1)}][\varepsilon_k^{(j+1)} - \varepsilon_k^{(j)}] = 1 \quad (1.39a)$$

for indices k, j in the range $k = 0, 1, 2, \ldots$ and $j \geq -[k/2]$. The artificial initialization conditions are

$$\varepsilon_{-1}^{(j)} = 0, \quad j = 0, 1, 2, \ldots,$$

and

$$\varepsilon_{2k}^{(-k-1)} = 0, \quad k = 0, 1, 2, \ldots. \quad (1.39b)$$

The usual initialization condition, using values derived from an interpolating polynomial, is

$$\varepsilon_0^{(j)} = r^{[j/0]}(z). \quad (1.39c)$$

Elements of the ε-table (see Section 3.3, Table 1, extended above the diagonal in Section 3.8, Table 2) are identified with values of rational interpolants by the formula

$$\varepsilon_{2k}^{(j)} = r^{[k+j/k]}(z), \quad k = 0, 1, 2, \ldots, j \geq -k. \quad (1.39d)$$

Proof of the results (1.39a)–(1.39d) is based on (1.34) and closely follows the corresponding argument of Section 3.4. The block structure is analyzed by Beckermann and Carstensen [1992].

The foregoing summary of methods of rational interpolation is biased toward methods which reduce to Padé methods in the confluent limit. Possibly the neatest of the continued-fraction algorithms for rational interpolation at distinct points is the t-g algorithm of Claessens [1976a,b]. The algorithm of Werner [1979], which leads to a Thiele–Werner interpolating rational fraction, has many advantages. It is based on principles similar to those of the generalized Viskovatov algorithm (see Section 4.2); it is a reliable algorithm, it has a confluent limit, and it is easily adapted to ensure numerical stability [Graves-Morris, 1980a]. The original papers on rational interpolation [Cauchy, 1821; Jacobi, 1846; Thacher and Tukey, 1960; Stoer, 1961; Wetterling, 1963; Larkin, 1967] lead to the methods developed in the review of Werner and Schaback [1972].

The problems of normality and degeneracy are reviewed by Meinguet [1970], and we refer to Gallucci and Jones [1976], Claessens [1978a, b], and Gutknecht [1989] for details of normality and degeneracy in the context of the Newton–Padé table. 'Newton–Padé approximation' is the name usually reserved for constructions based on Newton series (which have an ordered point set).

Most of the analytical theorems about Padé approximants stated in this book have extensions to N-point Padé approximation. A prototype of these generalizations is

Theorem 7.1.2 [Saff, 1972]. *Let E be a closed, bounded set in the complex z-plane whose complement K, with $\infty \in K$, is connected. Further, we suppose that a sequence of interpolation points*

$$\beta_1^{(0)},$$
$$\beta_1^{(1)}, \beta_2^{(1)},$$
$$\vdots$$
$$\beta_1^{(n)}, \beta_2^{(n)}, \ldots, \beta_{n+1}^{(n)}$$
$$\cdots\cdots$$

is defined, with no limit point in K, such that

$$\lim_{n\to\infty} \left| \prod_{i=1}^{n+1} (z - \beta_i^{(n)}) \right|^{1/n} = \mathrm{cap}\,(E) \exp\{G(z)\}$$

uniformly in z on each compact subset of K, where $\mathrm{cap}\,(E)$ is defined in Section 6.6. For each $\sigma > 1$, let Γ_σ denote the locus $G(z) = \ln \sigma$, and let

E_σ denote the interior of Γ_σ. Let $f(z)$ be analytic in E and meromorphic with precisely M poles, counting multiplicities, in E_ρ for some $\rho > 1$. Then, for all L sufficiently large, there exists a unique rational function $r^{[L/M]}(z)$ which interpolates $f(z)$ at $\beta_1^{(L+M)}, \beta_2^{(L+M)}, \ldots, \beta_{L+M+1}^{(L+M)}$. Each $r^{[L+M]}(z)$ has M finite poles which converge to the poles of $f(z)$ in E_ρ. Furthermore, $r^{[L/M]}(z) \to f(z)$ uniformly on any compact subset of E_ρ not containing any pole of $f(z)$, and

$$\lim_{n \to \infty} (\sup_{z \in E_\rho} |f(z) - r^{[L/M]}(z)|)^{1/n} \leq \frac{1}{\rho}.$$

Remarks. de Montessus's theorem, as expressed by Theorem 6.2.2, is a corollary of Saff's theorem, in which we take $E = \{z: |z| \leq R/\rho\}$, and the limits $\rho \to \infty$ and $\text{cap}(E) \to 0$. We refer (historically) to Warner [1976] for a discussion of how the interpolation points $\beta_i^{(n)}$ may be consistently chosen within E, for a set-theoretic generalization of Saff's theorem, and for the interesting connection with Runge's phenomenon in polynomial interpolation. We simply note that the Green's function $G(z)$ is well defined by the limit above if we take $\{\beta_i^{(n-1)}, i = 1, 2, \ldots, n\}$ to be the zeros of the Tchebycheff polynomial of order n for the region E. The proof of Theorem 7.1.2 follows Saff's method as described in Section 6.3 almost exactly.

Corresponding divergence theorems were proved by Wallin [1984]; interesting earlier related results were described by Walsh [1964a, b, 1965a, b], Karlsson [1976], and Gončar and Guillermo-Lopez [1978].

Bounding properties of rational interpolants for Stieltjes functions are given by Baker [1969] and Barnsley [1973].

By way of applications, we mention two important areas out of the many applications of rational interpolation. There has always been considerable interest in Padé-type methods as algorithms for finding zeros of functions. There is no difficulty in finding high-order, interpolatory methods [Merz, 1968; Zinn-Justin, 1970; Larkin, 1981], but in this context the principal difficulties are connected with minimizing the number of operations associated with the slowest path the algorithm may take, and avoidance of branching decisions based on noise [Garside, Jarratt, and Mack, 1968; Grant and Hitchens, 1971; Jarratt, 1970; Bus and Dekker, 1975].

Rational interpolation is a popular means of deferred extrapolation to the limit $h \to 0$ in the context of the numerical solution of ordinary differential equations using a finite grid of spacing h. This technique was pioneered by Bulirsch and Stoer [1964] and Gragg [1965]. Rational forms of the solution are used by Lambert and Shaw [1965, 1966] and Luke, Fair, and Wimp [1975].

This completes our summary of the principal algorithms for N-point Padé approximation. A simple application of N-point Padé approximation to the acceleration of convergence is given in Section 10.1. Next, we turn our attention to the special case of $N = 2$, as described by (1.1), (1.2), and (1.3). These have been extensively investigated within the framework of continued fractions, and are often called general T-fractions after Thron [1948].

We return to the theme of Section 4.5 and assume that we are given a pair of formal power series for a function $f(z)$:

$$L_0 = \sum_{i=0}^{\infty} c_i z^i = c_0 + c_1 z + c_2 z^2 + \cdots, \tag{1.40}$$

$$L_\infty = -\sum_{i=1}^{\infty} c_{-i} z^{-i} = -c_{-1} z^{-1} - c_{-2} z^{-2} - c_{-3} z^{-3} - \cdots, \tag{1.41}$$

so that L_0 is the ordinary Maclaurin series of $f(z)$, and L_∞ is the Taylor expansion of $f(z)$ about $z = \infty$. Such a function would naturally be approximated by

$$r^{[M-1/M]}(z) = \frac{a_0 + a_1 z + \cdots + a_{M-1} z^{M-1}}{b_0 + b_1 z + \cdots + b_M z^M}. \tag{1.42}$$

We will not be concerned with degenerate cases, for example, where $c_{-1} = 0$, because the representation (1.42) would become unnatural. The simplest example of (1.42) is the case with $M = 1$, for which $a_0:b_0:b_1$ are found from

$$a_0/b_0 = c_0, \quad a_0/b_1 = -c_{-1},$$

giving

$$r^{[0/1]}(z) = \frac{c_0}{1 - zc_{-1}/c_0}.$$

More generally, the Maclaurin and Taylor expansions of $r^{[M-1/M]}(z)$ are required to agree with those of (1.40), (1.41) to orders z^{M-1}, z^{-M}, inclusively and respectively:

$$r^{[M-1/M]}(z) - \sum_{i=0}^{\infty} c_i z^i = O_+(z^M), \tag{1.43}$$

$$r^{[M-1/M]}(z) + \sum_{i=1}^{\infty} c_{-i} z^{-i} = O_-(z^{-M-1}), \tag{1.44}$$

where the notation O_+, O_- is an order notation for power series of the types (1.40), (1.41), respectively.

7.1 Multipoint Padé approximants

After cross-multiplication, we require that

$$a_0 = b_0 c_0,$$
$$a_1 = b_0 c_1 + b_1 c_0,$$
$$\vdots$$
$$a_{M-1} = b_0 c_{M-1} + \cdots + b_{M-1} c_0$$

are the L_0 equalities, and then for L_∞:

$$-a_{M-1} = b_M c_{-1},$$
$$-a_{M-2} = b_M c_{-2} + b_{M-1} c_{-1},$$
$$\vdots$$
$$-a_0 = b_M c_{-M} + \cdots + b_1 c_{-1}.$$

By addition, the numerator variables $a_0, a_1, \ldots, a_{M-1}$ are eliminated and one obtains the equations

$$\begin{bmatrix} c_{-M} & c_{-M+1} & \cdots & c_0 \\ c_{-M+1} & c_{-M+2} & \cdots & c_1 \\ \vdots & \vdots & & \vdots \\ c_{-1} & c_0 & \cdots & c_{M-1} \end{bmatrix} \begin{bmatrix} b_M \\ b_{M-1} \\ \vdots \\ b_0 \end{bmatrix} = 0. \qquad (1.45)$$

This is the homogeneous system of M equations in $M-1$ unknowns which was solved in Section 1.1, with the result that

$$\sum_{i=0}^{M} b_i z^i = Q^{[-1/M]}(z) = \det \begin{bmatrix} c_{-M} & \cdots & c_{-1} & c_0 \\ \vdots & \ddots & \ddots & \vdots \\ c_{-1} & c_0 & \cdots & c_{M-1} \\ z^M & z^{M-1} & \cdots & 1 \end{bmatrix} \qquad (1.46)$$

is the denominator polynomial of $r^{[M-1/M]}(z)$ with determinantal normalization. Notice that it is normally the case that $c_j \neq 0$ for $j < 0$ in this context. The unusual sign convention of (1.41) is chosen so as to produce the standard form (1.46), for which there are useful identities. As the following theorem shows, the representation (1.46) is essentially unique for most cases of interest.

Theorem 7.1.3 (Uniqueness). *If the matrix in (1.45) has full rank M, the representation (1.46) is nontrivial and essentially unique.*

Proof. The hypothesis ensures that the determinant in (1.46) is not identically zero, and the representation is nontrivial. Suppose that

$A(z)/B(z)$, $P(z)/Q(z)$ are both representations of (1.42) with

$$\deg\{A(z)\} \leq M - 1, \quad \deg\{B(z)\} = M, \quad \deg\{P(z)\} = M - 1,$$
$$\deg\{Q(z)\} = M,$$

and that both fractions match L_0, L_∞ according to (1.43) and (1.44). Then (1.43) implies that $\Delta(z)$ defined by

$$\Delta(z) = \{A(z)Q(z) - B(z)P(z)\}z^{-M}$$

is a polynomial of degree $M - 1$ at most. From (1.44),

$$\frac{\Delta(z)}{Q(z)B(z)} = O_-(z^{-2M-1}).$$

Hence $\Delta(z) = 0$ and $r^{[M-1/M]}(z)$ is unique. Were there to exist a representation $A(z)/B(z)$ with $\deg B(z) < M$, the coefficients of $B(z)$ would form an annihilating column vector for the matrix in (1.45), contradicting the rank hypothesis. Therefore the representation (1.46) of $Q^{[-1/M]}(z)$ is nontrivial and the corresponding fraction $r^{[M-1/M]}(z)$ is unique.

Next, we show how a T-fraction can be constructed whose Mth convergent equals $r^{[M-1/M]}(z)$ and so satisfies the accuracy-through-order properties (1.43), (1.44) with the given expansions L_0, L_∞. Following the approach of Section 4.5, T-fractions are defined as expressions of the form

$$T(z) = \frac{F_1}{1 + G_1 z} + \frac{F_2 z}{1 + G_2 z} + \frac{F_3 z}{1 + G_3 z} + \cdots \quad (1.47)$$

with all F_i, $G_i \neq 0$. Suppose that D is the domain of convergence of the convergents of (1.47). $T(z)$ is thereby defined for $z \in D$; otherwise the expression (1.47) is purely formal. T-fractions have the remarkable property of being able to approximate series such as (1.40), (1.41) order by order: the values of F_1, G_1, F_2, G_2, ... do not depend on the terms of higher orders $-M$, $M - 1$ occurring later in the series L_∞, L_0, respectively. This result follows from the proof of Theorem 7.1.4. As an example, we note that the first two convergents of $T(z)$ have the properties that

$$T_1(z) = \frac{F_1}{1 + G_1 z} = F_1 + O_+(z) \quad (1.48)$$

$$= \frac{F_1}{G_1 z} + O_-(z^{-2}) \quad (1.49)$$

and that

$$T_2(z) = \frac{F_1(1 + G_2 z)}{(1 + G_1 z)(1 + G_2 z) + F_2 z} = F_1 - F_1(G_1 + F_2)z + O_+(z^2) \tag{1.50}$$

$$= \frac{F_1}{G_1 z} - \frac{(G_2 + F_2)F_1}{G_1^2 G_2 z^2} + O_-(z^{-3}). \tag{1.51}$$

Notice that the O_+ expansions of (1.48) and (1.50) are consistent, as are the O_- expansions of (1.49), (1.51).

The recursion relations associated with $T(z)$ are

$$P_i(z) = (1 + G_i z)P_{i-1}(z) + F_i z P_{i-2}(z), \tag{1.52}$$

$$Q_i(z) = (1 + G_i z)Q_{i-1}(z) + F_i z Q_{i-2}(z) \tag{1.53}$$

for $i = 2, 3, \ldots$, with the initialization

$$P_0(z) = 0, \quad Q_0(z) = 1, \quad P_1(z) = F_1, \quad Q_1(z) = 1 + G_1 z. \tag{1.54}$$

From (1.52), (1.53), it follows that

$$P_M(0) = F_1, \quad Q_M(0) = 1 \tag{1.55}$$

and that the leading coefficients of $P_M(z)$, $Q_M(z)$ are given by

$$\dot{Q}_M = \prod_{i=1}^{M} G_i, \quad \dot{P}_M = F_1 \prod_{i=2}^{M} G_i \tag{1.56}$$

for $M = 1, 2, 3, \ldots$. Notice that the ratios $\dot{P}_M / \dot{Q}_M = F_1/G_1$ and $P_M(0)/Q_M(0) = F_1$ do not depend on M, but reflect the leading behavior of $T(z)$, when convergent, at $z = \infty, 0$, respectively.

The standard construction of the elements F_i, G_i from the data $\{c_i\}_{-\infty}^{\infty}$ is the F–G algorithm. However, to establish the theoretical framework, a rather indirect construction is given by (1.61) as part of the following theorem.

Theorem 7.1.4. *A T-fraction of the form (1.47) can be constructed whose Mth convergent equals $r^{[M-1/M]}(z)$, and whose Maclaurin and Taylor expansions agree with L_0, L_∞, respectively, provided that each previous convergent $P_n(z)/Q_n(z)$ has the property that the accuracy-through-order properties (1.43), (1.44) hold precisely, in the sense that*

$$f(z) - P_n(z)/Q_n(z) = \alpha_n z^n + O_+(z^{n+1}) \tag{1.57}$$

$$= \beta_n z^{-n-1} + O_-(z^{-n-2}) \tag{1.58}$$

hold with $\alpha_n \neq 0$, $\beta_n \neq 0$, for $n = 0, 1, \ldots, M - 1$.

Proof. For the case $n = 0$, we have taken $P_0(z) = 0$, $Q_0(z) = 1$, and so the accuracy-through-order results

$$P_n(z)/Q_n(z) - f(z) = O_+(z^n), \; O_-(z^{-n-1}) \quad (1.59)$$

hold trivially for $n = 0$. For consistency, take $r^{[-1/0]}(z) = 0$. From (1.40), (1.41),

$$f(z) - P_0(z)/Q_0(z) = c_0 + O_+(z)$$
$$= -c_{-1}z^{-1} + O_-(z^{-2}).$$

The hypotheses (1.56), (1.57) therefore imply that $c_0 \neq 0$, $c_{-1} \neq 0$. Hence the definitions $F_1 = c_0$, $G_1 = -c_0/c_{-1}$ can be made, so that (1.59) holds with $n = 1$, and

$$r^{[0/1]}(z) = \frac{c_0}{1 - c_0 z/c_{-1}}.$$

Next, make the inductive hypothesis that the convergents

$$P_n(z)/Q_n(z) = r^{[n-1/n]}(z) \quad (1.60)$$

exist for $n = M - 2$, $M - 1$. This has been established for the cases $n = 0, 1$. Combining this hypothesis with that of (1.57), (1.58), we have $\alpha_{M-1}, \alpha_{M-2}, \beta_{M-1}, \beta_{M-2} \neq 0$. This permits us to define F_M and G_M by

$$F_M = -\alpha_{M-1}/\alpha_{M-2}, \quad G_M = -F_M \beta_{M-2}/(G_{M-1}\beta_{M-1}) \quad (1.61)$$

with $F_M \neq 0$, $G_M \neq 0$. From (1.55), (1.56), it follows that $Q_M(0) = 1$, $\deg\{Q_M(z)\} = M$. From the recursion relation and (1.61),

$$Q_M(z)f(z) - P_M(z) = (1 + G_M z)[Q_{M-1}(z)f(z) - P_{M-1}(z)]$$
$$+ F_M z[Q_{M-2}(z)f(z) - P_{M-2}(z)]$$
$$= (\alpha_{M-1} + F_M \alpha_{M-2})z^{M-1} + O_+(z^M)$$
$$= O_+(z^M);$$

likewise

$$Q_M(z)f(z) - P_M(z) = (G_M \beta_{M-1}\dot{Q}_{M-1} + F_M \beta_{M-2}\dot{Q}_{M-2}) + O_-(z^{-1})$$
$$= O_-(z^{-1}).$$

Therefore (1.59) holds with $n = M$, and the theorem is proved.

Corollary 1. *Given that $f(z)$ possesses a T-fraction expansion (1.47), equations (1.57), (1.58) take the explicit forms*

$$f(z) - P_n(z)/Q_n(z) = (-z)^n \prod_{i=1}^{n+1} F_i + O_+(z^{n+1}) \quad (1.62)$$

7.1 Multipoint Padé approximants

$$= -(-z)^{-n-1} G_{n+1} \prod_{i=1}^{n+1} \left(\frac{F_i}{G_i}\right)^2 + O_-(z^{-n-2}). \quad (1.63)$$

Proof. Both these results follow from (1.56)–(1.58) and (1.61), together with the various initializations.

Corollary 2. *For each M, $P_M(z)$, $Q_M(z)$ have no nontrivial common factors.*

Proof. Suppose otherwise. Then $P_M(z)/Q_M(z)$ is reducible and $P_M(z)/Q_M(z) = P_{M-i}(z)/Q_{M-i}(z)$ for some $i > 0$. Hence (1.57), (1.58) do not hold for the case $n = M - i$, which is impossible. Therefore there are no common factors of $P_M(z)$, $Q_M(z)$.

Once it has been established that the expansions L_0, L_∞ possess a T-fraction expansion, certain other properties of their two-point Padé approximants follow directly. First, the uniqueness theorem establishes that

$$r^{[M-1/M]}(z) = P_M(z)/Q_M(z), \quad M = 0, 1, 2, \ldots, \quad (1.64)$$

for each M.

By Corollary 2, $r^{[M-1/M]}(z)$ as expressed by (1.64) is irreducible, and therefore

$$Q_M(z) = Q^{[-1/M]}(z)/Q^{[-1/M]}(0) = Q^{[-1/M]}(z)/C(-1/M). \quad (1.65)$$

There is a converse of Theorem 7.1.4 which is:

Theorem 7.1.5. *If polynomials $p_n(z)$, $q_n(z)$, of exact degrees $n - 1$, n, respectively, exist such that the order properties*

$$f(z) - p_n(z)/q_n(z) = O_+(z^n), \quad (1.66)$$

$$= O_-(z^{-n-1}) \quad (1.67)$$

hold precisely, i.e., without being stronger, for $n = 0, 1, 2, \ldots$, then constants γ_n exist such that

$$p_n(z) = \gamma_n P_n(z), \quad q_n(z) = \gamma_n Q_n(z) \quad (1.68)$$

for all n.

Proof. For every M, the polynomials $p_M(z)$ and $q_M(z)$ have no common factor. For otherwise, polynomials $\tilde{p}_{M-1}(z)$, $\tilde{q}_{M-1}(z)$ of degrees $M - 2$, $M - 1$, respectively, exist for which $\tilde{p}_{M-1}(z)/\tilde{q}_{M-1}(z) = p_M(z)/q_M(z)$.

Thus the uniqueness theorem implies that

$$\tilde{p}_{M-1}(z)/\tilde{q}_{M-1}(z) = p_{M-1}(z)/q_{M-1}(z)$$

and then (1.67) cannot hold precisely for the case $n = M - 1$. Therefore each pair $p_M(z)$, $q_M(z)$ has no common factors.

The uniqueness theorem also implies that

$$p_M(z)/q_M(z) = P_M(z)/Q_M(z).$$

Corollary 2 of Theorem 7.1.4 states that $P_M(z)$ and $Q_M(z)$ have no common factors. Thus every zero of $q_M(z)$ is a zero of $Q_M(z)$, and

$$q_M(z) = \gamma_M Q_M(z)$$

for some constant γ_M, and the result (1.68) follows at once.

We proceed now on the assumptions that all $C(L/M) \neq 0$ and that all the required T-fractions do exist. Our aim is to establish the F–G algorithm for the construction of T-fractions in a normal table, given the expansions L_0, L_∞ in (1.40), (1.41). The method used is based on identifying the recursion (1.53) for the denominators of the T-fraction with the equivalent recursion for the denominators (1.46) of Padé approximants.

Using Frobenius

$$\begin{bmatrix} & * \\ * & * \end{bmatrix} \text{ and } \begin{bmatrix} * & * \\ & * \end{bmatrix}$$

identities as given by (3.5.11), we obtain the

$$\begin{bmatrix} * \\ * \\ * \end{bmatrix}$$

identity as

$$C(L/M + 1)C(L + 1/M + 1)zQ^{[L/M-1]}(z)$$
$$- \{C(L + 1/M)C(L/M + 1) - zC(L + 1/M + 1)C(L/M)\}Q^{[L/M]}(z)$$
$$+ C(L + 1/M)C(L/M)Q^{[L/M+1]}(z) = 0. \qquad (1.69)$$

Set $L = -1$ and substitute for $Q_M(z)$ using (1.65): equation (1.69) becomes

$$Q_M(z) = \left\{1 - z\frac{C(0/M)C(-1/M - 1)}{C(0/M - 1)C(-1/M)}\right\}Q_{M-1}(z)$$
$$- z\frac{C(0/M)C(-1/M - 2)}{C(0/M - 1)C(-1/M - 1)}Q_{M-2}(z). \qquad (1.70)$$

Equation (1.70) is readily identified with (1.53), yielding

$$F_M = -\frac{C(0/M)C(-1/M-2)}{C(0/M-1)C(-1/M-1)}, \quad G_M = -\frac{C(0/M)C(-1/M-1)}{C(0/M-1)C(-1/M)} \quad (1.71)$$

for all M, except that

$$F_1 = c_0, \quad G_1 = -c_0/c_{-1}, \quad F_2 = \frac{c_1}{c_0} - \frac{c_0}{c_{-1}}.$$

From (1.71), one can establish the main algorithm for computing the elements of $T(z)$. One conclusion to be drawn from Theorems 7.1.4, 7.1.5 is that T-fraction developments are only satisfactory numerically in cases which are not nearly degenerate, and in these cases the following QD type algorithm should work satisfactorily.

The F–G algorithm [McCabe and Murphy, 1976].

Initialization. Complete the first *three* columns of the F–G table by setting

$$F_1^{(k)} = c_k, \quad G_1^{(k)} = -c_k/c_{k-1}, \quad F_2^{(k)} = G_1^{(k+1)} - G_1^{(k)} \quad (1.72)$$

for all values of k.

Recursion. The next G-column is constructed using the

identity:

$$G_{j+1}^{(k+1)} = F_{j+1}^{(k+1)} G_j^{(k)} / F_{j+1}^{(k)}, \quad (1.73)$$

and the next F-column is constructed using the

identity:

$$F_{j+1}^{(k)} = F_j^{(k+1)} + G_j^{(k+1)} - G_j^{(k)}. \quad (1.74)$$

Termination. The elements required for $T(z)$ are those in the principal row of the F–G table, that is,

$$F_j = F_j^{(0)}, \quad G_j = G_j^{(0)}, \quad j = 1, 2, 3, \ldots.$$

Proof. The validity of the algorithm must be established. This is done by proving that

$$f^{(k)}(z) = \frac{F_1^{(k)}}{1 + G_1^{(k)}z} + \frac{F_2^{(k)}z}{1 + G_2^{(k)}z} + \frac{F_2^{(k)}z}{1 + G_3^{(k)}z} + \cdots, \quad (1.75)$$

where $f^{(k)}(z)$ is defined by the expansions

$$f^{(k)}(z) = c_k + c_{k+1}z + c_{k+2}z^2 + \cdots, \quad z \to 0, \quad (1.76)$$

$$= c_{-k-1}z^{-1} + c_{-k-2}z^{-2} + c_{-k-3}z^{-3} + \cdots, \quad z \to \infty. \quad (1.77)$$

The proof of the initialization is straightforward. For the recursion, it follows from (1.71), (1.76), and (1.77) that the elements of (1.75) are given by

$$F_j^{(k)} = -\frac{C(k/j)C(k-1/j-2)}{C(k/j-1)C(k-1/j-1)}, \quad G_j^{(k)} = -\frac{C(k/j)C(k-1/j-1)}{C(k/j-1)C(k-1/j)}. \quad (1.78)$$

From this specification (1.78), we find that

$$\frac{G_j^{(k)}}{F_{j+1}^{(k)}} = \frac{G_{j+1}^{(k+1)}}{F_{j+1}^{(k+1)}} = \frac{C(k/j)^2}{C(k/j-1)C(k/j+1)},$$

and (1.73) is thereby proved for $j \geq 1$. Sylvester's identity is used to show that

$$F_{j+1}^{(k)} + G_k^{(j)} = F_j^{(k+1)} + G_j^{(k+1)} = -\frac{C(k-1/j-1)C(k+1/j)}{C(k/j)C(k/j-1)},$$

and (1.74) is thereby proved for $j \geq 2$.

The F–G table consists of entries arranged as

$\vdots$	$\vdots$	$\vdots$	$\vdots$	$\vdots$	
$F_1^{(-1)}$	$G_1^{(-1)}$	$F_2^{(-1)}$	$G_2^{(-1)}$	$F_3^{(-1)}$	$\cdots$
$F_1^{(0)}$	$G_1^{(0)}$	$F_2^{(0)}$	$G_2^{(0)}$	$F_3^{(0)}$	$\cdots$ ← principal row
$F_1^{(1)}$	$G_1^{(1)}$	$F_2^{(1)}$	$G_2^{(1)}$	$F_3^{(1)}$	$\cdots$
$\vdots$	$\vdots$	$\vdots$	$\vdots$	$\vdots$	

The first three columns are determined by the initialization (1.72) and the remaining columns by the recursions (1.73), (1.74), allowing the calculation of the elements of the principal row.

The next result [Dijkstra, 1977; Wynn, 1962a] is a good example of a T-fraction expansion. It is

$$\frac{{}_1F_1(a, b+1; z)}{{}_1F_1(a, b; z)} = \frac{b}{b+z} - \frac{z(b+1-a)}{b+1+z} - \cdots - \frac{z(b+n-a)}{b+n+z} - \cdots$$

with $a, b \geq 0$ and $z \geq 0$. It embodies the Maclaurin expansion about $z = 0$ and the asymptotic expansion about $z = +\infty$ of the left-hand side. The special case with $b = a$ is

$$\frac{{}_1F_1(a, a+1; z)}{e^z} = \frac{a}{a+z} - \frac{z}{a+1+z} - \cdots - \frac{nz}{a+n+z} - \cdots.$$

Using Kummer's transformation, this becomes

$$_1F_1(1, a+1; z) = \frac{a}{a-z} + \frac{z}{a+1+z} + \cdots + \frac{nz}{a+n+z} + \cdots. \tag{1.79}$$

The original restriction to $z \geq 0$ may now be dropped, because the right-hand side of (1.79) is convergent throughout the z-plane. This formula is used to provide a T-fraction development of the error function, the incomplete gamma function, and the generalized Dawson integral. We define [Dijkstra, 1977]

$$F(p, x) = e^{-x^p} \int_0^x e^{t^p} dt, \quad x \geq 0, p > 0,$$

to be the generalized Dawson integral; Dawson's integral is $D(x) = F(2, x)$, and its relation to the error function is expressed by

$$\operatorname{erf}(z) = \frac{2i}{\sqrt{\pi}} e^{-z^2} D(-iz). \tag{1.80}$$

Then we find (see Section 4.6)

$$\gamma(a, z) = z^a e^{-z} a^{-1} {}_1F_1(1, 1+a; z), \tag{1.81}$$

$$F(p, x) = x {}_1F_1(1, a+1; -z), \quad \text{with } z = x^p \text{ and } a = p^{-1}, \tag{1.82}$$

$$D(z) = z {}_1F_1(1, \tfrac{3}{2}; -z^2), \tag{1.83}$$

and

$$\operatorname{erf}(z) = \frac{2ze^{-z^2}}{\pi} {}_1F_1(1, \tfrac{3}{2}; z^2). \tag{1.84}$$

T-fraction developments of all these functions then follow from (1.79) and are valid for all z.

The block structure of the two-point Padé table and degenerate cases are treated by Cooper, Magnus, and McCabe [1986], and the extension to the case in which the c_i do not commute is given by Draux [1983, 1987].

For further details on the subject of T-fractions, we refer to other original articles by Murphy [1971], McCabe [1974, 1975], and Drew and Murphy [1977], to the reviews by Jones and Thron [1988] and Jones, Njåstad, and Thron [1986], to the editions of Jones, Thron, and Waadeland [1982], Thron [1986], and Jacobsen [1989], and to the books by Jones and Thron [1980] and Lorentzen and Waadeland [1992].

7.2 Baker–Gammel approximants

A Padé–Legendre series is a series

$$g(z) = \sum_{n=0}^{\infty} c_n P_n(z) = \sum_{n=0}^{\infty} f_n P_n(-z) \tag{2.1}$$

where $\{P_n(z), n = 0, 1, 2, \ldots\}$ are Legendre polynomials and $f_n = (-)^n c_n$. We discuss rational approximants for series such as (2.1) toward the end of Section 7.4 whereas in this section we discuss an alternative class of approximating functions suitable for (2.1) and similar series. Except in degenerate cases, the coefficients c_n can be expressed as

$$c_n = \sum_{i=1}^{M} \alpha_i h_i^n \quad \text{for } n = 0, 1, 2, \ldots, 2M - 1. \tag{2.2}$$

If we suppose that (2.2) is approximately true for $n = 2M, 2M + 1, \ldots$, which is to say that we suppose that the coefficients c_n behave in a generalized geometric manner, then we see that $g(z)$ is approximated by

$$G^{[M-1/M]}(z) = \sum_{i=1}^{M} \alpha_i (1 - 2h_i z + h_i^2)^{-1/2}. \tag{2.3}$$

Equation (2.3) is based on a generating function for the Legendre polynomials and provides a nonrational approximation for $g(z)$. To determine the parameters $\{\alpha_i, h_i, i = 1, 2, \ldots, M\}$ occurring in (2.3), we deduce from (2.2) that

$$f(z) \equiv \sum_{n=0}^{\infty} c_n z^n = \sum_{i=1}^{M} \frac{\alpha_i}{1 - h_i z} + O(z^{2M}) \tag{2.4}$$

is formally valid, so h_i^{-1} are the poles of the $[M - 1/M]$ Padé approximant of $f(z)$ and $-\alpha_i/h_i$ are the corresponding residues. Thus the approximation (2.3) is easily constructed provided that the $[M - 1/M]$ Padé approximant of $f(z)$ defined by (2.3) exists.

The ideas underlying the approach expressed by (2.1)–(2.4) evolved from the papers of Gammel, Rousseau, and Saylor [1967], Baker [1967], and Common [1969a, b]. A rather different approach, based on the theory of positive functionals and different hypotheses from those of this section, leading to the inequalities of Theorem 7.2.2, was derived by M. Riesz [1923]; Shohat extended this approach to the case of Hamburger series [Akhiezer, 1965]. Some of the bounds obtained in this section may be compared with those obtained by the methods of Section 5.2.

Baker–Gammel approximants exploit and extend these ideas by considering functions having a formal expansion

$$g(z) = \sum_{m=0}^{\infty} f_m k_m(z)$$

which generalizes (2.1), and the integral representation

$$g(z) = \int_0^{\infty} k(z,u)\, d\phi(u), \qquad (2.5)$$

where $k(z, u)$ is a kernel, yet to be specified, and $d\phi(u)$ is a Stieltjes measure.

For the special case that $k(z, u) = (1 + uz)^{-1}$, (2.5) becomes a Stieltjes series, and the methods of Chapter 5 should be used. More generally, we begin by constructing Padé approximants to $f(z)$ defined by

$$f(z) \equiv \sum_{n=0}^{\infty} c_n z^n \equiv \int_0^{\infty} \frac{d\phi(u)}{1 + zu} = [M + J/M] + O(z^{2M+J+1}), \qquad (2.6)$$

which is a Stieltjes series. Baker–Gammel approximants for $g(z)$ in (2.5) are defined by

$$G^{[M+J/M]}(z) = \sum_{j=0}^{J} \beta_j k_j(z) + \sum_{i=1}^{M} \alpha_i k(z, u_i), \qquad (2.7)$$

where $k(z, u)$ is defined so that

$$k_j(z) = \frac{1}{j!} \left(\frac{\partial}{\partial u}\right)^j k(z, u) \bigg|_{u=0}, \qquad j = 0, 1, \ldots. \qquad (2.8)$$

The values of β_j, $j = 0, 1, \ldots, J$, and u_i, α_i, $i = 1, 2, \ldots, M$, remain to be specified; if $J = -1$ the polynomial term of (2.7) is absent.

If we ignore, for the moment, all questions of convergence in (2.9) and (2.10), (2.8) leads to

$$k(z, u) = \sum_{m=0}^{\infty} u^m k_m(z). \qquad (2.9)$$

Then (2.5), (2.6), and (2.9) show that $g(z)$ has the series expansion

$$g(z) = \sum_{m=0}^{\infty} f_m k_m(z), \qquad (2.10)$$

where each f_m is represented by

$$f_m = (-)^m c_m = \int_0^{\infty} u^m \, d\phi(u), \quad m = 0, 1, \ldots. \qquad (2.11)$$

From (2.10), (2.7), and (2.9), we see that the coefficients of $k_m(z)$ in $g(z)$ and $G^{[M+J/M]}(z)$ are equal for $m = 0, 1, \ldots, 2M + J$ provided

$$f_m = \beta_m + \sum_{i=1}^{M} \alpha_i u_i^m, \quad m = 0, 1, \ldots, J, \qquad (2.12)$$

and

$$f_m = \sum_{i=1}^{M} \alpha_i u_i^m, \quad m = J+1, J+2, \ldots, 2M + J. \qquad (2.13)$$

From (2.12) and (2.13), we find formally that

$$\sum_{m=0}^{\infty} f_m(-z)^m = \sum_{j=0}^{J} \beta_j(-z)^j + \sum_{i=1}^{M} \frac{\alpha_i}{1 + zu_i} + O(z^{2M+J+1}). \qquad (2.14)$$

Hence we deduce that the parameters β_j, α_i, u_i occurring in (2.7) are defined by (2.6) with the identification

$$[M + J/M]_f(z) = \sum_{j=0}^{J} \beta_j(-z)^j + \sum_{i=1}^{M} \frac{\alpha_i}{1 + zu_i}. \qquad (2.15)$$

This establishes that existence of the $[M + J/M]$ Padé approximant for $f(z)$ defined by (2.6) is sufficient formally to derive a Baker–Gammel approximant for $g(z)$ defined by (2.5). The justification of the former procedure lies in the following theorem.

Theorem 7.2.1. *Let $k(z, u)$ be analytic on the positive real u-axis and at least within a distance Δ from it for all $z \in \mathscr{C}$, a compact region in the z-plane. Further, let $|k(z, u)|$ be uniformly bounded by $[\ln u]^{-(1+\mu)}$ with $\mu > 0$ as $u \to +\infty$. Then*

$$G^{[M+J/M]}(z) \to g(z) \quad \text{for } z \in \mathscr{C}, \quad \text{as } M \to \infty.$$

Proof. With a Cauchy representation for $k(z, u)$, (2.5) becomes

$$g(z) = \int_0^{\infty} \frac{d\phi(u)}{2\pi i} \int_{\Gamma_1} \frac{k(z, w)}{w - u} \, dw,$$

7.2 Baker–Gammel approximants

where Γ_1 is a circular contour with center u and radius Δ. Using the assumed analyticity properties of $k(z, w)$, we deduce that

$$g(z) = \frac{1}{2\pi i} \int_0^\infty d\phi(u) \int_\Gamma \frac{k(z, w)}{w - u} dw,$$

where Γ is a contour at distance Δ from the positive real w-axis, as shown in Figure 7.2.1.

Hence,

$$g(z) = \frac{1}{2\pi i} \int_\Gamma \frac{k(z, w)}{w} dw \int_0^\infty \frac{d\phi(u)}{1 - u/w}$$

$$= \frac{1}{2\pi i} \int_\Gamma k(z, w) f(-1/w) \frac{dw}{w}. \quad (2.16)$$

Baker–Gammel approximants are formed by replacing the function f in (2.16) with its Padé approximant (2.15), so that

$$G^{[M+J/M]}(z) = \frac{1}{2\pi i} \int_\Gamma k(z, w) \left[\sum_{j=0}^J \beta_j w^{-j} + \sum_{i=1}^M \frac{\alpha_i}{1 - u_i/w} \right] \frac{dw}{w}. \quad (2.17)$$

Because $f(z)$ is a Stieltjes series, the poles at $w = u_i$ in (2.17) lie inside Γ, and we deduce from (2.8) that

$$G^{[M+J/M]}(z) = \sum_{j=0}^J \beta_j k_j(z) + \sum_{i=1}^M \alpha_i k(z, u_i).$$

Thus the heuristic arguments connecting (2.6), (2.7), and (2.15) are rigorously justified under the conditions of the theorem. The error formula, which follows from (2.16) and (2.17), is

$$g(z) - G^{[M+J/M]}(z)$$

$$= \frac{1}{2\pi i} \int_\Gamma \frac{k(z, w)}{w} \left\{ f\left(\frac{-1}{w}\right) - \sum_{j=0}^J \beta_j w^{-j} - \sum_{i=1}^M \frac{\alpha_i}{1 - u_i/w} \right\} dw. \quad (2.18)$$

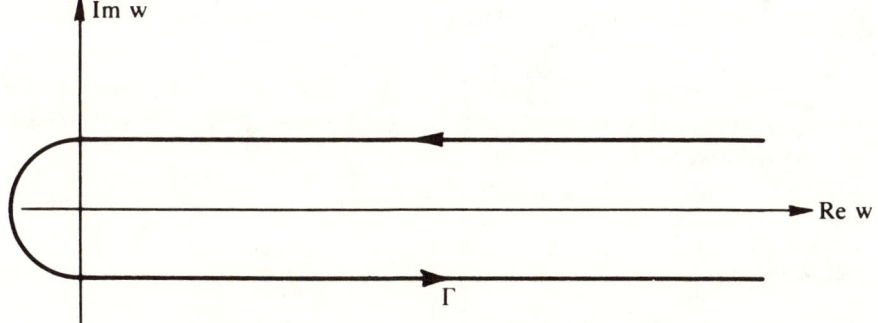

Figure 7.2.1. The contour Γ in the complex w-plane.

Using Theorems 5.2.6 and 5.3.1, we deduce that the term in brackets in (2.18) tends uniformly to zero as $M \to \infty$ for $w \in \Gamma$; furthermore it is $O(w^{-1})$ as $w \to \infty$. Hence (2.18) tends to zero as $M \to \infty$ for $z \in \mathscr{C}$, and convergence of the Baker–Gammel approximants is established.

By imposing a further condition on the kernel $k(z, u)$, these approximants become upper and lower bounds for $g(z)$, as is shown in the next theorem. But first we require a lemma.

Lemma. *With the definitions and notation of* (2.17),

$$\int_\Gamma w^{m-1} \left\{ [M + J + 1/M + 1]_f\left(\frac{-1}{w}\right) - [M + J/M]_f\left(\frac{-1}{w}\right) \right\} dw = 0$$

$$\text{for } m = 0, 1, 2, \ldots, 2M + J. \quad (2.19)$$

Proof. We notice first that the Padé accuracy-through-order conditions guarantee that the integral in (2.19) is well defined at $w = \infty$. Hence Γ may be deformed to a contour Γ' enclosing $w = 0$ and the poles of the approximants. Using (2.12), (2.13), and (2.14), we see that, for $m = 0, 1, \ldots, 2M + J$,

$$\frac{1}{2\pi i} \int_{\Gamma'} w^{m-1} [M + J/M]_f\left(\frac{-1}{w}\right) dw$$

$$= \frac{1}{2\pi i} \int_{\Gamma'} w^{m-1} \left\{ \sum_{j=0}^{J} \beta_j w^{-j} + \sum_{i=1}^{M} \frac{\alpha_i}{1 - u_i/w} \right\} dw$$

$$= f_m. \quad (2.20)$$

The other term of (2.19) is treated similarly, and the lemma is proved.

Theorem 7.2.2. *The inequalities*

$$(-)^{J+1}\{G^{[M+J+1/M+1]}(x) - G^{[M+J/M]}(x)\} \geq 0, \quad (2.21)$$

$$(-)^{J+1}\{G^{[M+J/M]}(x) - G^{[M+J+1/M-1]}(x)\} \geq 0, \quad (2.22)$$

$$G^{[M/M]}(x) \geq g(x) \geq G^{[M-1/M]}(x) \quad (2.23)$$

for all x real and positive, all $M \geq 1$, and $J \geq -1$, hold if and only if $k(x, u)$ satisfies

$$\left(-\frac{\partial}{\partial u}\right)^j k(x, u) \geq 0 \quad (2.24)$$

for all real nonnegative x and u and all $j = 0, 1, 2, \ldots$.

Proof. First, we show that (2.21) follows from (2.24). Consider a function $K(w)$ which is analytic within a distance Δ from the real w-axis. Let

7.2 Baker–Gammel approximants

$$I_K = \frac{1}{2\pi i} \int_\Gamma \frac{K(w)}{w} \left\{ [M+J+1/M+1]_f\left(\frac{-1}{w}\right) - [M+J/M]_f\left(\frac{-1}{w}\right) \right\} dw.$$

(2.25)

Since $f(z)$ is a Stieltjes series, the zeros of

$$w^{2M+1} Q^{[M+J+1/M+1]}(-1/w) Q^{[M+J/M]}(-1/w)$$

occur at points $w = w_i$, $i = 1, 2, \ldots, 2M+1$, lying on the real w-axis. Using Newton's formula (1.7), we may express $K(w)$ as

$$K(w) = \sum_{i=1}^{2M+1} K[w_1, w_2, \ldots, w_i] \prod_{k=1}^{i-1}(w - w_k)$$
$$+ K[w_1, w_2, \ldots, w_{2M+1}, w] \prod_{k=1}^{2M+1}(w - w_k).$$

Upon substitution of this into (2.25), the lemma shows that

$$I_K = \frac{1}{2\pi i} \int_\Gamma \frac{K[w_1, w_2, \ldots, w_{2M+1}, w]}{w}$$
$$\times \prod_{k=1}^{2M+1}(w - w_k) \left\{ [M+J+1/M+1]_f\left(\frac{-1}{w}\right) - [M+J/M]_f\left(\frac{-1}{w}\right) \right\} dw.$$

Using (3.5.18), it follows that

$$I_K = \frac{(-)^{J+1}[C(M+J+1/M+1)]^2}{2\pi i}$$
$$\times \int_\Gamma \frac{w^{-2M-J-2} \prod_{k=1}^{2M+1}(w-w_k) K[w_1, w_2, \ldots, w_{2M+1}, w]}{Q^{[M+J+1/M+1]}\left(\frac{-1}{w}\right) Q^{[M+J/M]}\left(\frac{-1}{w}\right)} dw$$

$$= \frac{(-)^{J+1}}{2\pi i} \frac{C(M+J+1/M+1)}{C(M+J/M)}$$
$$\times \int_\Gamma \frac{w^{-J-1} \prod_{k=1}^{2M+1}\left(1 - \frac{w_k}{w}\right) K[w_1, w_2, \ldots, w_{2M+1}, w]}{B^{[M+J+1/M+1]}\left(\frac{-1}{w}\right) B^{[M+J/M]}\left(\frac{-1}{w}\right)} dw$$

$$= \frac{(-)^{J+1}}{2\pi i} \frac{C(M+J+1/M+1)}{C(M+J/M)} \int_\Gamma w^{-J-1} K[w_1, w_2, \ldots, w_{2M+1}, w] \, dw.$$

Since $f(z)$ is a Stieltjes series, $C(M+J+1/M+1)/C(M+J/M)$ is positive if J is odd and negative otherwise [see (4.3.6)]. Using (1.5), (1.7), and Rolle's theorem, we have

$$\frac{(2M+J+1)!}{2\pi i}\int_\Gamma w^{-J-1} K[w_1, w_2, \ldots, w_{2M+1}, w]\, dw = \left(\frac{d}{du}\right)^{2M+J+1} K(u)\bigg|_{u=\bar{w}},$$

where $\bar{w}$ lies on the real w-axis. Now we invoke (2.17), (2.25) to deduce that

$$(-)^{J+1}\{G^{[M+J+1/M+1]}(x) - G^{[M+J/M]}(x)\} = p\left(\frac{-\partial}{\partial u}\right)^{2M+J+1} k(x, u)\bigg|_{u=\bar{w}},$$

where p is positive. Hence we see that (2.24) is a sufficient condition to prove (2.21).

The other inequalities (2.22), (2.23) are proved similarly, and the convergence of Theorem 7.2.1 justifies the appearance of $g(x)$ in (2.23) as stated. The converse is based on the linearity of (2.5) in $d\phi(u)$, and we refer to Baker [1970] for the details.

Example 1

$$k(z, u) = \frac{1}{1+zu}. \tag{2.26}$$

The conditions of Theorem 7.2.1 and (2.24) are satisfied. The Baker–Gammel approximants are given by (2.7) as

$$G^{[M+J/M]}(z) = \sum_{j=0}^{J} \beta_j(-z)^j + \sum_{i=1}^{M} \frac{\alpha_i}{1+zu_i}.$$

We see that (2.26) reduces to the special case of $[M+J/M]$ Padé approximants to a Stieltjes series $g(z)$ given by (2.5).

Example 2 (Exponential approximants).

$$k(z, u) = e^{-zu}. \tag{2.27}$$

The conditions of Theorem 7.2.1 and (2.24) are satisfied. The Baker–Gammel approximants are given by (2.7) as

$$G^{[M+J/M]}(z) = \sum_{j=0}^{J} \beta_j \frac{(-z)^j}{j!} + \sum_{i=1}^{M} \alpha_i e^{-z\sigma_i}.$$

These approximants are especially useful when the given function is exponentially damped.

Example 3 (Multipole approximants).

$$k(z, u) = \left[1 + \frac{zu}{n}\right]^{-n}. \tag{2.28}$$

The conditions of Theorem 7.2.1 and (2.24) are satisfied for $n > 0$. The Baker–Gammel approximants are

$$G^{[M+J/M]}(z) = \sum_{j=0}^{J} \beta_j \frac{\Gamma(n+j)}{\Gamma(n)} \left(\frac{-z}{n}\right)^j + \sum_{i=1}^{M} \frac{\alpha_i}{(1 + zu_i/n)^n}.$$

This class of functions interpolates between Example 1 ($n = 1$) and Example 2 ($n = \infty$).

Example 4 (Padé–Legendre approximants).

$$k(z, u) = (1 + 2uz + u^2)^{-1/2}. \tag{2.29}$$

The conditions of Theorem 7.2.1 are satisfied. To establish (2.24) in this case, we let $x > 1$ and $u > 0$. Then the two roots of $1 + 2ux + u^2 = 0$ occur at u_1, u_2 such that $-\infty < u_1 < u_2 < 0$. Cauchy's theorem leads to an integral representation of the kernel as

$$k(z, u) = \frac{1}{\pi} \int_{u_1}^{u_2} \frac{dw}{w - u} \frac{1}{\sqrt{-(1 + 2wz + w^2)}},$$

and the condition (2.24) is established. Thus (2.29) is a Baker–Gammel kernel. The approximants

$$G^{[M+J/M]}(x) = \sum_{j=0}^{J} \beta_j P_j(x) + \sum_{i=1}^{M} \alpha_i (1 + 2u_i x + u_i^2)^{-1/2}$$

give converging upper and lower bounds for $x > 1$ from Theorem 7.2.2.

Clearly, the preceding analysis of the Padé–Legendre series (2.1) generated by the kernel (2.29) extends to a wider class of orthogonal polynomials, but no general theory has yet been established.

Example 5 (Padé–Borel approximation). Consider the kernel

$$k(z, u) = \int_0^\infty \frac{e^{-t} dt}{1 + uzt^p} \quad \text{for } p > 0,$$

which has the asymptotic expansion

$$k(z, u) \simeq \sum_{j=0}^{\infty} (-uz)^j (pj)!,$$

and satisfies the requirements of Theorem 7.2.1 and (2.2.4). Equation (2.10) takes the form

$$g(z) = \sum_{j=0}^{\infty} c_j(pj)! z^j$$

in this case, and so we need to consider Padé approximants for

$$f(z) = \sum_{j=0}^{\infty} c_j z^j.$$

The difference between these power series of $g(z)$ and $f(z)$ is that a 'convergence factor' of $1/(pj)!$ has appeared. The Padé–Borel approximants of $g(z)$ take the form

$$g^{[M+J/M]}(z) = \sum_{j=0}^{J} \beta_j (pj)! z^j + \int_0^{\infty} e^{-t} \sum_{i=1}^{M} \frac{\alpha_i}{1 + u_i z t^p} dt$$

$$= \int_0^{\infty} e^{-t} f^{[M+J/M]}(zt^p) dt.$$

The Padé–Borel method was first used by Graffi, Grecchi, and Simon [1970] in an application to the anharmonic oscillator (see Bender and Wu [1968] and Bender and Orszag [1978]); the approach we adopt here is based on that of Baker [1975a, p. 287]. A relaxed survey of applications in perturbation theory of these and related techniques is given by Simon [1982].

A recent extension of the ideas of this section is appropriate when approximants are needed for the series (2.10)

$$g(z) = \sum_{m=0}^{\infty} f_m P_m(z) \tag{2.30}$$

and information is available a priori about the asymptotic form of the coefficients f_m [Baker and Gubernatis, 1981]. The scheme of Baker–Gammel approximants can easily be modified to incorporate this extra information. For example, a number of coefficients of the series

$$g(z) = \sum_{m=0}^{\infty} f_m P_m(z) \tag{2.31}$$

might be known, and additionally it might be known that numbers $\{\gamma_n\}$ are given such that

$$f_m \simeq \gamma_m \quad \text{as } m \to \infty. \tag{2.32}$$

In fact, it turns out that it is only necessary to know the asymptotic form of the ratio

$$R_m = f_{m+1}f_{m-1}/f_m^2, \qquad (2.33)$$

as will become clear shortly. Note the similarity of the hypothesis (2.32) to that of Levin's method of Section 10.1.

We present a method for obtaining asymptotic Baker–Gammel approximants, omitting the preconditions for existence and convergence. Equations (2.12), (2.13) are replaced by the hypothesis that, for $J \geq -1$, the coefficients $\{f_m\}$ can be expressed as

$$f_m = \gamma_m \left(\beta_m + \sum_{i=1}^{M} \alpha_i u_i^m \right), \qquad m = 0, 1, \ldots, J, \qquad (2.34)$$

$$f_m = \gamma_m \left(\sum_{i=1}^{M} \alpha_i u_i^m \right), \qquad m = J+1, J+2, \ldots, 2M+J. \qquad (2.35)$$

We then suppose that (2.35) is approximately true for $m = 2M, 2M+1$, ... We also see that multiplying each γ_m by a factor of αh^m for $m = 0, 1, 2, \ldots, 2M-1$ leaves the representation (2.35) invariant, showing that the asymptotic form of the ratio R_n in (2.33) is all that need be specified to define the asymptotic approximants.

Equation (2.9) is replaced by

$$k^{(a)}(z, u) = \sum_{m=0}^{\infty} u^m \gamma_m k_m(z). \qquad (2.36)$$

It is understood that $k^{(a)}(z, u)$ is defined by analytic continuation in u where necessary. Then we find that the approximants for $g(z)$, defined by the coefficients $\{f_m, m = 0, 1, \ldots, 2M+J\}$ of (2.30) and (2.35)–(2.37) are

$$g^{[M+J/M]}(z) = \sum_{i=1}^{M} \alpha_i k^{(a)}(z, u_i) + \sum_{m=0}^{J} \gamma_m \beta_m k_m(z). \qquad (2.37)$$

Exploitation of the information expressed by (2.32) follows the spirit of Levin's method of Section 10.1, and this method also bears resemblance to that of Common and Stacey [1979a, b, c].

Example 6. Suppose that approximants are required for the series

$$g(z) = \sum_{m=0}^{\infty} f_m P_m(z),$$

and it is known that

$$f_m \simeq 6(m+1)3^m \qquad \text{as } m \to \infty.$$

We define $\gamma_m = m + 1$ for $n = 0, 1, 2, \ldots$, and deduce from (2.36) that the kernel appropriate for (2.37) is

$$k^{(a)}(z, u) = (1 - uz)(1 - 2uz + u^2)^{-3/2}.$$

Example 7. The Le Roy function is given by

$$L_\zeta(x) = \int_0^\infty \exp[-(t + xt^\zeta)] \, dt.$$

Note that $k(z, u) = L_\zeta(zu)$ satisfies the conditions of Theorem 7.2.1 and Equation (2.24) for $\zeta \geq 0$. Investigate the special cases $\zeta = 0$, $\zeta = 1$.

7.3 Series analysis

The fundamental motive for studying Padé approximants is the extraction of information about a function from the first few terms of its power series. Before the advent of Padé approximants in theoretical physics, various methods with the collective name of series analysis were used. We start at the beginning with the ratio method, take the Padé method for granted, proceed to Gammel–Guttmann–Gaunt–Joyce approximants[†] (abbreviated G³J approximants), and finally come to quadratic approximants.

The methods we are about to discuss depend on having either some knowledge of what is a 'reasonable functional form' or else of the 'general behavior of the power-series coefficients.' The methods of this section are working techniques for applied scientists. They work in several contexts, and it is instructive to review, very briefly, one context so as to see the kind of intuitive information which proves useful.

In statistical mechanics there are a large number of problems for which the first few terms of the power series may be obtained exactly (they are integers), while the exact solution is unobtainable. The three-dimensional Ising model is a good example. The first few terms of the series of expansion of a thermodynamic quantity, such as the magnetic susceptibility, are generated, and the series is then analyzed to determine the behavior of the thermodynamic quantity in question. In statistical mechanics the emphasis is on phase transitions, such as the abrupt onset of ferromagnetic order from the disordered paramagnetic state at the critical temperature. It was expected that the magnetic susceptibility $f(z)$, a function of $z = T^{-1}$, the inverse temperature, would be given by

$$f(z) \approx A(1 - \mu z)^{-\gamma}, \tag{3.1}$$

[†] These approximants are so named because John Gammel and David Gaunt talked at the Canterbury Conference about their work, and discovered that day that A. J. Guttmann and G. S. Joyce were publishing a letter [1972] to the same effect. They are also called integral curve or integral approximants. See also Section 8.6.

where $1/\mu$ is the 'critical point,' γ is the 'critical exponent,' and $\approx$ means that the functional form is reasonably good near $\mu z = 1$ and no more. It is known that

$$f(z) = c_0 + c_1 z + c_2 z^2 + c_3 z^3 + \cdots, \qquad (3.2)$$

and the problem can be expressed as one of fitting

$$c_n \simeq A \frac{(\gamma + 1) \cdots (\gamma + n)}{n!} \mu^n, \quad n = 0, 1, 2, \ldots, N,$$

for the best values of A, μ, and γ. The ratio method [Domb and Sykes, 1961] consists of fitting the ratios

$$r_n \equiv \frac{c_n}{c_{n-1}} \approx \mu \left(1 + \frac{\gamma}{n}\right) \qquad (3.3)$$

by suitable means, such a graph of r_n versus n^{-1}. The ratio method is simple but limited in scope. It was replaced by the logarithmic derivative method [Baker, 1961], which consists of noting that (3.1) is equivalent to

$$g(z) \equiv \frac{d}{dz} \ln f(z) \approx \frac{\gamma \mu}{1 - \mu z}.$$

Thus a rational fraction is expected to be the dominant part of $g(z)$, and the values of γ and μ are easily obtained from the pole and residue of the Padé approximants of $g(z)$. This method remains satisfactory because the presumed functional form has been fully exploited but does not constrain the approximation method. Properly used, this method is simple, reliable, and elegant (cf. Section 2.2).

G^3J approximants appear as a generalization of the statement that (3.3) is tantamount to the requirement that c_i obey a recursion relation

$$nc_n - \mu(n + \gamma)c_{n-1} \approx 0$$

which may be rewritten as

$$A_{01} n c_n + [A_{11}(n - 1) + A_{10}] c_{n-1} \approx 0$$

with $A_{01} = 1$, $A_{11} = -\mu$, and $A_{10} = -\mu(1 + \gamma)$. This is generalized to become the Mth-order recursion relation [Guttmann and Joyce, 1972]

$$R_{2M}(c_n) \equiv \sum_{i=0}^{\min(M,n)} \{A_{i2}(n - i)^2 + A_{i1}(n - i) + A_{i0}\} c_{n-i} = 0 \quad (3.4)$$

which is valid for $n = 1, 2, 3, \ldots$ with the conditions $A_{02} = 1$, $A_{00} = 0$. The unknown coefficients

$$\{A_{01}; A_{i2}, A_{i1}, A_{i,0}, i = 1, 2, \ldots, M\} \qquad (3.5)$$

are determined by the $3M + 1$ linear equations

$$R_{2M}(c_n) = 0, \quad n = 1, 2, \ldots, 3M + 1. \tag{3.6}$$

Thus (3.6) determines (3.5), and (3.5) determines an entire sequence of coefficients c_i which are given by (3.2) to order $3M + 1$ and define an approximating function

$$\psi_M(z) = \sum_{i=0}^{\infty} c_i z^i$$

which agrees with the given expansion to order z^{3M+1}. It may be verified that $\psi_M(z)$ satisfies the ordinary homogeneous differential equation

$$Q(z)\frac{d^2\psi_M}{dz^2} + R(z)\frac{d\psi_M}{dz} + S(z)\psi_M = 0, \tag{3.7}$$

where

$$Q(z) = z\sum_{i=0}^{M} A_{i2} z^i,$$

$$R(z) = \sum_{i=0}^{M} (A_{i2} + A_{i1}) z^i,$$

and

$$S(z) = \sum_{i=0}^{M-1} A_{i+1,0} z^i.$$

If $Q(z)$ vanishes either once or twice at $z = z_0$ — by which we mean that

$$Q(z) = C_1(z - z_0) + O((z - z_0)^2), \quad C_1 \neq 0,$$

or

$$Q(z) = C_2(z - z_0)^2 + O((z - z_0)^3), \quad C_2 \neq 0$$

—and $S(z_0) \neq 0$, then z_0 is called a regular singular point of the differential equation. Depending upon the behavior of $R(z)$ at $z = z_0$, the following functional forms are possible for one of the solutions of (3.7):

$$\psi_M(z) = (z - z_0)^p \phi_1(z) + \phi_2(z) \tag{3.8a}$$

or

$$\psi_M(z) = \ln(z - z_0)\phi_1(z) + \phi_2(z) \tag{3.8b}$$

or

$$\psi_M(z) = (z - z_0)^p \ln(z - z_0)\phi_1(z) + \phi_2(z) \tag{3.8c}$$

or

$$\psi_M(z) = \ln(z - z_0)\phi_1(z) + (z - z_0)^{-p}\phi_2(z) + \phi_3(z), \tag{3.8d}$$

where $\phi_1(z)$, $\phi_2(z)$, and $\phi_3(z)$ are analytic at $z = z_0$, p is an arbitrary positive integer, and ρ is nonintegral. If $Q_M(z_1) \neq 0$, then z_1 is a regular point of the differential equation and $\psi_M(z)$ is analytic at $z = z_1$. Equations (3.8a)–(3.8d) show the types of singularity which can be produced accurately by the G³J scheme defined by (3.4). But the scheme is clearly much more general in its scope, and the inhomogeneous equation

$$Q(z)\frac{d^2\psi_M}{dz^2} + R(z)\frac{d\psi_M}{dz} + S(z)\psi_M = T(z)$$

has also been used in a similar framework. The G³J approximants satisfy the accuracy-through-order criterion, but are not necessarily rational functions. Their importance is that they allow intuition about the functional form—in other words, its presumed analytic structure near the critical point—to become an integral part of the formulation of the solution.

It is rarely clear in advance which values should be specified for the allowed degrees of the polynomials $Q(z)$, $R(z)$, $S(z)$, and $T(z)$. However, there is one equation of this type for which there is a satisfactory answer [Hunter and Baker, 1979]: the equation

$$R_N(z)\frac{d\psi}{dz} + S_M(z)\psi = T_L(z),$$

where $R_N(z)$, $S_M(z)$, and $T_L(z)$ are polynomials of orders $N = M + 2$, M, and $L = M$, respectively. As degenerate special cases, we note that if $R_N(z) = 0$, the approximant is a diagonal Padé approximant, and if $T_L(z) = 0$, the approximant is a D-log Padé approximant. The important feature of this equation is that it preserves the homographic-invariance property. Suppose that we use $3M + 4$ terms of the given series $f(z) = \sum_{i=0}^{\infty} c_i z^i$ to derive a G³J approximant which is $\psi_f(z)$. If we make a change of variable $z = w/(a + bz)$, we define

$$g(w) = f(z) = \sum_{i=0}^{\infty} g_i w^i,$$

for which we can obtain a new G³J approximant which is $\psi_g(w)$. The invariance theorem states that the approximant is invariant under the change of variable:

$$\psi_g(w) = \psi_f(z).$$

Ideas similar to the previous ones motivated the introduction of quadratic approximants [Shafer, 1974]. They are a useful outgrowth of

Padé's general problem of finding n polynomials $A_1(z)$, $A_2(z)$, ..., $A_n(z)$ of degrees $\mu_1, \mu_2, \ldots, \mu_n$ which satisfy

$$A_1(z)f_1(z) + A_2(z)f_2(z) + \cdots + A_n(z)f_n(z) = O(z^{\mu_1+\mu_2+\cdots+\mu_n+n-1}),$$

where $f_i(z)$, $i = 1, 2, \ldots, n$, are given functions; see Section 8.5. If $f(z)$ is given and if $Q(z)$, $R(z)$, and $S(z)$ are polynomials of orders q, r, and s which satisfy

$$Q(z)[f(z)]^2 + 2R(z)f(z) + S(z) = O(z^{q+r+s+2}), \qquad (3.9)$$

then the quadratic equation

$$Q(z)[\psi(z)]^2 + 2R(z)\psi(z) + S(z) = 0 \qquad (3.10)$$

is easily solved, and a solution is

$$\psi(z) = \frac{-R(z) + \sqrt{[R(z)]^2 - S(z)Q(z)}}{Q(z)} \qquad (3.11a)$$

$$= \frac{-S(z)}{R(z) + \sqrt{[R(z)]^2 - S(z)Q(z)}}. \qquad (3.11b)$$

Since $Q(z)$, $R(z)$, and $S(z)$ are polynomials, $\psi(z)$ is analytic except for a finite number of poles and branch points of square-root type. The correct branch of $\psi(z)$ must also be assigned; normally this branch is clear from the context, as in the following.

Example

$$f(z) = \tan^{-1} z.$$

Since $\tan^{-1} z$ is an odd function which vanishes at $z = 0$, (3.9) takes the form

$$(1 + \alpha z^2)[f(z)]^2 + \beta z f(z) + \gamma z^2 = O(z^8). \qquad (3.12)$$

The given expansion is

$$f(z) = z - \tfrac{1}{3}z^3 + \tfrac{1}{5}z^5 - \tfrac{1}{7}z^7 + O(z^8);$$

therefore

$$[f(z)]^2 = z^2 - \tfrac{2}{3}z^4 + \tfrac{23}{45}z^6 + O(z^8),$$

and substitution in (3.12) gives three simultaneous linear equations for α, β, and γ with the solution

$$\alpha = \tfrac{5}{3}, \quad \beta = 3, \quad \text{and } \gamma = -4.$$

From (3.11b), the quadratic approximant of type $[q, r, s] = [2, 2, 2]$ is

$$\psi(z) = \frac{8z}{3 + (25 + \tfrac{80}{3}z^2)^{1/2}} = \tan^{-1}(z) + O(z^8).$$

This approximation has several attractive features. For x real, $\tan^{-1} x$ is well defined and $\tan^{-1}(\infty) = \pi/2 = 1.57\ldots$ is approximated by $\psi(\infty) = 8\sqrt{\frac{3}{80}} = 1.55\ldots$. For $z = iy$ purely imaginary,

$$\tan^{-1}(iy) = \tfrac{1}{2}\ln\frac{1+iy}{1-iy} = i\tanh^{-1} y,$$

and therefore

$$\tanh^{-1} y = \frac{1}{2i}\ln\frac{1+iy}{1-iy} = \frac{8y}{3 + (25 - \frac{80}{3}y^2)^{1/2}} + O(y^8),$$

showing the branch points of the quadratic approximants at $y^2 = \frac{15}{16}$ which model those of $\tanh^{-1} y$ at $y^2 = 1$. Thus the global qualities of the quadratic approximants are apparent: we have accurate approximation on the entire real axis and an accurately located singularity on the imaginary axis. Of course, $\tan^{-1} z$ is related to a Stieltjes function, and so the standard Padé method is known to converge systematically in this case; see (4.6.5).

There are some series for which the Padé method is an inappropriate choice. Various examples of power series have already been given, such as noisy series, and series based on random numbers. Another example, in this case an ordinary series, is $\sum_{i=0}^{\infty} c_i$ where $f(z) = \sum_{i=0}^{\infty} c_i z^i$ has a branch point at $z = 1$ and yet $f(1)$ is well defined. One expects the poles of the Padé approximants to accumulate at $z = 1$ and convergence to be slow. For the present, we simplistically regard the Padé method for series summation as the hypothesis that the series is generated by N geometric components

$$c_n = \sum_{i=1}^{N}\alpha_i(r_i)^n, \quad n = 0, 1, 2, \ldots,$$

where each $|r_i| < 1$ and the partial sums define the sequences

$$S_n = \sum_{i=1}^{N}\alpha_i\frac{1 - r_i^{n+1}}{1 - r_i} = S_\infty - \sum_{i=1}^{N}\left(\frac{\alpha_i r_i}{1 - r_i}\right)r_i^n.$$

In this case, the $[L/M]$ approximant is exact provided $L, M \geq N$. If this hypothesis is inapplicable, one may consider $G^3 J$ approximants for $f(z) = \sum_{i=0}^{\infty} c_i z^i$ and evaluate at $z = 1$, especially if the locations of the singularities of $f(z)$ are known. A quite different approach is based on the hypothesis that S_n is a smooth function of the variable $w = 1/n$ on the interval $0 \leq w \leq 1$. The problem is then to interpolate the value of S_n at $n = \infty$. A standard approach is to use rational interpolation (N-point Padé approximation) to interpolate to $w = 0$; any reliable method from

Section 7.1 is usually very satisfactory in this context. This theme is taken up in Section 10.1, where some comparisons are made. All these methods and many others are surveyed and reviewed by Guttmann [1989].

7.4 Padé–Laurent, Padé–Fourier, and Padé–Tchebycheff approximants

In most of this book, power series of the form $f(z) = \sum_{i=0}^{\infty} c_i z^i$ are considered, and mostly those series having a nontrivial circle of convergence $|z| < R$. A natural generalization of this functional form is the Laurent series

$$f(z) = \sum_{i=-\infty}^{\infty} c_i z^i \qquad (4.1)$$

which is defined in the annulus $r < |z| < R$. We may think of the terms with $i > 0$ as arising from the singularities of $f(z)$ in $|z| \geq R$, as is the case with Maclaurin series, and the terms with $i < 0$ as arising from the singularities in $|z| \leq r$. The series coefficients in (4.1) are given by

$$c_j = \frac{1}{2\pi i} \int_\Gamma \frac{f(z)}{z^{j+1}} dz \qquad (4.2)$$

for all j, where Γ is a suitable contour enclosing the origin and within the annulus $r < |z| < R$ (see Copson [1948] for details). A widely quoted example of a Laurent series is [Titchmarsh, 1939]

$$f(z) = e^{(a/2)(z+z^{-1})} = \sum_{j=-\infty}^{\infty} c_j z^j.$$

Using (4.2) and the contour $|z| = 1$ for Γ, we find that

$$c_j = \frac{1}{2\pi} \int_0^{2\pi} \cos(j\theta - a \sin\theta) \, d\theta = I_j(a),$$

where $I_j(x)$ is the modified Bessel function of the first kind. In cases such as this, the series coefficients seem to form a natural continuum over $-\infty < j < \infty$. In this section, we consider rational approximations based on using as many c_j of negative index as of positive index. The methods are therefore more naturally applicable to cases in which the constant term c_0 determines a natural split of the Laurent expansion.

Given the expansion (4.1), Padé–Laurent approximants of type $[L/M]$ are formed in the following way. Partial denominator coefficients $\beta_0^+, \beta_1^+, \ldots, \beta_M^+$ are calculated as solutions of the homogeneous system

$$\sum_{j=0}^{M} c_{i-j} \beta_j^+ = 0, \quad i = L+1, L+2, \ldots, L+M, \qquad (4.3)$$

and $\beta_0^-, \beta_1^-, \ldots, \beta_m^-$ as solutions of

$$\sum_{j=0}^{M} c_{j-i}\beta_j^- = 0, \quad i = L+1, L+2, \ldots, L+M. \tag{4.4}$$

Note that these equations differ from (1.1.6) only for the case $L < M - 1$ in that the coefficient matrices in (4.3), (4.4) are normally full matrices. From the solutions of (4.3), (4.4), two partial denominator polynomials

$$\beta^\pm(z) = \sum_{j=0}^{M} \beta_j^\pm z^j \tag{4.5}$$

are formed.

We may now interpret (4.3) and (4.4) as

$$[f(z)\beta^+(z)]_{L+1}^{L+M} = 0, \quad [f(z^{-1})\beta^-(z)]_{L+1}^{L+M} = 0$$

using Nuttall's notation; we may express the polynomials as

$$\beta^+(z) = \det \begin{vmatrix} c_{L-M+1} & \cdots & c_{L+1} \\ \vdots & & \vdots \\ c_L & \cdots & c_{L+M} \\ z^M & \cdots & 1 \end{vmatrix},$$

$$\beta^-(z) = \det \begin{vmatrix} c_{M-L-1} & \cdots & c_{-L-1} \\ \vdots & & \vdots \\ c_{-L} & \cdots & c_{-L-M} \\ z^M & \cdots & 1 \end{vmatrix} \tag{4.6}$$

as was done in (1.1.8). Again, we note that $\beta^+(z) \neq Q^{[L/M]}(z)$ if $L < M - 1$.

The series (4.1) is split into two parts:

$$f^+(z) = \sum_{i=0}^{\infty} {}'c_i z^i, \quad f^-(z^{-1}) = \sum_{i=0}^{\infty} {}'c_{-i} z^{-i}, \tag{4.7}$$

where the prime denotes that the term involving c_0 is to be halved. Thus the constant c_0 is shared between $f^+(z)$ and $f^-(z)$, and (4.7) implies that

$$f(z) = f^+(z) + f^-(z^{-1}). \tag{4.8}$$

Partial numerator coefficients are defined by

$$\alpha_i^+ = \sum_{j=0}^{i} {}'c_{i-j}\beta_j^+, \quad \alpha_i^- = \sum_{j=0}^{i} {}'c_{j-i}\beta_j^- \quad \text{for } 0 \leq i \leq l, \tag{4.9}$$

where $l = \max(L, M)$ and it is understood that $\beta_j^\pm = 0$ for $j > M$. The partial numerators are in turn defined by

$$\alpha^+(z) = \sum_{i=0}^{l} \alpha_i^+ z^i, \quad \alpha^-(z) = \sum_{i=0}^{l} \alpha_i^- z^i \tag{4.10}$$

with the interpretation that

$$\alpha^+(z) = [\beta^+(z)f^+(z)]_0^l, \quad \alpha^-(z) = [\beta^-(z)f^-(z)]_0^l. \tag{4.11}$$

If, using the determinantal normalization (4.6), either $\beta^+(0) = 0$ or $\beta^-(0) = 0$, then the approximants are degenerate. Let us assume that this is not the case, so that nondegenerate solutions exist for (4.3), (4.4). Then a Padé–Laurent numerator and denominator of type $[L/M]$ are defined [Gragg and Johnson, 1974] by

$$p(z) = \beta^-(z^{-1})\alpha^+(z) + \beta^+(z)\alpha^-(z^{-1}), \tag{4.12}$$

$$q(z) = \beta^+(z)\beta^-(z^{-1}) \tag{4.13}$$

and the associated Padé–Laurent approximant is

$$r(z) = \frac{\alpha^+(z)}{\beta^+(z)} + \frac{\alpha^-(z^{-1})}{\beta^-(z^{-1})} = \frac{p(z)}{q(z)}. \tag{4.14}$$

From (4.13), we see that the Laurent degree of $q(z)$ is M, but from (4.12) it appears that the Laurent degree of $p(z)$ is l. This would be unsatisfactory if $L < M$, but the following lemma shows that there is, in fact, no problem.

Lemma 1. *The Laurent degree of $p(z)$ does not exceed L.*

Proof. Let p_k denote the coefficient of z^k in $p(z)$. Then, for cases in which $L + 1 \leq k \leq M$, we find from (4.12) that

$$p_k = \sum_{i=k}^{M} \alpha_i^+ \beta_{i-k}^- + \sum_{j=k}^{M} \alpha_{j-k}^- \beta_j^+$$

and hence, from (4.9), and then by algebra,

$$p_k = \sum_{i=k}^{M} \beta_{i-k}^- \sum_{j=0}^{i} {}' c_{i-j} \beta_j^+ + \sum_{j=k}^{M} \beta_j^+ \sum_{i=0}^{j-k} {}' c_{i+k-j} \beta_i^-$$

$$= \sum_{i=k}^{M} \beta_{i-k}^- \sum_{j=0}^{i} {}' c_{i-j} \beta_j^+ + \sum_{i=k}^{M} \beta_{i-k}^- \sum_{j=i}^{M} {}' c_{i-j} \beta_j^+$$

$$= \sum_{i=k}^{M} \beta_{i-k}^- \sum_{j=0}^{M} c_{i-j} \beta_j^+.$$

Using (4.3), we find that $p_k = 0$.

Therefore, it turns out that $p(z)$ and $q(z)$ are normally of Laurent degrees L, M in all cases.

7.4 Padé–Laurent, Padé–Fourier, and Padé–Tchebycheff approximants

Example 1. Find the Padé–Laurent approximation of type [1/2] for

$$f(z) = b\sum_{i=0}^{\infty} h^i z^i + c\sum_{i=0}^{\infty} k^{-i} z^{-i}, \quad \text{for } h < 1, k > 1.$$

Solution. The split, according to (4.7), is

$$f^+(z) = \frac{b+c}{2} + b\sum_{i=1}^{\infty} h^i z^i = \frac{b+c}{2} + \frac{bhz}{1-hz},$$

$$f^-(z) = \frac{b+c}{2} + c\sum_{i=1}^{\infty} k^{-i} z^{-i} = \frac{b+c}{2} + \frac{cz^{-1}}{k-z^{-1}}.$$

Thus $f^+(z)$ is analytic in $|z| < h^{-1}$, $f^-(z)$ is analytic in $|z| > k^{-1}$, and $f(z)$ is analytic in the annulus $k^{-1} < |z| < h^{-1}$ and on $|z| = 1$ in particular.

From (4.6), we find

$$\beta^+(z) = \det \begin{vmatrix} b+c & bh & bh^2 \\ bh & bh^2 & bh^3 \\ z^2 & z & 1 \end{vmatrix} = \beta^+(0)(1-zh)$$

and hence, from (4.11),

$$\alpha^+(z) = \frac{\beta^+(0)}{2}\{(b+c) + (b-c)hz\}.$$

Similarly, we find that

$$\beta^-(z) = \det \begin{vmatrix} b+c & c/k & c/k^2 \\ c/k & c/k^2 & c/k^3 \\ z^2 & z & 1 \end{vmatrix} = \beta^-(0)(1-z/k),$$

$$\alpha^-(z) = \frac{\beta^-(0)}{2}\{(b+c) + (c-b)z/k\}.$$

Hence, from (4.13),

$$q(z) = \beta^+(0)\beta^-(0)(1-zh)(1-k^{-1}z^{-1})$$

and, from (4.12), (4.14),

$$r(z) = b + c + \frac{bhz}{1-hz} + \frac{cz^{-1}}{k-z^{-1}}.$$

In fact, $r(z)$ reproduces $f(z)$ exactly. It is not a normal approximant, since $q(z)$ is of Laurent degree one. It is not reducible, since it is readily found that it is different from the Padé–Laurent approximant of type [1/1].

Equation (4.3) determining the β_i^+, and hence $\beta^+(z)$, is very similar to the equivalent Padé equations (1.1.6). The difference, of course, is that the coefficients $c_{-1}, c_{-2}, \ldots, c_{L-M+1}$ are involved when subdiagonal approximants with $L < M - 1$ are considered. In fact, all the recursion relations derived for the Padé denominator polynomials, such as those given in Section 3.5, continue to hold for $\beta^+(z)$, and similar relationships also hold for $\beta^-(z)$. All the recursion relations hold for the polynomials $\beta^+(z)$, $\alpha^+(z)$ and the partial approximants $\alpha^+(z)/\beta^+(z)$ when every actual or implied index $L \geq M - 1$, because these quantities are identical to the equivalent Padé polynomials and approximants.

Padé–Fourier approximation

In numerous branches of science, one encounters Fourier series of the form

$$F(\theta) = \sum_{k=0}^{\infty} {}' a_k \cos k\theta + \sum_{k=1}^{\infty} b_k \sin k\theta, \tag{4.15}$$

where $-\pi \leq \theta \leq \pi$ and a_k, b_k are real or complex numbers. The series (4.15) can be re-expressed as a power series of trigonometric functions,

$$F(\theta) = \sum_{k=0}^{\infty} \tilde{a}_k \cos^k \theta + \sin \theta \sum_{k=1}^{\infty} \tilde{b}_k \cos^k \theta,$$

but (4.15) is usually the more convenient form. We will construct rational trigonometric functions which are intended to approximate $F(\theta)$ more accurately when only a finite number of $\{a_k, b_k\}$ are available.

With the identification $z = e^{i\theta}$, the method given previously in this section can be used without difficulty. We define $c_0 = \tfrac{1}{2} a_0$ and

$$c_k = \tfrac{1}{2}(a_k - ib_k), \quad c_{-k} = \tfrac{1}{2}(a_k + ib_k), \quad k = 1, 2, 3, \ldots.$$

Then, using (4.15), we introduce $f(z)$ as

$$F(\theta) = \sum_{k=-\infty}^{\infty} c_k e^{ikz} = f(z). \tag{4.16}$$

Then $\beta^+(z)$, $\beta^-(z)$ are defined by (4.5) using (4.3), and $\alpha^+(z)$, $\alpha^-(z)$ are defined by (4.10) using (4.9). The Padé–Fourier approximant of type $[L/M]$ is then given by (4.14) as

$$F^{\text{PF}}(\theta) = r(e^{i\theta}) = \frac{\alpha^+(e^{i\theta})}{\beta^+(e^{i\theta})} + \frac{\alpha^-(e^{-i\theta})}{\beta^-(e^{-i\theta})}. \tag{4.17}$$

The case in which $F(\theta)$ is real valued, which happens when $a_k, b_k \in \mathbb{R}$, is of importance. Then

$$c_{-k} = c_k^*, \quad k = 0, 1, 2, \ldots,$$

and the superscripts $\pm$ can be suppressed if we take

$$\alpha^+(z) = \alpha(z), \quad \alpha^-(z) = \alpha^*(z),$$

where the asterisk denotes the functional complex conjugate, which in this case means the complex conjugates of the Maclaurin coefficients in (4.5), (4.10). Then, from (4.17),

$$F^{\text{PF}}(\theta) = \frac{\sum_{j=0}^{l} \alpha_j z^j}{\sum_{j=0}^{M} \beta_j z^j} + \frac{\sum_{j=0}^{l} \alpha_j^* z^{-j}}{\sum_{j=0}^{M} \beta_j^* z^{-j}}$$

$$= 2 \frac{\sum_{j=0}^{l} \sum_{k=0}^{M} \{\text{Re}(\alpha_j \beta_k^*) \cos(j-k)\theta - \text{Im}(\alpha_j \beta_k^*) \sin(j-k)\theta\}}{\sum_{j=0}^{M} \sum_{k=0}^{M} \{\text{Re}(\beta_j \beta_k^*) \cos(j-k)\theta - \text{Im}(\beta_j \beta_k^*) \sin(j-k)\theta\}},$$

(4.18)

which is real valued. As was shown in Lemma 1, this expression (4.18) reduces to a trigonometric rational polynomial of type $[L/M]$ in all cases.

Padé–Tchebycheff approximation

Formal expansions of the form

$$F(x) = \sum_{j=0}^{\infty} {}' c_j T_j(x) \tag{4.19}$$

arise naturally in the context of near-best polynomial approximations for $x \in [-1, 1]$. Here, $T_j(x)$ is the Tchebycheff polynomial of order j, defined by

$$T_j(x) = \cos(j \cos^{-1} x)$$

so that $T_0(x) = 1$, $T_1(x) = x$, ... The higher-order polynomials can be calculated using the recursion relation

$$T_{j+1}(x) - 2x T_j(x) + T_{j-1}(x) = 0, \quad j = 1, 2, 3, \ldots.$$

In fact, the section

$$p_n(x) = \sum_{j=0}^{n} {}' c_j T_j(x)$$

of (4.19) is the best L_2 polynomial approximation of degree n for a continuous function $F(x)$ on the interval $[-1, 1]$ with respect to weight function $w(x) = (1 - x^2)^{-\frac{1}{2}}$. In this case, the coefficients are given by

$$c_k = \int_{-1}^{1} F(x) T_k(x) \frac{dx}{\sqrt{1 - x^2}}, \quad k = 0, 1, \ldots, n. \tag{4.20}$$

It is remarkable that the best L_2 polynomial approximations $p_n(x)$ are scarcely worse than the best polynomial approximations $p_n^*(x)$ of moderate degree n, because Powell [1967] has shown that

$$\|p_n - F\| \le 5 \|p_n^* - F\|, \quad n \le 500.$$

In this subsection, we describe rational approximations of $F(x)$ on $[-1, 1]$, given a certain number of coefficients of its Tchebycheff expansion (4.19).

The standard substitutions for analyzing (4.19) are

$$x = \cos \theta, \quad z = e^{i\theta} = x + i\sqrt{1 - x^2}, \tag{4.21}$$

where the last equality defines the continuation away from $x \in [-1, 1]$ for complex x and $z(x)$. The inverse transformation is $x = \frac{1}{2}(z + z^{-1})$, and

$$T_j(x) = \cos j\theta = \tfrac{1}{2}(z^j + z^{-j}), \quad j = 0, 1, 2, \ldots. \tag{4.22}$$

The Laurent expansion associated with (4.19) is

$$f(z) = 2F(x) = \sum_{j=-\infty}^{\infty} c_j z^j, \tag{4.23}$$

where $c_j = c_{-j}$ for all $j \ge 0$. Then (4.3) and (4.5) determine one partial denominator polynomial

$$\beta(z) = \beta^+(z) = \beta^-(z),$$

and (4.9), (4.10) determine its corresponding partial numerator polynomial $\alpha(z)$. The corresponding Padé–Tchebycheff approximant is given in Laurent form by (4.14) as

$$r(z) = \frac{\alpha(z)}{\beta(z)} + \frac{\alpha(z^{-1})}{\beta(z^{-1})}$$

$$= \frac{\sum_{j=0}^{l} \alpha_j z^j \sum_{k=0}^{M} \beta_k z^{-k} + \sum_{j=0}^{l} \alpha_j z^{-j} \sum_{k=0}^{M} \beta_k z^k}{\sum_{j=0}^{M} \beta_j z^j \sum_{k=0}^{M} \beta_k z^{-k}}$$

$$= \frac{\sum_{j=0}^{l}\sum_{k=0}^{M}\alpha_j\beta_k(z^{j-k}+z^{k-j})}{\sum_{j=0}^{M}\beta_j^2+\sum_{j=1}^{M}\sum_{k=0}^{j-1}\beta_j\beta_k(z^{j-k}+z^{k-j})}. \qquad (4.24)$$

If we transform to the x-variable, the numerator of (4.24) is

$$P^{\text{PT}}(x) = p(z) = \sum_{j=0}^{l}\sum_{k=0}^{M}\alpha_j\beta_k(z^{j-k}+z^{k-j})$$

$$= 2\sum_{j=0}^{l}\sum_{k=0}^{M}\alpha_j\beta_k T_{|j-k|}(x), \qquad (4.25)$$

the denominator is

$$Q^{\text{PT}}(x) = q(z) = \sum_{j=0}^{M}\beta_j^2 + \sum_{j=1}^{M}\sum_{k=0}^{j-1}\beta_j\beta_k(z^{j-k}+z^{k-j})$$

$$= \sum_{j=0}^{M}\beta_j^2 + \sum_{j=1}^{M}\sum_{k=0}^{j-1}\beta_j\beta_k T_{j-k}(x)$$

$$= 2\sum_{j=0}^{M}{'}T_j(x)\left\{\sum_{k=j}^{M}\beta_k\beta_{k-j}\right\}, \qquad (4.26)$$

and the approximant is

$$R^{\text{PT}}(x) = P^{\text{PT}}(x)/Q^{\text{PT}}(x). \qquad (4.27)$$

In Lemma 1, it is proved that $P^{\text{PT}}(x)$ contains Tchebycheff polynomials of order no greater than L, so $R^{\text{PT}}(x)$ is the Padé–Tchebycheff rational approximant of type $[L/M]$ for $F(x)$ in all cases where (4.3) is soluble.

In fact, the Padé–Tchebycheff approximants derived by this method are precisely the Baker–Gammel approximants of Section 7.2 for the Tchebycheff series (4.19) for $L \geqslant M - 1$. It is instructive to see why this is so. Recall that $J = L - M$. As in (2.13), we use the representation of c_j as a sum of M geometric components with ratios $-u_1, -u_2, \ldots, -u_M$:

$$c_j = \sum_{i=1}^{M}\gamma_i(-u_i)^j, \quad j = J+1, \ldots, 2M+J. \qquad (4.28)$$

Then the partial denominator (4.5) is given by

$$\beta(z) = \kappa\prod_{i=1}^{M}(1+zu_i), \qquad (4.29)$$

where κ is a normalizing constant. The Laurent denominator defined by (4.13) is

$$q(z) = \kappa^2 \prod_{i=1}^{M}(1 + zu_i)(1 + z^{-1}u_i);$$

by (4.26) we obtain the Padé–Tchebycheff denominator as

$$Q^{PT}(x) = q(z) = \kappa^2 \prod_{i=1}^{M}(1 + 2u_i x + u_i^2). \tag{4.30}$$

To use Baker–Gammel approximants, (4.19) is identified with (2.10) using $f_j = (-1)^j c_j$, so that

$$k_j(x) = (-1)^j T_j(x).$$

Then the kernel required in (2.9) is effectively a generating function of the Tchebycheff polynomials:

$$k(x, u) = \sum_{j=0}^{\infty}(-u)^j T_j(x) = \frac{1}{1 + 2ux + u^2}.$$

The denominator of the Baker–Gammel approximants is derived from (2.17b) as

$$Q^{BG}(x) = \prod_{i=1}^{M}(1 + 2u_i x + u_i^2) \tag{4.31}$$

which equals $Q^{PT}(x)$ in (4.30) up to a normalizing constant. Equality of the numerator polynomials follows by uniqueness of the Laurent expansions obtained after multiplication of the Padé–Laurent and Baker–Gammel approximants by their common denominator.

Now we turn to consider the role of Padé–Tchebycheff approximants as near-best rational approximants of a real-valued function $F(x)$ on the interval $[-1, 1]$. The representation (4.28) expresses the coefficients c_j in terms of geometric ratios $(-u_i)$. If these ratios satisfy $|u_i| < 1$, then the roots of $\beta(z)$ lie outside the unit circle, as would be expected in the 'natural' representation of $f_+(z)$ explained in Example 1. The roots of $\beta(z^{-1})$ occur inside the unit circle, and the roots of $q(z)$ are paired.

If $|u_i| = 1$ for any i, the Padé–Tchebycheff approximant is of little use for near-best approximation, because $Q^{PT}(x) = 0$ when $x = \cos(\arg u_i)$. Provided $|u_i| \neq 1$ for all i,

$$Q^{PT}(x) \neq 0 \quad \text{for } -1 \leq x \leq 1.$$

Because the error function defined by

$$E(x) = F(x) - P^{PT}(x)/Q^{PT}(x)$$
$$= [f_+(z) - \alpha(z)/\beta(z)] + [f_-(z) - \alpha(z^{-1})/\beta(z^{-1})]$$

has a Laurent series with a gap in the powers $\{z^j\}_{j=-L-M}^{L+M}$ it can be expressed as

$$E(x) = \sum_{j=L+M+1}^{\infty} e_j T_j(x).$$

Such an error function has at least $L + M + 1$ sign changes over $[-1, 1]$ (see Cheney [1966], p. 110), and this property of alternation is a primary requirement for an error function. For example, Figure 7.4.1 shows the error of the Padé–Tchebycheff approximant of type $[2/2]$ for $\exp(x)$ on $[-1, 1]$, in which there are five changes of sign of the error.

All these considerations indicate that the Padé–Tchebycheff approximant is an accurate approximation procedure for approximation on an interval, and as such it is suitable for initializing an iterative procedure for minimax rational approximation.

The methods described in this section are primarily due to Gragg and Johnson [1974], Gragg [1977]; the Padé–Tchebycheff method was introduced independently by Clenshaw and Lord [1974] from a complementary viewpoint. There are other approaches to rational approximation of

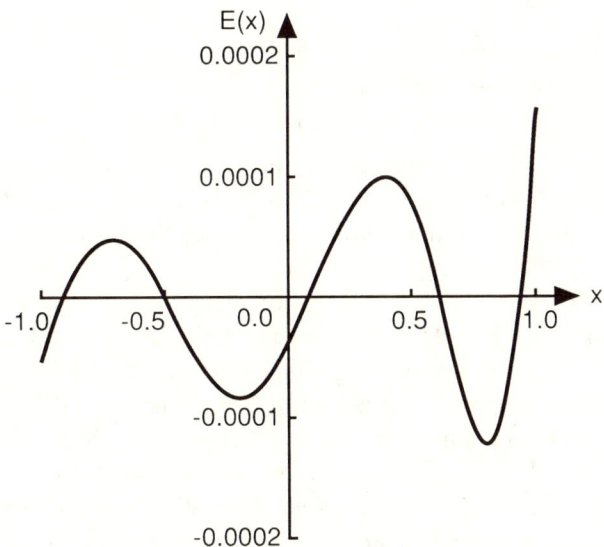

Figure 7.4.1. $\exp(x) - [2/2]^{PT}(x)$.

series of Tchebycheff polynomials, and these methods can also be adapted to series of general orthogonal polynomials. To effect the approximation, the fundamental requirement is that

$$\sum_{i=0}^{\infty}{'} c_i T_i(z) \approx \frac{\sum_{i=0}^{L}{'} a_i T_i(z)}{\sum_{j=0}^{M}{'} b_j T_j(z)}. \qquad (4.32)$$

To give meaning to this approximate equality, we cross-multiply and use the multiplication law

$$T_i(z) T_j(z) = \tfrac{1}{2}[T_{i+j}(z) + T_{|i-j|}(z)]$$

to derive

$$\left[\sum_{j=0}^{M}{'} b_j T_j(z)\right]\left[\sum_{i=0}^{\infty}{'} c_i T_i(z)\right] = \frac{1}{2}\sum_{i=0}^{\infty}{'}\left[\sum_{j=0}^{M}{'} b_j(c_{i+j} + c_{|i-j|})\right] T_i(z)$$

$$\approx \sum_{i=0}^{L}{'} a_i T_i(z). \qquad (4.33)$$

We interpret the latter approximate equality by equating coefficients of $T_i(z)$ for $i = 0, 1, \ldots, L + M$, yielding

$$\tfrac{1}{2}\sum_{j=0}^{M}{'} b_j(c_{i+j} + c_{|i-j|}) = 0, \quad i = L+1, \ldots, L+M, \qquad (4.34)$$

$$\tfrac{1}{2}\sum_{j=0}^{M}{'} b_j(c_{i+j} + c_{|i-j|}) = a_i, \quad i = 0, 1, \ldots, L. \qquad (4.35)$$

The equations (4.34) determine $\{b_j\}$ using the given coefficients, and then the equations (4.35) determine $\{a_i\}$. This appears to be a satisfactory way of calculating the approximants until one notices that the data coefficients occurring in (4.34), (4.35) are $c_0, c_1, \ldots, c_{L+2M}$. An unfortunate consequence of the multiplication law of orthogonal polynomials is that the formation of an $[L/M]$ approximant needs $L + 2M + 1$ coefficients of the given series. Obviously, this approximation scheme is uneconomic. Padé–Tchebycheff approximants defined in this way [Maehly, 1956; Holdeman, 1969] are sometimes called 'cross-multiplied' approximants [Fleischer, 1973b], to emphasize their derivation from (4.33). The simplest, but rather unsatisfactory, solution to the dilemma which occurs when only $c_0, c_1, \ldots, c_{L+M}$ are specified is to use (4.34), (4.35) with $c_{L+M+1}, c_{L+M+2}, \ldots, c_{L+2M}$ reset to zero.

To avoid these difficulties, one may revert to (4.32) and expand the denominator using

$$\left[\sum_{j=0}^{M}{}'b_j T_j(z)\right]^{-1} = \sum_{j=0}^{\infty}{}'\beta_j T_j(z). \qquad (4.36)$$

With the orthogonality property

$$\int_{-1}^{1} T_i(x) T_j(x) \frac{dx}{\sqrt{1-x^2}} = \frac{\pi}{2}(\delta_{ij} + \delta_{i0}\delta_{0j})$$

the coefficients β_i in (4.36) are given by

$$\beta_i = \sqrt{\frac{2}{\pi}} \int_{-1}^{1} \frac{T_i(x)}{\sum_{j=0}^{M}{}'b_j T_j(x)} \frac{dx}{\sqrt{1-x^2}}, \quad i = 0, 1, 2, \ldots. \qquad (4.37)$$

Substituting (4.36) into (4.32) and equating coefficients of $T_i(x)$, we find that

$$\frac{1}{2}\sum_{j=0}^{L}{}'a_j(\beta_{i+j} + \beta_{|i-j|}) = c_i, \quad i = 0, 1, \ldots, L+M.$$

Substituting from (4.37) for β_{i+j} and $\beta_{|i-j|}$, we obtain a horrific system of nonlinear equations for $a_0, \ldots, a_L, b_1, \ldots, b_M$. The approximants so defined are called 'properly expanded' approximants [Fleisher, 1973b] and are seldom used.

Comparisons of the various methods have been made by Chisholm and Common [1981], who also give detailed attention to the case in which both x and c_j in (4.19) are complex. By using methods of potential theory, Gončar, Rakhmanov, and Suetin [1992] have proved that, for Stieltjes functions, the Padé–Tchebycheff approximants converge substantially more rapidly than the cross-multiplied approximants.

The block structure of the Padé–Tchebycheff table is explained by Geddes [1981] and Trefethen and Gutknecht [1985, 1987].

7.5 Laurent–Padé approximation and Toeplitz systems

Laurent–Padé approximation refers to a natural extension of the formalism of Padé approximation (as given in Chapter 1) when the given series is a Laurent series rather than a Maclaurin series. Indeed, this approach was adopted in the previous section for the construction of the partial denominator $\beta^+(z)$ for Padé–Laurent approximation. However, for Laurent–Padé approximation the numerator is itself a Laurent series, as is clear from (5.6)–(5.8). We begin with the expansion

$$f(z) = \sum_{i=-\Omega}^{\infty} c_i z^i \qquad (5.1)$$

which is given. The integer Ω is taken to be sufficiently large, meaning that the number of terms with negative powers is to be bounded. Coefficients $q_0, q_1, \ldots, q_M$ of a denominator polynomial $q^{[L/M]}(z)$ are defined by

$$\begin{bmatrix} c_{L+1} & c_L & \cdots & c_{L-M+2} & c_{L-M+1} \\ c_{L+2} & c_{L+1} & \cdots & c_{L-M+3} & c_{L-M+2} \\ \vdots & & \ddots & \vdots & \vdots \\ c_{L+M} & c_{L+M-1} & \cdots & c_{L+1} & c_L \end{bmatrix} \begin{bmatrix} q_0 \\ q_1 \\ \vdots \\ q_M \end{bmatrix} = 0. \quad (5.2)$$

Notice that the ordering of the columns of the coefficient matrix in (5.2) is the reverse of that of (1.1.6), so the coefficient matrix has Toeplitz structure. The equations (5.2) are important for the construction of digital filters, and ARMA models of time series; in these contexts they are known as the Yule–Walker equations [e.g., Jackson, 1986]. Any solution of these equations serves to define a denominator polynomial

$$q(z) = \sum_{i=0}^{M} q_i z^i. \quad (5.3)$$

If the coefficient matrix has full rank M, the Laurent–Padé denominator with the standard determinantal normalization is uniquely defined as

$$q^{[L/M]}(z) = \det \begin{vmatrix} c_{L+1} & \cdots & c_{L-M+1} \\ \vdots & & \vdots \\ c_{L+M} & \cdots & c_L \\ 1 & \cdots & z^M \end{vmatrix}. \quad (5.4)$$

For cases in which $c_j = 0$ for $j < 0$, Laurent–Padé approximation reduces to ordinary Padé approximation. By comparing (5.4) with (1.1.8), the denominators in this case are related by

$$q^{[L/M]}(z) = (-1)^{\frac{1}{2}M(M+1)} Q^{[L/M]}(z).$$

Parallel to the notation of (3.6.10), we define the Toeplitz matrix $T_{L;M}$ by

$$T_{L;M} = \begin{bmatrix} c_L & \cdots & c_{L-M+1} \\ \vdots & \ddots & \vdots \\ c_{L+M-1} & \cdots & c_L \end{bmatrix} \quad (5.5)$$

and its determinant by $T(L/M)$. From (5.5),

$$T(L/M) = (-1)^{\frac{1}{2}(M-1)M} C(L/M).$$

Analogously to the Padé case, the denominator polynomial $q^{[L/M]}(z)$ is said to be nondegenerate if $T(L/M) \neq 0$.

7.5 Laurent–Padé approximation and Toeplitz systems

The Yule–Walker equations (5.2) arise from the requirement that

$$[f(z)q(z)]_{L+1}^{L+M} = 0,$$

using (5.3) and Nuttall's notation. The corresponding Laurent–Padé numerator is the Laurent polynomial

$$p(z) = [f(z)q(z)]_{-\Omega}^{L}. \tag{5.6}$$

In particular, the definition

$$p^{[L/M]}(z) = [f(z)q^{[L/M]}(z)]_{-\Omega}^{L} \tag{5.7}$$

accords with the determinantal normalization and, more explicitly,

$$p^{[L/M]}(z) = \det \begin{vmatrix} c_{L+1} & c_L & \cdots & c_{L-M+1} \\ \vdots & \ddots & \ddots & \vdots \\ c_{L+M} & \cdots & c_{L+1} & c_L \\ \sum_{i=-\Omega}^{L} c_i z^i & \cdots & \cdots & \sum_{i=-\Omega}^{L-M} c_i z^{i+M} \end{vmatrix}. \tag{5.8}$$

The Laurent–Padé form of type $[L/M]$ is given uniquely by

$$r^{[L/M]}(z) = p(z)/q(z) = p^{[L/M]}(z)/q^{[L/M]}(z), \tag{5.9}$$

using (5.3) and (5.6), or (5.4) and (5.8). Uniqueness of this rational form (5.9) follows in much the same way as for Padé approximants, as was explained in Section 1.4. If $q(0) \neq 0$ or $q^{[L/M]}(0) \neq 0$, $r^{[L/M]}(z)$ is called a Laurent–Padé approximant, and

$$f(z) - r^{[L/M]}(z) = O_+(z^{L+M+1}). \tag{5.10}$$

From (1.1.8), (1.1.9), (1.1.12), (5.2), (5.8), and (5.10) we have the relationship

$$r^{[L/M]}(z) = z^{-\Omega}[L + \Omega/M]_{(z^\Omega f)}(z)$$

between the Laurent–Padé and ordinary Padé approximants.

The Laurent–Padé table consists of all the approximants and forms $r^{[L/M]}(z)$ which can be defined for $f(z)$ as given by (5.1). Note that it is more extensive than the Padé table, because negative values of L are included. Following the analysis of Section 1.4, if all $T(L/M) \neq 0$, $\deg\{p^{[L/M]}(z)\} = L$, $\deg\{q^{[L/M]}(z)\} = M$ and their leading coefficients are given by

$$\dot{p}^{[L/M]} = (-1)^M T(L/M + 1), \quad \dot{q}^{[L/M]} = T(L + 1/M). \tag{5.11}$$

In such cases, the Laurent–Padé table is called normal.

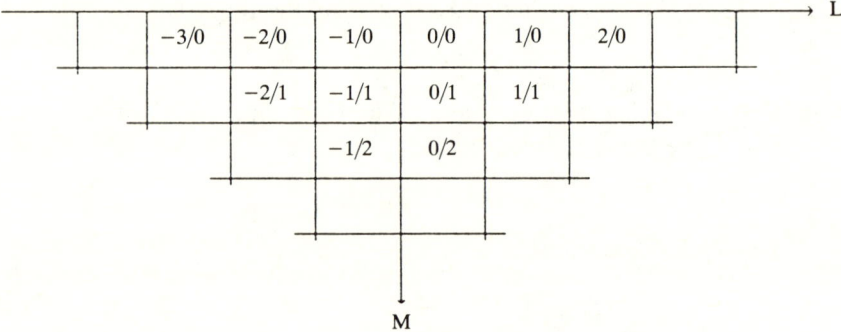

Figure 7.5.1. A section of the Laurent–Padé table.

The case in which the coefficients c_n are defined by

$$c_n = \frac{1}{2\pi}\int_{-\pi}^{\pi} e^{-in\theta}v(\theta)\,d\theta, \tag{5.12}$$

where $v(\theta)$ is a positive integrable weight function, is particularly important in the theory of polynomials orthogonal on the unit circle, and in applications to signal processing. In the latter context, the weight function $v(\theta)$ is often called the power spectral density. From (5.12), the matrices $T_{0;n}$ are seen to be Hermitian, and from the theory of quadratic forms [see (5.1.13) et seq or Bellman, 1970] their Toeplitz determinants are positive:

$$T(0/n+1) = \det \begin{vmatrix} c_0 & \cdots & c_{-n} \\ \vdots & & \vdots \\ c_n & \cdots & c_0 \end{vmatrix} > 0. \tag{5.13}$$

The fundamental properties of the connection between the Laurent–Padé denominators for $L = -1$ and orthogonality on the circle is encapsulated in:

Theorem 7.5.1. *With the definition* (5.12), *the polynomials defined by*

$$\phi_n(z) = T(0/n)^{-1/2}T(0/n+1)^{-1/2}q^{[-1/n]}(z)$$

$$= T(0/n)^{-1/2}T(0/n+1)^{-1/2}\begin{vmatrix} c_0 & c_{-1} & \cdots\cdots & c_{-n} \\ \vdots & \ddots & & \vdots \\ c_{n-1} & \cdots\cdots & c_0 & c_{-1} \\ 1 & \cdots & \cdots & z^n \end{vmatrix} \tag{5.14}$$

form a system of polynomials that are orthogonal on the unit circle, where orthogonality is defined in the sense that

$$\frac{1}{2\pi}\int_{-\pi}^{\pi} \phi_n(z)[\phi_m(z)]^* v(\theta)\,d\theta = \delta_{n,m}. \tag{5.15a}$$

7.5 Laurent–Padé approximation and Toeplitz systems

Proof. If $m \neq n$, taking $m < n$ loses no generality. For $i = 0, 1, \ldots, n - 1$, consider

$$I_i = \frac{1}{2\pi} \int_{-\pi}^{\pi} \phi_n(z) z^{-i} v(\theta) \, d\theta. \tag{5.15b}$$

With the substitution for $\phi_n(z)$ from (5.14), the entry in the last row of column j of the determinantal form of I_i (where $0 \leq j \leq n$) is

$$\frac{1}{2\pi} \int_{-\pi}^{\pi} z^{j-i} v(\theta) \, d\theta = c_{i-j}.$$

This is also the entry in row i; therefore $I_i = 0$ for $i = 0, 1, \ldots, n - 1$, and orthogonality is established for $m < n$. For the case in which $m = n$, (5.14) and (5.15b) give

$$I_n = T(0/n)^{-\frac{1}{2}} T(0/n + 1)^{\frac{1}{2}};$$

the leading coefficient of $\phi_n(z)$ follows from (5.14) as

$$\dot{\phi}_n = T(0/n)^{\frac{1}{2}} T(0/n + 1)^{-\frac{1}{2}}.$$

Combining these two formulas, and using orthogonality,

$$\frac{1}{2\pi} \int_{-\pi}^{\pi} \phi_n(z) [\phi_n(z)]^* v(\theta) \, d\theta = \dot{\phi}_n I_n^* = 1.$$

A most important property of the polynomials $\phi_n(z)$ is that their zeros lie in $|z| < 1$. For this, and many other fundamental properties of these polynomials, we refer to Szegö [1967, Chapter 11]. Their role in Laurent–Padé approximation is developed by Bultheel [1987], and the extent of their widespread significance is indicated by Kailath, Vieira, and Morf [1978].

Applications in real-time signal processing have motivated considerable interest in obtaining numerical solutions of the Toeplitz system

$$T_{0;M} \mathbf{x} = \begin{bmatrix} c_0 & \cdots & c_{-M+1} \\ \vdots & & \vdots \\ c_{M-1} & \cdots & c_0 \end{bmatrix} \mathbf{x} = \mathbf{b} \tag{5.16}$$

for arbitrary right-hand sides $\mathbf{b}$. By reversing the order of the equations or the variables, (5.16) becomes a Hankel system. The emphasis in this section on Toeplitz systems arises from the interest in extending the system (5.16) of dimension M to $M + 1, M + 2, \ldots$. This extension is easily seen to introduce different coefficients from a similar extension of the Hankel system (3.6.1), (3.6.30). Nevertheless, by row reversal, the formulas of Theorems 3.6.2, 3.6.3 are easily adapted to provide explicit inverses for the Toeplitz system. For example, (3.6.15) becomes

$$T_{L;M}^{-1} = \frac{1}{r_0 b_0} \left\{ \begin{bmatrix} b_0 & & 0 \\ \vdots & \ddots & \\ b_{M-1} & \cdots & b_0 \end{bmatrix} \begin{bmatrix} q_{M-1} & \cdots & q_0 \\ & \ddots & \vdots \\ 0 & & q_{M-1} \end{bmatrix} \right.$$

$$\left. - \begin{bmatrix} 0 & & 0 \\ q_0 & \ddots & \\ \vdots & \ddots & \\ q_{M-2} & \cdots & q_0 & 0 \end{bmatrix} \begin{bmatrix} 0 & & 0 \\ b_M & \cdots & b_1 \\ & \ddots & \vdots \\ 0 & & b_M \end{bmatrix} \right\},$$

(5.17)

where $B(z) = \sum_{i=0}^{M} b_i z^i$ is the denominator of a nondegenerate Laurent–Padé polynomial of type $[L/M]$ and $Q(z) = \sum_{i=0}^{M-1} q_i z^i$ is the denominator of a Laurent–Padé form of type $[L-1/M-1]$ whose remainder is given by

$$f(z)Q(z) - P(z) = r_0 z^{L+M-1} + O(z^{L+M}). \qquad (5.18)$$

Notice that the formula (5.17), like (3.6.15), holds for any convenient normalization of $B(z)$ and $Q(z)$.

A useful result due to Trench [1964] follows directly from (5.17) as

$$[T_{L;M}]_{i+1,j+1}^{-1} - [T_{L;M}]_{i,j}^{-1} = (b_{i+1} q_{M-j-2} - q_i b_{M-1-j})(r_0 b_0)^{-1}, \qquad (5.19)$$

where $[T_{L;M}]_{ij}$ is indexed by $i, j = 0, 1, \ldots, M-1$. The elements of the first row and column of $T_{L;M}^{-1}$ are given simply by (5.17), and the importance of the formula (5.17) is that it allows the remaining elements of $T_{L;M}^{-1}$ to be completed sequentially. See also Bareiss [1969].

In the remainder of this section, we will be concerned with how the quantities $\{b_i\}$, $\{q_j\}$ in (5.17) can be obtained. We begin with two algorithms which are based on the sequential construction of the approximants, or their denominators, in two neighboring column sequences of the Laurent–Padé table, assuming that this table is normal. The Toeplitz system (5.16) is expressed in standard form in (5.16) with $L = 0$. Thus the analysis of this case is based on the use of the columns of approximants of types $[0/m]$, $[-1/m]$ in the Laurent–Padé table.

The methodology of Section 3.5 holds for Laurent–Padé approximants without significant change, and the Frobenius identities hold in the Laurent–Padé table as stated in (3.5.11). The

$$\begin{bmatrix} * & * \\ & * \end{bmatrix} \text{ and } \begin{bmatrix} * & * \\ & * \end{bmatrix}$$

identities justify the linked progression for fixed L and increasing M

$$q^{[L-1/M]}(z) = -\hat{\alpha} q^{[L/M-1]}(z) + z q^{[L-1/M-1]}(z), \qquad (5.20)$$

$$p^{[L-1/M]}(z) = -\hat{\alpha}p^{[L/M-1]}(z) + zp^{[L-1/M-1]}(z), \qquad (5.21)$$

$$q^{[L/M]}(z) = q^{[L/M-1]}(z) - \alpha z q^{[L-1/M-1]}(z), \qquad (5.22)$$

$$p^{[L/M]}(z) = p^{[L/M-1]}(z) - \alpha z p^{[L-1/M-1]}(z). \qquad (5.23)$$

The normalization determined by (5.20)–(5.23) is that with $q^{[L/M]}(0) = 1$, $\dot{q}^{[L-1/M]}(z) = 1$, again for fixed L but all M, where the dot denotes the leading coefficient. Note that this normalization differs from that of (3.5.11) and (5.4). The constant $\hat{\alpha} = \hat{\alpha}(L, M)$ in (5.20), (5.21) is determined by the requirement that the coefficient of z^L in $p^{[L-1/M]}(z)$ is zero, and therefore $\hat{\alpha}$ is given by

$$0 = -\hat{\alpha}\dot{p}^{[L/M-1]} + \dot{p}^{[L-1/M-1]}. \qquad (5.24)$$

To obtain $\alpha = \alpha(L, M)$, the error function is introduced generally by

$$f(z)q^{[L/M]}(z) - p^{[L/M]}(z) = z^{L+M+1}e^{[L/M]}(z), \qquad (5.25)$$

and $e^{[L/M]}(z)$ is determined by its Maclaurin-series coefficients:

$$e_i^{[L/M]} = \sum_{j=0}^{M} c_{L+M+i-j+1} q_j^{[L/M]}, \quad i = 0, 1, 2, \ldots. \qquad (5.26)$$

To ensure that the $[L/M]$ approximant has degree of precision $L + M$, α in (5.22) must satisfy

$$e_0^{[L/M-1]} - \alpha e_0^{[L-1/M-1]} = 0. \qquad (5.27)$$

The constants α, $\hat{\alpha}$ determined by (5.24), (5.27) are called the Schur–Szegö parameters of the iteration (5.20)–(5.23).

Schur's algorithm consists of using (the coefficients of) the polynomials $q^{[L-1/M-1]}(z)$, $q^{[L/M-1]}(z)$, the Laurent polynomials $p^{[L-1/M-1]}(z)$, $p^{[L/M-1]}(z)$, and the residuals $e_0^{[L-1/M-1]}$ and $e_0^{[L/M-1]}$ given by (5.26) first to calculate α, $\hat{\alpha}$ in (5.24), (5.27) and then the next set of polynomials as expressed by (5.20)–(5.23).

Partly because the formula (5.17) does not require evaluation of the numerator Laurent polynomials, the previous algorithm can be made much more concise. We suppose that (the coefficients of) $q^{[L/M-1]}(z)$ and $q^{[L-1/M-1]}(z)$ have been computed, and that we are to compute $q^{[L-1/M]}(z)$ and $q^{[L/M]}(z)$ using (5.20) and (5.21). This calculation requires the values of α, $\hat{\alpha}$. In order to calculate $\hat{\alpha}$, $\dot{p}^{[L-1/M-1]}$ is calculated from

$$\dot{p}^{[L-1/M-1]} = \sum_{j=0}^{M-1} c_{L-1-j} q_j^{[L-1/M-1]}, \qquad (5.28)$$

$\dot{p}^{[L/M-1]}$ will have been precomputed from (5.32), and then

$$\hat{\alpha} = \dot{p}^{[L-1/M-1]}/\dot{p}^{[L/M-1]}. \tag{5.29}$$

In order to calculate α, $e_0^{[L/M-1]}$ is calculated from

$$e_0^{[L/M-1]} = \sum_{j=0}^{M-1} c_{L+M-j} q_j^{[L/M-1]}, \tag{5.30}$$

$e_0^{[L-1/M-1]}$ will have been precomputed from (5.33), and then

$$\alpha = e_0^{[L/M-1]}/e_0^{[L-1/M-1]}. \tag{5.31}$$

It is then straightforward to calculate $q^{[L-1/M]}(z)$ and $q^{[L/M]}(z)$ from (5.20) and (5.22). Before we continue the iteration, the quantities $\dot{p}^{[L/M]}$ and $e_0^{[L-1/M]}$ are required. From (5.23), and then from (5.24),

$$\dot{p}^{[L/M]} = \dot{p}^{[L/M-1]} - \alpha \dot{p}^{[L-1/M-1]}$$
$$= \dot{p}^{[L/M-1]}(1 - \alpha\hat{\alpha}). \tag{5.32}$$

From (5.20), and then from (5.27),

$$e_0^{[L-1/M]} = -\hat{\alpha} e_0^{[L/M-1]} + e_0^{[L-1/M-1]}$$
$$= e_0^{[L-1/M-1]}(1 - \alpha\hat{\alpha}). \tag{5.33}$$

Even the calculation (5.33) is unnecessary. Because $e_0^{[L-1/0]} = \dot{p}^{[L/0]} = c_L$, it follows from (5.32) and (5.33) that

$$e_0^{[L-1/M]} = \dot{p}^{[L/M]},$$

as is also evident from (1.1.9), (1.1.11).

Example. The calculation of $q^{[0/2]}(z)$ and $q^{[-1/2]}(z)$ using the modified Levinson algorithm.

Initialization

$$q^{[-1/0]}(z) = 1, \quad q^{[0/0]}(z) = 1,$$
$$\dot{p}^{[-1/0]} = c_{-1}, \quad \dot{p}^{[0/0]} = c_0,$$
$$e_0^{[-1/0]} = c_0, \quad e_0^{[0/0]} = c_1.$$

Stage $M = 1$.
From (5.29) and (5.31),

$$\hat{\alpha} = c_{-1}/c_0, \quad \alpha = c_1/c_0.$$

From (5.20), (5.22),
$$q^{[-1/1]}(z) = -(c_{-1}/c_0) + z, \quad q^{[0/1]}(z) = 1 - (c_1/c_0)z.$$
Use (5.32), (5.33) to get
$$\dot{p}^{[0/1]} = e_0^{[-1/1]} = c_0(1 - c_1 c_{-1}/c_0^2).$$

Stage M = 2.
From (5.28), (5.30),
$$\dot{p}^{[-1/1]} = c_{-2} - c_{-1}^2/c_0, \quad e_0^{[0/1]} = c_2 - c_1^2/c_0.$$
From (5.29) and (5.31),
$$\hat{\alpha} = \frac{c_0 c_{-2} - c_{-1}^2}{c_0^2 - c_1 c_{-1}}, \quad \alpha = \frac{c_2 c_0 - c_1^2}{c_0^2 - c_1 c_{-1}}.$$
From (5.20),
$$q^{[-1/2]}(z) = \frac{c_{-1}^2 - c_0 c_{-2}}{c_0^2 - c_1 c_{-1}}\left(1 - \frac{c_1 z}{c_0}\right) + z\left(z - \frac{c_{-1}}{c_0}\right).$$
From (5.22),
$$q^{[0/2]}(z) = 1 - \frac{c_1 z}{c_0} - z\frac{c_2 c_0 - c_1^2}{c_0^2 - c_1 c_{-1}}\left(z - \frac{c_{-1}}{c_0}\right).$$

If we allow for the different normalization, these results can be checked against (5.4).

The algorithm described based on using (5.29), (5.31), (5.20), (5.22), and (5.32) is Levinson's algorithm, with the modifications due to Trench [1964] and Zohar [1974]. This algorithm and Schur's algorithm are expressed more algorithmically by Bultheel [1987, Chapter 3]; see also Bultheel [1980c].

Schur's algorithm and the modified Levinson algorithm are entirely satisfactory under conditions such as that of (5.12) where positivity guarantees normality of the Laurent–Padé table. In other cases, it may be that some approximants in the bicolumnar sequence of types [0/m], [−1/m] are nonexistent or close to nonnormal. The occurrence of nearly degenerate approximants in a sequence is a well-known source of numerical error in the iterative calculation of subsequent members of the sequence. The use of nearly degenerate approximants can be avoided by using a technique similar to that of Section 3.6, where Hankel problems were solved using a progression along a paradiagonal sequence of the Padé table. The following theorem forms the basis of a stable numerical procedure for Toeplitz problems. It uses a column sequence in the Laurent–Padé table, with a fixed value of L; to solve the problem (5.16), take $L = 0$.

Theorem 7.5.2 (A progression theorem) [Gutknecht, 1993]. *Polynomials $a(z), b(z), u(z), v(z)$ can be found with*

$$\deg\{a(z)\} \leq k - 1, \quad \deg\{b(z)\} \leq k, \quad \deg\{u(z)\} \leq k - 2,$$
$$\deg\{v(z)\} \leq k - 1 \qquad (5.34)$$

and such that

$$\begin{bmatrix} p^{[L/M+k]}(z) & p^{[L/M+k-1]}(z) \\ q^{[L/M+k]}(z) & q^{[L/M+k-1]}(z) \end{bmatrix} = \begin{bmatrix} p^{[L/M]}(z) & p^{[L/M-1]}(z) \\ q^{[L/M]}(z) & q^{[L/M-1]}(z) \end{bmatrix}$$
$$\times \begin{bmatrix} b(z) & v(z) \\ za(z) & zu(z) \end{bmatrix}, \qquad (5.35)$$

where $p^{[l/m]}(z), q^{[l/m]}(z)$ denote numerator Laurent polynomial and denominator polynomial of the Laurent–Padé forms of type $[l/m]$, without any particular normalization convention.

If $p^{[L/M]}(z)/q^{[L/M]}(z)$ is a nondegenerate Laurent–Padé approximant (i.e., $q^{[L/M]}(0) \neq 0$) and $b(0) \neq 0$, then $p^{[L/M+k]}(z)/q^{[L/M+k]}(z)$ is a nondegenerate Laurent–Padé approximant.

The coefficients of the polynomials $a(z), b(z)$ are determined by

$$\begin{bmatrix} p_L^{[L/M-1]} & \cdots & p_{L-k+1}^{[L/M-1]} & p_L^{[L/M]} & \cdots & p_{L-k+1}^{[L/M]} \\ & \ddots & \vdots & & \ddots & \vdots \\ 0 & & p_L^{[L/M-1]} & 0 & & p_L^{[L/M]} \\ \hline e_0^{[L/M-1]} & & 0 & 0 & & 0 \\ \vdots & \ddots & & e_0^{[L/M]} & 0 & \\ & & & \vdots & \ddots & \ddots \\ e_{k-1}^{[L/M-1]} & \cdots & e_0^{[L/M-1]} & e_{k-2}^{[L/M]} & \cdots & e_0^{[L/M]} & 0 \end{bmatrix} \begin{bmatrix} a_0 \\ \vdots \\ a_{k-1} \\ \hline b_1 \\ \vdots \\ b_k \end{bmatrix}$$

$$= -b_0 \begin{bmatrix} 0 \\ \vdots \\ 0 \\ \hline e_0^{[L/M]} \\ \vdots \\ e_{k-1}^{[L/M]} \end{bmatrix} \qquad (5.36)$$

7.5 Laurent–Padé approximation and Toeplitz systems

and the coefficients of $u(z)$, $v(z)$ by the homogeneous equations

$$\begin{bmatrix} p_L^{[L/M-1]} & \cdots & p_{L-k+2}^{[L/M-1]} & 0 & p_L^{[L/M]} & \cdots & p_{L-k+2}^{[L/M]} \\ & \ddots & \vdots & & & \ddots & \vdots \\ 0 & & p_L^{[L/M-1]} & 0 & 0 & & p_L^{[L/M]} \\ \hdashline e_0^{[L/M-1]} & & 0 & e_0^{[L/M]} & 0 & & 0 \\ \vdots & \ddots & & \vdots & & \ddots & \\ e_{k-2}^{[L/M-1]} & \cdots & e_0^{[L/M-1]} & e_{k-2}^{[L/M]} & \cdots & e_0^{[L/M]} & 0 \end{bmatrix} \begin{bmatrix} u_0 \\ \vdots \\ u_{k-2} \\ \hdashline v_0 \\ \vdots \\ v_{k-1} \end{bmatrix}$$

$$= \mathbf{0}, \quad (5.37)$$

where the coefficients $e_i^{[L/M]}$ are defined in (5.26).

Proof. Regard (5.35) as defining the quantities on its left-hand side. From the equations

$$\begin{aligned} q^{[L/M+k]}(z) &= q^{[L/M]}(z)b(z) + zq^{[L/M-1]}(z)a(z), \\ q^{[L/M+k-1]}(z) &= q^{[L/M]}(z)v(z) + zq^{[L/M-1]}(z)u(z), \end{aligned} \quad (5.38)$$

it is readily verified that

$$\deg\{q^{[L/M+k]}(z)\} \leq M + k, \quad \deg\{q^{[L/M+k-1]}(z)\} \leq M + k - 1. \quad (5.39)$$

A similar result does not apply to the degrees of the numerator Laurent polynomials: the first k equations of the set (5.36), and the first $k-1$ equations of the set (5.37), respectively, ensure that

$$\deg\{p^{[L/M+k]}(z)\} \leq L, \quad \deg\{p^{[L/M+k-1]}(z)\} \leq L. \quad (5.40)$$

From (5.35), it follows that

$$f(z)q^{[L/M+k]}(z) - p^{[L/M+k]}(z) = z^{L+M+1}[e^{[L/M]}(z)b(z) + e^{[L/M-1]}(z)a(z)].$$

The last k equations of the set (5.36) ensure that

$$[e^{[L/M]}(z)b(z) + e^{[L/M-1]}(z)a(z)]_0^{k-1} = 0 \quad (5.41)$$

and then

$$f(z)q^{[L/M+k]}(z) - p^{[L/M+k]}(z) = O_+(z^{L+M+k+1}). \quad (5.42)$$

The results (5.39), (5.40), and (5.42) establish $p^{[L/M+k]}(z)/q^{[L/M+k]}(z)$ as a Laurent–Padé form of type $[L/M+k]$. The proof for $p^{[L/M+k-1]}(z)/q^{[L/M+k-1]}(z)$ is similar. Finally, if (5.36) has a solution with $b_0 \neq 0$, and

$q^{[L/M]}(0) \neq 0$, then (5.38) implies that

$$q^{[L/M+k]}(0) \neq 0.$$

The important conclusion of this theorem is that existence of both the Laurent–Padé approximant of type $[L/M]$ and nonsingularity of the linear system (5.36) is sufficient for the construction of a nonsingular Laurent–Padé approximant of type $[L/M + k]$, using the progression specified by (5.35).

The existence of a Laurent–Padé approximant $a(z)/b(z)$ of type $[0/M]$ is precisely what is required for the solution of the linear Toeplitz system (5.16), using the representation (5.17). But (5.17) also requires the values of $q_0, \ldots, q_{M-1}$ which are the coefficients of the denominator polynomial of the Laurent–Padé approximant of type $[-1/M - 1]$. This denominator $Q(z)$ is readily obtained using the Frobenius

$$\begin{pmatrix} * & * \\ & * \end{pmatrix}$$

identity (5.22), up to an irrelevant normalizing constant. The corresponding value of r_0 follows from (5.18), and the representation (5.17) is thereby completed. This procedure constitutes a stable recursive procedure for calculating the inverses of Toeplitz matrices and the solutions of Toeplitz systems.

It is worth noting that the rational functions $a(z)/b(z)$ and $u(z)/v(z)$ defined and used in Theorem 7.5.2 are in fact two-point Padé approximants. For example, from (5.41), we see that $a(z)/b(z)$ matches $-e^{[L/M]}(z)/e^{[L/M-1]}(z)$ at the origin, and the condition (5.40) ensures that $a(z)/b(z)$ matches $-z^{-1}p^{[L/M]}(z)/p^{[L/M-1]}(z)$ at infinity.

An alternative but similar procedure is based on a related result.

Theorem 7.5.3 (A progression theorem) [Gutknecht, 1993]. *Polynomials $a(z)$, $b(z)$, $u(z)$, $v(z)$ can be found with $\deg\{a(z)\} \leq k - 1$, $\deg\{b(z)\} \leq k$, $\deg\{u(z)\} \leq k - 2$, $\deg\{v(z)\} \leq k - 1$ and such that*

$$\begin{bmatrix} p^{[L/M+k]}(z) & p^{[L-1/M+k-1]}(z) \\ q^{[L/M+k]}(z) & q^{[L-1/M+k-1]}(z) \end{bmatrix}$$
$$= \begin{bmatrix} p^{[L/M]}(z) & p^{[L-1/M-1]}(z) \\ q^{[L/M]}(z) & q^{[L-1/M-1]}(z) \end{bmatrix} \begin{bmatrix} b(z) & v(z) \\ z^2 a(z) & z^2 u(z) \end{bmatrix}. \quad (5.43)$$

If $q^{[L/M]}(0) \neq 0$ and $b(0) \neq 0$, then $p^{[L/M+k]}(z)/q^{[L/M+k]}(z)$ is a nondegenerate Laurent–Padé approximant.

The coefficients of the polynomials $a(z)$, $b(z)$ are determined by

7.5 Laurent–Padé approximation and Toeplitz systems

$$\begin{bmatrix} p_{L-1}^{[L-1/M-1]} & \cdots & p_{L-k}^{[L-1/M-1]} & \vline & p_L^{[L/M]} & \cdots & p_{L-k-1}^{[L/M]} \\ & \ddots & \vdots & \vline & & \ddots & \vdots \\ 0 & & p_{L-1}^{[L-1/M-1]} & \vline & 0 & & p_L^{[L/M]} \\ \hline e_0^{[L-1/M-1]} & & 0 & \vline & 0 & & 0 \\ & \ddots & & \vline & e_0^{[L/M]} & 0 & \\ \vdots & & & \vline & \vdots & \ddots & \ddots \\ e_{k-1}^{[L-1/M-1]} & \cdots & e_0^{[L-1/M-1]} & \vline & e_{k-2}^{[L/M]} & \cdots & e_0^{[L/M]} & 0 \end{bmatrix}$$

$$\times \begin{bmatrix} a_0 \\ \vdots \\ a_{k-1} \\ \hline b_1 \\ \vdots \\ b_k \end{bmatrix} = -b_0 \begin{bmatrix} 0 \\ \vdots \\ 0 \\ \hline e_0^{[L/M]} \\ \vdots \\ e_{k-1}^{[L/M]} \end{bmatrix} \qquad (5.44)$$

and those of $u(z)$, $v(z)$ by the homogeneous equations

$$\begin{bmatrix} p_{L-2}^{[L-1/M-1]} & \cdots & & p_{L-k}^{[L-1/M-1]} & \vline & p_L^{[L/M]} & \cdots & & p_{L-k+1}^{[L/M]} \\ p_{L-1}^{[L-1/M-1]} & & & \vdots & \vline & & & \ddots & \vdots \\ 0 & & \ddots & p_{L-1}^{[L-1/M-1]} & \vline & 0 & & & p_L^{[L/M]} \\ \hline e_0^{[L-1/M-1]} & 0 & & 0 & \vline & e_0^{[L/M]} & 0 & 0 & 0 \\ \vdots & \ddots & \ddots & & \vline & \vdots & \ddots & \ddots & \ddots \\ e_{k-3}^{[L-1/M-1]} & \cdots & e_0^{[L-1/M-1]} & 0 & \vline & e_{k-3}^{[L/M]} & \cdots & e_0^{[L/M]} & 0 & 0 \end{bmatrix}$$

$$\times \begin{bmatrix} u_0 \\ \vdots \\ u_{k-2} \\ \hline v_0 \\ \vdots \\ v_{k-1} \end{bmatrix} = 0.$$

Proof. The proof is similar to that of Theorem 7.5.2.

The progression described by this theorem is advantageous for use in calculating the representation (5.17) because it gives the necessary parameters $b_0, \ldots, b_M, q_0, \ldots, q_{M-1}$ directly.

The important feature of the computational procedures arising from Gutknecht's theorems is their capacity to avoid the use of nearly degenerate approximants and the associated ill-conditioning of the Levinson-type algorithms at intermediate steps in applications without positivity. For $k = 1, 2, \ldots$, equations (5.36), (5.44) should be solved as a homogeneous system, e.g., by the QR algorithm, until a solution is found for which b_0 is (comparatively) not close to zero [Stewart, 1993]. In this way, the construction process progresses by k steps. Procedures of this kind are also called look-ahead algorithms [Freund, Golub, and Nachtigal, 1992].

7.6 Multivariable approximants

A natural problem is the generalization of Padé approximants to more than one variable. It turns out that the problems associated with many variables have the same kind of solution as the two-variable problems, and we confine our attention to this case for ease of exposition. First we consider various schemes for the formation of two-variable rational approximants, and then we consider more general classes of approximating functions. Let us assume that we are given the coefficients of the series expansion

$$f(x, y) = \sum_{i=0}^{\infty} \sum_{j=0}^{\infty} c_{ij} x^i y^j. \tag{6.1}$$

The problem consists of defining lattice spaces $\mathcal{N}$ and $\mathcal{D}$ and polynomials

$$A(x, y) = \sum_{i,j \in \mathcal{N}} a_{ij} x^i y^j \tag{6.2}$$

and

$$B(x, y) = \sum_{i,j \in \mathcal{D}} b_{ij} x^i y^j, \tag{6.3}$$

so that

$$f(x, y) = \frac{A(x, y)}{B(x, y)} + \sum_{i=0}^{\infty} \sum_{j=0}^{\infty} e_{ij} x^i y^j, \tag{6.4}$$

where as many coefficients e_{ij} as possible are zero. We have taken the numerator and denominator coefficients to lie in lattice spaces $\mathcal{N}$ and $\mathcal{D}$, and we require that $e_{ij} = 0$ for $i, j \in \mathcal{E}$, the equality lattice space.

Taking $b_{00} = 1$ as part of the definition, this scheme is normally determinate if

$$\dim(\mathcal{E}) = \dim(\mathcal{N}) + \dim(\mathcal{D}) - 1.$$

This analysis provides the foundation for a variety of approximation schemes. These schemes are only useful if their properties are known, and the most systematic developments are known as the Canterbury approximants (or generalized Chisholm approximants). These approximants have many properties; they satisfy accuracy-through-order conditions and reduce to Padé approximants if either x or y is zero. The original and simplest of the Canterbury approximants is the Chisholm approximant [Chisholm, 1973], defined by writing (6.2) and (6.3) as

$$A^{[L/L]}(x, y) = \sum_{i=0}^{L} \sum_{j=0}^{L} a_{ij} x^i y^j, \tag{6.5}$$

$$B^{[L/L]}(x, y) = \sum_{i=0}^{L} \sum_{j=0}^{L} b_{ij} x^i y^j, \quad b_{00} = 1. \tag{6.6}$$

The lattice spaces $\mathcal{N}$ and $\mathcal{D}$ corresponding to (6.5) and (6.6) are shown in Figure 7.6.1. Then one has equality at order $x^\alpha y^\beta$ if

$$\sum_{i=0}^{\alpha} \sum_{j=0}^{\beta} b_{ij} c_{\alpha-i, \beta-j} = a_{\alpha\beta} \quad \text{for } (\alpha, \beta) \in \mathcal{N} \tag{6.7}$$

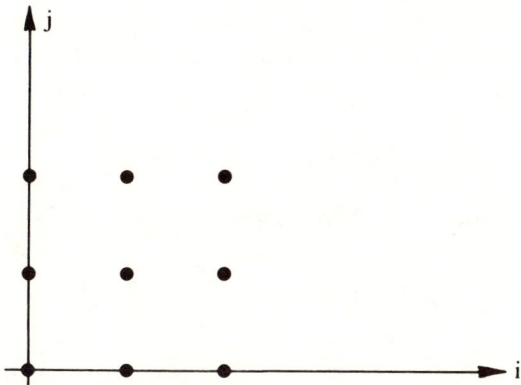

Figure 7.6.1. The lattice space $\mathcal{D}$ required for the [2/2] Chisholm approximant. In this case, the lattice spaces $\mathcal{N}$ and $\mathcal{D}$ are identical.

and

$$\sum_{i=0}^{\min(\alpha,L)} \sum_{j=0}^{\min(\beta,L)} b_{ij} c_{\alpha-i,\beta-j} = 0 \quad \text{for } (\alpha, \beta) \in \mathscr{E}, (\alpha, \beta) \notin \mathscr{N}. \quad (6.8)$$

The numerator coefficients are determined by (6.7) once the denominator coefficients are determined. The b_{ij} are determined by (6.8), and the lattice space $\mathscr{E}$ is most simply described by an example. In Figure 7.6.2, we show the space for the [2/2] Chisholm approximant.

To determine b_{10} and b_{20}, and obtain the [2/2] Padé approximant when $y = 0$, accuracy to orders x^3 and x^4 is required, and no higher. In fact, six of the eight equations for the b_{ij} are obtained by requiring accuracy through orders x^3, x^4, x^3y, y^3, y^4, and y^3x, as indicated by * in Figure 7.6.2. Two further equations are needed. One is obtained by writing down the equations (6.8) for orders x^4y and xy^4 and adding them, and the other by doing the same for orders x^3y^2 and x^2y^3. These two equations are called symmetrized equations. Thus Chisholm approximants always satisfy accuracy-through-order conditions for orders up to $x^{2L-\alpha}y^\alpha$, $\alpha = 0, 1, \ldots, 2L$. For symmetric functions, the symmetrizing process becomes a formality, and the approximants are accurate through orders x^{2L}, y^{2L}, and $x^{2L+1-\alpha}y^\alpha$, $\alpha = 1, 2, \ldots, 2L$. Provided that the necessary approximants exist, the scheme has the following properties [Chisholm, 1973; Common and Graves-Morris, 1974]:

Chisholm approximants reduce to diagonal Padé approximants if either variable is zero.

They satisfy restricted homographic invariance: let

$$x = \frac{Au}{1 + Bu}, \quad y = \frac{Av}{1 + Cv}, \quad A \neq 0.$$

If

$$f(x, y) = g(u, v),$$

then

$$[L/L]_f(x, y) = [L/L]_g(u, v).$$

They satisfy duality: if $f(0, 0) \neq 0$ and

$$g(x, y) = \frac{1}{f(x, y)},$$

then

$$[L/L]_g(x, y) = \frac{1}{[L/L]_f(x, y)}.$$

7.6 Multivariable approximants

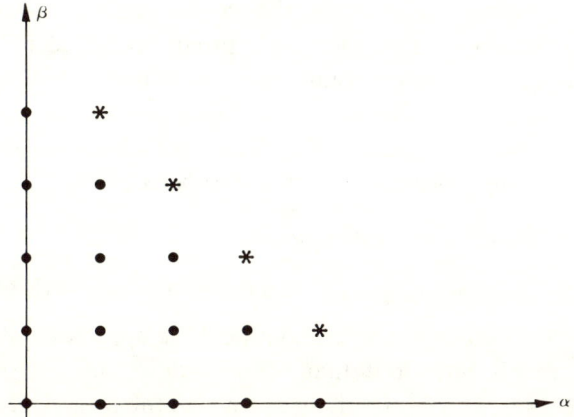

Figure 7.6.2. The lattice space $\mathscr{E}$ for a [2/2] Chisholm approximant. A star denotes a lattice point corresponding to an equation to be symmetrized.

They preserve unitarity: if $f(x, y)f^*(x, y) = 1$, then $[L/L]_f(x, y) \times [L/L]_f^*(x, y) = 1$.

They satisfy the factorization rule: if $f(x, y)$ is a product function, the approximant factorizes to a product of Padé approximants. Expressed in formulas, this rule states that if $f(x, y) = g(x)h(y)$, then $[L/L]_f(x, y) = [L/L]_g(x)[L/L]_h(y)$.

Formation of Chisholm approximants commutes with bilinear transformations of the function: let

$$g(x, y) = \frac{A + Bf(x, y)}{C + Df(x, y)}, \quad C + Df(0, 0) \neq 0.$$

Then

$$[L/L]_g(x, y) = \frac{A + B[L/L]_f(x, y)}{C + D[L/L]_f(x, y)}.$$

The proofs of all these properties are based on the accuracy-through-order principle. In particular, the homographic-invariance and factorization properties seem to be essential ingredients of a useful scheme.

The general system of Hughes Jones approximants [Hughes Jones, 1976] is defined by

$$A^{[L/M]}(x, y) = \sum_{i=0}^{L_1} \sum_{j=0}^{L_2} a_{ij} x^i y^j,$$

$$B^{[L/M]}(x, y) = \sum_{i=0}^{M_1} \sum_{j=0}^{M_2} b_{ij} x^i y^j, \quad b_{00} = 1.$$

The equality lattice space $\mathscr{E}$ is shown in Figures 7.6.3, 7.6.4. If $\min(M_1, M_2) \leq \min(L_1, L_2)$, the simple situation of Figure 7.6.3 applies; otherwise, we have the more complicated situation of Figure 7.6.4. The logic of these figures can be understood in terms of the prong method [Hughes Jones and Makinson, 1974]. The coefficients b_{ij} are calculated sequentially in prongs; prongs are defined to be vectors

$$\mathbf{b}^{(0)} = (b_{10}, b_{20}, \ldots, b_{m_10}; b_{01}, b_{02}, \ldots, b_{0m_2}),$$

$$\mathbf{b}^{(1)} = (b_{21}, b_{31}, \ldots, b_{m_11}; b_{12}, b_{13}, \ldots, b_{1m_2}; b_{11}), \text{ etc.}$$

Calculating $\mathbf{b}^{(0)}$ is equivalent to calculating Padé approximants for $f(x, 0)$ and $f(0, y)$, and it turns out that the evaluation of $\mathbf{b}^{(j)}$ only requires values of $\mathbf{b}^{(i)}$, $i = 0, 1, \ldots, j - 1$. In summary, the prong method reduces the calculation of b_{ij} to linear algebra with a lower triangular block coefficient matrix.

To define the Canterbury approximants completely, there is the possibility of weighting the symmetrizing equations before adding them. Naturally, for symmetric functions with $c_{ij} = c_{ji}$, the weights are equal. There are two different weighting schemes which treat the variables symmetrically, and each has its own advantages. One weighting scheme [Chisholm and Hughes Jones, 1975] gives full homographic invariance under the changes of variable

$$x = \frac{Au}{1 + Bu}, \quad y = \frac{Cu}{1 + Du}. \tag{6.9}$$

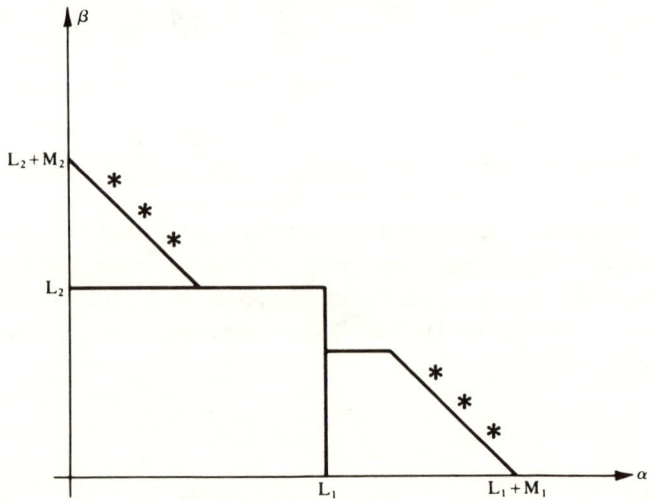

Figure 7.6.3. Schematic drawing of the lattice space $\mathscr{E}$ for a Canterbury approximant with $M_2 < M_1 < \min(L_1, L_2)$ and three symmetrized equations.

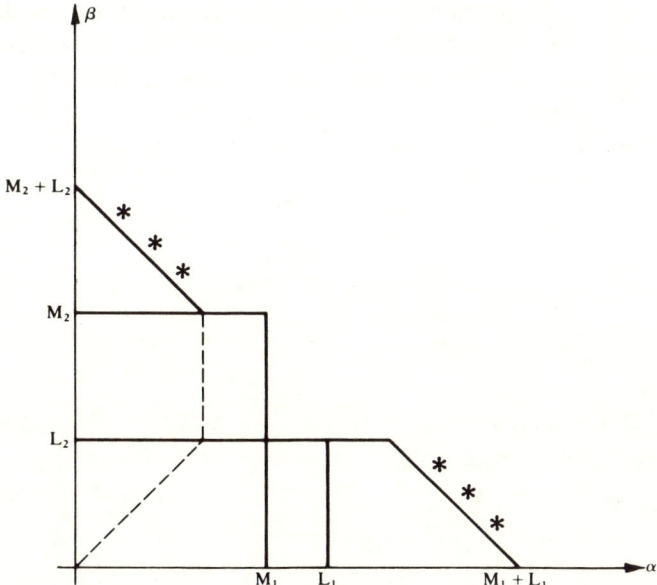

Figure 7.6.4. Schematic drawing of the lattice space $\mathscr{E}$ for a Canterbury approximant with $M_1 < L_1$ and $L_2 < \min(M_1, M_2)$ and three symmetrized equations.

The other scheme [Graves-Morris and Roberts, 1975] guards against accidental degeneracy by maximizing the numerical stability of the system, and has homographic invariance provided $|A| = |C|$ in (6.9). Fortunately, both schemes lead to similar results in trials.

As stated previously, the preceding ideas may be generalized systematically by geometrical methods and by the prong method to multivariable approximants at the price of algebraic complexity only [Chisholm and McEwan, 1974; Hughes Jones, 1976].

Some convergence theorems of the 'de Montessus' type have been proven for this class of approximants by Cuyt [1990, 1992]. In the case of single-variable Padé approximants, a function $f(z)$ meromorphic in a disk $|z| < R$ can be characterized by the requirement that there is a polynomial $B(z)$ (6.2.4) which vanishes at each pole z_i of $f(z)$ in the disk, and nowhere else in the disk, in such a way that the value of $\lim_{z \to z_i} B(z)f(z) \neq 0, \infty$ at every pole in the disk. Then de Montessus Theorem 6.2.1 implies that if the degree of the denominator is M and equals the degree of $B(z)$, and if the degree of the numerator goes to infinity, then the denominator converges to $B(z)$ and the Padé numerator converges to $B(z)f(z)$ in the disk. In the multivariate case, the situation is more complex. We expect that the zeros will lie on $(2l - 2)$-dimensional surfaces in the case of l complex variables. We just discuss the case of two variables, as we believe it manifests the general procedure for a multivariable case. We define a meromorphic function of two variables in a similar manner to that above. We say that a function $f(x, y)$ is meromorphic in the polydisk $B = \{(x, y), |x| < R_1, |y| < R_2\}$, if there exists a polynomial

$$R_\mathcal{M}(x, y) = \sum_{(d,e) \in \mathcal{M} \subset \mathcal{I}^2} r_{de} x^d y^e = \sum_{i=0}^{m} r_{d_i e_i} x^{d_i} y^{e_i}$$

such that $R_\mathcal{M}(x, y) f(x, y)$ is analytic in B, where $\mathcal{I}$ is the set of all nonnegative integers.

Theorem 7.6.1 [Cuyt, 1990]. *Let $f(x, y)$ be a function which is meromorphic in the polydisk B. Further assume that $R_\mathcal{M}(0, 0) \neq 0$ so that necessarily $(0, 0) \in \mathcal{M}$. Let there exist m zeros $(x_h, y_h) \in B$ of $R_\mathcal{M}(x, y)$ satisfying*

$$\lim_{(x,y) \to (x_h, y_h)} f(x, y) R_\mathcal{M}(x, y) \neq 0, \quad h = 1, \ldots, m,$$

and

$$\begin{vmatrix} x_1^{d_1} y_1^{e_1} & \cdots & x_1^{d_m} y_1^{e_m} \\ \vdots & \ddots & \vdots \\ x_m^{d_1} y_m^{e_1} & \cdots & x_m^{d_m} y_m^{e_m} \end{vmatrix} \neq 0.$$

The index sets $\mathcal{N}$, $\mathcal{D}$, and $\mathcal{E}$ are specified in (6.1)–(6.3) and we require that $\mathcal{D} = \mathcal{M}$ so the degree of the multivariate Padé denominator matches that of $R_\mathcal{M}(x, y)$ exactly. Let

$$[\mathcal{N}/\mathcal{D}]_\mathcal{E}(x, y) = \frac{A^{[\mathcal{N}/\mathcal{D}]}(x, y)}{B^{[\mathcal{N}/\mathcal{D}]}(x, y)}$$

denote the multivariable Padé approximant of type $[\mathcal{N}/\mathcal{D}]$ formed from equation set $\mathcal{E}$. Then the $[\mathcal{N}/\mathcal{D}]_\mathcal{E}$ converge to $f(x, y)$ uniformly on compact subsets of $B \cap \{R(x, y) \neq 0\}$, and the denominator B of (6.3) converges to $R_\mathcal{M}(x, y)$, provided

$r_\tau = \max[\lambda, (i, \lambda - i) \in \mathcal{E}$ for each $0 \leq i \leq \lambda] \to \infty$,

$r_T = \max[\lambda, x^i y^{\lambda - i} \in \{A^{[\mathcal{N}/\mathcal{D}]}(x, y) B^{[\mathcal{N}/\mathcal{D}]}(x, y)\}$, for each $0 \leq i \leq \lambda] \to \infty$

as the sets $\mathcal{N}$ and $\mathcal{E}$ grow along a column in the Padé table.

The proof uses the extension to two variables of Hermite's polynomial interpolation formula (6.3.1) and the corresponding error formula (6.3.2). The error formula generalizes to

$$f(x, y) - [N/\mathcal{D}]_{\mathcal{E}} = \left(\frac{1}{2\pi i}\right)^2 \frac{1}{B(x, y)R_{\mathcal{M}}(x, y)} \sum_{\mathcal{I}^2 \setminus \mathcal{E}} x^i y^j$$

$$\times \int_{|t|=R_1} \int_{|u|=R_2} \frac{f(t, u)B(t, u)R_{\mathcal{M}}(t, u)}{t^{i+1} u^{j+1}} \, dt \, du,$$

where we will drop the superscript on $B^{[N/\mathcal{D}]}(x, y)$ when no confusion will occur. Since the numerator of the integrand is regular in B, the terms in the series are of the order of $H(x/R_1)^i(y/R_2)^j$, where H is some constant independent of i and j, and hence the above sum converges geometrically in B provided that we are not at a zero of $B(x, y)$ or $R_{\mathcal{M}}(x, y)$ as by the conditions in the theorem the order of contact r_τ tends to infinity. It is claimed that the remainder of the proof follows the method of Saff's proof of de Montessus's theorem as discussed in Section 6.3. It is here that the determinantal condition enters. Its purpose is to ensure that the polynomial $R_{\mathcal{M}}(x, y)$ is uniquely implied by $f(x, y)$ and the polydisk B. The construction of these interpolants is reviewed by Cuyt and Verdonk [1994]. A rather different theorem of the de Montessus type is given by Graves-Morris [1977]. It is based on some technical assumptions about the locations of the zeros of $B(x, 0)$ and $B(0, y)$, and establishes convergence in a polydisk rather smaller than B.

There are two other systems of N-variable rational approximants with distinctive merits. Both lead to higher-order polynomials in numerator and denominator than the accuracy-through-order criterion requires.

First, one may partition the series for $f(x, y)$ by defining [Hillion, 1977a; Watson, 1974]

$$F(x, y; \lambda) = \sum_{i=0}^{\infty} \lambda^i \sum_{k=0}^{i} x^k y^{i-k} c_{k, i-k}.$$

Then $f(x, y) = F(x, y; 1)$, and $[L/M]$ Padé approximants in λ to $F(x, y; \lambda)$ define rational approximants to $f(x, y)$ at $\lambda = 1$. These are very convenient approximants in the presence of many variables, and reduce to $[L/M]$ Padé approximants to $f(x, 0)$ and $f(0, y)$ on the axes. Furthermore, their values may be obtained by the ε-algorithm. They have the disadvantage of requiring a relatively specific set of coefficients for their formation.

Second, one may treat $f(x, y)$ as if it defines a moment problem [Alabiso and Butera, 1975]. Let us suppose

$$f(x, y) = \int \frac{\rho(u, t) \, du \, dt}{1 + ux + ty} = \sum_{m,n} \binom{m+n}{m} f_{m,n} x^m y^n, \qquad (6.10)$$

and that $|g\rangle \in \mathcal{H}$, where $\mathcal{H}$ is a Hilbert space with two commuting operators A, B and such that

$$f_{m,n} = \langle g|A^m B^n|g\rangle. \tag{6.11}$$

These equations (6.10), (6.11) are formally equivalent to the condition that

$$f(x, y) = \langle g|\frac{1}{1 + Ax + By}|g\rangle. \tag{6.12}$$

The vector

$$|\psi^{(N)}\rangle = \sum_{p=0}^{N}\sum_{q=0}^{p}\psi_{pq}^{(N)} A^{p-q} B^q|g\rangle \tag{6.13}$$

is an approximation to

$$|\psi\rangle = \frac{1}{1 + Ax + By}|g\rangle \tag{6.14}$$

provided

$$\langle h_{rs}|1 + Ax + By|\psi^{(N)}\rangle = \langle h_{rs}|g\rangle \tag{6.15}$$

for

$$\langle h_{rs}| = \langle g|A^{r-s} B^s$$

with

$$r = 0, 1, \ldots, N \quad \text{and} \quad s = 0, 1, \ldots, r.$$

Equation (6.15) provides $\frac{1}{2}N(N + 1)$ linear equations for $\psi_{pq}^{(N)}$, namely,

$$\sum_{p=0}^{N}\sum_{q=0}^{p}\psi_{pq}^{(N)}(f_{r+p,s+q} + xf_{r+p+1,s+q} + yf_{r+p,s+q+1}) = f_{r,s}.$$

This equation determines $|\psi^{(N)}\rangle$ from (6.13) and a rational approximant given by $\langle g|\psi^{(N)}\rangle$ following (6.12). The general result is clear from the example of $N = 1$, for which

$$\langle f|\psi^{(1)}\rangle$$

$$= (f_{00} f_{10} f_0)\begin{bmatrix} f_{00} - xf_{10} - yf_{01} & f_{10} - xf_{20} - yf_{11} & f_{01} - xf_{11} - yf_{02} \\ f_{10} - xf_{20} - yf_{11} & f_{20} - xf_{30} - yf_{21} & f_{11} - xf_{21} - yf_{12} \\ f_{01} - xf_{11} - yf_{02} & f_{11} - xf_{21} - yf_{12} & f_{02} - xf_{12} - yf_{03} \end{bmatrix}^{-1}$$

$$\times \begin{bmatrix} f_{00} \\ f_{10} \\ f_{01} \end{bmatrix}.$$

7.6 Multivariable approximants

This system of approximants converges for strict Stieltjes functions in two variables, but the accuracy-through-order principle and the property of reduction to Padé approximants on the axes are forfeited.

Probably the best of the multivariable approximants described in this section is the one which fits the context of the original problem most closely. For this reason we now consider alternative approximants, not necessarily rational, which may be formed from the coefficients c_{ij} and in some sense approximate

$$f(x, y) = \sum_{i=0}^{\infty} \sum_{j=0}^{\infty} c_{ij} x^i y^j. \tag{6.1}$$

In the context of critical phenomena, we seek a generalization of G^{3}J approximants (Section 7.3) and D-log Padé approximants (2.2.7) for functions having branch points as well as poles. As an example, f might be the specific heat of a substance, depending on temperature and pressure. In this case we set $x = T^{-1}$ and $y = p$ in (6.1), and investigate the properties of f near the critical point. It is conjectured [Pfeuty, Jasnow, and Fisher, 1974] that such functions might behave as

$$f(x, y) \approx (x_c - x)^{-\gamma} Z\left(\frac{y_c - y}{(x_c - x)^\phi}\right) \tag{6.16}$$

near the critical point, where γ and ϕ are exponents, and $Z(z)$ is an unknown scaling function. Fisher approximants [Fisher, 1977; Fisher and Kerr, 1977] are designed for problems of the kind just outlined. From the power-series coefficients given, one considers the truncated Maclaurin polynomial

$$f_T(x, y) = \sum_{i,j \in \mathfrak{S}} c_{ij} x^i y^j.$$

Its derivatives $\partial f_T(x, y)/\partial x$ and $\partial f_T(x, y)/\partial y$ follow immediately. In the absence of degeneracy, one may construct polynomials $P_\mathfrak{L}(x, y)$, $Q_\mathfrak{M}(x, y)$, and $R_\mathfrak{N}(x, y)$ which satisfy

$$P_\mathfrak{L}(x, y) f_T(x, y) = Q_\mathfrak{M}(x, y) \frac{\partial f_T(x, y)}{\partial x} + R_\mathfrak{N}(x, y) \frac{\partial f_T(x, y)}{\partial y}$$

$$+ \text{ high-order terms.} \tag{6.17}$$

The lattice space $\mathfrak{S}$ contains the data coefficients, the lattice spaces $\mathfrak{L}$, $\mathfrak{M}$, $\mathfrak{N}$ define the orders of the polynomials $P_\mathfrak{L}(x, y)$, $Q_\mathfrak{M}(x, y)$, and $R_\mathfrak{N}(x, y)$, and the scheme is normally determinate if

$$\dim(\mathfrak{S}) = \dim(\mathfrak{L}) + \dim(\mathfrak{M}) + \dim(\mathfrak{N}) - 1. \tag{6.18}$$

This follows because (6.17) defines a homogeneous linear system of equations for the coefficients of $P_\mathfrak{L}(x, y)$, $Q_\mathfrak{M}(x, y)$, and $R_\mathfrak{N}(x, y)$. The precise nature of the lattice spaces $\mathfrak{L}$, $\mathfrak{M}$, and $\mathfrak{N}$ may be chosen to suit the problem at hand. Using these polynomials $P_\mathfrak{L}(x, y)$, $Q_\mathfrak{M}(x, y)$, and $R_\mathfrak{N}(x, y)$, the Fisher approximant of $f(x, y)$ is defined to be a solution of the partial differential equation

$$P_\mathfrak{L}(x, y)F(x, y) = Q_\mathfrak{M}(x, y)\frac{\partial F(x, y)}{\partial x} + R_\mathfrak{N}(x, y)\frac{\partial F(x, y)}{\partial y}. \quad (6.19)$$

In order that $F(x, y)$ may be uniquely defined, a boundary condition must be specified, e.g., the function $F(0, y)$ must be known. $F(0, y)$ might be estimated by a G^3J approximant if its exact value is not available.

Example. The given series coefficients are derived from

$$F(x, y) = \sum_{i=0}^{\infty}\sum_{j=0}^{\infty} c_{ij} x^i y^j = (1 - 2x + y)^{-\gamma} + \theta(1 - x + 2y)^{-2\gamma}, \quad (6.20)$$

where θ, γ are constants. With the choice of lattices, $\mathfrak{L}$, $\mathfrak{M}$, and $\mathfrak{N}$ implied by

$$P_\mathfrak{L}(x, y) = 6\gamma, \quad Q_\mathfrak{M}(x, y) = 5 - 7x + 2y, \quad R_\mathfrak{N}(x, y) = 4 - 2x - 2y, \quad (6.21)$$

$f(x, y)$ is an exact solution of (6.19). With the boundary condition

$$F(0, y) = (1 + y)^{-\gamma} + \theta(1 + 2y)^{-\gamma},$$

(6.20) is the unique solution of (6.19) and (6.21).

Any solution of (6.19) is normally necessarily singular at the point where $Q_\mathfrak{M}(x, y) = R_\mathfrak{N}(x, y) = 0$, and this would represent the critical point. For example, any solution of (6.19), (6.21) is normally singular at $x = 1$, $y = 1$. Solutions of (6.19) with the properties shown in Figure 7.6.5 require a more sophisticated demonstration. Fisher approximants are especially important because they can reproduce scaling properties [Fisher and Kerr, 1977].

Fisher approximants are linear in the sense that if F_1, F_2 are solutions of (6.19) for specified $P_\mathfrak{L}(x, y)$, $Q_\mathfrak{M}(x, y)$ and $R_\mathfrak{N}(x, y)$, then $F_1 + F_2$ is also a solution. A class of nonlinear two-variable approximants are the branching approximants which arise as generalizations of Shafer's quadratic approximants. Given the power series (6.1), or at least a sufficient subset $\mathfrak{S}$ of its coefficients, the coefficients of the power series of $[f(x, y)]^2$ may be constructed. With preassigned lattice spaces $\mathfrak{L}$, $\mathfrak{M}$, and

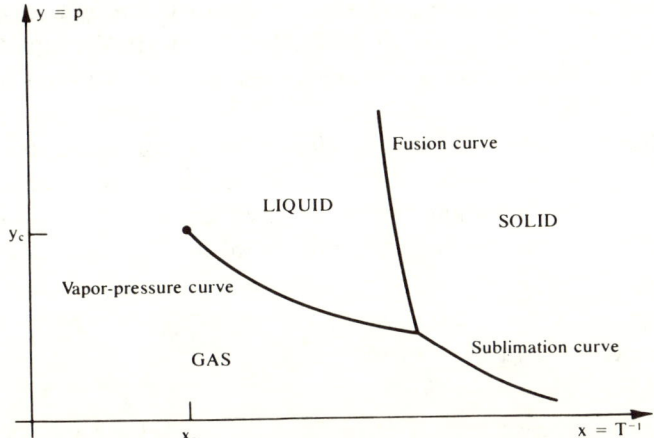

Figure 7.6.5. Singularity structure of the specific heat of a fluid in the (x, y)-plane as a function of $x = T^{-1}$ and $y = p$. The critical point (x_c, y_c) and the vapor-pressure line representing the liquid–gas transition are shown.

$\mathfrak{N}$, one may normally construct polynomials $Q_{\mathfrak{L}}(x, y)$, $R_{\mathfrak{M}}(x, y)$, and $S_{\mathfrak{N}}(x, y)$ such that

$$Q_{\mathfrak{L}}(x, y)[f(x, y)]^2 + 2R_{\mathfrak{M}}(x, y)f(x, y) + S_{\mathfrak{N}}(x, y) = 0 \quad \text{to high order}$$

provided (6.18) is satisfied. As in (3.11a), the quadratic approximant is defined [Chisholm, 1977b] by

$$\psi(x, y) = \frac{-R_{\mathfrak{M}}(x, y) + \sqrt{[R_{\mathfrak{M}}(x, y)]^2 - S_{\mathfrak{N}}(x, y)Q_{\mathfrak{L}}(x, y)}}{Q_{\mathfrak{L}}(x, y)}. \quad (6.22)$$

The correct branch of $\psi(x, y)$ to represent $f(x, y)$ must also be assigned; this branch is normally clear from the context. The quadratic approximant scheme may be generalized both to higher-order algebraic equations and to different functional equations. We reemphasize that the best N-variable approximant to use is the one which fits the context of the original problem most closely.

For further details about Canterbury approximants, we refer to Lutterodt [1974], to the review by Chisholm [1977a], and to Graves-Morris, Hughes Jones, and Makinson [1974], Chisholm and Graves-Morris [1975], Roberts, Griffiths, and Wood [1975], Chisholm and Roberts, [1976], Graves-Morris [1977], Roberts [1977], and Graves-Morris and Hughes Jones [1976]. These approximants have found application in multidimensional systems theory by Bose and Basu [1980]; see also [Bose, 1982].

For further details of other kinds of multivariable approximants, we refer to Levin [1976], Fisher [1977], and Baker and Moussa [1978]; to

Genz [1977]; to Karlsson and Wallin [1977]; to Barnsley and Robinson [1978]; and to Chisholm [1977b, 1978a, b], Short [1978], and Brezinski [1978b]. Fisher approximants are developed further by Fisher and Styer [1982], Styer [1983], Styer and Fisher [1983], and Liu and Fisher [1989].

A traditional approach to two-dimensional continued fractions is described by Skorobogatko [1983], but more recently interest has properly focused on continued fractions in which the values of the entries c_{ii} in (6.1) determine the pivotal elements in the sequential construction of a fraction such as

$$C(x, y) = c_{00} + \sum_{i=1}^{\infty} c_{i0} x^i + \sum_{j=1}^{\infty} c_{0j} y^j$$

$$+ \cfrac{xy}{c_{00}^{(1)} + \sum_{i=1}^{\infty} c_{i0}^{(1)} x^i + \sum_{j=1}^{\infty} c_{0j}^{(1)} y^j} + \cfrac{xy}{c_{00}^{(2)} + \sum_{i=1}^{\infty} c_{i0}^{(2)} x^i + \sum_{j=1}^{\infty} c_{0j}^{(2)} y^j + } \cdots$$

[Murphy and O'Donohoe, 1978; Kuchminskaya, 1978; Cuyt, 1983]. The development of this approach has been reviewed by Kuchminskaya and Siemasko [1987] and Cuyt and Verdonk [1988a]. Cuyt [1994] gives a Q.D. algorithm which enables the denominator polynomial $B^{[N/\mathscr{D}]}(x, y)$ of Theorem 7.6.1 to be constructed.

8

Multiseries approximants

In the bulk of this book we have been concerned with the approximation of functions described by a single power series, at one or more points. The case of several points yields a theory which is formally similar to that for one point in the Jacobi approach [Baker, 1975a, Chapter 8]. In this chapter we will consider cases where there are multiple series. In some cases, such as the matrix Padé and Hermite–Padé, there will be a finite number of series describing the function. In other cases, such as the multivariate approximants, there will be an infinite number of such series.

8.1 Simultaneous Padé approximants

The approximation of the ratios $f_1(z)/f_0(z)$, $f_2(z)/f_0(z)$, ..., $f_d(z)/f_0(z)$ of several power series $f_0(z)$, $f_1(z)$, ..., $f_d(z)$ taken together is called simultaneous Padé approximation. For many purposes, it is useful to take $f_0(z) = 1$ and the object of approximation to be

$$\mathbf{f}(z) = (f_1(z), f_2(z), \ldots, f_d(z)). \tag{1.1}$$

As this notation suggests, simultaneous Padé approximation is a form of vector Padé approximation, but we reserve the term vector Padé approximation here for the methods of Section 8.4. Nevertheless, it is useful to think of simultaneous Padé approximation primarily as approximation of the vector function $\mathbf{f}(z) \in \mathbb{C}^d[[z]]$, as given by (1.1).

The problem of simultaneous Padé approximation is often referred to as the German polynomial approximation problem, because a Gothic typeface was popular for the approximating polynomials.

To begin with, some algorithms are introduced which show how simultaneous Padé approximants are applied to the approximation of vector power series given numerically. Next we review some row convergence theorems which are analogous to de Montessus's [1902] theorem, and then we consider theorems which apply to systems of Stieltjes functions and are generalizations of Markov's theorem. The study of simultaneous Padé approximants was enlivened when Mahler [1968] published his results which make intricate and unexpected connections between simultaneous and Hermite–Padé approximants. This topic involves greater mathematical generality and it is introduced toward the end of this section.

By way of introduction, suppose that the power series

$$\left.\begin{aligned} f_1(z) &= c_0^{(1)} + c_1^{(1)}z + c_2^{(1)}z^2 + \cdots + c_{N_1}^{(1)}z^{N_1} + \cdots, \\ f_2(z) &= c_0^{(2)} + c_1^{(2)}z + c_2^{(2)}z^2 + \cdots + c_{N_2}^{(2)}z^{N_2} + \cdots, \\ &\vdots \\ f_d(z) &= c_0^{(d)} + c_1^{(d)}z + c_2^{(d)}z^2 + \cdots + c_{N_d}^{(d)}z^{N_d} + \cdots \end{aligned}\right\} \quad (1.2)$$

are given, up to and including the coefficients of z^{N_j} in $f_j(z)$. Let

$$\mathbf{c}_i = (c_i^{(1)}, c_i^{(2)}, \ldots, c_i^{(d)})$$

denote the coefficient of z^i in $\mathbf{f}(z)$, so that

$$(\mathbf{c}_i)_j = c_i^{(j)}, \quad j = 1, 2, \ldots, d.$$

An essential ingredient of the definition of simultaneous Padé approximants is an index set

$$I = (m_1, m_2, \ldots, m_d). \quad (1.3)$$

The approximants have numerator polynomials denoted by

$$\mathbf{p}(z) = (p_1(z), p_2(z), \ldots, p_d(z))$$

whose degrees are specified by

$$\deg\{p_j(z)\} \le L_j = N_j - m_j, \quad \text{for } j = 1, 2, \ldots, d. \quad (1.4)$$

The denominator polynomial $q(z)$ of the simultaneous Padé approximant is of degree M at most, as defined by

$$\deg\{q(z)\} \le M = m_1 + m_2 + \cdots m_d. \quad (1.5)$$

The system of equations

$$q(z)f_j(z) - p_j(z) = O(z^{N_j+1}), \quad j = 1, 2, \ldots, d, \quad (1.6)$$

is interpreted as a system of $\sum_{j=1}^{d}(N_j + 1)$ homogeneous equations for the $\sum_{j=1}^{d}(L_j + 1)$ unknown coefficients of $\mathbf{p}(z)$ and the $M + 1$ unknown coefficients of $q(z)$. Note that

8.1 Simultaneous Padé approximants

$$M + \sum_{j=1}^{d}(L_j + 1) = \sum_{j=1}^{d}(N_j + 1), \tag{1.7}$$

so this system (1.6) always has a solution for a simultaneous Padé form $(\mathbf{p}(z), q(z))$ which satisfies the requirements above.

Using Nuttall's notation, the defining equations (1.6) become

$$[q(z)f_j(z)]_{L_j+1}^{N_j} = 0, \quad j = 1, 2, \ldots, d, \tag{1.8}$$

which is a compact means of expressing M homogeneous equations for the $M + 1$ coefficients of $q(z)$. If these equations are nonsingular, their solution is given by Cramer's rule and we obtain

$$q(z) = \det \begin{vmatrix} c_{N_1-M-m_1+1}^{(1)} & c_{N_1-M-m_1+2}^{(1)} & \cdots & c_{N_1-m_1+1}^{(1)} \\ \vdots & \vdots & & \vdots \\ c_{N_1-M}^{(1)} & c_{N_1-M+1}^{(1)} & \cdots & c_{N_1}^{(1)} \\ \hline \vdots & \vdots & & \vdots \\ \hline c_{N_d-M-m_d+1}^{(d)} & c_{N_d-M-m_d+2}^{(d)} & \cdots & c_{N_d-m_d+1}^{(d)} \\ \vdots & \vdots & & \vdots \\ c_{N_d-M}^{(d)} & c_{N_d-M+1}^{(d)} & \cdots & c_{N_d}^{(d)} \\ z^M & z^{M-1} & \cdots & 1 \end{vmatrix}. \tag{1.9}$$

The numerator polynomials follow as the Maclaurin sections

$$p_j(z) = [f_j(z)q(z)]_0^{L_j}, \quad j = 1, 2, \ldots, d. \tag{1.10}$$

If, in the determinantal representation (1.9), $q(0) \neq 0$, it then follows that (1.8) represents a homogeneous linear system of equations of full rank M, and the system has a one-parameter family of solutions (i.e., its solution is one dimensional). Consequently, if $q(0) \neq 0$ in (1.9), then

$$\mathbf{r}(z) = \mathbf{p}(z)/q(z) \tag{1.11}$$

is a uniquely determined rational fraction which is the simultaneous Padé approximant of $\mathbf{f}(z)$ of type $[\mathbf{L}/M]$ with index set $\{m_1, m_2, \ldots, m_d\}$.

The condition that $q(0) \neq 0$ with $q(z)$ given by (1.9) is a sufficient condition for uniqueness of (1.11), and it is necessarily a stronger condition than its equivalent for the scalar ($d = 1$) case. For example, with

$$f_1(z) = 1, \quad N_1 = 2, \quad m_1 = 1, \quad L_1 = 1,$$
$$f_2(z) = 1 + z, \quad N_2 = 1, \quad m_2 = 0, \quad L_2 = 1,$$

we may take $q(z) = 1 + q_1 z$ for any q_1 and satisfy (1.8). Then

$$\mathbf{r}(z) = \left(1, \frac{1 + (q_1 + 1)z}{1 + q_1 z}\right), \qquad (1.12)$$

which is not unique.

If numerical values are given for the coefficients $\mathbf{c}_i$, the first practical problem is the specification of a suitable sequence of simultaneous Padé approximants which accelerates the convergence of the Maclaurin sections of $\mathbf{f}(z)$. For this purpose Van Iseghem [1986, 1987a, b, 1989] defines a sequence indexed by the denominator degree $M = 0, 1, 2, \ldots$, and integers k, ρ defined by

$$\rho = \left[\frac{M}{d}\right], \quad k = M - d\rho, \qquad (1.13)$$

so that $k = M \bmod (d)$ and $0 \leq k < d$. The degrees are then specified by $L_j = M - 1$ for all j and

$$m_j = \rho + 1, \; N_j = M + \rho \quad \text{for } j = 1, 2, \ldots, k, \text{ and}$$

$$m_j = \rho, \quad N_j = M + \rho - 1 \text{ for } j = k + 1, k + 2, \ldots, d. \qquad (1.14)$$

Note that either set of values of j in (1.14) may be empty, and that

$$\sum_{j=1}^{d} m_j = k(\rho + 1) + (d - k)\rho = M$$

as required by (1.5). With the specification (1.14), equation (1.9) becomes

$$q(z) = \pm \det \begin{vmatrix} \mathbf{c}_0 & \cdots & \mathbf{c}_M \\ \vdots & & \vdots \\ \mathbf{c}_{\rho-1} & \cdots & \mathbf{c}_{\rho+M-1} \\ \mathbf{c}_\rho^{(1)} & \cdots & \mathbf{c}_{\rho+M}^{(1)} \\ \vdots & & \vdots \\ \mathbf{c}_\rho^{(k-1)} & \cdots & \mathbf{c}_{\rho+M}^{(k-1)} \\ z^M & \cdots & 1 \end{vmatrix}, \qquad (1.15)$$

where the $\mathbf{c}_j$ have been taken as column vectors. The numerator $\mathbf{p}(z)$ is given by (1.10) and the approximant by (1.11). Because all the coefficients starting with $\mathbf{c}_0$ are used in (1.15), the sequence of simultaneous Padé approximants generated this way makes the most of the data available. These approximants are called vector Padé approximants by their originators, Brezinski and Van Iseghem [1994] and Van Iseghem [1985, 1986, 1987a, b, 1989]. In these papers a generalization of the ε-algorithm is given (which involves the use of some auxiliary $d \times d$

determinants) and this allows sequential numerical calculation of this sequence of approximants, as well as a generalization of the Q.D. algorithm for coefficient calculations.

The index set $I = (m_1, m_2, \ldots, m_d)$ was introduced in (1.3) as the essential specification of the degree reduction from N_j to L_j along each axis of $\mathbb{C}^d$. A development of this approach is to specify instead a set of independent real direction vectors $\{\mathbf{w}_i\}_{i=1}^l$ along each of which degree reduction by m_i is required. In this context, it is assumed that $N_i = N$ for all i, and that $\mathbf{c}_0, \mathbf{c}_1, \ldots, \mathbf{c}_N$ are the data. Then the denominator is defined by

$$q(z) = \det \begin{vmatrix} \mathbf{c}_{N-M-m_1+1} \cdot \mathbf{w}_1 & \cdots & \mathbf{c}_{N-m_1+1} \cdot \mathbf{w}_1 \\ \vdots & & \vdots \\ \mathbf{c}_{N-M} \cdot \mathbf{w}_1 & \cdots & \mathbf{c}_N \cdot \mathbf{w}_1 \\ \hline \vdots & & \vdots \\ \hline \mathbf{c}_{N-M-m_l+1} \cdot \mathbf{w}_l & \cdots & \mathbf{c}_{N-m_l+1} \cdot \mathbf{w}_l \\ \vdots & & \vdots \\ \mathbf{c}_{N-M} \cdot \mathbf{w}_l & \cdots & \mathbf{c}_N \cdot \mathbf{w}_l \\ z^M & \cdots & 1 \end{vmatrix} \quad (1.16)$$

and

$$\mathbf{p}(z) = [\mathbf{f}(z)q(z)]_0^N. \quad (1.17)$$

It is readily verified that (1.16) and (1.17) lead to

$$\deg\{\mathbf{p}(z) \cdot \mathbf{w}_i\} \leq N - m_i$$

as the specification requires and the approximant $\mathbf{r}(z) = \mathbf{p}(z)/q(z)$ is called a directed simultaneous Padé approximant. The procedure above amounts to simultaneous Padé approximation after making a change of basis for $\mathbf{f}(z)$ in $\mathbb{C}^d$.

The particular form of (1.16) with $l = 1$, $\mathbf{w}_1 = \mathbf{w}$, $M_1 = M$ is more familiar in the context of Brezinski's topological ε-algorithm. Given a sequence $\{\mathbf{s}_j\}_{j=0}^\infty$ of real or complex vectors, this algorithm is initialized with

$$\boldsymbol{\varepsilon}_{-1}^{(j)} = \mathbf{0}, \quad \boldsymbol{\varepsilon}_0^{(j)} = \mathbf{s}_j, \quad j = 0, 1, 2, \ldots,$$

and the recursive steps are

$$\boldsymbol{\varepsilon}_{2k+1}^{(j)} = \boldsymbol{\varepsilon}_{2k-1}^{(j+1)} + \frac{\mathbf{w}}{(\boldsymbol{\varepsilon}_{2k}^{(j+1)} - \boldsymbol{\varepsilon}_{2k}^{(j)}) \cdot \mathbf{w}}, \quad (1.18)$$

$$\varepsilon_{2k+2}^{(j)} = \varepsilon_{2k}^{(j+1)} + \frac{\varepsilon_{2k}^{(j+1)} - \varepsilon_{2k}^{(j)}}{(\varepsilon_{2k+1}^{(j+1)} - \varepsilon_{2k+1}^{(j)}) \cdot (\varepsilon_{2k}^{(j+1)} - \varepsilon_{2k}^{(j)})} \quad (1.19)$$

for $j, k = 0, 1, 2, \ldots$. It turns out [Brezinski, 1980, p. 179] that

$$\varepsilon_{2k}^{(j)} = \det \begin{vmatrix} \mathbf{w} \cdot \mathbf{c}_{j+1} & \cdots & \mathbf{w} \cdot \mathbf{c}_{j+k+1} \\ \vdots & & \vdots \\ \mathbf{w} \cdot \mathbf{c}_{j+k} & \cdots & \mathbf{w} \cdot \mathbf{c}_{j+2k} \\ \mathbf{s}_j & \cdots & \mathbf{s}_{j+k} \end{vmatrix}$$

$$\div \det \begin{vmatrix} \mathbf{w} \cdot \mathbf{c}_{j+1} & \cdots & \mathbf{w} \cdot \mathbf{c}_{j+k+1} \\ \vdots & & \vdots \\ \mathbf{w} \cdot \mathbf{c}_{j+k} & \cdots & \mathbf{w} \cdot \mathbf{c}_{j+2k} \\ 1 & \cdots & 1 \end{vmatrix},$$

(1.20)

where

$$\mathbf{c}_0 = \mathbf{s}_0 \text{ and}$$

$$\mathbf{c}_j = \mathbf{s}_j - \mathbf{s}_{j-1} = \Delta \mathbf{s}_{j-1} \quad \text{for } j = 1, 2, 3, \ldots.$$

The connection with (1.16) is made in the same way as was done in Section 1.3.

The Jacobi–Perron algorithm [Perron, 1957] is probably the best known algorithm for calculating simultaneous Padé approximants recursively, given the absence of degeneracy, and it can even lead to convergence theorems [Aptekharev, 1985]. Several algorithms of Kronecker type have been found [Graves-Morris and Wilkins, 1987], and some progress in dealing with the problems of degeneracy has been reported by de Bruin [1974]. For most practical purposes, direct solution of the system of linear equations (1.8) for the denominator coefficients has the same advantages as in the scalar case discussed in Section 2.1. For large-scale problems, sequential methods analogous to those given in Section 3.6 are given by Labahn [1992] as part of a major extension of Mahler's theorem. The theoretical framework is comprehensively analyzed by Van Barel and Bultheel [1992a].

Next, we turn to the theory of convergence of simultaneous Padé approximants. This topic is of importance both as an aspect of approximation theory, and as an indicator of the areas in which numerical estimates formed using simultaneous Padé approximants of a vector function are better or worse than those found by Padé approximation of the individual component functions. By inspection of the determinant (1.9), we see that the functions must be significantly different from each other: for example, if the functions $f_i(z)$ for which $m_i > 0$ are linearly dependent, $q(z)$

vanishes identically. These considerations led to the following hypothesis, on which a useful analogue of de Montessus's theorem is based.

Definition. Let each of the functions $f_1(z)$, $f_2(z)$, ..., $f_d(z)$ be meromorphic in the disk $D_R = \{z: |z| < R\}$. and let nonnegative integers $m_1, m_2, \ldots, m_d$ be given as in (1.3). Then the functions $f_i(z)$ are *polewise independent with respect to* $m_1, m_2, \ldots, m_d$ *in* D_R if there do not exist polynomials $\pi_1(z), \pi_2(z), \ldots, \pi_d(z)$, at least one of which is nonnull, satisfying

$$\partial\{\pi_i(z)\} \leq m_i - 1, \quad \text{if } m_i > 0,$$

$$\pi_i(z) = 0, \quad \text{if } m_1 = 0,$$

and such that

$$\Phi(z) = \sum_{i=1}^{d} \pi_i(z) f_i(z)$$

is analytic throughout D_R.

The theorem of de Montessus de Ballore applies to the case in which the degree of the denominator precisely matches the number of poles (counting multiplicities) of the given function. The hypothesis of polewise independence is the appropriate analogue for simultaneous Padé approximation. The conditions of the definition imply that each $f_i(z)$ must have poles of total multiplicity of at least m_i in D_R, whereas, if $m_i = 0$, implying that the power-series coefficients of $f_i(z)$ do not appear in (1.9), $f_i(z)$ may be analytic in D_R.

Theorem 8.1.1 [Graves-Morris and Saff, 1984]. *Suppose that each of the d functions $f_1(z), f_2(z), \ldots, f_d(z)$ is analytic in the disk D_R except for possible poles at the M not necessarily distinct points $z_1, z_2, \ldots, z_M$ in D_R, none of which is the origin, and that the $f_i(z)$ are polewise independent with respect to $m_1, m_2, \ldots, m_d$ in D_R as in the previous definition.*

Then, for each integer N sufficiently large, there exist polynomials $q_N(z)$, $\{p_{j,N}(z)\}_{j=1}^{d}$ satisfying (1.4), (1.5), and (1.6) where all $N_j = N$. The denominators (with monic normalization) converge as

$$\lim_{N \to \infty} q_N(z) = \prod_{i=1}^{M} (z - z_i) = q(z) \qquad (1.21)$$

and the approximants converge as

$$\lim_{N \to \infty} p_{j,N}(z)/q_N(z) = f_j(z) \quad \text{for } j = 1, 2, \ldots, d \text{ and } z \in D_R^-, \qquad (1.22)$$

where
$$D_R^- = D_R \setminus \{z_i\}_{i=1}^M.$$

The rates of convergence are governed by

$$\limsup_{N \to \infty} \|q_N - q\|_K^{1/N} \leq \max_{1 \leq i \leq M} \left|\frac{z_i}{R}\right| < 1, \quad (1.23)$$

where K is any compact subset of $\mathbb{C}$, *and*

$$\limsup_{N \to \infty} \|f_j - p_{j,N}/q_N\|_E^{1/N} \leq \frac{\|z\|_E}{R} < 1 \quad (1.24)$$

for $j = 1, 2, \ldots, d$, *where E is any compact subset of* D_R^-.

The proof of this theorem is based on careful estimates of $q_N(z_i)$, just as in method 3 of Section 6.3. The theorem itself is an extension of a result of Mall [1934]; in turn, Theorem 8.1.1 is readily extended to cover the formulation of the directed simultaneous Padé approximants (1.16), (1.17), and the case of multipoint Padé approximants [Graves-Morris and Saff, 1984]. Corresponding divergence results are given by Graves-Morris and Saff [1991b].

Another area which naturally lends itself to convergence theorems is that of simultaneous approximation of Stieltjes functions. Generalizing directly from the definitions and analysis of Section 5.3, consider the system of d functions

$$f_j(z) = \int_{\Delta_j} \frac{d\sigma_j(t)}{1 + zt} = \sum_{i=0}^{\infty} c_i^{(j)} z^i, \quad j = 1, 2, \ldots, d, \quad (1.25)$$

where $\Delta_1, \Delta_2, \ldots, \Delta_d$ are disjoint intervals of the real axis $\mathbb{R}$. The sequence of simultaneous Padé approximants $\mathbf{r}^{(\mathbf{m})}(z)$ specified by $I = (m_1, m_2, \ldots, m_d)$ and

$$L_j = M - 1, \ N_j = M + m_j - 1, \ M = \sum_{j=1}^d m_j \quad (1.26)$$

turns out to be a natural sequence to consider for the approximation of $\mathbf{f}(z)$. Following (1.6)–(1.11), but with the approximants labeled by a superscript $\mathbf{m}$, they satisfy

$$r_j^{(\mathbf{m})}(z) = p_j^{(\mathbf{m})}(z)/q^{(\mathbf{m})}(z) = f_j(z) + O(z^{M+m_j}), \ j = 1, 2, \ldots, d. \quad (1.27)$$

Following the approach of Section 5.4, make the substitution $w = -z^{-1}$, and then

$$F_j(w) = w^{-1} f_j(-w^{-1}) = \int_{\Delta_j} \frac{d\sigma_j(t)}{w - t}, \quad j = 1, 2, \ldots, d, \quad (1.28)$$

8.1 Simultaneous Padé approximants

$$Q^{(\mathbf{m})}(w) = w^M q^{(\mathbf{m})}(-w^{-1})/q^{(\mathbf{m})}(0), \tag{1.29}$$

$$\mathbf{P}^{(\mathbf{m})}(w) = w^{M-1} \mathbf{p}^{(\mathbf{m})}(-w^{-1})/q^{(\mathbf{m})}(0). \tag{1.30}$$

In this context, it is always the case that $q^{(\mathbf{m})}(0) \neq 0$, and hence $Q^{(\mathbf{m})}(w)$ is a monic polynomial. From (1.28)–(1.30) one obtains

$$P_j^{(\mathbf{m})}(w)/Q^{(\mathbf{m})}(w) = F_j(w) + O(w^{-M-m_j-1}). \tag{1.31}$$

From (1.9) and (1.29),

$$Q^{(\mathbf{m})}(w) = (-1)^M \det \begin{vmatrix} c_0^{(1)} & c_1^{(1)} & \cdots & c_M^{(1)} \\ \vdots & \vdots & & \vdots \\ c_{m_1-1}^{(1)} & c_{m_1}^{(1)} & \cdots & c_{M+m_1-1}^{(1)} \\ \hline \vdots & \vdots & & \vdots \\ \hline c_0^{(d)} & c_1^{(d)} & \cdots & c_M^{(d)} \\ \vdots & \vdots & & \vdots \\ c_{m_d-1}^{(d)} & c_{m_d}^{(d)} & \cdots & c_{M+m_d-1}^{(d)} \\ 1 & -w & \cdots & (-w)^M \end{vmatrix} / q^{(\mathbf{m})}(0) \tag{1.32}$$

and hence we have the important orthogonality property that, for each j,

$$\int_{\Delta_j} Q^{(\mathbf{m})}(t) t^i \, d\sigma_j(t) = 0, \quad i = 0, 1, \ldots, m_j - 1. \tag{1.33}$$

It is useful to define

$$E_j^{(\mathbf{m})}(w) = \frac{1}{Q^{(\mathbf{m})}(w)} \int_{\Delta_j} \frac{Q^{(\mathbf{m})}(t)}{w-t} \, d\sigma_j(t), \quad j = 1, 2, \ldots, d, \tag{1.34}$$

which we shall soon prove to be the error functions of the approximation scheme. Let $R(t)$ be any polynomial of degree m_j at most. Then

$$E_j^{(\mathbf{m})}(w) = \frac{1}{Q^{(\mathbf{m})}(w)} \int_{\Delta_j} \frac{Q^{(\mathbf{m})}(t)}{w-t} \, d\sigma_j(t)$$

$$- \frac{1}{Q^{(\mathbf{m})}(w)R(w)} \int_{\Delta_j} \frac{R(w)-R(t)}{w-t} Q^{(\mathbf{m})}(t) \, d\sigma_j(t) \tag{1.35}$$

because (1.33) implies that the second integral on the right-hand side of (1.35) is zero. By merging these integrals, we obtain

$$E_j^{(\mathbf{m})}(w) = \frac{1}{Q^{(\mathbf{m})}(w)R(w)} \int_{\Delta_j} \frac{Q^{(\mathbf{m})}(t)R(t)}{w-t} \, d\sigma_j(t). \tag{1.36}$$

By taking $R(w) = w^{m_j}$, for example, it follows from (1.36) that

$$E_j^{(\mathbf{m})}(w) = O(w^{-M-m_j-1}). \tag{1.37}$$

From (1.28) and (1.34), we obtain

$$\mathbf{F}(w) - \mathbf{E}(w) = \frac{\mathbf{P}^{(m)}(w)}{Q^{(m)}(w)}, \qquad (1.38)$$

where the components of $\mathbf{P}^{(m)}(w)$ are

$$P_j^{(m)}(w) = \int_{\Delta_j} \frac{Q^{(m)}(w) - Q^{(m)}(t)}{w - t} \, d\sigma_j(t). \qquad (1.39)$$

Notice that (1.39) defines $\mathbf{P}_j^{(m)}(w)$ as a polynomial of degree $M - 1$ at most. Then $R^{(m)}(w)$ defined by

$$\mathbf{R}^{(m)}(w) = \mathbf{P}^{(m)}(w)/Q^{(m)}(w) \qquad (1.40)$$

is a vector rational approximant of type $[M - 1/M]$ in the variable w, with denominator given by (1.32) and numerator by (1.39); the factor $q^{(m)}(0)$ in (1.32) is irrelevant in (1.40). From (1.37), (1.38), and (1.40),

$$R_j^{(m)}(w) = F_j(w) + O(w^{-M-m_j-1}) \qquad (1.41)$$

and $\mathbf{R}^{(m)}(w)$ is the simultaneous Padé approximant of type $[M - 1/M]$ with index set I, for $\mathbf{F}(w)$ based on an expansion about $w = \infty$. Note that it is the particular degree conditions (1.26) which give rise to the elegant formula (1.39) for $\mathbf{P}^{(m)}(w)$.

In the setting described by (1.25)–(1.41), Gončar and Rakhmanov [1981] established criteria governing convergence or divergence of certain sequences of simultaneous Padé approximants. The sequences they consider are those for which

$$m_i/M \to \gamma_i \quad \text{as } M \to \infty. \qquad (1.42)$$

Two of their principal conclusions are (i) that the domains of convergence or divergence depend on both $\gamma = (\gamma_1, \gamma_2, \ldots, \gamma_d)$ and on the geometry of the support of $d\sigma(t)$ but not on the magnitudes of $d\sigma(t)$ themselves and (ii) that even for seemingly benign geometries there are significant domains of divergence of the approximants for (say) $\gamma = (\frac{1}{2}, \frac{1}{2})$ in the $d = 2$ case.

The key to their analysis is the interpretation of (1.33) as a minimality property. The proof of method 2 used in Section 5.4 shows that precisely m_j of the zeros of $Q^{(m)}(w)$ lie in Δ_j; these zeros are called $\{t_{j,k}\}_{k=1}^{m_j}$. Define monic polynomials by

$$Q_j^{(m)}(t) = \prod_{k=1}^{m_j} (t - t_{j,k}), \qquad (1.43)$$

$$\hat{Q}_j^{(m)}(t) = \prod_{\substack{i \neq j \\ i=1}}^{d} Q_i^{(m)}(t) = Q^{(m)}(t)/Q_j^{(m)}(t). \qquad (1.44)$$

Then (1.33) becomes

$$\int_{\Delta_j} [Q_j^{(m)}(t)R(t)]|\hat{Q}_j^{(m)}(t)|\,d\sigma_j(t) = 0. \tag{1.45}$$

Because $\hat{Q}_j^{(m)}(t)$ has no zero in Δ_j, (1.45) can be interpreted as the Tchebycheff-like minimality result

$$\min_Q \int_{\Delta_j} Q^2(t)|\hat{Q}_j^{(m)}(t)|\,d\sigma_j(t) = \int [Q_j^{(m)}(t)]^2|\hat{Q}_j^{(m)}(t)|\,d\sigma_j(t) = \lambda_j^{(m)} \text{ (say)}, \tag{1.46}$$

for $j = 1, 2, \ldots, d$, where Q is any monic polynomial of degree m_j precisely.

We associate with a polynomial $p_k(t)$ of degree k a measure $dv_{p_k}(t)$ and a potential $V_{p_k}(w)$ given by

$$V_{p_k}(w) = \frac{1}{k}\ln\frac{1}{|p_k(w)|} = \int \ln\frac{1}{|w-t|}\,dv_{p_k}(t),$$

[Landkof, 1972]. Because (1.46) represents a minimality property, Gončar and Rakhmanov could establish the existence of unique limiting distributions $dv_j(t)$ for which

$$V_j(w) = \int_{\Delta_j} \ln\frac{1}{|w-t|}\,dv_j(t) \tag{1.47}$$

and then

$$\lim_{M\to\infty} |Q_j^{(m)}(w)|^{1/m_j} = e^{-V_j(w)}, \quad w \in \mathbb{C}\backslash\Delta_j, \tag{1.48}$$

$$\lim_{M\to\infty} |Q^{(m)}(w)Q_j^{(m)}(w)|^{1/M} = e^{-W_j(w)}, \quad w \in \mathbb{C}\backslash \cup \Delta_j, \tag{1.49}$$

and

$$\lim_{M\to\infty} (\lambda_j^{(m)})^{1/M} = e^{-w_j}. \tag{1.50}$$

The exponent $W_j(w)$ in (1.49) is related to $V_j(w)$ by

$$W_j(w) = V_j(w) + \sum_{i=1}^{d}\gamma_i V_i(w). \tag{1.51}$$

Gončar and Rakhmanov proved that, as $M \to \infty$, all the d integrals in (1.46) can be minimized self-consistently if the limiting measures $dv_j(t)$ in (1.47) are chosen to minimize the energy integral

$$J(v) = \sum_i\sum_j \int_{\Delta_i}\int_{\Delta_j} \ln\frac{1}{|t-u|}\,dv_i(t)\,dv_j(u) + \sum_j\int_{\Delta_j}\int_{\Delta_j}\ln\frac{1}{|t-u|}\,dv_j(t)\,dv_j(u), \tag{1.52}$$

subject to the charges $\gamma_j = \int_{\Delta_j} dv_j(t)$ being assigned to the interval Δ_j.

Note that the measures $d\sigma_j(t)$ do not occur in (1.52), and that the support of the measures $dv_j(t)$ is determined by the geometry of the intervals Δ_j.

By taking $R(w) = Q_j^{(m)}(w)$ in (1.36) and using (1.46), (1.49), it follows that

$$\lim_{M \to \infty} |F_j(w) - R_j^{(m)}(w)|^{1/M} = e^{W_j(w) - w_j}. \tag{1.53}$$

Thus there is a domain $\{D^+ : |W_j(z)| < w_j\}$ of convergence and a domain $\{D^- : |W_j(z)| > w_j\}$ of divergence of $R_j^{(m)}(w)$.

Despite the fact that the zeros of $Q^{(m)}(w)$ are located in the intervals Δ_j, Gončar and Rakhmanov found examples with $d = 2$, $\gamma_1 = \gamma_2 = \frac{1}{2}$, for which the domain of divergence D^- is larger than $\cup \Delta_j$. Kalyagin [1979] proved convergence in $\mathbb{C} \setminus \cup \Delta_j$ when $\gamma_1 = \gamma_2 = \frac{1}{2}$ and Δ_1, Δ_2 adjoin and are of equal length. If

$$|\Delta_1| \geq |\Delta_2|, \tag{1.54}$$

Gončar and Rakhmanov proved that $R_2^{(m)}(w) \to F_2(w)$ for all $w \in \mathbb{C} \setminus (\Delta_1 \cup \Delta_2)$, but the weaker condition (1.54) is insufficient for convergence in $R_1^{(m)}(w)$ in the same domain. In fact, they proved divergence of $R_1^{(m)}(w)$ for the case of $\gamma_1 = \gamma_2 = \frac{1}{2}$ in the ovular shapes D_1^- for the two geometries shown in Figure 8.1.1.

The orthogonality conditions (1.33) are often called biorthogonality conditions. Numerous instructive examples of systems of biorthogonal polynomials $Q^{(m)}(w)$ are given by Iserles and Nørsett [1988]. Together with (1.39), they constitute useful examples of simultaneous Padé approximants [Iserles and Saff, 1987]. Other explicit examples of systems of simultaneous Padé approximants are reviewed by de Bruin [1984], and new results for hypergeometric and q-hypergeometric functions are given by de Bruin [1984, 1988] and de Bruin, Driver, and Lubinsky [1992]. We refer to Stahl [1988, 1989] for a discussion of convergence results for Nikisin systems, in which the functions of the system are Stieltjes functions whose measures have common support.

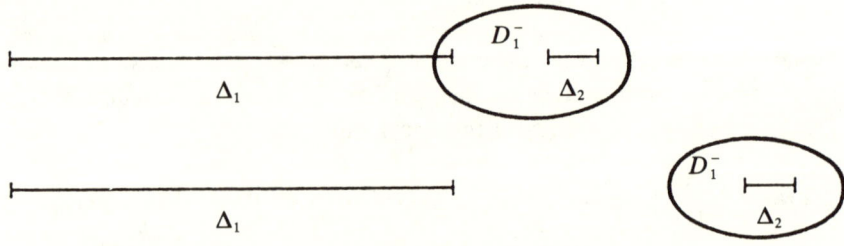

Figure 8.1.1. Ovular shapes of divergence with $|\Delta_1| > |\Delta_2|$.

Mahler discovered remarkable connections between simultaneous Padé approximants and Hermite–Padé approximants. The latter approximants are properly introduced in Section 8.5, but this section may be read first. As was mentioned at the beginning, Mahler's more abstract setting for simultaneous Padé approximants is based on $d+1$ power series $f_0(z)$, $f_1(z), \ldots, f_d(z)$ which are given. Then simultaneous Padé polynomials $P_0^{(\mathbf{m})}(z)$, $P_1^{(\mathbf{m})}(z), \ldots, P_d^{(\mathbf{m})}(z)$ are determined from the coefficient equations arising from the requirement that

$$f_0(z)P_j^{(\mathbf{m})}(z) - f_j(z)P_0^{(\mathbf{m})}(z) = O(z^{N+1}), \quad j = 1, 2, \ldots, d. \quad (1.55)$$

The index vector $\mathbf{m} = (m_0, m_1, \ldots, m_d)$ and the common order of correspondence N are specified as data too. The degrees of the polynomials in (1.55) must satisfy

$$\deg\{P_i^{(\mathbf{m})}(z)\} \leq N - m_i, \quad i = 0, 1, \ldots, d, \quad (1.56)$$

and

$$N = \sum_{i=0}^{d} m_i. \quad (1.57)$$

The system (1.55) represents $(N+1)d$ homogeneous linear equations for $(N+1)d - 1$ unknown coefficients. Under a rank condition similar to that given earlier in the section, the system (1.55) has a one-dimensional family of solutions for $P_0^{(\mathbf{m})}(z), \ldots, P_d^{(\mathbf{m})}(z)$. The connection with (1.3)–(1.7) is made with

$$M = N - m_0, \quad N_j = N, \quad q(z) = P_0^{(\mathbf{m})}(z), \quad f_0(z) = 1.$$

Hermite–Padé polynomials $Q_0^{(\mathbf{m})}(z)$, $Q_1^{(\mathbf{m})}(z), \ldots, Q_d^{(\mathbf{m})}(z)$ are determined from the coefficient equations arising from the requirement that

$$f_0(z)Q_0^{(\mathbf{m})}(z) + f_1(z)Q_1^{(\mathbf{m})}(z) + \cdots + f_d(z)Q_d^{(\mathbf{m})}(z) = O(z^{N-1}). \quad (1.58)$$

The index vector $\mathbf{m}$ and the order N are the same as above, and the degree conditions are

$$\deg\{Q_i^{(\mathbf{m})}(z)\} \leq m_i - 1. \quad (1.59)$$

We emphasize that the degree conditions (1.56), (1.59) are useful in this context, but that (1.59) is *not* the convention generally used, nor the one used in Section 8.5. Under rank conditions such as those given in Section 8.5, the coefficient equations arising from (1.59) have a one-dimensional family of solutions for $Q_0^{(\mathbf{m})}(z), \ldots, Q_d^{(\mathbf{m})}(z)$. The following result typifies those discovered by Mahler [1968].

Theorem 8.1.2. *Let* $\mathbf{e}_0, \mathbf{e}_1, \ldots, \mathbf{e}_d$ *denote unit vectors in* $\mathbb{N}^{d+1}$, *and let* $f_0(0) \neq 0$. *Then any solutions of* (1.55)–(1.59) *satisfy*

$$\begin{bmatrix} P_0^{(\mathbf{m}-\mathbf{e}_0)}(z) & \cdots & P_d^{(\mathbf{m}-\mathbf{e}_0)}(z) \\ \vdots & & \vdots \\ P_0^{(\mathbf{m}-\mathbf{e}_d)}(z) & \cdots & P_d^{(\mathbf{m}-\mathbf{e}_d)}(z) \end{bmatrix} \begin{bmatrix} Q_0^{(\mathbf{m}+\mathbf{e}_0)}(z) & \cdots & Q_0^{(\mathbf{m}+\mathbf{e}_d)}(z) \\ \vdots & & \vdots \\ Q_d^{(\mathbf{m}+\mathbf{e}_0)}(z) & \cdots & Q_d^{(\mathbf{m}+\mathbf{e}_d)}(z) \end{bmatrix} = z^N \cdot D,$$

(1.60)

where D *is a constant diagonal matrix.*

Proof. Let M denote the matrix product on the left-hand side of (1.60). Then

$$M_{ij} = \sum_{k=0}^{d} P_k^{(\mathbf{m}-\mathbf{e}_i)}(z) Q_k^{(\mathbf{m}+\mathbf{e}_j)}(z).$$

Using (1.55) and then (1.58),

$$f_0(z) M_{ij} = f_0(z) P_0^{(\mathbf{m}-\mathbf{e}_i)}(z) Q_0^{(\mathbf{m}+\mathbf{e}_j)}(z) + P_0^{(\mathbf{m}-\mathbf{e}_i)}(z) \sum_{k=1}^{d} f_k(z) Q_k^{(\mathbf{m}+\mathbf{e}_j)}(z)$$

$$+ O(z^N)$$

$$= O(z^N). \quad (1.61)$$

Hence $M_{ij} = O(z^N)$. The specifications (1.56), (1.59) imply that

$$\deg \{P_k^{(\mathbf{m}-\mathbf{e}_i)}(z)\} \leq N - m_i - 1 + \delta_{ik},$$

$$\deg \{Q_k^{(\mathbf{m}+\mathbf{e}_j)}(z)\} \leq m_i - 1 + \delta_{kj},$$

and therefore

$$\deg \{M_{ij}\} \leq \max_k \{\deg \{P_k^{(\mathbf{m}-\mathbf{e}_i)}(z)\} + \deg \{Q_k^{(\mathbf{m}+\mathbf{e}_j)}(z)\}\}$$

$$\left. \begin{array}{ll} \leq N - 1 & \text{if } i \neq j, \text{ and} \\ \deg \{M_{ij}\} \leq N & \text{if } i = j. \end{array} \right\} \quad (1.62)$$

The constraints (1.61), (1.62) show that D is a constant diagonal matrix.

A useful example of this theorem and of the approximants is taken from Labahn [1992]. For the case of $m_0 = 2$, $m_1 = 2$, $m_2 = 1$,

$$f_0(z) = 1,$$

$$f_1(z) = 2z^2 - 2z^3 + z^4 - 2z^5 + O(z^6),$$

$$f_2(z) = z - z^3 - z^4 + O(z^6),$$

the identity (1.60) takes the form

$$\begin{bmatrix} z^3 + \frac{z^2}{4} + \frac{z}{2} - \frac{1}{2} & z^2 & -\frac{z^3}{4} + \frac{z^2}{2} + \frac{z}{2} \\ \frac{3z^2}{8} + \frac{z}{4} - \frac{1}{4} & z^3 - \frac{z^2}{2} & \frac{5z^3}{8} + \frac{z^2}{4} - \frac{z}{4} \\ -\frac{z^2}{4} - \frac{z}{2} - \frac{1}{2} & -z^2 & z^4 + \frac{z^3}{4} - \frac{z^2}{2} - \frac{z}{2} \end{bmatrix}$$

$$\times \begin{bmatrix} z^2 - \frac{z}{4} & -z & \frac{z}{4} \\ -\frac{3z}{8} - \frac{1}{2} & z^2 + \frac{z}{2} & -\frac{5z}{8} - \frac{1}{2} \\ \frac{1}{4} & 1 & z - \frac{1}{4} \end{bmatrix} = z^5 I.$$

A remarkable generalization of Theorem 8.1.2 along the lines of (3.6.6) is also given by Labahn [1992], which allows reliable recursive calculation of both simultaneous and Hermite–Padé approximants. These two approximation schemes have been formally unified by Beckermann and Labahn [1992, 1994], and computational algorithms have been given by Cabay and Labahn [1992], Cabay, Labahn, and Beckermann [1992], and Van Barel and Bultheel [1991].

8.2 Operator Padé approximants

Most of this book is about Padé approximants for $\sum_{i=0}^{\infty} c_i z^i$, where c_i are real or complex numbers. In fact, many of the properties of Padé approximants previously discussed have an interpretation when c_i are operators on a Hilbert or Banach space [Pindor, 1979b]. For example, the c_i might be infinite L_2 matrices. The essential features of operator Padé approximants are adequately explained by treating the case where the c_i are constant $p \times p$ square matrices. We begin with

$$f(z) = c_0 + c_1 z + c_2 z^2 + \cdots, \quad c_i \in \mathbb{C}^{p \times p}, \tag{2.1}$$

and so $f(z)$ is a $p \times p$ matrix whose elements are functions, or power series to be precise. In general, we cannot assume that multiplication of c_i and c_j is commutative, and so the order of multiplication in this context becomes significant. Generalization of the determinantal formulas (1.1.8, 1.1.9) has not proved useful, and we adopt a different start.

In nondegenerate cases, Viskovatov's method provides a natural way of introducing rational approximants for (2.1). This series may be formally inverted by setting

$$d_0 = c_0^{-1}, \qquad (2.2)$$

$$d_j = -\left[\sum_{i=0}^{j-1} d_i c_i\right] c_0^{-1}, \quad j = 1, 2, 3, \ldots, \qquad (2.3)$$

which yields

$$\left(\sum_{j=0}^{\infty} d_j z^j\right)\left(\sum_{i=0}^{\infty} c_i z^i\right) = I \qquad (2.4)$$

in the sense of equality of coefficients in (2.4) at every order. Operations such as (2.2) require that c_0 be invertible, and for this property it is essential that the matrices c_i are square. Moreover, existence of c_0^{-1} is also a sufficient condition for the complete construction of $(\sum_{i=0}^{\infty} c_i z^i)^{-1}$.

Following the procedure of Section 4.2, we expand

$$f(z) = c_0 + c_1 z (1 + c_1^{(1)} z + c_2^{(1)} z^2 + \cdots) \qquad (2.5)$$

$$= c_0 + \frac{c_1 z}{I} + \frac{c_1^{(1)} z}{I + c_1^{(2)} z + c_2^{(2)} z^2 + \cdots}$$

$$= c_0 + \frac{c_1 z}{I} + \frac{c_1^{(1)} z}{I} + \frac{c_1^{(2)} z}{I} + \frac{c_1^{(3)} z}{I} + \cdots, \qquad (2.6)$$

where the right-handed multiplication and division rule introduced in (2.5) must be respected in (2.6). For example, (2.6) can be rewritten as

$$f(z) = c_0 + c_1 z \{I + c_1^{(1)} z [I + c_1^{(2)} z (I + \cdots)^{-1}]^{-1}\}^{-1} \qquad (2.7)$$

to emphasize the right-handed convention, but at the expense of clarity. Likewise, a corresponding series with a left-handed convention can be introduced by re-expressing (2.5) as

$$f(z) = c_0 + (I + \tilde{c}_1^{(1)} z + \tilde{c}_2^{(1)} z^2 + \cdots) c_1 z$$

$$= c_0 + \frac{c_1 z}{I} + \frac{\tilde{c}_1^{(1)} z}{I} + \frac{\tilde{c}_1^{(2)} z}{I} + \cdots \qquad (2.8)$$

$$= c_0 + \{[(\cdots + I)^{-1} \tilde{c}_1^{(2)} z + I]^{-1} \tilde{c}_1^{(1)} z + I\}^{-1} c_1 z. \qquad (2.9)$$

Thus, to any matrix continued fraction of the form (2.6) or (2.8) should be conjoined a statement of whether it is a left- or right-handed representation. Unless otherwise stated, right-handed representations are used.

A natural generalization of the representation (2.6) is

$$f(z) = b_0 + \frac{a_1 z}{b_1} + \frac{a_2 z}{b_2} + \cdots + \frac{a_i z}{b_i} + \cdots, \qquad (2.10)$$

where a_i, b_i are constant $p \times p$ matrices. Equation (2.10) is the matrix analogue of the regular C-fractions discussed in Chapter 4. Following the methods of that chapter, we introduce polynomial matrices $A_i(z)$, $B_i(z)$ by

$$\left. \begin{array}{ll} A_0(z) = b_0, & B_0(z) = I, \\ A_1(z) = b_0 b_1 + a_1 z, & B_1(z) = b_1, \end{array} \right\} \qquad (2.11)$$

and sequentially for $i = 2, 3, 4, \ldots$

$$A_i(z) = A_{i-1}(z) b_i + A_{i-2}(z) a_i z, \qquad (2.12)$$

$$B_i(z) = B_{i-1}(z) b_i + B_{i-2}(z) a_i z. \qquad (2.13)$$

The purpose of (2.11)–(2.13) is the construction of $A_i(z) B_i(z)^{-1}$ as the $i+1$th convergent of (2.10), which may be proved by the methods of Section 4.1. The necessity of observing the conventions for right-handed approximants is self-evident from the ordering of the products in (2.12), (2.13).

In fact, (2.10) is a good starting point for the introduction of operator Padé approximants, provided one makes the rather strong hypothesis that all the b_i are invertible, as would be natural within a continued-fraction framework. From (2.13),

$$B_i(0) = b_0 b_1 \cdots b_i \qquad (2.14)$$

and so $B_i(0)$ is invertible, and $A_i(z) B_i(z)^{-1}$ is a matrix Padé approximant of type

$$\left[\left[\frac{i+1}{2} \right] \middle/ \left[\frac{i}{2} \right] \right]$$

whenever the b_i are invertible. Otherwise, we have the weaker result that

$$f(z) B_i(z) - A_i(z) = O(z^{i+1}), \quad i = 0, 1, 2, \ldots, \qquad (2.15)$$

which follows directly from the definitions (2.12), (2.13).

For the moment, the conclusion that we draw based on the development of (2.5)–(2.13) is that right-handed matrix Padé approximants for (2.1) normally take the form $A(z) B(z)^{-1}$, where $A(z)$ and $B(z)$ are $p \times p$ polynomial matrices.

The main aim of this section is the construction of polynomial matrices $A^{[L/M]}(z)$, $B^{[L/M]}(z)$ which satisfy conditions of the form

$$\deg\{A^{[L/M]}(z)\} \leq L, \quad \deg\{B^{[L/M]}(z)\} \leq M, \tag{2.16}$$

$$f(z)B^{[L/M]}(z) - A^{[L/M]}(z) = O(z^{L+M+1}), \tag{2.17}$$

$$B^{[L/M]}(0) = I. \tag{2.18}$$

Definition 8.2.1. If (2.16)–(2.18) are satisfied, $A^{[L/M]}(z)B^{[L/M]}(z)^{-1}$ is called the right-handed matrix Padé approximant for $f(z)$.

Definition 8.2.2. If (2.16)–(2.17) are satisfied by matrix polynomials $P^{[L/M]}(z)$, $Q^{[L/M]}(z)$, and $Q^{[L/M]}(z)$ is invertible, $P^{[L/M]}(z)Q^{[L/M]}(z)^{-1}$ is called a right-handed matrix Padé form for $f(z)$.

Notice that the latter definition is phrased to reflect the fact that, in degenerate cases, matrix Padé forms are not necessarily unique. Some authors replace the condition (2.18) by the requirement that $B^{[L/M]}(0)$ be an invertible matrix, and this also achieves the accuracy-through-order property

$$f(z) - A^{[L/M]}(z)B^{[L/M]}(z)^{-1} = O(z^{L+M+1}) \tag{2.19}$$

which is required for matrix Padé approximants.

Theorem 8.2.1. *If both left- and right-handed matrix Padé forms exist, they are identical.*

Proof. Denote these forms by $^LQ^{[L/M]}(z)^{-1}\,^LP^{[L/M]}(z)$ and $^RP^{[L/M]}(z)\,^RQ^{[L/M]}(z)^{-1}$, respectively. Multiply (2.17) by $^LQ^{[L/M]}(z)$ on the left, its left-handed equivalent by $^RQ^{[L/M]}(z)$ on the right, and subtract. Then

$$^LQ^{[L/M]}(z)\,^RP^{[L/M]}(z) - {}^LP^{[L/M]}(z)\,^RQ^{[L/M]}(z) = O(z^{L+M+1})$$

$$= 0 \tag{2.20}$$

because there are no terms of order z^{L+M+1} on the left-hand side. The theorem follows by matrix division.

This somewhat paradoxical theorem is easily appreciated by noticing that $\frac{1}{2} \times 6$ and $9 \times \frac{1}{3}$ are different representations of the same number, and Example 1 will demonstrate the equality. The most direct method for calculating a matrix Padé approximant of type $[L/M]$ is analogous to that of Section 2.1. For a right-handed approximant, we use the Baker definition $b_0 = I$ and solve

$$\begin{bmatrix} c_{L-M+1} & \cdots & c_L \\ \vdots & & \vdots \\ c_L & \cdots & c_{L+M-1} \end{bmatrix} \begin{bmatrix} b_M \\ \vdots \\ b_1 \end{bmatrix} = - \begin{bmatrix} c_{L+1} \\ \vdots \\ c_{L+M} \end{bmatrix}, \quad (2.21)$$

which is a block system of equations, representing p sets of simultaneous linear equations with the same $Mp \times Mp$ coefficient matrix. Notice that the b_i used in (2.21) differ from those in (2.10)–(2.14). Again, it is useful to define

$$C(L/M) = \begin{vmatrix} c_{L-M+1} & \cdots & c_L \\ \vdots & & \vdots \\ c_L & \cdots & c_{L+M-1} \end{vmatrix} \quad (2.22)$$

as an $Mp \times Mp$ determinant.

Theorem 8.2.2. *If $C(L/M) \neq 0$, then both left- and right-handed matrix Padé approximants of type $[L/M]$ exist.*

Proof. The condition that $C(L/M) \neq 0$ suffices for (2.21) to represent p sets of nonsingular linear equations, with unique solutions which constitute the matrices $b_1, b_2, \ldots, b_M$. These matrices form

$$B^{[L/M]}(z) = \sum_{i=0}^{M} b_i z^i \quad (2.23)$$

uniquely, with $b_0 = I$. Similarly, the left-handed denominator is formed uniquely.

Theorem 8.2.3. *If both left- and right-handed matrix Padé approximants of type $[L/M]$ exist, they are equal.*

Proof. This is an immediate consequence of Theorem 8.2.1.

Example 1. Find matrix Padé approximants of type $[1/1]$ for

$$f(z) = \begin{bmatrix} 1 + z + z^2 & 0 \\ -z & 1 + z - z^2 \end{bmatrix}$$

$$= \begin{bmatrix} 1 & 0 \\ 0 & 1 \end{bmatrix} + \begin{bmatrix} 1 & 0 \\ -1 & 1 \end{bmatrix} z + \begin{bmatrix} 1 & 0 \\ 0 & -1 \end{bmatrix} z^2.$$

Solution. For the right-handed approximant, we expect

$$c_0 + c_1 z + c_2 z^2 = (a_0 + a_1 z)(I + b_1 z)^{-1} + O(z^3)$$

and we need to solve

$$c_0 = a_0,$$
$$c_1 + c_0 b_1 = a_1,$$
$$c_2 + c_1 b_1 = 0,$$

with the result that

$$^R[1/1]_f = \begin{bmatrix} 1 & 0 \\ -2z & 1+2z \end{bmatrix} \begin{bmatrix} 1-z & 0 \\ -z & 1+z \end{bmatrix}^{-1}.$$

For the left-handed approximant, we expect

$$c_0 + c_1 z + c_2 z^2 = (I + \tilde{b}_1 z)^{-1}(\tilde{a}_0 + \tilde{a}_1 z) + O(z^3),$$

and we need to solve

$$c_0 = \tilde{a}_0,$$
$$c_1 + \tilde{b}_1 c_0 = \tilde{a}_1,$$
$$c_2 + \tilde{b}_1 c_1 = 0,$$

with the result that

$$^L[1/1]_f = \begin{bmatrix} 1-z & 0 \\ z & 1+z \end{bmatrix}^{-1} \begin{bmatrix} 1 & 0 \\ 0 & 1+2z \end{bmatrix}.$$

Note that in this example,

$$^R B^{[1/1]}(z) \neq {}^L B^{[1/1]}(z).$$

We construct explicitly the exact inverses of $^R Q(z)$ and $^L Q(z)$ and obtain

$$^R[1/1] = \begin{bmatrix} 1 & 0 \\ -2z & 1+2z \end{bmatrix} \begin{bmatrix} \dfrac{1+z}{1-z^2} & 0 \\ \dfrac{z}{1-z^2} & \dfrac{1-z}{1-z^2} \end{bmatrix} = \begin{bmatrix} \dfrac{1}{1-z} & 0 \\ \dfrac{-z}{1-z^2} & \dfrac{1+2z}{1+z} \end{bmatrix},$$

$$^L[1/1] = \begin{bmatrix} \dfrac{1+z}{1-z^2} & 0 \\ \dfrac{-z}{1-z^2} & \dfrac{1-z}{1-z^2} \end{bmatrix} \begin{bmatrix} 1 & 0 \\ 0 & 1+2z \end{bmatrix} = \begin{bmatrix} \dfrac{1}{1-z} & 0 \\ \dfrac{-z}{1-z^2} & \dfrac{1+2z}{1+z} \end{bmatrix},$$

showing that in this case the left- and right-handed Padé approximants are identical, and confirming that each equals the given matrix through order z^2.

Having discussed the nondegenerate case in which $C(L/M) \neq 0$, we proceed to consider the cases of degeneracy. In fact, there are more forms of degeneracy of matrix Padé approximants than there are in the scalar case. If $C(L/M) = 0$, solutions of the matrix approximation problem are best obtained by considering the space X of solutions $\mathbf{x} = (x_0, x_1, \ldots, x_M) \in \mathbb{C}^{(M+1)p}$ of

$$\begin{bmatrix} c_{L-M+1} & \cdots & c_{L+1} \\ \vdots & & \vdots \\ c_L & \cdots & c_{L+M} \end{bmatrix} \begin{bmatrix} x_M \\ \vdots \\ x_1 \\ x_0 \end{bmatrix} = \mathbf{0}. \qquad (2.24)$$

The rank of the coefficient matrix cannot exceed Mp, and so the system (2.24) has at least p linearly independent solutions. If the number of linearly independent solutions exceeds p, not only is the matrix denominator not specified uniquely, but the matrix Padé form is not necessarily unique either, in contrast to the scalar case. Difficulties can arise even when the system has precisely p linearly independent solutions, as the following ingenious example of Labahn and Cabay [1989] demonstrates.

Example 2. Find matrix Padé forms of type [2/3] for

$$f(z) = \begin{bmatrix} 1 & 0 \\ 0 & 1 \end{bmatrix} + \begin{bmatrix} 1 & 0 \\ 0 & 1 \end{bmatrix} z^2 + \begin{bmatrix} 2 & 0 \\ 0 & 1 \end{bmatrix} z^4 + \begin{bmatrix} -1 & 0 \\ -1 & 0 \end{bmatrix} z^5 + \cdots . \qquad (2.25)$$

Solution. Equation (2.24), together with its solution, is

$$\begin{bmatrix} 1 & 0 & 0 & 0 & 1 & 0 & 0 & 0 \\ 0 & 1 & 0 & 0 & 0 & 1 & 0 & 0 \\ 0 & 0 & 1 & 0 & 0 & 0 & 2 & 0 \\ 0 & 0 & 0 & 1 & 0 & 0 & 0 & 1 \\ 1 & 0 & 0 & 0 & 2 & 0 & -1 & 0 \\ 0 & 1 & 0 & 0 & 0 & 1 & -1 & 0 \end{bmatrix} \begin{bmatrix} 0 & 0 \\ 0 & -1 \\ 0 & 0 \\ -1 & 0 \\ 0 & 0 \\ 0 & 1 \\ 0 & 0 \\ 1 & 0 \end{bmatrix} = \begin{bmatrix} 0 & 0 \\ 0 & 0 \\ 0 & 0 \\ 0 & 0 \\ 0 & 0 \\ 0 & 0 \end{bmatrix}. \qquad (2.26)$$

It is easy to check that the coefficient matrix has its full rank of 6 by converting it to row echelon form. It follows that the two solutions given span the solution space, and we may take

$$Q^{[2/3]}(z) = \begin{bmatrix} 0 & 0 \\ 1 - z^2 & z - z^3 \end{bmatrix}$$

as the 'denominator' polynomial, and hence we obtain

$$f(z)\begin{bmatrix} 0 & 0 \\ 1-z^2 & z-z^3 \end{bmatrix} - \begin{bmatrix} 0 & 0 \\ 1 & z \end{bmatrix} = O(z^6).$$

We notice that $Q^{[2/3]}(0)$ is not invertible, nor is it zero. Thus (2.25) does not have a right-handed matrix Padé approximant, although it does have a left-handed one. We also note that $Q^{[2/3]}(z)$ has one independent column, allowing for polynomial multipliers, whereas (2.26) admits two linearly independent solutions.

The previous example, in which the denominator polynomial is neither zero nor invertible, makes it plain that computation of a staircase sequence or a paradiagonal sequence cannot be a straightforward extension of Viskovatov's algorithm or Werner's algorithm. However, the scale of the problem for large values of p makes a reliable and iterative procedure of calculation all the more desirable, and we next describe the look-ahead method due to Labahn and Cabay [1989].

We consider the paradiagonal sequence of types $[m + J/m]$, for $m = 0$, 1, 2, ... if $J \geq 0$ or for $m = -J, -J+1, \ldots$ if $J < 0$. If a particular approximant of type $[L/M]$ is required, set $J = L - M$ as usual. Theorem 8.2.2 states that nonsingular approximants are identified by $C(m + J/m) \neq 0$. The iterative computation of a sequence of matrix approximants is based on the use of the sequence for which $C(m_i + J/m_i) \neq 0$, $i = 1, 2, 3, \ldots$, and is based on the same principles as the fast algorithm explained in Section 3.6. We begin with two theorems which justify the procedures.

Theorem 8.2.4. *If a matrix Padé approximant of type $[L/M]$ is nonsingular, the right-handed matrix Padé form of its predecessor of type $[L - 1/M - 1]$ exists and is essentially unique.*

Proof. By hypothesis, $C(L/M) \neq 0$, and hence the $pM \times pM$ matrix

$$\mathbf{M} = \begin{bmatrix} c_{L-M+1} & \cdots & c_L \\ \vdots & & \vdots \\ c_L & \cdots & c_{L+M-1} \end{bmatrix} \quad (2.27)$$

has rank pM. By deleting its last block row, it follows that the $p(M-1) \times pM$ matrix

$$\widetilde{\mathbf{M}} = \begin{bmatrix} c_{L-M+1} & \cdots & c_L \\ \vdots & & \vdots \\ c_{L-1} & \cdots & c_{L+M-2} \end{bmatrix} \quad (2.28)$$

has rank $p(M - 1)$, and the dimension of its solution space $\Xi = \{\mathbf{x} \colon \tilde{M}\mathbf{x} = \mathbf{0}, \mathbf{x} \in \mathbb{C}^{pM}\}$ is p. Then, for example, any orthogonal basis $\mathbf{x}^{(1)}, \mathbf{x}^{(2)}, \ldots, \mathbf{x}^{(p)}$ of Ξ may be used to construct $Q^{[L-1/M-1]}(z)$ column by column, and hence the corresponding right-handed matrix Padé form. The column vectors $\hat{\mathbf{x}}^{(1)}, \hat{\mathbf{x}}^{(2)}, \ldots, \hat{\mathbf{x}}^{(p)}$ representing any other solution $\hat{Q}^{[L-1/M-1]}(z)$ must, by Definition 8.2.2, have dimension precisely p, and hence

$$\hat{Q}^{[L-1/M-1]}(z) = Q^{[L-1/M-1]}(z) \cdot C,$$

where C is a nonsingular constant $p \times p$ matrix. To see this, let $X, Y \in \mathbb{C}^{pM \times p}$ denote the solutions of $\tilde{M}X = 0$ from which $Q^{[L-1/M-1]}(z)$ and $\hat{Q}^{[L-1/M-1]}(z)$ respectively are formed. Both X and Y have rank p and span the same solution space Ξ, and so a nonsingular matrix C exists such that $Y = XC$. By reversing the order of the block rows of this equation, we see that C is the matrix required.

Theorem 8.2.5. *Let $m, k \geq 1$ be two integers that are given, together with the matrix power series $f(z)$ of (2.1).*

Assume that $P^{[m+J-1/m-1]}(z)$, $Q^{[m+J-1/m-1]}(z)$ are right-handed numerator and denominator matrix Padé polynomials of type $[m + J - 1/m - 1]$ for the power series $f(z)$, and similarly for $P^{[m+J/m]}(z)$ and $Q^{[m+J/m]}(z)$, but also with

$$C(m + J/m) \neq 0. \tag{2.29}$$

With these polynomials, two 'remainder series' are defined by

$$d(z) = \{f(z)Q^{[m+J-1/m-1]}(z) - P^{[m+J-1/m-1]}(z)\}/z^{2m+J-1}, \tag{2.30}$$

$$e(z) = \{f(z)Q^{[m+J/m]}(z) - P^{[m+J/m]}(z)\}/z^{2m+J+1}. \tag{2.31}$$

Then $d(0)$ is invertible. Assume that $A^{[k/k+1]}(z)$, $B^{[k/k+1]}(z)$ are nondegenerate right-handed numerator and denominator matrix Padé polynomials of type $[k/k + 1]$ for the rational power series $\tilde{e}(z) = [d(z)]^{-1}e(z)$.

Then the right-handed matrix Padé polynomials of types $[m + k + J/m + k]$ and $[m + k + J + 1/m + k + 1]$ for $f(z)$ are given by

$$\begin{bmatrix} P^{[m+k+J/m+k]}(z) & P^{[m+k+J+1/m+k+1]}(z) \\ Q^{[m+k+J/m+k]}(z) & Q^{[m+k+J+1/m+k+1]}(z) \end{bmatrix}$$

$$= \begin{bmatrix} P^{[m+J/m]}(z) & P^{[m+J-1/m-1]}(z) \\ Q^{[m+J/m]}(z) & Q^{[m+J-1/m-1]}(z) \end{bmatrix} \begin{bmatrix} B^{[k-1/k]}(z) & B^{[k/k+1]}(z) \\ -z^2 A^{[k-1/k]}(z) & -z^2 A^{[k/k+1]}(z) \end{bmatrix},$$

$$\tag{2.32}$$

and the matrix approximant of type $[m + k + J + 1/m + k + 1]$ is nondegenerate.

Proof. Using the proof of the previous theorem, the hypothesis that $C(m + J/m) \neq 0$ implies that the system of equations

$$\begin{bmatrix} c_{J+1} & \cdots & c_{m+J} \\ \vdots & & \vdots \\ c_{m+J} & \cdots & c_{2m+J-1} \end{bmatrix} \begin{bmatrix} q_{m-1} \\ \vdots \\ q_0 \end{bmatrix} = \begin{bmatrix} 0 \\ 0 \\ \vdots \\ 0 \\ y \end{bmatrix}, \quad (2.33)$$

which is a homogeneous system for $q_0, q_1, \ldots, q_{m-1}$ and in turn determines y, has an essentially unique solution. The same hypothesis (2.29) implies that the coefficient matrix M of (2.33) is nonsingular. If the $p \times p$ matrix y were singular, a linear combination of its columns would vanish. Let this same linear combination of the columns of $(q_m, \ldots, q_0)^T$ be called $\mathbf{w}$, and then $\mathbf{Mw} = \mathbf{0}$. This contradicts the fact that M is nonsingular, and hence y is nonsingular.

A consequence of (2.33) is that

$$f(z)(q_0 + q_1 z + \cdots + q_{m-1} z^{m-1}) = p_0 + p_1 z + \cdots + p_{m+J-1} z^{m+J-1} + y z^{2m+J-1} + O(z^{2m+J}).$$

Thus $d(0) = y$, and $d(0)$ is invertible.

Using the methods of Theorem 3.6.1, we find that $P^{[m+k+J+1/m+k+2]}(z)$ and $Q^{[m+k+J+1/m+k+1]}(z)$ defined by (2.32) satisfy the appropriate degree conditions and accuracy-through-order conditions (2.16)–(2.17) indicated.

Also, from (2.32),

$$Q^{[m+k+J+1/m+k+1]}(0) = Q^{[m+J/m]}(0) B^{[k/k+1]}(0).$$

The hypothesis (2.29) guarantees that $Q^{[m+J/m]}(0)$ is invertible, and therefore $Q^{[m+k+J+1/m+k+1]}(0)$ and $B^{[k/k+1]}(0)$ are singular and nonsingular together. In particular, if the approximant $[k/k+1]_f(z)$ is nondegenerate, so is $[m+k+J+1/m+k+1]_f(z)$.

The iterative method of computation of the matrix Padé approximants in the paradiagonal sequence of type $[m + J/m - 1]$ is based on locating the ith approximant of type $[m_i + J/m_i]$ which is nonsingular in accord with (2.29), finding the least value of k, say k_i, for which the matrix Padé approximants of type $[k/k+1]$ for $[d(z)]^{-1}e(z)$ are nonsingular, and using (2.32) to construct the next nonsingular matrix Padé approximant in the sequence. A natural requirement for this method to work efficiently is a facility for calculating right-handed polynomial bigradients of the type $[k-1/k]$ and $[k/k+1]$ for $[d(z)]^{-1}e(z)$ without forming $d(z)^{-1}$ explicitly. Following the method of Section 2.4 as far as possible to form the matrix Padé polynomials of type $[k/k + 1]$, we must find a solution of a system of the form

$$\begin{bmatrix} d_0 & & & & & 0 & e_0 & & & & 0 \\ d_1 & d_0 & & & & & e_1 & e_0 & & & \\ & d_1 & \ddots & & & & & e_1 & \ddots & & \\ & & \ddots & d_0 & & & & & \ddots & & \\ \vdots & \vdots & & d_1 & d_0 & & \vdots & & & \ddots & \\ & & & & \vdots & & & & & & e_0 \\ d_{2k+1} & d_{2k} & \cdots & d_{k+2} & d_{k+1} & & e_{2k+1} & e_{2k} & \cdots & & e_k \end{bmatrix} \begin{bmatrix} a_0 \\ a_1 \\ \vdots \\ a_k \\ -b_0 \\ -b_1 \\ \vdots \\ -b_{k+1} \end{bmatrix}$$

$$= \mathbf{0} \quad (2.34)$$

with $b_0 = I$ and for the least value of k. Thus we take k to have the smallest value for which the determinant of the first $2k$ block columns of the coefficient matrix of (2.34) is nonsingular. Since d_0 is invertible, k is taken to have the least value k_i for which $C_e(k/k + 1) \neq 0$. The solution of (2.34) and that of its predecessor of type $[k - 1/k]$ are then used in conjunction with (2.32) to form the next matrix Padé approximants in the sequence, which are of types $[m_i + k_i + J/m_i + k_i]$ and $[m_i + k_i + J + 1/m_i + k_i + 1]$. It is also relatively straightforward to adapt the techniques of Section 3.6 to introduce a robust strategy to avoid the approximate equality $C_e(k/k + 1) \approx 0$.

We conclude this section with some identities and results which have proved useful in scattering theory.

Theorem 8.2.6. *Consider a transformation of the series coefficients defined by*

$$c_i \rightarrow c_i' = v c_i v^{-1},$$

where v is any constant $d \times d$ invertible matrix. This transformation defines an automorphism of the algebra of coefficient matrices, and specifically it defines

$$g(z) = \sum_{i=0}^{\infty} c_i' z^i = v \cdot f(z) \cdot v^{-1}.$$

Then $[L/M]_g(z) = v \cdot [L/M]_f(z) \cdot v^{-1}$, *provided either approximant exists.*

Theorem 8.2.7. *If $f(0) = c_0$ is invertible and*

$$g(z) = \{f(z)\}^{-1},$$

then

$$[L/M]_g(z) = \{[M/L]_f(z)\}^{-1},$$

provided either approximant exists.

Theorem 8.2.8. *Let L_1, L_2, L_3, L_4 be constant $d \times d$ matrices, and let*

$$g(z) = (L_1 f(z) + L_2)(L_3 f(z) + L_4)^{-1}.$$

Then

$$[M/M]_g = \{L_1[M/M]_f + L_2\}\{L_3[M/M]_f + L_4\}^{-1}$$

provided $L_3 f(0) + L_4$ is invertible and $[M/M]_f$ exists.

Theorem 8.2.9 [Gammel and McDonald, 1966]. *Let $S(z)$ be unitary, viz.,*

$$S(z)S^\dagger(z) = I, \qquad (2.35)$$

where † denotes the Hermitian conjugate, i.e., transposition followed by complex conjugation. Provided the $[M/M]$ matrix Padé approximants exist, then they are unitary:

$$[M/M]_S(z)[M/M]_S^\dagger(z) = I. \qquad (2.36)$$

Theorem 8.2.10. *Let $w = \alpha z/(\beta z + \gamma)$ and $g(w) = f(z)$. Then $[N/N]_g(w) = [N/N]_f(z)$ provided $\alpha, \gamma \neq 0$ and either approximant exists.*

Proofs. The proofs are all based in the accuracy-through-order method and are formally identical to those of the scalar case.

An interesting view of matrix Padé approximants derives from a comparison with the consequences of a variational approach. Define a right-handed functional by

$$\mathcal{R} = \mathcal{R}(\mu_0, \mu_1, \ldots, \mu_N) = \sum_{k=0}^{N} c_k \mu_k z^k,$$

a left-handed functional by

$$\mathcal{L} = \mathcal{L}(\lambda_0, \lambda_1, \ldots, \lambda_N) = \sum_{j=0}^{N} \lambda_j c_j z^j,$$

and a bilateral Hankel functional by

$$\mathcal{H} = \mathcal{H}(\lambda_0, \lambda_1, \ldots, \lambda_N; \mu_0, \mu_1, \ldots, \mu_N)$$
$$= \sum_{j=0}^{N} \sum_{k=0}^{N} \lambda_j z^j (c_{j+k} - z c_{j+k+1}) z^k \mu_k.$$

The parameters λ_j, μ_k are arbitrary $d \times d$ matrices.

Theorem 8.2.11. *The variational functions are defined by*

$$[F_1(\lambda_0, \ldots, \lambda_N; \mu_0, \ldots, \mu_N)] = \mathcal{R}\mathcal{H}^{-1}\mathcal{L},$$

$$[F_2(\lambda_0, \ldots, \lambda_N; \mu_0, \ldots, \mu_N)] = \mathcal{R} - \mathcal{H} + \mathcal{L}.$$

These two functionals are identical at their respective stationary points and are then the $[M-1/M]$ matrix Padé approximants of $\sum_{i=0}^{\infty} c_i z^i$.

Space permits neither the proof of this theorem nor a full incursion into the elegant formalism of variational methods with matrix coefficients. In Section 9.8, we derive matrix Padé approximants from the ordinary Schwinger principle in the context of potential theory.

The convergence theory of matrix Padé approximants is somewhat incomplete, and especially so if the matrices are infinite-dimensional matrices. However, if $f(z)$ is an R-operator corresponding to the Stieltjes series with infinite matrices as the coefficients c_i, convergence theorems exist [Narcowich and Allen, 1975].

The approach to Padé approximants adopted in this section follows that of Wynn [1962b, 1963], then that of Bessis [1973], and finally that of Labahn and Cabay [1989]. The matrix analogues of the Gohberg–Heinig formulas, such as (3.6.15), are given by Labahn, Choi, and Cabay [1990]. Applications of operator Padé approximants are reviewed by Pindor and Turchetti [1982] in the context of potential scattering, and Pindor [1984] develops the theme to include operator continued fractions applied to the analysis of anharmonic oscillators. Applications to operator Stieltjes series and network models have been investigated by Basu and Bose [1983].

An extensive theory of Padé approximation for series whose coefficients lie in an abstract Banach space has been developed by Cuyt [1984], with application to the solution of nonlinear integral equations [Cuyt, 1982].

We conclude with an explanation based on the following theorem of why direct extension of the scalar theory to the operator case is hazardous.

Theorem 8.2.12. *If $C(L/M) \neq 0$ in the representation (2.22), the poles of the $[L/M]$-type matrix Padé approximant of $f(z)$ occur at the zeros of the (scalar) polynomial*

$$Q(z) = \det \begin{vmatrix} c_{L-M+1} & c_{L-M+2} & \cdots & c_{L+1} \\ c_{L-M+2} & c_{L-M+3} & \cdots & c_{L+2} \\ \vdots & \vdots & & \vdots \\ c_L & c_{L+1} & \cdots & c_{L+M} \\ z^M \cdot I & z^{M-1} \cdot I & \cdots & I \end{vmatrix}, \quad (2.37)$$

where I denotes the $d \times d$ identity matrix.

Proof. We write the equations (2.21) schematically as

$$Cb = c$$

in which b and c are matrices having Md rows and d columns. Hence the denominators of the approximant may be expressed as

$$^R B^{[L/M]}(z) = I + \sum_{i=1}^{M} b_i z^i$$

$$= I + (zI, z^2 I, \ldots, z^M I) C^{-1} c.$$

We use the identity [Beckenbach and Bellman, 1965]

$$\det X = \det C \det (F - DC^{-1} E)$$

for any matrix X which has a block decomposition

$$X = \begin{bmatrix} C & D \\ E & F \end{bmatrix}$$

in which C is invertible to obtain

$$\det {}^R B^{[L/M]}(z) = Q(z)/\det C.$$

If we use the formula

$$X^{-1} = [\text{adjugate } X]^T / \det X \qquad (2.38)$$

for the inverse of $X = {}^R B^{[L/M]}(z)$, we may express the matrix Padé approximant in (2.19) as

$$^R[L/M](z) = P(z)/Q(z) \qquad (2.39)$$

where $P(z)$ is a $d \times d$ matrix polynomial, and the theorem is proved.

When $C(L/M) \neq 0$, (2.37) shows that $Q(z)$ is normally a polynomial of degree Md. Using (2.38) and (2.39), we find that $P(z)$ is normally a matrix polynomial of degree $L + M(d-1)$. If $d > 2$, this number is greater than the number of terms needed to define the matrix Padé approximant. For this reason, we do not specify matrix Padé approximants by a representation of the type (2.37), but by using a matrix denominator as in (2.19).

8.3 Rectangular matrix Padé approximants for minimal partial-realization problems

This section concerns the case in which the coefficients c_k of the given power series are rectangular but not necessarily square matrices. The

8.3 Rectangular matrix Padé approximants for minimal partial-realization problems

principal motivation for this work came from linear systems theory [Kailath, 1980]; the problem was to design multiple-input, multiple-output (MIMO) systems (see Figure 8.3.2) with given Markov parameters c_k. To be general, we suppose that

$$F(z) = c_0 + c_1 z + c_2 z^2 + \cdots + c_{L+M} z^{L+M} + \cdots, \quad c_k \in \mathbb{R}^{p \times m} \text{ (or } \mathbb{C}^{p \times m}),$$
(3.1)

is given, meaning that each c_k is a real or complex matrix having p rows and m columns. The complex case is an entirely straightforward generalization of the real case, and so we need not refer to it explicitly. At first, it might seem natural to seek polynomial matrices $P(z) \in \mathbb{R}^{p \times m}[z]$, $Q(z) \in \mathbb{R}^{m \times m}[z]$ such that

$$\partial\{P(z)\} \leq L, \quad \partial\{Q(z)\} \leq M \qquad (3.2)$$

and

$$F(z)Q(z) = P(z) + O(z^{L+M+1}). \qquad (3.3)$$

However, this result requires the $(L + M + 1)mp$ parameters of the first $L + M + 1$ terms of $F(z)$ to be represented using $(L + 1)mp$ parameters for $P(z)$ and $(M + 1)m^2$ parameters for $Q(z)$. Of these, the m^2 parameters of $Q(0)$ are normally superfluous. Thus, in general, we need to have $m \geq p$ to use the right-handed approximation scheme of (3.1)–(3.3), and the left-handed form

$$^{(L)}Q(z)F(z) = {}^{(L)}P(z) + O(z^{L+M+1}) \qquad (3.4)$$

would be appropriate if $m \leq p$. Either way, there is a mismatch between the number of equations and the number of parameters to be found if $p \neq m$. Therefore, a specification of a unique rectangular matrix Padé approximant is necessarily quite different from that for operator Padé approximants. In this section, we describe methods of solution of some rather imprecisely posed problems, and note the important characteristics of the solutions obtained. We begin with two examples.

Example 1. Rectangular matrix Padé approximants for

$$F(z) = \begin{bmatrix} z & -1+z & 1 \\ 1+z & 1 & 1+z \end{bmatrix} + O(z^2)$$

$$= \begin{bmatrix} 0 & -1 & 1 \\ 1 & 1 & 1 \end{bmatrix} + z \begin{bmatrix} 1 & 1 & 0 \\ 1 & 0 & 1 \end{bmatrix} + \cdots \qquad (3.5)$$

are found using the method described below. We obtain

$$Q(z) = \begin{bmatrix} 1 & -\frac{1}{2} & -1 \\ 0 & 1+\frac{z}{2} & 1 \\ -z & 0 & 1 \end{bmatrix},$$

and then (3.3) takes the form

$$F(z)Q(z) = \begin{bmatrix} z & -1+z & 1 \\ 1+z & 1 & 1+z \end{bmatrix} \begin{bmatrix} 1 & -\frac{1}{2} & -1 \\ 0 & 1+\frac{z}{2} & 1 \\ -z & 0 & 1 \end{bmatrix}$$

$$+ O(z^2)$$

$$= \begin{bmatrix} 0 & -1 & 0 \\ 1 & \frac{1}{2} & 1 \end{bmatrix} + O(z^2). \tag{3.6}$$

Using the method below for $F^T(z)$, i.e., the transposed problem, we derive

$$^{(L)}Q(z)F(z) = \begin{bmatrix} -1-z & 1 \\ 1 & -z \end{bmatrix} \begin{bmatrix} z & -1+z & 1 \\ 1+z & 1 & 1+z \end{bmatrix} + O(z^2)$$

$$= \begin{bmatrix} 1 & 2 & 0 \\ 0 & -1 & 1-z \end{bmatrix} + O(z^2) \tag{3.7}$$

as a best possible result resembling the form (3.3).

We note that $(^{(L)}P(z))_{23}$ in (3.7) is of degree one, and conclude that a specification of the form (3.1)–(3.3) is inadequate for rectangular (as distinct from square) matrix Padé approximation.

Returning to the fundamental problem as expressed in (3.3), we seek to solve the equations

$$\begin{bmatrix} c_{L-M+1} & c_{L-M+2} & \cdots & c_{L+1} \\ c_{L-M+2} & c_{L-M+3} & \cdots & c_{L+2} \\ \vdots & \vdots & & \vdots \\ c_L & c_{L+1} & \cdots & c_{L+M} \end{bmatrix} \begin{bmatrix} Q_M \\ Q_{M-1} \\ \vdots \\ Q_0 \end{bmatrix} = \begin{bmatrix} 0 \\ 0 \\ \vdots \\ 0 \end{bmatrix} \tag{3.8}$$

which take $M \times (M+1)$ block form. All that is obvious, when $m \geq p$, is that at least one nonnull column-vector solution of (3.8) exists. In this way, a nontrivial matrix polynomial

$$Q(z) = \sum_{i=0}^{M} Q_i z^i \tag{3.9}$$

is found which satisfies (3.3); $P(z)$ is obtained by cross-multiplication.

8.3 Rectangular matrix Padé approximants for minimal partial-realization problems

Because, in the case $p = m = 2$, for example, it may happen that

$$Q(0) = \begin{bmatrix} 1 & 0 \\ 0 & 0 \end{bmatrix}, \tag{3.10}$$

it may be that we have not made much progress toward our goal of approximating $F(z)$ as

$$F(z) = P(z)Q(z)^{-1} + O(z^{L+M+1}). \tag{3.11}$$

As we shall see later, specification of the approximation problem and uniqueness of its solution is phrased in terms of given Kronecker indices, or related degree conditions.

The first method we describe in this section is that of Kung [1977] and Kailath [1980]. It yields the equivalent of a Padé approximant of type $[L/M]$ for $f(z)$, in the sense that the offset $L - M$ is specified. The columns of $Q(z)$ are generated sequentially, each having minimal degree. The second method we describe is a progressive algorithm. It constructs the equivalent of a staircase sequence of Padé approximants for $f(z)$, and is due to Dickinson, Morf, and Kailath [1974]. Finally, we explain the application in circuit synthesis for which these methods were originally devised.

The method of Kung and Kailath. Given the matrix series (3.1) and the offset $L - M$, the main step is the formation of the incomplete $2M \times 2M$ block Hankel matrix

$$H = \begin{bmatrix} c_{L-M+1} & c_{L-M+2} & c_{L-M+3} & \cdots & c_{L+M} \\ c_{L-M+2} & c_{L-M+3} & c_{L-M+4} & \cdots & \times \\ \vdots & \vdots & \vdots & & \vdots \\ c_{L+M-1} & c_{L+M} & \times & \cdots & \times \\ c_{L+M} & \times & \times & \cdots & \times \end{bmatrix} \tag{3.12}$$

in which the ×'s are unspecified, because their values are immaterial. Decompose H into its constituent columns:

$$H = (\mathbf{h}_1, \mathbf{h}_2, \ldots, \mathbf{h}_{2Mm}). \tag{3.13}$$

Let $\mathbf{h}_{k_1+1}$ be the first of these which is linearly dependent on its predecessors, in the sense that

$$\mathbf{h}_{k_1+1} = \sum_{k=1}^{k_1} \mu_k^{(1)} \mathbf{h}_k \tag{3.14}$$

for some convenient values of the ×'s, and let

$$\kappa_1 = \left[\frac{k_1}{m} \right], \quad j_1 = k_1 - m\kappa_1 + 1. \tag{3.15}$$

An alternative way of expressing this choice of k_1 is to require that k_1 is the least integer for which (3.14) holds when the vectors $\mathbf{h}_1, \mathbf{h}_2, \ldots, \mathbf{h}_{k_1+1}$ are truncated to their first $2M - \kappa_1$ block rows. The ×'s are called nice elements, because they can now be chosen so that (3.14) remains true for columns of block length $2M$. If there is freedom of choice of the $\mu_k^{(1)}$ in (3.14), we arrange it that as many as possible of $\mu_k^{(1)}$ are zero, i.e., λ_1 is the largest integer for which

$$\mu_1^{(1)} = \mu_2^{(1)} = \cdots = \mu_{\lambda_1}^{(1)} = 0. \tag{3.16}$$

We define an auxiliary vector $\boldsymbol{\mu}^{(1)} \in \mathbb{C}^{2Mm}$ by

$$\boldsymbol{\mu}^{(1)} = (\mu_1^{(1)}, \mu_2^{(1)}, \ldots, \mu_{k_1}^{(1)}, -1, 0, 0, \ldots, 0)^T,$$

and from its blocks we form auxiliary vectors $\mathbf{q}_0^{(1)}, \mathbf{q}_1^{(1)}, \ldots, \mathbf{q}_{\kappa_1}^{(1)}$ as

$$\mathbf{q}_{\kappa_1}^{(1)} = \begin{bmatrix} -\mu_1^{(1)} \\ -\mu_2^{(1)} \\ \cdot \\ \cdot \\ \cdot \\ -\mu_m^{(1)} \end{bmatrix}, \quad \mathbf{q}_{\kappa_1-1}^{(1)} = \begin{bmatrix} -\mu_{m+1}^{(1)} \\ -\mu_{m+2}^{(1)} \\ \cdot \\ \cdot \\ \cdot \\ -\mu_{2m}^{(1)} \end{bmatrix}, \ldots, \mathbf{q}_0^{(1)} = \begin{bmatrix} -\mu_{m\kappa_1+1}^{(1)} \\ \vdots \\ -\mu_{k_1}^{(1)} \\ 1 \\ 0 \\ \vdots \\ 0 \end{bmatrix}, \tag{3.17}$$

and then we define $\mathbf{Q}^{(1)}(z)$ by

$$\mathbf{Q}^{(1)}(z) = \sum_{i=0}^{\kappa_1} \mathbf{q}_i^{(1)} z^i. \tag{3.18}$$

Vectors such as $\mathbf{Q}^{(1)}(z)$ are often referred to as solution vectors. The index k_1 in (3.14) cannot be found if, for example, $\mathbf{h}_1 = \mathbf{h}_2 = \cdots = \mathbf{h}_{2Mm-1} = \mathbf{0}$. If k_1 cannot be found by the specification in (3.14) above, we define $\kappa_1 = 2M$, $j_1 = m$, and

$$\mathbf{Q}^{(1)}(z) = \mathbf{q}_0^{(1)} = (0, 0, \ldots, 0, 1)^T.$$

From this construction, we obtain $\mathbf{Q}^{(1)}(z)$ as an m component vector of polynomials of degree κ_1 at most. Any freedom of choice in (3.14) was used in (3.16) to ensure that $\deg\{\mathbf{Q}^{(1)}(z)\}$ is minimal, and this choice is especially relevant when $m > p$. Moreover, the components of $\mathbf{Q}^{(1)}(z)$ satisfy

$$(\mathbf{Q}^{(1)}(0))_{j_1} = 1, \tag{3.19}$$

and

$$(\mathbf{Q}^{(1)}(0))_j = 0, \quad j = j_1 + 1, j_1 + 2, \ldots, m, \tag{3.20}$$

8.3 Rectangular matrix Padé approximants for minimal partial-realization problems

when $j_1 < m$. The vector polynomial $\mathbf{Q}^{(1)}(z)$ constitutes the first column of $Q(z)$. The choices expressed by (3.14) and (3.16) ensure that $\mathbf{Q}^{(1)}(z)$ is of minimal degree.

By restructuring the index k in (3.14) as $k = im + j$ and using (3.17), we have

$$\mathbf{h}_{k_1+1} = -\sum_{i=0}^{\kappa_1-1}\sum_{j=1}^{m}\mathbf{h}_{im+j}(\mathbf{q}^{(1)}_{\kappa_1-i})_j - \sum_{j=1}^{j_1-1}\mathbf{h}_{\kappa_1 m+j}(\mathbf{q}^{(1)}_0)_j \tag{3.21}$$

which simplifies to

$$\sum_{i=0}^{\kappa_1}\sum_{j=1}^{m}\mathbf{h}_{im+j}(\mathbf{q}^{(1)}_{\kappa_1-i})_j = \mathbf{0}. \tag{3.22}$$

In component form, this is

$$\sum_{i=0}^{\kappa_1}\sum_{j=1}^{m}(\mathbf{h}_{im+j})_l(\mathbf{q}^{(1)}_{\kappa_1-i})_j = 0, \tag{3.23}$$

for $l = 1, 2, \ldots, (2M - \kappa_1)p$. We express l in terms of a block index $s = 1, 2, \ldots, 2M - \kappa_1$ and remainder $r \in [1, p]$ by $l = (s - 1)p + r$, and then

$$(\mathbf{h}_{im+j})_l = (c_{L-M+i+s})_{rj}. \tag{3.24}$$

Thus (3.23) becomes

$$\sum_{i=0}^{\kappa_1}\sum_{j=1}^{m}(c_{L-M+i+s})_{rj}(\mathbf{q}^{(1)}_{\kappa_1-i})_j = 0, \quad s = 1, 2, \ldots, 2M - \kappa_1. \tag{3.25}$$

Equation (3.25) expresses the fact that the coefficient of $z^{\kappa_1+L-M+s}$ in the vector polynomial $\mathbf{X}(z)$ given by

$$X_r(z) = \sum_{j=1}^{m} F_{rj}(z) Q_j^{(1)}(z), \quad r = 1, 2, \ldots, p,$$

vanishes. Hence we can define $\mathbf{P}^{(1)}(z)$ with the property that

$$\sum_{j=1}^{m} F_{rj}(z) Q_j^{(1)}(z) = P_r^{(1)}(z) + O(z^{L+M+1}), \tag{3.26}$$

where

$$\partial\{\mathbf{P}^{(1)}(z)\} \leq L - M + \kappa_1. \tag{3.27}$$

From (3.23) it follows that all columns of index j greater than $k_1 + 1$ and satisfying $j = k_1 + 1 \mod (m)$ are linearly dependent on their predecessors, using the same multipliers $\mu_k^{(1)}$, and the ×'s (the nice elements) remain immaterial. We do not use any such consequential linear dependencies to form polynomials, and we let $\mathbf{h}_{k_2+1}$ be the next column of H, with index $k_2 > k_1$ and $k_2 \neq k_1 \mod(m)$, which is linearly dependent on its predecessors. Then

$$\mathbf{h}_{k_2+1} = \sum_{k=1}^{k_2} \mu_k^{(2)} \mathbf{h}_k. \tag{3.28}$$

We proceed to define κ_2 as in (3.15), $\mathbf{Q}^{(2)}(z)$ as in (3.18) etc., and $\mathbf{P}^{(2)}(z)$ as in (3.26). We continue with this process step by step, and require that each k_i satisfy

$$k_i \neq k_1, k_2, \ldots, k_{i-1} \mod(m). \tag{3.29}$$

We end up with a matrix

$$Q(z) = (\mathbf{Q}^{(1)}(z), \mathbf{Q}^{(2)}(z), \ldots, \mathbf{Q}^{(m)}(z)). \tag{3.30}$$

The condition (3.29) and the structure in (3.19), (3.20) ensure that $Q(0)$ is similar to a unit upper-triangular matrix by permutation of its columns. Thus $Q(0)$ is nonsingular. Similarly we form

$$P(z) = (\mathbf{P}^{(1)}(z), \mathbf{P}^{(2)}(z), \ldots, \mathbf{P}^{(m)}(z)). \tag{3.31}$$

Hence we obtain

$$F(z) = P(z)Q(z)^{-1} + O(z^{L+M+1}) \tag{3.32}$$

as the first major result of this section. The jth column of $P(z)$ is of degree $L - M + \kappa_j$ at most, the jth column of $Q(z)$ is of degree κ_j at most, and each κ_j lies in the range

$$0 \leq \kappa_j \leq 2M. \tag{3.33}$$

The numbers $\kappa_1, \kappa_2, \ldots, \kappa_m$ are called the Kronecker column indices.

If we place the numbers $1, 2, \ldots, (2M+1)m$ one by one into the rows of a table with m columns and $2M + 1$ rows, and circle the numbers $k_1 + 1, k_2 + 1, \ldots, k_m + 1$, it is called a crate table. The condition (3.29) ensures that each circle occurs in precisely one column, and the analogues of (3.15) show that κ_j is the index (in the range 0 to $2M$) of the row in which $k_j + 1$ occurs.

The same process of construction of the partial-realization matrix Padé approximant can be applied to the transpose of $F(z)$; alternatively row operations may be used. Either way, a left-handed matrix Padé approximant of $F(z)$ is obtained, having properties equivalent to (3.19), (3.20) as well as

$$^{(L)}Q(z)^{-1\,(L)}P(z) = F(z) + O(z^{L+M+1}). \tag{3.34}$$

8.3 Rectangular matrix Padé approximants for minimal partial-realization problems

Its corresponding Kronecker indices are called Kronecker row indices and are denoted by $\kappa'_1, \kappa'_2, \ldots, \kappa'_m$.

Example 2. Rectangular matrix Padé approximants with offset $L - M = -1$ are required for

$$F(z) = \begin{bmatrix} 1 & 1 & 1 \\ 1 & 1 & 1 \\ 1 & 0 & 0 \end{bmatrix} + z \begin{bmatrix} 2 & 1 & 3 \\ 2 & 1 & 3 \\ 0 & 0 & 1 \end{bmatrix} + O(z^2). \qquad (3.35)$$

Remark. In this example, the specification means that the left- and right-handed matrix Padé approximants of type [0/1] are required. In fact, the techniques of the previous section could be used, because the matrices in (3.35) are square.

Solution. To form the right-handed approximant, we use

$$H = \begin{bmatrix} 1 & 1 & 1 & 2 & 1 & 3 \\ 1 & 1 & 1 & 2 & 1 & 3 \\ 1 & 0 & 0 & 0 & 0 & 1 \\ 2 & 1 & 3 & \times & \times & \times \\ 2 & 1 & 3 & \times & \times & \times \\ 0 & 0 & 1 & \times & \times & \times \end{bmatrix}. \qquad (3.36)$$

The first three columns of H are linearly independent, and we find the multipliers used in (3.14) et seq. to be

$$-\boldsymbol{\mu}^{(1)} = (0, 0, -2, 1, 0, 0)^T \quad \text{with } k_1 + 1 = 4,$$

$$-\boldsymbol{\mu}^{(2)} = (0, 0, 0, -\tfrac{1}{2}, 1, 0)^T \quad \text{with } k_2 + 1 = 5,$$

$$-\boldsymbol{\mu}^{(3)} = (-1, 0, 0, -1, 0, 1)^T \quad \text{with } k_3 + 1 = 6.$$

The crate table is

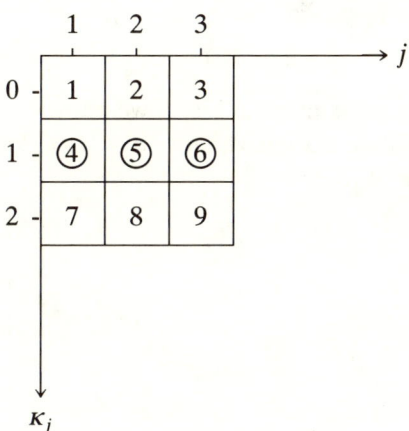

and the corresponding Kronecker column indices are

$$\kappa_1 = \kappa_2 = \kappa_3 = 1. \tag{3.37}$$

From these multipliers we form the denominator matrix as in (3.18):

$$Q(z) = \begin{bmatrix} 1 & -\frac{1}{2} & -1-z \\ 0 & 1 & 0 \\ -2z & 0 & 1 \end{bmatrix}. \tag{3.38}$$

By cross-multiplication, as in (3.26), we find

$$P(z) = \begin{bmatrix} 1 & \frac{1}{2} & 0 \\ 1 & \frac{1}{2} & 0 \\ 1 & -\frac{1}{2} & -1 \end{bmatrix}. \tag{3.39}$$

Hence the [0/1] type right-handed approximant is given by

$$P(z)Q(z)^{-1} = \begin{bmatrix} 1 & 1-z-z^2 & 1+z \\ 1 & 1-z-z^2 & 1+z \\ 1-2z & z^2 & z \end{bmatrix} / (1-2z-2z^2)I \tag{3.40}$$

in which each element is a rational function, and the elements have common poles at the zeros of

$$q(z) = 1 - 2z - 2z^2.$$

However, for application in which the elements of $P(z)$ and $Q(z)$ are related to the values of circuit elements, it is the forms (3.38), (3.39) which are more useful than (3.40), and we express our result as

$$F(z) = \begin{bmatrix} 1+2z & 1+z & 1+3z \\ 1+2z & 1+z & 1+3z \\ 1 & 0 & z \end{bmatrix} + \cdots$$

$$= \begin{bmatrix} 1 & \frac{1}{2} & 0 \\ 1 & \frac{1}{2} & 0 \\ 1 & -\frac{1}{2} & 0 \end{bmatrix} \begin{bmatrix} 1 & -\frac{1}{2} & -1-z \\ 0 & 1 & 0 \\ -2z & 0 & 1 \end{bmatrix}^{-1} + O(z^2). \tag{3.41}$$

For the left-handed approximant of $F(z)$, we copy the previous procedure and find the right-handed approximant to $F^T(z)$. For this, the column table is the transpose of (3.36),

$$H^T = \begin{bmatrix} 1 & 1 & 1 & 2 & 2 & 0 \\ 1 & 1 & 0 & 1 & 1 & 0 \\ 1 & 1 & 0 & 3 & 3 & 1 \\ 2 & 2 & 0 & \times & \times & \times \\ 1 & 1 & 0 & \times & \times & \times \\ 3 & 3 & 1 & \times & \times & \times \end{bmatrix}. \tag{3.42}$$

The multipliers are found to be

8.3 Rectangular matrix Padé approximants for minimal partial-realization problems

$$-\boldsymbol{\mu}'^{(1)} = (-1, 1, 0, 0, 0, 0)^T \quad \text{with } k_1' + 1 = 2,$$
$$-\boldsymbol{\mu}'^{(2)} = (\tfrac{1}{2}, 0, \tfrac{1}{2}, -\tfrac{1}{2}, 0, 1)^T \quad \text{with } k_2' + 1 = 6,$$

and we set

$$-\boldsymbol{\mu}'^{(3)} = (0, 0, 0, 1, 0, 0)^T \quad \text{with } k_3' + 1 = 7.$$

The last of these, $\boldsymbol{\mu}'^{(3)}$, exemplifies the situation when we have 'run out' of useful columns. The Kronecker indices are

$$\kappa_1' = 0, \quad \kappa_2' = 1, \quad \kappa_3' = 2 \tag{3.43}$$

and the crate table is

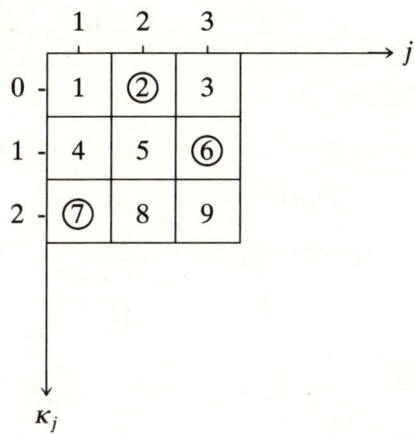

We find

$$^{(L)}Q^T(z) = \begin{bmatrix} -1 & -\tfrac{1}{2} + \tfrac{1}{2}z & 1 \\ 1 & 0 & 0 \\ 0 & 1 + \tfrac{1}{2}z & 0 \end{bmatrix} \tag{3.44}$$

and

$$^{(L)}Q(z)^{-1} \cdot {}^{(L)}P(z) = \begin{bmatrix} -1 & 1 & 0 \\ -\tfrac{1}{2} + \tfrac{1}{2}z & 0 & 1 + \tfrac{1}{2}z \\ 1 & 0 & 0 \end{bmatrix}^{-1}$$

$$\times \begin{bmatrix} 0 & 0 & 0 \\ \tfrac{1}{2} & -\tfrac{1}{2} & -\tfrac{1}{2} \\ 1 + 2z & 1 + z & 1 + 3z \end{bmatrix}$$

$$= \begin{bmatrix} 1 + 2z & 1 + z & 1 + 3z \\ 1 + 2z & 1 + z & 1 + 3z \\ \dfrac{2 + z - 2z^2}{2 + z} & \dfrac{-z^2}{2 + z} & \dfrac{2z - 3z^2}{2 + z} \end{bmatrix}. \tag{3.45}$$

Notice that the degree of the last row of $^{(L)}P(z)$ is unity, reflecting the fact that $\kappa'_3 = 2$ and that the left-handed approximant has precisely one pole. It is evident that the right-handed partial realization approximant (3.40) differs from its left-handed counterpart (3.45). Nevertheless, we have secured the required accuracy-through-order property

$$F(z) = P(z)Q(z)^{-1} + O(z)^2 = {}^{(L)}Q(z)^{-1}{}^{(L)}P(z) + O(z^2).$$

Example 1 [continued from (3.5)]. To solve for the right-handed approximant, we form

$$H = \begin{bmatrix} 0 & -1 & 1 & 1 & 1 & 0 \\ 1 & 1 & 1 & 1 & 0 & 1 \\ 1 & 1 & 0 & \times & \times & \times \\ 1 & 0 & 1 & \times & \times & \times \end{bmatrix}.$$

The first three columns are linearly independent; we find

$$-\boldsymbol{\mu}^{(1)} = (0, 0, -1, 1, 0, 0)^T,$$
$$-\boldsymbol{\mu}^{(2)} = (0, \tfrac{1}{2}, 0, -\tfrac{1}{2}, 1, 0)^T,$$
$$-\boldsymbol{\mu}^{(3)} = (0, 0, 0, -1, 1, 1)^T,$$

and from these we form

$$Q(z) = \begin{bmatrix} 1 & -\tfrac{1}{2} & -1 \\ 0 & 1 + \tfrac{1}{2}z & 1 \\ -z & -1 & 1 \end{bmatrix}$$

leading to the result (3.6). However, we could have used

$$-\boldsymbol{\mu}^{(2)} = (0, \tfrac{1}{2}, -\tfrac{1}{2}, 0, 1, 0)$$

which would lead to a different denominator matrix polynomial.

For the left-handed approximant, we consider

$$F^T(z) = \begin{bmatrix} z & 1 + z \\ -1 + z & 1 \\ 1 & 1 + z \end{bmatrix}$$

and form

$$H^T = \begin{bmatrix} 0 & 1 & 1 & 1 \\ -1 & 1 & 1 & 0 \\ 1 & 1 & 0 & 1 \\ 1 & 1 & \times & \times \\ 1 & 0 & \times & \times \\ 0 & 1 & \times & \times \end{bmatrix}.$$

8.3 Rectangular matrix Padé approximants for minimal partial-realization problems

We find that $\mathbf{h}'_4 = \mathbf{h}'_1 + \mathbf{h}'_3$, but $\mathbf{h}'_3$ cannot be expressed as a linear combination of the other column vectors of H^T. Consequently, we form

$$-\boldsymbol{\mu}'^{(1)} = (-1, 0, -1, 1)^T$$

and

$$\mathbf{Q}^{(1)}(z) = \begin{bmatrix} -1 - z \\ 1 \end{bmatrix}.$$

There is no ideal choice for $\mathbf{Q}^{(2)}(z)$, and (3.7) gives one best possible result out of many.

Further examples are given by Kailath [1980], Van Barel and Bultheel [1987a], and Van Barel [1989].

Kronecker's theorem. *The Kronecker row and column indices for the $p \times m$ partial-realization matrix Padé approximant problem satisfy*

$$\sum_{i=1}^{m} \kappa_i = \sum_{i=1}^{p} \kappa'_i. \tag{3.46}$$

Proof. First, the nice elements of H are assigned values compatible with its Hankel structure, so that H is fully specified and $\text{rank}(H)$ is given meaning. Consider the column vectors $\{\mathbf{h}_j\}_{j=m+1}^{2Mm}$ in turn. If, for any α and any integer $l \geq 0$, $j = k_\alpha + lm + 1$, then the nice elements of $\mathbf{h}_j$ are determined by

$$\mathbf{h}_j = \sum_{k=1}^{k_\alpha} \mu_k^{(\alpha)} \mathbf{h}_{k+lm}.$$

Otherwise, $j \neq k_\alpha + lm + 1$ for any α and any $l \geq 0$, and then any nice elements of $\mathbf{h}_j$ that are not determined by the Hankel structure are set to zero. In this way, the linear dependencies (3.21), (3.28) are extended to the full matrix H.

A Kronecker column index of value κ_i implies that precisely $2M - \kappa_i$ of the vectors $\{\mathbf{h}_j\}_{j=1}^{2Mm}$ are linearly dependent, and this result does not duplicate similar results for other column vectors [see (3.29)]. Hence

$$\text{rank } H = \text{rank } \{\mathbf{h}_j\}_{j=1}^{2Mm} = 2Mm - \sum_{i=1}^{m}(2M - \kappa_i) = \sum_{i=1}^{m} \kappa_i.$$

Second, the above argument does not apply to rows of the full matrix H with the above choice of nice elements, because the equivalent $2M - \kappa'_i$ linear dependencies for subrows associated with Kronecker row index κ'_i do not necessarily extend to the whole rows; specification of the nice entries cannot introduce linear dependency, but may only remove it.

Hence

$$\text{rank}\,(H) = \text{rank}\,\{(H^T)_j\}_{j=1}^{2Mp} \geq 2Mp - \sum_{i=1}^{p}(2M - \kappa_i') = \sum_{i=1}^{p}\kappa_i',$$

where $(H^T)_j$ denotes the jth column of H^T, and hence $\sum_{i=1}^{p}\kappa_i' \leq \sum_{i=1}^{m}\kappa_i$.

By applying the previous argument to H^T, we find that $\sum_{i=1}^{m}\kappa_i \leq \sum_{i=1}^{p}\kappa_i'$, and therefore (3.46) is established.

This theorem [see Bosgra and van der Weiden, 1980] is borne out by Example 2. Note that the default specification $\kappa_i = 2M$ or $\kappa_i' = 2M$ allows this theorem, which is basically a result about rank, to hold even when only trivial left-handed (or right-handed) solution vectors can be constructed.

The left- and right-handed approximants found by the previous method have the properties that

$$^{(L)}Q(z)^{-1}{}^{(L)}P(z) = F(z) + O(z^{L+M+1}),$$

$$P(z)Q(z)^{-1} = F(z) + O(z^{L+M+1}).$$

In consequence, we have the following uniqueness result.

Theorem. *Suppose that the column-vector polynomials constituting $Q(z)$, $P(z)$ satisfy*

$$\deg\{\mathbf{Q}^{(i)}(z)\} \leq \alpha_i,$$

$$\deg\{\mathbf{P}^{(i)}(z)\} \leq L - M + \alpha_i,$$

and the row-vector polynomials constituting $^{(L)}Q(z)$, $^{(L)}P(z)$ satisfy

$$\deg\{\mathbf{Q}^{(i)T}(z)\} \leq \alpha_i',$$

$$\deg\{\mathbf{P}^{(i)T}(z)\} \leq L - M + \alpha_i'.$$

Then, provided $\max\{\alpha_i\} + \max\{\alpha_i'\} \leq 2M$,

$$^{(L)}Q(z)^{-1} \cdot {}^{(L)}P(z) = P(z)Q(z)^{-1}.$$

Proof. Because $Q(0)$, $^{(L)}Q(0)$ are nonsingular, this result follows by cross-multiplication.

To treat large systems and to further the mathematical development of rectangular matrix Padé approximation, we are also interested in a recursive solution of the minimal partial-realization problem. Indeed, the method we describe next has the same iteration procedure as the original method given by Dickinson, Morf, and Kailath [1974] and Anderson,

8.3 Rectangular matrix Padé approximants for minimal partial-realization problems

Brasch, and Lopresti [1975] for the problem, and we explain it as a logical extension of the Berlekamp–Massey algorithm, but with the initialization of Van Barel and Bultheel [1987a] and Van Barel [1989].

We begin with the transfer function defined in s-form by

$$F(s) = c_0 + c_1 s^{-1} + c_2 s^{-2} + \cdots, \quad c_k \in \mathbb{R}^{p \times m}, \quad (3.47)$$

following engineering conventions, but note that (3.47) is slightly notationally inconsistent with (3.1). As with (3.1), the treatment is exactly parallel for $c_k \in \mathbb{C}^{p \times m}$. Using the connections explained in Section 4.4, it is straightforward to substitute $z = s^{-1}$ to convert (3.47) to (3.1) and to re-express the formalism in that way.

As in the scalar case, the method is based on the elimination of residuals, but in this case the residuals lie in $\mathbb{R}^p$ rather than $\mathbb{R}$. Therefore, either explicitly or implicitly, we have to use a basis for $\mathbb{R}^p$ and perform the elimination on as many of the p elements of this basis as is necessary.

Specification. We find a sequence $\{(P_n(s), Q_n(s))\}_{n=0}^\infty$ of polynomial matrices $P_n(s) \in \mathbb{R}^{p \times m}[s]$, $Q_n(s) \in \mathbb{R}^{m \times m}[s]$ of minimal degrees such that

$$F(s) \mathbf{Q}_n^{(j)}(s) - \mathbf{P}_n^{(j)}(s) = O(s^{\alpha_n^{(j)} - n - 1}). \quad (3.48)$$

Here, $\alpha_n^{(j)} = \deg\{\mathbf{Q}_n^{(j)}(s)\}$, where $\mathbf{Q}_n^{(j)}(s)$ is the jth column of $Q_n(s)$; the degree of a square matrix means the degree of its determinant.

Again, from a mathematical viewpoint, the specification of 'minimal degrees' above is slightly incomplete. The exact form of the polynomial matrices derived depends significantly on the initialization adopted and is determined by the action of the algorithm. We refer to Bultheel and Van Barel [1987c] and Van Barel [1989] for the modifications needed when an offset between the degrees of $Q_n(s)$ and $P_n(s)$ is required. As in the z-formulation, the columns $\mathbf{Q}_n^{(j)}(s)$ are still referred to as solution vectors, even though their role as described in this section explains why they are also called annihilating column vectors.

Construction. The iterative stage of the construction is designed to eliminate the residual vectors progressively. We define the residual vector $\mathbf{R}_n^{(j)}(s)$ by

$$F(s) \mathbf{Q}_n^{(j)}(s) = \mathbf{P}_n^{(j)}(s) + \mathbf{R}_n^{(j)}(s) s^{\alpha_n^{(j)} - n - 1}. \quad (3.49)$$

By construction, $\{\mathbf{R}_n^{(j)}(\infty)\}_{j=1}^m$ are well defined, and it is these m leading residuals which are, if necessary, eliminated at each stage n. There are a number of different ways to initialize the process; with a view to the V-matrix formulation, we use the decomposition

$$F(s) I_m = c_0 + R_0(s) \quad (3.50)$$

at stage $n = 0$ and artificially at stage $n = -1$ we use

$$F(s)O = I_p + R_{-1}(s), \tag{3.51}$$

where I_p, I_m are unit p-dimensional and m-dimensional matrices, respectively, and the O in (3.51) denotes an $m \times p$ null matrix. This procedure may be viewed as initialization with approximants of types [0/−1] in (3.51) and [0/0] in (3.50). These formulas are reflected in the actual initializations.

$$Q_0(s) = I_m, \quad P_0(s) = c_0, \quad X_0(s) = 0, \quad Y_0(s) = I_p.$$

For reasons to be explained next, the latter two matrices are assigned potential degrees 1 at stage $n = 0$, reflecting their introduction as constants at stage -1. Just as in (4.4.38), (4.4.1)–(4.4.44), auxiliary vectors are required to eliminate the leading residuals from (3.49). Each auxiliary vector $\mathbf{X}_n^{(j)}(s)$ originated as a column vector of $Q_{n'}(s)$ for some $n' < n$. We suppose that, at some stage $n' < n$, it was $\mathbf{Q}_{n'}^{(j')}$ which was transferred to $X_{n'}(s)$ and thence to $X_n(s)$. Thus to each auxiliary vector $\mathbf{X}_n^{(j)}(s)$ we associate a corresponding residual vector function $\mathbf{S}_n^{(j)}(s)$ defined by

$$F(s)\mathbf{X}_n^{(j)}(s) = \mathbf{Y}_n^{(j)}(s) + \mathbf{S}_n^{(j)}(s)s^{\alpha_{n'}^{(j')}-n'-1}, \tag{3.52}$$

where $\mathbf{Y}_n^{(j)}(s)$ is the polynomial part of the left-hand side, and $\mathbf{S}_n^{(j)}(\infty)$ is well defined, as a property inherited from (3.49). We also associate with $\mathbf{X}_n^{(j)}(s)$, $\mathbf{Y}_n^{(j)}(s)$, and $\mathbf{S}_n^{(j)}(s)$ a potential degree defined by

$$\pi_n^{(j)} = n + \alpha_{n'}^{(j')} - n'. \tag{3.53}$$

This quantity depends on the stage n which the algorithm has reached, and so is incremented by unity at the end of each stage of the numerical algorithm. From (3.52), (3.53), we have

$$F(s)\mathbf{X}_n^{(j)}(s) = \mathbf{Y}_n^{(j)}(s) + \mathbf{S}_n^{(j)}(s)s^{\pi_n^{(j)}-n-1}. \tag{3.54}$$

By comparing (3.54) with (3.49), we see why $\pi_n^{(j)}$ is called the potential degree of the auxiliary quantities.

At each stage n of the iteration, the column indices $j = 1, 2, \ldots, m$ are considered in turn, and the leading residuals $\mathbf{R}_n^{(j)}(\infty)$ are formed, using a formula based on (3.49). If $\mathbf{R}_n^{(j)}(\infty) = \mathbf{0}$, there is no work to be done and we proceed to the next value of j. Otherwise, it is always possible to express

$$\mathbf{R}_n^{(j)}(\infty) + \sum_{l=1}^{p} \mathbf{S}_n^{(l)}(\infty)\mu_l = \mathbf{0} \tag{3.55}$$

using nontrivial and uniquely determined parameters μ_l, because at every stage the vectors $\{\mathbf{S}_n^{(l)}(\infty)\}_{l=1}^{p}$ form a basis of $\mathbb{R}^p$. Let $\mathbf{S}_n^{(\hat{j})}(\infty)$ be the

8.3 Rectangular matrix Padé approximants for minimal partial-realization problems

vector of maximum potential degree $\pi_n^{(\hat{j})}(\infty)$ of the vectors used nontrivially in (3.55). We suppose that (3.55) has been constructed so that $\pi_n^{(\hat{j})}$ is a minimum over all choices of $\{\mu_l\}$.

If $\pi_n^{(\hat{j})} \leq \alpha_n^{(j)}$, a minor change analogous to (4.4.43), (4.4.44) is all that is needed to eliminate the jth leading residual at stage n. We define

$$\mathbf{Q}_{n+1}^{(j)}(s) = \mathbf{Q}_n^{(j)}(s) + \sum_{l=1}^{p} \mathbf{X}_n^{(l)}(s)\mu_l s^{\alpha_n^{(j)} - \pi_n^{(l)}}, \qquad (3.56)$$

$$\mathbf{P}_{n+1}^{(j)}(s) = \mathbf{P}_n^{(j)}(s) + \sum_{l=1}^{p} \mathbf{Y}_n^{(l)}(s)\mu_l s^{\alpha_n^{(j)} - \pi_n^{(l)}}, \qquad (3.57)$$

and note that the right-hand sides of (3.56), (3.57) are polynomial vectors chosen to reduce the order of the residual without increasing the degree of the solution vector.

If $\pi_n^{(\hat{j})} > \alpha_n^{(j)}$, it is not possible to reduce the order of the residual without increasing the order of the solution vector. We make the major change analogous to (4.4.41), (4.4.42), and define

$$\mathbf{Q}_{n+1}^{(j)}(s) = s^{\pi_n^{(\hat{j})} - \alpha_n^{(j)}} \mathbf{Q}_n^{(j)}(s) + \sum_{l=1}^{p} \mathbf{X}_n^{(l)}(s)\mu_l s^{\pi_n^{(\hat{j})} - \pi_n^{(l)}}, \qquad (3.58)$$

$$\mathbf{P}_{n+1}^{(j)}(s) = s^{\pi_n^{(\hat{j})} - \alpha_n^{(j)}} \mathbf{P}_n^{(j)}(s) + \sum_{l=1}^{p} \mathbf{Y}_n^{(l)}(s)\mu_l s^{\pi_n^{(\hat{j})} - \pi_n^{(l)}}. \qquad (3.59)$$

The vectors $\mathbf{X}_n^{(\hat{j})}(s)$, $\mathbf{Y}_n^{(\hat{j})}(s)$, and $\mathbf{S}_n^{(\hat{j})}(s)$ are replaced immediately with $\mathbf{Q}_n^{(j)}(s)$, $\mathbf{P}_n^{(j)}(s)$, and $\mathbf{R}_n^{(j)}(s)$. In this sense, the former quantities are updated as the algorithm progresses and these symbols do not have an absolute mathematical meaning. At this stage, their potential degree is made equal to their actual degree, as indicated by (3.53). From (3.55), it is clear that the vectors of the updated space spanned by $\{\mathbf{S}_n^{(l)}(\infty)\}$ form a basis of $\mathbb{R}^p$, and that the property expressed by (3.55) is established inductively.

At the end of each stage n, the potential degrees are incremented by one, and their vectors $\mathbf{X}_n^{(l)}(s)$, $\mathbf{Y}_n^{(l)}(s)$, and $\mathbf{S}_n^{(l)}(s)$ are transferred by reindexing to stage $n + 1$.

Example [Dickinson, Morf, and Kailath, 1974]. Find the right-hand minimal partial realization for the transfer function

$$F(s) = \begin{bmatrix} -1 & 1 \\ 0 & 0 \end{bmatrix} s^{-1} + \begin{bmatrix} 0 & 1 \\ 0 & 0 \end{bmatrix} s^{-2} + \begin{bmatrix} 1 & 1 \\ -1 & 1 \end{bmatrix} s^{-3} + \begin{bmatrix} 1 & 2 \\ 0 & 1 \end{bmatrix} s^{-4}$$
$$+ \begin{bmatrix} 1 & 4 \\ 1 & 1 \end{bmatrix} s^{-5} + \begin{bmatrix} 2 & 7 \\ 1 & 2 \end{bmatrix} s^{-6} + \cdots.$$

Solution

Initialization

$$Q_0(s) = \begin{bmatrix} 1 & 0 \\ 0 & 1 \end{bmatrix}, \quad P_0(s) = \begin{bmatrix} 0 & 0 \\ 0 & 0 \end{bmatrix}, \quad X_0(s) = \begin{bmatrix} 0 & 0 \\ 0 & 0 \end{bmatrix},$$

$$Y_0 = \begin{bmatrix} 1 & 0 \\ 0 & 1 \end{bmatrix}, \quad R_0(s) = \begin{bmatrix} -1 & 1 \\ 0 & 1 \end{bmatrix} + \cdots, \quad S_0(s) = \begin{bmatrix} -1 & 0 \\ 0 & -1 \end{bmatrix},$$

and the potential degrees of each $\mathbf{X}_0^{(j)}$ at stage $n = 0$ are $\pi_0^{(j)} = 1$.

Iteration. The residuals $\mathbf{R}_n^{(j)}(\infty)$ are calculated from (3.49) using a formula of the form (4.4.9) at each stage (n, j) of the algorithm.

Stage $n = 0$.
First,

$$\mathbf{R}_0^{(1)}(\infty) = \begin{bmatrix} -1 \\ 0 \end{bmatrix} \quad \text{and} \quad \mathbf{R}_0^{(1)}(\infty) - \mathbf{S}_0^{(1)}(\infty) = \mathbf{0}.$$

The relevant degrees are $\alpha_0^{(1)} = 0$, $\pi_0^{(1)} = 1$, and so we make a major change of the form (3.56), (3.57):

$$\mathbf{Q}_1^{(1)}(s) = s\mathbf{Q}_0^{(1)}(s) - \mathbf{X}_0^{(1)}(s) = \begin{bmatrix} s \\ 0 \end{bmatrix},$$

$$\mathbf{P}_1^{(1)}(s) = s\mathbf{P}_0^{(1)}(s) - \mathbf{Y}_0^{(1)}(s) = \begin{bmatrix} -1 \\ 0 \end{bmatrix}.$$

Then $\mathbf{Q}_0^{(1)}(s)$, $\mathbf{P}_0^{(1)}(s)$, and $\mathbf{R}_0^{(1)}(\infty)$ replace $\mathbf{X}_0^{(1)}(s)$, $\mathbf{Y}_0^{(1)}(s)$, and $\mathbf{S}_0^{(1)}(\infty)$, respectively, so currently

$$S_0(\infty) = \begin{bmatrix} -1 & 0 \\ 0 & -1 \end{bmatrix}, \quad X_0(s) = \begin{bmatrix} 1 & 0 \\ 0 & 0 \end{bmatrix}, \quad Y_0(s) = \begin{bmatrix} 0 & 0 \\ 0 & 1 \end{bmatrix}$$

with potential degrees 0, 1 for their columns.
Second,

$$\mathbf{R}_0^{(2)}(\infty) = \begin{bmatrix} 1 \\ 0 \end{bmatrix} \quad \text{and} \quad \mathbf{R}_0^{(2)}(\infty) + \mathbf{S}_0^{(1)}(\infty) = \mathbf{0}.$$

The relevant degrees are $\alpha_0^{(2)} = 0$ and $\pi_0^{(1)} = 0$, and so we make a minor change of the form (3.54), (3.55):

$$\mathbf{Q}_1^{(2)}(s) = \mathbf{Q}_0^{(2)}(s) + \mathbf{X}_0^{(1)}(s) = \begin{bmatrix} 1 \\ 1 \end{bmatrix},$$

$$\mathbf{P}_1^{(2)}(s) = \mathbf{P}_0^{(2)}(s) + \mathbf{Y}_0^{(1)}(s) = \begin{bmatrix} 0 \\ 0 \end{bmatrix}.$$

8.3 Rectangular matrix Padé approximants for minimal partial-realization problems

We finish stage $n = 0$ with

$$Q_1(s) = \begin{bmatrix} s & 1 \\ 0 & 1 \end{bmatrix}, \quad P_1(s) = \begin{bmatrix} -1 & 0 \\ 0 & 0 \end{bmatrix}, \quad X_1(s) = \begin{bmatrix} 1 & 0 \\ 0 & 0 \end{bmatrix},$$

$$Y_1(s) = \begin{bmatrix} 0 & 0 \\ 0 & 1 \end{bmatrix}, \quad S_1(\infty) = \begin{bmatrix} -1 & 0 \\ 0 & -1 \end{bmatrix}, \quad \pi_1^{(1)} = 1, \quad \pi_1^{(2)} = 2.$$

Stage $n = 1$.
First,

$$\mathbf{R}_1^{(1)}(\infty) = \begin{bmatrix} 0 \\ 0 \end{bmatrix},$$

and therefore

$$\mathbf{Q}_2^{(1)}(s) = \mathbf{Q}_1^{(1)}(s) = \begin{bmatrix} s \\ 0 \end{bmatrix}.$$

Second,

$$\mathbf{R}_1^{(2)}(\infty) = \begin{bmatrix} 1 \\ 0 \end{bmatrix} \quad \text{and } \mathbf{R}_1^{(2)}(\infty) + \mathbf{S}_1^{(1)}(\infty) = \mathbf{0}.$$

The relevant degrees are $\alpha_1^{(2)} = 0$, $\pi_1^{(1)} = 1$, and so we make a major change of the form (3.56), (3.57):

$$\mathbf{Q}_2^{(2)}(s) = s\mathbf{Q}_1^{(2)}(s) + \mathbf{X}_1^{(1)}(s) = \begin{bmatrix} s + 1 \\ s \end{bmatrix},$$

$$\mathbf{P}_2^{(2)}(s) = s\mathbf{P}_1^{(2)}(s) + \mathbf{Y}_1^{(1)}(s) = \begin{bmatrix} 0 \\ 0 \end{bmatrix}.$$

The vectors $\mathbf{Q}_1^{(2)}(s)$, $\mathbf{P}_1^{(2)}(s)$, and $\mathbf{R}_1^{(2)}(s)$ replace $\mathbf{X}_1^{(1)}(s)$, $\mathbf{Y}_1^{(1)}(s)$, and $\mathbf{S}_1^{(1)}(s)$, respectively, with potential degree zero at stage $n = 1$.

We finish stage $n = 1$ with

$$Q_2(s) = \begin{bmatrix} s & s+1 \\ 0 & s \end{bmatrix}, \quad P_2(s) = \begin{bmatrix} -1 & 0 \\ 0 & 0 \end{bmatrix}, \quad X_2(s) = \begin{bmatrix} 1 & 0 \\ 1 & 0 \end{bmatrix},$$

$$Y_2(s) = \begin{bmatrix} 0 & 0 \\ 0 & 1 \end{bmatrix}, \quad S_2(\infty) = \begin{bmatrix} 1 & 0 \\ 0 & -1 \end{bmatrix}, \quad \pi_2^{(1)} = 1, \quad \pi_2^{(2)} = 3.$$

Stage $n = 2$.
First,

$$\mathbf{R}_2^{(1)}(\infty) = \begin{bmatrix} 1 \\ -1 \end{bmatrix},$$

and

$$\mathbf{R}_2^{(1)}(\infty) - \mathbf{S}_2^{(1)}(\infty) - \mathbf{S}_2^{(2)}(\infty) = \mathbf{0}.$$

The relevant degrees are $\alpha_2^{(1)} = 1$, $\pi_2^{(1)} = 1$, $\pi_2^{(2)} = 3$ and so we make a major change of the form (3.56), (3.57):

$$\mathbf{Q}_3^{(1)}(s) = s^2 \mathbf{Q}_2^{(1)}(s) - s^2 \mathbf{X}_2^{(1)}(s) - \mathbf{X}_2^{(2)}(s) = \begin{bmatrix} s^3 - s^2 \\ -s^2 \end{bmatrix},$$

$$\mathbf{P}_3^{(1)}(s) = s^2 \mathbf{P}_2^{(1)}(s) - s^2 \mathbf{Y}_2^{(1)}(s) - \mathbf{Y}_2^{(2)}(s) = \begin{bmatrix} -s^2 \\ -1 \end{bmatrix},$$

and $\mathbf{Q}_2^{(1)}(s)$, $\mathbf{P}_2^{(1)}(s)$, and $\mathbf{R}_2^{(1)}(\infty)$ replace $\mathbf{X}_2^{(2)}(s)$, $\mathbf{Y}_2^{(2)}(s)$, and $\mathbf{S}_2^{(2)}(\infty)$, respectively, so that currently

$$S_2(\infty) = \begin{bmatrix} 1 & 1 \\ 0 & -1 \end{bmatrix}, \quad X_2(s) = \begin{bmatrix} 1 & s \\ 1 & 0 \end{bmatrix}, \quad Y_2(s) = \begin{bmatrix} 0 & -1 \\ 0 & 0 \end{bmatrix},$$

with potential degrees 1, 1 for their columns.

Second,

$$\mathbf{R}_2^{(2)}(\infty) = \begin{bmatrix} 2 \\ 0 \end{bmatrix}$$

and

$$\mathbf{R}_2^{(2)}(\infty) - 2\mathbf{S}_2^{(1)}(\infty) = \mathbf{0}.$$

The relevant degrees are $\alpha_2^{(2)} = 1$, $\pi_2^{(1)} = 1$ and so we make the minor change

$$\mathbf{Q}_3^{(2)}(s) = \mathbf{Q}_2^{(2)}(s) - 2\mathbf{X}_2^{(1)}(s) = \begin{bmatrix} s - 1 \\ s - 2 \end{bmatrix},$$

$$\mathbf{P}_3^{(2)}(s) = \mathbf{P}_2^{(2)}(s) - 2\mathbf{Y}_2^{(1)}(s) = \begin{bmatrix} 0 \\ 0 \end{bmatrix}.$$

We finish stage $n = 2$ with

$$Q_3(s) = \begin{bmatrix} s^3 - s^2 & s - 1 \\ -s^2 & s - 2 \end{bmatrix}, \quad P_3(s) = \begin{bmatrix} -s^2 & 0 \\ -1 & 0 \end{bmatrix}, \quad X_3(s) = \begin{bmatrix} 1 & s \\ 1 & 0 \end{bmatrix},$$

$$Y_3(s) = \begin{bmatrix} 0 & -1 \\ 0 & 0 \end{bmatrix}, \quad S_3(\infty) = \begin{bmatrix} 1 & 1 \\ 0 & -1 \end{bmatrix}, \quad \pi_3^{(1)} = 2, \quad \pi_2^{(2)} = 2.$$

Stage $n = 3$.
First,

$$\mathbf{R}_3^{(1)}(\infty) = \begin{bmatrix} -1 \\ 0 \end{bmatrix}$$

and

$$\mathbf{R}_3^{(1)}(\infty) + \mathbf{S}_3^{(1)}(\infty) = \mathbf{0},$$

8.3 Rectangular matrix Padé approximants for minimal partial-realization problems

leading to the minor change

$$Q_4^{(1)}(s) = Q_3^{(1)}(s) + sX_3^{(1)}(s) = \begin{bmatrix} s^3 - s^2 + s \\ -s^2 + s \end{bmatrix},$$

$$P_4^{(1)}(s) = P_3^{(1)}(s) + sY_3^{(1)}(s) = \begin{bmatrix} -s^2 \\ -1 \end{bmatrix}.$$

Second,

$$R_3^{(2)}(\infty) = 0,$$

and we finish with

$$Q_4(s) = \begin{bmatrix} s^3 - s^2 + s & s - 1 \\ -s^2 + s & s - 2 \end{bmatrix}, \quad P_4(s) = \begin{bmatrix} -s^2 & 0 \\ -1 & 0 \end{bmatrix}, \quad X_4 = \begin{bmatrix} 1 & s \\ 1 & 0 \end{bmatrix}$$

$$Y_4(s) = \begin{bmatrix} 0 & -1 \\ 0 & 0 \end{bmatrix}, \quad S_4(\infty) = \begin{bmatrix} 1 & 1 \\ 0 & -1 \end{bmatrix}, \quad \pi_4^{(1)} = 3, \quad \pi_4^{(2)} = 3.$$

At stages 4 and 5 the residuals are zero, and we find that

$$F(s) \simeq P_4(s)[Q_4(s)]^{-1} = \begin{bmatrix} -s^2 & 0 \\ -1 & 0 \end{bmatrix} \begin{bmatrix} s^3 - s^2 + s & s - 1 \\ -s^2 + s & s - 2 \end{bmatrix}^{-1} \quad (3.60)$$

provides the partial realization required.

It will be noted that the results obtained by Van Barel and Bultheel [1987a] and Dickinson, Morf, and Kailath [1974], namely,

$$Q_4^{\text{BVB}}(s) = \begin{bmatrix} s^3 & s - 1 \\ -s & s - 2 \end{bmatrix}, \quad Q_4^{\text{DMK}}(s) = \begin{bmatrix} -s^3 - s^2 + s & s - 1 \\ -s^2 + 3s & s - 2 \end{bmatrix},$$

differ from $Q_4(s)$ in (3.60) by an unimportant unimodular transformation and the reason for the difference is that different initialization procedures have been used.

The further mathematical development of this topic has been based on V-matrices, introduced in Section 4.4, which concisely represent the formulas (3.56)–(3.59). We avoid the notational ambiguity of updates by using $Q_{n,j}(s)$, $P_{n,j}(s)$, $X_{n,j}(s)$, and $Y_{n,j}(s)$ to denote the denominator and numerator polynomials and their auxiliaries at the stage (n, j) of the iteration. Then the updates all take the form

$$\begin{bmatrix} Y_{n,j}(s) & P_{n,j}(s) \\ X_{n,j}(s) & Q_{n,j}(s) \end{bmatrix} V_{n,j}(s) = \begin{bmatrix} Y_{n,j+1}(s) & P_{n,j+1}(s) \\ X_{n,j+1}(s) & Q_{n,j+1}(s) \end{bmatrix} \quad (3.61)$$

where the minor change of (3.56), (3.57) is represented by

$$V_{n,j}^{\text{minor}}(s) = \begin{bmatrix} 1 & & & & & & & \mu_1 s^{\alpha_n^{(j)}-\pi_n^{(1)}} & & & \\ & 1 & & & & & & \cdot & & & \\ & & \cdot & & 0 & & & \cdot & & 0 & \\ & & & \cdot & & & & \cdot & & & \\ & 0 & & & 1 & & & & & & \\ & & & & & 1 & & \mu_p s^{\alpha_n^{(p)}-\pi_n^{(p)}} & & & \\ \hline & & & & & & 1 & & & & \\ & & & & & & & 1 & & 0 & \\ & & 0 & & & & & & \cdot & & \\ & & & & & & & 0 & & 1 & \\ & & & & & & & & & & 1 \end{bmatrix}$$

$$\uparrow$$
column $p+j$

(3.62)

and the major change of (3.58), (3.59) is represented by

$V_{n,j}^{\text{major}}(s) =$

$$\begin{bmatrix} 1 & & & & & & & & & \mu_1 s^{\pi_n^{(\hat{j})}-\pi_n^{(1)}} & & & \\ & 1 & & & & & 0 & & & \cdot & & & \\ & & \cdot & & & & & & & \cdot & & & \\ & & & 1 & & & & & & \cdot & & & \\ & & & & 0 & & & & & 0 & \mu_{\hat{j}} & & 0 \\ & & & & 1 & & & & & \cdot & & & \\ & 0 & & & & \cdot & & & & \cdot & & & \\ & & & & & & 1 & & & \mu_p s^{\pi_n^{(\hat{j})}-\pi_n^{(p)}} & & & \\ \hline & & 0 & & & & & 1 & & & & & \\ & & \cdot & & & & & & \cdot & & & 0 & \\ & & 0 & & & & & & & 1 & & & \\ & 0 & 1 & & 0 & & & & & & s^{\pi_n^{(\hat{j})}-\alpha_n^{(j)}} & & \\ & & 0 & & & & & & & & & 1 & \\ & & \cdot & & & & & & & 0 & & & \cdot \\ & & 0 & & & & & & & & & & 1 \end{bmatrix}$$

$$\uparrow \qquad\qquad \uparrow$$
column $\hat{j}$ column $p+j$

(3.63)

8.3 Rectangular matrix Padé approximants for minimal partial-realization problems

Each update is induced by a unimodular matrix (3.62) or (3.63). Moreover, the initialization is

$$\begin{bmatrix} Y_{0,1}(s) & P_{0,1}(s) \\ X_{0,1}(s) & Q_{0,1}(s) \end{bmatrix} = \begin{bmatrix} I_p & c_0 \\ 0 & I_m \end{bmatrix}$$

which is also unimodular, and hence we find that

$$\begin{bmatrix} Y_{n,j}(s) & P_{n,j}(s) \\ X_{n,j}(s) & Q_{n,j}(s) \end{bmatrix} = \prod_{\substack{n'<n \\ j'<j}} V_{n',j'}(s)$$

(with an ordered product) is a unimodular representation of the partial realization. It has been exploited by Antoulas [1986] and Van Barel and Bultheel [1987b] for the recursive construction of linear systems in more detail than we give here. Our main conclusion is that (3.56)–(3.59) are a set of equations to be used for the recursive computation of rectangular matrix Padé approximants, and their action is represented schematically by (3.62), (3.63).

The two methods given in this section can be compared.

The denominator $Q_n(s)$ is said to be column reduced (see Kailath [1980, p. 413]) with column degrees $\alpha_n^{(1)}, \ldots, \alpha_n^{(m)}$. By postmultiplying (3.48) by $\text{diag}\{s^{-\alpha_n^{(1)}}, \ldots, s^{-\alpha_n^{(m)}}\}$, it follows that

$$F(s) = P_n(s)Q_n(s)^{-1} + O(s^{-n-1}).$$

If we run the algorithm until $n = 2M$ and set

$$P(z) = P_n(z^{-1}) \, \text{diag}\{z^{\alpha_n^{(1)}}, \ldots, z^{\alpha_n^{(m)}}\},$$

$$Q(z) = Q_n(z^{-1}) \, \text{diag}\{z^{\alpha_n^{(1)}}, \ldots, z^{\alpha_n^{(m)}}\},$$

we obtain a solution for (3.32) in the case $L = M$. However, if $L = M - 1$, it is necessary to reinitialize (3.47) with $F(s^{-1}) \to s^{-1}F(s^{-1})$ so that $c_0 = 0$ (as in the previous example), and then to replace $P(z)$ by

$$P(z) = z^{-1} P_n(z^{-1}) \, \text{diag}\{z^{\alpha_n^{(1)}}, \ldots, z^{\alpha_n^{(m)}}\}.$$

Up to a permutation, the degrees $\alpha_n^{(i)}$ are the Kronecker column indices. A more direct connection between these results and (3.30), (3.31) depends on generalizing the Kailath–Kung method to include all equivalent osculatory matrix rational functions of the required type [Van Barel and Bultheel, 1987c; Van Barel, 1989].

The principal motive for deriving the methods of this section was the derivation of the circuit elements on an m-input, p-output system having certain given Markov parameters. The single-input, single-output system has already been described in Section 4.4 as an application of ordinary Padé approximation.

The single-input, p-output, system would involve $c_1, c_2, \ldots, c_N$ as the Markov parameters to be realized, with $c_i \in \mathbb{R}^p$. We would then begin with

$$f(z) = c_1 s^{-1} + c_2 s^{-2} + \cdots + c_N s^{-N} + \cdots$$

as the transfer function to be realized. The approximant required can be derived by the methods of this section, or using the explicit formulas for simultaneous Padé approximation (see Section 8.1). For example, for the case of $N = 3$, $p = 2$,

$$Q(s) = \begin{vmatrix} c_{3,1} & c_{2,1} & c_{1,1} \\ c_{3,2} & c_{2,2} & c_{1,2} \\ s^2 & s & 1 \end{vmatrix} \begin{vmatrix} c_{2,1} & c_{1,1} \\ c_{2,2} & c_{1,2} \end{vmatrix}^{-1} = q_0 + q_1 s + s^2, \quad (3.64)$$

where $c_{i,j}$ denotes the jth component of $c_i \in \mathbb{R}^2$.

The corresponding vector numerator is found by cross-multiplication to be

$$P(s) = p_0 + p_1 s = c_2 + q_1 c_1 + c_1 s. \quad (3.65)$$

Equations (3.64), (3.65) supply the parameters which deliver the outputs $c_1, c_2, c_3, \ldots$ from the circuit of Figure 8.3.1.

For the m-input, p-output system, the required Markov parameters form the expansion

$$F(s) = c_0 + c_1 s^{-1} + c_2 s^2 + \cdots + c_N s^{-N} + \cdots, \qquad c_i \in \mathbb{R}^{p \times m}.$$

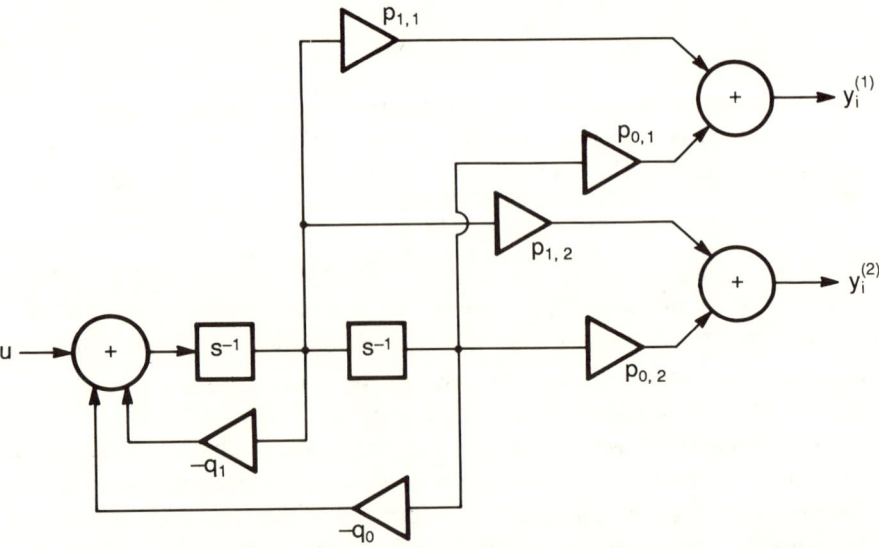

Figure 8.3.1. Controller canonical form of a single-input, two-output system.

8.3 Rectangular matrix Padé approximants for minimal partial-realization problems

We will assume that one of the methods of this section has been used to construct the rectangular matrix Padé approximant $P(s)Q(s)^{-1}$ with accuracy through order N. Then, from (3.48), for example, we have

$$F(s)\mathbf{Q}^{(j)}(s) - \mathbf{P}^{(j)}(s) = O(s^{\alpha^{(j)}-N-1}),$$

where the superscript j denotes the jth column of the originating matrix, and $\alpha^{(j)} = \deg(\mathbf{Q}^{(j)}(s))$. Let $\mathbf{Q}_{hc}^{(j)}$ denote the vector of coefficients of $s^{\alpha^{(j)}}$ in $\mathbf{Q}^{(j)}(s)$, where hc stands for highest coefficient. Combining these vectors, we form Q_{hc} as the matrix of the highest coefficients of $Q(s)$. Assuming that Q_{hc}^{-1} exists [its equivalent was proved nonsingular in (3.32)], we may then form

$$\tilde{Q}(s) = Q(s)Q_{hc}^{-1}, \quad \tilde{P}(s) = P(s)Q_{hc}^{-1},$$

so $\tilde{Q}_{hc} = I_m$. The coefficients of $\tilde{P}(s)$, $\tilde{Q}(s)$ are used to form the circuit elements shown schematically in Figure 8.3.2.

For further explanation of the construction of the circuits and other engineering issues, we refer to Kailath [1980]. The question of stable rational interpolation is properly addressed by Antoulas and Anderson [1989]. Some fast algorithms for SIMO systems are given by Graves-Morris and Wilkins [1987]. The whole area is comprehensively covered in

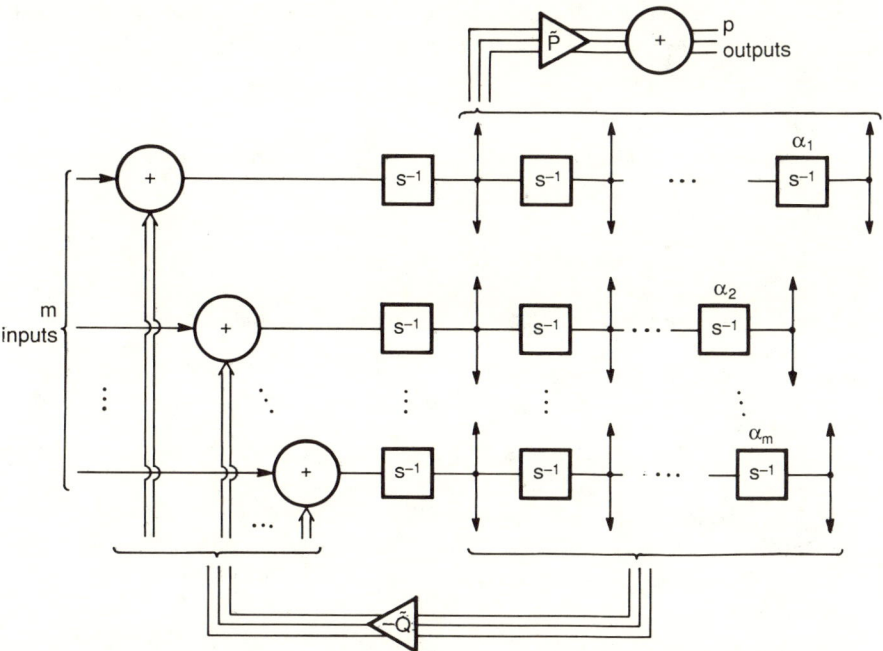

Figure 8.3.2. Schematic controller form of the realization with $\tilde{P}(s)\tilde{Q}(s)^{-1}$ for an m-input, p-output system, with α_i delays in the ith line.

a series of reports by Van Barel and Bultheel [1987a, b, c], which appear in Van Barel's thesis [1989], and the papers of Van Barel and Bultheel [1989a, b]. Some fundamental questions of existence and uniqueness are treated by Xu and Bultheel [1990]. As an alternative to the approach used in this section, the construction of the approximants can be based on an antidiagonal sequence, as is done by Kronecker's algorithm (see Section 3.7) in the scalar case. With this approach, it is natural to introduce the V-matrices at the beginning [Bultheel and Van Barel, 1990b].

8.4 Vector Padé approximants

In just the same way as the convergence of scalar sequences can be accelerated using the ε-algorithm, so can the convergence of vector sequences be similarly accelerated. We begin this section with a statement of the vector ε-algorithm, and give an example of its application. Next we define vector Padé approximants and show their connection with and origination from continued fractions formed with vector elements. Then we show how matrix Padé approximants are associated with vector series, and that these approximants satisfy Wynn's identity. The connection with the vector ε-algorithm is thereby established. We conclude with a specification, based on generalized inverses, of rectangular matrix Padé approximants. In Section 8.4.1, the generalization to functional Padé approximants is explained.

Suppose that we are given the vector sequence

$$\mathbf{S} = \{\mathbf{s}_0, \mathbf{s}_1, \mathbf{s}_2, \ldots : \mathbf{s}_i \in \mathbb{C}^d\} \tag{4.1}$$

and that

$$\mathbf{s}_i \to \mathbf{s} \quad \text{as } i \to \infty, \tag{4.2}$$

but that the convergence of (4.2) is too slow to be useful. The vector ε-algorithm [Wynn, 1962b, 1964] is initialized with

$$\boldsymbol{\varepsilon}_0^{(j)} = \mathbf{s}_j, \quad j = 0, 1, 2, \ldots, \tag{4.3a}$$

$$\boldsymbol{\varepsilon}_{-1}^{(j)} = \mathbf{0}, \quad j = 1, 2, 3, \ldots, \tag{4.3b}$$

where (4.3a) enters the data and (4.3b) is an artificial initialization of the left-hand column of the vector ε-table. The recursion is

$$\boldsymbol{\varepsilon}_{k+1}^{(j)} = [\boldsymbol{\varepsilon}_k^{(j+1)} - \boldsymbol{\varepsilon}_k^{(j)}]^{-1} + \boldsymbol{\varepsilon}_{k-1}^{(j+1)} \tag{4.4}$$

with the same diagrammatic interpretation (3.3.1) as for the scalar case. The vector inverse is interpreted as the Moore–Penrose generalized inverse defined by

8.4 Vector Padé approximants

$$\mathbf{u}^{-1} = \mathbf{u}^*/|\mathbf{u}|^2 \tag{4.5}$$

in which * denotes complex conjugation.

Example 1. The vector sequence $\{\mathbf{s}_i\}_{i=0}^{\infty}$ is taken from Gauss–Seidel iteration. In the following model problem [Wynn, 1962b], we find the solution of

$$A\mathbf{x} = \mathbf{b}, \tag{4.6}$$

where A and $\mathbf{b}$ are given by

$$A = \begin{bmatrix} 5 & 7 & 6 & 5 \\ 7 & 10 & 8 & 7 \\ 6 & 8 & 10 & 9 \\ 5 & 7 & 9 & 10 \end{bmatrix}, \quad \mathbf{b} = \begin{bmatrix} 23 \\ 32 \\ 33 \\ 31 \end{bmatrix}. \tag{4.7}$$

The example is purely illustrative, because realistic applications involve matrices of high dimension. To generate an iterative solution of (4.6), A is decomposed as

$$A = B + C \tag{4.8}$$

where

$$B = \begin{bmatrix} 5 & 0 & 0 & 0 \\ 7 & 10 & 0 & 0 \\ 6 & 8 & 10 & 0 \\ 5 & 7 & 9 & 10 \end{bmatrix}, \quad C = \begin{bmatrix} 0 & 7 & 6 & 5 \\ 0 & 0 & 8 & 7 \\ 0 & 0 & 0 & 9 \\ 0 & 0 & 0 & 0 \end{bmatrix}. \tag{4.9}$$

The recursion generating the sequence of Gauss–Seidel iterates is

$$\mathbf{s}_{i+1} = E\mathbf{s}_i + \mathbf{d} \tag{4.10}$$

where

$$E = -B^{-1}C, \quad \mathbf{d} = B^{-1}\mathbf{b}, \quad \text{and } \mathbf{s}_0 = \mathbf{0}.$$

In fact the eigenvalues of the iteration matrix $E = -B^{-1}C$ are

$$\lambda_1 = 0.9969, \quad \lambda_2 = 0.8373, \quad \lambda_3 = 0.6038, \quad \lambda_4 = 0.$$

It will be noticed that the exact solution of equation (4.6) appears in the $k = 3$ column of entries in Table 8.4.1, as $\boldsymbol{\varepsilon}_6^{(j)}$, $j = 1, 2, 3, \ldots$. This is a direct consequence of the fact that $\lambda_1, \lambda_2, \lambda_3 \neq 0$ and $\lambda_4 = 0$, but to understand why, the analytic theory must be developed, and we return to this example after McLeod's theorem has been proved.

Starting from $\mathbf{S}$ as defined in (4.1), we define

$$\mathbf{c}_0 = \mathbf{s}_0 \quad \text{and } \mathbf{c}_i = \Delta\mathbf{s}_{i-1}, \quad i = 1, 2, 3, \ldots, \tag{4.11}$$

and proceed with Definition 8.4.1.

Table 8.4.1. *Entries in even index columns of the vector ε-table for Example 1. For brevity, only six vectors of the initializing sequence are shown.*

0.000				→ k
0.000				
0.000				
0.000				$\varepsilon_{2k}^{(j)}$
4.600	3.781			
−0.020	−0.018		j	
0.556	0.738		↓	
0.314	0.454			
3.647	2.290	2.310		
−0.017	0.059	0.078		
0.843	1.037	0.978		
0.530	0.980	1.010		
3.083	2.310	2.328	2.324	
−0.003	0.078	0.201	0.203	
0.976	0.978	0.664	0.666	
0.682	1.010	1.197	1.196	
2.751	2.329	2.324	1.000	1.000
0.016	0.093	0.203	1.000	1.000
1.023	0.929	0.666	1.000	1.000
0.793	1.034	1.196	1.000	1.000
2.557	2.347	2.320	1.000	
0.036	0.107	0.206	1.000	indeterminate
1.023	0.886	0.667	1.000	
0.875	1.054	1.196	1.000	
$k = 0$	$k = 1$	$k = 2$	$k = 3$	$k = 4$

Definition 8.4.1 (**A vector Padé approximant of type** $[n/2k]$). A vector-valued function is formally defined by

$$\mathbf{f}(z) = \mathbf{c}_0 + \mathbf{c}_1 z + \mathbf{c}_2 z^2 + \cdots \qquad (4.12)$$

with $\mathbf{c}_i \in \mathbb{C}^d$. If a vector polynomial $\mathbf{p}(z)$ and a real scalar polynomial $q(z)$ exist satisfying the axioms

(i) $\qquad\qquad\qquad \partial\{\mathbf{p}\} \leq n, \quad \partial\{q\} = 2k, \qquad (4.13)$

(ii) $\qquad\qquad\qquad q(z) \,|\, \mathbf{p}(z) \cdot \mathbf{p}^*(z), \qquad (4.14)$

(iii) $\qquad\qquad\quad \mathbf{f}(z) - \mathbf{p}(z)/q(z) = O(z^{n+1}), \qquad (4.15)$

(iv) $\qquad\qquad\qquad q(0) \neq 0,$

then
$$\mathbf{r}(z) = \mathbf{p}(z)/q(z) \qquad (4.16)$$
is a vector Padé approximant of type $[n/2k]$.

The factorization condition (4.14) is unusual; it could be re-expressed in terms of the existence of another real polynomial $Q(z)$ of degree $2n - 2k$ at most for which
$$q(z)Q(z) = \mathbf{p}(z) \cdot \mathbf{p}^*(z). \qquad (4.17)$$
Notice also that the degree condition (4.13) on $q(z)$ is crucial. Were it to be the inequality $\partial\{q\} \leq 2k$, it would be possible to satisfy all the axioms using $q(z) = 1$, $\mathbf{p}(z) = [\mathbf{f}(z)]_0^n$, which would be unproductive. Thus the condition on $q(z)$ of degree equality is essential for the construction of a genuinely rational approximant. The above definition immediately raises the questions of existence, uniqueness, and degeneracy.

***Theorem* 8.4.1 (Uniqueness theorem) [Graves-Morris and Jenkins, 1986].** *If a vector Padé approximant $\mathbf{r}^{[n/2k]}(z)$ of type $[n/2k]$ exists, then it is unique.*

Proof. Suppose that $(\mathbf{p}(z), q(z))$, $(\tilde{\mathbf{p}}(z), \tilde{q}(z))$ define two different vector Padé approximants, separately satisfying the axioms (4.13)–(4.16). Consider the vector polynomial $\boldsymbol{\pi}(z)$ defined by
$$\boldsymbol{\pi}(z) = \mathbf{p}(z)\tilde{q}(z) - \tilde{\mathbf{p}}(z)q(z). \qquad (4.18)$$
Then
$$\deg\{\boldsymbol{\pi}(z)\} \leq n + 2k, \quad \boldsymbol{\pi}(z) = O(z^{n+1}). \qquad (4.19)$$
From (4.14) it follows that
$$\boldsymbol{\pi}(z) \cdot \boldsymbol{\pi}^*(z) = q(z)\tilde{q}(z)\tilde{Q}(z)$$
for some polynomial $\tilde{Q}(z)$. From (4.19), we find that
$$\deg\{\tilde{Q}(z)\} \leq 2n, \quad \tilde{Q}(z) = O(z^{2n+2})$$
which are contradictory unless $\tilde{Q}(z) = 0$. Hence $\boldsymbol{\pi}(z) = \mathbf{0}$ and $\mathbf{r}^{[n/2k]}(z)$ is uniquely defined.

Next, we answer the question of existence rather directly.

***Construction* 1 (The determinantal formula).** Given $\{\mathbf{c}_0, \mathbf{c}_1, \ldots, \mathbf{c}_n: \mathbf{c}_j \in \mathbb{C}^d\}$, the vector Padé approximant of type $[n/2k]$ for $\mathbf{f}(z)$ can be *constructed in terms of its denominator polynomial*

$$q(z) = \begin{vmatrix} 0 & M_{01} & M_{02} & \cdots & M_{0,2k} \\ -M_{01} & 0 & M_{12} & \cdots & M_{1,2k} \\ -M_{02} & -M_{12} & 0 & \cdots & M_{2,2k} \\ \vdots & \vdots & \vdots & & \vdots \\ z^{2k} & z^{2k-1} & z^{2k-2} & \cdots & 1 \end{vmatrix} \quad (4.20)$$

provided $q(0) \neq 0$ *and* $\deg\{q(z)\} = 2k$. *For* $j > i$, *the elements* M_{ij} *of* $q(z)$ *are defined by*

$$M_{ij} = \sum_{l=0}^{j-i-1} \mathbf{c}_{l+i+n-2k+1} \cdot \mathbf{c}^*_{j-l+n-2k}, \quad (4.21)$$

and the vector numerator of the approximant is given by

$$\mathbf{p}(z) = [\mathbf{f}(z)q(z)]_0^n. \quad (4.22)$$

Example 2. The vector Padé approximant of type $[n/2]$ has denominator polynomial given by

$$q(z) = \begin{vmatrix} 0 & |\mathbf{c}_{n-1}|^2 & 2\,\mathrm{Re}\,(\mathbf{c}^*_{n-1} \cdot \mathbf{c}_n) \\ -|\mathbf{c}_{n-1}|^2 & 0 & |\mathbf{c}_n|^2 \\ z^2 & z & 1 \end{vmatrix}$$

$$= |\mathbf{c}_{n-1}|^2(\mathbf{c}_{n-1} - z\mathbf{c}_n) \cdot (\mathbf{c}^*_{n-1} - z\mathbf{c}^*_n). \quad (4.23)$$

The numerator is given, as before, by

$$\mathbf{p}(z) = [\mathbf{f}(z)q(z)]_0^n. \quad (4.24)$$

Proof of (4.21). Define the nth Maclaurin section of $\mathbf{f}(z)$ by

$$\mathbf{F}(z) = [\mathbf{f}(z)]_0^n.$$

Then

$$\mathbf{p}(z) = [\mathbf{F}(z)q(z)]_0^n$$

and define $\theta(z)$ by

$$\theta = (\mathbf{p}^* - \mathbf{F}^*q) \cdot (\mathbf{p} - \mathbf{F}q) = \mathbf{p}^* \cdot \mathbf{p} - q\mathbf{F}^* \cdot \mathbf{p} - q\mathbf{F} \cdot \mathbf{p}^* + q^2\mathbf{F}^* \cdot \mathbf{F}. \quad (4.25)$$

Thus $\theta(z)$ is a polynomial which is divisible by $q(z)$ and

$$\theta(z) = O(z^{2n+2}).$$

With the notation of (4.17),

$$Q(z) - \mathbf{F}^*(z) \cdot \mathbf{p}(z) - \mathbf{F}(z) \cdot \mathbf{p}^*(z) + q(z)\mathbf{F}^*(z) \cdot \mathbf{F}(z) = O(z^{2n+2}),$$

and therefore

$$[-\mathbf{F}^*(z) \cdot \mathbf{p}(z) - \mathbf{F}(z) \cdot \mathbf{p}^*(z) + q(z)\mathbf{F}^*(z) \cdot \mathbf{F}(z)]_{2n-2k+1}^{2n+1} = 0.$$

We display the involvement of q by writing this as

$$[-\mathbf{F}^* \cdot [\mathbf{F}q]_0^n - \mathbf{F} \cdot [\mathbf{F}^*q]_0^n + q\mathbf{F}^* \cdot \mathbf{F}]_{2n-2k+1}^{2n+1} = 0. \tag{4.26}$$

Equation (4.26) represents $2k+1$ linear equations for the unknowns q_0, $q_1, \ldots, q_{2k}$, of $q(z) = \sum_{i=0}^{2k} q_i z^i$. The coefficients of q_i in the jth equation of (4.26) turn out to be the elements M_{ij} of (4.21). The proof of this result is messy and is omitted. In fact, the entries M_{ij} form an antisymmetric matrix of odd order, which has zero determinant. This property is essential, because it guarantees that a nontrivial solution for $q_0, q_1, \ldots, q_{2k}$ exists.

With the conditions $q(0) \neq 0$, $\partial\{q(z)\} = 2k$, it is now elementary to check that $(\mathbf{p}(z), q(z))$ defined by (4.20)–(4.22) satisfy axioms (i), (ii), and (iii) of (4.13)–(4.15). Thus $\mathbf{r}(z)$ exists as the unique vector Padé approximant of type $[n/2k]$ for $\mathbf{f}(z)$.

From a numerical viewpoint, it may well be better to view (4.20) as arising from a linear system of equations for the cofficients q_i of $q(z) = \sum_{i=0}^{2k} q_i z^i$. If either $q_0 = 0$ or $q_{2k} = 0$, we describe the vector Padé approximant as degenerate. A full treatment of such cases is given by Graves-Morris and Jenkins [1989] and Graves-Morris and Roberts [1994], but notice that axiom (iv), $q(0) \neq 0$, is not used in the proof of uniqueness of $\mathbf{p}(z)/q(z)$.

Analogously to the Padé table in Section 1.1, the vector rational fractions constructed using (4.20)–(4.22) can be laid out in a table:

Table 8.4.2. *The table of vector Padé approximants.*

$\mathbf{r}^{[0/0]}(z)$	$\mathbf{r}^{[1/0]}(z)$	$\mathbf{r}^{[2/0]}(z)$	$\cdots$
$\mathbf{r}^{[1/2]}(z)$	$\mathbf{r}^{[2/2]}(z)$	$\mathbf{r}^{[3/2]}(z)$	$\cdots$
$\vdots$	$\vdots$	$\vdots$	$\ddots$

Each vector entry of Table 8.4.2 corresponds to its scalar counterpart in the same location in the ordinary Padé table, even though the indices denoting the allowed degrees of the polynomials differ in all rows except the first. To explain this correspondence with the scalar case, we take $f(z)$ to be the scalar function defined in (1.1.1), and let $(p^{\text{PA}}(z), q^{\text{PA}}(z))$

denote the Padé polynomials of type $[L/M]$ for $f(z)$, where $q^{PA}(z)$ is defined by the determinantal formula (1.1.8) and $p^{PA}(z)$ is defined by $p^{PA}(z) = [f(z)q^{PA}(z)]_0^L$. We embed the function $f(z)$ in $\mathbb{C}^d$ by taking $\mathbf{u}$ to be an arbitrary fixed unit vector in $\mathbb{C}^d$ and defining

$$\mathbf{f}(z) = \mathbf{u}f(z). \tag{4.27}$$

If we also define

$$q^{VPA}(z) = q^{PA}(z)q^{PA*}(z), \tag{4.28}$$

$$\mathbf{p}^{VPA}(z) = \mathbf{u}p^{PA}(z)q^{PA*}(z), \tag{4.29}$$

then $(\mathbf{p}^{VPA}(z), q^{VPA}(z))$ satisfy the axioms (4.13)–(4.16) above. The uniqueness theorem implies that

$$\mathbf{r}^{VPA}(z) = \mathbf{p}^{VPA}(z)/q^{VPA}(z)$$

is the vector Padé approximant of type $[L + M/2M]$ for $\mathbf{f}(z)$.

The correspondence established in (4.28), (4.29) between scalar and vector Padé approximants raises the question of how $q^{VPA}(z)$ defined by (4.20) factorizes when $\mathbf{f}(z)$ takes the special form (4.27). Woodcock and Graves-Morris [1993, 1995] find that

$$q^{VPA}(z) = q^{PA}(z)q^{PA*}(z)q^{PA}(0)q^{PA*}(0), \tag{4.30}$$

where $q^{VPA}(z)$ is the denominator of the vector Padé approximant of type $[L + M/2M]$ as given by (4.20) with $n = L + M$ and $k = M$, and $q^{PA}(z)$ is the denominator of the Padé approximant of type $[L/M]$ as given by (1.1.8).

Let us begin again with a continued fraction

$$\mathbf{F}(z) = \mathbf{b}_0 + \cfrac{z}{\mathbf{b}_1 +} \cfrac{z}{\mathbf{b}_2 +} \cdots \cfrac{z}{+ \mathbf{b}_n +} \cdots, \tag{4.31}$$

formally defined in terms of vector elements $\mathbf{b}_0, \mathbf{b}_1, \mathbf{b}_2, \ldots \in \mathbb{C}^d$, as a natural alternative starting point to (4.12) [Wynn, 1968a; Graves-Morris, 1983]. The fraction $\mathbf{F}(z)$ corresponding to a power series $\mathbf{f}(z)$ as given by (4.12) can be constructed by Viskovatov's algorithm. The method is the same as that of Section 4.2; vector-valued power series are inverted (reciprocated) using formulas of the form $[\mathbf{f}(z)]^{-1} = \mathbf{f}^*(z)[\mathbf{f}(z) \cdot \mathbf{f}^*(z)]^{-1}$. The nth convergent of $\mathbf{F}(z)$ is defined by

$$\mathbf{F}^{(n)}(z) = \mathbf{b}_0 + \cfrac{z}{\mathbf{b}_1 +} \cfrac{z}{\mathbf{b}_2 +} \cdots \cfrac{z}{+ \mathbf{b}_n} \tag{4.32}$$

and it is easily evaluated by backward recursion for $z \in \mathbb{R}$. Standard forward three-term recursions are meaningless, because they would involve incompatible matrix multiplications. However, Roberts [1992] has devised new forward recursion formulas which can be practically implemented in this case.

We show next how evaluation of $\mathbf{F}^{(n)}(z)$ by backward recursion yields the representation (4.33) which is compatible with the formulation (4.13)–(4.16). We make the inductive hypothesis that

$$\mathbf{S}_k(z) = \cfrac{1}{\mathbf{b}_k + } \cfrac{z}{\mathbf{b}_{k+1} + } \cdots \cfrac{z}{+ \mathbf{b}_n} = \mathbf{p}_k(z)/q_k(z), \qquad (4.33)$$

where $\mathbf{p}_k(z)$ is a vector polynomial and $q_k(z)$ is a real scalar polynomial, and that there exists another real scalar polynomial $Q_k(z)$ such that

$$\mathbf{p}_k(z) \cdot \mathbf{p}_k^*(z) = q_k(z)Q_k(z). \qquad (4.34)$$

By definition, and for $z \in \mathbb{R}$, we have

$$\mathbf{S}_{k-1}(z) = [\mathbf{b}_{k-1} + z\mathbf{S}_k(z)]^{-1}$$
$$= q_k(z)[\mathbf{b}_{k-1}q_k(z) + z\mathbf{p}_k(z)]^{-1}.$$

We are led to define

$$\mathbf{p}_{k-1}(z) = \mathbf{b}_{k-1}^* q_k(z) + z\mathbf{p}_k^*(z), \qquad (4.35)$$

$$q_{k-1}(z) = \mathbf{p}_{k-1}(z) \cdot \mathbf{p}_{k-1}^*(z)/q_k(z)$$
$$= |\mathbf{b}_{k-1}|^2 q_k(z) + z\mathbf{p}_k(z) \cdot \mathbf{b}_{k-1}^* + z\mathbf{p}_k^*(z) \cdot \mathbf{b}_{k-1} + z^2 Q_k(z) \qquad (4.36)$$

and by this construction,

$$\mathbf{S}_{k-1}(z) = \mathbf{p}_{k-1}(z)/q_{k-1}(z). \qquad (4.37)$$

Notice that $q_{k-1}(z)$ is a real polynomial with the property that

$$\mathbf{p}_{k-1}(z) \cdot \mathbf{p}_{k-1}^*(z) = q_{k-1}(z)q_k(z).$$

Thus the representation (4.33) with the factorization property (4.34) is established inductively, and the origin of the factorization axiom (4.14) is explained. By degree counting, one may verify that the sequence of vector Padé approximants generated by (4.32) for $n = 0, 1, 2, \ldots$ is a staircase sequence:

$$\mathbf{F}^{(n)}(z) = \mathbf{r}^{[2k/2k]}(z), \quad \text{if } n = 2k = \text{even,}$$

or

$$\mathbf{F}^{(n)}(z) = \mathbf{r}^{[2k+1/2k]}(z), \quad \text{if } n = 2k + 1 = \text{odd.}$$

The connection between vector and matrix Padé approximants is made by constructing an appropriate Clifford algebra, $\mathcal{A}$, based on a 2^n-dimensional matrix representation along the following lines, using sequential block decomposition. Begin with the definition of $E_1, E_2, \ldots, E_n$ by

$$E_1 = \begin{bmatrix} I & 0 \\ 0 & -I \end{bmatrix}, \tag{4.38}$$

where I denotes a block unit matrix, and

$$E_j = \begin{bmatrix} 0 & X_j^{(2)} \\ X_j^{(2)} & 0 \end{bmatrix}, \quad j = 2, 3, \ldots, n. \tag{4.39}$$

For $k = 2, 3, \ldots, n$ and $i = k, k+1, \ldots, n$, square matrices $X_i^{(k)}$ are defined by

$$X_k^{(k)} = \begin{bmatrix} I & 0 \\ 0 & -I \end{bmatrix}, \quad k = 2, 3, \ldots, n-1, \tag{4.40}$$

or recursively by

$$X_j^{(k)} = \begin{bmatrix} 0 & X_j^{(k+1)} \\ X_j^{(k+1)} & 0 \end{bmatrix}, \quad j = k+1, k+2, \ldots, n, \tag{4.41}$$

or ultimately by

$$X_n^{(n)} = \begin{bmatrix} 1 & 0 \\ 0 & -1 \end{bmatrix}. \tag{4.42}$$

Following McLeod [1971], we note that (4.40)–(4.42) imply that

$$X_k^{(k)} X_j^{(k)} + X_j^{(k)} X_k^{(k)} = 2I\delta_{jk}, \quad 2 \leq k \leq j \leq n, \tag{4.43}$$

from which it follows that the E_i have dimension 2^n and

$$E_i E_j + E_j E_i = 2\delta_{ij} I. \tag{4.44}$$

The subalgebra, consisting of elements of the form λI, where λ is a scalar, is called $\mathcal{S}$.

If we are given a vector $\mathbf{x} \in \mathbb{R}^d$ defined by

$$\mathbf{x} = (x_1, x_2, \ldots, x_d), \tag{4.45}$$

with Euclidean norm

$$\|\mathbf{x}\| = \left\{ \sum_{i=1}^d x_i^2 \right\}^{1/2}, \tag{4.46}$$

we take $n = d$ in (4.38)–(4.43) and associate with $\mathbf{x}$ an element $x \in \mathcal{A}$ by

$$x = \sum_{i=1}^d x_i E_i. \tag{4.47}$$

Let all elements of the form (4.47) constitute the real linear subspace $\mathcal{V}_R \subset \mathcal{A}$. From (4.43), (4.46), and (4.47),

8.4 Vector Padé approximants

$$x^2 = \|\mathbf{x}\|^2 I \tag{4.48}$$

and we are led to define the modulus of x by

$$|x| = \|\mathbf{x}\|. \tag{4.49}$$

Thus we have an elementary isomorphism between

(i) the Euclidean space $\mathbb{R}^d$ with representative element $\mathbf{x}$ given by (4.45) and norm by (4.46), and
(ii) the vector space $\mathcal{V}_R$ with representative element x given by (4.47) and its modulus by (4.49).

The Moore–Penrose generalized inverse of $\mathbf{x} \in \mathbb{R}^d$ is defined by (4.5) as

$$\mathbf{x}^{-1} = \mathbf{x}/\|\mathbf{x}\|^2.$$

For $x \in \mathcal{V}_R$, we may use (4.48) in the form

$$x^{-1} = x/\|\mathbf{x}\|^2, \tag{4.50}$$

and therefore the isomorphism is compatible with Moore–Penrose inverses.

Instead of beginning with $\mathbf{x} \in \mathbb{R}^d$, suppose that we are given

$$\mathbf{z} = (z_1, z_2, \ldots, z_d) \in \mathbb{C}^d. \tag{4.51}$$

Its Moore–Penrose inverse is defined by (4.5) as

$$\mathbf{z}^{-1} = \mathbf{z}^*/\|\mathbf{z}\|^2,$$

where the star denotes complex conjugate, and

$$\|\mathbf{z}\| = \left(\sum_{i=1}^{d}|z_i|^2\right)^{1/2}. \tag{4.52}$$

Using (4.38)–(4.42), we take $n = 2d + 1$ and we define

$$J = E_n. \tag{4.53}$$

Define also

$$B_i = E_i, \quad C_i = JE_{d+i}, \quad i = 1, 2, \ldots, d. \tag{4.54}$$

By expressing the vector $\mathbf{z} \equiv (\mathbf{x} + i\mathbf{y}) \in \mathbb{C}^d$ in terms of its real and imaginary parts $\mathbf{x}, \mathbf{y} \in \mathbb{R}^d$, we associate

$$z = \sum_{i=1}^{d} x_i B_i + \sum_{i=1}^{d} y_i C_i. \tag{4.55}$$

Let $\mathcal{V}_C$ denote the subspace of $\mathcal{A}$ formed by elements of the form (4.55).

The corresponding associate of $\mathbf{z}^*$ is

$$\tilde{z} = \sum_{i=1}^{d} x_i B_i - \sum_{i=1}^{d} y_i C_i \qquad (4.56)$$

and we find that

$$z\tilde{z} = \left[\sum_{i=1}^{d}(x_i^2 + y_i^2)\right] I = \|\mathbf{z}\|^2 I. \qquad (4.57)$$

Thus we have another elementary isomorphism between

(iii) the linear space $\mathbb{C}^d$ with representative element $\mathbf{z}$ given by (4.51) and its norm by (4.52), and
(iv) the linear space $\mathcal{V}_C$ with representative element z given by (4.55) and its modulus defined by $|z| = \|\mathbf{z}\|$.

From (4.57), we see that

$$z^{-1} = \tilde{z}/|z|^2 \qquad (4.58)$$

and hence the inverses of elements in $\mathcal{V}_C$ lie in $\mathcal{V}_C$ and correspond to Moore–Penrose inverses in $\mathbb{C}^d$.

We use the tilde operation introduced in (4.56) for the reverse automorphism of the algebra A generated by $E_0, E_1, \ldots, E_n$. We define

$$\widetilde{E_{j_1} E_{j_2} \cdots E_{j_k}} = E_{j_k} \cdots E_{j_2} E_{j_1} \qquad (4.59)$$

with its natural extension to all elements of $\mathcal{A}$. For example, if $u = E_1 + E_2 E_3 + E_4 E_5 E_6$, then

$$\tilde{u} = E_1 + E_3 E_2 + E_6 E_5 E_4 = E_1 - E_2 E_3 - E_4 E_5 E_6.$$

We emphasize that the association (4.55) is invariant under real linear transformations $z \to \alpha z$, $\alpha \in \mathbb{R}$, but not under complex transformations with $\alpha \in \mathbb{C}$. This fact is reinforced by using the tilde notation rather than a bar. For the representation (4.38)–(4.42), it is interesting to note that the tilde operation has the same effect as matrix transposition. It should be noted that $\mathcal{A}$ is not a division algebra, because

$$(I + E_1)(I - E_1) = 0,$$

showing that $I + E_1$ is not invertible. However, all nonzero elements of the subspaces $\mathcal{V}_R$ or $\mathcal{V}_C$ are invertible, and all constructs of the vector ε-algorithm have images of this form.

We will be concerned with operations on formal power series of the form

8.4 Vector Padé approximants

$$f(z) = \sum_{i=0}^{\infty} c_i z^i, \quad c_i \in \mathcal{V}_R \text{ or } \mathcal{V}_C. \tag{4.60}$$

We may re-express (4.60) as

$$f(z) = \sum_{i=1}^{d} \alpha_i(z) E_i + \sum_{i=1}^{d} \beta_i(z) J E_{d+i}, \tag{4.61}$$

where $\alpha_i(z) = \sum_{k=0}^{\infty} \alpha_k^{(i)} z^k$ and $\beta_i(z) = \sum_{k=0}^{\infty} \beta_k^{(i)} z^k$ are power series in z. Formally, we have

$$f(z)\tilde{f}(z) = I \sum_{i=1}^{d} ([\alpha_i(z)]^2 + [\beta_i(z)]^2)$$

which is a scalar in $\mathcal{A}$. Provided $f(0) \neq 0$, there exists a formal power series for $[f(z)\tilde{f}(z)]^{-1}$, and we may usefully express

$$[f(z)]^{-1} = \tilde{f}(z)[f(z)\tilde{f}(z)]^{-1}.$$

This form makes it plausible that application of Viskovatov's algorithm, as described in Section 8.2, to formal matrix power series of the form (4.60) produces a continued fraction whose elements are the matrix images of the vector elements in (4.32). This is indeed the case [Graves-Morris and Roberts, 1994]. For present purposes, Kronecker's algorithm provides a simpler reliable algorithm for the construction of a matrix Padé approximant in the particular case of $c_i \in \mathcal{V}$, where $\mathcal{V}$ denotes $\mathcal{V}_R$ or $\mathcal{V}_C$. The procedure and the results are directly analogous to those of Section 3.7.

Kronecker's algorithm for matrix Padé approximants is initialized with

$$a_0(z) = c_0 + c_1 z + \cdots + c_n z^n, \tag{4.62}$$

$$b_0(z) = I \tag{4.63}$$

and artificially with

$$a_{-1}(z) = I z^{n+1}, \quad b_{-1}(z) = 0. \tag{4.64}$$

With our conventions,

$$\partial\{b_0\} = 0, \quad \partial\{b_{-1}\} = -1. \tag{4.65}$$

With the initialization (4.62)–(4.64), it is trivial that

$$a_0(z)\tilde{a}_{-1}(z) \in \mathcal{V}, \quad b_0(z)\tilde{b}_{-1}(z) \in \mathcal{V}. \tag{4.66}$$

The iterative step of Kronecker's algorithm consists of defining

$$\rho_j(z) = \underset{z \to \infty}{\text{principal part}} \{[a_j(z)]^{-1} a_{j-1}(z)\}, \tag{4.67}$$

$$a_{j+1}(z) = a_j(z)\rho_j(z) - a_{j-1}(z), \qquad (4.68)$$

$$b_{j+1}(z) = b_j(z)\rho_j(z) - b_{j-1}(z), \qquad (4.69)$$

for $j = 0, 1, 2, \ldots$. As an example to explain the notation of (4.67), if $c_n \neq 0$,

$$\begin{aligned}\rho_0(z) &= \text{PP}\{z^{n+1}(c_0 + c_1 z + \cdots + c_n z^n)^{-1}\} \\ &= \text{PP}\{z c_n^{-1}(I + c_{n-1} c_n^{-1} z^{-1})^{-1}\} \\ &= z c_n^{-1} - c_n^{-1} c_{n-1} c_n^{-1}.\end{aligned}$$

The algorithm is terminated at the index ω for which we find

$$a_{\omega+1}(z) = 0. \qquad (4.70)$$

The formulas (4.67), (4.68) signify that $\rho_j(z)$ is defined as the matrix polynomial quotient of $a_j(z)$ divided by $a_{j-1}(z)$, and that $a_{j+1}(z)$ is the remainder. Thus (4.68) represents application of the actual Euclidean algorithm.

The left-handed versions of the initializations and termination are the same as the right-handed forms, but the equivalents of (4.67)–(4.69) are

$$\breve{\rho}_j(z) = \text{principal part } \{\breve{a}_{j-1}(z)[\breve{a}_j(z)]^{-1}\}, \qquad (4.71)$$

$$\breve{a}_{j+1}(z) = \breve{\rho}_j(z)\breve{a}_j(z) - \breve{a}_{j-1}(z), \qquad (4.72)$$

$$\breve{b}_{j+1}(z) = \breve{\rho}_j(z)\breve{b}_j(z) - \breve{b}_{j-1}(z), \qquad (4.73)$$

where the $\breve{}$ denotes the left-handed form (denoted by ^{L} in Section 8.2). We denote the leading coefficient of a polynomial $p(z)$ by using a dot, so that $\dot{p} \neq 0$ unless $p(z) \equiv 0$.

Theorem 8.4.2. *With the specification above, $\{\dot{a}_j\}_{j=0}^{\omega}$ are invertible, so $\rho_j(z)$ is well defined by (4.67). Also*

$$\tilde{a}_j(z)a_{j-1}(z) \in \mathscr{V}, \quad \tilde{b}_j(z)b_{j-1}(z) \in \mathscr{V}, \quad \rho_j(z) \in \mathscr{V}, \\ \tilde{b}_j(z)b_j(z) \in \mathscr{S} \qquad (4.74)$$

for $j = 0, 1, \ldots, \omega$. Similar results hold for the left-handed polynomial matrices and

$$\breve{\rho}_j(z) \equiv \rho_j(z). \qquad (4.75)$$

Proof. The results are trivial for the cases $j = 0, 1$. We make the inductive hypothesis that $\dot{a}_j$ is invertible. Then $\rho_j(z)$ and $a_{j+1}(z)$ are well defined by (4.67) and (4.68), respectively. From (4.68), the identity

$$[a_j(z)]^{-1}a_{j+1}(z) = \rho_j(z) - [a_j(z)]^{-1}a_{j-1}(z) \qquad (4.76)$$

holds as an identity among matrix rational fractions. Both terms on the right-hand side of (4.76) lie in $\mathscr{V}$, so the left one does too. Hence $\dot{a}_j^{-1}\dot{a}_{j+1} \in \mathscr{V}$ and $\dot{a}_{j+1}$ is invertible. By first establishing that $[a_j(z)]^{-1}a_{j-1}(z) = \breve{a}_{j-1}(z)[\breve{a}_j(z)]^{-1}$, $j = 1, 2, \ldots$, we can prove (4.75). The proofs of the rest of (4.74) and the left-handed equivalent of (4.74) are straightforward.

The previous theorem is the key which establishes the feasibility of the Kronecker algorithm in the present context. The following theorems are virtually identical to those for the more familiar scalar case. The differences are that we need to distinguish the left- and right-handed representations of the matrix approximants, and to establish invertibility.

Theorem 8.4.3. *The matrix polynomials $a_j(z)$, $b_j(z)$ have the accuracy-through-order property*

$$f(z)b_j(z) - a_j(z) = O(z^{n+1}), \quad j = 0, 1, \ldots, \omega. \tag{4.77}$$

If $b_j(0) \neq 0$,

$$f(z) = a_j(z)[b_j(z)]^{-1} + O(z^{n+1}) \tag{4.78}$$

and $a_j(z)[b_j(z)]^{-1}$ is the unique right-handed matrix Padé approximant of $f(z)$. Similar results hold for the left-handed polynomials.

Proof. The result (4.77) is easily proved by an induction based on (4.78) and (4.79); we omit the details. To establish (4.78), let β be the coefficient of the least power of z in $b_{j-1}(z)$. From (4.74), $\tilde{b}(0)\beta \in \mathscr{V}$. Therefore $b_j(0)$ is invertible unless it is zero. Uniqueness follows from Theorem 8.2.1.

Theorem 8.4.4. *For $k = 1, 2, \ldots, \omega$,*

$$\partial a_k = n + 1 - \sum_{j=0}^{k} \partial \rho_j, \tag{4.79}$$

$$\partial b_k = \sum_{j=0}^{k-1} \partial \rho_j, \tag{4.80}$$

$$\partial \rho_j \geq 1. \tag{4.81}$$

The degrees of $\{b_k(z)\}$ are monotonically increasing and the degrees of $\{a_k(z)\}$ are monotonically decreasing with k, and the algorithm necessarily terminates according to (4.70). Similar results hold for the left-handed polynomials.

Proof. By construction,

$$\partial \rho_j = \partial a_{j-1} - \partial a_j = \partial b_{j+1} - \partial b_j, \qquad (4.82)$$

and the remainder of the proof is straightforward.

Theorem 8.4.5. *The leading coefficients of $a_j(z)$ and $b_j(z)$ are*

$$\dot{a}_j = [\dot{\rho}_j \dot{\rho}_{j-1} \cdots \dot{\rho}_0]^{-1}, \qquad (4.83)$$

$$\dot{b}_j = \dot{\rho}_0 \dot{\rho}_1 \cdots \dot{\rho}_{j-1}, \qquad (4.84)$$

which are invertible for $j = 1, 2, \ldots, \omega$.

Proof. The proof is trivial from (4.68), (4.69).

Theorem 8.4.6. *For $j = 0, 1, \ldots, \omega$,*

$$\breve{a}_{j-1}(z) b_j(z) - \breve{b}_{j-1}(z) a_j(z) = \breve{b}_j(z) a_{j-1}(z) - \breve{a}_j(z) b_{j-1}(z) = z^{n+1} I. \qquad (4.85)$$

Proof. The cases $j = 0$ and $j = 1$ follow trivially from (4.62)–(4.63). The other cases in (4.85) follow by a two-step induction, using (4.68), (4.69), (4.72), and (4.73). It is then clear that the matrix fractions $\{a_k(z)[b_k(z)]^{-1}\}$ are distinct.

Theorem 8.4.7. *Consider the jth matrix Padé polynomials and define l, m and v by*

$$l = \partial a_j, \quad m = \partial b_j, \quad v = n - l - m + 1,$$

so that

$$v = \partial \rho_j. \qquad (4.86)$$

If $b_j(0) \neq 0$, $(a_j(z), b_j(z))$ are the minimal matrix Padé polynomials of type $[l/m]$. In addition, if $v > 1$, $(a_j(z), b_j(z))$ form the minimal element of a nontrivial block of dimension at least v in the table of matrix Padé approximants.

Proof. The result (4.86) follows from (4.79), (4.80). Suppose that $a_j(z)$ and $b_j(z)$ share a common right-handed invertible factor $d(z)$. Since $b_j(0) \neq 0$, $d(0) \neq 0$ and therefore z does not divide $d(z)$. Any common right-handed matrix polynomial factor of $a_j(z)$ and $b_j(z)$ divides the left-hand side of (4.85). Therefore $d(z)$ divides the right-hand side too, and it can only be a constant. Thus $a_j(z)$, $b_j(z)$ have no common right-handed matrix polynomial factor. The remainder of the proof is straightforward.

The conclusions of Theorems 8.4.4, 8.4.7 are depicted in Figure 3.7.4, because the result is identical to that of the scalar case.

Next, we show that the matrix Padé approximants constructed for the series (4.60) using Kronecker's algorithm satisfy Wynn's compass identity. Suppose that $C(z)$ having the right- and left-handed representations

$$C(z) = p_C(z)[q_C(z)]^{-1} = [\breve{q}_C(z)]^{-1}\breve{p}_C(z) \qquad (4.87)$$

is a standard matrix Padé approximant of precise type $[l/m]$, whose block (trivially) has dimension $r = 1$. We use the usual compass point notation, $N(z)$, $W(z)$, $E(z)$, and $S(z)$ to denote its neighbors, and we take minimal degrees for the numerator and denominator polynomials, so

$$\partial p_N = l - \alpha, \quad \partial p_W \leq l - \beta - 1,$$
$$\partial q_N \leq m - \alpha - 1, \quad \partial q_W = m - \beta,$$

where either $\alpha = 0$ or $\beta = 0$, so that $N(z)$ and $W(z)$ are distinct.

By degree counting and using properties of accuracy through order only, we find that

NC: $\quad \breve{p}_N(z)q_C(z) - \breve{q}_N(z)p_C(z) = z^{l+m-\alpha}\breve{p}_N\dot{q}_C,$ (4.88a)

WC: $\quad \breve{p}_C(z)q_W(z) - \breve{q}_C(z)p_W(z) = z^{l+m-\beta}\breve{p}_C\dot{q}_W,$ (4.88b)

NW: $\quad \breve{p}_N(z)q_W(z) - \breve{q}_N(z)p_W(z) = z^{l+m-\alpha-\beta}\breve{p}_N\dot{q}_W.$ (4.88c)

From (4.84), we see that $\dot{q}_\Omega$ is invertible for $\Omega = N, S, E, W,$ or C. Likewise, each $\breve{q}_\Omega$ is invertible. We use (4.88) to get

$$[N(z) - C(z)]^{-1} - [W(z) - C(z)]^{-1}$$
$$= [N(z) - C(z)]^{-1}(W(z) - N(z))[W(z) - C(z)]^{-1}$$
$$= [\breve{q}_N^{-1}\breve{p}_N - p_Cq_C^{-1}]^{-1}(p_Wq_W^{-1} - \breve{q}_N^{-1}\breve{p}_N)[p_Wq_W^{-1} - \breve{q}_C^{-1}\breve{p}_C]^{-1}$$
$$= q_C[\breve{p}_Nq_C - \breve{q}_Np_C]^{-1}(\breve{q}_Np_W - \breve{p}_Nq_W)[\breve{q}_Cp_W - \breve{p}_Cq_W]^{-1}\breve{q}_C$$
$$= z^{-l-m}q_C(z)\dot{q}_C^{-1}\breve{p}_C^{-1}\breve{q}_C(z).$$

Similarly, we have

$$\partial p_E = l + r, \quad \partial q_S = m + r,$$

EC: $\quad \breve{p}_E(z)q_C(z) - \breve{q}_E(z)p_C(z) = \breve{p}_E\dot{q}_C z^{l+m+r},$

SC: $\quad \breve{p}_C(z)q_S(z) - \breve{q}_C(z)p_S(z) = \breve{p}_C\dot{q}_S z^{l+m+r},$

ES: $\quad \breve{p}_E(z)q_S(z) - \breve{q}_E(z)p_S(z) = \breve{p}_E\dot{q}_S z^{l+m+2r},$

where $r = 1$, from which we obtain

$$[E(z) - C(z)]^{-1} - [S(z) - C(z)]^{-1} = z^{-l-m}q_C(z)\dot{q}_C^{-1}\breve{p}_C^{-1}\breve{q}_C(z). \quad (4.89)$$

Combining (4.87) and (4.89), we have Wynn's identity

$$[N(z) - C(z)]^{-1} - [W(z) - C(z)]^{-1} = [E(z) - C(z)]^{-1}$$
$$- [S(z) - C(z)]^{-1}. \quad (4.90)$$

Thus far, we have seen how vector-valued power series of the form (4.12),

$$\mathbf{f}(z) = c_0 + c_1 z + c_2 z^2 + \cdots,$$

are to be associated with matrix-valued power series of the form (4.60) using a Clifford algebra. By taking $n = L + M$ in (4.62), Kronecker's algorithm reliably constructs the matrix Padé approximant of type $[L/M]$ if it exists. Moreover, these matrix Padé approximants have been shown to satisfy Wynn's compass identity. The next stage is to make the connection between these matrix Padé approximants and the vector Padé approximants defined in (4.12)–(4.16).

Let $(a(z), b(z))$ denote the polynomial matrices constructed by Kronecker's algorithm and representing any nondegenerate matrix Padé approximant of type $[n - k/k]$ for $f(z)$, having $\deg\{b(z)\} = k$. From Theorem 8.4.2, $b(z)\tilde{b}(z) \in \mathscr{S}$, and hence

$$q(z)I = b(z)\tilde{b}(z) \quad (4.91)$$

for some ordinary polynomial $q(z)$ of degree $2k$ with $q(0) \neq 0$. Furthermore, $a(z)\tilde{b}(z) \in \mathscr{V}_R$, or $\mathscr{V}_C$, and hence an ordinary vector polynomial $\mathbf{p}(z) \in \mathbb{R}^d[z]$ or $\mathbb{C}^d[z]$ of degree n at most follows from (4.61). Because

$$[a(z)\tilde{b}(z)][\overline{a(z)\tilde{b}(z)}] = q(z)a(z)\tilde{a}(z),$$

it follows that

$$q(z)Q(z) = \mathbf{p}(z) \cdot \mathbf{p}^*(z) \quad (4.92)$$

for some polynomial $Q(z)$. In this way, the matrix polynomials $(a(z), b(z))$ derived using Kronecker's algorithm yield vector and scalar polynomials $(\mathbf{p}(z), q(z))$ which satisfy the axioms (4.12)–(4.17) of a vector Padé approximant. The uniqueness Theorems 8.2.1 and 8.4.1 for matrix Padé approximants and vector Padé approximants readily establish that the connection is general and does not depend on the method of derivation of the approximants. In short, we have seen that the isomorphic correspondence

$$\mathbf{r}^{[n/2k]}(z) \leftrightarrow r^{[n-k/k]}(z)$$

between the entries $\mathbf{r}^{[n/2k]}(z)$ in Table 8.4.2 and the matrix Padé approximants of type $[n - k/k]$.

Equation (4.90) shows that matrix Padé approximants satisfy Wynn's compass identity, and therefore so too do the corresponding vector Padé approximants. In fact, there is the remarkable result that vector Padé approximants satisfy Cordellier's identity. This was proved geometrically by Cordellier [1974] and in more detail in his thesis [1989]; analytical proofs have been given by Graves-Morris and Jenkins [1986, 1989] and Graves-Morris and Roberts [1994].

This section began with a description of the vector ε-algorithm. Because we have established that vector Padé approximants satisfy Wynn's compass identity, the connection between the entries of the vector ε-table and vector Padé approximants is easily established recursively: the connection is

$$\varepsilon_{2k}^{(n-2k)} = \mathbf{r}^{[n/2k]}(1), \quad n = 2k, 2k+1, \ldots. \tag{4.93}$$

We return to the theme of the vector ε-algorithm with McLeod's theorem, which shows that the vector ε-algorithm sums certain geometric vector sequences exactly. We begin with

Lemma 1. *Given a recursion for* $\{\mathbf{s}_i\}_{i=0}^{\infty}$ *of the form*

$$\sum_{i=0}^{k} \beta_i \mathbf{s}_{i+j} = \mathbf{0}, \quad j = 0, 1, 2, \ldots,$$

in which $\sum_{i=0}^{k} \beta_k = 0$, *the* $\mathbf{s}_i$ *satisfy a (possibly inhomogeneous) recursion relation of lower order.*

Proof. By hypothesis, we see that $z - 1$ is a factor of $\sum_{i=0}^{k} \beta_i z^i$, and hence coefficients γ_i can be defined by

$$\sum_{i=0}^{k} \beta_i z^i = (z-1) \sum_{i=0}^{k-1} \gamma_i z^i.$$

Define the sequence $\{\mathbf{t}_j\}_{j=0}^{\infty}$ by

$$\mathbf{t}_j = \sum_{i=0}^{k-1} \gamma_i \mathbf{s}_{i+j}.$$

Then, for $j \geq 0$,

$$\mathbf{t}_{j+1} - \mathbf{t}_j = \sum_{i=0}^{k-1} \gamma_i \mathbf{s}_{i+j+1} - \sum_{i=0}^{k-1} \gamma_i \mathbf{s}_{i+j}$$

$$= \gamma_{k-1} \mathbf{s}_{j+k} + \sum_{i=1}^{k-1} (\gamma_{i-1} - \gamma_i) \mathbf{s}_{i+j} - \gamma_0 \mathbf{s}_j$$

$$= \sum_{i=0}^{k} \beta_i \mathbf{s}_{j+i} = \mathbf{0}.$$

Therefore $\mathbf{t}_j = \mathbf{t}$ = constant, and

$$\sum_{i=0}^{k-1} \gamma_i \mathbf{s}_{i+j} = \mathbf{t}, \quad j = 0, 1, 2, \ldots.$$

Theorem 8.4.8 [McLeod, 1971; Graves-Morris, 1983, 1992]. *Suppose that the vector sequence* $\mathbf{S}$ *of* (4.1) *satisfies a nontrivial recursion relation*

$$\sum_{i=0}^{k} \beta_i \mathbf{s}_{i+j} = \left[\sum_{i=0}^{k} \beta_i\right] \mathbf{s}, \quad j = 0, 1, 2, \ldots, \quad (4.94)$$

with $\beta_i \in \mathbb{C}$. *Then the vector* ε-*algorithm of* (4.3), (4.4) *leads to*

$$\varepsilon_{2k}^{(j)} = \mathbf{s}, \quad j = 0, 1, 2, \ldots, \quad (4.95)$$

provided that zero divisors are not encountered in the construction.

Proof. By subtracting the forms of (4.94) for two consecutive values of j, it follows [see (4.11)] that

$$\sum_{i=0}^{k} \beta_i \mathbf{c}_{i+j} = \mathbf{0}, \quad j = 1, 2, \ldots. \quad (4.96a)$$

Define

$$\beta(z) = \beta_0 z^k + \beta_1 z^{k-1} + \cdots + \beta_k,$$

and consider [see (4.12)]

$$\beta(z)\mathbf{f}(z) = \left(\sum_{i=0}^{k} \beta_i z^{k-i}\right)\left(\sum_{j=0}^{\infty} \mathbf{c}_j z^j\right)$$

$$= \sum_{i=0}^{k} \sum_{n=-i}^{\infty} \beta_i \mathbf{c}_{n+i} z^{n+k}$$

$$= z^k \sum_{i=0}^{k} \sum_{n=-i}^{0} \beta_i \mathbf{c}_{n+i} z^n$$

$$= \boldsymbol{\alpha}(z), \quad (4.96b)$$

thereby defining $\boldsymbol{\alpha}(z)$ as a vector polynomial of degree k at most. Thus $\mathbf{f}(z)$ is a vector-valued meromorphic function:

$$\mathbf{f}(z) = \mathbf{p}(z)/q(z),$$

where

$$\mathbf{p}(z) = \boldsymbol{\alpha}(z)\beta^*(z), \quad q(z) = \beta(z)\beta^*(z),$$

and $(\mathbf{p}(z), q(z))$ constitutes a vector Padé approximant representation of $\mathbf{f}(z)$ of type $[2k/2k]$ satisfying the axioms (4.13)–(4.16).

From (4.96a) and (4.96b), we have

$$\sum_{i=0}^{k} \beta_i \mathbf{s}_{i+j} = \sum_{i=0}^{k} \beta_i \sum_{n=-i}^{j} \mathbf{c}_{n+i} = \sum_{i=0}^{k} \beta_i \sum_{n=-i}^{0} \mathbf{c}_{n+i} = \boldsymbol{\alpha}(1). \quad (4.96c)$$

Suppose, for the moment, that $\beta(1) \neq 0$. Then, by (4.93), by uniqueness, and by (4.96c) and (4.94),

$$\varepsilon_{2k}^{(j)} = \mathbf{r}^{[j+2k/2k]}(1) = \boldsymbol{\alpha}(1)/\beta(1) = \mathbf{s}, \quad j = 0, 1, 2, \ldots,$$

which establishes (4.95).

It remains to consider the case in which $\beta(1) = 0$. If so, let κ be the least integer, necessarily less than k, for which

$$\sum_{i=0}^{\kappa} \theta_i \mathbf{s}_{i+j} = \mathbf{0}, \quad j = 0, 1, 2, \ldots,$$

for some constants $\theta_0, \theta_1, \ldots, \theta_\kappa$, yet $\sum_{i=0}^{\kappa} \theta_i \neq 0$. We can follow the previous argument with $\{\theta_i\}_{i=0}^{\kappa}$ replacing $\{\beta_i\}_{i=0}^{k}$, because Lemma 1 justifies the hypothesis equivalent to (4.96a), and conclude that

$$\varepsilon_{2\kappa}^{(j)} = \mathbf{s}, \quad j = 1, 2, 3, \ldots.$$

This result implies that

$$\varepsilon_{2\kappa+1}^{(j)} = \infty, \quad j = 1, 2, 3, \ldots,$$

contrary to hypothesis. Therefore $\beta(1) \neq 0$ and (4.95) is established under the conditions of the theorem.

Example 3. The sequence of vectors **S** is defined by

$$\mathbf{s}_j = \mathbf{u}_0 + \mathbf{u}_1 \sum_{i=0}^{j} \alpha_1^i + \mathbf{u}_2 \sum_{i=0}^{j} \alpha_2^i + \cdots + \mathbf{u}_k \sum_{i=0}^{j} \alpha_k^i, \quad j = 0, 1, 2, \ldots,$$

with $\mathbf{u}_i \in \mathbb{C}^d$, $\alpha_i \in \mathbb{C}$, $|\alpha_i| < 1$. **S** has a limit given by

$$\mathbf{s} = \mathbf{u}_0 + \frac{\mathbf{u}_1}{1 - \alpha_1} + \frac{\mathbf{u}_2}{1 - \alpha_2} + \cdots + \frac{\mathbf{u}_k}{1 - \alpha_k}.$$

Then $\{\mathbf{s}_j\}$ satisfy the recursion

$$\beta_0 \mathbf{s}_j + \beta_1 \mathbf{s}_{j+1} + \cdots + \beta_k \mathbf{s}_{j+k} = \mathbf{s} \sum_{j=0}^{k} \beta_j$$

provided $\alpha_1, \alpha_2, \ldots, \alpha_k$ are roots of

$$\beta_0 + \beta_1 r + \beta_2 r^2 + \cdots + \beta_k r^k = 0.$$

If, say $\beta_0 \neq 0$, then the recursion (4.94) is nontrivial, and then, provided zero divisors are not encountered in the construction, McLeod's theorem states that $\varepsilon_{2k}^{(j)} = \mathbf{s}$. Were any of the $|\alpha_i| > 1$, this result would be reinterpreted with $\mathbf{s}$ redefined as an (unstable) fixed point of the recursion (4.94). McLeod's theorem is also directly applicable to Example 1. The iteration matrix E of (4.10) satisfies its characteristic equation

$$E(E - \lambda_1 I)(E - \lambda_2 I)(E - \lambda_3 I) = 0.$$

By applying this result to the sequence $\{\mathbf{s}_i\}_{i=0}^{\infty}$ generated by (4.10) and (4.11), it is readily found that $\{\mathbf{s}_i\}_{i=1}^{\infty}$ satisfies a four-term inhomogeneous recursion relation. Zero divisors are not found in the implementation of the ε-algorithm as far as the column $k = 2$ of Table 8.4.1, and therefore the exact answer, being the fixed point of the recursion (4.10), is found in the entries $\varepsilon_6^{(j)}$, $j = 1, 2, 3, \ldots$.

Now consider another example of the same kind, but in which convergence of the individual columns of the ε-table is the topic of interest.

Example 4 (Acceleration of convergence of SOR iterates). The following diagram represents a system of linear equations $A\mathbf{x} = \mathbf{b}$ in an example given by Varga [1962]. It originates in the solution by spatial discretization of an equation for neutron diffusion in a two-dimensional model reactor. We refer to the appendix of his text for the actual numerical values of the entries of A, which is symmetric and sparse. Diagrammatically, it is expressed as

$$\begin{bmatrix} \times & \times & & & \times & & & & & & & & & & & & \\ \times & \times & \times & & & \times & & & & & & & & & & & \\ & \times & \times & \times & & & \times & & & & & & & & & & \\ & & \times & \times & \times & & & \times & & & & & & & & & \\ \times & & & \times & \times & \times & & & \times & & & & & & & & \\ & \times & & & \times & \times & \times & & & \times & & & & & & & \\ & & \times & & & \times & \times & \times & & & \times & & & & & & \\ & & & \times & & & \times & \times & \times & & & \times & & & & & \\ & & & & \times & & & \times & \times & \times & & & \times & & & & \\ & & & & & \times & & & \times & \times & \times & & & \times & & & \\ & & & & & & \times & & & \times & \times & \times & & & \times & & \\ & & & & & & & \times & & & \times & \times & \times & & & \times & \\ & & & & & & & & \times & & & \times & \times & \times & & & \\ & & & & & & & & & \times & & & \times & \times & \times & & \\ & & & & & & & & & & \times & & & \times & \times & \times & \\ & & & & & & & & & & & \times & & & \times & \times & \times \\ & & & & & & & & & & & & \times & & & \times & \times \end{bmatrix} \begin{bmatrix} \times \\ \times \\ \times \\ \times \\ \times \\ \times \\ \times \\ \times \\ \times \\ \times \\ \times \\ \times \\ \times \\ \times \\ \times \\ \times \\ \times \end{bmatrix} = \mathbf{0}.$$

(4.97)

The right-hand side would not be zero in a realistic application. It is taken as zero here so as to demonstrate the speed of convergence. The system is decomposed as

$$(D - L - U)\mathbf{x} = \mathbf{0}$$

in Varga's notation, with D as the matrix of diagonal elements of A, and $-L$, $-U$ as its upper and lower triangular parts. The system is solved iteratively by

$$\mathbf{s}^{(j+1)} = L_\omega \mathbf{s}^{(j)}, \quad j = 0, 1, 2, \ldots, \tag{4.98}$$

and the SOR iteration matrix is defined by

$$L_\omega = (D - \omega L)^{-1}\{\omega U + (1 - \omega)D\}. \tag{4.99}$$

The parameter ω is called the overrelaxation parameter.

Method 1 (Solution by successive overrelaxation). The parameter ω is chosen so as to minimize $\rho(L_\omega)$, the spectral radius of L_ω. In this way, the number of iterations required in (4.98) is minimized. In Varga's example

$$\omega = 1.92 \ldots$$

and 139 iterations are required to reduce an initial vector $\mathbf{s}_0 = (1, 1, 1, \ldots, 1)$ to $\mathbf{s}_{139}$ with tolerance 10^{-4}, i.e., $\|\mathbf{s}_{139}\|_\infty \leq 10^{-4}$.

Method 2 (Solution using vector epsilon algorithm). The overrelaxation parameter ω is chosen so that

(i) L_ω is allowed to have one large eigenvalue λ_1, and
(ii) the remaining 15 eigenvalues have the smallest possible maximum modulus.

This situation was achieved by taking $\omega = 1.52$, for which

$$\lambda_1 = 0.994 \ldots, \tag{4.100}$$

$$|\lambda_i| \leq 0.52 \ldots, \quad i = 2, 3, \ldots, 16. \tag{4.101}$$

In fact, equality is found in the modulus bound given by (4.101), and (4.101) could be expressed as $|\lambda_i| = \omega - 1$. After stating Theorem 8.4.9, we explain why this modulus, rather than that of (4.100), governs the rate of convergence of $\boldsymbol{\varepsilon}_2^{(j)}$. We find that the column of index 2 of the vector ε-table converges to tolerance 10^{-4} in 26 iterations. Moreover, the speed of convergence is relatively insensitive to the value of ω.

The following theorem governs the convergence of certain rows of the vector Padé table. Its corollary with z fixed at $z = 1$ governs convergence of columns of the vector ε-table, as used in the previous example.

Theorem 8.4.9 (A row convergence theorem) [Graves-Morris and Saff, 1988]. *Let $f(z)$ be a vector function which may be expressed as*

$$\mathbf{f}(z) = \mathbf{g}(z)/Q(z), \tag{4.102}$$

where

(i) $Q(z)$ is a monic real polynomial of degree k with roots $\{z_1, \ldots, z_k\}$, and $0 < |z_i| < \rho$, $i = 1, 2, \ldots, k$; \hfill (4.103)
(ii) $\mathbf{g}(z)$ is analytic in $|z| < \rho$; \hfill (4.104)
(iii) $\mathbf{g}(z_i) \cdot \mathbf{g}^*(z_i) \neq 0$, $i = 1, 2, \ldots, k$.

Thus $\mathbf{f}(z)$ is analytic in the deleted disk

$$D_\rho^- = \{z : |z| < \rho\} - \bigcup_{i=1}^{k} \{z_i\}.$$

Let $(\mathbf{P}_n(z), Q_n(z))$ be vector Padé polynomials of type $[n/2k]$ in which $Q_n(z)$ has leading coefficient unity. Then

$$\lim_{n \to \infty} \mathbf{P}_n(z)/Q_n(z) = \mathbf{f}(z), \quad z \in D_\rho^-,$$

and the rate of convergence is governed by

$$\limsup_{n \to \infty} \|\mathbf{f} - \mathbf{P}_n/Q_n\|_K^{1/n} \leq \mu/\rho. \tag{4.105}$$

The denominators converge according to

$$\lim_{n \to \infty} Q_n(z) = Q^2(z) \tag{4.106}$$

and the rate is given by

$$\limsup_{n \to \infty} \|Q_n - Q^2\|_E^{1/n} \leq \max_{1 \leq i \leq k} |z_i|/\rho, \tag{4.107}$$

for any fixed $\mu \in (0, \rho)$. In (4.105) K is any compact subset of $D_\rho^- \cap \{z : |z| < \mu\}$, and in (4.107) E is any compact subset of $\mathbb{C}$.

Although rather cumbersome, the conditions of this theorem are valid for the important case in which $\mathbf{f}$ is a real symmetric function, i.e., $\mathbf{f}(x) \in \mathbb{R}^d$ for $x \in \mathbb{R}$. Indeed, with the correspondence (4.93), this theorem applies directly in the context of Example 4. In terms of actual numbers, the data generated by (4.98) define the generating function (4.12), which has the representation (4.102). In (4.103) $z_1 = \lambda_1^{-1} = 1.006$ and in (4.104) $\rho = |\lambda_2|^{-1} = 1.92$. With $\mu = 1$, equation (4.105) indicates a convergence rate of $O(0.52)^j$ for $\boldsymbol{\varepsilon}_2^{(j)}$, where $\boldsymbol{\varepsilon}_2^{(j)}$ is the result obtained after j iterations of the process described in Method 2. This expected rate of convergence was obtained in practice, which explains why only 26 matrix iterations are required when convergence acceleration is used appropriately.

Despite the elegance of the convergence result (4.105), it does not imply that the residues of the approximants (which are vectors) converge to the residue of $\mathbf{f}(z)$ at a pole z_i. In fact, numerical experiments and a simple analysis [Coope and Graves-Morris, 1993] indicate that these residues do not converge. One resolution of this difficulty is to use hybrid vector Padé approximants.

In several numerical experiments it was found that the positions of the zeros of the denominators of the vector Padé approximants could provide remarkably accurate estimates of the position of the pole of the underlying generating function. The contexts were those in which it was known a priori that $\mathbf{f}(z)$ possessed real poles at $z_1, z_2, \ldots, z_k$ in the terminology of Theorem 8.4.8. This theorem states that the denominators $q^{[n/2k]}(z)$ have zeros $\zeta_1, \zeta_1^*, \zeta_2, \zeta_2^*, \ldots, \zeta_k, \zeta_k^*$ which converge as

$$\xi_i \to z_i, \quad i = 1, 2, \ldots, k, \quad \text{as } n \to \infty.$$

It was found empirically in several examples that

$$\xi_i = \operatorname{Re} \zeta_i \tag{4.108}$$

provided much more accurate estimates of z_i. It was found that the error in $\operatorname{Im} \zeta_i$ was much greater than that of $\operatorname{Re} \zeta_i$, and the choice (4.108) gave roughly twice as many significant figures for ξ_i when $\mathbf{S}$ is generated by a symmetric operator [Roberts, 1996]. Thus, for reconstructing functions $\mathbf{f}(z)$ whose poles are real, we define

$$q^h(z) = \prod_{i=1}^{k}(z - \xi_i) \tag{4.109}$$

as the hybrid denominator, with corresponding vector polynomial numerator

$$\mathbf{p}^h(z) = [\mathbf{f}(z)q^h(z)]_0^n. \tag{4.110}$$

An example which demonstrates the exceptional precision obtained using hybrid vector Padé approximants

$$\mathbf{r}^h(z) = \mathbf{p}^h(z)/q^h(z) \tag{4.111}$$

is given in the next subsection.

Another prime motive for introducing hybrid vector Padé approximants as defined by (4.108)–(4.111) is that the choice (4.109) removes the superfluous zero doubling in the denominator polynomial which is so evident from (4.23), (4.30). This zero doubling leads to a serious residue problem at the poles of vector Padé approximants, as is exemplified by Coope and Graves-Morris [1993] in the case $d = \infty$. Hybrid vector Padé approximants of type $[n/1]$ for a function $\mathbf{f}(z)$ with coefficients $\mathbf{c}_i \in \mathbb{R}^d$ take an interesting explicit form. From (4.23), and with a convenient normalization, we obtain

$$q^{\mathrm{H}}(z) = \mathbf{c}_n \cdot (\mathbf{c}_{n-1} - z\mathbf{c}_n), \qquad (4.112)$$

and the actual approximant follows from (4.110), (4.111). The corresponding vector accelerator is obtained by setting $z = 1$ and defining $\mathbf{t}_n = \mathbf{p}^{\mathrm{H}}(1)/q^{\mathrm{H}}(1)$. We obtain the hybrid vector accelerator

$$\mathbf{t}_n^{\mathrm{H}} = \mathbf{s}_{n-1} - \Delta\mathbf{s}_{n-1} \frac{\Delta\mathbf{s}_{n-2} \cdot \Delta\mathbf{s}_{n-1}}{\Delta\mathbf{s}_{n-1} \cdot \Delta^2\mathbf{s}_{n-2}}, \quad n = 2, 3, 4, \ldots. \qquad (4.113)$$

A similar formula, namely, the Aitken-like vector accelerator

$$\mathbf{t}_n^{\mathrm{A}} = \mathbf{s}_{n-1} - \Delta\mathbf{s}_{n-1} \frac{(\Delta\mathbf{s}_{n-2})^2}{\Delta\mathbf{s}_{n-2} \cdot \Delta^2\mathbf{s}_{n-2}}, \quad n = 2, 3, 4, \ldots, \qquad (4.114)$$

was given by Graves-Morris [1992] based on the same principles. These formulas (4.113), (4.114) are different, but they are compatible in the sense that they are derived in different ways using the same approximation. In general, the aim of the hybrid methods is to locate approximate double roots of $q(z)$, which are given by (4.20) or (4.23), and then to construct a new denominator polynomial with single roots at the same points, as in (4.108), (4.109). To find an approximate double root of $q(z)$ given by (4.23), we may use either the formula

$$|\mathbf{c}_n|^2 |\mathbf{c}_{n-1} - z\mathbf{c}_n|^2 = [\mathbf{c}_n \cdot (\mathbf{c}_{n-1} - z\mathbf{c}_n)]^2 + D \qquad (4.115)$$

or the formula

$$|\mathbf{c}_{n-1}|^2 |\mathbf{c}_{n-1} - z\mathbf{c}_n|^2 = [\mathbf{c}_{n-1} \cdot (\mathbf{c}_{n-1} - z\mathbf{c}_n)]^2 + z^2 D, \qquad (4.116)$$

where $z \in \mathbb{R}$ and

$$D := |\mathbf{c}_n|^2 |\mathbf{c}_{n-1}^2| - (\mathbf{c}_n \cdot \mathbf{c}_{n-1})^2 \geq 0.$$

The hybrid formulas (4.112), (4.113) arise from (4.115) under the hypothesis that $D \approx 0$. The Aitken formula (4.114) arises from (4.116), also under the hypothesis that $D \approx 0$.

Formula (4.114) is a remarkably natural generalization of Aitken's Δ^2 algorithm [Graves-Morris, 1992]. It generates a sequence $\{\mathbf{t}_n^{\mathrm{A}}\}_{n=2}^{\infty}$, which is the same as the 'first column' generated by the minimal-polynomial extrapolation algorithm (MPE) of Cabay and Jackson [1976], although the principles involved in the respective derivations are quite different. When applied to the vector sequence generated by (4.98) and Method 2, only 20 iterations for convergence to the prescribed tolerance are required. The performance of this algorithm, regarded as a general method for accelerating the convergence of SOR estimates using (4.98) with a suboptimal ω, has been found to be very acceptable when applied to a number of standard examples [da Cunha and Hopkins, 1994]. Proof that this algorithm converges under reasonable hypotheses also follows from Theorem 8.4.9, or more directly from theorems of Sidi [1986, 1994]. The general principle of using 'more iterations and less extrapolation' is supported by the work of Gander, Golub, and Gruntz [1989], who report and explain a number of connections between various algorithms for acceleration of convergence of vector sequences. Polynomial acceleration of SOR estimates using a suboptimal value of ω was first proposed by Dancis [1991a, b].

In summary, we note that acceleration of convergence of a sequence of vectors is conveniently implemented using the ε-algorithm. This procedure is supported by the theory of vector Padé approximation. However, it suffers from the pole doubling and residue failure problems, and so hybrid vector Padé approximants are introduced. To lowest order, extrapolation methods for real vector sequences based on hybrid approximants take the forms (4.113), (4.114). They generalize Aitken's Δ^2 procedure, and appear to give good numerical precision when properly applied. However, for all but the lowest orders, the hybrid approximants are awkward to compute.

The question of how Padé approximants should be defined for series of the form

$$f(z) = c_0 + c_1 z + c_2 z^2 + \cdots, \; c_i \in \mathbb{C}^{d \times m},$$

in which the c_i are rectangular matrices with m in the range $1 \leq m \leq d$ and which reduce to vector Padé approximants when $m = 1$ and to matrix Padé approximants when $m = d$ has remained open for some time. An approach based on the requirements that

$$q(z) \in \mathbb{R}^{m \times m}, \quad \partial\{q\} = 2k,$$
$$p(z) = [f(z)q(z)]_0^n \in \mathbb{C}^{d \times m}$$

and the existence of a matrix polynomial $P(z) \in \mathbb{R}^{d \times d}$ for which

$$p^T(z)p(z)q^{-1}(z)p^T(z)p(z) = p^T(z)P(z)p(z) \qquad (4.117)$$

seems promising [Graves-Morris, 1994].

8.4.1 Functional Padé approximants

Lastly, we come to functional Padé approximants. They arise as the $d \to \infty$ limit of vector Padé approximants. The power series that is given has coefficients $c_i(x)$ which themselves are functions of x. In this context λ is a more natural name for the expansion variable than z. The data series takes the form

$$f(x, \lambda) = c_0(x) + \lambda c_1(x) + \lambda^2 c_2(x) + \cdots \qquad (4.118)$$

with $\lambda \in \mathbb{C}$ and, for example, $c_i(x) \in L_2(a, b)$, where (a, b) is a real interval which is a natural domain for the $c_i(x)$.

The formalism is precisely analogous to that given already in this section. For example, the functional Padé approximant of type $[n/2k]$ for (4.118) takes the form

$$r(x, \lambda) = p(x, \lambda)/q(\lambda), \qquad (4.119)$$

where

$$q(\lambda) = \begin{vmatrix} 0 & M_{0,1} & \cdots & M_{0,2k-1} & M_{0,2k} \\ -M_{0,1} & 0 & \cdots & M_{1,2k-1} & M_{1,2k} \\ \vdots & \vdots & & \vdots & \vdots \\ -M_{0,2k-1} & -M_{1,2k-1} & \cdots & 0 & M_{2k-1,2k} \\ \lambda^{2k} & \lambda^{2k-1} & \cdots & \lambda & 1 \end{vmatrix}, \qquad (4.120)$$

and

$$M_{ij} = \sum_{l=0}^{j-i-1} \int_a^b c_{l+i+n-2k+1}(x)[c_{j-l+n-2k}(x)]^* \, dx \qquad (4.121)$$

for $i = 0, 1, \ldots, 2k$ and $j = i+1, i+2, \ldots, 2k$, and

$$p(x, \lambda) = [q(\lambda)f(x, \lambda)]_0^n. \qquad (4.122)$$

A weight function can be incorporated into (4.121) if appropriate. These approximants find a natural application in linear integral equations, in which the Fredholm denominator does not depend on the variable x. This situation is faithfully represented by the functional Padé denominator given by (4.120). An example will serve to illustrate the use of functional Padé approximants.

Example 5. The functional Padé approximant of type [2/2] and the Aitken-type hybrid functional Padé approximant of type [2/1] for the Neumann series of

$$\phi(x) = 1 + \lambda \int_0^1 \{1 + |x - y|\} \phi(y) \, dy \qquad (4.123)$$

are required.

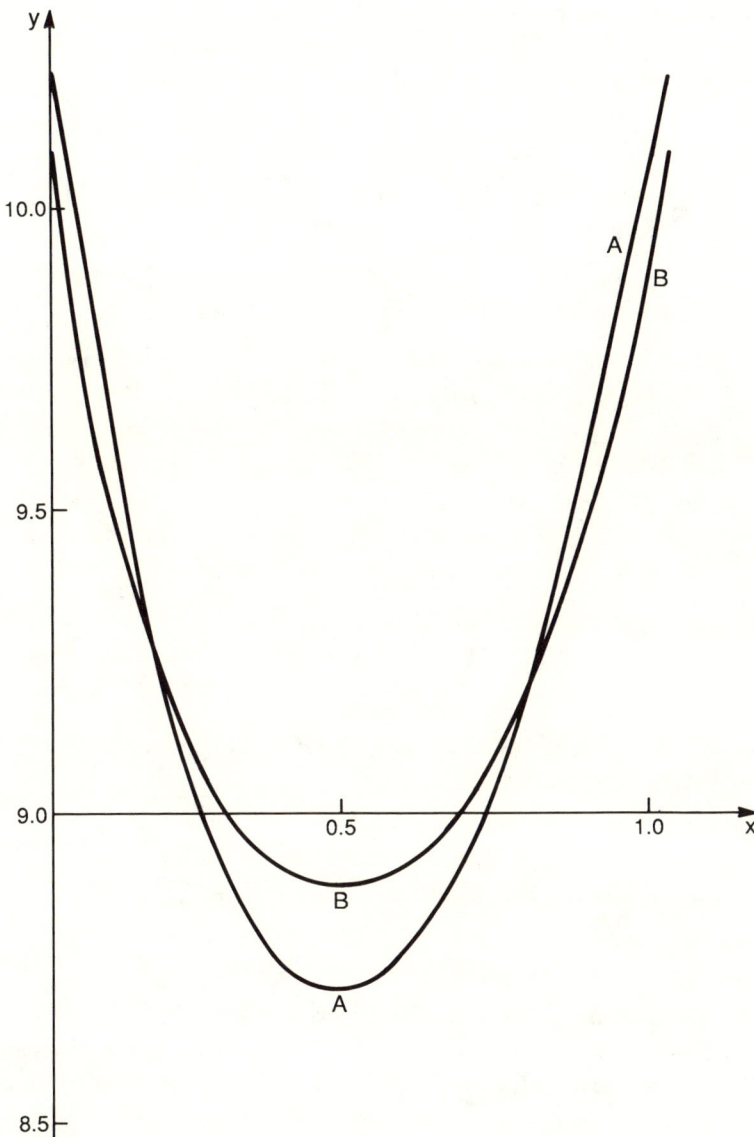

Figure 8.4.1. Curve A shows the exact solution of (4.123) for $\lambda = 2/3$, which is imperceptibly different from the Aitken-type hybrid solution $r^A(x, 2/3)$ of type [2/1] given by (4.125). Curve B is the [2/2]-type function-valued Padé approximant $r(x, 2/3)$ given by (4.119).

Solution. The Neumann series is

$$f(x, \lambda) = 1 + \left[\frac{5}{4} + \left(x - \frac{1}{2}\right)^2\right]\lambda$$
$$+ \left[\frac{161}{96} + \frac{5}{4}\left(x - \frac{1}{2}\right)^2 + \frac{1}{6}\left(x - \frac{1}{2}\right)^4\right]\lambda^2 + \cdots,$$

as is found by iteration of (4.123). From (4.121),

$$q(\lambda) = \begin{vmatrix} 0 & \int_0^1 [c_1(x)]^2 \, dx & 2\int_0^1 c_2(x)c_1(x) \, dx \\ -\int_0^1 [c_1(x)]^2 \, dx & 0 & \int_0^1 [c_2(x)]^2 \, dx \\ \lambda^2 & \lambda & 1 \end{vmatrix}$$

and the functional Padé approximant is

$r(x, \lambda) =$

$$\frac{1 + [-1.425 + (x - \tfrac{1}{2})^2]\lambda + [0.122 - 1.425(x - \tfrac{1}{2})^2 + 0.167(x - \tfrac{1}{2})^4]\lambda^2}{1 - 2.675\lambda + 1.788\lambda^2}$$

(4.124)

Analogously to (4.109)–(4.111) and (4.116), the Aitken-type hybrid functional Padé approximant is

$r^A(x, \lambda) =$

$$\frac{1 + [-0.087 + (x - \tfrac{1}{2})^2]\lambda + [-0.006 - 0.087(x - \tfrac{1}{2})^2 + 0.167(x - \tfrac{1}{2})^4]\lambda^2}{1 - 1.337\lambda}.$$

(4.125)

These two approximants are shown in Figure 8.4.1.

This and other examples of functional Padé approximants are described more fully by Graves-Morris [1990] and Graves-Morris and Thukral [1992].

8.5 Hermite–Padé polynomials

Hermite–Padé approximants were introduced by Hermite [1893] and Padé [1892, 1894b] as a generalization of Padé approximants, which they thought of as the approximants to continued fractions. As a preliminary to the discussion of the special cases of Hermite–Padé approximants to be reviewed in the next section, we will discuss in this section the auxiliary Hermite–Padé polynomials and some of their properties. To this end, let us be given $k + 2$ formal power series $f_{-1}(z)$, $f_0(z)$, ..., $f_k(z)$. This slightly strange way of denoting them is used for future convenience in the next section. We assume, without loss of significant generality, that at least one of the f's does not vanish at $z = 0$. By the symmetry of the problem, we may choose it to be $f_{-1}(z)$ without making any essential difference. For given $(m_{-1}, m_0, \ldots, m_k)$ we now define, through the accuracy-through-order principle, the polynomials $Q_{j,m_j}(z)$ by

$$\sum_{j=-1}^{k} Q_{j,m_j}(z) f_j(z) = O(z^{s+1}), \tag{5.1}$$

where $Q_{j,m_j}(z)$ is of degree m_j at most. We also define

$$M = \sum_{j=0}^{k} (m_j + 1) - 1, \quad s = M + m_{-1}. \tag{5.2}$$

If we set $m_j = -1$, it means that $Q_j(z) \equiv 0$. Note is taken that the Q's defined in this manner are determined only up to an overall constant factor. This point will be further discussed later. If all the solutions of (5.1) differ from each other just by a constant factor, then the solution is one dimensional, and we will properly call it 'essentially unique.' It is also the case that if we divide (5.1) by $f_{-1}(z)$ then this operation represents only a reorganization of the linear equations for the coefficients of the Q's corresponding to (5.1). It does not change the solutions so defined for the Hermite–Padé polynomials. Consequently, for analytical purposes we may as well set $f_{-1}(z) = 1$, as we will do explicitly in the next section. Using the notation

$$f_j(z) = \sum_{n=0}^{\infty} f_n^{(j)} z^n, \tag{5.3}$$

we can write out explicit solutions for the Q's in terms of determinants. For example, if the solution of (5.1) is essentially unique, we can use Cramer's rule to solve (5.1) for the $(m_j + 1)$ coefficients of $Q_{j,m_j}(z)$ and we obtain

$$Q_{j,m_j}(z) =$$

$$\det \begin{vmatrix} f_0^{(-1)} & 0 & \cdots & 0 & & f_0^{(j)} & 0 & \cdots & 0 & \\ f_1^{(-1)} & f_0^{(-1)} & \cdots & 0 & & f_1^{(j)} & f_0^{(j)} & \cdots & 0 & \\ \vdots & \vdots & \ddots & \vdots & & \vdots & \vdots & \ddots & \vdots & \\ f_{m_{-1}}^{(-1)} & f_{m_{-1}-1}^{(-1)} & \cdots & f_0^{(-1)} & \cdots & f_{m_j}^{(j)} & f_{m_j-1}^{(j)} & \cdots & f_0^{(j)} & \cdots \\ \vdots & \vdots & & \vdots & & \vdots & \vdots & & \vdots & \\ f_s^{(-1)} & f_{s-1}^{(-1)} & \cdots & f_{s-m_{-1}}^{(-1)} & & f_s^{(j)} & f_{s-1}^{(j)} & \cdots & f_{s-m_j}^{(j)} & \\ 0 & 0 & \cdots & 0 & \cdots & 1 & z & \cdots & z^{m_j} & \cdots \end{vmatrix},$$

$$\tag{5.4}$$

where the last row is zero everywhere except for the jth block as indicated. It may happen that the determinantal solution fails because the equations admit multiple solutions. There must exist nontrivial solutions, because there is one more unknown than there are equations in (5.1). Let us define an $(s + 1) \times (s + 2)$ matrix by

$$\Delta_{\vec{m}} = \begin{bmatrix} f_0^{(-1)} & 0 & \cdots & 0 & f_0^{(j)} & 0 & \cdots & 0 \\ f_1^{(-1)} & f_0^{(-1)} & \cdots & 0 & f_1^{(j)} & f_0^{(j)} & \cdots & 0 \\ \vdots & \vdots & \ddots & \vdots & \vdots & \vdots & \ddots & \vdots \\ f_{m_{-1}}^{(-1)} & f_{m_{-1}-1}^{(-1)} & \cdots & f_0^{(-1)} & \cdots & f_{m_j}^{(j)} & f_{m_j-1}^{(j)} & \cdots & f_0^{(j)} & \cdots \\ \vdots & \vdots & & \vdots & & \vdots & \vdots & & \vdots \\ f_s^{(-1)} & f_{s-1}^{(-1)} & \cdots & f_{s-m_{-1}}^{(-1)} & & f_s^{(j)} & f_{s-1}^{(j)} & \cdots & f_{s-m_j}^{(j)} \end{bmatrix}$$

(5.5)

We note that the solution $Q_{j,m_j}(z)$ given by (5.4) is essentially unique and of precise type $(m_{-1}, \ldots, m_k)$ if and only if $\operatorname{rank}(\Delta_{\vec{m}}) = s + 1$. In the case of multiple solutions we can prove the following result.

***Theorem* 8.5.1 (Multi-niqueness) [Baker and Graves-Morris, 1990].** *There exist at most $k + 1$ linearly independent solutions $(Q_{-1,m_{-1}}^{(i)}, \ldots, Q_{k,m_k}^{(i)})$ for the Hermite–Padé polynomials of type $(m_{-1}, \ldots, m_k)$ to $f_{-1}(z), \ldots, f_k(z)$.*

Remark. It is certainly the case that the multiplicity of solutions that are independent in the ordinary sense is not limited by $k + 1$, but following the usual rules it equals the nullity of $\Delta_{\vec{m}}$ in (5.5). The linear independence governed by Theorem 8.5.1 refers to polynomial factors. For example, if both zQ_{j,m_j} and Q_{j,m_j} are solutions of (5.1), they are linearly dependent in the sense of multi-niqueness, in which polynomial multipliers are allowed, but linearly independent in the standard sense of algebraic multiplicity over the coefficient space $\mathbb{C}^{s+2}$.

Proof. Suppose that there exist $k + 2$ solutions of the equations for the Hermite–Padé polynomials. Then

$$\sum_{j=-1}^{k} f_j(z) Q_{j,m_j}^{(i)}(z) = r^{(i)}(z), \quad i = 1, \ldots, k + 2, \qquad (5.6)$$

where for all i,

$$r_L^{(i)}(z) = O(z^{s+1}).$$

Regarding (5.6) as a set of linear equations in the $f_j(z)$, $j = 0, \ldots, k$, we have the condition

$$f_{-1}(z) \det \begin{vmatrix} Q_{k,m_k}^{(1)} & \cdots & Q_{0,m_0}^{(1)} & Q_{-1,m_{-1}}^{(1)} \\ \vdots & & \vdots & \vdots \\ Q_{k,m_k}^{(k+2)} & \cdots & Q_{0,m_0}^{(k+2)} & Q_{-1,m_{-1}}^{(k+2)} \end{vmatrix}$$

$$= \det \begin{vmatrix} Q_{k,m_k}^{(1)} & \cdots & Q_{0,m_0}^{(1)} & r^{(1)} \\ \vdots & & \vdots & \vdots \\ Q_{k,m_k}^{(k+2)} & \cdots & Q_{0,m_0}^{(k+2)} & r^{(k+2)} \end{vmatrix} \quad (5.7)$$

for their internal consistency. Since every term in the determinant on the left-hand side of (5.7) is a polynomial of degree $\sum_{j=-1}^{k} m_j \leq s - k$ and the determinant on the right-hand side is $O(z^{s+1})$, the left-hand side must be identically zero and therefore any $k+2$ solutions must be linearly dependent.

Note that the case $k = 0$ corresponds directly to the uniqueness theorem [Baker, 1975a, Theorem 1.1] for Padé approximants.

8.5.1 Minimality definitions and uniqueness

It is useful to adopt a definition of Hermite–Padé polynomials which makes them essentially unique. The prime motivation for requiring this uniqueness is the study of the convergence properties of the approximants. The definition which we will adopt will be motivated by the consideration of the special cases of the integral and algebraic approximants in the next section. In this context, the term $Q_{k,m_k}(z)f_k(z)$ in (5.6) is special, because it is naturally the leading term. Since the number of singular points in the integral or algebraic approximants is determined by the number of zeros of $Q_{k,m_k}(z)$, there is an advantage in reducing the number of arbitrary or spurious zeros in this polynomial. In the cases where the solution is not uniquely defined by (5.1), we feel that as it is very likely that the extra, arbitrary zeros so introduced would not be relevant to the underlying structure of the function being approximated, they should be eliminated. Accordingly, our definition of minimality is motivated by a natural hierarchy of $f_k(z), f_{k-1}(z), \ldots, f_{-1}(z)$.

Definition 8.5.1 (Minimality) [Baker and Graves-Morris, 1990]. A solution for the Hermite–Padé polynomials of type $(m_{-1}, \ldots, m_k)$ to $f(z)$ is called minimal if it is of the lowest degree in the following sense. First there exists no other solution of type $(m_{-1}, \ldots, m_k)$ for which the actual degree of Q_{k,m_k} is smaller. If there exist solutions of type $(m_{-1}, \ldots, m_k)$ for which $Q_{k,m_k} \equiv 0$, then we minimize the degree of $Q_{k-1,m_{k-1}}$, $Q_{k-2,m_{k-2}}$, etc. to find the minimal solution.

***Theorem* 8.5.2 (Uniqueness) [Baker and Graves-Morris, 1990].** *For any specified type* $(m_{-1}, \ldots, m_k)$ *there exist Hermite–Padé polynomials* $(Q_{-1,m_{-1}}, \ldots, Q_{k,m_k})$ *to* $f(z)$ *which are minimal and essentially unique.*

Proof. First, we have already shown that a solution exists. Second, suppose that there exist two solutions of minimal degree $Q^{(1)}_{j,m_j}(z)$ and $Q^{(2)}_{j,m_j}(z)$, both of true degree $n_j \leq m_j$, and $j \leq k$ is the largest index for which the $Q^{(i)}_{j,m_j} \not\equiv 0$. Then

$$\alpha(Q^{(1)}_{-1}, \ldots, Q^{(1)}_{k,m_k}) + (1 - \alpha)(Q^{(2)}_{-1,m_{-1}}, \ldots, Q^{(2)}_{k,m_k}) \quad (5.8)$$

is also a solution of minimal degree. Choose α such that the coefficient of z^{n_j} in that solution is zero. But this result implies that n_j is not the true minimal degree, contrary to assumption, which is a contradiction. Therefore there is a one-dimensional minimal solution of type $(m_{-1}, \ldots, m_k)$. As such it is essentially unique.

An alternative definition which leads to a unique set of Hermite–Padé polynomials was introduced by Paszkowski [1990]. He refers to these as 'optimal'. Their principal value is described by him as theoretical rather than numerical. In what follows, we will assume that the f's are polynomially independent. That is to say, there does not exist any finite set of polynomials $P_j(z)$, $j = -1, \ldots, k$, such that $\sum_{j=-1}^{k} P_j(z)f_j(z) \equiv 0$. This restriction is equivalent to the requirement of not approximating a rational function in a discussion of Padé approximants. The reason for the restriction in that context is that a finite-order Padé approximant trivially gives the exact solution of the approximation problem at all higher orders. The reason here is the same: a finite-order set of Hermite–Padé polynomials would exactly satisfy (5.1). We also introduce the notation that $\rho(m_{-1}, \ldots, m_k; \mu)$ is the rank of the $\mu \times (s + 2)$ matrix obtained by either truncating or extending the number of rows of f's in (5.5) to get μ rows. We can ignore the case in which $\rho(m_1, \ldots, m_k; \infty) \leq s + 1$, because it would follow that (5.1) has a finite-order exact solution, as explained previously. Paszkowski shows that there exists an integer $\omega = \omega(m_{-1}, \ldots, m_k)$ such that

$$\omega \geq s + 1, \quad \rho(m_{-1}, \ldots, m_k; \omega) = s + 1,$$

$$\rho(m_{-1}, \ldots, m_k; \omega + 1) = s + 2. \quad (5.9)$$

For the normal case where the solution is essentially unique, $\omega = s + 1$. The implication of this result is that the right-hand side of (5.1) could be replaced by $O(z^\omega)$ and so the polynomial defining equations would be oversatisfied by $\omega - s - 1$ orders in z. The notions of oversatisfaction (r) and the deficiency index $(\omega_{L,M})$ played an essential role in Padé's [1892,

1900] analysis of his block theorem (Theorem 1.4.3), and other matters discussed in Section 1.4.

Definition 8.5.2 (Optimal Hermite–Padé polynomials) [Paszkowski, 1990]. It follows from the properties (5.9) of ω that there exist Hermite–Padé polynomials $\mathcal{Q}_{j,m_j}(z)$, $-1 \leq j \leq k$, of type $(m_{-1}, \ldots, m_k)$ which oversatisfy (5.1) by $\omega - s - 1$ powers of z. If, for at least one j, $\mathcal{Q}_{j,m_j}(0) \neq 0$, then the $\mathcal{Q}$'s are called the optimal Hermite–Padé polynomials.

Corollary 1 [Paszkowski, 1990]. The optimal Hermite–Padé polynomials of type $(m_{-1}, \ldots, m_k)$ are essentially unique.

Proof. By the properties (5.9) of $\omega(m_{-1}, \ldots, m_k)$ there must exist a subset of $s + 1$ linearly independent rows of the ω rows of the matrix of rank $s + 1$ which was described before (5.9). It follows directly that the solution for this subset of the equations is essentially unique, and by the properties of rank this solution satisfies the rest of the equations as well.

Because Paszkowski requires that, for at least one j, $Q_{j,m_j}(0) \neq 0$, sometimes the optimal Hermite–Padé polynomials do not exist. In this respect, the situation is analogous to that of nonexistent Padé approximants, q.v. Section 1.4. It is the case that Definition 8.5.1 ensures that the Hermite–Padé polynomials have no common factors of the form $a + bz$ with $a \neq 0$. This result can easily be seen, since if there were a common factor in the supposed minimal set of polynomials, then we could just divide it out without changing the order in z in the right-hand-side of (5.1) and obtain a new solution for which the true degree of $Q_{k,m_k}(z)$ was one less. The same does not hold for factors of z. In order to impose the requirement that, for at least one j, $Q_{j,m_j}(0) \neq 0$, the required division by z would change the order in z of the right-hand-side of (5.1). We will discuss the elimination of such factors in Lemma 3.

Definition 8.5.2 requires the use of higher-order information to break the degeneracy of the fundamental defining equations. In this way, a solution is found which is essentially unique. But because this definition requires the knowledge of an indefinite number of series coefficients, it is not so useful numerically. However, it is related in spirit to Definition 8.5.1 in that it seeks to get the maximum representation out of the smallest possible set of polynomial coefficients.

There is yet another definition of minimal Hermite–Padé polynomials. It is due to Beckermann [1990], but it does not necessarily lead to a unique specification. We shall call it Beckermann minimality. To simplify our writing we will introduce the notation $\vec{m} = (m_{-1}, \ldots, m_k)$ and $|\vec{m}| = \sum_{j=-1}^{k} m_j$. Note that $|\vec{m}| \geq -k - 2$ rather than 0 because the values $m_j = -1$ are possible. The definition involves a partial ordering relation: $\vec{a} \leq \vec{c}$ means that every component of $\vec{a}$ is less than or equal to the corresponding component of $\vec{c}$. The relation $\vec{a} = \vec{c}$ means equality of each component. We can then define the further ordering relation $\vec{a} < \vec{c}$, which means that $\vec{a} \leq \vec{c}$ and $\vec{a} \neq \vec{c}$. Notice that if $a_i > c_i$, $c_j > a_j$, $i \neq j$, then the pair of elements $\vec{a}$, $\vec{c}$ are not comparable by means of these partial ordering relations, as shown in Figure 8.5.1.

***Definition 8.5.3* (Minimality) [Beckermann, 1990].** A nontrivial solution of (5.1) for the Hermite–Padé polynomials is called a Beckermann minimal solution if, among all the solutions (5.1), there is no other nontrivial solution whose degree is less according to the above given partial ordering relation.

***Definition 8.5.4* (Minimal region) [Beckermann, 1990].** The Beckermann minimal region associated with a Beckermann minimal solution is the set of all points in index space for which the given solution is a Beckermann minimal solution.

The solutions which are minimal according to Definition 8.5.2 are also Beckermann minimal solutions, but not necessarily vice versa. Thus with a point in index space, there may be associated several Beckermann minimal solutions. Furthermore, in the next example we see that it is not necessarily true that the solution of (5.1) for the point in index space occupied by a Beckermann minimal solution (or a minimal solution for that matter) is necessarily essentially unique. In addition, 'minimal

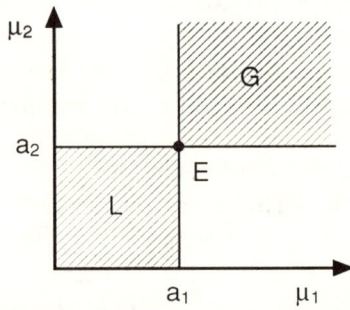

Figure 8.5.1. At E, $\vec{\mu} = \vec{a}$. In region L, $\vec{\mu} \leq \vec{a}$, and in region G, $\vec{\mu} \geq \vec{a}$. Elsewhere, $\vec{\mu}$ and $\vec{a}$ are not comparable.

8.5 Hermite–Padé polynomials

regions' can be associated with each solution. As an example, consider the case

$$f_{-1}(z) = 1 + 3z^2 + 6z^3 + 9z^4 + z^5,$$
$$f_0(z) = 1 + z + 2z^2 + 4z^3 + 5z^4,$$
$$f_1(z) = 1.$$

The Beckermann minimal solution of type $(1, 1, 1)$ is

$$\vec{Q}_{(1,1,1)}(z) = (-1, 1 + z, -2z),$$

which is essentially unique. Notice that this solution is actually of degree $(0, 1, 1)$. Since it oversatisfies the defining equations, it is also a solution, which turns out to be essentially unique, of types $(0, 2, 1)$ and $(0, 1, 2)$. Consequently, it is a Beckermann minimal solution of those types as well. This result illustrates the concept of a Beckermann minimal region. Independent solutions of type $(0, 1, 1)$ are taken to be

$$\vec{Q}_A(z) = (0, -1 + 2z, 1 - z)$$

and

$$\vec{Q}_B(z) = (2, -3, 1 + 3z),$$

which may be combined to give

$$\vec{Q}_{(1,1,1)}(z) = \tfrac{1}{2}[\vec{Q}_A(z) - \vec{Q}_B(z)],$$
$$\vec{Q}_{(0,1,1)}(z) = 3\vec{Q}_A(z) + \vec{Q}_B(z),$$

This result illustrates that a Beckermann minimal solution is not necessarily an essentially unique solution for the point in index space designated by its true degree, and it is not even necessarily a minimal solution at that point, as the above given $\vec{Q}_{0,1,1}(z)$ is a solution of actual degree $(0, 1, 0)$ [but it is not a solution of type $(1, 1, 1)$]. This illustration leads us to the next development.

8.5.2 Table structure results

Let us now introduce the notation that $\vec{e}_j$ is a unit vector in index space in the jth dimension.

Lemma 1 (Local properties of a Beckermann [1990] minimal solution).
(i) Let $\vec{\mathfrak{Q}} = (\mathfrak{Q}_{-1,m_{-1}}(z), \ldots, \mathfrak{Q}_{k,m_k}(z))$ be a minimal solution of type $\vec{m} = (m_{-1}, \ldots, m_k)$. Further let us define t, the actual degree of

oversatisfaction of (5.1), by

$$\sum_{j=-1}^{k} \mathcal{Q}_{j,m_j}(z) f_j(z) = O(z^{s+t+1}), \tag{5.10}$$

where $s = |\vec{m}| + k$ and the order in (5.10) holds precisely. Then for any j, $\vec{\mathcal{Q}}$ is an $\vec{m} + \vec{e}_j$ Beckermann minimal solution if and only if $t > 0$. (ii) If $\vec{\mathcal{Q}}$ is an $\vec{m} + \vec{e}_j$ Beckermann minimal solution of true degree $\vec{n} = (n_{-1}, \ldots, n_k)$, then for any $l \neq j$, $\vec{\mathcal{Q}}$ is also an $\vec{m} + \vec{e}_l$ Beckermann minimal solution if and only if $\vec{n} \leq \vec{m}$.

Proof. By Definition 8.5.3, $\vec{n}_\mathcal{Q} \leq \vec{m}$, where we are using $\vec{n}$ to represent the true degree and the subscript to denote which solution. With respect to the result (i), we know that $\vec{n}_\mathcal{Q} < \vec{m} + \vec{e}_j$, so $\vec{\mathcal{Q}}$ is a solution of (5.1) of type $\vec{m} + \vec{e}_j$ if and only if $t > 0$. If $\vec{\mathcal{Q}}$ is not a Beckermann minimal solution of type $\vec{m} + \vec{e}_j$, then there must exist a solution $\vec{\mathcal{P}}$ such that $\vec{n}_\mathcal{P} \leq \vec{n}_\mathcal{Q}$. If this is the case, then $\vec{\mathcal{P}}$ is also a Beckermann minimal solution of type $\vec{m}$ which is a contradiction. One can prove result (ii) in the same way because, by the accuracy-through-order principle, $\vec{\mathcal{Q}}$ is also an $\vec{m} + \vec{e}_l$ solution if and only if $\vec{n} \leq \vec{m} = \min\{\vec{m} + \vec{e}_j, \vec{m} + \vec{e}_l\}$.

We also need the definition of an index set which has the shape of a (possibly) truncated simplex in index space. Let $\mathcal{T}(\vec{m}, \delta, d)$ be the set of all points $\vec{m} + \vec{v}$ such that $\vec{v} \geq \vec{0}$ and $0 \leq \delta \leq |\vec{v}| \leq d$, as shown schematically in Figure 8.5.2. On the basis of these definitions and the lemma, we can prove that a Beckermann minimal region takes the shape of a (possibly truncated) simplex.

***Theorem* 8.5.3 (Geometrical figures induced by a minimal solution) [Beckermann, 1990].** Let $\vec{\mathcal{Q}}$ be a Beckermann minimal solution of true degree $\vec{n}$ and of type $\vec{m}$, and let t be given by (5.10). If we define

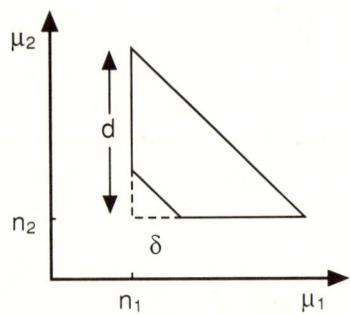

Figure 8.5.2. A schematic diagram of $\mathcal{T}(\vec{m}, \delta, d)$.

8.5 Hermite–Padé polynomials

$d = s + t + 2 - |\vec{n} + \sum_{j=-1}^{k} \vec{e}_j|$, *then there exists a δ, $0 \leq \delta \leq d$, such that the minimal region of $\vec{\mathcal{Q}}$ is $\mathcal{T}(\vec{n}, \delta, d)$. If $\delta > 0$, there exist at least two polynomially independent solutions of (5.1) for all points in index space of type $\vec{a} \in \mathcal{T}(\vec{n}, 0, \delta - 1)$.*

Proof. According to (5.10), $\vec{\mathcal{Q}}$ is a solution for every point in $\mathcal{T}(\vec{n}, 0, d)$ and by hypothesis a Beckermann minimal solution for at least one point $\vec{m}$ in this region. Therefore by Definition 8.5.3, there must exist a $\vec{c}$ of minimal norm such that $\vec{\mathcal{Q}}$ is a Beckermann minimal solution of type $\vec{n} + \vec{c}$ and we set $\delta = |\vec{c}|$. By a step-by-step construction based on Lemma 1, $\mathcal{T}(\vec{n}, \delta, d)$ is in the Beckermann minimal region for $\vec{\mathcal{Q}}$, as required. If $\delta = 0$, the whole theorem is proved. Otherwise, $\delta > 0$, and there must exist an l such that $\vec{\mathcal{Q}}$ is not a Beckermann minimal solution for $\vec{n} + \vec{c} - \vec{e}_l$. But since $\vec{\mathcal{Q}}$ is an $\vec{n} + \vec{c} - \vec{e}_l$ solution of (5.1), there must also be a Beckermann minimal solution $\vec{\mathcal{P}}$ of type $\vec{n} + \vec{c} - \vec{e}_l$ with $n_{\mathcal{P}} \leq n_{\mathcal{Q}}$. Therefore, $\vec{\mathcal{P}}$ is a solution at least in the set $\mathcal{T}(\vec{n}, 0, \delta - 1)$ with a smaller degree than $\vec{\mathcal{Q}}$, so $\vec{\mathcal{Q}}$ is a Beckermann minimal solution nowhere in $\mathcal{T}(\vec{n}, 0, \delta - 1)$.

In the previous illustration, there is an essentially unique Beckermann minimal solution for $\vec{\mathcal{Q}}_{(1,1,1)}(z)$. This uniqueness need not be the case in general. The minimal region associated with that Beckerman minimal solution is $\mathcal{T}((0, 1, 1), 1, 1)$. For this case $\delta = 1$; we see explicitly the existence of a region in which the solutions are not essentially unique. These 'predecessor' structures, as we have seen in the proof of the theorem, are minimal regions associated with a different type than the type $\vec{m}$ which we started with, and so knowledge of the structure of the Beckermann minimal regions for one point in index space can have implications for some of the related points.

There are some other things which can be deduced from the structure of the Beckermann minimal solutions at one point. Baker and Graves-Morris [1990] pointed out that if the degree of oversatisfaction of any solution $t > k + 1$, which may of course include a Beckermann minimal solution, then $z\vec{\mathcal{Q}}$ is also a solution, but of type $\vec{m} + \sum_{j=-1}^{k} \vec{e}_j$. This process can be repeated until the degree of oversatisfaction is reduced so that $t \leq k + 1$. Note that the preceding statements tacitly assume that $\vec{n} \geq \vec{0}$. If such is not the case, then $k + 1$ is replaced by the number of non-identically zero polynomials minus unity. Beckermann [1990] adds the result that if $\vec{\mathcal{Q}}$ is a Beckermann minimal solution at least in $\mathcal{T}(\vec{n}, \delta, t)$, then there exists a Beckermann minimal solution which contains $\mathcal{T}(\vec{n} + \sum_{j=-1}^{k} \vec{e}_j, t - k - 1, t - k - 1)$ in its Beckermann minimal region.

Suppose that there are at least two Beckermann minimal solutions, $\vec{\mathcal{Q}}$ and $\vec{\mathcal{P}}$. Suppose further that, in the notation of Theorem 8.5.3 above, $d_{\mathcal{Q}} + |\vec{n}_{\mathcal{Q}} + \sum_{j=-1}^{k} \vec{e}_j| = d_{\mathcal{P}} + |\vec{n}_{\mathcal{P}} + \sum_{j=-1}^{k} \vec{e}_j|$, and that the intersection $\mathcal{T}(\vec{n}_{\mathcal{Q}}, d_{\mathcal{Q}}, d_{\mathcal{Q}}) \cap \mathcal{T}(\vec{n}_{\mathcal{P}}, d_{\mathcal{P}}, d_{\mathcal{P}})$ is not empty. With the notation $\vec{m}_\mu = \max\{\vec{m}_{\mathcal{Q}}, \vec{m}_{\mathcal{P}}\}$, $\delta_\mu = \max\{\delta_{\mathcal{Q}} - |\vec{m}_{\mathcal{Q}} - \vec{m}_\mu|, \delta_{\mathcal{P}} - |\vec{m}_{\mathcal{P}} - \vec{m}_\mu|\}$, and $d_\mu = \min\{d_{\mathcal{Q}} - |\vec{m}_{\mathcal{Q}} - \vec{m}_\mu|, d_{\mathcal{P}} - |\vec{m}_{\mathcal{P}} - \vec{m}_\mu|\}$, Beckermann [1990] proved for any j such that $\vec{e}_j \cdot \vec{m}_{\mathcal{Q}} = \vec{e}_j \cdot \vec{m}_{\mathcal{P}} = \vec{e}_j \cdot \vec{m}_\mu \geq 0$, and any $i \leq d_\mu$ that there exists a Beckermann minimal solution $\vec{\mathcal{R}}$ for which

$$\mathcal{T}(\vec{m}_\mu, i, i) \subset \mathcal{T}(\vec{m}_\mu - \vec{e}_j, i+1, i+1)$$
$$\subset \text{Beckermann minimal region of } \vec{\mathcal{R}}, \quad (5.11)$$

where the first inclusion is strict. The proof of these results revolves around the idea that a linear combination of two polynomially independent solutions can be made which will satisfy (5.1) to one order higher in z and the geometry of the truncated simplices $\mathcal{T}$.

By selecting at most $k+1$ Beckermann minimal solutions, Beckermann [1990] showed that one can obtain a basis for all the solutions of (5.1) of a given type, in accordance with Theorem 8.5.1.

Rather than pursue the complex structure of Beckermann minimal solutions and their pendant minimal regions, we will now focus on the structure of the table of minimal Hermite–Padé polynomials. As these minimal polynomials are also Beckermann minimal polynomials, the more straightforward aspects of the general structure follow from the above results. We expect that for isolated instances of oversatisfaction of the polynomial defining equations (5.1) there will be a minimal region of the form $\mathcal{T}(\vec{m}, 0, d)$, plus the appropriate extensions with $z\vec{\mathcal{Q}}$, $z^2\vec{\mathcal{Q}}$, etc., as described above. The δ of Theorem 8.5.3 is zero, because otherwise the polynomially independent, multiple solutions which must occur are, as we will see in some detail below, incompatible with an isolated structure.

In the case of Padé approximants (see Section 1.4), the structure of the table of solutions is related to the structure of an underlying set of determinants. The situation is the same here, but regrettably more complex. In order to make this connection, we will introduce a family of determinants, and give relations among them. First, however, we will prove a lemma.

Lemma 2 (Nullity lemma) [Baker and Graves-Morris, 1993a]. *The change in the nullity of the coefficient matrix $\Delta_{\vec{m}}$ (see (5.5)) of the defining equations for the Hermite–Padé polynomials changes by at most unity as any one of the m_i is changed by unity.*

8.5 Hermite–Padé polynomials

Proof. We observe that the increase of any of the m_i by unity changes the coefficient matrix $\Delta_{\vec{m}}$, which we will denote here by $\mathcal{A}$, by inserting a column of coefficients right after the column which ends with $f^{(i)}_{s-m_i}$ and adding a row of coefficients after the row which begins with $f^{(-1)}_s$. For convenience, and without loss of generality, we illustrate with the case $i = k$. We will first extend $\mathcal{A}$ to $\bar{\mathcal{A}}$ by bordering it on the right and bottom by zeros. Symbolically, the change from m_k to $m_k + 1$ yields $\mathcal{A} \to \bar{\mathcal{A}} + \mathcal{B}$ with

$$\mathcal{B} = \begin{bmatrix} \mathbb{O} & a \\ b & c \end{bmatrix}, \tag{5.12}$$

where $\mathbb{O}$ is an $(s + 1) \times (s + 2)$ block of zeros, a is an $(s + 1)$-dimensional column vector, b is an $(s + 2)$-dimensional row vector, and c is a number. We note that the ranks of $\mathcal{A}$ and $\bar{\mathcal{A}}$ are the same, and then we may conclude that

$$\operatorname{rank}(\mathcal{A}) \leq \operatorname{rank}(\bar{\mathcal{A}} + \mathcal{B}) \leq \operatorname{rank}[\mathcal{A}, a] + 1 \leq \operatorname{rank}(\mathcal{A}) + 2, \tag{5.13}$$

where the first inequality follows because, if $\mathcal{A}$ is of rank $r \leq s + 1$, then the $r \times r$ critical minor which determines the rank will still be there in $\bar{\mathcal{A}} + \mathcal{B}$. The second inequality follows from the reduction of one row and the final by the reduction of one column. Using this bound and the relation nullity equals dimension minus rank, we can rewrite (5.13) as

$$\operatorname{nullity}(\mathcal{A}) - 1 \leq \operatorname{nullity}(\bar{\mathcal{A}} + \mathcal{B}) \leq \operatorname{nullity}(\mathcal{A}) + 1, \tag{5.14}$$

which provides the conclusion of the lemma.

We now introduce the determinants for $j = 0, 1, \ldots$

$$D^{(j)}(\vec{m}) =$$

$$\det \begin{vmatrix} f^{(-1)}_0 & 0 & \cdots & 0 & & f^{(i)}_0 & \cdots & 0 & \\ f^{(-1)}_1 & f^{(-1)}_0 & \cdots & 0 & & f^{(i)}_1 & \cdots & 0 & \\ \vdots & \vdots & \ddots & \vdots & & \vdots & \ddots & \vdots & \\ f^{(-1)}_{m_{-1}} & f^{(-1)}_{m_{-1}-1} & \cdots & f^{(-1)}_0 & \cdots & f^{(i)}_{m_i} & \cdots & f^{(i)}_0 & \cdots \\ \vdots & \vdots & & \vdots & & \vdots & & \vdots & \\ f^{(-1)}_s & f^{(-1)}_{s-1} & \cdots & f^{(-1)}_{s-m_{-1}} & & f^{(i)}_s & \cdots & f^{(i)}_{s-m_i} & \\ f^{(-1)}_{s+1+j} & f^{(-1)}_{s+j} & \cdots & f^{(-1)}_{s-m_{-1}+1+j} & & f^{(i)}_{s+1+j} & \cdots & f^{(i)}_{s-m_i+1+j} & \end{vmatrix}. \tag{5.15}$$

When the solutions for the Hermite–Padé polynomials are normalized as in (5.4), then we may give the series expansion

$$\sum_{i=-1}^{k} Q_{i,m_i}(z) f_i(z) = \sum_{j=0}^{\infty} D^{(j)}(\vec{m}) z^{s+j+1} \tag{5.16}$$

for the error term. The first coefficient $D^{(0)}(\vec{m})$ has extra significance in the case of Hermite–Padé approximation of type $\vec{m} + \vec{e}_i$, because, using the normalization (5.4), it so happens that $D^{(0)}(\vec{m})$ is the coefficient of z^{m_i+1} in $Q_{i,m_i+1}(z)$ for all values of i. Thus zeros in the table of values of $D^{(0)}(\vec{m})$ characterize relationships among the Hermite–Padé polynomials in much the same way as the determinants of the C-table characterize the table of Padé approximants. In particular, the vanishing of $D^{(j)}(\vec{m})$ for $j = 0, \ldots, J$ can signify that there is oversatisfaction by $t = J$ in (5.10).

A number of authors have contributed determinantal identities in recent years, particularly Della Dora [1980], Della Dora and Di Crescenzo [1980], Paszkowski [1982], and Baker and Graves-Morris [1993a]. We will give only a few of them. First we give a generalization due to Baker and Graves-Morris [1993a] of the 'generalized Frobenius formula' of Della Dora [1980]. Our argument is of the same type as that which leads to Sylvester's identity (Theorem 1.4.1), but extended. Suppose we denote by $\mathcal{M}$ the matrix in (5.15) whose determinant is $D^{(0)}(\vec{m})$. We notice that the elements of the row we add are independent of μ when we go from $\vec{m}$ to $\vec{m} + \vec{e}_\mu$, except for one element in the column added. Let us further denote by $\mathcal{A}$ the matrix whose determinant is $D^{(j)}(\vec{m} + \sum_{l=1}^{n} \vec{e}_{\mu_l})$. We illustrate the required manipulations for the case $n = 3$,

$$\pm \det \mathcal{A} \det \mathcal{M} = \det \begin{vmatrix} \mathcal{M} & r & q & p & 0 \\ h & g & f & e & 0 \\ d & c & b & a & 0 \\ v & u & t & s & v \\ 0 & 0 & 0 & 0 & \mathcal{M} \end{vmatrix}$$

$$= \det \begin{vmatrix} \mathcal{M} & r & q & p & 0 \\ h & g & f & e & 0 \\ d & c & b & a & 0 \\ v & u & t & s & v \\ \mathcal{M} & r & q & p & \mathcal{M} \end{vmatrix}$$

$$= \det \begin{vmatrix} \mathcal{M} & r & q & p & 0 \\ h & g & f & e & 0 \\ d & c & b & a & 0 \\ 0 & u & t & s & v \\ 0 & r & q & p & \mathcal{M} \end{vmatrix}, \qquad (5.17)$$

where the entries are understood to be two-dimensional arrays of the appropriate size. For example, r is an $(s + 2) \times 1$ array, h is a $1 \times (s + 2)$ array and g is a 1×1 array. The three rows and columns bordering the $\mathcal{M}$ in the upper left-hand corner are what must be added to $\mathcal{M}$ to produce $\mathcal{A}$, with some permutation of its columns. If we now use the Laplace expansion of the last determinant in (5.17), we get that

$$\pm \det \mathcal{A} \det \mathcal{M}$$

$$= \det \begin{vmatrix} \mathcal{M} & r & q \\ h & g & f \\ d & c & b \end{vmatrix} \det \begin{vmatrix} s & v \\ p & \mathcal{M} \end{vmatrix} - \det \begin{vmatrix} \mathcal{M} & r & p \\ h & g & e \\ d & c & a \end{vmatrix} \det \begin{vmatrix} t & v \\ q & \mathcal{M} \end{vmatrix}$$

$$+ \det \begin{vmatrix} \mathcal{M} & q & p \\ h & f & e \\ d & b & a \end{vmatrix} \det \begin{vmatrix} u & v \\ r & \mathcal{M} \end{vmatrix}, \qquad (5.18)$$

where only the nonzero entries of the $s+4$, $s+3$ split have been recorded. These can now be identified as $D^{(j)}$ determinants, and we can write (5.17)–(5.18) as

$$D^{(j)}\left(\vec{m} + \sum_{l=1}^{n} \vec{e}_{\mu_l}\right) D^{(0)}(\vec{m})$$

$$= \sum_{l=1}^{n} (-1)^{n-l} D^{(0)}\left(\vec{m} - \vec{e}_{\mu_l} + \sum_{\lambda=1}^{n} \vec{e}_{\mu_\lambda}\right) D^{(j+n-1)}(\vec{m} + \vec{e}_{\mu_l}), \quad (5.19)$$

where the signs are as stated when the μ_l are taken to be in ascending order. The computation of the signs comes simply from keeping track of the number of row and column permutations required to put the various matrices involved in their original form. The special case $n = k+1$ is Della Dora's result mentioned above and the case $n = 1$ is trivial.

Paszkowski [1982] has derived a large number of interesting identities which interconnect the elements of the $D^{(0)}(\vec{m})$ table. Baker and Graves-Morris [1993a] have extended them to the $D^{(j)}(\vec{m})$ case. The first such case that we report is

$$D^{(j)}(\vec{m} + \vec{e}_\mu + \vec{e}_\lambda) D^{(0)}(\vec{m} + \vec{e}_\theta) = D^{(j)}(\vec{m} + \vec{e}_\theta + \vec{e}_\mu) D^{(0)}(\vec{m} + \vec{e}_\lambda)$$
$$- D^{(j)}(\vec{m} + \vec{e}_\theta + \vec{e}_\lambda) D^{(0)}(\vec{m} + \vec{e}_\mu), \qquad (5.20)$$

where $\theta < \mu < \lambda$ and θ, μ, λ can range from -1 to k. Notice that when $j = 0$, (5.20) only involves elements of the $D^{(0)}(\vec{m})$ table. This relation can be derived from Schweins's little-known identity. If $\mathcal{X}$ has l rows and $l + 1$ columns, then if we delete as appropriate, row l and columns 1, 2, or $l+1$, Schweins's identity is, in the notation of Theorem (1.4.1),

$$\det |\mathcal{X}_{\varnothing;l+1}| \det |\mathcal{X}_{l;1,2}| = \det |\mathcal{X}_{\varnothing;2}| \det |\mathcal{X}_{l;1,l+1}| - \det |\mathcal{X}_{\varnothing;1}| \det |\mathcal{X}_{l;2,l+1}|, \qquad (5.21)$$

where $\varnothing$ is the empty set. This identity can be derived from Sylvester's identity by adjoining a final row to $\mathscr{X}$ of the form $(0, \ldots, 0, 1)$ as is done in Section 3.4.

Paszkowski gives his result for general n as in (5.19) which can be proven by the same sort of arguments which lead to that result. In turn it is easily generalized to general j, so we get

$$\sum_{l=1}^{n} (-1)^l D^{(j)}(\vec{m} - \vec{e}_{\mu_l}) D^{(0)}\left(\vec{m} + \vec{e}_{\mu_l} - \sum_{\lambda=1}^{n} \vec{e}_{\mu_\lambda}\right) = 0, \qquad (5.22)$$

where the signs are as stated when the μ_l are taken to be in ascending order. Again when $j = 0$ the identity interrelates the elements of the $D^{(0)}(\vec{m})$ table. Paszkowski [1982] also generalizes the 'star identity' (1.4.10) to be

$$[D^{(0)}(\vec{m})]^2 = \sum_{i=-1}^{k} D^{(0)}(\vec{m} - \vec{e}_i) D^{(0)}(\vec{m} + \vec{e}_i), \quad \vec{m} \geq \vec{0}. \qquad (5.23)$$

We give a simpler proof. Let us first define the row vectors

$$\mathbf{b}_\lambda = (f_\lambda^{(-1)}, f_{\lambda-1}^{(-1)}, \ldots, f_{\lambda-m_{-1}}^{(-1)}, \ldots, f_\lambda^{(k)}, f_{\lambda-1}^{(k)}, \ldots, f_{\lambda-m_k}^{(k)}),$$

and then

$$D^{(0)}(\vec{m}) = \det \begin{vmatrix} \Delta_{\vec{m}} \\ \mathbf{b}_s \end{vmatrix}. \qquad (5.24)$$

Now consider the determinant

$$\mathscr{D} = \begin{vmatrix} \Delta_{\vec{m}} & \vdots & \varpi \\ \mathbf{b}_{s+1} & \vdots & \Delta_{\vec{m}} \\ \mathbf{b}_{s+2} & \vdots & \mathbf{b}_{s+1} \\ \mathcal{O} & \vdots & \Delta_{\vec{m}} \end{vmatrix} = (-1)^{s+1} [D^{(0)}(\vec{m})]^2, \qquad (5.25)$$

where ϖ is a row vector of $s + 1$ zeros and $\mathcal{O}$ is an $(s + 1) \times (s + 2)$ array of zeros. The equality follows by first using elementary row operations to convert the upper right block of $\mathscr{D}$ to a block of zeros, and then by using the Laplace expansion of the determinant with an $s + 2$, $s + 2$ split. The sign follows by counting the number of exchanges necessary to move $\mathbf{b}_{s+1}$ to the correct place. Now because of the structure of $\Delta_{\vec{m}}$, $\mathbf{b}_{s+1}$, $\mathbf{b}_{s+2}$ we can create by elementary column operations zeros in all of the top $s + 3$ entries in every one of the columns in the right half of $\mathscr{D}$ except for those which are needed to complete one of the $D^{(0)}(\vec{m} + \vec{e}_j)$. By means of the Laplace expansion, using this time an $s + 3$, $s + 1$ split, we get

$$\mathscr{D} = (-1)^{s+1} \sum_{j=-1}^{k} D^{(0)}(\vec{m} - \vec{e}_j) D^{(0)}(\vec{m} + \vec{e}_j), \qquad (5.26)$$

where again the sign is obtained by counting the number of exchanges necessary to put the columns in the correct places. Equations (5.25) and (5.26) together yield the desired result (5.23).

We now turn our attention to the structure of the Hermite–Padé polynomial table and its relation to the determinants, when the minimal Definition 8.5.2 is used. Della Dora [1980] has pointed out by means of a counterexample that the simple, square-block structure for the Padé approximants does not generalize to a cube, etc. His example is as follows. Consider the three functions $f_1(z) = f(z)$, $f_0(z) = f^2(z)$, $f_{-1}(z) = 1$, where

$$f(z) = 1 + \beta z^2 + \gamma z^3 + \delta z^4 + \varepsilon z^5 + \cdots.$$

Direct calculation shows that

$$D^{(j)}((0, 0, 0)) = 0 \quad \forall j \geq 0, \quad \text{and}$$

$$D^{(0)}((1, 0, 0)) = D^{(0)}((0, 1, 0)) = D^{(0)}((0, 0, 1)) = 0,$$

but

$$D^{(0)}((1, 1, 1)) = \beta^4.$$

Since we can choose $\beta \neq 0$, this example provides a counterexample to the idea that these zeros must occur only in cubical blocks.

As we saw above, if any (minimal) solution oversatisfies the polynomial defining equations (5.10), then, by Theorem 8.5.3, it is also a solution in $\mathcal{T}(\vec{n}, 0, d)$, where the notation is as in Theorem 8.5.3. However, it is not necessarily a minimal solution throughout. If the degree of oversatisfaction of the polynomial defining equations is high enough, then, as remarked above, there are the associated $z\vec{\mathcal{D}}$, $z^2\vec{\mathcal{D}}$ solutions. We make an additional observation, which follows from accuracy through order and is independent of uniqueness and minimality: when some of the Hermite–Padé polynomials have their true degree less than their nominal degree, then there is a solution of the defining equations for the type equal to the true degree type which oversatisfies the defining equations sufficiently so as to include the set of Hermite–Padé polynomials with which we started. Therefore, if any of the Hermite–Padé polynomials have their true degree less than their nominal degree, the Hermite–Padé polynomials must lie in a simplex.

We will give next a series of theorems which relate the structure of the determinant table to that of the Hermite–Padé table.

Theorem 8.5.4 (Rule of the simplex) **[Baker and Graves-Morris, 1993a].** *If for some λ $(-1 \leq \lambda \leq k)$,*

(a) $$D^{(0)}(\vec{m} - \vec{\delta}_\lambda) \neq 0, \quad \text{and}$$

(b) $$D^{(j)}(\vec{m}) = 0, \quad j = 0, \ldots, J, \qquad (5.27)$$

then the solution of type $\vec{m}$ is essentially unique, it fills the simplex $\mathcal{T}(\vec{m}, 0, J)$, and

$$D^{(n)}(\vec{m} + \vec{v}) = 0 \quad \forall n, \vec{v} \text{ such that } \vec{v} \geq \vec{0}, |\vec{v}| + n \leq J. \quad (5.28)$$

Proof. Condition (a) implies that the coefficient of z^{m_λ} in $Q_{\lambda, m_\lambda}(z)$ is not zero in the determinantal normalization (5.4). This is sufficient to ensure that rank $(\Delta_{\vec{m}}) = s + 1$ in (5.5) and so the solution of type $\vec{m}$ is essentially unique. By the remarks made above, that solution must fill the simplex $\mathcal{T}(\vec{m}, 0, J)$. It remains to prove (5.28). Let $\mathcal{A}$ denote the matrix whose determinant is $D^{(0)}(\vec{m} - \vec{\delta}_\lambda)$. We next add a row and a column to make

$$D^{(j)}(\vec{m}) = \pm \det \begin{vmatrix} \mathcal{A} & a \\ b_j & c_j \end{vmatrix} = \det \begin{vmatrix} \mathcal{A} & a \\ 0 & 0 \end{vmatrix} = 0, \quad j \leq J, \quad (5.29)$$

where the entries are compatibly sized arrays: a is an $(s + 1)$-dimensional column vector, each b_j is an $(s + 1)$-dimensional row vector, and each c_j is a number. The final determinantal form in (5.29) represents the effect of elementary row operations, made possible because $\mathcal{A}$ is nonsingular, to impose condition (b). Any of the determinants of (5.28) can, using this procedure, be put in the form

$$D^{(n)}(\vec{m} + \vec{v}) = \det \begin{vmatrix} \mathcal{A} & a & u \\ 0 & 0 & w \end{vmatrix}, \quad (5.30)$$

where w is a $(\theta + 1) \times \theta$ matrix with $\theta = |\vec{v}|$. The determinant of a matrix of this structure is necessarily zero, as it has nullity at least unity.

There is an additional determinantal theorem which adds further assumptions, and plays in some sense the role of Padé's block theorem (Theorem 1.4.3).

Theorem 8.5.5 (Isolated simplex) [Baker and Graves-Morris, 1993a]. *In addition to the assumptions of Theorem 8.5.4, we further assume that condition (a) holds as stated but for all λ. Then for each μ and for all $\vec{v} \geq \vec{0}$ such that $v_\mu = 0$,*

$$D^{(J+1-|\vec{v}|)}(\vec{m} + \vec{v}) \neq 0, \quad \text{for } |\vec{v}| \leq J + 1, \quad (5.31)$$

if and only if

(a) $\quad D^{(0)}(\vec{m} + \vec{v} - \vec{e}_\mu) \neq 0, \quad \text{for } |\vec{v}| \leq J + 1, \quad \text{and}$

(b) $\quad D^{(0)}(\vec{m} + \vec{v}) \neq 0, \quad \text{for } |\vec{v}| = J + 1. \quad (5.32)$

Proof. In order to demonstrate this theorem, we need to use the identity (5.19) for $n = 2$ in the form

$$D^{(J+1-|\vec{v}|)}(\vec{m} + \vec{v})D^{(0)}(\vec{m} + \vec{v} - \vec{e}_\mu - \vec{e}_\lambda)$$
$$= -D^{(0)}(\vec{m} + \vec{v} - \vec{e}_\mu)D^{(J+2-|\vec{v}|)}(\vec{m} + \vec{v} - \vec{e}_\lambda)$$
$$+ D^{(0)}(\vec{m} + \vec{v} - \vec{e}_\lambda)D^{(J+2-|\vec{v}|)}(\vec{m} + \vec{v} - \vec{e}_\mu). \quad (5.33)$$

First let us show that (5.31) implies (5.32). The second condition in (5.32) is a special case of (5.31). To show the first condition, we can start with $\vec{v} = \vec{e}_\lambda$, where $\vec{e}_\lambda$ is any direction in the surface of the hyperplane, i.e., perpendicular to a particular $\vec{e}_\mu$. Since the left-hand side of (5.33) is nonzero by hypothesis, as is also the second factor of the first term on the right-hand side, and since the first factor of the second term on the right-hand side is zero by Theorem 8.5.4, it follows that the second factor on the right-hand side is nonzero. This result is that the $D^{(0)}$ for one of the nearest neighbors of the simplex is nonzero. We can repeat this argument in a systematic manner to run over the whole face of the simplex (perpendicular to $\vec{e}_\mu$). We can then extend it to run over all the faces of the simplex. To show the converse that (5.32) implies (5.31), we start with $\vec{v} = (J + 1)\vec{e}_\lambda$. As (5.32) implies that both the left-hand side and the first factor of the first term on the right-hand side of (5.33) are nonzero, and the second term vanishes by Theorem 8.5.4, we can conclude that the second factor of the first term on the right-hand side of (5.33) is nonzero. We can repeat this argument, and use it to demonstrate completely that (5.32) implies condition (5.31). This completes the demonstration of the theorem.

A simplex satisfying conditions (a), (b), etc. for each μ in Theorem 8.5.4 is an isolated simplex. In other words either all the nearest neighbors (on the hypercubic lattice in index space, but not, of course, those in the simplex itself) of its face planes have nonzero determinants $D^{(0)}$, or, equivalently, the $D^{(j)} \neq 0$ for appropriate j for each point in the face planes and the face-plane 'bases'. We do not expect that the 'base' of the whole simplex, i.e., the surface of the simplex which is not perpendicular to any coordinate axis, necessarily to be completely covered with nonzero determinants because if $J \geq k + 2$, then, as remarked above, the simplex extensions will generate multiple solutions to the Hermite–Padé polynomial defining equations which will imply the occurrence of zero determinants. Note that in the case described by this theorem, both minimality and Beckermann minimality yield uniquely the vertex solution for the isolated simplex throughout the whole simplex.

Corollary 2 [Baker and Graves-Morris, 1993a]. *If we have for some $\lambda \neq \mu$,*

$$D^{(0)}(\vec{m} + j\vec{e}_\lambda) = 0, \quad j = 0, \ldots, J,$$
$$D^{(0)}(\vec{m} + (J+1)\vec{e}_\lambda) \neq 0, \tag{5.34}$$
$$D^{(0)}(\vec{m} + j\vec{e}_\lambda - \vec{e}_\mu) \neq 0, \quad j = 0, \ldots, J+1,$$

then

$$D^{(j)}(\vec{m}) = 0, \quad j = 0, \ldots, J,$$
$$D^{(J+1-j)}(\vec{m} + j\vec{e}_\lambda) \neq 0, \quad j = 0, \ldots, J+1. \tag{5.35}$$

Proof. This corollary follows directly from the argument used for the converse case in Theorem 8.5.5.

The consequence of this corollary is that the assumption of a line of zero determinants adjoined by nonzero determinants is sufficient to establish by Theorem 8.5.4 the existence of a whole simplex of size J. The necessary added argument is that $D^{(J+1)}(\vec{m}) \neq 0$ is enough to imply that the solution of type $\vec{m}$ is essentially unique. The assumptions of Corollary 2 are stated in terms of the determinants $D^{(0)}$ alone and do not involve the $D^{(j)}$'s, whereas the assumptions of Theorem 8.5.4 are essentially local in index space, but do involve the $D^{(j)}$'s. As we have just noted, the assumption (a) of Theorem 8.5.4 could have been replaced by $D^{(J+1)}(\vec{m}) \neq 0$, with the same conclusion.

An important property which characterizes blocks in the Padé table is the occurrence of only a single approximant. In terms of the Hermite–Padé polynomials, it is that there is just one linearly independent solution in the sense of polynomial coefficients. We now give a theorem for Hermite–Padé polynomials.

Theorem 8.5.6 (Isolated simplex uniqueness) [Baker and Graves-Morris, 1993a]. *There is at most one linearly independent (allowing z dependent factors) solution for the Hermite–Padé polynomials in an isolated simplex.*

Proof. Since an isolated simplex is bordered by nonzero determinants by Theorem 8.5.5, Lemma 2 implies that the nullity of the defining equations for the Hermite–Padé polynomials cannot exceed the distance from the faces of the simplex (plus unity). Thus the equations in the faces have nullity unity, and so the solutions there are essentially unique; however, by Theorem 8.5.4, the solution is just that solution at the vertex of the simplex. At a distance unity from at least one face, and at least unity from all other faces, the nullity can be two. Here we can construct another solution which is just $z\vec{\mathfrak{D}}$, which together with $\vec{\mathfrak{D}}$ exhausts the available number of solutions. We can continue this argument up to the maximum possible distance from all faces and we find that by using powers of z times $\vec{\mathfrak{D}}$, we always exhaust the number of solutions that can exist by the nullity lemma. Thus all solutions of the defining equations are given by a polynomial times the solution for the vertex of the simplex. Hence it can be taken as the only linearly independent solution.

The case of isolated simplices is one of rather simple structure. This case, plus the simplex extension, is all that can occur for Padé approximants (case $k = 0$). At this point it is worthwhile to give an example to show that for $k > 0$, simplices can intersect. Consider the case where

$$f_{-1}(z) = 1, \quad f_0(z) = \frac{1}{1-z}, \quad f_1(z) = f_0'(z) = \frac{1}{(1-z)^2}. \quad (5.36)$$

It is easy to check that

$$(1-z)f_0(z) - f_{-1}(z) = 0, \quad (5.37)$$

yielding $(-1, 1-z, 0)$ as the solution of type $(0, 1, -1)$, and by differentiation that

$$(1-z)f_0'(z) - f_0(z) = (1-z)f_1(z) - f_0(z) = 0, \quad (5.38)$$

yielding $(0, -1, 1-z)$ as the solution of type $(-1, 0, 1)$. In fact, both the $(0, 1, -1)$- and the $(-1, 0, 1)$-type Hermite–Padé polynomials form the vertices of simplices of infinite size. It is clear that neither vertex lies in the other simplex and that the simplices overlap, for example, at $(0, 1, 1)$. That particular case has the solution

$$(Q_{-1}, Q_0, Q_1) = a(-1, 1-z, 0) + b(0, -1, 1-z), \quad (5.39)$$

and the minimal solution for this case is $b = 0$, as given by (5.37). However, for the types of the form $(-1, m, 2)$, with $m > 0$, the minimal solution is always that of (5.38). In this case we always find that the minimal solution is that of one or the other of the simplex vertices.

The conditions of Theorem 8.5.5 may seem to be unduly strong in view of the large number of determinants assumed to be nonzero. Let us review them. Let us start with the idea of a simplex containing just one linearly independent solution in the sense of polynomials. There are basically just two types of simplices: those for which the vertex solution is essentially unique and those for which it is not ($\delta > 0$ in the notation of Theorem 8.5.3). Those simplices for which it is not lie in the intersection of other simplices, for example. Manifestly we need the first kind for our current considerations. As the simplex of Theorem 8.5.4 could refer to only a portion of a larger simplex, in line with our idea, we require that we have the actual vertex. Thus as observed above, every Hermite–Padé polynomial needs to be of full nominal degree. Therefore this condition plus essential uniqueness implies that

$$D^{(0)}(\vec{m} - \vec{e}_\mu) \neq 0, \quad \forall \mu. \tag{5.40}$$

The question does not arise for $J = 0$, so now suppose that one of the conditions (5.32),

$$D^{(0)}(\vec{m} + \vec{v} - \vec{e}_\mu) \neq 0, \quad 1 \leq |\vec{v}| \leq J, \quad v_\mu = 0, \tag{5.41}$$

is violated for $\vec{v} = \vec{v}_0$, $\mu = \mu_0$, and $J \geq 1$. We may also suppose by (5.40) that we may examine one such violation for which there exists a $\lambda \neq \mu_0$ such that $D^{(0)}(\vec{m} + \vec{v}_0 - \vec{e}_\lambda - \vec{e}_{\mu_0}) \neq 0$, so that the integral approximants of type $(\vec{m} + \vec{v}_0 - \vec{e}_{\mu_0})$ are essentially unique and oversatisfy by at least one order in z. Thus polynomials of type $(\vec{m} + \vec{v}_0 - \vec{e}_{\mu_0})$ are also a solution of type $(\vec{m} + \vec{v}_0)$ and they are linearly independent of those of type $(\vec{m})$, as the μth polynomial of type $(\vec{m} - \vec{e}_\mu)$ is lower by at least one in degree and the λth polynomial is of full degree by the remark after (5.16). If the degree of oversatisfaction of the Hermite–Padé polynomial determining equation is exactly J orders in z, then in light of (5.40) and (5.16) we must have the conditions (5.27b) and the condition

$$D^{(J+1)}(\vec{m}) \neq 0. \tag{5.42}$$

If these conditions hold then we can conclude that condition (5.31) of Theorem 8.5.5 holds by the use of identity (5.33), but only for $|\vec{v}| \leq J$ instead of $J + 1$. The remaining conditions in Theorem 8.5.5 that we have not discussed yet are

$$D^{(0)}(\vec{m} + \vec{v} - \vec{e}_\mu) \neq 0, \quad D^{(0)}(\vec{m} + \vec{v}) \neq 0, \quad \text{provided } |\vec{v}| = J + 1, v_\mu = 0, \tag{5.43}$$

and they serve a dual role. First, as we showed in Theorem 8.5.5, they imply (together with the other conditions discussed) (5.42) and the rest of (5.31). They are, however, not necessary for the conclusion of that theorem if we replace them with (5.42). If we do assume (5.42) then we can conclude that both of the determinants of (5.43) are either zero or nonzero together. Second, the conditions (5.43) do serve to isolate the base plane of the simplex and so to create an isolated simplex. If the degree of oversatisfaction for the simplex is large enough, then as remarked earlier in this section, it will necessarily have the appropriate associated simplex extension. We can now summarize our conclusions by

Corollary 3. *It is both necessary and sufficient that* (5.27b), (5.40)–(5.42) *hold for there to be exactly one linearly independent solution (in the sense of polynomials) for the Hermite–Padé polynomials in a simplex at whose vertex the Hermite–Padé polynomials oversatisfy their defining equations by exactly J orders in z.*

8.5.3 Recursion relations

There is another area of application of the determinantal identities. This application is in the area of recursion relations between the Hermite–Padé polynomials of contiguous types. There are several known families of recursion relations (Della Dora [1980], Della Dora and Di Crescenzo [1980], Paszkowski [1982, 1987], Baker and Graves-Morris [1993a], and Beckermann [1993]) for the Hermite–Padé polynomials. To date, there are two basic types of expositions of the recursion relations: those which express all the coefficients in terms of determinants of the same general size as those required to solve for the Hermite–Padé polynomials directly and those which do not. From the practical point of view, the ones which require the evaluation of large determinants are not very useful and we will give only a brief discussion of a few of them to make a connection with the above work. We will concentrate mainly on methods which avoid the computation of large determinants. Generally in our discussion of the recursion relations, we will assume that all the Hermite–Padé polynomials considered are 'essentially unique'. For some results on cases where this property fails, see Beckermann [1993].

The recursion relations we shall discuss can be roughly grouped into two types. The first type consists of those which do not depend on the series coefficients explicitly, but only through the coefficients of the 'prior' Hermite–Padé polynomials. This distinction is somewhat artificial, but we find it useful. Our first example can be thought of as belonging to either type and we use it to introduce the methods that we shall employ. The example is the recursion relation for

$$Q_{i,\vec{m}+\sum_{l=1}^{n}\vec{e}_{\mu_l}}(z) D^{(0)}(\vec{m}) = \sum_{l=0}^{n}(-1)^{n-l} D^{(0)}\left(\vec{m} + \sum_{\lambda=1}^{n}\vec{e}_{\mu_\lambda} - \vec{e}_{\mu_l}\right) Q_{i,\vec{m}+\vec{e}_{\mu_l}}(z)$$

(5.44)

where the subscript i tells which of the $i = -1, 0, \ldots, k$ integral polynomials is involved and the quantity $\vec{m}$ gives the type of the Hermite–Padé polynomial, i.e., the position in the Hermite–Padé table. This change of notation makes the presentation of the results on the recursion relations more convenient. The sign in (5.44) is correct for the

μ's taken in ascending order, $\mu_{l+1} > \mu_l$. The numbering refers to the order in which the corresponding series appear in the determinants $D^{(0)}(\vec{m})$. The relation (5.44) follows by analogy from (5.19) when we observe that it doesn't really matter what we put in for the row corresponding to the subscript j. We have merely used the last line of (5.4) to create $Q_{n,\vec{m}}(z)$ instead of $D^{(j)}(\vec{m})$. By the use of the identities (5.19) and (5.44) we can first fill in the table of $D^{(0)}(\vec{m})$ and then the table of the Q's. One possible pattern would be to fill in by standard methods two, two-dimensional, boundary planes with a common direction. Then (5.19) can be used to fill in the line next to the line of intersection, and then to continue to fill in a whole plane next to one of the original planes. This process can be continued to then generate a further plane, and so on to get the entire $D^{(0)}$ table. The same method can then be used with (5.44) to fill in the entire Q table.

For expository purposes, we will now concentrate on the special case $n = 2$ of (5.44), but see (5.60) below. The key thing to notice about (5.44) is that the determinants $D^{(0)}(\vec{m})$ are just numbers independent of z and i. In order for the relation (5.44) to hold in the case $i = \mu_1$, the first and second terms must combine to cancel the highest power of z, in order to agree with the required order in $Q_{i,\vec{m}+\vec{e}_{\mu_2}}(z)$. Thus the ratio of the determinants is determined by the polynomial coefficients in the first two terms. Only the overall scale factor is undetermined, but the resulting polynomials can then be normalized by whatever rule ($Q_k(0) = 1.0$, for example) that one finds suitable for one's current purposes. In order to illustrate our other method, we will rederive the $n = 2$ case of (5.44) in a different way. First we introduce the notation

$$\|\vec{m}\| \equiv \sum_{i=-1}^{k} (m_i + 1) - 1 = |\vec{m}| + k + 1, \quad \vec{m} \geq \vec{0}. \tag{5.45}$$

Now consider the defining equations

$$\sum_{j=-1}^{k} Q_{j,\vec{m}+\vec{e}_\mu}(z) f_j(z) = O(z^{\|\vec{m}\|+2}), \tag{5.46}$$

$$\sum_{j=-1}^{k} Q_{j,\vec{m}+\vec{e}_\lambda}(z) f_j(z) = O(z^{\|\vec{m}\|+2}). \tag{5.47}$$

In (5.46), for example, the degree of Q_j is $\vec{e}_j \cdot (\vec{m} + \vec{e}_\mu)$, and similarly for (5.47). We can now determine the corresponding Hermite–Padé polynomials for the position $\vec{m} + \vec{e}_\mu + \vec{e}_\lambda$ in the approximant table by taking a linear combination of (5.46) and (5.47) as

$$\sum_{j=-1}^{k} [aQ_{j,\vec{m}+\vec{e}_\mu}(z) + bQ_{j,\vec{m}+\vec{e}_\lambda}(z)] f_j(z) = O(z^{\|\vec{m}\|+2}). \tag{5.48}$$

8.5 Hermite–Padé polynomials

So far the equation (5.48) is one order short of determining the desired Hermite–Padé polynomials; however, we can use the freedom of our choice of a and b to rectify this problem. The coefficient of $z^{\|\vec{m}\|+2}$ in (5.48) is selected to be zero by requiring

$$\sum_{j=0}^{k} \sum_{i=0}^{m_j+\vec{e}_j \cdot (\vec{e}_\mu+\vec{e}_\lambda)} [aQ_{j,\vec{m}+\vec{e}_\mu,i} + bQ_{j,\vec{m}+\vec{e}_\lambda,i}] f^{(i)}_{\|\vec{m}\|+2-i} = 0, \tag{5.49}$$

where we have used the notation

$$Q_{j,\vec{m}+\vec{e}_\mu}(z) = \sum_{i=0}^{m_j+\vec{e}_j \cdot \vec{e}_\mu} Q_{j,\vec{m}+\vec{e}_\mu,i} z^i. \tag{5.50}$$

Equation (5.49) determines the ratio of a to b. The overall scale factor determines normalization, and again we select that in a manner suitable for our current purposes. The equation (5.44) uses the 'determinant normalization.' As (5.48) and (5.49) together are the defining equations for the Hermite–Padé polynomials of the type $\vec{m} + \vec{e}_\mu + \vec{e}_\lambda$, we can conclude that with an a and b which satisfy (5.49),

$$Q_{j,\vec{m}+\vec{e}_\mu+\vec{e}_\lambda}(z) = aQ_{j,\vec{m}+\vec{e}_\mu}(z) + bQ_{j,\vec{m}+\vec{e}_\lambda}(z). \tag{5.51}$$

It is to be noted that the first method described above, i.e., the one which does not involve solving an equation in which the defining series coefficients appear explicitly, works when the degree of contact with the defining series at the origin of the solved for Hermite–Padé polynomials is less than or equal to that for the known Hermite–Padé polynomials. In the second method, the degree of contact is increased in the sought Hermite–Padé polynomials and here equations are involved that depend explicitly on the defining series coefficients.

One observes that, from the practical point of view, either method described above uses information currently available in the recursion process and the computations are much quicker than the evaluation of the determinants.

Now we consider another identity that does not involve explicitly the defining series coefficients. It follows from (5.20) in the same way that (5.44) followed from (5.19).

$$Q_{j,\vec{m}+\vec{e}_\mu+\vec{e}_\lambda}(z) D^{(0)}(\vec{m} + \vec{e}_\theta)$$
$$= Q_{j,\vec{m}+\vec{e}_\theta+\vec{e}_\mu}(z) D^{(0)}(\vec{m} + \vec{e}_\lambda) - Q_{j,\vec{m}+\vec{e}_\theta+\vec{e}_\lambda}(z) D^{(0)}(\vec{m} + \vec{e}_\mu). \tag{5.52}$$

In this identity, the degree of contact of every set of Hermite–Padé polynomials is the same. It is also symmetric between the θ, μ, and λ directions. As it stands we can rewrite it as

$$Q_{j,\vec{m}+\vec{e}_\mu+\vec{e}_\lambda}(z) = aQ_{j,\vec{m}+\vec{e}_\theta+\vec{e}_\mu}(z) + bQ_{j,\vec{m}+\vec{e}_\theta+\vec{e}_\lambda}(z), \tag{5.53}$$

and the ratio between a and b can be determined by canceling the highest power of z in the Q_θ's on the right-hand side. Again the normalization can be fixed by whatever rule is convenient for present purposes. This identity is convenient to fill in a hyperplane of constant degree of contact with the defining series from its boundary values. For example, the three-series, Hermite–Padé polynomials can be filled in from the Padé approximants alone to $f_0(x)/f_{-1}(z)$, $f_1(z)/f_{-1}(z)$, and $f_1(z)/f_0(z)$ by means of this identity.

Another interesting identity of this type [Paszkowski, 1982] is

$$zQ_{j,\vec{m}}(z) = \sum_{i=-1}^{k} A_i Q_{j,\vec{m}+\vec{e}_i}(z). \tag{5.54}$$

The A_i are determined by equating the coefficients of the highest power of z on the left-hand side to that on the right-hand side for $i = j$. The Hermite–Padé polynomial defining equations for each term separately give zero for orders $z^0, \ldots, z^{\|\vec{m}\|+1}$. Hence, (5.54) will serve to give any one of the set of (unnormalized) Hermite–Padé polynomials on the right-hand side in terms of the remainder plus the one on the left-hand side. Note the interesting fact that as the left-hand side is divisible by z so too must the right-hand side be for each value of n. This fact is a consequence of the structure of the Hermite–Padé polynomial defining equations.

A number of the next class of identities have been given explicitly by Della Dora [1980] for the case $k = 1$. They involve the embeddings of a linear chain of $k + 3$ vertices (connected by $k + 2$ bonds) in the index space describing the Hermite–Padé polynomials. That is to say, each vertex of this linear chain corresponds to a set of Hermite–Padé polynomials and the bonds mean that location vectors for successive vertices differ by precisely unity in one and only one direction. These identities do not by any means exhaust the possibilities. In this class alone the number of types of possible identities, for which each successive vertex moves into a hyperplane of contact of one order higher with the defining series, obeys the relationship $N(k, m) = N(k-1, m)m + N(k-1, m-1)$, where m is the dimension of the subspace spanned by the identity. For $k = 0$ we have $N(0, 1) = 1$, $N(0, 2) = 1$. The sums over all possible m of $N(k, m)$ for $k = 0, 1, \ldots, 6$ are 2, 5, 15, 52, 203, 877, and 3839. Manifestly, we will only discuss a minute fraction of these possibilities.

First we consider the 'diagonal' type of identity. It is

$$Q_{j,\vec{m}+\sum_{i=-1}^{k}\vec{e}_i}(z) = azQ_{j,\vec{m}}(z) + A_{-1}Q_{j,\vec{m}+\vec{e}_{-1}}(z) + A_0 Q_{j,\vec{m}+\vec{e}_{-1}+\vec{e}_0}(z)$$
$$+ \cdots + A_{k-1}Q_{j,\vec{m}+\sum_{i=-1}^{k-1}\vec{e}_i}(z). \tag{5.55}$$

Any permutation of the order of adding directions is equally valid. In particular, the cyclic permutations will be required to advance step-by-step along the diagonal in the Hermite–Padé approximant table. The degree of contact on the right-hand side of (5.55) is equal to or greater than that for the second term on the right-hand side, because the first term has a factor of z. In like manner to (5.48) we impose the requirements that the coefficients of $z^{\|\vec{m}\|+2}$, ..., $z^{\|\vec{m}\|+2+k}$ on the right-hand side vanish in the Hermite–Padé polynomial defining equations. As the coefficients of orders z^0, ..., $z^{\|\vec{m}\|+1}$ vanish by construction, these additional equations ensure that the Hermite–Padé polynomial defining equations are satisfied for the Hermite–Padé polynomials on the left-hand side of (5.55). There is also the correct number of equations to determine, except for the overall normalization, the multipliers a, A_{-1}, A_0, ..., A_{k-1}. There results a set of $k+1$ linear equations (of Hessenberg type; see Golub and Van Loan, [1989]) to be solved for these multipliers, the solution of which is computationally much simpler than the direct solution of the complete set of Hermite–Padé polynomial defining equations. This mode of derivation also establishes the existence of these identities, without the prior knowledge which we have for $k = 0$, 1 that they exist, because we have shown that the derived polynomials directly satisfy the Hermite–Padé polynomial defining equations. We feel that this is a powerful method to find identities among the set of Hermite–Padé polynomials. For small values of k, such as 0 or 1, for instance, it is quite practical to solve the equations directly and give explicit expressions for the multipliers.

There are other 'space types' in this class of identities. That is to say, other embeddings of the $k+2$ step linear chain in the index-space lattice that lead to interesting identities. For example, there is the case where a single index alone increases. Here the result is

$$Q_{j,\vec{m}+(k+2)\vec{e}_\mu}(z) = azQ_{j,\vec{m}}(z) + \sum_{i=1}^{k+1}(A_i + B_i z)Q_{j,\vec{m}+i\vec{e}_\mu}(z). \quad (5.56)$$

In this case there are $2k+3$ constants on the right-hand side to be determined. Again, the polynomial defining equations of orders z^0, ..., $z^{\|\vec{m}\|+1}$ are satisfied by construction. We must impose the additional $k+1$ accuracy-through-order equations, for the orders $z^{\|\vec{m}\|+2}$, ..., $z^{\|\vec{m}\|+2+k}$. In addition in this case, in order to keep the left-hand side of (5.56) at its stated order, we require that the highest power of z cancel between the terms on the right-hand side for the cases $j = -1, 0, \ldots, k$ except for $j = \mu$. This condition leads to $k+1$ simultaneous linear equations among the B_i and a, which are thereby determined, except for overall normalization. The A_i can then be determined from the B_i and the accuracy-through-order equations mentioned above. This case therefore requires the solution of two sets of $k+1$ equations rather than one as in the previous case. It is nevertheless still computationally much simpler than the direct solution of the complete set of Hermite–Padé polynomial defining equations.

Identity (5.56) is an example of what we called 'reduced-dimension' identities. It has the property that all the Hermite–Padé polynomials lie in a subspace of the index space that is of less than full dimension. In the case of (5.56) it is a one-dimensional subspace. Another example is the identity

$$Q_{j,\vec{m}+\vec{e}_\lambda+(k+1)\vec{e}_\mu}(z)$$
$$= azQ_{j,\vec{m}}(z) + A_{k+1}Q_{j,\vec{m}+(k+1)\vec{e}_\mu}(z) + \sum_{i=1}^{k}(A_i + B_iz)Q_{j,\vec{m}+i\vec{e}_\mu}(z). \quad (5.57)$$

The difference here from (5.56) is that $B_{k+1} \equiv 0$ and the number of equations needed to cancel extra powers of z on the right-hand side is reduced by one (the case $n = \lambda$). Other than that, the discussion is identical. The structure of (5.57) will be seen to be intermediate between that of (5.55) and (5.56). We give one further such identity, again in a two-dimensional subspace. It is the analogue of the 'staircase' identities of Padé approximant theory. First let us define

$$\vec{r}_m = \begin{cases} \vec{r} + \tfrac{1}{2}m(\vec{e}_\mu + \vec{e}_\lambda), & m \text{ even} \\ \vec{r} + \vec{e}_\mu + \tfrac{1}{2}(m-1)(\vec{e}_\mu + \vec{e}_\lambda), & m \text{ odd}. \end{cases} \quad (5.58)$$

Then the identity is

$$Q_{j,\vec{r}_{k+2}}(z) = azQ_{j,\vec{r}}(z) + A_{k+1}Q_{j,\vec{r}_{k+1}}(z) + \sum_{i=1}^{k}(A_i + B_iz)Q_{j,\vec{r}_i}(z). \quad (5.59)$$

Here the structure is the same as in (5.57), and the method of derivation of the coefficients is also the same.

There is one item to be noted at this point, which is that in the above derivations, it is assumed that all the polynomials are nonzero. For instance, in (5.57) if $m_\lambda = -1$ so that the Hermite–Padé polynomials $Q_\lambda \equiv 0$ on the right-hand side of the equation, then manifestly, we cannot produce a nonzero one on the left-hand side of the equation. We see here the 'reduction effect.' Namely, the identity in this case involves only Hermite–Padé polynomials in a smaller subspace, so the right-hand side vanishes identically for such a case, for all j. This prevents, for example, one from using the Padé polynomials of type $[N/N]$ to compute in only a few steps the Hermite–Padé polynomials of the type $(N, N, 0)$ or the type $(N, N, 1)$.

The corresponding identity to (5.54) 'in the other direction' is known in the determinantal form (Della Dora and Di Crescenzo, [1980], Baker and Graves-Morris [1993a]). It is not a 'linear chain identity,' but one of the plethora of other possibilities. Here it is now necessary for us to involve $k + 3$ different sets of Hermite–Padé polynomials. Thus we can get a family of identities. They are

$$Q_{j,\vec{m}}(z) = \sum_{l=1}^{n} A_l Q_{j,\vec{m}+\vec{e}_{\mu_l}-\sum_{i=1}^{n}\vec{e}_{\mu_i}}(z). \quad (5.60)$$

In this case the A_l are determined, except for the overall normalization, by equating to zero the coefficients in the Hermite–Padé polynomial determining equations for the orders $z^{\|\vec{m}\|-n+1}, \ldots, z^{\|\vec{m}\|}$. Identity (5.44) is the special case $n = 2$ of (5.60).

8.5.4 Existence of sequences and the modified minimality definition

As we have remarked above, the minimality definition (Definition 8.5.1) ensures that the Hermite–Padé polynomials have no common factors of $(a + bz)$ with $a \neq 0$. There may, however, be common factors of z. We can see this result in the following example. Let

$$f_{-1}(z) = 1, \quad f_0(z) = 1 - 2z + z^2 - \tfrac{2}{3}z^3 + \tfrac{1}{12}z^4 - z^5 + \cdots,$$
$$f_1(z) = f_0'(z) = -2 + 2z - z^2 + \tfrac{1}{3}z^3 - 5z^4 + \cdots. \quad (5.61)$$

Then, observing that

$$f_1(z) + f_0(z) + f_{-1}(z) = O(z^4), \quad (5.62)$$

we can show that this equation gives the minimal Hermite–Padé polynomials of types $(0, 0, 0)$, $(0, 0, 1)$, $(0, 1, 0)$, $(1, 0, 0)$, $(0, 1, 1)$, $(1, 0, 1)$, and $(1, 1, 0)$. Direct computation shows, however, that the minimal solution for type $(1, 1, 1)$ is given by

$$zf_1(z) + zf_0(z) + zf_{-1}(z) = O(z^5), \quad (5.63)$$

which illustrates the possibility that such a case can occur. Since the division by a factor of z, and conceivably many such factors, can affect the degree of contact at the origin of the polynomial defining equations, there may be adverse consequences with regard to the convergence properties of the approximants. The problem we shall now address is whether we can or cannot add the restriction that the minimal solution satisfy $\vec{\mathcal{Q}}(0) \neq \vec{0}$, in order to ensure that the accuracy-through-order conditions hold for any so restricted sequence of minimal, Hermite–Padé polynomials. To this end we now prove a key lemma.

***Lemma* 3 [Baker and Graves-Morris, 1994a].** *If a minimal solution for the Hermite–Padé polynomials of type $\vec{m}$ is divisible by z^μ but not by $z^{\mu+1}$ for some $\mu \in \{0, 1, 2, \dots\}$, then the minimal solution for the Hermite–Padé polynomials of type $\vec{m} - \vec{v}$ is not divisible by z, if*

$$|\vec{v}| = \mu, \quad \vec{v} \geq \vec{0}. \tag{5.64}$$

Proof. If $\mu = 0$, the conclusion holds by hypothesis, and so we restrict our attention to the cases $\mu \geq 1$. Suppose that z^μ divides the minimal polynomials $Q_{j,\vec{m}}(z)$ defined by (5.1), and that $z^{\mu+1}$ does not for at least one j in the range $-1 \leq j \leq k$. Let us denote

$$\hat{Q}_{j,\vec{m}}(z) = z^{-\mu} Q_{j,\vec{m}}(z), \quad -1 \leq j \leq k. \tag{5.65}$$

Then the polynomial defining equations of the type $\vec{m} - \mu \sum_{j=-1}^{k} \vec{e}_j$ are oversatisfied by $(k+1)\mu$ orders in z by the $\hat{Q}(z)$, as

$$\sum_{j=1}^{k} \hat{Q}_{j,\vec{m}}(z) f_j(z) = O(z^{[s-(k+2)\mu]+(k+1)\mu+1}). \tag{5.66}$$

By (5.66), we see that the $\hat{Q}_{j,\vec{m}}(z)$ form a solution in $\mathcal{T}(\vec{m} - \mu\sum_{j=-1}^{k}\vec{e}_j, 0, (k+1)\mu)$. Note that this region includes the set of types defined by (5.64). Let us now suppose that the $\vec{v}$ of (5.64) is $\vec{v} = \mu \vec{e}_J$. The argument for any other $\vec{v}$ defined by (5.64) is not intrinsically different. The remarks above show that the $\hat{Q}_{j,\vec{m}}(z)$ form a solution of the type $\vec{m} - \mu\vec{e}_J$. If the highest powers of z in the $Q_{j,\vec{m}}(z)$ are v_j, then the highest powers of z in the $\hat{Q}_{j,\vec{m}}(z)$ are $\hat{v}_j = v_j - \mu$ (unless that particular component polynomial $Q_{j,\vec{m}} \equiv 0$). Either the $\hat{Q}_{j,\vec{m}}(z)$ form the unique minimal solution for type $\vec{m} - \mu\vec{e}_J$, or they do not. If they do, then by hypothesis z does not divide all the $\hat{Q}_{j,\vec{m}}(z)$ and the lemma is proven. Therefore suppose that they do not. By uniqueness, the highest power of z in the minimal solution $\tilde{Q}_{k,\vec{m}-\mu\vec{e}_J}(z)$ must be less than or equal to $\hat{v}_k - 1 = v_k - \mu - 1$. Since the solutions $\tilde{Q}$ and $\hat{Q}$ differ in the true degrees of $\tilde{Q}_{k,\vec{m}-\mu\vec{e}_J}(z)$ and $\hat{Q}_{k,\vec{m}}(z)$, and since the $\hat{Q}_{j,\vec{m}}(z)$ have no common factors, by construction, we have two different solutions of type $\vec{m} - \mu\vec{e}_J$. Therefore there exists a linear combination of them $\check{Q}$ which satisfies the polynomial defining equations to one higher order and so this linear combination is a solution of type $\vec{m} - (\mu - 1)\vec{e}_J$, with the highest power of z in $\check{Q}_{k,\vec{m}-\mu\vec{e}_J}(z)$ equal to at most $v_k - \mu$. In addition, $z\hat{Q}$ is a solution of type $\vec{m} - (\mu - 1)\vec{e}_J$, with the highest power of z in its $Q_k(z)$ equal to $v_k - (\mu - 1)$. Again we can construct a linear combination of these two solutions which satisfies the polynomial defining equations to one higher order and so it is a solution of type $\vec{m} - (\mu - 2)\vec{e}_J$ with the highest power of z in its $Q_k(z)$ equal to $v_k - (\mu - 2)$. This argument can be repeated until we obtain a solution of type $\vec{m}$ where the highest degree of z in its $Q_k(z)$ is $\nu_k - 1$. However, ν_k was the minimum possible so we have a contradiction. Therefore the 'simplex vertex solution' $\hat{Q}_{j,\vec{m}}(z)$ must have been the unique minimal solution of type $\vec{m} - \mu \vec{e}_j$. As this same set of arguments can also be repeated for any $\vec{v}$ allowed by (5.66), we conclude that the lemma is true.

At this point it is convenient to define a certain class of sequences.

Definition 8.5.5. A sequence $\{\vec{m}^{(i)}\}_{i=0}^{\infty}$ of Hermite–Padé polynomials is said to be a nearest-neighbor sequence if, for all i and a corresponding $j(i)$, $m_{j(i)}^{i+1} = m_{j(i)}^{(i)} + 1$ and, for all other $j \neq j(i)$, $m_j^{(i+1)} = m_j^{(i)}$.

Theorem 8.5.7 [Baker and Graves-Morris, 1994a]. *Any nearest-neighbor sequence of Hermite–Padé polynomials contains at least an infinite subsequence $\{\vec{Q}_{\vec{m}^{(i)}}\}_{i=0}^{\infty}$ of minimal Hermite–Padé polynomials such that $\vec{Q}_{\vec{m}^{(i)}}(0) \neq \vec{0}$, and such that the actual order of contact goes to infinity with i.*

Proof. In the nearest-neighbor sequence of Hermite–Padé polynomials originally given, the value of s of (5.2) goes to infinity by construction. Next, we note that we may assume, without loss of generality, that for some j, $-1 \leq j \leq k$, $f_j(0) \neq 0$, and for all the other j's, $f_j \sim z^{l(j)}$ for some finite, nonnegative l. Otherwise the problem is equivalent to one with a smaller value of k. In any solution with a sufficiently high degree of contact at the origin, it must be that there are at least two polynomials which are not identically zero, or else it would be impossible to satisfy the polynomial defining equations in a nontrivial manner. Therefore we can conclude that there exists a highest power $\mu(i) \geq 0$ of z which divides the minimal solution of type $\vec{m}^{(i)}$ where $\mu(i)$ is less than or equal to the second largest $m_j^{(i)}$. If $\mu(i) = 0$, we include type $\vec{m}^{(i)}$ in our sequence. Otherwise, by Lemma 8.5.16, we can conclude that associated with type $\vec{m}^{(i)}$ there are unique minimal solutions of types $\vec{m}^{(i)} - \vec{v}$ which are not divisible by z, where $|\vec{v}| = \mu(i)$ and $\vec{v} \geq \vec{0}$. However, this set of types is the complete set of possible μth backward precursor types allowed by Definition 8.5.5 for a nearest-neighbor sequence. Therefore one of these types is type $\vec{m}^{(i-\mu(i))}$. As all the members of this set of solutions are the same and are equal to $z^{-\mu(i)}\vec{Q}_{\vec{m}^{(i)}}(z)$, we lose nothing by discarding the case $\vec{m}^{(i)}$ in these circumstances. Finally, we remark that if $s(i)$ is the value of s in (5.1) for member i of the sequence, then $\mu(i) \leq \frac{1}{2}s(i)$ and so the order of contact of type $\vec{m}^{(i-\mu(i))}$ is greater than or equal to $\frac{1}{2}s(i)$ and so it goes to infinity with i, as does $s(i)$.

There is an elementary corollary which follows directly from this theorem, but it is useful to note it explicitly. First, however, we give a definition.

Definition 8.5.6. A near-partially diagonal sequence of Hermite–Padé polynomials will be any nearest-neighbor sequence where, for $j \in \mathcal{N} \subset \mathcal{I} = \{-1 \leq j \leq k\}$, $\lim_{i \to \infty} m_j^{(i)}/i$ is independent of j and for any $j \in \mathcal{I} \setminus \mathcal{N}$, the $m_j^{(i)}$ are independent of i.

Note that the case $\mathcal{N} = \mathcal{I}$ is called a 'near-diagonal sequence,' and the case where $\mathcal{N}$ has just one element is called an 'axis sequence' which includes a 'horizontal sequence' as a special case.

Corollary 4. *There exists at least an infinite subsequence of minimal, near-partially diagonal Hermite–Padé polynomials to given functional elements $f_i(z)$ such that $\vec{Q}_{\vec{m}}(0) \neq \vec{0}$.*

Proof. This is a special case of Theorem 8.5.7.

We see by these results that the subset of all minimal Hermite–Padé polynomials which have the property that $\vec{Q}_{\vec{m}}(0) \neq 0$ is as complete, in so far as the derived approximants will be concerned, as is the complete table of minimal Hermite–Padé approximants. Furthermore, in this subset there are the infinite subsequences necessary for the study of convergence properties and here that study is easier since the accuracy-through-order properties are valid.

8.6 Integral and algebraic approximants

In order to put the discussion of both integral and algebraic approximants in context, we need first to consider multiform (multivalued) functions of a complex variable. One of the features which distinguishes these approximants from Padé approximants is their capacity to be multiform functions. The usual description of multiform complex functions is in terms of Riemann surfaces. A good reference here is Springer [1957]. The multivaluedness is handled by use of a finite or possibly an infinite number of copies of the complex plane. These copies are connected by jumps from one copy to another across a cut. The cut begins and ends at a branch point. Note that polar and essential singularities are of a different character in that although the function may be undefined at such a point, it remains single valued. By the process of analytic continuation, a single functional element, by which we mean, for example, a convergent

Taylor's series expansion about a regular point z_0, can define an analytic function throughout a wider region of the complex plane, and if all possible paths are considered (denumerably many will do), a multiform function may result. There can result only denumerably many different coverings of the plane (i.e., the number of Riemann sheets is at most denumerable). We define the complete analytic function as the totality of points, including the point at infinity, which prove to be regular in the course of analytic continuation, each covered by its corresponding functional value. A simple illustration of a cut is given in Figure 8.6.1 for the case where there are just two copies of the complex plane or Riemann sheets.

8.6.1 Monodromy theory

There is another method of organizing the structure of a multiform complete analytic function of a complex variable. It is also due to Riemann. (For a reference here, see Baker [1990].) A very simple function, for example, $\ln z$, may have an infinite number of Riemann sheets, and yet the values on the different sheets may be trivially related to each other. In this example,

$$\ln z|_n = \ln |z| + i[\tan^{-1}(y/x) + 2\pi n], \quad \text{where } z = x + iy; \quad (6.1)$$

$\tan^{-1}$ is defined to yield values in the range $-\pi$ to π by taking note of the signs of x and y, and n denotes the sheet number. To take advantage of this simplicity, Riemann proposed to classify functions by the number of independent coverings of the complex plane. In the case of $\ln z$, there are two independent coverings, because the solution (6.1) may be represented by the linear combination $\ln z|_n = n \ln z|_1 - (n-1) \ln z|_0$. In the case of $\sqrt{z}$ there is just one such covering, as the two roots $\pm\sqrt{z}$ are linearly dependent on each other. In case there are an infinite number of linearly independent coverings, there must be an infinite number of Riemann sheets, but as we have just seen it can well be that an infinite number of sheets can correspond to only a finite number of linearly

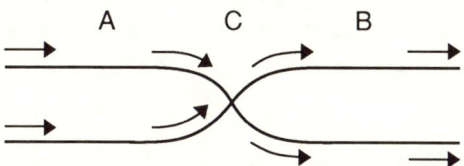

Figure 8.6.1. A cut in the complex plane (edge view) showing the cross connections of two Riemann sheets.

independent coverings. Thus we lose nothing in complexity in this classification with respect to the usual classification in terms of sheets, and may gain. Thus it makes sense to classify all complete analytic functions in terms of the number of linearly independent coverings m. Furthermore, there can be n branch points, also a finite or an infinite number. Suppose that m and n are finite for some complete analytic function. Start with some point, z_0, which is regular on all m of the coverings. Next define a functional element at z_0 for each of these coverings, $1 \leq f \leq m$, and extend it to be a complete analytic function y_j. Suppose now we encircle one of the n branch points i. Since there are only m linearly independent possibilities, we must find that after encircling the branch point and returning to the same value of z that

$$y_j \mapsto \sum_{k=1}^{m} M^{(i)}_{j,k} y_k, \quad \text{or} \quad \mathbf{y} \mapsto M^{(i)} \mathbf{y}. \tag{6.2}$$

These matrices M are called the monodromy matrices and have some useful properties. For example,

Lemma 1. *If $f(z)$ is a complete analytic function with m linearly independent coverings and n branch points in the complete complex plane, then there exists an order of the branch points such that,*

$$M^{(1)} M^{(2)} \cdots M^{(n)} = I, \tag{6.3}$$

where I is the identity matrix.

Proof. By Cauchy's theorem, $\oint (d\mathbf{y}/dz)\, dz = 0$ around any closed contour which does not contain any branch point. Thus $\mathbf{y}(z)$ returns to its original value when z encircles that contour. Expand this contour so that it passes each of the branch points one-by-one. The passing operation is illustrated in Figure 8.6.2 and in fact the contour does not actually cross any of the branch points so the value of the integral remains unchanged. Continue this operation until all the branch points have been passed. We now have one large contour with n small circles about the n branch points attached. We may reinterpret the behavior along the contour. After encircling the first singularity, $\mathbf{y}$ becomes $M^{(1)} \mathbf{y}$, and then after the second, $M^{(2)} M^{(1)} \mathbf{y}$, and so forth. But since we started with no change in $\mathbf{y}$, we have generated the identity matrix in this way, and (6.3) follows by reversing the numbering.

Because of (6.3), it follows that every M has an inverse, as the product of determinants equals the determinant of the product and since $\det I \neq 0$, none of the M's can be singular and hence each has a unique

8.6 Integral and algebraic approximants

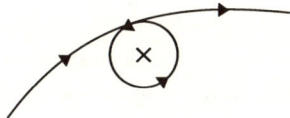

Figure 8.6.2. The contour distortion which allows the original contour to be separated into a contour beyond the singularity and one which encircles the singularity.

inverse. Hence the M's form a group $\mathcal{M}$, the so-called monodromy group. As an example, let us choose [Baker, 1990]

$$y_1(z) = (1-z)^{-1.5} + (1+z)^{-1.25},$$
$$y_2(z) = -(1-z)^{-1.5} + (1+z)^{-1.25}. \quad (6.4)$$

It is simple to see that y_2 is obtained from y_1 by encircling the singular point $z = 1$. It is straightforward to derive the corresponding monodromy matrices

$$M^{(1)} = \begin{bmatrix} 0 & 1 \\ 1 & 0 \end{bmatrix},$$

$$M^{(-1)} = \begin{bmatrix} \dfrac{1-i}{2} & \dfrac{-1-i}{2} \\ \dfrac{-1-i}{2} & \dfrac{1-i}{2} \end{bmatrix}, \quad (6.5)$$

$$M^{(\infty)} = \begin{bmatrix} \dfrac{-1+i}{2} & \dfrac{1+i}{2} \\ \dfrac{1+i}{2} & \dfrac{-1+i}{2} \end{bmatrix},$$

and to verify that $M^{(1)} M^{(-1)} M^{(\infty)} = I$. Note is made that at a pole, or an essential singularity the monodromy matrix corresponding to such a singularity is just I.

Riemann proved an important theorem in this area. In order to state it we need some notation. First let us define a class of complete analytic functions. We denote it by

$$Q \begin{bmatrix} a_1, & \cdots, & a_n \\ M^{(1)}, & \cdots, & M^{(n)} \end{bmatrix}. \quad (6.6)$$

It consists of all m-systems $\mathbf{y}$ with the properties that they are vectors whose components are m linearly independent coverings of the complex plane which are generated from a single functional element and that they have the singular points $a_1, \ldots, a_n$ and no others and have the corresponding given set of monodromy matrices $M^{(i)}$. In addition each of the y_k must satisfy the condition that there are no singularities of infinite order (in the sense of Poincaré). By this restriction we mean that for each i, there exist finite A_i and r_i such that

$$\limsup_{z \to a_i} |z - a_i|^{r_i} \left| \frac{y_k'}{y_k} \right| \leq A_i, \quad \forall k. \tag{6.7}$$

Theorem 8.6.1 (Riemann's monodromy theorem). *For any $m + 1$ m-systems which belong to the same class Q (in (6.6)), there exists a linear, homogeneous relation with polynomial coefficients in z,*

$$\sum_{j=1}^{m+1} A_j(z) \mathbf{y}_j(z) = \mathbf{0}. \tag{6.8}$$

Sketch of proof. This proof more or less directly follows that of Riemann [1953]. Suppose that the $\mathbf{y}_j$ are the $m + 1$ m-systems of the theorem. Consider a singular point a_k. Since all the different m-systems have the same monodromy matrix $M^{(k)}$ at that point, and it is nonsingular, we can introduce a change of basis

$$u_{j,l}(z) = \sum_{\eta=1}^{m} U_{k;l,\eta} y_{j,\eta}, \tag{6.9}$$

where the $y_{j,\eta}$ and $u_{j,l}$ are the vector components of $\mathbf{y}_j$ and $\mathbf{u}_j$, and U_k is a constant matrix such that

$$\Lambda_k = U_k M^{(k)} U_k^{-1}, \tag{6.10}$$

where Λ is a diagonal matrix. In this representation, we have

$$\Lambda_k u_{j,l} = \lambda_{k;l} u_{j,l}, \tag{6.11}$$

where the $\lambda_{k;l}$ are the eigenvalues of $M^{(k)}$. We will omit the extra complications that occur if the eigenvalues are degenerate. The solution for the u's is then

$$\mathbf{u}_{j,l}(z) = (z - a_k)^{v_{k;j,l}} \mathbf{h}_{k;j,l}(z), \quad v_{k;j,l} = \frac{\log \lambda_{k;l}}{2\pi i} + L_{k;j,l}, \tag{6.12}$$

where the L's are whole numbers and $h_{k;j,l}(z)$ is single valued in the neighborhood of a_k. Again we will not give the extra detail required if $r_k \neq 1$ and we will assume that $h_{k;j,l}(a_k) \neq 0, \infty$.

Next let us consider the equations

$$\sum_{j=1}^{m+1} c_j \mathbf{y}_j = \mathbf{0}. \tag{6.13}$$

These equations determine, up to an overall factor, the c_j in terms of the $\mathbf{y}_j$ because, by hypothesis, the different vector components of the $\mathbf{y}$'s are linearly independent. Thus, by Cramer's rule,

$$c_{j_0} = \Delta_{j_0}(z) = \det|y_{j,i}|, \quad \begin{array}{l} i = 1, \ldots, m, \\ j = 1, \ldots, m+1, j \neq j_0. \end{array} \tag{6.14}$$

In terms of our change of basis, we may rewrite this determinant as

$$\Delta_{j_0}(z) = \det|U_k| \det|u_{j,l}(z)|, \quad \begin{array}{l} l = 1, \ldots, m, \\ j = 1, \ldots, m+1, \quad j \neq j_0. \end{array} \tag{6.15}$$

If we define $v_{k;l} = \min_j(v_{k;l,j})$, then from our representation of the $u_{j,l}$ (6.12) we see that

$$\left[\prod_{l=1}^{m}(z - a_k)^{-v_{k;l}}\right]\Delta_{j_0}(z) \tag{6.16}$$

is regular in the neighborhood of $z = a_k$. Thus we can conclude that

$$\mathcal{P}_{j_0}(z) = \left[\prod_{k=1}^{n}\prod_{l=1}^{m}(z - a_k)^{-v_{k;l}}\right]\Delta_{j_0}(z) \tag{6.17}$$

is regular at $z = a_1, \ldots, a_n$ and hence in the entire (finite) z-plane by construction. Since (6.3) holds, and the product of the determinants is the determinant of the product, it must be that $\sum_{k=1}^{n}\sum_{l=1}^{m} v_{k;j,l}$ is a whole number for each j. As the difference between any $v_{k;j,l}$ and $v_{k;l}$ is a whole number by construction, the same conclusion holds when the $v_{k;l}$ replace the $v_{k;j,l}$. Consequently the behavior $\mathcal{P}_{j_0}(z)$ as $z \to \infty$ can only be $z^{\mathcal{L}}$ for some integer $\mathcal{L}$. The only such functions are polynomials. Thus all the $\mathcal{P}_j(z)$ can be brought into polynomial form.

Corollary 1. *If the m-system* $\mathbf{y}$ *belongs to the class Q defined in* (6.6), (6.7), *then*

$$\sum_{j=0}^{m} A_j(z) y_k^{(j)}(z) = 0, \quad k = 1, 2, \ldots, m, \tag{6.18}$$

where $y_k^{(j)}(z)$ *denotes the jth derivative of the kth vector component of* $\mathbf{y}$.

Proof. It follows easily by the differentiation of (6.2) that all the derivatives belong to the same class Q as does $\mathbf{y}$. Thus this corollary follows directly from Theorem 8.6.1.

We notice how the case of an algebraic equation fits into this system. Suppose that $f(z)$ satisfies

$$\sum_{j=0}^{m} B_j(z)[f(z)]^j = 0, \qquad (6.19)$$

where the $B_j(z)$ are given polynomials. We know from the fundamental theorem of algebra that at each point z there are exactly m solutions for $f(z)$. It is straightforward to establish the regularity of the solution $f(z)$. If we differentiate (6.19) we obtain

$$\left(\sum_{j=1}^{m} j B_j(z)[f(z)]^{j-1}\right) \frac{df(z)}{dz} = -\sum_{j=0}^{m} B_j'(z)[f(z)]^j. \qquad (6.20)$$

The procedure is to solve (6.19) for a solution $f(z_0)$ at some point z_0 where the coefficient in (6.20) of $(df(z)/dz)$ does not vanish. Then by standard methods, (6.20) normally provides for the construction of a differentiable, bounded function $f(z)$ and hence an analytic one. The same can be done for the other $m - 1$ solutions of (6.19) at z_0 and hence we can construct m complete analytic functions from the solution of (6.19). The singularities occur when both (6.19) holds and the coefficient of $(df(z)/dz)$ in (6.20) vanishes. In order for both of these conditions to hold, the $[(2m - 1) \times (2m - 1)]$ Sylvester resultant must vanish, i.e.,

$$\Delta_1 = \det \begin{vmatrix} B_0 & B_1 & \cdots & 0 \\ 0 & B_0 & \cdots & 0 \\ \vdots & \vdots & \ddots & \vdots \\ 0 & 0 & \cdots & B_m \\ B_1 & 2B_2 & \cdots & 0 \\ 0 & B_1 & \cdots & 0 \\ \vdots & \vdots & \ddots & \vdots \\ 0 & 0 & \cdots & mB_m \end{vmatrix} = 0. \qquad (6.21)$$

Note that $B_m(z)$ is a factor of Δ_1 and that its zeros correspond to the divergent ($f \to \infty$) singularities of $f(z)$. Since the $B_j(z)$ are polynomials, the Sylvester resultant [see Theorem (2.4.2)] is also a polynomial in z, and thus has a finite number of zeros. Hence the number of singular points of $f(z)$ (on any branch) must be finite and so we conclude that the solutions of (6.19) form an m-system $\mathbf{f}(z)$ and belong to a class Q for some finite n. Actually we do not need to restrict the $B_j(z)$ to be polynomials, but only the number of zeros of (6.21) to be finite, which is an awkward condition.

By the above results and by noting that the singularities which can occur are rational powers of $z - a_k$ whose denominators can only be chosen from $1, 2, \ldots, m$, we can conclude that the solution of algebraic equations with polynomial coefficients are a subclass of the general class of m-systems which we have considered in our sketch of part of monodromy theory.

8.6.2 Definitions and the accuracy-through-order principle

In the case of Padé approximants, we have approximated a given function by that member of the class of meromorphic functions which has a restricted number of poles and zeros. We will now introduce, in the same spirit, methods of approximation in a wider function class. These approximants will be special cases of the Hermite–Padé approximants which were discussed in the previous section. First we define *algebraic approximants*. These were introduced by Padé [1892] and an investigation of the quadratic case was made by Shafer [1974]. Let a function $f(z)$ be given in terms of its Maclaurin expansion. As in the previous section, we define the algebraic polynomials by the accuracy-through-order principle by means of

$$\sum_{j=-1}^{k} Q_{j,m_j}(z)[f(z)]^{j+1} = O(z^{s+1}), \tag{6.22}$$

where the $Q_{j,m_j}(z)$ are polynomials of degree at most m_j and s is given by (5.2). Again we will use the convention in this section that if $m_j = -1$, then $Q_{j,-1}(z) \equiv 0$. We will find it convenient later in this section to treat m_{-1} separately and so introduce the notation $\mathbf{m} = (m_0, \ldots, m_k)$. The algebraic approximant is denoted by $\langle m_{-1}/m_0, \ldots, m_k \rangle = \langle m_{-1}/\mathbf{m} \rangle$ and is defined by the solution $y(z)$ of

$$\sum_{j=-1}^{k} Q_{j,m_j}(z)[y(z)]^{j+1} = 0, \quad \text{where } y(0) = f(0). \tag{6.23}$$

Of course, we can impose additional boundary conditions at $z = 0$, if necessary, to break the degeneracy of the possible solutions. See McInnes [1992] for an additional discussion of this point. As we have seen above, these solutions form a subset of the set of complete analytic functions which have m linearly independent coverings of the complex plane and have a finite number of singularities.

Next we will consider the *integral approximants*. Early studies of this area were made by Guttmann and Joyce [1972, 1973], Gammel and Gaunt [Gammel, 1973b], Fisher and Au-Yang [1979], and Hunter and Baker [1979]. First we define the *integral polynomials* by means of the accuracy-through-order principle. Let a function $f(z)$ be given through its Maclaurin expansion. We define the $Q_{j,m_j}(z)$ by

$$Q_{-1,m_{-1}}(z) + \sum_{j=0}^{k} Q_{j,m_j}(z) f^{(j)}(z) = O(z^{s+1}), \tag{6.24}$$

where again the $Q_{j,m_j}(z)$ are polynomials of degree at most m_j and s is as given by (5.2). Note is taken that the Q's as defined in (6.22) and (6.24) are determined up to an overall constant factor. The integral approximant to the given $f(z)$ is denoted by $[m_{-1}/m_0, \ldots, m_k] = [m_{-1}/\mathbf{m}]$ and is the solution $y(z)$ of

$$Q_{-1,m_{-1}}(z) + \sum_{j=0}^{k} Q_{j,m_j}(z) y^{(j)}(z) = 0, \; y^{(j)}(0) = f^{(j)}(0), \; j = 0, \ldots, k-1, \quad (6.25)$$

in the usual case. In exceptional cases the manner of prescribing the boundary conditions may vary. See Baker and Graves-Morris [1990] for a further discussion of this point. Computer programs to calculate the integral approximants (called differential approximants by a number of authors) are given in the excellent review article by A. J. Guttmann [1989].

In order to complete the definition of these approximants, attention must be paid to the possibility that neither (6.22) nor (6.24) uniquely defines the polynomials. As we remarked in Section 8.5, the points where $Q_{k,m_k}(z)$ vanishes correspond to singularities. Consequently, in the case where the polynomials are not uniquely determined it would be desirable, in our view, to eliminate any arbitrary or spurious singularities that are introduced by this lack of uniqueness. Consequently we complete these definitions by using the minimal polynomials of Definition 8.5.1, and so uniqueness is ensured by Theorem 8.5.2. In addition, we will impose the condition that $\vec{Q}_{\vec{m}}(0) \neq \vec{0}$. By Theorem 8.5.7 we may do so and, although some types $\vec{m}$ may then fail to exist, the table of approximants is as complete, insofar as the derived approximants are concerned, as was the table of approximants derived from the minimal polynomials. Some of the repeats have, however, been dropped by this restriction.

The first property of Padé approximants which we discussed in Section 1.1 is that of accuracy through order. In that case, the requirement that the denominator polynomial does not vanish at the origin is sufficient to show that the Padé approximant agrees with the defining series to the same order in z as the accuracy of the polynomial defining equations. For integral approximants the same results obtain, but for algebraic approximants, as we shall see, the results are roughly true as well, but the situation is a bit more complex. This question has been addressed by Baker and Graves-Morris [1990] for the case of integral approximants, and by McInnes [1992] and Baker [1994] for algebraic approximants. In the first case we will make some improvements on their results.

***Theorem* 8.6.2 (Accuracy through order, integral approximants).** *If the error in the integral polynomial defining equation is $O(z^M)$ then the*

8.6 Integral and algebraic approximants

integral approximant differs from the defining series by at worst the same error.

Proof. Consider first (6.25). By our definitions, there must be a $0 \leq j_0 \leq k$ such that $Q_{j_0,m_{j_0}}(0) \neq 0$, as it can not be that only $Q_{-1,m_{-1}}(0) \neq 0$, since the terms in the defining series are all finite. Define γ as

$$\gamma = \max(k - \alpha_k, k - 1 - \alpha_{k-1}, \ldots, -\alpha_0), \quad (6.26)$$

where $Q_{j,m_j}(z) \sim z^{\alpha_j}$ as $z \to 0$, and of course we must have $\alpha_j \leq m_j$. As remarked above, $\alpha_{j_0} = 0$. Therefore we can conclude

$$\gamma \geq j_0 \geq 0. \quad (6.27)$$

Now select η as the smallest integer such that

$$z^{\gamma-k} Q_{k,m_k}(z)|_{z=0} = \cdots = z^{\gamma-k+\eta-1} Q_{k-\eta+1,m_k}(z)|_{z=0} = 0,$$
$$z^{\gamma-k+\eta} Q_{k-\eta,m_k}(z)|_{z=0} \neq 0, \quad (6.28)$$

where no negative powers of z can occur in (6.28) by construction. If we substitute a power-series expansion $y = \sum_{j=0}^{\infty} y_j z^j$ in (6.25) and multiply by z^γ we get, displaying the error for convenience,

$$z^\gamma Q_{-1,m_{-1}}(z) + Q_{0,m_0}(z) \sum_{j=0}^{\infty} y_j z^{j+\gamma} + \cdots$$
$$+ Q_{k,m_k}(z) \sum_{j=0}^{\infty} y_j [j]_k z^{j+\gamma-k} = 0 + O(z^{s+t+\gamma+1}), \quad (6.29)$$

where s is as in (5.2), t is the degree of oversatisfaction as in (5.10), and $[j]_k \equiv j(j-1) \cdots (j-k+1)$ and $[j]_0 \equiv 1$. The coefficient of $z^{\gamma+j}$ involves in a linear manner the series coefficients $y_0, \ldots, y_j$. The coefficient of y_j is

$$C(j) = z^{\gamma-k+\eta} Q_{k-\eta,m_{k-\eta}}(z)|_{z=0} [j]_{k-\eta} + \cdots + z^\gamma Q_{0,m_0}(z)|_{z=0} [j]_0, \quad (6.30)$$

as the other terms vanish by the definition of η. The leading term does not vanish by our construction, so $C(j)$ is a polynomial in j of degree $k - \eta$ exactly. If $\eta = 0$ only the first term survives and $C(j) \sim [j]_k$. In this case $C(j)$ vanishes for $j = 0, \ldots, k$ precisely. Then the boundary conditions listed in (6.25) may be used to determine the y_j for those values of j. The rest of the values of y_j follow uniquely by recursion through y_s by (6.29) and (6.30). Since the original series is a solution, we conclude that $y_j = f_j$ for $j = 0, \ldots, s+t$ which is the accuracy-through-order result.

It cannot be that $\eta = k$ because Ince [1944] has shown that there are no regular solutions for this case. But the original series is a regular solution, which would be a contradiction. For the cases $0 < \eta < k$, $C(j)$ is a polynomial of exact degree $k - r$. There can be at most $k - r$ positive integer zeros. It is for these values of j that a boundary condition $y_j = f_j$ is required to determine the y_j. This is the unusual case, but no more than k boundary conditions are required to determine the solution. When these are specified, the recursion relations determine the rest of the y_j uniquely and they must therefore agree with the f_j for $j = 0, \ldots, s + t$, and so the accuracy-through-order results hold in this case as well.

An illustration of the occurrence of nonstandard boundary conditions referred to in the above theorem is given by the seemingly innocuous equation

$$zf'' - 9f' + zf = 2z,$$

which has the indicial equation

$$\rho(\rho - 10) = 0.$$

The $[1/1; 0; 1]$ requires series-expansion terms through order z^7 to determine all its coefficients, but these terms are insufficient to determine the complete solution. To do so one must specify not only f_0, but also f_{10}. However, the boundary condition f_0 alone is sufficient to ensure the accuracy-through-order result for the $[1/1; 0; 1]$. In an alternative example, if the 9 of this example is replaced by a 5, then one would need to provide the value of f_6 as well as f_0 to achieve the accuracy-through-order result. The required number of boundary conditions remains fixed at the expected two, however.

In order to analyze the problem of accuracy through order for the algebraic approximants, we need to introduce the concept of *deficiency*. While our minimal definition for the algebraic polynomials and Theorem 8.5.7 ensure that we need only look at cases for which $\vec{Q}_{\vec{m}}(0) \neq \vec{0}$, there is another quantity which is important to consider in the algebraic case, i.e., the coefficient of df/dz in (6.20).

Definition 8.6.1 (Deficiency, algebraic polynomials). The deficiency d of the algebraic polynomials to $f(z)$ is given by

$$\sum_{j=0}^{k}(j+1)Q_{j,m_j}(z)[f(z)]^j \underset{z \to 0}{\sim} z^d. \qquad (6.31)$$

Theorem 8.6.3 (Accuracy-through-order, algebraic approximants) [Baker, 1994]. *If the error in the algebraic polynomial defining equation is $O(z^M)$ and the deficiency of the algebraic polynomials is d, then the algebraic*

8.6 Integral and algebraic approximants

approximant differs from the defining series by an error at worst $O(z^{M-d})$, when we specify that $y^{(j)}(0) = f^{(j)}(0)$ for $j = 0, \ldots, d$.

Proof. First, if we differentiate the approximant defining equation

$$\sum_{j=0}^{k} Q_{j,m_j}(z)[y(z)]^j = O(z^{s+t+1})$$

(with error displayed for convenience), we get

$$\left(\sum_{j=0}^{k}(j+1)Q_{j,m_j}(z)[y(z)]^j\right)\frac{dy(z)}{dz} = -\sum_{j=-1}^{k} Q'_{j,m_j}(z)[y(z)]^{j+1} + O(z^{s+t}),$$

(6.32)

where s is defined by (5.2) and t is the degree of oversatisfaction (5.10). Suppose that the deficiency is $d = 0$; then by letting $z \to 0$ in (6.32) we can compute y_1 from $y_0 = f_0$, where $y(z) = \sum_{j=0}^{\infty} y_j z^j$. If we now compute the coefficient of z^j in (6.32) we find that it involves y_{j+1} linearly with its coefficient proportional to the limit in (6.31) which is nonzero in this case. Thus we can solve in a unique recursive manner for all the coefficients y_j. By the defining equation, the f_j are also a solution within the error, so we get $y_j = f_j$ for $j = 0, \ldots, s+t$ which establishes the accuracy-through-order results for this case. If the deficiency $d > 0$, we examine (6.32) which is also satisfied by $f(z)$ as well as $y(z)$, and since $f'(0)$ is finite, the left-hand side vanishes like z^d and so too must the right-hand side. Here we have used the initial conditions $y^{(j)}(0) = f^{(j)}(0)$ for $j = 0, \ldots, d$. We next observe that in the defining equation (6.23) the coefficients of z^j through z^{2j-1} are linear in y_j, since y_j^2 cannot appear at lower order than z^{2j}. Consider the case of deficiency d. The coefficient of z^{2d+1} must be linear in $y_{d+1}, \ldots, y_{2d+1}$. The coefficient in (6.23) of $z^{2d+1} y_{d+n}$ for $n = 1, \ldots, d+1$ can be seen to be the coefficient of z^{d+1-n} in the quantity multiplying dy/dz on the left-hand side of (6.32). This result follows by noting that this coefficient is the partial derivative of (6.23) with respect to $y(z)$ and so will give the coefficient of the linear terms. Thus, because the deficiency is d, only y_{d+1} has a nonzero coefficient. This coefficient necessarily involves only $y_0, \ldots, y_d$, which we have set equal to the corresponding f_j's by the initial conditions. As we examine the higher orders we find that the coefficient of z^{d+n} is linear in y_n with a nonzero coefficient for $n \geq d+1$, and that, as in the case $n = d+1$ just discussed, the coefficients in this term of y_{n+l} all vanish for any $l > 0$. Thus we may, in a unique and recursive manner, compute all the y_j for $j = d+1, d+2, \ldots, s+t-d$. Since, as we have remarked, the f_j's satisfy the equation, we have proven the theorem, unless $d > s+t$. But it cannot be that $d > s$, since we have a minimal solution. For if it were so, then the quantity in (6.31) multiplied by $f(z)$ would be a different solution set (of the same order and the same polynomial degrees) to the polynomial defining equations. A linear combination of these two polynomial sets could then be formed which would give a new solution set of order $k - 1$ instead of k, which is a contradiction to our selection of a minimal set of algebraic polynomials.

In order for the algebraic approximants to be generally useful, we need to show that their degree of contact with the defining series goes to infinity with $\mathcal{M}$.

Theorem 8.6.4 (Degree of contact bound, algebraic approximants) [Baker, 1994]. *Given an essentially unique, minimal algebraic approximant whose polynomials satisfy $\vec{Q}_{\vec{m}}(0) \neq \vec{0}$ and for which the error in the polynomial defining equations is $O(z^{\mathcal{M}})$, the degree of contact c of the algebraic approximant with its defining series is bounded by*

$$c \geq \left[\frac{\mathcal{M} - 1}{\mu}\right], \tag{6.33}$$

where $[x]$ denotes the greatest integer less than or equal to x, and $\mu \geq 1$ is the smallest value of ν for which the coefficient of $[f(z) - f(0)]^\nu$ in (6.34) does not vanish for $z = 0$. The value of (6.33) for $\mu = k + 1$ always gives a bound.

Proof. Let us expand $f(z)$ in the polynomial defining equation (6.22) about the given value of $f(0)$. We get

$$\sum_{\nu=0}^{k+1} \frac{[f(z) - f(0)]^\nu}{\nu!} \left(\sum_{j=\nu-1}^{k} \frac{Q_{j,m_j}(z)(j + 1)![f(0)]^{j+1-\nu}}{(j + 1 - \nu)!} \right) = O(z^{s+t+1}). \tag{6.34}$$

Now we note that the $\nu = 0$ term vanishes at $z = 0$ as is required by the polynomial defining equations. If the coefficient of the second term does not vanish at $z = 0$, then the deficiency is zero. It is the case, when we denote the coefficient of $[f(z) - f(0)]^\nu$ by $\mathcal{Q}_\nu(z)$, that $\mathcal{Q}_\mu(0) \neq 0$ for at least one $\nu = \mu$. This result can be seen as follows. First consider the case $\nu = k + 1$. If $\mathcal{Q}_{k+1}(0) \neq 0$, we have the desired result. Otherwise, consider the case $\nu = k$. In this circumstance, $\mathcal{Q}_k(0) = 0$ if and only if $Q_{k-1,m_{k-1}}(0) = 0$. We can continue in this manner through all the $Q_\nu(0)$'s (with $k + 1 \geq \nu > 0$). We must eventually find one which does not vanish, since by hypothesis, $\vec{Q}_{\vec{m}}(0) \neq \vec{0}$. In order to focus on the computation of c, it is simplest to consider the determination of f_1. Let μ be the smallest value of ν for which $\mathcal{Q}_\nu(0) \neq 0$. From (6.34) there will be a nonvanishing term, proportional to f_1^μ, and potentially other terms of order z^μ involving f_1 but no other series coefficients, except f_0. Let us now examine the series in z term by term. The first term (beyond the $v = 0$ term) is linear in f_1 and if the coefficient does not vanish, we can solve for f_1 directly. If it does vanish, then the second term will be a quadratic equation in f_1 and there will be no mixed terms involving f_1 or terms involving f_j with $j > 1$ because their coefficients vanish as a consequence of the vanishing of the first term. The examination continues in this manner until we either find an equation for f_1, or the form of the next order term in z involves f_1 alone. We are, however, guaranteed that we will find an equation for f_1 with at least one nonzero coefficient, not beyond the z^μ term. It will be of μth order in f_1 and corresponds to μ roots which are coincident at the origin. If the appropriate root of this equation for f_1 is not a multiple root, then we can solve for all the rest of the coefficients f_j, $j > 1$, because they will first appear linearly and if $R(f) = 0$ is the equation for f_1 then their coefficient will always be $(dR(f)/df)|_{f_1}$ which is nonzero for this case. By analysis of this case through some rather tedious but straightforward algebra, one can see that there will always be an equation with at least one nonvanishing coefficient to determine f_j; if not, we certainly always have the term $\mathcal{Q}_\mu(0)f_j^\mu z^{j\mu}$. Thus the accuracy-through-order equations are always sufficient to determine at least $[(\mathcal{M} - 1)/\mu]$ coefficients.

An illustration of these theorems is given by the function

$$f(z) = -1 + z - \frac{1}{2}z^2 + \frac{3}{8}z^3 - \frac{5}{16}z^4 + \frac{35}{128}z^5 - \frac{63}{256}z^6 + O(z^7),$$

which satisfies

$$(1 + z)[f(z)]^2 + 2(1 + z)f(z) + 1 + z - z^2 = O(z^7).$$

Condition (6.32) becomes in this case

$$2(1 + z)(f(z) + 1)\frac{df(z)}{dz} = -[f(z)]^2 - 2f(z) - 1 + 2z + O(z^6). \quad (6.35)$$

We see by direct substitution of the defining series in (6.35) that, as z divides (6.35), the deficiency is $d = 1$, thereby reducing the degree of contact by unity. This deficiency is associated with the coincidence of two solutions at $z = 0$, and the corresponding vanishing of the first two terms in (6.34). Those solutions for $f(z)$ are

$$f(z) = -1 \pm \sqrt{\frac{z^2}{1 + z}} + O(z^6),$$

with the + sign selected by the imposition of the additional boundary condition $f'(0) = 1$ beyond $f(0) = -1$.

8.6.3 Equivalence properties

One fundamental property for Padé approximants is the equivalence between the occurrence of an infinite block in the Padé table and the rational nature of the function being approximated. This property is described by Theorem 1.4.4. Next we give the corresponding results for the integral and the algebraic approximants.

Theorem 8.6.5 **(Equivalence theorem for integral approximants) [Baker and Graves-Morris, 1990].** *The statement* (i) $f(z)$ *is a functional element at* $z = 0$ *which satisfies*

$$Q_{-1,m_{-1}}(z) + \sum_{j=0}^{k} Q_{j,m_j}(z) f^{(j)}(z) = 0, \tag{6.36}$$

where $\vec{Q}_{\vec{m}}$ *is of true nominal degree and essentially unique, is equivalent to* (ii) *there exists a functional element* $g(z)$ *at* $z = 0$ *for which*

$$\vec{\mathcal{Q}}_{\vec{M}} = \vec{Q}_{\vec{m}}, \quad \forall \vec{M} \geq \vec{m}, \tag{6.37}$$

where $\vec{Q}_m$ *is minimal and* $\vec{\mathcal{Q}}_{\vec{M}}$ *is a Beckermann minimal solution of type* $\vec{M}$ *for* $g(z)$, *the inequality is in the sense of Definition 8.5.3, and* $g(z) = f(z)$.

Proof. First, (i) implies (ii) since, if we pick $g(z) = f(z)$, then $\vec{Q}_{\vec{m}}$ is a solution of type $\vec{M}$ for any $\vec{M} \geq \vec{m}$ and as by hypothesis $\vec{Q}_{\vec{m}}$ is minimal, it is at least a Beckermann minimal solution of type $\vec{M}$.

Second, we must consider whether (ii) implies (i). Suppose that (ii) holds; therefore we must have

$$Q_{-1,m_{-1}}(z) + \sum_{j=0}^{k} Q_{j,m_j}(z) g^{(j)}(z) = O(z^t), \quad \forall t < \infty. \tag{6.38}$$

By the minimality of $\vec{Q}_{\vec{m}}(z)$, there are no common factors of the type $1 + az$ and division by any factor of z^j leaves (6.38) unchanged except that $m_i \to m_i - j$, as it holds for all $t < \infty$. Since $g(z)$ is a functional element, the left-hand side of (6.38) is regular at $z = 0$ and so by (6.38) must be identically zero. The remaining step to complete the proof is to show that $g(z) = f(z)$. This result follows from the accuracy-through-order theorem (Theorem 8.6.2) and the principle of analytic continuation.

Theorem 8.6.6 (Equivalence theorem for algebraic approximants) [Baker, 1995]. *The statement* (i) $f(z)$ *is a functional element at* $z = 0$ *which satisfies*

$$\sum_{j=-1}^{k} Q_{j,m_j}(z)[f(z)]^{j+1} = 0, \qquad (6.39)$$

where $\vec{Q}_{\vec{m}}$ *is of true nominal degree and essentially unique, is equivalent to* (ii) *there exists a functional element* $g(z)$ *at* $z = 0$ *for which,*

$$\vec{\mathcal{Q}}_{\vec{M}} = \vec{Q}_{\vec{m}}, \quad \forall \vec{M} \geq \vec{m}, \qquad (6.40)$$

where $\vec{Q}_{\vec{m}}$ *is minimal and* $\vec{\mathcal{Q}}_{\vec{M}}$ *is a Beckermann minimal solution of type* $\vec{M}$ *for* $g(z)$, *the inequality is in the sense of Definition* 8.5.3, *and* $g(z) = f(z)$.

Proof. The line of argument follows identically that given for Theorem 8.6.5, except that Theorems 8.6.3 and 8.6.4 are relied on instead of Theorem 8.6.2.

8.6.4 Invariance properties

The invariance properties which a particular class of Padé approximants possesses go a long way toward explaining the generality of its convergence properties. These properties were discussed in Section 1.5. There are analogous properties for the algebraic approximants and, to some extent, for the integral approximants as well.

Theorem 8.6.7 (Homographic invariance under argument transformations for algebraic approximants) [Baker 1984a]. *Let* $f(z) = \sum_{i=0}^{\infty} c_i z^i$. *Define an origin-preserving linear fractional transformation*

$$z = \frac{\alpha w}{1 + \beta w}, \qquad (6.41)$$

and thereby a new function $g(w) = f(z)$. *If there exists an essentially unique minimal algebraic approximant* $y_f(z)$ *for* $f(z)$ *of the type* $\vec{m} = (m, m, \ldots, m)$, *then there exists an essentially unique minimal algebraic approximant* $y_g(w)$ *for* $g(w)$ *and* $y_f(z) = y_g(w)$.

Proof. Let us use the transformation (6.41) in the defining equation for the algebraic polynomials (6.22). It gives

$$\sum_{j=-1}^{k} Q_{j,\vec{m}}\left(\frac{\alpha w}{1 + \beta w}\right)\left[f\left(\frac{\alpha w}{1 + \beta w}\right)\right]^{j+1} = O(w^{s+1}). \qquad (6.42)$$

It is easy to see that we get form invariance if and only if all the $m_j = m$ (or $m_j = -1$), as

$$\sum_{j=-1}^{k} (1 + \beta w)^m Q_{j,\vec{m}}\left(\frac{\alpha w}{1 + \beta w}\right)\left[f\left(\frac{\alpha w}{1 + \beta w}\right)\right]^{j+1} = O(w^{s+1}) \quad (6.43)$$

reduces the coefficients to polynomial form:

$$\sum_{j=-1}^{k} P_{j,\vec{m}}(w)\left[f\left(\frac{\alpha w}{1 + \beta w}\right)\right]^{j+1} = O(w^{s+1}). \quad (6.44)$$

If the original $\vec{Q_{\vec{m}}}$ is an essentially unique minimal solution, then it transforms into an essentially unique minimal solution for the polynomials. The reason is soon found by examining (5.4) as the transformation is effectively by elementary row and column operations, which are invertible, thus preserving the original uniqueness. The transformed (or the original) polynomials may not be of full degree because, if α/β is a root of one of the $Q_{j,\vec{m}}(z)$, then the coefficient of w^{m_j} in $P_{j,\vec{m}}(w)$ will vanish. The agreement of the approximants follows by construction because the map of the solution is the solution of the mapped equations.

For algebraic equations, we also get value-transformation invariance:

***Theorem* 8.6.8 (Homographic invariance under value transformations for algebraic approximants)** [Baker, 1984a]. *Given a function* $f(z) = \sum_{i=0}^{\infty} c_i z^i$, *we define*

$$g(z) = \frac{a + bf(z)}{c + df(z)}, \quad \text{where } ad - cb \neq 0. \quad (6.45)$$

If $y_f(z)$ is an essentially unique minimal algebraic approximant to $f(z)$ and $y_g(z)$ is the algebraic approximant to $g(z)$, both of type $\vec{m} = (m, \ldots, m)$, then $y_g(z)$ is the essentially unique minimal algebraic approximant to $g(z)$ of type $\vec{m}$ and is given by

$$y_g(z) = \frac{a + by_f(z)}{c + dy_f(z)}, \quad (6.46)$$

provided $c + df(0) \neq 0$.

Proof. If we substitute (6.45) directly in the polynomial defining equation (6.22) for $g(z)$ and multiply by $(c + df(z))^{k+1}$ we get

$$\sum_{j=-1}^{k} Q_{j,\vec{m}}(z)[c + df(z)]^{k+1-j}[a + bf(z)]^{j+1} = O(z^{s+1}), \quad (6.47)$$

or

$$\sum_{j=-1}^{k} \mathcal{Q}_{j,\vec{m}}(z)[f(z)]^{j+1} = O(z^{s+1}), \qquad (6.48)$$

so the defining equations are form invariant. Note that (6.47)–(6.48) yield an explicit linear expression for the $\mathcal{Q}$'s in terms of the Q's, and that the type $\vec{m}$ of the approximating polynomials is unchanged. Since $ad - bc \neq 0$ by hypothesis, we can invert (6.45) to give $f(z)$ as a linear fractional transformation of $g(z)$, and use this expression to convert (6.48) back into (6.22) for $g(z)$. Since we thus get an explicit linear expression for the Q's in terms of the $\mathcal{Q}$'s, the determinant of the transformation is not zero and hence the essential uniqueness of the y_f implies that of the y_g.

Remark. If $a = 0$ in this theorem, then $m_{-1} = -1$ is also allowed.

We next turn our attention to the invariance properties of the integral approximants as defined by (6.24)–(6.25). First we consider the problem of argument invariance. Note that by (6.41)

$$\frac{df(z)}{dz} = \left(\frac{df(z)}{dw}\right)\left(\frac{dw}{dz}\right) = (1 + \beta w)^2 \left(\frac{df(z)}{dw}\right)\bigg/\alpha. \qquad (6.49)$$

Thus we find that the structure

$$P_2(z)\left(\frac{df}{dz}\right) = \mathcal{P}_2(w)\left(\frac{df}{dw}\right) \qquad (6.50)$$

is form invariant, as was noticed by Hunter and Baker [1979]. The subscripts 2 in (6.50) indicate quadratic polynomials. If we substitute (6.49) into (6.24) with $k = 1$, we get that

$$Q_{-1,(n,n,n+2)}(z) + Q_{0,(n,n,n+2)}(z)f(z) + Q_{1,(n,n,n+2)}(z)f'(z) = O(z^{s+1})$$
$$(6.51)$$

is form invariant when we multiply by $(1 + \beta w)^n$. For second-order integral approximants, a new feature appears. By (6.50), the form-invariant structure is

$$P_n(z) + R_n(z)f(z) + S_{n+2}(z)f'(z) + T_{n+2}(z)\frac{d}{dz}\left[U_2(z)\frac{df(z)}{dz}\right] = O(z^{4n+9}), \qquad (6.52)$$

or

$$Q_{-1,(n,n,n+3,n+4)}(z) + Q_{0,(n,n,n+3,n+4)}(z)f(z)$$
$$+ Q_{1,(n,n,n+3,n+4)}(z)f'(z) + Q_{2,(n,n,n+3,n+4)}(z)f''(z) = O(z^{s+1}), \qquad (6.53)$$

where (6.53) is subject to the restriction, in the notation of (5.50), that $Q_{1(n,n,n+3,n+4),n+3} = 2Q_{2,(n,n,n+3,n+4),n+4}$.

An example of this result is well known in classical analysis and is the Riemann–Papperitz equation

$$\frac{d^2u}{dz^2} + \left\{ \frac{1-\alpha-\alpha'}{z-a} + \frac{1-\beta-\beta'}{z-b} + \frac{1-\gamma-\gamma'}{z-c} \right\} \frac{du}{dz}$$

$$+ \left\{ \frac{\alpha\alpha'(a-b)(a-c)}{z-a} + \frac{\beta\beta'(b-a)(b-c)}{z-b} + \frac{\gamma\gamma'(c-a)(c-b)}{z-c} \right\}$$

$$\times \frac{u}{(z-a)(z-b)(z-c)} = 0, \quad \alpha + \alpha' + \beta + \beta' + \gamma + \gamma' = 1,$$

which is of type $(-1, 2, 5, 6)$ as expected for $n = 2$, with no inhomogeneous term. The restriction is equivalent to that found for (6.53). A linear fractional transformation of the independent variable of the Riemann–Papperitz equation just maps the a, b, c in the same manner as it does z, but otherwise leaves the equation unchanged.

The general structure for a kth order integral appoximant that has form invariance is of type $(n$ or $-1, n, n+k, n+k+1, \ldots, n+2k)$. The number of necessary restrictions is one for $Q_{k-1,\vec{m}}(z)$, two for $Q_{k-2,\vec{m}}(z)$, and so on up to $k-1$ restrictions for $Q_{1,\vec{m}}(z)$. Baker, Oitmaa, and Velgakis [1988] have computed a number of sample cases and found in general no noticeable numerical advantage to using this extended-in-size, but restricted, polynomial structure dictated by argument invariance.

Now we look at the question of adding value invariance [Baker, 1984a] and begin our discussion with first-order integral approximants. It is straightforward to note that from (6.46),

$$g'(z) = \frac{(cb - ad)f'(z)}{(c + df(z))^2}. \tag{6.54}$$

If (6.54) is substituted into the polynomial defining equations (6.24) derived from $g(z)$ we get nonlinear terms. It is easy to see that the set of integral approximants (beyond the case $k = 0$ which are the Padé approximants) does not possess the property of value invariance, but note in passing that the generalized Riccati equation [Baker, Benofy, and Enting, 1986],

$$P_{m+2k}(z)[f'(z)]^k + \sum_{j=-1}^{2k-1} Q_{j,\vec{m}}(z)[f(z)]^{j+1} = O(z^{(2k+2)(m+2)-3}), \tag{6.55}$$

is form invariant. (See also Section 10.7.)

8.6.5 Separation properties

In the case of Padé approximants, we found by Theorem 6.2.2 that horizontal sequences converge for the class of meromorphic functions in compact subsets excluding the singular points of certain disks centered on the origin. For the case of integral approximants, we can find a function class for which it will turn out that we can prove a similar convergence theorem. First we need to define the concept of local monodromic dimension.

Definition 8.6.2. Given a convergent Maclaurin series $f(z)$ and a disk $\mathcal{D} = \{z, |z| \leq R\}$, we say that $f(z)$ has local monodromic dimension m if analytic continuation along all paths in $\mathcal{D}$ generates exactly m linearly independent coverings of $\mathcal{D}$.

Theorem 8.6.9 (Disk monodromy theorem) [Baker, Oitmaa, and Velgakis, 1988]. *Let $f(z)$ be a functional element at $z = 0$ of local monodromic dimension m in disk $\mathcal{D} = \{z, |z| \leq R\}$. Let there be exactly $n < \infty$ singular points a_k of finite order (see (6.7)) in $\mathcal{D}$, and $|a_k| < R$. Then there exist $\rho_j(z)$ such that*

$$\sum_{j=0}^{m} \rho_j(z) f^{(j)}(z) = 0, \qquad (6.56)$$

where ρ_m is a polynomial of finite degree, and $\rho_j(z)$, $j = 0, \ldots, m - 1$, are analytic in $\mathcal{D}$.

Proof. The proof of this theorem follows very closely that of Theorem 8.6.1. The difference comes after (6.17) where the conclusion here is that $\mathcal{P}_{j_0}(z)$ is regular in $\mathcal{D}$. Since $\mathcal{P}_m(z)$ can only have a finite number of zeros in $\mathcal{D}$ by the standard theory of functions of a complex variable, we can factor it into a polynomial which vanishes only at those zeros in $\mathcal{D}$ and another analytic function which has no zeros in $\mathcal{D}$. When we divide the equation by this other function, the conclusion of the theorem follows.

We are now in a position to define the appropriate function class as those functions which have the separation property and finite local monodromic dimension.

Definition 8.6.3 (Separation property) [Baker, Oitmaa, and Velgakis, 1988]. If a function $f(z)$, possibly multivalued, can be written as $f(z) = f_i(z) + f_o(z)$, where for a disk $\mathcal{D} = \{z, |z| \leq R\}$, $f_o(z)$ is analytic for all $z \in \mathcal{D}$ and every analytic continuation of $f_i(z)$ is regular in the finite complex plane outside of $\mathcal{D}$ (i.e., $\mathbb{C}\backslash(\mathcal{D} \cup \infty)$), then $f(z)$ has the separation property.

An elementary example is the class of meromorphic functions, for which convergence can be shown for horizontal sequences of Padé approximants. (See Section 6.2.) Another example is afforded by $f_o(z) = e^z$ and $f_i = \sqrt{(1 - 2z)(1 - 3z)(1 - 4z)}$ with $R = 1$. To see some of the consequences of the separation property, first consider $f_i(z)$. By Riemann's monodromy theorem, Theorem 8.6.1, there must exist polynomials $A_j(z)$ such that

$$\sum_{j=0}^{m} A_j(z)f_i^{(j)}(z) = 0, \qquad (6.57)$$

when $f_i(z)$ has local monodromic dimension m with respect to $\mathcal{D}$. It may be that the A_j's generated by the proof of Theorem 8.6.1 are not minimal, but by arguments parallel to those of Theorem 8.5.2 there does exist an essentially unique minimal set of polynomials of the type generated by the proof of Theorem 8.6.1, and we intend that the $A_j(z)$ be that set. When we add $f_o(z)$ to $f_i(z)$ to get $f(z)$, we get

$$\sum_{j=0}^{m} A_j(z)f^{(j)}(z) = \phi(z), \qquad (6.58)$$

where $\phi(z)$ is regular in $\mathcal{D}$ with the A's an essentially unique minimal solution as noted above since $f_o(z)$ is regular in $\mathcal{D}$. The proof of the convergence results mentioned above will be treated in the appropriate sequence below.

8.6.6 Convergence theory

One of the most important questions to be answered in any consideration of convergence is whether the sequence converges to the correct answer. We can answer this question in the affirmative for both integral and algebraic approximants under certain conditions.

Theorem 8.6.10 (Uniqueness of convergence for integral and algebraic approximants) [Baker, 1984b; Baker and Graves-Morris, 1995]. *Let s be defined as in (5.2). If there exists a sequence of integral or algebraic approximants y_λ of type $\vec{m}_\lambda$ to a functional element defined by its convergent Maclaurin series where $s \to \infty$ with λ such that, in a region $\mathcal{R}$, possibly multiply connected and which contains the origin as an interior point, the quantities*

$$\left| \frac{Q_{j,\vec{m}_\lambda}(z)}{Q_{m,\vec{m}_\lambda}(z)} \right| \qquad (6.59)$$

are uniformly bounded in $\mathfrak{R}$ and λ, and in addition for the case of algebraic approximants, the Sylvester resultant

$$\Delta_1 = \det \begin{vmatrix} Q_{-1} & Q_0 & \cdots & 0 \\ 0 & Q_{-1} & \cdots & 0 \\ \vdots & \vdots & \ddots & \vdots \\ 0 & 0 & \cdots & Q_k \\ Q_0 & 2Q_1 & \cdots & 0 \\ 0 & Q_0 & \cdots & 0 \\ \vdots & \vdots & \ddots & \vdots \\ 0 & 0 & \cdots & (k+1)Q_k \end{vmatrix} \neq 0, \qquad (6.60)$$

uniformly in $\mathfrak{R}$ and λ, then the sequence converges to the complete analytic function defined by the original Maclaurin series in the interior of $\mathfrak{R}$. The $Q_{j,\vec{m}}$ are defined by (6.22) or (6.24) and the corresponding approximants y by (6.23) or (6.25).

Sketch of proof. By the standard estimation theorems for the theory of differential equations [Kamke, 1951], the boundedness conditions (6.59) imply that the integral approximants are uniformly bounded throughout $\mathfrak{R}$ for all λ. Since the $Q_{j,\vec{m}}(z)$ are polynomials, they are regular and therefore the approximant y_λ is regular but not necessarily single valued, as the region may be multiply connected. For the algebraic approximants the boundedness condition (6.59) implies that all the solutions of (6.23) are bounded. Since by (6.60) and by the analysis of (6.19)–(6.21) there are no branch points in $\mathfrak{R}$, the approximant $y_\lambda(z)$ is regular but not necessarily single valued, as the region may be multiply connected. Since the origin is an interior point of $\mathfrak{R}$, there exists a disk about the origin in which all the y_λ are regular. In a disk of half that diameter about the origin, it is easy to show, using uniform boundedness and the fact that it follows from Theorem 8.6.2, or Theorem 8.6.3 with the observation that (6.60) implies that the deficiency is zero, that there exists a λ_0 such that the Maclaurin series of each y_λ, for all $\lambda > \lambda_0$, agrees with that of the original Maclaurin series to any required degree of accuracy, and that $\lim_{\lambda \to \infty} y_\lambda = f(z)$ in that disk. By Vitali's convergence continuation theorem we can conclude that

$$\lim_{\lambda \to \infty} y_\lambda = f(z), \qquad (6.61)$$

in the interior of $\mathfrak{R}$.

We will now consider the results which can be obtained for 'horizontal sequences' (cf. the remark after Definition 8.5.19). A basic theorem here is

***Theorem 8.6.11* (Pointwise convergence) [Baker, 1984b; Baker, Oitmaa, and Velgakis, 1988; Baker and Graves-Morris, 1995].** *Let us be given a functional element $f(z)$ defined by its Maclaurin series.* (i) *Let $f(z)$ satisfy*

$$\sum_{j=0}^{k} P_j(z) f^{(j)}(z) + \phi(z) = 0, \qquad (6.62)$$

where the $P_j(z)$ are polynomials of degree v_j, respectively, $\phi(z)$ is regular in $\mathcal{D} = \{z, |z| \leq R\}$, and all the zeros of $P_k(z)$ are in $\mathcal{D}$, and let the left-hand side of (6.62) be irreducible. Then

$$\lim_{L \to \infty} [L/v] = f(z), \qquad (6.63)$$

in compact subsets of the interior of $\mathcal{D}$ not containing a singular point of $f(z)$. (ii) *Let $f(z)$ satisfy*

$$\sum_{j=0}^{k} P_j(z) [f(z)]^{j+1} + \phi(z) = 0, \qquad (6.64)$$

where the $P_j(z)$ are polynomials of degree v_j respectively, $\phi(z)$ is regular in $\mathcal{D} = \{z, |z| \leq R\}$, and all the zeros $P_k(z)$ are in $\mathcal{D}$, and let the left-hand side of (6.64) be irreducible. Then

$$\lim_{L \to \infty} \langle L/v \rangle = f(z), \qquad (6.65)$$

in compact subsets of the interior of $\mathcal{D}$ not containing a singular point of $f(z)$.

Sketch of proof. The proof depends on a detailed analysis of the exact structure of the coefficients and a detailed analysis of the explicit solution for the approximants. The relevant polynomials are given by (5.4). If we use elementary column operations and the equations (6.62) and (6.64), respectively, we can reduce the Q_j's to a form where the column containing z^{v_j} is replaced by

$$p_{v_j} \begin{bmatrix} \phi_{L+1-v_j} \\ \vdots \\ \phi_{s-v_j} \\ P_j(z)/p_{v_j} \end{bmatrix}. \qquad (6.66)$$

Here p_{v_j} is the coefficient of z^{v_j} in $P_j(z)$. Since, by (6.62), there are v_k singularities of $[L/v]$ in $\mathcal{D}$, or alternatively at least v_k singularities of $\langle L/v \rangle$ in $\mathcal{D}$ since $P_k(z)$ divides the Sylvester resultant (6.60) for (6.64), we can write [Darboux, 1878]

$$f_n = \sum_{j=1}^{} \sum_{k}^{} \binom{-\gamma_j + k}{n} A_{j,k} z_j^n + g_n, \qquad (6.67)$$

where the z_j's are the singular points of $f(z)$ in $\mathscr{D}$ and g_n is the contribution from $\phi(z)$ which has a radius of convergence greater than R. In the computation of the subdeterminants, there will remain, since (6.62) and (6.64) are irreducible by hypothesis, singular terms which cannot be canceled out completely, so the worst divisor that we have to deal with in the solution for $P_j(z)$ is of the form

$$\frac{1}{n^{\text{fixed power}} R_{v_k}^n}, \tag{6.68}$$

where R_{v_k} is the largest absolute value of the singularities of $f(z)$ in $\mathscr{D}$. Since g_n vanishes faster than R^{-n} and $R_{v_k} < R$ by hypothesis, the error term in the $Q_{j,\mathbf{m}}(z)$ goes to zero geometrically. (The details of this argument are like those of Wilson's theorem, Theorem 11.2 in Baker [1975a].)

It now remains to show that $Q_{-1,L}(z)$ converges to $\phi(z)$ as $L \to \infty$. From Cramer's rule, we can easily deduce that

$$Q_{-1,L}(z) \propto$$

$$\det \begin{vmatrix} f_{L+1}^{(0)} & \cdots & f_{L+1-m_0}^{(0)} & \cdots & f_{L+1}^{(k)} & \cdots & f_{L+1-m_k}^{(k)} \\ \vdots & \ddots & \vdots & & \vdots & \ddots & \vdots \\ f_{L+M}^{(0)} & \cdots & f_{L+M-m_0}^{(0)} & \cdots & f_{L+M}^{(k)} & \cdots & f_{L+M-m_k}^{(k)} \\ \sum_{j=0}^{L} f_j^{(0)} z^j & \cdots & \sum_{j=m_0}^{L} f_{j-m_0}^{(0)} z^j & \cdots & \sum_{j=0}^{L} f_j^{(k)} z^j & \cdots & \sum_{j=m_k}^{L} f_{j-m_k}^{(k)} z^j \end{vmatrix}.$$

(6.69)

We also replace $f_n^{(j)}$ by the series coefficients of $[f(z)]^{j+1}$ in the algebraic case. By use of the coefficients of the $P_j(z)$, in a manner similar to that used to produce (6.66) above, we can reduce (6.69) to a form where its last column is replaced by

$$p_{m_k}^{-1} \begin{bmatrix} \phi_{L+1-m_k} \\ \vdots \\ \phi_{L+M-m_k} \\ p_{m_k} \sum_{j=0}^{L} \phi_j z^j \end{bmatrix}, \tag{6.70}$$

where p_{m_k} is the coefficient of z^{m_k} in $P_k(z)$. If we normalize $Q_{-1,L}(z)$ by dividing (6.69) by the determinant of the matrix U, where U is formed by deleting the last row and column in (6.69), then we can write

$$Q_{-1,L}(z) = \mathbf{f}^T U^{-1} \boldsymbol{\phi} + \sum_{j=0}^{L} \phi_j z^j \tag{6.71}$$

by standard methods of matrix theory, where

$$\mathbf{f}^T = \left(\sum_{j=0}^{L} f_j^{(0)} z^j, \ldots, \sum_{j=m_0}^{L} f_{j-m_0}^{(0)} z^j, \ldots, \sum_{j=0}^{L} f_j^{(k)} z^j, \ldots, \sum_{j=m_k}^{L} f_{j-m_k}^{(k)} z^j \right),$$

$$\boldsymbol{\phi} = (\phi_{L+1-m_k}, \ldots, \phi_{L+M-m_k}). \tag{6.72}$$

By Cauchy's inequalities, we know that $|\phi_j| \leq K/R^j$, for some fixed finite K. Let us now examine $\mathbf{f}^T U^{-1}$. By defining $f_j^{(i)} = 0$ for $j < 0$, we can write

$$\mathbf{f}^T U^{-1} = \sum_{j=0}^{L} z^j \mathbf{f}_j^T U^{-1}, \tag{6.73}$$

where

$$\mathbf{f}_j^T = (f_j^{(0)}, \ldots, f_{j-m_0}^{(0)}, \ldots, f_j^{(k)}, \ldots, f_{j-m_k}^{(k)}). \tag{6.74}$$

Now $|\mathbf{f}_j|$ is of the order of r_1^j where $r_j = |a_j|$ are the roots of $P_k(z)$ arranged in order of distance from the origin for the integral case and the location of the singularities for the algebraic case. If the singularity nearest to the origin cannot be separated from the other $m_k - 1$ singularities, then in the limit as $L \to \infty$ we can solve for all M polynomial coefficients from the details of the behavior at $z \approx a_1$ via its contribution to the expansion (6.67). The binomial coefficients $\binom{\gamma}{n} \sim n^{-(\gamma+1)}$ and are expandable in powers of n^{-1} for large n. The rest of the properties, such as exact representations at the other singular points, will be automatically built in. This case is the usual case for algebraic approximants and is a consequence of the hypotheses of this theorem. Otherwise, (6.64) factors into analytically separate factors, the solutions of which cannot be reached by analytic continuation of the solutions from the other factors. Hence the equation would not be irreducible in our sense. Our analysis of this case also holds in the integral-approximant case for a_1 being a confluent singular point or a logarithmic singularity, but the details are somewhat lengthy. In the current case, we estimate that $U^{-1} = O(L^{\text{fixed power}} r_1^L)$, so we infer from (6.71) and (6.73) that

$$Q_{-1,L}(z) = O([\text{constant} + (z/R)^L] r_1^L L^{\text{fixed power}} R^{-L}) + \text{constant},$$
$$= O(L^{\text{fixed power}} [(z/R)^L + (r_1/R)^L]) + \text{constant}, \tag{6.75}$$

which shows that $Q_{-1,L}(z)$ is bounded uniformly in $\mathcal{D}$ and L for the current case. The other case which we need to consider in detail is that where all the m_k singular points are separable from each other. Consider the reduced problem in which the singularity is at a_{m_k}. (To simplify our presentation we will assume that $r_{m_k} > r_{m_k-1}$.) For this reduced problem, there is necessarily an essentially unique differential multiplier of type

8.6 Integral and algebraic approximants

$\vec{\mu} < \vec{m}$ in the sense of Definition 8.5.3. In particular $\mu_k < m_k$ because there is one fewer singularity and the differential-multiplier coefficient $Q_{k,\mu_k}(a_{m_k}) \neq 0$. If we define ψ analogously to ϕ of (6.72) but consisting of the contribution to the right-hand side of the reduced problem of the m_kth singularity, then we may write

$$U\mathbf{y} = \phi + \psi, \tag{6.76}$$

where $\mathbf{y}$ is the vector of coefficients for the reduced problem of the type $\vec{\mu} = \vec{m} - \vec{e}_k$. The corresponding $\hat{Q}_{-1,L}(z)$ for the reduced problem is given by

$$\hat{Q}_{-1,L}(z) - \sum_{j=0}^{L}(\phi_j + \psi_j)z^j = \mathbf{f}^T U^{-1}(\phi + \psi),$$

$$= \mathbf{f}^T \cdot \mathbf{y} = O\left(\sum_{j=0}^{L} z^j(\psi_j + \phi_j)\right). \tag{6.77}$$

This result shows that $O(\mathbf{f}^T U^{-1})$ is independent of the contributions of the first $m_k - 1$ singularities, so we get for the full problem, in this case,

$$Q_{-1,L}(z) = O([\text{constant} + (z/r_{m_k})^L]r_{m_k}^L L^{\text{fixed power}} R^{-L}) + \text{constant},$$

$$= O(L^{\text{fixed power}}[(z/R)^L + (r_{m_k}/R)^L]) + \text{constant}, \tag{6.78}$$

which again leaves $Q_{-1,L}(z)$ uniformly bounded for all L and $z \in \mathcal{D}$. There are all sorts of special cases, but it always works out that the uncanceled part of $\mathbf{f}_j$ is balanced against the magnitude of U^{-1}, leaving the z^j and the R^{-L} from the ϕ_L term to force the vanishing of the large L contributions.

Finally, the convergence of the series expansion for $Q_{-1,L}(z)$ over the disk $|z| < \frac{1}{2}r_1$, plus the uniform boundedness just established, and Vitali's convergence continuation theorem, show that $Q_{-1,L}(z) \underset{L \to \infty}{\rightrightarrows} \phi(z)$, which are the results necessary to complete the proof of the theorem.

Baker and Lubinsky [1987] have studied extensively the convergence of horizontal sequences of both integral and algebraic approximants to meromorphic functions. In order to state their results in a systematic manner, they always impose the following conditions. Let $f(z)$ be analytic at $z = 0$ and meromorphic in $\mathcal{D} = \{z, |z| < R\}$, $0 < R \leq \infty$, with l distinct poles $z_1, \ldots, z_l$ of multiplicities $p_1, \ldots, p_l$, respectively. They define the auxiliary quantities

$$p = \sum_{j=1}^{l} p_j, \quad S_1(z) = \prod_{j=1}^{l}(z - z_j), \quad S(z) = \prod_{j=1}^{l}(z - z_j)^{p_j}. \tag{6.79}$$

In addition, define m as after (6.22) and M as in (5.2), but impose the condition $M = p + kl$ for integral approximants and $M = (k + 1)p$ for algebraic approximants. One of the keys to their results is in the following lemma. Before stating it we remember that we have proven the existence of the relevant integral and algebraic polynomials in Section 8.5 and the uniqueness of the corresponding minimal polynomials of every type (L, m).

Lemma 2. *Assume that $f(z)$ and* m *satisfy the Baker–Lubinsky conditions. Let us normalize the $Q_{j,\vec{m}}(z)$ for $j = 0, \ldots, k$ such that all of their coefficients have absolute magnitude at most unity, with equality for at least one coefficient. Let $\mathcal{S}$ be any infinite sequence of integers. Then we can always find a subsequence $\mathcal{S}'$ such that*

$$\lim_{\substack{L \to \infty \\ L \in \mathcal{S}'}} Q_{j,\vec{m}}(z) = Q^*_{j,\text{m}}(z), \quad j = 0, 1, \ldots, k, \tag{6.80}$$

uniformly on compact subsets of the complex plane $\mathbb{C}$. Here $\vec{Q}^* = (Q^*_{0,\text{m}}(z), \ldots, Q^*_{k,\text{m}}(z))$ is a differential or algebraic multiplier, respectively, for $f(z)$ in $\mathcal{D}$. Furthermore,

$$\lim_{\substack{L \to \infty \\ L \in \mathcal{S}'}} Q_{-1,\vec{m}}(z) = Q^*_{-1,\text{m}}(z) \tag{6.81}$$

uniformly in compact subsets of $\mathcal{D}$, where $Q^*_{-1,\text{m}}(z)$ is analytic in $\mathcal{D}$ and

$$Q^*_{-1,\text{m}}(z) + \sum_{j=0}^{k} f^{(j)}(z) Q^*_{j,\text{m}}(z) = 0 \tag{6.82a}$$

for integral polynomials, or for the algebraic case

$$Q^*_{-1,\text{m}}(z) + \sum_{j=0}^{k} [f(z)]^{j+1} Q^*_{j,\text{m}}(z) = 0. \tag{6.82b}$$

Proof. Since the $Q_{j,\vec{m}}$, $j = 0, \ldots, k$, have just $M + 1$ coefficients which are all bounded in absolute value by unity, by standard arguments on compact sets we may choose a subsequence for which the first coefficient converges, and from that, a subsequence for which the second coefficient converges, and so forth. Thus (6.80) holds. The limiting polynomials $\vec{Q}^*_\text{m}$ have degree at most m and $\vec{Q}^*_\text{m} \neq \vec{0}$ because at least one coefficient has absolute value unity.

Let us define $T(z) = S_1^k(z) S(z)$ for the integral case and $T(z) = S^{k+1}(z)$ for the algebraic case. These are polynomials of degrees $p + kl$ and $p(k+1)$, respectively. Also let $f_{-1}(z) = 1$ and for $j = 0, \ldots, k$,

8.6 Integral and algebraic approximants

$f_j(z) = f^{(j)}(z)$ for the integral case and $f_j(z) = [f(z)]^{j+1}$ for the algebraic case. Consider

$$T(z)\left\{Q_{-1,\vec{m}}(z) + \sum_{j=0}^{k} f_j(z)Q_{j,\vec{m}}(z)\right\}/z^{L+M+1} \qquad (6.83a)$$

which is analytic in $\mathcal{D}$, and by Cauchy's formula for $r < R$ can be re-expressed as

$$= \oint_{|t|=r} dt \frac{T(t)\{Q_{-1,\vec{m}}(t) + \sum_{j=0}^{k} f_j(t)Q_{j,\vec{m}}(t)\}}{t^{L+M+1}(t-z)} \qquad (6.83b)$$

$$= \oint_{|t|=r} dt \frac{T(t)\sum_{j=0}^{k} f_j(t)Q_{j,\vec{m}}(t)}{t^{L+M+1}(t-z)}, \qquad (6.83c)$$

since, for fixed z,

$$T(t)\{Q_{-1,\vec{m}}(t)/[t^{L+M+1}(t-z)]\} = O(t^{-2}), \quad |t| \to \infty, \qquad (6.84)$$

by the Baker–Lubinsky conditions and the left-hand side is analytic in t for all $|t| > r$. By distorting the contour in (6.83c) to a circle arbitrarily near to $|t| = R$ and by the use of the maximum-modulus principle on (6.83a) we conclude that, for $|z| < r' < r < R$,

$$\limsup_{\substack{L \to \infty \\ L \in \mathcal{S}'}} \left| T(z) \sum_{j=-1}^{k} f_j(z)Q_{j,\vec{m}}(z) \right|^{1/L} \leq \frac{r'}{R}. \qquad (6.85)$$

This result, together with (6.80), shows that

$$\lim_{\substack{L \to \infty \\ L \in \mathcal{S}'}} Q^*_{-1,\vec{m}}(z) + \sum_{j=0}^{k} f_{j+1}(z)Q^*_{j,m}(z) = 0 \qquad (6.86)$$

in compact subsets of $\mathcal{D}\backslash\{z_1, \ldots, z_l\}$. However, $Q_{-1,\vec{m}}(z)$ is a sequence of polynomials and therefore analytic functions which are bounded in $|z| \leq r' < R$, with the convergence properties just described. Therefore they converge uniformly in compact subsets of $\mathcal{D}$ to a limit $Q^*_{-1,m}(z)$ which is analytic in $\mathcal{D}$. Therefore, the remaining sum in (6.82) must be analytic as well and since $\vec{Q}_m \neq \vec{0}$, $\vec{Q}_m(z)$ is an integral or an algebraic multiplier of type $\mathbf{m}$ for $f(z)$ in $\mathcal{D}$.

In Corollary 4 of Section 8.5 we proved that there always exists a well-behaved horizontal sequence of Hermite–Padé polynomials. In the following theorem, we prove that with the Baker–Lubinsky conditions, not only does (after at most a finite number of exceptions) the whole sequence exist, but it also converges.

Theorem 8.6.12 **(Existence and convergence of approximants) [Baker and Lubinsky, 1987].** *Assume that the Baker–Lubinsky conditions hold; then there exists a differential or an algebraic multiplier $\vec{Q}^*_{\mathbf{m}}$ of type* **m**. *If in addition this multiplier is unique, then, for $L = m_{-1}$ large enough, there exist essentially unique integral or algebraic polynomials $\vec{Q}_{\vec{m}}$ of type $\vec{m}$. With suitable normalization,*

$$\lim_{L \to \infty} Q_{j,\vec{m}}(z) = Q^*_{j,\mathbf{m}}(z), \quad j = 0, \ldots, k, \qquad (6.87a)$$

in $\mathbb{C}$ and

$$\lim_{L \to \infty} Q_{-1,\vec{m}}(z) = Q^*_{-1,\mathbf{m}}(z) \qquad (6.87b)$$

uniformly on compact sets of $\mathcal{D} = \{z, |z| < R\}$. $Q^*_{-1,\mathbf{m}}(z)$ is analytic in $\mathcal{D}$ and satisfies

$$Q^*_{-1,\mathbf{m}}(z) + \sum_{j=0}^{k} f_j(z) Q^*_{j,\mathbf{m}}(z) = 0, \quad z \in \mathcal{D}, \qquad (6.88)$$

where $f_{-1}(z) = 1$ and $f_j(z) = f^{(j)}(z)$ in the integral case and $f_j(z) = [f(z)]^{j+1}$ in the algebraic case.

Proof. The existence and convergence of the Q's follow from Lemma 2. Next we need to show that the Hermite–Padé polynomials are essentially unique. Suppose that for an infinite set of L's there exist two linearly independent sets of Hermite–Padé polynomials, $\vec{Q}^{\#}_{\vec{m}}$ and $\vec{Q}_{\vec{m}}$, both of type $(L, \mathbf{m})$. It follows easily from the defining equations that $\vec{Q}^{\#}_{\mathbf{m}}$ and $\vec{Q}_{\mathbf{m}}$ are linearly independent as well. By (6.80) we may normalize $\vec{Q}^{\#}_{\mathbf{m}}$ and $\vec{Q}_{\mathbf{m}}$ so that as $L \to \infty$ (at least in some subsequence $\mathcal{S}''$) both $\vec{Q}^{\#}_{\mathbf{m}}$ and $\vec{Q}_{\mathbf{m}}$ converge to the unique differential multiplier. Choose a j such that $Q_{j,\mathbf{m}}(z) \neq 0$ and an integer s such that the coefficient of z^s is nonzero. Then for L sufficiently large, the coefficients of z^s in $Q_{j,\vec{m}}(z)$ and $Q^{\#}_{j,\vec{m}}(z)$ both converge to the same nonzero number. Hence we can select an α_L for L large enough so that the coefficient of z^s in $\alpha_L Q^{\#}_{j,\vec{m}}(z) + (1 - \alpha_L) Q_{j,\vec{m}}(z)$ is zero and hence converges to zero. By linear independence $\alpha_L \vec{Q}^{\#}_{\vec{m}}(z) + (1 - \alpha_L) \vec{Q}_{\vec{m}}(z)$ is also a sequence of Hermite–Padé polynomials. But by Lemma 2 any sequence $\mathcal{S}$ must contain at least a subsequence which converges to the unique multiplier. Since that multiplier has a nonzero coefficient of z^s, there is a contradiction. Therefore, with at most a finite number of exceptions, the whole sequence exists and it converges to the unique multiplier, and so the Hermite–Padé polynomials are essentially unique for L sufficiently large.

Definition 8.6.4. A differential or algebraic multiplier is said to be pole matching if all the zeros of $Q^*_{j,\mathfrak{m}}(z)$ in (6.80) are also zeros of $S(z)$ in (6.79).

Theorem 8.6.13 (Convergence theorem) [Baker and Lubinsky, 1987].
Assume that the Baker–Lubinsky conditions hold and that the differential or algebraic multiplier is unique and pole matching for $f(z)$ in $\mathfrak{D} = \{z, |z| < R\}$. In the algebraic case we assume in addition that the Sylvester resultant (6.60) does not vanish for $z = 0$. Then, for L large enough, $[L/\mathfrak{m}]$ or $\langle L/\mathfrak{m}\rangle$ exists and is uniquely defined in a neighborhood of the origin. It may be analytically continued to a single-valued, analytic function in any open set in $\mathfrak{D}$ whose closure does not contain any of the poles z_i of $f(z)$. Also

$$\lim_{L\to\infty}[L/\mathfrak{m}] = f(z), \text{ or } \lim_{L\to\infty}\langle L/\mathfrak{m}\rangle = f(z), \qquad (6.89)$$

uniformly on compact sets of $\mathfrak{D}\setminus\{z_i\}$. If we normalize so that $Q^*_{k,\mathfrak{m}}(z)$ is a monic polynomial, then in any compact subset $\mathcal{K} \subset \mathbb{C}$,

$$\limsup_{L\to\infty}\|Q_{j,\vec{m}} - Q^*_{j,\mathfrak{m}}\|_{\mathcal{K}}^{1/L} \leq (\max_i\{|z_i|\})/R, \quad j = 0,\ldots,k, \qquad (6.90)$$

where we use the sup norm. When $\mathcal{K} \subset \mathfrak{D}$,

$$\limsup_{L\to\infty}\|Q_{-1,\vec{m}} - Q^*_{-1,\mathfrak{m}}\|_{\mathcal{K}}^{1/L} \leq \|z\|_{\mathcal{K}}^{1/L}/R. \qquad (6.91)$$

Further, if $\mathcal{K}$ contains no poles of $f(z)$, and in the algebraic case no zeros of the Sylvester resultant $\Delta^*_1(z)$ constructed from the $Q^*_{j,\mathfrak{m}}(z)$, then

$$\limsup_{L\to\infty}\|[L/\mathfrak{m}] - f(z)\|_{\mathcal{K}}^{1/L} \leq \|z\|_{\mathcal{K}}/R, \quad \text{or}$$

$$\limsup_{L\to\infty}\|\langle L/\mathfrak{m}\rangle - f(z)\|_{\mathcal{K}}^{1/L} \leq \|z\|_{\mathcal{K}}/R. \qquad (6.92)$$

The proof of this theorem follows from Lemma 2 and Theorems 8.6.10 and 8.6.12, plus error-estimate calculations derived from the results in the proof of Lemma 2.

Baker and Lubinsky [1987] have derived conditions for the existence of unique multipliers. Specifically they prove that there is always a unique differential multiplier of type

$$\mathfrak{m} = (p, l-1, \ldots, l-1) \text{ or} \qquad (6.93a)$$

$$\mathfrak{m} = (p-1, \underbrace{l-1, \ldots, l-1}_{n-s}, \underbrace{-1, \ldots, -1}_{t-1}, tl, \underbrace{l-1, \ldots, l-1}_{k-n}),$$

$$(6.93b)$$

where $1 \leq t \leq n \leq k$ and l is defined in (6.79). In the special case $n = k$, the differential multiplier has also been proven to be pole matching and $Q_{k,\mathfrak{m}}(z) = [S_1(z)]^l$. In the algebraic case they have shown that there is always a unique algebraic multiplier of the type

$$\mathfrak{m} = (\underbrace{p-1, \ldots, p-1}_{n-t+1}, \underbrace{-1, \ldots, -1}_{t-1}, tp, \underbrace{p-1, \ldots, p-1}_{k-n},)$$

(6.94)

where $1 \leq t \leq n \leq k$. Again in the special case $n = k$ the algebraic multiplier is pole matching and $Q_{k,\mathfrak{m}}(z) = [S(z)]^t$.

These results parallel the results of de Montessus for Padé approximants to meromorphic functions given in Sections 6.2–6.4 and show that these more general approximants retain the convergence properties of the Padé approximants in this case.

Theorem 8.6.14 (Separation property convergence theorem) [Baker, Oitmaa, and Velgakis, 1988]. *Let $f(z)$ have the separation property (Definition 8.6.3) with respect to a disk $\mathcal{D}$, a finite number of singular points a_i in the interior of $\mathcal{D}$, but none on the boundary of $\mathcal{D}$, and $|a_i| > 0$ for all i. Assume further that all these singular points plus the point at infinity are of finite order (6.7) for $f_i(z)$. Let $f_i(z)$ be of exact monodromic dimension m. Then (i) there exists a unique differential multiplier whose coefficients are polynomials of degrees $\mathfrak{m}$ which is minimal in the sense of Definition 8.5.1, and (ii) the integral approximants $[L/\mathfrak{m}]$ converge to $f(z)$ on simply connected, compact subsets of $\mathcal{D}\setminus\{a_i\}$ which contain the origin.*

Proof. By (6.57) there must exist a differential multiplier, and, by the same type of arguments as those which lead to Theorem 8.5.2, there must be an essentially unique minimal differential multiplier. Thus we can conclude (i). The second part of the theorem follows directly from Theorem 8.6.11.

Baker and Graves-Morris [1991, 1993b] have established convergence results for the following case, which is perhaps the simplest one, which does not have the separation property. Let

$$(1-z)f'(z) + G(z)f(z) = H(z), \qquad (6.95)$$

where $H(z)$ and $G(z) = \sum_{i=1}^{\infty} G_i(1-z)^i$ are analytic (and not just polynomials) in the disk $\mathcal{D} = \{z, |z| < \rho\}$ for some $\rho > 1$. We assume that $\gamma = -G_i$ is not an integer. The solution of (6.95) is

$$f(z) = A(z)(1-z)^{-\gamma} + B(z) \qquad (6.96)$$

by the standard theory of differential equations, where $A(z)$ and $B(z)$ are analytic in $\mathcal{D}$. We further assume that $A(1) \neq 0$. If we write out explicitly the integral polynomials of type $(L - 1/M - 1, 1)$ and their defining equation as

$$[\lambda^{(L)}(1 - z) + \alpha^{(L)}]f'(z) + g^{(L)}(z)f(z) = h^{(L)}(z) + O(z^{L+M+1}), \quad (6.97)$$

where

$$g^{(L)}(z) = \sum_{i=0}^{M-1} g_i^{(L)}(1 - z)^i, \quad h^{(L)}(z) = \sum_{i=0}^{L-1} h_i^{(L)} z^i, \quad (6.98)$$

then the approximant $[L - 1/M - 1; 1] = y(z)$ is the solution of

$$[\lambda^{(L)}(1 - z) + \alpha^{(L)}]y'(z) + g^{(L)}(z)y(z) = h^{(L)}(z). \quad (6.99)$$

Theorem 8.6.15 (First-order, integral polynomial theorem) [Baker and Graves-Morris, 1991]. *Under the above given assumptions, for sufficiently large L, and normalizing $\lambda^{(L)} = 1$, we have*

$$\alpha^{(L)} = O(L^{-M-1}),$$

$$g_i^{(L)} = G_i + O(L^{-M+i}), \quad i = 0, 1, \ldots, M - 1, \quad (6.100)$$

and therefore

$$\lim_{L \to \infty} g^{(L)}(z) = \sum_{i=1}^{M-1} G_i(1 - z)^i,$$

$$\lim_{L \to \infty} h^{(L)}(z) = H(z) - \left\{ G(z) - \sum_{i=0}^{M-1} (1 - z)^i G_i \right\} f(z). \quad (6.101)$$

The proof of this theorem is fairly lengthy. It follows the spirit of the proof of Theorem 8.6.11. The determinantal solution is set out explicitly and the series terms are expressed by means of Darboux's result (6.67). The polynomial defining equations are reduced to an upper triangular form by an appropriate series of row operations and then they are solved for the error terms by back substitution. These results combined with a considerable number of algebraic manipulations and an appeal to the Bolzano–Weierstrass theorem lead to the conclusion of the theorem.

Theorem 8.6.16 (First-order, integral approximant theorem) [Baker and Graves-Morris, 1991, 1993b]. *Let $f(z)$ satisfy the above given conditions. Then*

$$\lim_{M \to \infty} \lim_{L \to \infty} [L/M; 1] = f(z) \quad (6.102)$$

on compact subsets of $|z| < 1$ and also on any compact set of a (finitely numbered) Riemann surface accessible from the disk $|z| \leq \rho$ with support in $\{|z| < 1\} \cap \{|1 - z| < \rho - 1\}$. For L large enough,

$$\gamma^{(L)} = \gamma + O(L^{-M}). \tag{6.103}$$

Further, for any given z in $1 < |z| < \rho$, and for any fixed $M \geq 0$,

$$\lim_{L \to \infty} |f(z) - [L - 1/M - 1; 1]| = \infty. \tag{6.104}$$

Sketch of proof. The proof of the convergence in $|z| < 1$ is a straightforward consequence of Theorem 8.6.10 and the estimates of Theorem 8.6.15. To get the convergence on higher Riemann sheets, the explicit solution is written out and separated into an analytic part and an analytic part times $(1 + \alpha^{(L)} - z)^{\gamma^{(L)}}$. The analytic part of the solution is no trouble as the area of concern is in $|z| < 1$. The singular part, when expanded about $z = 1$, converges inside the radius of convergence of $G(z)$. The estimates of Theorem 8.6.15 are sufficient to permit all the various errors to be forced to zero. Result (6.103) is given by (6.100).

The proof of the divergence outside the region of convergence depends on the Lagrange–Hermite interpolation formula (6.3.1), which gives an integral representation of the first terms of a Maclaurin expansion of an analytic function. With this formula we can develop an integral expression for the error term in (6.97) as we did in (6.3.2) for Padé approximants. By careful analysis and use of the value of the beta-function integral, we can show that there is a dominant term which overpowers all the other terms together and show that it diverges in the region indicated.

In addition Baker and Graves-Morris [1991] were able to give the corresponding estimates for a second-order differential equation in place of (6.95). When estimates of this type are available for a differential equation of any order, then they are sufficient to show that

$$\lim_{L \to \infty} [L/\text{m}] = f(z), \quad |z| < 1, \quad \text{m fixed and finite}. \tag{6.105}$$

These results go beyond those for the separation-property theorem in that, in the present cases, only ρ_m is a polynomial and the rest of the ρ's are analytic functions in the disk, whereas in Theorem 8.6.14 all the ρ's are polynomials. These results (even if the estimates could be given in general, which plausibly is possible) do not yet answer the convergence questions when the degree of ρ_m is larger than unity.

In some sense, Theorem 8.6.14 is the best possible. We next give a theorem which extends some of the results of the previous theorem and shows that lack of the separation property reduces the region of convergence.

Theorem 8.6.17 (**Separation property divergence theorem**) [**Baker and Graves-Morris, 1995**]. *Let $f(z)$ have local monodromic dimension k plus an analytic background in disk $\mathcal{D} = \{z, |z| \leq R\}$, and have n singularities a_j of finite order in the interior of $\mathcal{D}$. Suppose that at least one singularity a_v in $\mathcal{D}$ is nonseparable. Then there exists at least a convergent sequence of the integral polynomials. However, any sequence of type $[L/\mathfrak{m}]$ with integral approximants and fixed $\mathfrak{m}$ (order k), which converges in the neighborhood of zero, diverges in $\mathcal{D}$ if $|z| > |a_v|$.*

Proof. Let us normalize the $Q_{j,m_j}(z)$, $j = 0, \ldots, k$, such that the largest coefficient in the polynomials in absolute value in this set of polynomials is exactly unity. Thus all these coefficients lie in the unit disk. By standard compactness arguments, there must exist a subsequence of the L's such that every coefficient converges. Let us call these limiting polynomials $Q^*_{j,m_j}(z)$, $j = 0, \ldots, k$. Since by Theorem 8.6.9 and an argument similar to the proof of Theorem 8.5.2, there exists an essentially unique, minimal differential multiplier for this problem, and since one of the singularities a_v is nonseparable, then at least one of the ρ_j of (6.56) is not a polynomial. If in fact they are all polynomials (remembering that by minimality all the zeros of $\rho_k(z)$ are in $\mathcal{D}$), then we could deduce the separation property by the direct integration (6.56). Thus this sequence cannot converge to the differential multiplier. Therefore, for the subsequence above,

$$\lim \sum_{j=0}^{k} Q_{j,m_j}(z) f^{(j)}(z) = \sum_{j=0}^{k} Q^*_{j,m_j}(z) f^{(j)}(z) = g(z)$$

is not analytic in $\mathcal{D}$. In particular, it must be that the point a_v is a singularity of $g(z)$. Since each coefficient of $Q_{-1,v}(z)$ converges to the corresponding coefficient of $g(z)$, it must be that there is an eventual violation of the Cauchy inequalities derivable from the assumption that the $Q_{-1,\lambda}(z)$ converge in any disk $|z| \leq |a_v| + \varepsilon$ for any $\varepsilon > 0$. Thus we can deduce the divergence of the limit of the $Q_{-1,\lambda}(z)$ outside of $|z| \leq |a_v|$. This divergence, by the integration of (6.25), is sufficient to cause the divergence of the limit of the integral approximants for $|z| > |a_v|$. But this result holds for any convergent sequence. Hence we have proven the theorem.

***Theorem 8.6.18* [Baker and Graves-Morris, 1995].** *Let $f(z)$ have local monodromic dimension k plus an analytic background in disk $\mathcal{D} = \{z, |z| \leq R\}$, and have n singularities a_j of finite order in the interior of $\mathcal{D}$. Consider any sequence $[m_{-1,\lambda}/\mathfrak{m}_\lambda]$ for which, for one value of $j = \theta$, the $m_{\theta,\lambda}$ are uniformly less than the corresponding degree in the essentially unique differential multiplier for $f(z)$ in $\mathcal{D}$. The other m's may go to ∞, but we assume that $m_{j,\lambda+1} \geq m_{j,\lambda}$ for all j and λ. Then there exists at least a convergent subsequence. For any such subsequence, the error must diverge somewhere in $\mathcal{D}\setminus\{a_j\}$.*

Proof. We choose to normalize the $Q_{j,m_{j,\lambda}}(z)$ so that

$$\max\left(|(R/2)^i q_{\theta,i,\lambda}|, i = 0, \ldots, m_\theta, \sum_{i=0}^{m_{j,\lambda}}|(R/2)^i q_{j,i,\lambda}|, j = -1, \ldots, k, j \neq \theta\right)$$

$$= 1. \quad (6.106)$$

Since the number of polynomial coefficients is denumerable, and $m_{j,\lambda+1} \geq m_{j,\lambda}$, we may select a subsequence in which the first coefficient converges, as it is bounded in absolute value. From this subsequence we can select another in which the second coefficient converges and so forth for all the coefficients. If the limiting form of any of the polynomials is an infinite sequence, it must by (6.106) converge in $|z| < R/2$. The limit of the integral polynomials cannot vanish everywhere in $\mathcal{D}$ because of the normalization. Also the normalization condition and the maximum-modulus theorem imply that the resultant limits of the Q's are bounded on compact subsets of $|z| < R/2$, $z \notin \{a_j\}$. Since all the Q's and $f(z)$ are bounded and analytic inside $|z| < \frac{1}{2}|a_1|$, the resulting forms are analytic in that disk. However, since the differential multiplier is essentially unique, and by hypothesis the form of the sequence cannot match it, the sequence limit

$$\lim_{\lambda \to \infty}\left[\sum_{j=0}^{k} Q_{j,m_{j,\lambda}}(z) f^{(j)}(z) + Q_{-1,L_\lambda}(z)\right] = g(z) \quad (6.107)$$

cannot be analytic in all of $\mathcal{D}\setminus\{a_j\}$ and so forces some of the Q's to diverge somewhere, and also the corresponding integral approximants. Note is made that the defect in degree cannot be cured by division to remove one of the factors of $P_\theta(z)$ and so escape the degree restriction, as it must cause at least one of the integral polynomials to diverge somewhere in $\mathcal{D}$, since by hypothesis we began with a minimal multiplier.

In this subsection we have treated so far the convergence problems for horizontal sequences and sequences in which the degree of at least one of

8.6 Integral and algebraic approximants

the polynomials remains finite. We will now examine the case of diagonal or near-diagonal sequences in which the degrees of all the coefficients go to infinity together. The results here are generally more powerful and the region of convergence is no longer restricted to a disk, but the mode of convergence is no longer pointwise. First we give a theorem which helps set the stage for the theory by establishing a limitation.

Theorem 8.6.19 (**Baker's nonconvergence theorem**) [Baker, 1988b]. *A sequence of kth-order integral or algebraic approximants of type $\vec{m}$ cannot converge simultaneously on $k + 2$ linearly independent coverings of the complex plane to an $f(z)$ which has a monodromic dimension greater than $k + 1$. (Note that a uniform function plus a function of monodromic dimension greater than k belongs to this class.)*

Proof. On any Riemann sheet we may write

$$Q_{-1,m_{-1}}(z) + \sum_{j=0}^{k} Q_{j,m_j}(z) f^{(j)}(z^{(i)}) = R_{\vec{m}}(z^{(i)}), \qquad (6.108)$$

where $f^{(j)}(z)$ denotes the usual derivative in the integral case and $[f(z)]^{j+1}$ in the algebraic case, $\pi(z^{(i)}) = z$ is the projection onto the complex plane of the ith Riemann surface, and $R_{\vec{m}}(z)$ is the remainder. If we solve (6.108), using $k + 2$ coverings $f(z^{(i)})$, $i = 1, \ldots, k + 2$, for the $\vec{Q}_{\vec{m}}(z)$ in terms of the $R_{\vec{m}}(z^{(i)})$, we get

$$Q_{k,m_k}(z) = \frac{\det \begin{vmatrix} R_{\vec{m}}(z^{(1)}) & f^{(k-1)}(z^{(1)}) & \cdots & f(z^{(1)}) & 1 \\ \vdots & \vdots & & \vdots & \vdots \\ R_{\vec{m}}(z^{(k+2)}) & f^{(k-1)}(z^{(k+2)}) & \cdots & f(z^{(k+2)}) & 1 \end{vmatrix}}{\det \begin{vmatrix} f^{(k)}(z^{(1)}) & \cdots & f(z^{(1)}) & 1 \\ \vdots & & \vdots & \vdots \\ f^{(k)}(z^{(k+2)}) & \cdots & f(z^{(k+2)}) & 1 \end{vmatrix}},$$

(6.109)

for example. The solutions for subscripts other than k are of similar form. Next observe that if we use elementary row operations to clear the ones out of the right-hand column of the denominator determinant, except for the upper right-hand entry, and then expand the determinant along the last column, we can reduce it to the form

$$\det \begin{vmatrix} g^{(k)}(z^{(1)}) & \cdots & g(z^{(1)}) \\ \vdots & & \vdots \\ g^{(k)}(z^{(k+1)}) & \cdots & g(z^{(k+1)}) \end{vmatrix}, \qquad (6.110)$$

where we have defined

$$g(z^{(i)}) = f(z^{(i+1)}) - f(z^{(i)}). \qquad (6.111)$$

The subtraction in (6.111) will eliminate a uniform (single-valued) term, and might reduce the monodromic dimension but to no less than k. This is sufficient to ensure that (6.110) is not zero. Notice that (6.110) is independent of the type descriptor $\vec{m}$, and of the polynomial which we are computing. Since, by our definitions, the polynomials are not all identically zero, it cannot be that all the residues $R_{\vec{m}}(z^{(i)})$, $i = 1, \ldots, k+2$, vanish simultaneously. The structure of (6.109) is of the form

$$A(z) = \frac{\mathbf{R}^T \cdot \mathbf{U}_j(z)}{Q_{j,m_j}(z)}, \qquad (6.112)$$

where A, $\mathbf{U}_j$ are independent of the type descriptor $\vec{m}$. As $\mathbf{R}$ can be resolved into components parallel and perpendicular to $\mathbf{U}_j$, the magnitude of $\mathbf{R}$ is at least as great as that part parallel to $\mathbf{U}_j$. Thus we can deduce a uniform lower bound on $|\mathbf{R}|$ as

$$|\mathbf{R}| \geq \max_j \left(\frac{|A(z)||Q_{j,m_j}(z)|}{|\mathbf{U}_j(z)|} \right), \qquad (6.113)$$

which gives the conclusion of the theorem.

Next let us try to give some context to the diagonal-sequence convergence problem. Our mode of presentation for this material is a somewhat abbreviated version of the summary of Baker [1990]. Divide (6.108) by $Q_{k,m_k}(z)$ and take the limit as all the $m_j \to \infty$. If there is convergence, then we get

$$f^{(k)}(z^{(i)}) + \sum_{j=-1}^{k-1} \left(\frac{Q_{j,m_j}(z)}{Q_{k,m_k}(z)} \right) f^{(j)}(z) = 0. \qquad (6.114)$$

The major thing to notice about (6.114) is that the ratios are single-valued functions of z. Thus the set of equations (6.114) can be solved for these ratios explicitly in terms of $f(z)$, and its derivatives for the integral case or its powers for the algebraic case, from $k+1$ linearly independent coverings of the complex plane. Once these ratios are determined, we face the problem, like the one solved by Stahl's extremal domain theorem, Theorem 6.6.5, of finding a region to determine a cut structure that leaves these functions single valued on the Riemann surface of $f(z)$. There is an additional problem, namely, in the case where there are more than $k+1$ independent coverings, which will certainly happen if the monodromy dimension exceeds $k+1$ (or exceeds k plus a uniform background), which $k+1$ coverings should be selected. After all, the integral or algebraic approximants depend only on the Taylor series coefficients at the origin of the first Riemann sheet and there is no explicit mention as to which sheets we should select. A further point to take note of is that (6.114) is a 'two-way' equation. That is to say, given the ratios (and the initial conditions) and provided we have not hit on some degenerate case, we may integrate (6.114) to give the values of $f(z)$ on $k + 1$ coverings, and given a set of $k + 1$ independent coverings we can determine the ratios.

In the context of the diagonal-sequence convergence problem we remark that as the polynomials are necessarily single valued, the cuts seem to appear in the complex plane instead of on the Riemann surface. What is meant by this remark is illustrated by the example

$$f(z) = \tfrac{1}{2}\sqrt{1 + z} + \tfrac{1}{2}\sqrt{1 - z}.$$

This function has three branch points at $z = +1, -1$, and ∞, is locally analytic elsewhere, has four Riemann sheets, and has monodromic dimension two. The first-order integral approximants with $k = 1$ cannot represent it exactly in any finite order of z. The complete, minimal, diagonal sequence here reduces to just the $[2n \pm 1/2n \pm 1; 2n]$ approximants with the plus signs for n odd and the minus signs for n even. Since the function $f(z)$ is even we must have a symmetrical treatment of the two singularities. The correct description is as follows: for the first Riemann sheet we get $\mathscr{S}_0 = \mathbb{C}\setminus[(-\infty \leqslant z \leqslant -1) \cup (1 \leqslant z \leqslant +\infty)]$, which is not so surprising as it is just a cut from $z = -1$ through $z = \infty$ and on to $z = +1$ and identical to that which would be obtained by the Padé-approximant method. We can generate with the integral approximants a second sheet $\mathscr{S}_1$ by an analytic continuation through the cut $(1, +\infty)$ and a third sheet $\mathscr{S}_2$ by an analytic continuation through the cut $(-\infty, -1)$. The domain of convergence is then

$$\mathscr{S} = \mathscr{S}_0 \cup (\mathscr{S}_1 \cap \{\mathrm{Re}\,(z) > 0\}) \cup (\mathscr{S}_2 \cap \{\mathrm{Re}\,(z) < 0\}).$$

The singularities, which we can thus expect to be reflected in the positions of the zeros of the integral polynomials, are those at $z = \pm 1, \infty$ and along the imaginary axis. Even though the imaginary axis is a line of singularities, it is nevertheless true that on the first Riemann sheet, the limits of the solution from the right and left half plane join each other analytically at the imaginary axis, in the limit as the $m_j = m \to \infty$. On the second sheet this coincidence does not happen. Although the ratios are single valued and locally analytic, they are discontinuous across the cut on the imaginary axis. We can easily solve (6.114) for limiting values of them and we get

$$\frac{Q_0^*(z)}{Q_1^*(z)} = \frac{\mathrm{sgn}\,(\mathrm{Re}\,(z))}{2[1 - \mathrm{sgn}\,(\mathrm{Re}\,(z))z]},$$

$$\frac{Q_{-1}^*(z)}{Q_1^*(z)} = \frac{\mathrm{sgn}\,(\mathrm{Re}\,(z))\sqrt{1 + \mathrm{sgn}\,(\mathrm{Re}(z))z}}{2(1 - z^2)}.$$

This illustration shows that some rather surprising things can happen in this case, such as a cut through the origin. Numerical investigation shows that the zeros of $Q_{k,m_k}(z)$ approach the origin rather slowly and with diminishing residues, so no practical problem was encountered. This example does show that the interpretation of the results in this case differs from that for Padé approximants.

The next theorem answers the question of how to choose the 'right set of coverings.' It relies on modern potential theory; see, for example, Landkof [1972]. In order to state the theorem, we need some notation. We will use $\hat{\mathbb{C}}$ to denote the extended complex plane including the point at infinity. The next few theorems are due to Stahl [1988]. He uses the $(1/z)$-plane to express his results because they are more naturally expressed this way. His functional elements are thus expansions in $1/z$ about $z = \infty$. The original or zeroth Riemann sheet has the point at infinity denoted as $\infty^{(0)}$ and is his original expansion point. We need to define a standard linear factor

$$H(z, x) = \begin{cases} z - x & \text{for} & |x| \leq 1, \\ (z - x)/|x| & \text{for} & 1 < |x| < \infty, \\ 1 & \text{for} & x = \infty. \end{cases} \quad (6.115)$$

For a measure with support in $\hat{\mathbb{C}}$, we define the potential

$$\rho(\mu; z) = \int \ln|H(z, x)| \, d\mu(x). \quad (6.116)$$

Let us next consider $\mathscr{D}$ as the set of all domains $D \subset \mathscr{R}$ (the Riemann surface of a given multiform function $f(z)$) which possess a (generalized) Green's function $g_D(z, w)$ and for which the capacity (see Section 6.6) $\text{cap}(\partial D) > 0$. For our purposes, the Green's function $g_D(z, x)$ has a logarithmic singularity as $z \to x$ with unit amplitude and $g(z, x)$ vanishes on $x \in \partial D$. We now have on $\mathscr{R}$ the Green's potentials

$$g(\mu; z) = g(\mu, D; z) = \int g_D(z, w) \, d\mu(w), \quad (6.117)$$

where μ is now a measure on $\mathscr{R}$ and $D \in \mathscr{D}$. The Green's function also satisfies $g_D(z, w) = 0$ for any $w \in \mathscr{R} \setminus D$. Finally, we define the potential

$$\rho_\mu(z) = \rho(\mu; \pi(z)), \quad (6.118)$$

which is called the lifting of the potential (6.116) to $\mathscr{R}$. The projection operator π is as in (6.108).

If we are now given a domain $D \in \mathscr{D}$ with $\infty^{(0)} \in D$, and a normalized positive measure ν on $\hat{\mathbb{C}}$, then we can define the two auxiliary functions

$$r(\nu, D; z) = (k + 1)g_D(z, \infty^{(0)}) - \int g_D(z, w) \, d\nu(\pi(z)), \quad (6.119)$$

$$d(v, D; z) = r(v, D; z) - \rho_v(z), \quad z \in \mathcal{R}. \tag{6.120}$$

Some of the properties of these functions are noteworthy. Namely,

$$r(z) = \rho_v(z) \text{ for } z \in \mathcal{R} \setminus D, \tag{6.121}$$

$$r(z) = k \ln |\pi(z)| + O(1) \text{ for } z \to \infty^{(0)}, \tag{6.122}$$

and $r(z) - \ln |\pi(z)|$ is harmonic in $D \setminus \infty^{(0)}$ and continuous almost everywhere on ∂D. At the point $\infty^{(0)}$ the function $d(z)$ has a logarithmic singularity with a coefficient $k+1$ and inside D the measure is repeated as many times as the domain D covers the support of v in $\hat{\mathbb{C}}$. Let us denote 'the support of' by $S(\cdot)$. If it is the case that $\pi(D) \cap S(v) = \emptyset$, then $d(z) = (k+1)g(z, \infty^{(0)})$.

***Theorem* 8.6.20 (Minimal domain theorem) [Stahl, 1988].** *If $f(z)$ is locally analytic in $\hat{\mathbb{C}} \setminus \mathcal{E}(f)$, where $\mathcal{E}(f)$ is a compact subset of $\hat{\mathbb{C}}$ of capacity* cap$(\mathcal{E}(f)) = 0$, *if $f(z)$ has monodromic dimension at least $k+1$, and if $\mathcal{R}$ is the Riemann surface of $f(z)$ (which must have at least $k+1$ Riemann sheets), then there exists a unique domain $\mathcal{S} \subset \mathcal{R}$ and a probability measure v whose support is in $\hat{\mathbb{C}}$ such that* (i) $\infty^{(0)} \in \mathcal{S}$ *and* (ii) *the Green's potential $r(v, \mathcal{S}; z)$ is harmonic in a neighborhood of ∂S.*

There are some significant consequences of this theorem which serve to characterize the resulting domain. First suppose that $\Sigma \in \mathcal{D}$ is a domain with a smooth boundary $\partial \Sigma$, and let $\partial/\partial n$ denote the normal derivative on $\partial \Sigma$ directed toward Σ. On $\partial \Sigma$ we can define a measure μ_d by

$$d\mu_d(z) = \frac{1}{2\pi} \frac{\partial}{\partial n} d(z) \, d\Sigma_z, \quad z \in \partial \Sigma. \tag{6.123}$$

It also follows from (ii) of the theorem that

$$\overline{\pi(\mathcal{S})} = S(v), \tag{6.124}$$

because any component of v outside of $\overline{\pi(\mathcal{S})}$ would not contribute to the Green's potential anyway, and

$$\pi^{-1}(v)|_{\mathcal{S}} = \mu_{d(v, \mathcal{S}; z)}. \tag{6.125}$$

Thus the domain $\mathcal{S}$ and the measure v introduced in the above theorem are determined in such a way that the normal derivative of the function $d(z)$ at a boundary point $z \in \partial \mathcal{S}$ is equal to the density of the measure v at the point $\pi(z) \in \hat{\mathbb{C}}$. To reiterate, the measure μ_d sits on the Riemann surface, the border between $\mathcal{S}_1$ and $\mathcal{S}_2$ in our previous illustration, and v is its projection on the complex plane, which is what we will see when we calculate the approximants.

Note is made that the $k = 1$ case is just Theorem 6.6.5. The deviation from the expected notation for the case of $k = 0$ is made because this set of results deals with the homogeneous cases of type $(-1, \mathfrak{m})$ for integral approximants and order $k - 1$ (with leading term $\{f(z)\}^k$) instead of our usual order k for the algebraic case. The uniqueness part of the theorem, which follows from the maximum principle for harmonic functions, is important because it answers the selection question that we raised at (6.114). Before we give the next theorem, we need the notation that a sequence $(-1, \mathfrak{m}_\lambda)$ is near to diagonal if

$$\lim_{\lambda \to \infty} \sum_{j=0}^{k} m_{j,\lambda} = \infty, \text{ and } \lim_{\lambda \to \infty} \frac{m_j}{\sum_{j=0}^{k} m_{j,\lambda}} = \frac{1}{k+1} \quad \forall j. \quad (6.126)$$

Theorem 8.6.21 (Convergence in capacity) [Stahl, 1988]. *Let $f(z)$ be analytic in a neighborhood of $\infty \in \hat{\mathbb{C}}$. Further let $f(z)$ have a monodromic dimension of at least $k + 1$, and let $\mathfrak{R}$ be the Riemann surface of $f(z)$ (which must have at least $k + 1$ sheets). Let $[-1/m_0; \ldots; m_k]$ be a near-to-diagonal sequence of integral approximants or $\langle m_0/m_1, \ldots, m_k \rangle$ be a sequence of near-to-diagonal algebraic approximants. Let $\mathscr{S} \subset \mathfrak{R}$ be the minimal domain of the previous theorem, and let $d(z)$ be as defined in (6.120). Then we have*

$$[-1/m_{0,\lambda}; \ldots; m_{k,\lambda}] \text{ or } \langle m_{0,\lambda}/m_{1,\lambda}, \ldots, m_{k,\lambda} \rangle \xrightarrow[\mathscr{S}]{\text{cap}} f(z) \text{ for } \lambda \to \infty.$$

(6.127)

More precisely,

$$|f(z) - \{[-1/m_{0,\lambda}; \ldots; m_{k,\lambda}] \text{ or } \langle m_{0,\lambda}/m_{1,\lambda}, \ldots, m_{k,\lambda} \rangle\}|^{(k+1)/|\mathfrak{m}_\lambda|}$$
$$\xrightarrow[\mathscr{S}]{\text{cap}} e^{-d(z)} \text{ for } \lambda \to \infty, \quad (6.128)$$

where $\xrightarrow[\mathscr{S}]{\text{cap}}$ means 'converges in capacity on compact subsets of $\mathscr{S}$'.

The proof of this theorem is omitted because of its length and complexity.

8.6.7 Singular index and amplitude computations

One important application of integral approximants is, of course, extrapolation which gives the value of a function at regular points and also along cuts in the complex plane. These cuts are sometimes beyond the scope of application of the Padé approximant, but as we saw in the previous section, can be well within the domain of convergence of the integral approximant. Another important application of Padé and integral approximants has been the deduction of the behavior at singular points. This information is of particular interest in the theory of critical phenomena. The integral approximant has a widely expanded range, relative to the Padé approximant, of singularities which it can represent exactly.

From the standard theory of differential equations, we know that singularities in the solution of (6.25) occur when $Q_{k,m_k}(z_0) = 0$. The simplest and most common case is when $Q_{k-1,m_{k-1}}(z_0) \neq 0$. In this case, the various solutions can take the form

$$y(z) = \sum_{j=0}^{\infty} g_j(z - z_0)^{\nu+j}, \qquad (6.129)$$

where $g_0 \neq 0$ and the allowed values of ν are given by the indicial equation

$$(\nu)(\nu - 1) \ldots (\nu - k + 1)Q'_{k,m_k}(z_0)$$
$$+ (\nu)(\nu - 1) \ldots (\nu - k + 2)Q_{k-1,m_{k-1}}(z_0) = 0. \qquad (6.130)$$

The solutions are

$$\nu = 0, \ldots, k - 2 \text{ and } \nu = k - 1 - \frac{Q_{k-1,m_{k-1}}(z_0)}{Q'_{k,m_k}(z_0)}, \qquad (6.131)$$

i.e., there is one singular solution and the rest are all regular. The form of the solution in the neighborhood of $z = z_0$ is

$$y(z) = (z - z_0)^{\nu}\phi(z) + \psi(z), \qquad (6.132)$$

where both ϕ and ψ are regular at $z = z_0$ and ν is the singular index. The value of $\phi(z_0)$ is the singular amplitude and the value of $\psi(z_0)$ is called the remainder. While the value of the singular index, as we just saw, is given directly from the integral polynomials without the need for integration of (6.25), the amplitude and the remainder do require integration. For the special case of $k = 1$, Hunter and Baker [1979] have shown that

$$\psi(z_0) = -\frac{Q_{-1,m_{-1}}(z_0)}{Q_{0,m_0}(z_0)}, \qquad (6.133)$$

which is also obtained without integration because of the rather special structure in this case. In more general cases we will adopt the methods of Rehr, Joyce, and Guttman [1980]. For a single isolated singularity, the singular index is given by the last value in (6.131). In order to compute $\phi(z_0)$ and $\psi(z_0)$, we first note that it is numerically unstable to integrate up to or out from a regular singular point. We will illustrate the method for the case $k = 2$. First, we compute the power-series expansions of the solutions of (6.25) about $z = z_0$,

$$y_\nu(z) = (z - z_0)^\nu \left[1 + \sum_{n=1}^\infty a_n(z - z_0)^n\right],$$

$$y_0(z) = 1 + \sum_{n=1}^\infty b_n(z - z_0)^n. \tag{6.134}$$

The initial coefficients of unity in (6.134) are the required initial conditions for these solutions at a regular singular point. For $k > 2$, we would need, in addition to $y_0(z)$, $y_1(z) \sim (z - z_0)$, ..., $y_{k-2}(z) \sim (z - z_0)^{k-2}$, and the corresponding expansion of the subsequent equations. Next, (6.25) is integrated from the origin to some convenient intermediate point, $z = \xi$, where the series in (6.134) converge rapidly, and the solutions

$$y(\xi) = A_\nu y_\nu(\xi) + A_0 y_0(\xi),$$
$$y'(\xi) = A_\nu y'_\nu(\xi) + A_0 y'_0(\xi), \tag{6.135}$$

are matched by solving for A_ν and A_0. It then follows that

$$\phi(z_0) = A_\nu, \quad \psi(z_0) = A_0. \tag{6.136}$$

These results give a straightforward solution in non-degenerate cases, but there are other cases which are also of interest. In the case of critical-point theory (see Section 2.2 and Baker [1990]), problems arise where the singularity of interest can be confluent. That is to say, we would expect the form of $y(z)$ to be

$$y(z) = \sum_{i=1}^\mu (z - z_0)^{\nu_i} \phi_i(z) + \psi(z). \tag{6.137}$$

The appropriate conditions on the defining differential equation (6.25) for such a singularity to occur are that the $Q_{j,m_j}(z)$ must factor as

$$(z - z_0)^\mu \pi_m(z) \frac{d^m y}{dz^m} + (z - z_0)^{\mu-1} \pi_{m-1}(z) \frac{d^{m-1} y}{dz^{m-1}}$$
$$+ \cdots + (z - z_0)^0 \pi_{m-\mu}(z) \frac{d^{m-\mu}}{dz^{m-\mu}} y + \cdots = 0. \tag{6.138}$$

Of course, in practice at any finite degree of approximation, we are unlikely to have reached the limiting result exemplified by (6.138). Even when we are close, we still expect that the factors $(z - z_0)^j$ will be replaced by $\prod_{i=1}^j (z - z_{j,i})$ where the $z_{j,i} \approx z_0$. Baker, Oitmaa, and Velgakis [1988] produce an elegant algorithm based on ideas from monodromy theory (Section 8.6.1). They argue that, from a distance, a cluster of zeros looks just about like a single multiple zero, and so if a contour integral is taken about the cluster, its properties are a good representation of the true singular structure. We will give here a procedure, which is usually reasonably accurate, to determine the properties at a confluent singularity. We illustrate it for the simple case of the confluence of two singularities, in a second-order case. Let us write (6.25) for this case as

$$P(z)\frac{d^2y}{dz^2} + Q(z)\frac{dy}{dz} + R(z)y + T(z) = 0, \quad (6.139)$$

and suppose that $P(z)$ has two roots $z_1 \approx z_2$ which represent the confluent singularity. The method is based on a quadratic approximation to $P(z)$. First we compute the 'local singular indices,' in analogy to (6.131),

$$\beta_1 = 1 - \frac{Q(z_1)}{P'(z_1)}, \quad \beta_2 = 1 - \frac{Q(z_2)}{P'(z_2)}, \quad (6.140)$$

and the auxiliary quantities

$$r_1 = \frac{R(z_1)}{\lim_{z \to z_1} \frac{P(z)}{(z-z_1)(z-z_2)}}, \quad r_2 = \frac{R(z_2)}{\lim_{z \to z_2} \frac{P(z)}{(z-z_1)(z-z_2)}}. \quad (6.141)$$

From these quantities, we compute

$$\beta = 1 - \beta_1 - \beta_2, \quad r = \tfrac{1}{2}(r_1 + r_2). \quad (6.142)$$

The two approximate singular indices are then given by the solution of an (approximate) indicial equation,

$$\gamma^2 + \beta\gamma + r = 0, \quad (6.143)$$

as

$$\gamma_\pm = -\tfrac{1}{2}[\beta \pm \sqrt{\beta^2 - 4r}]. \quad (6.144)$$

More elaborate approximate schemes can be devised along the same lines in case there are more than two confluent singularities.

We expect a representation in the neighborhood of the confluent singularity of the form

$$y(z) \approx A_+(z-z_c)^{\gamma_+} + A_-(z-z_c)^{\gamma_-} + A_0. \quad (6.145)$$

To complete a description of y in the neighborhood of z_0, we need the amplitudes. To obtain them, we need to first define $y_0(z) = y(z^{(0)})$, $y_1(z) = y(z^{(1)})$, and $y_2(z) = y(z^{(2)})$, where as in the previous section $z^{(i)}$

refers to z projected onto the ith Riemann sheet. Here we get to the next Riemann sheet by following a contour which encircles both z_1 and z_2. In order to compute the amplitudes, we decompose

$$y(z) = y_+(z) + y_-(z) + \hat{y}_0(z). \tag{6.146}$$

The relation between these y's and the others just defined by the integration of (6.139) is

$$y_+(z) = \frac{\exp(2\pi i \gamma_-)(y - y_1) - (y_1 - y_2)}{(\exp(2\pi i \gamma_-) - \exp(2\pi i \gamma_+))(1 - \exp(2\pi i \gamma_+))},$$

$$y_-(z) = \frac{\exp(2\pi i \gamma_+)(y - y_1) - (y_1 - y_2)}{(\exp(2\pi i \gamma_+) - \exp(2\pi i \gamma_-))(1 - \exp(2\pi i \gamma_-))}, \tag{6.147}$$

$$\hat{y}_0(z) = y_0(z) - y_+(z) - y_-(z).$$

Now asymptotically, as $z \to z_0$,

$$y_+(z) \sim \phi_+(z)(z - z_0)^{\gamma_+},$$
$$y_-(z) \sim \phi_-(z)(z - z_0)^{\gamma_-}, \tag{6.148}$$
$$\hat{y}_0(z) \sim \phi_0(z),$$

where the ϕ's are analytic near $z = z_0$. The best estimator for z_0 that can be simply provided in this system of approximation seems, from experience, to be either z_1 or z_2, according to which of the local indices β_1 or β_2 corresponds to the most divergent singularity, i.e., $\min(\operatorname{Re}(\beta_i))$.

The corresponding problem for the algebraic approximants has not been so thoroughly analyzed. As remarked at the end of sub-section 8.6.1, the only possible singular indices for the solution of (6.23) are rational numbers with the denominator less than or equal to $k + 1$. The size of the numerator can also be limited in terms of the degrees of the polynomials. The divergent singularities occur when $Q_{k,m_k}(z_0) = 0$. The analysis for the singular index of an isolated divergent singularity follows in terms of

$$(k + 1)\zeta = \alpha_k + \max(k\zeta - \alpha_{k-1}, (k - 1)\zeta - \alpha_{k-2}, \ldots, -\alpha_{-1}), \tag{6.149}$$

where $Q_{j,m_j}(z) \sim (z - z_0)^{\alpha_j}$ as $z \to z_0$ and, by our definitions, at least one $\alpha_{j_0} \neq 0$. Since the coefficient of ζ on the left-hand side is larger than any on the right-hand side, the left-hand side is larger than the right-hand side for sufficiently large ζ and smaller for sufficiently negative ζ. Thus since the left-hand side is piecewise linear, there is a unique solution ζ_0 of (6.149). The singular index at z_0 is then ζ_0, so

$$f(z) = \phi(z)(z - z_0)^{\zeta_0} + \psi(z), \tag{6.150}$$

where ϕ is regular in the neighborhood of z_0 and ψ is a subdominant term. The amplitude of the singularity follows from the equation obtained by directly substituting $A_0(z - z_0)^{\zeta_0}$ into (6.25) and equating the most divergent part to zero.

9

Connection with integral equations and quantum mechanics

9.1 The general method and finite-rank kernels

We will consider linear, inhomogeneous integral equations of the type

$$f(x) = g(x) + \lambda \int_a^b A(x, y)f(y)\,dy. \tag{1.1}$$

These are sometimes called Fredholm equations of the second kind. If (1.1) is solved iteratively, the Liouville–Neumann expansion results:

$$f(x) = g(x) + \lambda \int_a^b A(x, y)g(y)\,dy$$
$$+ \lambda^2 \int_a^b A(x, y_1) \int_a^b A(y_1, y)g(y)\,dy\,dy_1 + \cdots. \tag{1.2}$$

The expansion (1.2) formally satisfies (1.1), and is a solution of (1.1) if the series (1.2) is uniformly convergent. It is often the case that the quantity of interest is not $f(x)$ but rather

$$\langle h|f \rangle \equiv \int_a^b h^*(x)f(x)\,dx. \tag{1.3}$$

and then

$$\langle h|f \rangle = \int_a^b h^*(x)g(x)\,dx + \lambda \int_a^b \int_a^b h^*(x)A(x, y)g(y)\,dx\,dy$$
$$+ \lambda^2 \int_a^b \int_a^b \int_a^b h^*(x)A(x, y_1)A(y_1, y)g(y)\,dx\,dy_1\,dy + \cdots \tag{1.4}$$

for suitable convergence conditions. We defer detailed questions of convergence to Section 9.2. The series (1.4) is a power series in λ, which we write concisely as

$$\langle h|f \rangle = c_0 + \lambda c_1 + \lambda^2 c_2 + \cdots, \tag{1.5}$$

where

$$c_n = \int_a^b \int_a^b \cdots \int_a^b h^*(y_1) A(y_1, y_2) A(y_2, y_3) \cdots A(y_{n-1}, y_n) g(y_n) \, dy_1 \cdots dy_n.$$
(1.6)

Provided all the coefficients $\{c_n\}$ of (1.6) exist, we may sum the series (1.5) using the Padé approximants in the hope that the method converges. The basic application of Padé approximants in integral equations is very simple: use an iterative solution, and use Padé approximants to interpret it. In this chapter we will consider a variety of applications. In some instances the Padé method may be proved to converge, and we know of no circumstances where the Padé method, properly applied, has failed for integral equations in genuine applications.

We will now consider an important and very particular class of kernels which are called finite-rank kernels. They may be written as

$$A(x, y) = \sum_{j=1}^n X_j(x) Y_j(y). \tag{1.7}$$

We will suppose that n is the smallest integer allowing the representation (1.7), so as to obviate unnecessary later difficulties. Finite-rank kernels are important because they allow the basic integral equation (1.1) to be solved exactly, and this solution sheds a great deal of light on the nature of the solution with more general kernels.

We will find the explicit solution of (1.1) using the finite-rank kernel (1.6) and show that the Padé approximant solution is exact.

Substitute (1.7) in (1.1) to obtain

$$f(x) = g(x) + \lambda \sum_{j=1}^n X_j(x) \int_a^b Y_j(y) f(y) \, dy$$

$$= g(x) + \lambda \sum_{j=1}^n f_j X_j(x), \tag{1.8}$$

where

$$f_j = \int_a^b Y_j(y) f(y) \, dy, \quad j = 1, 2, \ldots, n, \tag{1.9}$$

Let

$$g_i = \int_a^b Y_i(x) g(x) \, dx, \quad i = 1, 2, \ldots, n, \tag{1.10}$$

$$h_i = \int_a^b X_i(x) h^*(x) \, dx, \quad i = 1, 2, \ldots, n, \tag{1.11}$$

and
$$d_{ij} = \int_a^b Y_i(y) X_j(y)\, dy, \quad i, j = 1, 2, \ldots, n. \tag{1.12}$$

Equations (1.9)–(1.12) define the basic vectors **f**, **g**, and **h** and the matrix D which are needed presently. By integrating (1.8) over $Y_i(x)$, we find

$$f_i = g_i + \lambda \sum_{j=1}^{n} d_{ij} f_j,$$

i.e.,
$$\mathbf{f} = \mathbf{g} + \lambda D \mathbf{f},$$

and so $\mathbf{f} = (I - \lambda D)^{-1} \mathbf{g}$, provided $I - \lambda D$ is invertible. By substituting this formula for f_i in (1.8), we find

$$f(x) = g(x) + \lambda \sum_{i=1}^{n} [(I - \lambda D)^{-1} \mathbf{g}]_i X_i(x), \tag{1.13}$$

and so

$$\langle h | f \rangle = \int_a^b h^*(x) g(x)\, dx + \lambda \sum_{i=1}^{n} \sum_{j=1}^{n} h_i [(I - \lambda D)^{-1}]_{ij} g_j. \tag{1.14}$$

Hence (1.14) provides a solution of the problem, using the definitions (1.10), (1.11), and (1.12). We notice that if the inverse matrix $(I - \lambda D)^{-1}$ is assumed to exist (and it certainly does if $|\lambda|$ is sufficiently small), then the power-series expansion of (1.14) is valid. In general, we notice that the solution (1.14) is a rational function of λ. We demonstrate this by writing

$$(I - \lambda D)^{-1} = \frac{\text{adj}\,(I - \lambda D)}{\det(I - \lambda D)},$$

where the adjugate matrix, adj A, is defined to be the transposed matrix of cofactors of A. Hence we see that (1.14) implies that

$$\langle h | f \rangle = \frac{N(\lambda)}{D(\lambda)}, \tag{1.15}$$

where $N(\lambda)$ and $D(\lambda)$ are normally polynomials of degree n. In particular, any $[L/M]$ Padé approximant to the power series of (1.14) with $L \geq n$ and $M \geq n$ is exact, from Theorem 1.4.4 [Chisholm, 1963]. We are also naturally led to consider diagonal Padé approximants for the solution of integral equations.

An important and useful conclusion to this section is that the diagonal Padé approximants to the solution $\langle h | f \rangle$ of the integral equation (1.1) converge to the exact solution, if the kernel is of finite rank.

More generally, for any $\alpha \in (0, 1)$, the $[L/M]$ Padé approximants to the expansion of $\langle h|f \rangle$, given by (1.4), converge to the solution $\langle h|f \rangle$ of the integral equation (1.1) with a finite-rank kernel, provided

$$\alpha < L/M < \alpha^{-1} \quad \text{and } \min(L, M) \to \infty.$$

Furthermore, for a kernel of rank n, the $[L/M]$ Padé approximants are exact provided $\min(L, M) \geq n$.

We defer questions of convergence to the next section; however, the answers to such questions with finite-rank kernels are scarcely difficult, given the existence of an explicit analytical solution.

9.2 Padé approximants and integral equations with compact kernels

In the previous section, we stated the Padé method and motivated it by the study of finite-rank kernels, following the historical approach. In this section we will summarize some of the results of the theory of integral equations and relate these to the theory of Padé approximants. The underlying theme comprises two parts: establishing that $\langle h|f \rangle$ of (1.4) is a meromorphic function of λ, and selecting the sequence of Padé approximants which converge to such a function. If special properties of the integral equation (1.1) imply that $\langle h|f \rangle$ is not merely meromorphic (for example, $\langle h|f \rangle$ might be a Stieltjes series), the methods of Section 9.4 may be applicable, and stronger conclusions then follow. Presently we consider the general integral equation of the type

$$f(x) = g(x) + \lambda \int_a^b A(x, y) f(y) \, dy, \tag{2.1}$$

and its solution in the form

$$\langle h|f \rangle \equiv \int_a^b h^*(x) f(x) \, dx \tag{2.2}$$

as at the start of this chapter. Let us recall the results of Hilbert-space theory. $\mathcal{H}$ is a linear space with elements $f, g, h, \ldots$ [not yet to be associated with f, g and h of (2.1) and (2.2)] including a null element 0. To every pair of elements there is an inner product $\langle f|g \rangle$ satisfying the usual rules, including $\langle f|g \rangle = \langle g|f \rangle^*$. Each element f has a norm $\|f\| = \langle f|f \rangle^{1/2}$ given by the inner product. Every Cauchy sequence of elements of $\mathcal{H}$ converges to a limit element which belongs to $\mathcal{H}$.

The first important linear space we use is the space $\mathscr{C}[a, b]$ of continuous functions defined on a closed finite interval $[a, b]$. The inner product is defined by

$$\langle h|g \rangle = \int_a^b h^*(x) g(x) \, dx. \tag{2.3}$$

$\mathscr{C}[a, b]$ is not a Hilbert space, because it is not complete. This is so because Cauchy sequences of continuous functions defined on $[a, b]$ may be found which converge to discontinuous functions, such as step functions, using the norm implied by (2.3). We will assume in this context that $g(x)$ of (2.1) belongs to $\mathscr{C}[a, b]$ and that $A(x, y)$ of (2.1) is continuous for all $x, y \in [a, b]$. Then given $\varepsilon > 0$, Weierstrass's approximation theorem states that a polynomial $S_N(x, y)$ of sufficiently high degree N, and depending on ε, exists such that

$$|S_N(x, y) - A(x, y)| < \varepsilon$$

for all $x, y \in [a, b]$. Then $T(x, y)$ is defined by

$$A(x, y) = S_N(x, y) + T(x, y) \tag{2.4}$$

with the property that $\|T\|_\infty < \varepsilon$. This property is sufficient to show that the operator

$$(1 - \lambda T)^{-1} = 1 + \lambda T + \lambda^2 T^2 + \cdots \tag{2.5}$$

exists for all $|\lambda| < \varepsilon^{-1}$. By this we mean that (2.5) converges to a continuous kernel and this maps any function from $\mathscr{C}[a, b]$ into a function of $\mathscr{C}[a, b]$. We use (2.4) and (2.5) in (2.1):

$$f = g + \lambda(S_N + T)f, \tag{2.6}$$

$$(1 - \lambda T)f = g + \lambda S_N f,$$

$$f = (1 - \lambda T)^{-1} g + \lambda (1 - \lambda T)^{-1} S_N f. \tag{2.7}$$

Equation (2.7) is an integral equation of finite rank, and so f is meromorphic in λ, with at most N poles in the circle $|\lambda| < \varepsilon^{-1}$. We deduce that $\langle h|f\rangle$ is meromorphic in λ.

The second space we need to consider is the Hilbert space $\mathscr{L}^2(a, b)$, the space of square-integrable functions defined on the open interval (a, b). For this space, we allow the possibility that $a = -\infty$, or $b = \infty$, or both. The inner product is the same as (2.3). If a and b are finite, $\mathscr{C}[a, b] \subset \mathscr{L}^2(a, b)$. To make use of this space, we will assume that

$$\int_a^b |g(x)|^2 \, dx < \infty \quad \text{and} \quad \int_a^b \int_a^b |A(x, y)|^2 \, dx \, dy < \infty$$

in this context. The singular value decomposition theorem [Smithies, 1958] states that, for any $\varepsilon > 0$, any $\mathscr{L}^2$ kernel $A(x, y)$ has an expansion

$$A(x, y) = \sum_{n=1}^{N} \frac{u_n(x) v_n(y)}{\lambda_n} + T(x, y) \tag{2.8}$$

where $\|T\| = \int_a^b \int_a^b |T(x,y)|^2 \, dx \, dy < \varepsilon$. The functions u and v belong to $\mathscr{L}^2(a,b)$ and are normalized eigenfunctions of $AA^\dagger$ and $A^\dagger A$, respectively. An analysis identical to that following (2.6) shows again that $\langle h|f \rangle$ of (2.2) is meromorphic in λ.

We have considered kernels in two different Hilbert spaces, and used them as operators in the sense that

$$Af = \int_a^b A(x,y) f(y) \, dy. \qquad (2.9)$$

Further, they are compact operators. An operator is defined to be compact (sometimes called completely continuous) if it transforms every bounded set of elements in the space into a sequence which contains a convergent subsequence. For example, we see that the identity operator and also v defined by $vg = v(x)g(x)$ with $v \neq 0$ are not compact operators. The concept of a compact operator is useful in both Hilbert- and Banach-space theory. A Banach space has all the axioms of a Hilbert space except that the norm is not required to be an inner product. In an arbitrary Banach space, it is not true that every compact operator can be approximated in norm by a kernel of finite rank, so an equation such as (2.8) is unavailable. But it is true that the resolvent of every compact operator on $\mathscr{B}$ is meromorphic in λ. Referring to (2.1) and (2.2), if $g \in \mathscr{B}$, $h \in \mathscr{B}^*$ (the dual of $\mathscr{B}$), and A is a compact operator on $\mathscr{B}$, then $\langle h|f \rangle$ is a meromorphic function of λ.

Three approaches of increasing sophistication have been given which suffice to prove that $\langle h|f \rangle$ is a meromorphic function of λ. Let us turn to consider the theorems and conjectures available to utilize this information.

If a kernel S_N of finite rank N exists which is a good approximation to A, one expects the $[N/N]$ Padé approximant to $\langle h|f \rangle$ to be a good approximation, and this is usually the case. Motivated by the analysis of the previous section, the folklore is that use of the diagonal sequence of Padé approximants for $\langle h|f \rangle$ is an efficient way of solving integral equations occurring in practice. The fact that the method works well in practice is understood as implying that ordinary kernels have (whether or not one can find them) accurate finite-rank approximations in the sense of the relevant norm.

Counterexample [Graves-Morris, 1978a]. There exist continuous functions $g(x)$, $h(x)$ and a continuous kernel $A(x,y)$ for which the sequence of diagonal approximants of (1.4) does not converge.

We use Legendre polynomials $P_n(x)$ to define $\{p_n(x) = (n + \tfrac{1}{2})^{1/2} P_n(x)$, $n = 0, 1, 2, \ldots\}$, which is an orthogonal sequence on $[-1, 1]$. We define $g(x) = p_0(x) = 1/\sqrt{2}$ and

$$A(x, y) = \sum_{n=0}^{\infty} \frac{p_{n+1}(x) p_n(y)}{(n + 1)^3}. \tag{2.10}$$

Because $|P_n(x)| \leq 1$ on $-1 \leq x \leq 1$, the sum in (2.10) is uniformly convergent (by the Weierstrass M-test) and so $A(x, y)$ is continuous on $-1 \leq x \leq 1$, $-1 \leq y \leq 1$. Hence the kernel (2.10) is compact on the Hilbert space $\mathscr{L}^2(-1, 1)$.

We define $h(x)$ by

$$h(x) = \sum_{n=0}^{\infty} h_n p_n(x) = \sum_{n=0}^{\infty} \{n!\}^3 c_n p_n(x), \tag{2.11}$$

where the coefficients c_n are specified as follows. Indices n_k are defined iteratively by

$$n_1 = 1, \quad n_{k+1} = 2n_k + 1;$$

we define $c_n = \alpha_k \lambda_k^{-1}$ if n is such that $n_k \leq n < n_{k+1}$; α_k is defined by

$$\alpha_k = \{(2n_k)!\}^{-4} \min \{|\lambda_k|^{n_k}, |\lambda_k|^{2n_k}\}, \tag{2.12}$$

and the sequence $\{\lambda_k\}$ is defined to be any set of points (allowing repetition) which is dense in the λ-plane. The choice (2.12) of α_k ensures that

$$|c_n| < (n!)^{-4}, \quad n = 1, 2, \ldots,$$

so that (2.11) is uniformly convergent on $-1 \leq x \leq 1$. We have constructed $g(x)$, $A(x, y)$, and $h(x)$ specifically so that the series

$$\langle h | f \rangle = \sum_{n=0}^{\infty} \lambda^n \langle h | A^n | g \rangle = \sum_{n=0}^{\infty} c_n \lambda^n$$

is essentially the same as the series constructed by Gammel (Section 6.7), the only minor difference being an extra factor $(2n_k)!^{-3}$ in (2.12) which is not in Gammel's example, but which is needed for reasons associated with the convergence of (2.10) and (2.11). Using Gammel's argument, it follows that the sequence of diagonal approximants of (1.4) diverges on a dense set of points in the whole λ-plane. This establishes the counterexample.

The conjecture of Baker, Gammel, and Wills [1961], in this context, is that at least an infinite subsequence of diagonal Padé approximants to any meromorphic function converges in a bounded region of the λ-plane, except on arbitrary open sets enclosing the poles of the function. There is as yet no proof of this conjecture, but there are strong indications of its validity in this application.

The theorem of Nuttall [1970b] of convergence in measure is a solid result, albeit not quite the result one might have expected. In this context, the theorem states that paradiagonal $[M + J/M]$ sequences converge to $\langle h|f \rangle$ in measure. This means, colloquially, that the area of the region of the complex λ-plane where any approximant of sufficiently high order is less accurate than a given error bound is small. Of course, the location of this region is unknown, and it depends on M.

Nuttall's theorem is strengthened by Pommerenke's modification [1973], the relevant portion of which replaces the notion of measure by capacity. This change has the effect of restricting the maximum diameter of any closed connected exceptional region of the λ-plane instead of restricting its measure. This result and the restrictions on the size of the exceptional region are explained in Section 6.6.

To summarize, we do not yet know the conditions ensuring that the general method we have just described using a paradiagonal sequence of Padé approximants converges, but we have every reason to believe that the 'bad' approximants of any sequence are rare.

Lastly, we consider the poles of the inverse operator $(1 - \lambda A)^{-1}$. Clearly, the Padé approximants of $\langle h|(1 - \lambda A)^{-1}|g \rangle$ do not converge at the poles in the usual way. One can introduce the chordal metric (convergence on the sphere) as described in Section 6.4 to include this case, and this is often desirable. If inspection of a paradiagonal sequence of Padé approximants reveals a stable, convergent sequence of poles, in practice these are always poles of the solution $\langle h|f \rangle$. Location of these poles enables us to find the values of λ which solve the homogeneous integral equation

$$f(x) = \lambda \int_a^b A(x, y) f(y) \, dy. \qquad (2.13)$$

The analysis of (2.6) and (2.7), showing that (2.1) has a unique solution, except at $\lambda = \lambda_n$, $n = 1, 2, 3, \ldots$, with

$$|\lambda_n| \geq \|A(x, y)\|^{-1},$$

implies that (2.13) has no solution except possibly at $\lambda = \lambda_n$, $n = 1, 2, 3, \ldots$. The Padé method of finding these values is to solve the inhomogeneous equation (2.1) with some arbitrary g, locate the poles of the paradiagonal sequence of approximants, and extrapolate their limiting points. In principle, g must be chosen so that it is not orthogonal to the relevant eigenfunction. It should also be remembered that operators of the type $(1 - \lambda A)^{-1}$ discussed in this and the previous section need not have any poles at all. For example, the integral equation

$$f(x) = g(x) + \lambda \int_a^x A(x, y) f(y) \, dy$$

is a Volterra integral equation, and if A is an $\mathscr{L}^2$ kernel, $\langle h|f \rangle$ is an entire function of λ. This statement means that its Maclaurin series converges for any value of λ. In such a case, one would not use the Padé method except for the explicit purpose of accelerating convergence.

The books by Smithies [1958], Tricomi [1957], and Riesz and Szökefalvi-Nagy [1955] are good general texts on integral equations. For further details of the connection with Padé approximants, we refer to Baker and Gammel [1961], Vorobyev [1965], and Graves-Morris [1973a].

9.3 Projection techniques

We discuss in this section the solution of the standard inhomogeneous linear integral equation which takes the form

$$f(x) = g(x) + \lambda \int_a^b A(x, y) f(y) \, dy. \tag{3.1}$$

It will be convenient to write (3.1) in the Hilbert-space formalism using the Dirac notation as

$$|f\rangle = |g\rangle + \lambda A |f\rangle. \tag{3.2}$$

We have in mind the Hilbert spaces of $\mathscr{L}^2$ functions on either finite or infinite ranges (a, b), and A is a linear operator on the Hilbert space. The theory we discuss is quite general, and applications in Banach spaces are sometimes appropriate. The basic problem we wish to treat is the evaluation of the overlap integral

$$\langle h|f\rangle = \int_a^b h^*(x) f(x) \, dx,$$

where $\langle h|$ is another vector from the Hilbert space, and $|f\rangle$ is the solution of (3.1, 3.2). By expanding the iterative solution of (3.1, 3.2), we find

$$|f\rangle = |g\rangle + \lambda A |g\rangle + \lambda^2 A^2 |g\rangle + \cdots, \tag{3.3}$$

and therefore

$$\langle h|f\rangle = \langle h|g\rangle + \lambda \langle h|A|g\rangle + \lambda^2 \langle h|A^2|g\rangle + \cdots \tag{3.4}$$

$$= c_0 + c_1 \lambda + c_2 \lambda^2 + \cdots, \tag{3.5}$$

where we have defined $c_i = \langle h|A^i|g\rangle$, $i = 0, 1, 2, \ldots$. These equalities are well defined if $|\lambda| \cdot \|A\| < 1$, and they are only formal if $|\lambda| \cdot \|A\| > 1$, It is clear from (3.4) that the basic vectors $\langle h|$ and $|g\rangle$ enter the problem more symmetrically than the original formulation suggests.

If $A(x, y)$ is a constant, (3.4) becomes a geometric series,

$$\langle h|f\rangle = \frac{c_0}{1 - \lambda A},$$

and the [0/1] Padé approximant is exact. More generally, for the [$N - 1/N$] Padé approximant, it follows from (1.3.6) that

$$[N - 1/N]_{\langle h|f\rangle} = \sum_{i,j=0}^{N-1} c_i [C^{-1}]_{ij} c_j, \qquad (3.6)$$

where the matrix

$$C_{ij} \equiv c_{i+j} - \lambda c_{i+j+1}. \qquad (3.7)$$

We will see in this section that we may interpret the [$N - 1/N$] Padé approximants to the Neumann series as the exact solution of an integral equation similar to the original equation (3.1), (3.2).

Let us consider the subspace of the original Hilbert space spanned by the N elements

$$|g\rangle, A|g\rangle, A^2|g\rangle, \ldots, A^{N-1}|g\rangle$$

and define

$$\left.\begin{array}{l} |\psi_i\rangle = A^{i-1}|g\rangle \\ \langle \psi_i'| = \langle h|A^{i-1} \end{array}\right\} \quad i = 1, 2, \ldots, N. \qquad (3.8)$$

Then we define S_N to be the space spanned by $|\psi_1\rangle, |\psi_2\rangle, \ldots, |\psi_N\rangle$, and S_N' to be space spanned by $\langle \psi_1'|, \langle \psi_2'|, \ldots, \langle \psi_N'|$. We define the matrix R by

$$R_{ij} = \langle \psi_i'|\psi_j\rangle = c_{i+j-2}, \quad i, j = 1, 2, \ldots, \qquad (3.9)$$

and the projection operator $\mathcal{P}_N$ by

$$\mathcal{P}_N = \sum_{i,j=1}^{N} |\psi_j\rangle (R^{-1})_{ij} \langle \psi_i'|. \qquad (3.10)$$

It is immediately clear by matrix multiplication that

$$\mathcal{P}_N^2 = \mathcal{P}_N,$$

so $\mathcal{P}_N$ is a projection operator. $\mathcal{P}_N$ projects Hilbert-space vectors into the space S_N. We see that

$$\mathcal{P}_N|\psi_i\rangle = |\psi_i\rangle \quad \text{for } i = 1, 2, \ldots, N,$$

and so $\mathcal{P}_N$ is represented by the identity on S_N. We define T_N to be the complement of S_N in the Hilbert space. In the general case, T_N contains $A^N|g\rangle$, $A^{N+1}|g\rangle$, and many other vectors. These are mapped by $\mathcal{P}_N$ into S_N, but unfortunately it is a difficult matter to put a bound on $\|\mathcal{P}_N\|$ in this case. $\mathcal{P}_N$ is not an orthogonal projection, but an oblique projection, and the following analogy may help to understand it. The horizontal plane of a sundial is the space S_N, and the vertical is T_N. The shadow of the gnomon always lies in S_N, and may have any length, depending on the elevation of the sun. In general it is clear that

$$\|\mathcal{P}_N\| \geq 1,$$

and $\mathcal{P}_N$ may be represented schematically by

$$\mathcal{P}_N \begin{bmatrix} S_N \\ T_N \end{bmatrix} = \begin{bmatrix} I & X \\ 0 & 0 \end{bmatrix} \begin{bmatrix} S_N \\ T_N \end{bmatrix}.$$

It turns out that the $[N-1/N]$ Padé approximant solution of (3.1), (3.2) is the exact solution of the integral equation

$$|f_N\rangle = |g\rangle + \lambda \mathcal{P}_N A |f_N\rangle. \tag{3.11}$$

Proof. Equation (3.11) implies that $|f_N\rangle \in S_N$, and therefore we may write

$$|f_N\rangle = \sum_{i=1}^{N} d_i |\psi_i\rangle. \tag{3.12}$$

To find the $\{d_i\}$, substitute (3.10) and (3.12) in (3.11) to give

$$\sum_{i=1}^{N} d_i |\psi_i\rangle = |g\rangle + \lambda \sum_{i,j,k=1}^{N} |\psi_i\rangle (R^{-1})_{ij} \langle \psi_j' | A | \psi_k \rangle d_k.$$

Forming the inner product with $\langle \psi_m'|$ for $m = 1, 2, \ldots, N$ yields N linear equations for the $\{d_i\}$,

$$\sum_{i=1}^{N} d_i R_{mi} = \langle \psi_m' | g \rangle + \lambda \sum_{k=1}^{N} \langle \psi_m' | A | \psi_k \rangle d_k.$$

From (3.8), (3.9),

$$\sum_{k=1}^{N} (c_{m+k-2} - \lambda c_{m+k-1}) d_k = c_{m-1},$$

and again from (3.8), (3.12)

$$\langle h | f_N \rangle = \sum_{i=1}^{N} d_i c_{i-1}$$

$$= \sum_{i,j=0}^{N-1} c_i (C^{-1})_{ij} c_j,$$

where the matrix C has elements given by (3.7). This derivation proves that the $[N - 1/N]$ Padé approximant to $\langle h|f \rangle$ given by (3.6) is the exact solution of (3.11).

Before moving on, we note that if $\mathcal{P}_N$ were an orthogonal projection, so that $\mathcal{P}_N T_N = 0$, we could prove that $\mathcal{P}_N A \to A$ as $N \to \infty$ and we would deduce that the Padé approximant solution to (3.1) converges. However, $\mathcal{P}_N$ is an oblique projection and we may only deduce convergence of the Padé approximants to (3.1) when constraints are placed on A, $|g\rangle$, and $\langle h|$ so that $\mathcal{P}$ is approximately orthogonal [Baker, 1975b].

We now review the previous method of construction of the Padé-approximant solution of (3.1) using the elements of S_N given by (3.8) in the light of Lanczos's biorthogonal algorithm (method of minimal iterations) [Lanczos, 1952, Golub and Van Loan, 1989, Chapter 9; Brezinski and Sadok, 1993]. This connection is due to Garibotti and Villani [1969a], who showed that the Padé denominator of $\langle h|f_N \rangle$ is an orthogonal polynomial developed in the theory. This algorithm develops the previous analysis by using bases of vectors in S_N and S'_N which are biorthogonal.

From the vectors

$$|\psi_i\rangle = A^{i-1}|g\rangle \in S_N, \quad i = 1, 2, \ldots, N, \quad (3.13)$$

and

$$|\psi'_j\rangle = (A^\dagger)^{j-1}|h\rangle \in S'_N, \quad j = 1, 2, \ldots, N,$$

we construct a new set of functions $|\phi_i\rangle$, $|\phi'_j\rangle$ which satisfy the orthogonality condition that

$$\langle \phi'_i | \phi_j \rangle = \omega_i \delta_{ij}, \quad i, j = 0, 1, \ldots, N - 1. \quad (3.14)$$

The $\{\omega_i\}$ are normalization constants only. In the event that $|g\rangle = |h\rangle$ and $A = A^\dagger$ (i.e., $g(x) = \{h(x)\}^*$ and $A(x, y) = \{A(y, x)\}^*$, it follows that S_N and S'_N are essentially the same space, and the analysis is much simpler, leading to a set of orthogonal vectors in S_N (or S'_N) in the usual sense. In the general case, the biorthogonal algorithm is most easily developed by defining polynomial operators sequentially by

$$p_0(A) = 1, \quad (3.15a)$$

$$p_1(A) = A - \alpha_0, \quad (3.15b)$$

$$p_{i+1}(A) = (A - \alpha_i) p_i(A) - \beta_i p_{i-1}(A) \quad \text{for } i = 1, 2, \ldots, \quad (3.15c)$$

and the constants α_i, β_i required in (3.15) are chosen so that

$$\alpha_i = \frac{\langle \phi_i' | A | \phi_i \rangle}{\langle \phi_i' | \phi_i \rangle} \quad \text{and} \quad \beta_i = \frac{\langle \phi_i' | \phi_i \rangle}{\langle \phi_{i-1}' | \phi_{i-1} \rangle} = \frac{\omega_i}{\omega_{i-1}}. \tag{3.16}$$

We may prove by induction that we may take

$$|\phi_i\rangle = p_i(A)|g\rangle$$

and (3.17)

$$|\phi_i'\rangle = p_i(A^\dagger)|h\rangle$$

for $i = 0, 1, 2, \ldots$ to obey the biorthogonality condition (3.14). The proof is straightforward, and we omit the details.

In the event that A is a finite-rank operator of rank N, the elements $A^N|g\rangle$, $A^{N+1}|g\rangle$, ... are linearly dependent on elements of S_N. From (3.15) we see that the leading term of $p_i(A)$ is A^i, guaranteeing that the space spanned by $|\phi_0\rangle$, $|\phi_1\rangle$, ..., $|\phi_{N-1}\rangle$ is S_N for all N. Hence we see that if A is a finite-rank operator, then $|\phi_N\rangle$ is a linear combination of $|\phi_0\rangle, \ldots, |\phi_{N-1}\rangle$ and it follows from the biorthogonality condition (3.14) that $|\phi_N\rangle = 0$. We conclude that if A has rank N, then $|\phi_N\rangle = 0$, and we prove similarly that $\langle \phi_N' | = 0$.

We may adopt the attitude in the general case that by truncating the orthogonalization process at stage N, for N large enough, we have taken sufficiently many dominant eigenfunctions, characterized by their relatively large eigenvalues, into account. At any rate, we will formally curtail the orthogonalization process by taking $|\phi_N\rangle = \langle \phi_N'| = 0$. This leads to the truncated integral equation

$$|f_N\rangle = |g\rangle + \lambda \mathcal{P}_N A |f_N\rangle, \tag{3.11}$$

which we now solve by using the expansion

$$|f_N\rangle = \sum_{j=0}^{N-1} e_j |\phi_j\rangle. \tag{3.18}$$

Substituting (3.18) in (3.11) and taking the inner product with $\langle \phi_i' |$ yields

$$e_i \omega_i = \omega_0 \delta_{i0} + \lambda \langle \phi_i' | A \sum_{j=0}^{N-1} e_j | \phi_j \rangle. \tag{3.19}$$

From (3.15c) we find that for $i = 1, 2, \ldots, N-2$

$$\langle \phi_i' | A = \langle h | p_i(A) A$$
$$= \langle h | p_{i+1}(A) + \alpha_i p_i(A) + \beta_i p_{i-1}(A)$$
$$= \langle \phi_{i+1}' | + \alpha_i \langle \phi_i' | + \beta_i \langle \phi_{i-1}' |,$$

and hence from (3.19)

$$e_i \omega_i = \lambda(e_{i+1}\omega_{i+1} + \alpha_i e_i \omega_i + \beta_i e_{i-1}\omega_{i-1}).$$

Dividing by ω_i and using (3.16),

$$e_i = \lambda(\beta_{i+1} e_{i+1} + \alpha_i e_i + e_{i-1}). \tag{3.20a}$$

The initial equation for $i = 0$ is

$$e_0 = 1 + \lambda(\beta_1 e_1 + \alpha_0 e_0), \tag{3.20b}$$

and the final equation for $i = N - 1$ is

$$e_{N-1} = \lambda(\alpha_{N-1} e_{N-1} + e_{N-2}). \tag{3.20c}$$

Equations (3.20a)–(3.20c) are N equations for N unknowns $e_0, e_1, \ldots, e_{N-1}$. To solve these equations, let us take $\mu = \lambda^{-1}$, so that they become recursion relations

$$\beta_1 e_1 + (\alpha_0 - \mu)e_0 + \mu = 0, \tag{3.21a}$$

$$\beta_{i+1} e_{i+1} + (\alpha_i - \mu)e_i + e_{i-1} = 0, \quad i = 1, 2, \ldots, N - 2, \tag{3.21b}$$

and

$$(\alpha_{N-1} - \mu)e_{N-1} + e_{N-2} = 0. \tag{3.21c}$$

To make (3.21b) take the form of (3.15c), substitute

$$e'_i = \omega_i e_i \tag{3.22}$$

and multiply (3.21b) by ω_i to give

$$e'_{i+1} + (\alpha_i - \mu)e'_i + e'_{i-1}\beta_i = 0 \tag{3.23a}$$

with the associated equations

$$e'_1 + (\alpha_0 - \mu)e'_0 + \omega_0 \mu = 0 \tag{3.23b}$$

and

$$(\alpha_{N-1} - \mu)e'_{N-1} + e'_{N-2}\beta_{N-1} = 0. \tag{3.23c}$$

The solution of the recursion relation (3.23a) with initial conditions $p_0(\mu) = 1$, $p_1(\mu) = \mu - \alpha_0$ are the polynomials $p_i(\mu)$. Let us define polynomials $\tilde{p}_i(\mu)$ to be the solutions of the recursion (3.23a) with initial conditions $\tilde{p}_0(\mu) = 0$, and $\tilde{p}_1(\mu) = \mu$. Then the solution of (3.23) is a linear combination of these,

$$e'_i = \gamma p_i(\mu) + \delta \tilde{p}_i(\mu), \tag{3.24}$$

provided (3.23b) and (3.23c) are satisfied. These give

$$\delta + \omega_0 = 0$$

and

$$\gamma p_N(\mu) + \delta \tilde{p}_N(\mu) = 0,$$

from which the values of δ and γ follow. Hence (3.24) and (3.22) and (3.18) give $|f_N\rangle$ explicitly, and we find

$$\langle h|f_N\rangle = e_0\omega_0 = e_0' = \gamma$$

$$= \frac{\tilde{p}_N(\mu)}{p_N(\mu)}\omega_0.$$

The significance of this result is that we see directly that the Padé-approximant solution of the integral equation has poles at the zeros of $p_N(1/\lambda)$ and zeros at the zeros of $\tilde{p}_N(1/\lambda)$. By writing the equations (3.21a, b) in the form

$$\beta_1\left(\frac{e_1}{e_0}\right) + (\alpha_0 - \mu) + \frac{\mu}{e_0} = 0$$

and

$$\beta_{i+1}\left(\frac{e_{i+1}}{e_i}\right) + \alpha_i - \mu + \frac{e_{i-1}}{e_i} = 0,$$

it follows that

$$e_0 = \frac{-\mu}{\alpha_0 - \mu +}\;\frac{\beta_1}{\alpha_1 - \mu +}\;\frac{\beta_2}{\alpha_2 - \mu +}\;\cdots\;\frac{\beta_{N-1}}{\alpha_{N-1} - \mu}$$

and hence that

$$\langle h|f_N\rangle = \frac{-\omega_0}{\lambda\alpha_0 - 1 +}\;\frac{\lambda\beta_1}{\lambda\alpha_1 - 1 +}\;\frac{\lambda\beta_2}{\lambda\alpha_2 - 1 +}\;\cdots\;\frac{\lambda\beta_{N-1}}{\lambda\alpha_{N-1} - 1}.$$

Thus we see that the biorthogonal algorithm enables the Padé approximant of the integral equation to be represented as a continued fraction, and the sequence of $[N - 1/N]$ Padé-approximant solutions appear as the Nth convergent of an infinite continued fraction. The coefficients α_i and μ_i of the continued fraction are given initially by (3.16), and the role of the expansion parameter λ is preserved.

9.4 Potential scattering

Many of the applications of Padé approximants have turned up in the theory of potential scattering, because potential-scattering theory provides a fine testing ground for the effectiveness of the Padé method on a profusion of interesting problems. It is also true that potential theory has proved a useful guide to the properties of quantum field theory. Indeed, because of the difficulty of calculating more than a few low-order terms of T-matrix elements in quantum field theory, Padé approximants may prove to be an essential tool for deriving reasonable extrapolations from a few terms of power series. We present here in this section the basic elements of scattering theory needed in our applications in quantum mechanics. We shall follow the notation and normalization of the book by Newton [1966, Chapter 7] where possible, because this book emphasizes K-matrix theory and Jost functions which are a necessary component of the theory of convergence of the Padé method; see also Graves-Morris [1973a].

We use Schrödinger's equation in the two-particle relative coordinate $\mathbf{r}$, and take $\hbar = 1$. The reduced mass of the system, $m = (m_1^{-1} + m_2^{-1})^{-1}$, is taken to be $\frac{1}{2}$, defining the mass scale, so the equation takes the form

$$H\psi = -\nabla^2 \psi(\mathbf{r}) + V(r)\psi(\mathbf{r}) = k^2 \psi(\mathbf{r}). \tag{4.1}$$

We assume that $V(r)$ is a short-range, spherically symmetric potential, which means that a positive μ exists for which $|e^{\mu r}V(r)|$ is bounded as $r \to \infty$. With these hypotheses, the equation (4.1) has scattering solutions of the type

$$\psi^{\text{asy}}(\mathbf{r}) \simeq e^{i\mathbf{k}\cdot\mathbf{r}} + f(\theta)\frac{e^{ikr}}{r} \quad \text{as } r \to \infty. \tag{4.2}$$

Figure 9.4.1. shows the wave fronts in a pictorial representation of the scattering process, with an incident plane wave and a scattered wave.

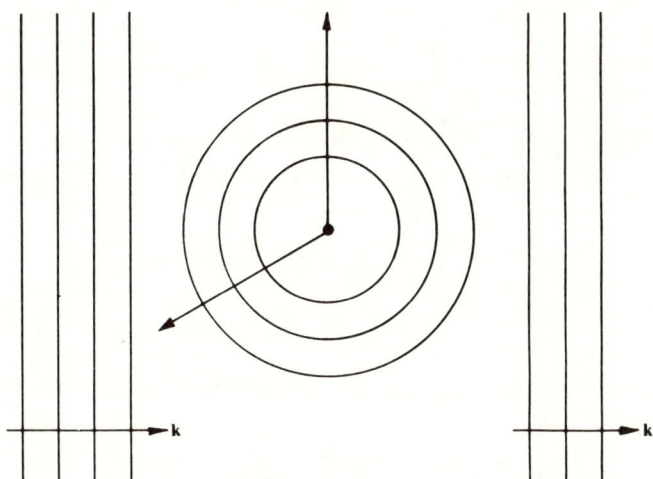

Figure 9.4.1. A pictorial representation of scattering, showing an incident wave of momentum **k** and an outgoing spherical wave.

One of the observable effects of the scattering process is the scattering cross-section, defined as the number of particles scattered per second per unit solid angle per unit incident flux, which is expressed in terms of the scattering amplitude $f(\theta)$ by the formula

$$\frac{d\sigma}{d\Omega} = |f(\theta)|^2.$$

To solve the Schrödinger equation, we write it as

$$(\nabla^2 + k^2)\psi(\mathbf{r}) = V(r)\psi(\mathbf{r})$$

and use the free Green's function, which is a solution of the equation

$$(\nabla^2 + k^2)G(\mathbf{r}, \mathbf{r}', k^2) = \delta(\mathbf{r} - \mathbf{r}') \tag{4.3}$$

representing propagation from the point $\mathbf{r} = \mathbf{r}'$. The solution we need is

$$G(\mathbf{r}, \mathbf{r}', k^2) = -\frac{1}{4\pi} \frac{e^{ik|\mathbf{r}-\mathbf{r}'|}}{|\mathbf{r} - \mathbf{r}'|}, \tag{4.4}$$

and the solution of Schrödinger's equation is given by the Born equation

$$\psi(\mathbf{r}) = e^{i\mathbf{k} \cdot \mathbf{r}} - \int \frac{e^{ik|\mathbf{r}-\mathbf{r}'|}}{4\pi|\mathbf{r} - \mathbf{r}'|} V(r')\psi(\mathbf{r}')\, dr'. \tag{4.5}$$

We consider the form of $\psi(\mathbf{r})$ at a representative point $\mathbf{r}$ far from the scattering center. Let $\hat{\mathbf{r}}$ be a unit vector in the direction of $\mathbf{r}$, and $\mathbf{k}'$ be the momentum associated with scattering in this direction, so that $\mathbf{k}' = k\hat{\mathbf{r}}$. Then the asymptotic form of $\psi(\mathbf{r})$ follows from the Born equation, and is

$$\psi^{\text{asy}}(\mathbf{r}) = e^{i\mathbf{k} \cdot \mathbf{r}} - \frac{e^{ikr}}{4\pi r} \int e^{-i\mathbf{k}' \cdot \mathbf{r}'} V(r')\psi(\mathbf{r}')\, dr' + O\left(\frac{1}{r^2}\right), \tag{4.6}$$

provided $V(r)$ is a short-range potential. This condition also implies that $|\psi(\mathbf{r})|$ is bounded as $r \to \infty$, and by comparing (4.2) and (4.6) we are led to define

$$T(\mathbf{k}', \mathbf{k}) = \langle \mathbf{k}'|T|\mathbf{k} \rangle = -4\pi f(\theta)$$

$$= \int e^{-i\mathbf{k}' \cdot \mathbf{r}} V(r)\psi(\mathbf{r})\, dr \tag{4.7}$$

as the scattering amplitude, and the momentum-space representation of the potential

$$V(\mathbf{k}', \mathbf{k}) = \int e^{-i\mathbf{k}' \cdot \mathbf{r}} V(r) e^{i\mathbf{k} \cdot \mathbf{r}}\, dr$$

$$= \langle \mathbf{k}'|V|\mathbf{k} \rangle. \tag{4.8}$$

9.4 Potential scattering

We can solve for the free Green's function by momentum-space methods, taking care with the boundary conditions, and find as in (4.4) that

$$G(\mathbf{r}, \mathbf{r}', k^2) = \frac{e^{ik|\mathbf{r}-\mathbf{r}'|}}{-4\pi|\mathbf{r}-\mathbf{r}'|} = \lim_{\varepsilon \downarrow 0} \int \frac{d\mathbf{q}}{(2\pi)^3} \frac{e^{i\mathbf{q}\cdot(\mathbf{r}-\mathbf{r}')}}{k^2 - q^2 + i\varepsilon}. \quad (4.9)$$

From (4.5), (4.7), and (4.9),

$$T(\mathbf{k}', \mathbf{k}) = V(\mathbf{k}', \mathbf{k}) + \frac{1}{(2\pi)^3} \int V(\mathbf{k}', \mathbf{q}) \frac{d\mathbf{q}}{k^2 - q^2 + i\varepsilon} T(\mathbf{q}, \mathbf{k}). \quad (4.10)$$

This equation is the Lippmann–Schwinger equation for the half-off-shell T-matrix, which is defined by (4.10) for all real $\mathbf{k}'$. We refer to Reed and Simon [1979, p. 98] for a discussion of how the Lippmann–Schwinger equation is incorporated in rigorous scattering theory. The on-shell T-matrix follows by taking $|\mathbf{k}'| = |\mathbf{k}|$. In three-body theory, it is useful to define a totally off-shell T-matrix by

$$T(\mathbf{k}', k^2, \mathbf{k}'') = V(\mathbf{k}', \mathbf{k}'') + \frac{1}{(2\pi)^3} \int V(\mathbf{k}', \mathbf{q}) \frac{d\mathbf{q}}{k^2 - q^2 + i\varepsilon} T(\mathbf{q}, k^2, \mathbf{k}''), \quad (4.11)$$

and we shall return to this in Section 9.8. Equations (4.10) and (4.11) may be written schematically as

$$T = V + VGT = (1 - VG)^{-1}V$$
$$= V + TGV = V(1 - GV)^{-1}$$
$$= V + VGV + VGVGV + \cdots, \quad (4.12)$$

but these formulas do no more than denote algebraic manipulations on (4.10) and (4.11). Provided the inverses exist and the series converge, the abbreviations are useful. For actual calculations, it is essential to reduce the dimension of (4.10) by expanding in partial waves, by means of the following projections:

$$t^{(l)}_{k',k} = \frac{1}{8\pi} \int_{-1}^{1} T(\mathbf{k}', \mathbf{k}) P_l(\cos \theta_{\mathbf{k}',\mathbf{k}}) \, d(\cos \theta_{\mathbf{k}',\mathbf{k}}), \quad (4.13a)$$

$$v^{(l)}_{k',k} = \frac{1}{8\pi} \int_{-1}^{1} V(\mathbf{k}', \mathbf{k}) P_l(\cos \theta_{\mathbf{k}',\mathbf{k}}) \, d(\cos \theta_{\mathbf{k}',\mathbf{k}}), \quad (4.13b)$$

so that

$$V(\mathbf{k}', \mathbf{q}) = -4\pi \sum_{l=0}^{\infty} (2l+1) P_l(\cos \gamma) v^{(l)}_{k',q} \quad (4.14a)$$

and

$$T(\mathbf{k}', \mathbf{k}) = -4\pi \sum_{l=0}^{\infty} (2l+1) P_l(\cos \beta) t^{(l)}_{k',k}. \quad (4.14b)$$

We have used $\gamma = \angle(\mathbf{k}', \mathbf{q})$ and $\beta = \angle(\mathbf{k}', \mathbf{k})$ in (4.14), and we need $\alpha = \angle(\mathbf{k}, \mathbf{q})$ as well. These angles are shown in Figure 9.4.2. We need the 'addition formula'

$$P_l(\cos \gamma) = P_l(\cos \alpha) P_l(\cos \beta)$$
$$+ 2 \sum_{m=1}^{l} \frac{(l-m)!}{(l+m)!} P_l^m(\cos \alpha) P_l^m(\cos \beta) \cos m\phi$$

to give

$$t_{k'k}^{(l)} = v_{k'k}^{(l)} - \frac{2}{\pi} \int_0^\infty v_{k'q}^{(l)} \frac{q^2 dq}{k^2 - q^2 + i\varepsilon} t_{qk}^{(l)}. \tag{4.15}$$

This equation is the partial-wave Lippmann–Schwinger equation [Goldberger and Watson, 1964, p. 918], given formally by

$$T = V + VGT, \tag{4.16}$$

just as in (4.12). To understand it better, let us consider the cases of exponential and Yukawa potentials. If V is an exponential potential,

$$V(r) = \lambda e^{-\mu r}.$$

With the momentum transfer Δ given by $\Delta^2 = (\mathbf{k} - \mathbf{k}')^2$, the potential is represented by

$$V(\mathbf{k}', \mathbf{k}) = \frac{8\pi \lambda \mu}{(\Delta^2 + \mu^2)^2}$$

in momentum space, and its S-wave component is given by

$$v_{k'k}^{(0)} = \frac{-2\lambda\mu}{[(k'+k)^2 + \mu^2][(k'-k)^2 + \mu^2]}. \tag{4.17}$$

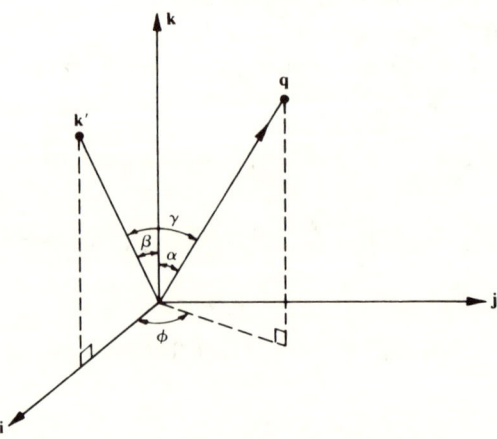

Figure 9.4.2. Angles between vectors $\mathbf{k}'$, $\mathbf{q}$, and $\mathbf{k}$, used to describe the scattering process.

This takes on an especially simple form on the mass shell, and in general allows exact calculations with (4.15). If V is a Yukawa potential,

$$V(r) = \lambda e^{-\mu r}/r,$$

then

$$V(\mathbf{k}', \mathbf{k}) = \frac{4\pi\lambda}{(\mathbf{k}' - \mathbf{k})^2 + \mu^2}$$

and

$$v^{(l)}_{k'k} = \frac{-\lambda}{2kk'} Q_l\left(\frac{k'^2 + k^2 + \mu^2}{2kk'}\right) \quad (4.18)$$

gives the partial-wave projection in terms of the lth second-kind Legendre function. These functions $v^{(l)}_{k'k}$ are well-behaved $\mathscr{L}^2$ functions which cause no difficulties in (4.15). But the partial-wave Lippmann–Schwinger equation (4.15) is a singular integral equation (in the $\mathscr{L}^2$ space), because the full kernel

$$S(k', q) = -\frac{2}{\pi} v^{(l)}_{k'q} \frac{q^2}{k^2 - q^2 + i\varepsilon}$$

does not satisfy the $\mathscr{L}^2$ condition

$$\iint |S(k', q)|^2 \, dk' \, dq < \infty,$$

and so Hilbert–Schmidt methods do not apply directly. However, the formal techniques of (4.12) may be valid for an iterative solution. In fact, the key requirement we need is that $t^{(l)}_{qk}$ be differentiable in q (or a Hölder continuity condition is sufficient) so that the right-hand side of (4.15) is well defined. Hence we are led to use a suitably restricted Banach space for $t^{(l)}_{qk}$. Then VG is a compact operator on the space, $(1 - VG)^{-1}$ is a well-defined operator on the space, and the usual theory of integral equations with compact kernels is applicable.

Let us define the partial-wave K-matrix by the equation

$$K_{pq} = v_{pq} - \frac{2}{\pi} \int K_{pk'} \frac{k'^2 \, dk'}{k^2 - k'^2} v_{k'q}. \quad (4.19)$$

The integral is a principal-value integral, and the superscript l for partial waves is suppressed. A Fourier transform shows that the Green's function analogous to (4.4) for the three-dimensional K-matrix equation is

$$G_r(\mathbf{r}, \mathbf{r}', k^2) = \frac{\cos k|\mathbf{r} - \mathbf{r}'|}{-4\pi|\mathbf{r} - \mathbf{r}'|} = \frac{1}{(2\pi)^3} \int e^{i\mathbf{k}' \cdot (\mathbf{r} - \mathbf{r}')} \frac{d\mathbf{k}'}{k^2 - k'^2}, \quad (4.20)$$

which is real, as is denoted by the subscript r. The wave functions in each partial wave are real and are standing waves, and so the K-matrix is sometimes called the reaction matrix, or R-matrix for short. Equation (4.19) may be written formally as

$$K = V + KG_r V = (1 + KG_r)V. \tag{4.21}$$

Operating on each side of (4.16) with $(1 + KG_r)$ on the left gives

$$(1 + KG_r)T = K + KGT$$

and

$$T = K + K(G - G_r)T.$$

This equation is interpreted as

$$t_{pk} = K_{pk} - \frac{2}{\pi}\int K_{pq}\left(\frac{q^2}{k^2 - q^2 + i\varepsilon} - \frac{q^2}{k^2 - q^2}\right)t_{qk}\,dq$$
$$= K_{pk} + ikK_{pk}t_{kk} \tag{4.22}$$

and the on-shell partial-wave T-matrix is given by

$$t_{kk} = K_{kk}(1 - ikK_{kk})^{-1}. \tag{4.23}$$

Writing $K_{kk}^{(l)} = (\tan \delta_l)/k$, we find from (4.23) that

$$t_{kk}^{(l)} = \frac{e^{i\delta_l}\sin\delta_l}{k}. \tag{4.24}$$

Furthermore, (4.6), (4.7), and (4.14b) imply that

$$\psi^{\text{asy}}(\mathbf{r}) = e^{i\mathbf{k}\cdot\mathbf{r}} + \frac{1}{r}\sum_{l=0}^{\infty}(2l + 1)e^{ikr+i\delta_l}\sin\delta_l P_l(\cos\theta), \tag{4.25}$$

and this equation shows why δ_l is called the phase shift of the lth partial wave.

We have approached the scattering problem by first finding the momentum-space representation and then going to partial waves. Frequently there are advantages in using partial waves ab initio, and then we will need the partial-wave Green's functions explicitly.

If the wave function is separated into partial waves,

$$\psi(\mathbf{r}) = \sum_{l=0}^{\infty}\sum_{m=-l}^{l} a_{lm}P_l^{|m|}(\cos\theta)e^{im\phi}R_l(r), \tag{4.26}$$

then the radial part $R_l(r)$ satisfies Schrödinger's equation in the form

$$\frac{1}{r}\frac{d^2}{dr^2}[rR_l(r)] - \frac{l(l+1)}{r^2}R_l(r) + k^2 R_l(r) = V(r)R_l(r), \tag{4.27}$$

and the substitution

$$\psi_l(r) = rR_l(r) \tag{4.28}$$

reduces (4.27) to

$$\frac{d^2\psi_l(r)}{dr^2} - \frac{l(l+1)}{r^2}\psi_l(r) + k^2\psi_l(r) = V(r)\psi_l(r). \tag{4.29}$$

If $V(r) = 0$, the noninteracting solutions of this equation obey

$$\frac{d^2\psi_l(r)}{dr^2} - \frac{l(l+1)}{r^2}\psi_l(r) + k^2\psi_l(r) = 0. \tag{4.30}$$

The independent solutions are $u_l(kr)$ and $v_l(kr)$, which are defined in terms of Bessel and spherical Bessel functions by

$$u_l(x) = \sqrt{\tfrac{1}{2}\pi x}\, J_{l+1/2}(x) = xj_l(x) \tag{4.31a}$$

and

$$v_l(x) = \sqrt{\tfrac{1}{2}\pi x}\, N_{l+1/2}(x) = xn_l(x). \tag{4.31b}$$

From (4.28) and (4.33a) we see that the radial part of an incident plane wave must be the solution of (4.29) vanishing at $r = 0$, which is $u_l(kr)$. In fact, an incident plane wave has the expansion

$$e^{i\mathbf{k}\cdot\mathbf{r}} = \frac{1}{kr}\sum_{l=0}^{\infty}(2l+1)P_l(\cos\theta_{\mathbf{k},\mathbf{r}})u_l(kr)i^l, \tag{4.32}$$

which may be verified by using $(d/d\cos\theta)^l$ and then $r = 0$. It is conventional to take account of the constant phase i^l explicitly in each partial wave, so that the incident wave is real.

We will need the following facts about $u_l(x)$ and $v_l(x)$, which are easily derived from well-known properties of Bessel functions:

$$u_l(x) \sim \frac{x^{l+1}}{(2l+1)!!} + O(x^{l+3}) \quad \text{as} \quad x \to 0, \tag{4.33a}$$

$$v_l(x) \sim -\frac{(2l-1)!!}{x^l} + O(x^{-l+2}) \quad \text{as} \quad x \to 0, \tag{4.33b}$$

$$u_l(x) \sim \sin(x - \tfrac{1}{2}\pi l) \quad \text{as} \quad x \to +\infty, \tag{4.34a}$$

$$v_l(x) \sim -\cos(x - \tfrac{1}{2}\pi l) \quad \text{as} \quad x \to +\infty. \tag{4.34b}$$

The Wronskian of $u_l(x)$ and $v_l(x)$ is defined by

$$W(u_l, v_l) = \begin{vmatrix} u_l(x) & v_l(x) \\ u_l'(x) & v_l'(x) \end{vmatrix} = u_l(x)v_l'(x) - v_l(x)u_l'(x).$$

Because u_l and v_l satisfy the same second-order ordinary differential equation, it follows that $W(u_l, v_l)$ is independent of x, and either (4.33) or (4.34) shows that

$$W(u_l, v_l) = 1. \tag{4.35}$$

We need the further formulas and definitions

$$w_l^{(+)}(x) = -e^{i\pi l}[v_l(x) - iu_l(x)] \tag{4.36}$$

$$\sim e^{ix + (1/2)i\pi l} \quad \text{as } x \to +\infty. \tag{4.37}$$

We see that $w_l^+(x)$ represents an outgoing, scattered wave and is irregular at $x = 0$. From (4.36),

$$W(u_l, w_l^+) = e^{i\pi(l+1)}. \tag{4.38}$$

As a simple example of these functions, we have the S-wave formulas for (4.31) and (4.36):

$$u_0(x) = \sin x, \quad v_0(x) = -\cos x, \quad \text{and } w_0^+(x) = e^{ix}.$$

The solution of the partial-wave Schrödinger equation (4.29) is found by Green's functions to be

$$\psi_l(r) = u_l(kr) + \int_0^\infty g(r, r', k^2) V(r') \psi_l(r') \, dr', \tag{4.39}$$

provided the Green's function is chosen so that

(i) $$\frac{\partial^2 g}{\partial r^2} - \frac{l(l+1)}{r^2} g + k^2 g = \delta(r - r'), \tag{4.40}$$

(ii) $$g(r, r', k^2) \propto e^{ikr} \quad \text{as } r \to \infty,$$

(iii) $$g(0, r') = 0,$$

so that $\psi_l(r)$ is regular at the origin and the interaction leads to scattered waves only. We deduce that

$$g(r, r', k^2) = f_1(r') u_l(kr) \quad \text{for } r < r',$$
$$g(r, r', k^2) = f_2(r') w_l^+(kr) \quad \text{for } r > r', \tag{4.41}$$

and to satisfy (4.40) at $r = r'$,

$$\left(\frac{\partial g}{\partial r}\right)_{r=r'+} - \left(\frac{\partial g}{\partial r}\right)_{r=r'-} = 1,$$

i.e.,

$$f_2(r) w_l^{+\prime}(kr) - f_1(r) u_l'(kr) = k^{-1}.$$

9.4 Potential scattering

From (4.38) we deduce that

$$f_1(r) = -e^{-i\pi l}k^{-1}w_l^+(kr),$$
$$f_2(r) = -e^{i\pi l}k^{-1}u_l(kr),$$

which allows (4.41) to be written compactly as

$$g^{(l)}(r, r', k^2) = -e^{-i\pi l}u_l(kr_<)w_l^+(kr_>)/k. \qquad (4.42)$$

As an example, the standard Green's function for scattering in S-waves is

$$g^{(0)}(r, r', k^2) = -k^{-1}\sin kr_< e^{ikr_>}.$$

We may deduce from (4.42) and (4.36) that the K-matrix (standing-wave) Green's function, which is real, is

$$g_r^{(l)}(r, r', k^2) = u_l(kr_<)v_l(kr_>)/k. \qquad (4.43)$$

As an example, the standing-wave Green's function for the reaction in S-waves is

$$g_r^{(0)}(r, r', k^2) = -k^{-1}\sin kr_< \cos kr_>.$$

We substitute (4.42) in (4.39) to obtain the asymptotic form

$$\psi_l^{\text{asy}}(r) = \sin(kr - \tfrac{1}{2}\pi l) - k^{-1}e^{ikr-(1/2)i\pi l}\int_0^\infty u_l(kr')V(r')\psi_l(r')\,dr'. \qquad (4.44)$$

Just as with (4.6), we are led to define

$$\tau_l = -k^{-1}\int_0^\infty u_l(kr')V(r')\psi_l(r')\,dr', \qquad (4.45)$$

and then

$$\psi_l^{\text{asy}}(r) = \frac{e^{ikr-(1/2)i\pi l}(1 + 2i\tau_l) - e^{-ikr+(1/2)i\pi l}}{2i}, \qquad (4.46)$$

in which the first and second terms represent the outgoing wave and incoming wave, respectively. Conservation of flux between these waves requires that

$$|1 + 2i\tau_l| = 1,$$

and it is usual to define

$$S_l = 1 + 2i\tau_l = e^{2i\delta_l} \quad \text{and} \quad \tau_l = e^{i\delta_l}\sin\delta_l. \qquad (4.47)$$

Of course, the phase shifts of (4.47) are precisely the same as those of (4.25) by virtue of (4.26), (4.28), (4.32), and (4.34a), which amount to (4.46).

To complete our discussion of partial-wave Green's functions, we need the Green's function for Jost solutions of the partial-wave Schrödinger equation (4.29). These are the solutions

$$f(k, r) \sim e^{ikr} \quad \text{and} \quad f(-k, r) \sim e^{-ikr} \tag{4.48}$$

representing outgoing and incoming waves at infinity. Each is an unphysical solution because it does not conserve flux. The solutions are given by

$$f_l(k, r) = e^{ikr} + \int_0^\infty \tilde{g}_l(r, r', k^2) V(r') f_l(k, r'). \tag{4.49b}$$

We need the Green's function satisfying

$$\tilde{g}(r, r', k^2) = 0 \quad \text{for } r > r', \tag{4.50a}$$

and a derivation similar to that of (4.42) but using (4.35) instead of (4.38) gives

$$\tilde{g}_l(r, r', k^2) = \frac{1}{k} \{ v_l(kr') u_l(kr) - u_l(kr') v_l(kr) \} \quad \text{for } r < r'. \tag{4.50b}$$

For the particular case of S-waves needed in (9.3) below,

$$\tilde{g}(r, r', k^2) = -k^{-1} \sin k(r - r') \quad \text{for } r < r',$$

$$\tilde{g}(r, r', k^2) = 0 \quad \text{for } r > r',$$

giving rise to the Volterra integral equations for the S-wave:

$$f(k, r) = e^{ikr} - k^{-1} \int_r^\infty \sin k(r - r') V(r') f(-k, r') \, dr'$$

and
$$\tag{4.51}$$

$$f(-k, r) = e^{-ikr} - k^{-1} \int_r^\infty \sin k(r - r') V(r') f(-k, r') \, dr'.$$

From these two independent solutions of Schrödinger's equation we may form the physical solution

$$\psi_l(r) = af(k, r) + bf(-k, r).$$

As $r \to \infty$, (4.46)–(4.48) show that $S_l = -(a/b)e^{i\pi l}$. At $r = 0$, $\psi_l(0) = 0$, and hence

$$S_l = \frac{f(-k, 0)}{f(k, 0)} e^{i\pi l}, \tag{4.52}$$

showing how the Jost functions determine the S-matrix.

Considerable attention has been paid to the solution of the Lippmann–Schwinger equation, (4.15), because it is a prototype equation for quantum scattering theory. The natural method of solution of the Lippmann–Schwinger equation using Padé approximants is to use a sequence of nearly diagonal approximants of the Liouville–Neumann series, as described in Section 9.2. The first numerical results for the Yukawa potential using this approach were given by Caser, Piquet, and Vermeulen [1969] and were most encouraging. We emphasize again (cf. Section 2.1 and Schwartz [1966]) that considerable numerical accuracy is required for the coefficients of the Liouville–Neumann series when Padé approximation is used. In this chapter, we discuss various developments of this direct approach to the Lippmann–Schwinger equation. It is known [Lovelace, 1964; Graves-Morris and Rennison, 1974] that the kernel of the Lippmann–Schwinger equation for Yukawa scattering is compact on a suitably chosen Banach space. In the light of the discussion of Section 9.2, it is therefore not surprising that the direct Padé method is successful in this application.

A variety of quite distinct Padé methods have been used to solve the Lippmann–Schwinger equation and related equations. In Section 9.7, we prove that the direct Padé method converges for single-sign ordinary potentials. In Section 9.8, we discuss variational Padé methods which normally improve the accuracy of the direct Padé method. In Section 9.9, we discuss singular potentials. An interesting approach to low-energy scattering problems is to start with the doubly off-shell partial-wave K-matrix Lippmann–Schwinger equation (4.19),

$$K_{pq}(k^2) = v_{pq} + \frac{2}{\pi} \int_0^\infty K_{pk'}(k^2) \frac{k'^2 \, dk'}{k'^2 - k^2} v_{k'q}.$$

The kernel, as a Hilbert-space kernel, is nonsingular for $k^2 \leq 0$, and the equation is easily solved by matrix inversion. The direct Padé method necessarily converges (see Section 9.7). The solution is evaluated for fixed p and q, and N different negative values of k^2. The solution is then extrapolated, using N-point rational interpolants, to the scattering region, $k^2 > 0$ [Schlessinger and Schwartz, 1966]. A similar extrapolation is also useful for the Bethe–Salpeter equation [Haymaker, 1968; Haymaker and Schlessinger, 1970].

The Faddeev equations describe quantum-mechanical three-body potential scattering. They are integral equations, involving double integrals over a kernel containing polar and logarithmic singularities, and are usually coupled integral equations in cases of physical interest. The Faddeev equations are singular in such a complicated way that solution by formation of Padé approximants to the Liouville–Neumann series is a very attractive method [Tjon, 1970, 1973, 1977].

For further details of the role of Padé approximants in particular approaches to the Lippmann–Schwinger equation, we refer to Ruijgrok [1968], Stern and Warburton [1972], and Warburton and Stern [1968].

9.5 Derivation of Padé approximants from variational principles

Variational principles have often turned out to lead to the most accurate methods of solving the Schrödinger equation in bound-state and scattering problems. Certainly it is true that quite drastic approximations to the wave functions lead to surprisingly accurate values of the binding energy in the application of the Rayleigh–Ritz method.

We first discuss bound-state problems in quantum mechanics, which are directly associated with homogeneous integral equations. These are to be distinguished carefully from scattering problems, associated with inhomogeneous integral equations, which we treat second and separately.

Bound-state problems usually involve the determination of discrete energy levels. The associated integral equations have solutions only for particular discrete values of a parameter occurring in the theory. Although it is not equivalent to Padé approximation, we consider first the Rayleigh–Ritz variational method. We will then use this method with a particular approach which yields Padé approximants as the variational solution. We consider a potential $V(r)$ which is at least partly attractive. The Rayleigh–Ritz principle shows that such a potential has at least one bound state, and we wish to determine the energy of the lowest-lying bound state.

Schrödinger's equation for the wave function ψ of a particle in a potential V may be written as

$$H\psi \equiv -\nabla^2\psi + V\psi \equiv H_0\psi + V\psi = E\psi \qquad (5.1)$$

following (4.1). The Rayleigh–Ritz principle is the statement that

$$[E] = \frac{\langle \psi_t | H | \psi_t \rangle}{\langle \psi_t | \psi_t \rangle}, \qquad (5.2)$$

which takes real values for any normalizable function $|\psi_t\rangle$, has a minimum value with respect to variations of the trial function $|\psi_t\rangle$. This minimum value of $[E]$ is the ground-state binding energy, and the associated trial function is the bound-state wave function. The proof of the principle is easy if the existence of a complete orthonormal set of eigenfunctions of H is assumed. The variational method consists of selecting M functions

$$\psi_i(\mathbf{x}), \quad i = 1, 2, \ldots, M, \qquad (5.3)$$

and forming from these a trial wave function

$$\psi_t(\mathbf{x}) = \sum_{i=1}^{M} \alpha_i \psi_i(\mathbf{x}). \qquad (5.4)$$

9.5 Derivation of Padé approximants from variational principles

As an example, we might choose

$$\psi_t(x) = \alpha_1 e^{-|x|} + \alpha_2 e^{-3|x|}.$$

Consequently, $[E]$ becomes a function of $\{\alpha_i\}$; the values of $\{\alpha_i\}$ which give a minimum value of $[E]$ determine the best approximation to the true wave function, and the actual minimum value of $[E]$ is the estimated binding energy. Another form of the variational method allows each $\psi_i(x)$ to depend on other parameters

$$\{\beta_{ij}\}, \quad j = 1, 2, \ldots, M_i,$$

which may also be varied so as to give a minimum of $[E]$. For example,

$$|\psi_t\rangle = \alpha_1 |x|^{\beta_{11}} e^{-\beta_{12}|x|} + \alpha_2 |x|^{\beta_{21}} e^{-3|x|}$$

may be varied with respect to the five variational parameters to yield a minimum value of $[E]$ and an approximation to $\psi(x)$.

The Rayleigh–Ritz method has been used for very precise calculations of the binding energy of atomic helium, where the significant forces are electromagnetic and are accurately known [Pekeris 1958, 1959]. The method has also been used for the calculation of the binding energy of triton using nuclear potentials, where the results are more of a test of reliability of the potentials [Delves and Phillips, 1969].

Having considered the basic Rayleigh–Ritz principle, we contrast this procedure with another variational principle, and see in what form this approach yields Padé approximants as the variational solution. The bound-state problem is reformulated to be the determination of the potential strength that gives a particular bound-state energy. The potential strength is measured by a parameter λ, and we take

$$V(r) = \lambda U(r) \tag{5.5}$$

and consider the functional

$$\left[-\frac{1}{\lambda}\right] = \frac{\langle \psi_t | U(H_0 - E)^{-1} U | \psi_t \rangle}{\langle \psi_t | U | \psi_t \rangle}. \tag{5.6}$$

First we notice that if $|\psi_t\rangle$ is a solution of the Schrödinger equation (5.1), then $[-1/\lambda] = -1/\lambda$. Let us define $G = (H_0 - E)^{-1}$ as the usual free Green's function and consider first-order variations in $[-1/\lambda]$,

$$\delta\left[\frac{-1}{\lambda}\right] = \frac{\langle \psi_t | UGU | \delta\psi_t \rangle}{\langle \psi_t | U | \psi_t \rangle} + \frac{\langle \delta\psi_t | UGU | \psi_t \rangle}{\langle \psi_t | U | \psi_t \rangle}$$
$$- \langle \psi_t | UGU | \psi_t \rangle \langle \delta\psi_t | U | \psi_t \rangle [\langle \psi_t | U | \psi_t \rangle]^{-2}$$
$$- \langle \psi_t | UGU | \psi_t \rangle \langle \psi_t | U | \delta\psi_t \rangle [\langle \psi_t | U | \psi_t \rangle]^{-2}. \tag{5.7}$$

Since the exact solution $|\psi\rangle$ satisfies

$$GV|\psi\rangle = -|\psi\rangle, \tag{5.8}$$

it follows that

$$\delta\left[\frac{-1}{\lambda}\right] = 0 \quad \text{to first order.}$$

Hence we say that the functional $[-1/\lambda]$ is stationary at

$$|\psi_t\rangle = |\psi\rangle. \tag{5.9}$$

Quite clearly, a great deal more needs to be done to give conditions to make the previous formulas rigorous. But provided $\langle\psi_t|U|\psi_t\rangle$ is sufficiently close to $\langle\psi|U|\psi\rangle$, the rigorous discussion adds little to the understanding. In conclusion, this variational method consists of finding a turning point of $[-1/\lambda]$ by varying the parameters of $|\psi_t\rangle$, and it gives the value of the coupling strength λ which produces a bound state of given energy.

We use the following set of trial functions [Nuttall, 1966, 1967, 1970a]:

$$|\psi_t\rangle = \sum_{i=0}^{M-1} d_i(GU)^i|\psi_0\rangle, \tag{5.10}$$

and the coefficients $d_0, \ldots, d_{N-1}$ are to be determined. Then

$$\left[\frac{-1}{\lambda}\right] = \frac{\sum_{i=0}^{M-1}\sum_{j=0}^{M-1} d_i d_j^* \langle\psi_0|U(GU)^{i+j+1}|\psi_0\rangle}{\sum_{i=0}^{M-1}\sum_{j=0}^{M-1} d_i d_j^* \langle\psi_0|U(GU)^{i+j}|\psi_0\rangle}. \tag{5.11}$$

We make $[-1/\lambda]$ stationary by requiring

$$\frac{\partial}{\partial d_i}\left[\frac{-1}{\lambda}\right] = 0 \quad \text{for } i = 0, 1, \ldots M-1 \tag{5.12}$$

and define

$$c_k = \langle\psi_0|U(GU)^k|\psi_0\rangle, \quad k = 0, 1, 2, \ldots. \tag{5.13}$$

Substituting from (5.11) into (5.12), we find the set of linear equations

$$\sum_{j=0}^{M-1} d_j^* c_{i+j+1} - \left[\frac{-1}{\lambda}\right] \sum_{j=0}^{M-1} d_j^* c_{i+j} = 0.$$

The elements c_k of (5.13) are real in the bound-state problem, and hence we see that $d_0, \ldots, d_{M-1}$ are determined by

9.5 Derivation of Padé approximants from variational principles

$$\begin{bmatrix} c_0 + [\lambda]c_1 & c_1 + [\lambda]c_2 & \cdots & c_{M-1} + [\lambda]c_M \\ c_1 + [\lambda]c_2 & c_2 + [\lambda]c_3 & \cdots & c_M + [\lambda]c_{M+1} \\ \vdots & \vdots & & \vdots \\ c_{M-1} + [\lambda]c_M & c_M + [\lambda]c_{M+1} & \cdots & c_{2M-2} + [\lambda]c_{2M-1} \end{bmatrix} \begin{bmatrix} d_0 \\ d_1 \\ \vdots \\ d_{M-1} \end{bmatrix}^* = 0. \quad (5.14)$$

These equations determine $d_0:d_1:\cdots:d_{N-1}$ and $[\lambda] \equiv [1/\lambda]_{st}^{-1}$ from the consistency condition that

$$\begin{vmatrix} c_0 + [\lambda]c_1 & c_1 + [\lambda]c_2 & \cdots & c_{M-1} + [\lambda]c_M \\ c_1 + [\lambda]c_2 & c_2 + [\lambda]c_3 & \cdots & c_M + [\lambda]c_{M+1} \\ \vdots & \vdots & & \vdots \\ c_{M-1} + [\lambda]c_M & c_M + [\lambda]c_{M+1} & \cdots & c_{2M-2} + [\lambda]c_{2M-1} \end{vmatrix} = 0. \quad (5.15)$$

We compare this method with the Padé method of solving the Schrödinger equation (5.1), which we write as a homogeneous integral equation

$$|\psi\rangle = -GV|\psi\rangle. \quad (5.16)$$

To solve (5.16) we introduce a function $|\psi_0\rangle$ and parameters λ and η. λ retains its interpretation through (5.5), and we seek a solution of

$$|\psi\rangle = \eta|\psi_0\rangle - \lambda GU|\psi\rangle \quad (5.17)$$

with $\eta = 0$. Provided $\eta \neq 0$, we find formally

$$|\psi\rangle = \eta\{|\psi_0\rangle - \lambda GU|\psi_0\rangle + (-\lambda GU)^2|\psi_0\rangle + \cdots\},$$

and

$$\langle \psi_0|U|\psi\rangle = \eta\{c_0 - \lambda c_1 + (-\lambda)^2 c_2 + \cdots\}.$$

Forming Padé approximants to the series $c_0 + \lambda' c_1 + \lambda'^2 c_2 + \cdots$, we have

$$Q^{[M-1/M]}(-\lambda)\langle\psi_0|U|\psi\rangle = \eta P^{[M-1/M]}(-\lambda) + O(\lambda^{2M}).$$

We see that a zero of $Q^{[M-1/M]}(-\lambda)$ is consistent with $\eta = 0$ within the approximation scheme. Hence the Padé method boils down to the condition that $Q^{[M-1/M]}(-\lambda) = 0$, which may be written [see (1.3.2)] as

$$\begin{vmatrix} c_0 + \lambda c_1 & c_1 + \lambda c_2 & \cdots & c_{M-1} + \lambda c_M \\ c_1 + \lambda c_2 & c_2 + \lambda c_3 & \cdots & c_M + \lambda c_{M+1} \\ \vdots & \vdots & & \vdots \\ c_{M-1} + \lambda c_M & c_M + \lambda c_{M+1} & \cdots & c_{2M-2} + \lambda c_{2M-1} \end{vmatrix} = 0, \quad (5.18)$$

agreeing with (5.15). We conclude that the Padé method and the specific variational method of (5.10) and (5.11) give identical results.

Let us proceed to the problems of scattering theory, where the Kohn method for Green's function and the Schwinger method for the T-matrix are widely used. Suppose we wish to calculate matrix elements of the full interacting Green's function

$$I = \langle h|(E - H)^{-1}|g\rangle. \quad (5.19)$$

Then a stationary expression for I is

$$[I] = \langle h|\psi_t\rangle + \langle \psi_t'|g\rangle - \langle \psi_t'|E - H|\psi_t\rangle. \quad (5.20)$$

$[I]$ is a bivariational functional, which must be made stationary with respect to independent variations in $|\psi_t\rangle$ and $\langle \psi_t'|$. This leads to

$$\delta[I] = \langle h|\delta\psi_t\rangle - \langle \psi_t'|E - H|\delta\psi_t\rangle + \langle \delta\psi_t'|g\rangle$$
$$- \langle \delta\psi_t'|E - H|\psi_t\rangle - \langle \delta\psi_t'|E - H|\delta\psi_t\rangle. \quad (5.21)$$

Hence we see that $[I]$ is stationary if

$$|\psi_t\rangle = (E - H)^{-1}|g\rangle \quad (5.22a)$$

and

$$\langle \psi_t'| = \langle h|(E - H)^{-1}, \quad (5.22b)$$

and at the stationary point

$$[I] = \langle h|(E - H)^{-1}|g\rangle,$$

agreeing with (5.19). If we make the choices [Nuttall, 1966, 1967, 1970a; Conn, 1974]

$$|\psi_t\rangle = \sum_{i=0}^{M-1} d_i (GU)^i G|g\rangle \quad (5.23a)$$

and

$$\langle \psi_t'| = \sum_{i=0}^{M-1} d_i' \langle h|(GU)^i G, \quad (5.23b)$$

we can then find the stationary points of $[I]$:

$$[I] = \sum_{i=0}^{M-1} (d_i + d_i') \langle h|(GU)^i G|g\rangle$$
$$- \sum_{i=0}^{M-1} \sum_{j=0}^{M-1} d_i' d_j \langle h|(GU)^i [1 - \lambda GU](GU)^j G|g\rangle,$$

and defining $c_i = \langle h|(GU)^i G|g\rangle$ for $i = 0, 1, \ldots$ reduces this expression to

9.5 Derivation of Padé approximants from variational principles

$$[I] = \sum_{i=0}^{M-1} (d_i + d_i')c_i - \sum_{i=0}^{M-1}\sum_{j=0}^{M-1} d_i' d_j (c_{i+j} - \lambda c_{i+j+1}),$$

$$\frac{\partial[I]}{\partial d_i} = 0 \quad \text{if } c_i = \sum_{j=0}^{M-1} d_j'(c_{i+j} - \lambda c_{i+j+1}),$$

$$\frac{\partial[I]}{\partial d_i'} = 0 \quad \text{if } c_i = \sum_{j=0}^{M-1} d_j(c_{i+j} - \lambda c_{i+j+1}).$$

These are $2M$ linear equations for $2M$ unknowns, d_j, d_j', $j = 0, 1, \ldots, M - 1$. We define [as in (3.7)] the matrix

$$C_{ij} = c_{i+j} - \lambda c_{i+j+1},$$

and then (5.20) reduces to

$$[I] = \sum_{i=0}^{M-1}\sum_{j=0}^{M-1} c_i (C^{-1})_{ij} c_j. \tag{5.24}$$

This equation is the variational solution for the interacting Green's-function matrix element using the variations of (5.23). It is also the Padé-approximant solution obtained from expansion of the full Green's function in terms of the free Green's function

$$I = \langle h | G + GVG + GVGVG + \cdots | g \rangle.$$

As usual, we form Padé approximants to the series

$$I = \langle h | G + \lambda GUG + \lambda^2 GUGUG + \cdots | g \rangle$$
$$= c_0 + \lambda c_1 + \lambda^2 c_2 + \cdots,$$

and (1.3.6) shows that

$$[M - 1/M]_I = \sum_{i=0}^{M-1}\sum_{j=0}^{M-1} c_i (C^{-1})_{ij} c_j,$$

confirming (5.24).

It is interesting to perform this calculation with any potential for which the full Green's function is known analytically. For the exponential potential

$$V(r) = \lambda \exp(-r/a),$$

the S-wave radial equation may be solved by making the substitution $x = \exp(-r/a)$. The full S-wave Green's function is [Basdevant and Lee, 1969b]

$$g(r, r', k^2) = \frac{-a\pi}{\sin \pi \nu} \frac{J_{-\nu}(\eta \zeta_>)}{J_{-\nu}(\eta)} [J_\nu(\eta) J_{-\nu}(\eta \zeta_<) - J_{-\nu}(\eta) J_\nu(\eta \zeta_<)],$$

where

$$\zeta_< = \exp(-r_</2a), \quad \zeta_> = \exp(-r_>/2a),$$
$$\eta = 2a\sqrt{-\lambda}, \quad \text{and } v = 2aik.$$

The relevant Bessel functions may be expanded in powers of η, leading to

$$g(r, r', k^2) = \sum_{i=0}^{\infty} \lambda^i c_i(r, r', k^2), \tag{5.25}$$

where the coefficients c_i are known functions of r, r', and k^2. The poles of g are then given approximately by the zeros of $Q^{[M-1/M]}(\lambda)$. These zeros should be independent of r, r' from the mathematical theory of integral equations, or because physically they represent couplings for bound states. Accordingly, one makes a sensible choice, such as $r = r' = a$, upon which we will improve in the next section. The condition

$$Q^{[M-1/M]}(\lambda) = 0 \tag{5.26}$$

then becomes an equation in the variables k, λ giving either the couplings for a bound state of given binding energy or, because we have an analytic solution, the binding energy for a given coupling.

The Schwinger variational method calculates the amplitude T given by

$$T = \langle h|V + V(E - H)^{-1}V|g\rangle \tag{5.27}$$

from the bivariational functional

$$[T] = \langle \psi'_t|V|g\rangle + \langle h|V|\psi_t\rangle - \langle \psi'_t|V - VGV|\psi_t\rangle. \tag{5.28}$$

Then we may prove, as with the Kohn principle, that $[T]$ is stationary with respect to all possible variations of $|\psi_t\rangle$ and $\langle \psi'_t|$ if $|\psi_t\rangle$ and $\langle \psi'_t|$ satisfy

$$\langle \psi'_t| = \langle h| + \langle \psi'_t|VG$$

and

$$|\psi_t\rangle = |g\rangle + GV|\psi_t\rangle. \tag{5.29}$$

In this case,

$$[T] = \langle h|(1 - VG)^{-1}V|g\rangle = \langle h|V + V(E - H)^{-1}V|g\rangle$$
$$= T,$$

proving that $[T]$ is a bivariational functional for the T-matrix. For the scattering problem, we have the formal expansion for the T-matrix elements

$$T = \langle h|V + VGV + VGVGV + \cdots |g\rangle,$$

where $\langle h|, |g\rangle$ are usually plane-wave states or their partial-wave components. We may write

$$T = \lambda\langle h|U|g\rangle + \lambda^2\langle h|UGU|g\rangle + \cdots$$
$$= \lambda c_0 + \lambda^2 c_1 + \cdots$$

and form Padé approximants to T/λ. The result for the $[M - 1/M]$ Padé approximant is precisely the same as for the Schwinger variational method using [Garibotti, 1972]

$$|\psi_t\rangle = \sum_{i=0}^{M-1} d_i(GV)^i|g\rangle$$

and

$$\langle \psi_t'| = \sum_{i=0}^{M-1} d_i\langle h|(VG)^i.$$

The analysis is very similar to that for the Kohn method and not worth repeating. We summarize by noting that for three particular variational methods, the solution obtained is the same as that of the $[M - 1/M]$ Padé approximant.

To complete the discussion, let us observe that we can derive all Padé approximants of type $[N + J - 1/N - J]$, for $J = 0, 1, 2, \ldots, N$, which lie above the diagonal by restricting the choice of trial vectors. For example, in (5.22), the choices

$$|\psi_t\rangle = \sum_{i=0}^{J-1} \lambda^i(GU)^iG|g\rangle + \sum_{i=J}^{M-1} d_i(GU)^iG|g\rangle$$

and

$$\langle \psi_t'| = \sum_{i=0}^{J-1} \lambda^i\langle h|(GU)^iG + \sum_{i=J}^{N-1} d_i'\langle h|(GU)^iG$$

allow a similar analysis, leading to $2(N - J)$ equations for $2(N - J)$ unknowns $\{d_i, d_i'\}$ and eventually to the $[N + J - 1/N - J]$ Padé approximant.

Reverting to the Rayleigh–Ritz principle at the beginning of the section, we consider a refinement which is a variational principle of Bessis [1976]. The Rayleigh–Ritz method and principle are summarized by the equations

$$[E] = \frac{\langle \psi_t|H|\psi_t\rangle}{\langle \psi_t|\psi_t\rangle},$$

$$E_0 = \min_{|\psi_t\rangle}[E].$$

If it so happens that some of the higher moments of the trial wave function $|\psi_t\rangle$ are available, namely

$$\mu_i = \langle \psi_t | H^i | \psi_t \rangle, \tag{5.30}$$

Bessis's method [1976] makes use of this information. We use a supertrial wave function

$$|\psi_s\rangle = \sum_{i=0}^{M-1} d_i H^i |\psi_t\rangle, \tag{5.31}$$

where $|\psi_t\rangle$ is the ordinary trial wave function and $\{d_i\}$ are variational parameters. The principle is now rewritten as

$$E_0 = \min_{|\psi_t\rangle} \min_{\{d_i\}} \frac{\langle \psi_s | H | \psi_s \rangle}{\langle \psi_s | \psi_s \rangle}, \tag{5.32}$$

indicating that the minimization with respect to $\{d_i\}$ is to be done first. Of course, the lowest bound-state energy E_0 would be given exactly if ever minimization over all trial wave functions $|\psi_t\rangle$ could be realized. But because in practice this is approximate, the minimization over $\{d_i\}$ may be worthwhile. Retaining the definition of $\{\mu_i\}$ given by (5.30), we find

$$[E] = \frac{\sum_{i=0}^{M-1}\sum_{j=0}^{M-1} d_i d_j^* \langle \psi_t | H^{i+j+1} | \psi_t \rangle}{\sum_{i=0}^{M-1}\sum_{j=0}^{M-1} d_i d_j^* \langle \psi_t | H^{i+j} | \psi_t \rangle}$$

$$= \frac{\sum_{i=0}^{M-1}\sum_{j=0}^{M-1} d_i d_j^* \mu_{i+j+1}}{\sum_{i=0}^{M-1}\sum_{j=0}^{M-1} d_i d_j^* \mu_{i+j}}. \tag{5.33}$$

We make $[E]$ stationary by requiring

$$\frac{\partial [E]}{\partial d_i} = 0 \quad \text{for } i = 0, 1, \ldots, M-1. \tag{5.34}$$

Substituting (5.33) into (5.34), we find the set of linear equations

$$\sum_{j=0}^{M-1} d_j^* \mu_{i+j+1} - [E] \sum_{j=0}^{M-1} d_j^* \mu_{i+j} = 0. \tag{5.35}$$

The moments $\{\mu_i\}$ are all real, and hence $[E]$ is determined by the consistency condition for the M homogeneous equations (5.35) for $d_0:d_1:d_2:\cdots:d_{M-1}$. This condition is

9.5 Derivation of Padé approximants from variational principles

$$\begin{vmatrix} \mu_1 - [E]\mu_0 & \mu_2 - [E]\mu_1 & \cdots & \mu_M - [E]\mu_{M-1} \\ \mu_2 - [E]\mu_1 & \mu_3 - [E]\mu_2 & \cdots & \mu_{M+1} - [E]\mu_M \\ \vdots & \vdots & & \vdots \\ \mu_M - [E]\mu_{M-1} & \mu_{M+1} - [E]\mu_M & \cdots & \mu_{2M-1} - [E]\mu_{2M-2} \end{vmatrix} = 0,$$

or alternatively

$$\begin{vmatrix} \mu_0 & \mu_1 & \cdots & \mu_M \\ \mu_1 & \mu_2 & \cdots & \mu_{M+1} \\ \vdots & \vdots & & \vdots \\ \mu_{M-1} & \mu_M & \cdots & \mu_{2M-1} \\ 1 & [E] & \cdots & [E]^M \end{vmatrix} = 0. \tag{5.36}$$

Bessis's principle is that a minimum value of the solution $[E]$ of (5.36) is to be found over a set of trial wave functions $|\psi_t\rangle$ which determine $\{\mu_i\}$ through (5.30).

This is precisely the scheme which would emerge by the simplest Padé method. To find the poles in the E-plane of the Green's-function matrix element

$$\langle \psi_t | (E - H)^{-1} | \psi_t \rangle,$$

it is natural to use $w = E^{-1}$ and to write

$$\langle \psi_t | (E - H)^{-1} | \psi_t \rangle = w \langle \psi_t | 1 + wH + w^2 H^2 + \cdots | \psi_t \rangle$$

$$= \sum_{i=0}^{\infty} w^{i+1} \mu_i. \tag{5.37}$$

We see that the solution (5.36) of the variational problem is the position of the pole on the left-hand real w-axis nearest the origin given by the $[M/M]$ Padé approximant to $\langle \psi_t | (E - H)^{-1} | \psi_t \rangle$ of (5.37). The spectrum of the exact full Green's function is shown in Figure 9.5.1.

Probably the most useful solution is that with $m = 2$, with a quadratic equation for $[E]$,

$$[E]^2 \begin{vmatrix} \mu_0 & \mu_1 \\ \mu_1 & \mu_2 \end{vmatrix} + [E] \begin{vmatrix} \mu_2 & \mu_0 \\ \mu_3 & \mu_1 \end{vmatrix} + \begin{vmatrix} \mu_1 & \mu_2 \\ \mu_2 & \mu_3 \end{vmatrix} = 0,$$

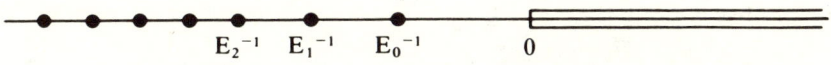

Figure 9.5.1. The spectrum of the full Green's function.

which has an explicit solution in terms of μ_0, μ_1, μ_2, μ_3. Even so, Bessis's principle requires the minimization over the parameters of $|\psi_t\rangle$ of a nonlinear and complicated function. However, the variational justification of it shows that it must be at least as accurate as the simple Rayleigh–Ritz principle.

In this section we have concentrated on the connection between Padé approximants and variational principles rather than the bounding properties which may be derived under suitable conditions. In this context, we refer to Nuttall [1966, 1967, 1970a, 1977], Bessis and Villani [1975], Giraud [1978], and Giraud and Turchetti [1978]. For further details of the connection with Brillouin–Wigner and Rayleigh–Schrödinger perturbation theories, we refer to the review of Killingbeck [1977], and to Amos [1978], Benofy and Gammel [1977], Young, Biedenharn, and Feenberg [1957], Schofield [1972], and Goldhammer and Feenberg [1956].

9.6 An error bound on Padé approximants from variational principles

Surprisingly, variational methods such as the Kohn method, for which the error takes a simple form, can lead to (a posteriori) error estimates for the Padé solution to integral equations. We revert to the formalism of Section 9.3 for integral equations of the type

$$|f\rangle = |g\rangle + \lambda A|f\rangle \tag{6.1}$$

and the Hermitian-conjugate equation

$$\langle f'| = \langle h| + \lambda \langle f'|A. \tag{6.2}$$

To solve (6.1) for the quantity $\langle h|f \rangle$, we consider the variational quantity

$$[J] = \langle h|\psi_t\rangle + \langle \psi_t'|g\rangle - \langle \psi_t'|1 - \lambda A|\psi_t\rangle. \tag{6.3}$$

The stationary value of $[J]$ occurs when

$$|\psi_t\rangle = |f\rangle \quad \text{and} \quad \langle \psi_t'| = \langle f'|, \tag{6.4}$$

under which circumstances we find, as in Section 9.5, that

$$[J]_{\text{st}} = J = \langle h|f\rangle = \langle f'|g\rangle. \tag{6.5}$$

When $|\psi_t\rangle$ and $\langle \psi_t'|$ are not set according to (6.4), the variation of $[J]$ is given by

$$\delta[J] = [J] - J. \tag{6.6}$$

Recall that if

$$|\psi_t\rangle = \sum_{i=0}^{N-1} d_i A^i |g\rangle \tag{6.7a}$$

9.6 An error bound on Padé approximants from variational principles

and

$$\langle \psi'_t | = \sum_{i=0}^{N-1} d'_i \langle h | A^i \quad (6.7b)$$

are the variational solutions of (6.3), then $[J]$ is the Padé approximant to $\langle h|f \rangle$ and $[\delta J]$ is the error of the Padé approximation. We define the differences

$$|\delta \psi_t \rangle = |\psi_t \rangle - |f \rangle \quad (6.8a)$$

and

$$\langle \delta \psi'_t | = \langle \psi'_t | - \langle f' |. \quad (6.8b)$$

From (6.3), (6.5), (6.6), and (6.8) we find, after cancellation, that

$$\delta[J] = -\langle \delta \psi'_t | 1 - \lambda A | \delta \psi_t \rangle. \quad (6.9)$$

This expression may be written symmetrically and be bounded by

$$|\delta[J]| \leq \frac{\|\langle \delta \psi'_t |(1 - \lambda A)\| \|(1 - \lambda A)|\delta \psi_t \rangle\|}{1 - |\lambda| \|A\|} \quad (6.10)$$

provided $|\lambda| < \|A\|^{-1}$.

We consider

$$(1 - \lambda A)|\delta \psi_t \rangle = (1 - \lambda A)(|\psi_t \rangle - |f \rangle)$$
$$= (1 - \lambda A)|\psi_t \rangle - |g \rangle, \quad (6.11)$$

using (6.1) and (6.8a). From (5.24) and the preceding equations,

$$|\psi_t\rangle = \frac{-1}{\det C} \begin{vmatrix} 0 & c_0 & \cdots & c_{N-1} \\ |g\rangle & c_0 - \lambda c_1 & \cdots & c_{N-1} - \lambda c_N \\ A|g\rangle & c_1 - \lambda c_2 & \cdots & c_N - \lambda c_{N+1} \\ \vdots & \vdots & & \vdots \\ A^{N-1}|g\rangle & c_{N-1} - \lambda c_N & \cdots & c_{2N-2} - \lambda c_{2N-1} \end{vmatrix},$$

where

$$\det C = \begin{vmatrix} c_0 - \lambda c_1 & \cdots & c_{N-1} - \lambda c_N \\ \vdots & & \vdots \\ c_{N-1} - \lambda c_N & \cdots & c_{2N-2} - \lambda c_{2N-1} \end{vmatrix} \quad (6.12)$$

and $c_i = \langle h | A^i | g \rangle$ for $i = 0, 1, 2, 3, \ldots$. It follows from (6.1a) and (6.11) that

$$(1 - \lambda A)|\psi_t\rangle - |g\rangle = (1 - \lambda A)|\delta \psi_t\rangle = \frac{-(-\lambda)^N}{\det C} |G\rangle, \quad (6.13)$$

where

$$|G\rangle = \begin{vmatrix} |g\rangle & c_0 & c_1 & \cdots & c_{N-1} \\ A|g\rangle & c_1 & c_2 & \cdots & c_N \\ \vdots & \vdots & \vdots & & \vdots \\ A^N|g\rangle & c_N & c_{N+1} & \cdots & c_{2N-1} \end{vmatrix}. \quad (6.14a)$$

We define similarly

$$\langle H| = \begin{vmatrix} \langle h| & c_0 & c_1 & \cdots & c_{N-1} \\ \langle h|A & c_1 & c_2 & \cdots & c_N \\ \vdots & \vdots & \vdots & & \vdots \\ \langle h|A^N & c_N & c_{N+1} & \cdots & c_{2N-1} \end{vmatrix}. \quad (6.14b)$$

Noting that $\det C = Q^{[N-1/N]}(\lambda)$ in (6.12), it follows from (6.10), (6.12), (6.13), and (6.14) that

$$|\delta[J]| \leq \frac{|\lambda|^{2N} \langle H|H\rangle^{1/2} \langle G|G\rangle^{1/2}}{(1 - \lambda\|A\|)Q^{[N-1/N]}(\lambda)^2} \quad (6.15)$$

provided that $|\lambda| < \|A\|^{-1}$. This formula is a bound on the error of the Padé approximants, which is calculated for

$$\delta[J] = \langle h|\chi_t\rangle - \langle h|f\rangle = \langle \chi_t'|g\rangle - \langle f'|g\rangle.$$

The construction of (6.15) does use additional information, such as values of $\langle g|(K^\dagger)^j K^k|g\rangle$, which is not used in the construction of the Padé approximants themselves. It is valid within the circle of convergence of the Neumann series of the integral equation. Most important, it is a formula comprising the accuracy-through-order concept, namely, the factor $|\lambda|^{2N}$ in (6.15). It can be used to find bounds for Padé approximants to particular functions which may be represented by matrix elements of the solution of integral equations

$$\langle h|f\rangle = \sum_{i=0}^{\infty} c_i \lambda^i$$

within the circle of convergence. It is also easy to generalize it to refer to $[N + J - 1/N - J]$ Padé approximants with $J = 0, 1, 2, \ldots, N$, [Barnsley and Baker, 1976].

For further details of what may be achieved using methods similar to those of this section, we refer to Barnsley and Robinson [1974a–c] and Robinson and Barnsley [1979].

9.7 Single-sign potentials in scattering theory etc.

We consider in this section the problems of scattering from purely repulsive or purely attractive regular potentials in quantum mechanics.

9.7 Single-sign potentials in scattering theory etc.

We can prove that the Padé approximants of the forward-scattering K-matrix and the partial-wave scattering amplitude converge. These results also hold for the Bethe-Salpeter equation in the one-particle-exchange approximation for equal-mass scattering. We start with potential theory and make the foregoing remarks more explicit.

The key requirement on the potential is that [Tani, 1966b; Masson, 1967a, b; Garibotti and Villani, 1969a, b]

$$V(r) = \lambda U(r) \quad \text{with } U(r) \geq 0 \quad \text{for all } r.$$

This enables $U^{1/2}(r)$ to be defined to be real and positive. In three-dimensional momentum space, it is represented by

$$U^{1/2}(p, q) = \int e^{-i(\mathbf{p}-\mathbf{q})\cdot \mathbf{r}} U^{1/2}(r) \, d\mathbf{r},$$

and in one dimension, in the lth partial wave,

$$U_l^{1/2}(p, q) = \int_0^\infty u_l(pr) U^{1/2}(r) u_l(qr) \, dr.$$

The K-matrix has been defined by (4.19):

$$K = V + VPGK$$
$$= V + KPGV$$
$$= \lambda U + \lambda UPGK. \tag{7.1}$$

This K-matrix has the Born–Neumann expansion

$$K = \lambda U + \lambda^2 UPGU + \lambda^3 UPGUPGU + \cdots. \tag{7.2}$$

Equation (7.1) is a formal integral equation. It is easily made explicit in momentum space in one and three dimensions. The expansion (7.2) suggests that we focus attention on the symmetric kernel $U^{1/2}PGU^{1/2}$. This kernel is a real and symmetric $\mathscr{L}^2$ kernel, and so Hilbert-Schmidt theory implies that

$$[U^{1/2}PGU^{1/2}] \doteq \sum_{i=0}^\infty \frac{\phi_i(\mathbf{p})\phi_i(\mathbf{q})}{\lambda_i}.$$

We have used the three-dimensional momentum-space interpretation; the λ_i are real, $\{\phi_i(\mathbf{p})\}$ are real orthogonal $\mathscr{L}^2$ functions, and convergence is in the $\mathscr{L}^2$ norm, i.e. in the mean. We may now sum the series (7.2) after leaving the first two terms and obtain

$$K = \lambda U + \lambda^2 UPGU + \lambda^3 U^{1/2} \left[\sum_{i=0}^\infty \phi_i \frac{1}{\lambda_i(\lambda_i - \lambda)} \phi_i \right] U^{1/2}. \tag{7.3}$$

It follows that the residue of the ith pole of $K(p, p)/\lambda$ is

$$\gamma_i = -\lambda_i \left[\int U^{1/2}(\mathbf{p}, \mathbf{q}) \phi_i(\mathbf{q}) \, d\mathbf{q} \right]^2,$$

which is real. Therefore $K(\mathbf{p}, \mathbf{p})/\lambda$ has only poles with positive residues on the negative real axis and poles with negative residues on the positive real axis. An analogous result holds in one dimension for $K_l(p, p)/\lambda$. In this sense, we say that the K-matrix for forward scattering and the on-shell partial-wave K-matrix are both Hamburger series, and we deduce that the sequence of $[M - 1/M]$ Padé approximants of both $K(\mathbf{p}, \mathbf{p})/\lambda$ and $K_l(p, p)/\lambda$ converge. There is a slight ambiguity about how to define Padé approximants to (7.2), because the series may be interpreted as having a vanishing first term. We choose this interpretation, that the first term is zero, preserving the fundamental idea that the $[L/M]$ Padé approximant to (7.2) agrees with (7.2) to order λ^{L+M} inclusive. We recall from (4.23) that the on-shell partial-wave T-matrix is a homographic transform of the K-matrix,

$$T_l(k, k) = \frac{K_l(k, k)}{1 - ikK_l(k, k)}.$$

The usual accuracy-through-order proof shows that

$$[L/M]_{T_l(\lambda)} = [L/M]_{K_l(\lambda)} \{1 - ik[L/M]_{K_l(\lambda)}\}^{-1}$$

provided $L \leq M$. In particular, we note that Padé approximants of type $[M - 1/M]$ for $K_l(\lambda)/\lambda$ suggested by the theory of Section 9.2 as the preferred choice, correspond to unitary $[M/M]$ approximants of the partial wave S-matrix. In conclusion, we have proved that the diagonal sequence of Padé approximants to the on-shell partial wave S-Matrix or T-matrix for single-sign potential scattering converges and is unitary. It has also been proved from (7.3) [Garibotti and Villani, 1969b; Baker, 1975b] that the numerators and denominators of the diagonal approximants of the S-matrix converge separately to Jost functions, as indicated by (4.52).

We obtain similar results for the Bethe-Salpeter equation by following a similar approach. We refer to the classic paper of Schwartz and Zemach [1966] for a fuller explanation of the scattering theory with the Bethe-Salpeter equation in the two-particle sector, using the Wick rotation, and we borrow their notation for the theory we need. The Wick rotation is a 90° rotation in the complex energy (p_0) plane which is valid for the Bethe-Salpeter equation below the three-particle production threshold and removes the two-particle propagator singularities from the equation. The wave function, which now involves a relative time coordinate,

9.7 Single-sign potentials in scattering theory etc.

satisfies the integral equation

$$\psi(x) = e^{i\mathbf{k}\cdot\mathbf{r}} + \int d^4x' G(x, x') I(x') \psi(x'), \tag{7.4}$$

where $I(x')$ denotes the interaction, $G(x, x')$ is the equal-mass two-particle Green's function, and $x = (\mathbf{r}, t)$. The formulas are

$$G(x, x') = \int \frac{d^4p}{(2\pi)^4} \frac{e^{ip(x-x')}}{[\mathbf{p}^2 - (p_0 + \omega)^2 + m^2 - i\varepsilon][\mathbf{p}^2 - (p_0 - \omega)^2 + m^2 - i\varepsilon]} \tag{7.5a}$$

$$= \frac{i}{16\pi\omega} \left\{ \frac{e^{ik|\mathbf{r}-\mathbf{r}'|}}{|\mathbf{r}-\mathbf{r}'|} - \left(\int_{-\infty}^{-\omega} + \int_{\omega}^{\infty}\right) \frac{d\beta}{\pi} e^{i\beta(t-t')} K_0(QR) \right\}, \tag{7.5b}$$

where

$$R = \sqrt{(\mathbf{r}-\mathbf{r}')^2 - (t-t')^2}, \quad Q = \sqrt{\beta^2 - k^2},$$

$$\omega = \sqrt{k^2 + m^2}, \quad \text{and } |x| = \sqrt{r^2 - t^2}.$$

The interaction is given by one-particle exchange:

$$I(x) = \frac{4\lambda\mu}{|x|} K_1(\mu|x|) \approx \frac{\lambda\sqrt{8\pi\mu}}{|x|^{3/2}} e^{-\mu|x|} \quad \text{for } |x| \to +\infty, \tag{7.6}$$

showing Yukawa exponential falloff for large $|x|$. A necessary constituent of the nonrelativistic and the relativistic analysis is a real Green's function, and we denote the standing-wave solutions associated with a real Green's function by a subscript r. The associated standing-wave function obeys

$$\psi_r(x) = e^{i\mathbf{k}\cdot\mathbf{r}} + \int d^4x' G_r(x, x') I(x') \psi_r(x'),$$

where

$$G_r(x, x') = \int \frac{d^4p}{(2\pi)^4} e^{ip(x-x')}$$
$$\times \{[\mathbf{p}^2 - (p_0 + \omega)^2 + m^2 - i\varepsilon]^{-1}$$
$$\times [\mathbf{p}^2 - (p_0 - \omega)^2 + m^2 - i\varepsilon]^{-1}$$
$$+ 2\pi^2 \delta^+[(p_0+\omega)^2 - m^2]\delta^+[(p_0-\omega)^2 - m^2]\} \tag{7.7a}$$

$$= G(x, x') + \frac{i}{32\pi\omega|\mathbf{r}-\mathbf{r}'|}(e^{-ik|\mathbf{r}-\mathbf{r}'|} - e^{ik|\mathbf{r}-\mathbf{r}'|})$$

$$= \frac{i}{16\pi\omega}\left\{\frac{\cos k|\mathbf{r}-\mathbf{r}'|}{|\mathbf{r}-\mathbf{r}'|} - \left(\int_{-\infty}^{-\omega} + \int_{\omega}^{\infty}\right)\frac{d\beta}{\pi}e^{i\beta(t-t')}K_0(QR)\right\}. \tag{7.7b}$$

The asymptotic form, as $r \to \infty$ along the direction of $\mathbf{k}'$, of $\psi(x)$ follows from (7.4), (7.5b) and is

$$\psi(x) \sim e^{i\mathbf{k}\cdot\mathbf{r}} + \frac{ie^{ikr}}{16\pi\omega r}\int d^4x' e^{-i\mathbf{k}'\cdot\mathbf{x}'} I(x')\psi(x').$$

This motivates the definition of the T-matrix, namely

$$T(\mathbf{k}', \mathbf{k}) = \frac{i}{16\pi\omega}\int d^4x\, e^{-i\mathbf{k}'\cdot\mathbf{r}} I(x)\psi(x). \tag{7.8a}$$

By analogy, Nuttall [1967] defines the R-matrix (the relativistic K-matrix or reaction matrix) as

$$R(\mathbf{k}', \mathbf{k}) = \frac{i}{16\pi\omega}\int d^4x\, e^{-i\mathbf{k}'\cdot\mathbf{r}} I(x)\psi_r(x). \tag{7.8b}$$

Equation (7.8a) gives the Born approximation to the T-matrix,

$$T^B(\mathbf{k}', \mathbf{k}) = \frac{\pi\lambda/\omega}{(\mathbf{k}-\mathbf{k}')^2 + \mu^2}.$$

We now make the Wick rotation, $t = \tau e^{-i\phi}$, in which ϕ increases from 0 to $\pi/2$ and τ is real. We define

$$\psi_r(\mathbf{r}, t) = \phi_r(\mathbf{r}, \tau)$$

and

$$I(\mathbf{r}, t) = V(\mathbf{r}, \tau).$$

It follows that $|x| = \sqrt{\mathbf{r}^2 + \tau^2}$ and $V(\mathbf{r}, \tau) = 4\lambda\mu K_1(\mu|x|)/|x|$. Thus we may define the positive quantity

$$U(\mathbf{r}, \tau) = 4\mu K_1(\mu|x|)|x|^{-1},$$

which has a well-defined square root. This overcomes the immediate problem of finding the analogue of the single-sign potential function. Next, we need a real integral equation with a symmetric kernel to determine $R(\mathbf{k}, \mathbf{k}')$. This follows from (7.7b) using

$$H_r(x, x') = \frac{1}{16\pi\omega}\left\{\frac{\cos k|\mathbf{r}-\mathbf{r}'|}{|\mathbf{r}-\mathbf{r}'|} - \left(\int_{-\infty}^{-\omega} + \int_{\omega}^{\infty}\right)\frac{d\beta}{\pi} e^{\beta(\tau-\tau')} K_0(QR)\right\},$$

which is real. Hence we find

$$\phi_r = e^{i\mathbf{k}\cdot\mathbf{r}} + \lambda H_r U \phi_r, \tag{7.9}$$

showing that ϕ_r obeys an integral equation with a real symmetric kernel. The inhomogeneous term of (7.9) is complex.

Equation (4.3) shows that the even-wave ($S, D, G, \ldots$) components of $e^{i\mathbf{k}\cdot\mathbf{r}}$ are real, and that the odd-wave ($P, F, H, \ldots$) components are purely imaginary. Equation (7.9) shows that ϕ_r shares this property. We express (7.8b) as

$$R(\mathbf{k}', \mathbf{k}) = \frac{1}{16\pi\omega} \int d\mathbf{r} \, d\tau \, e^{-i\mathbf{k}' \cdot \mathbf{r}} V(\mathbf{r}, \tau) \phi_r(\mathbf{r}, \tau). \tag{7.10}$$

The factor $e^{-i\mathbf{k}' \cdot \mathbf{r}}$ in (7.10) has an expansion of the form (4.3). We deduce that each partial-wave component of $R(\mathbf{k}', \mathbf{k})$ defined by (7.10) is real. Using a representation analogous to (7.3) and the discussion given previously, we see that the forward-scattering R-matrix, $R(\mathbf{k}, \mathbf{k})/\lambda$, is a Hamburger series in the coupling strength. Similarly, we deduce that the on-shell partial-wave projections of $R(\mathbf{k}', \mathbf{k})/\lambda$ are Hamburger series in the coupling strength, with a convergent set of $[M - 1/M]$ approximants. Further, the partial-wave T-matrix is given by the origin-preserving homographic transform

$$T_l(k, k) = \frac{R_l(k, k)}{1 - ikR_l(k, k)},$$

and so the diagonal sequence of Padé approximants to $T_l(k, k)$ also converges.

Partial-wave projections of (7.9) and (7.10) do not seem to have been used directly for solving the Bethe–Salpeter equation. These equations may be used with great success as ingredients for a Schwinger-type variational principle. For our purposes, they establish convergence of diagonal and subdiagonal sequences of Padé approximants to $T_l(k, k)$, and the momentum-space representation is used for calculations. This still leaves free the choices of whether to use a Wick rotation, subtraction methods, and the T or R matrix. The simplest conceptually is the unrotated, unsubtracted T-matrix equation, which explains the principles involved. Using (7.8a) and taking $k = (\mathbf{k}, 0)$, we define

$$T(k', k) = \frac{i}{16\pi\omega} \int d^4x \, e^{ik'x} I(x) \psi(x). \tag{7.11}$$

Substituting (7.4) and (7.5a) into (7.11) gives

$$T(k', k) = V(k', k)$$
$$+ \frac{16\pi\omega}{(2\pi)^4 i} \int \frac{d^4q \, V(k', q) T(q, k)}{[\mathbf{q}^2 - (q_0 + \omega)^2 + m^2 - i\varepsilon][\mathbf{q}^2 - (q_0 - \omega)^2 + m^2 - i\varepsilon]}, \tag{7.12}$$

where

$$V(k', k) = \frac{i}{16\pi\omega} \int d^4x \, e^{i(k-k')x} I(x).$$

The fully off-shell extension is given by using k'' instead of the on-shell energy momentum k in (7.12), which then becomes

$$T(k'; \omega; k'') = V(k', k'')$$
$$+ \frac{\omega}{\pi^3 i} \int \frac{d^4q\, V(k', q) T(q; \omega; k'')}{[q^2 - (q_0 + \omega)^2 + m^2 - i\varepsilon][q^2 - (q_0 - \omega)^2 + m^2 - i\varepsilon]}. \quad (7.13)$$

The definitions

$$k' = (\mathbf{p}', \omega'), \quad p' = |\mathbf{p}'|, \quad k'' = (\mathbf{p}'', \omega''), \quad p'' = |\mathbf{p}''|$$

explain the partial-wave decomposition

$$T(k'; \omega; k'') = \sum_{l=0}^{\infty} (2l+1) P_l(\cos\theta_{p',p''}) t_l(\omega', p'; \omega; \omega'', p''). \quad (7.14)$$

Following the method of (4.13), (4.14), and (4.15), we substitute (7.14) into (7.13), giving

$$t_l(\omega', p'; \omega; \omega'', p'') = v_l(\omega', p'; \omega'', p'')$$
$$+ \frac{4\omega}{\pi^2 i} \int \frac{q^2\, dq\, dq_0\, v_l(\omega', p'; q_0, q) t_l(q_0, q; \omega; \omega'', p'')}{[q^2 - (q_0 + \omega)^2 + m^2 - i\varepsilon][q^2 - (q_0 - \omega)^2 + m^2 - i\varepsilon]} \quad (7.15)$$

This equation is the partial-wave Bethe–Salpeter equation in momentum space, normalized so that the on-shell amplitude is

$$t_l(0, p; \sqrt{p^2 + m^2}; 0, p) = \frac{e^{i\delta} \sin\delta}{k} \quad (7.16)$$

in the elastic region. The inhomogeneous term of (7.13) is given by

$$V(k', k) = \frac{\lambda}{\pi\omega} \frac{1}{(k' - k)^2 + \mu^2 - i\varepsilon},$$

and so the inhomogeneous term of (7.15) is given by

$$v_l(\omega', p'; \omega'', p'') = \frac{\lambda}{\pi\omega} \frac{1}{2p'p''} Q_l\left(\frac{p'^2 + p''^2 + \mu^2 - (\omega' - \omega'')^2 - i\varepsilon}{2p'p''}\right). \quad (7.17)$$

Equations (7.15), (7.16), and (7.17) give the on-shell partial-wave t-matrix element, and we have already proved that the diagonal Padé sequence converges in the elastic region. We return to this in Section 9.8.

As a corollary to this section, concerned with exploitation of the kernel $U^{1/2}GU^{1/2}$, we include the results obtainable from the kernel $G^{1/2}UG^{1/2}$. No longer is U required to be single-signed, and the requirement falls on G. For the Lippmann-Schwinger equation (7.19), we may take

9.7 Single-sign potentials in scattering theory etc.

$$G(q; k^2; q) = -\frac{2}{\pi} \frac{q^2}{k^2 - q^2 + i\varepsilon} \delta(q - q'),$$

which has a well-defined square root (as an operator)

$$G^{1/2}(q; k^2; q') = \sqrt{\frac{2}{\pi}} \frac{q}{(-k^2 + q^2)^{1/2}} \delta(q - q')$$

provided k^2 is negative.

For the Bethe-Salpeter equation, the Wick-rotated kernel written in the form

$$H(x, x') = \frac{1}{16\pi^2 \omega} \int_{-\omega}^{\omega} d\beta e^{\beta(\tau - \tau')} K_0(QR)$$

is real and positive provided $\omega < m$, which allows $H^{1/2}$ to be defined. We now write the T-matrix formally as $T = V + VGV + VG^{1/2}(G^{1/2}VG^{1/2} + (G^{1/2}VG^{1/2})^2 + \cdots)G^{1/2}V$, knowing that $G^{1/2}VG^{1/2}$ is a real symmetric kernel for $k^2 < 0$. Thus T/λ is a Hamburger series, and therefore the diagonal sequence of Padé approximants of T converges. The implications are that the (unphysical) forward-scattering T-matrix and partial-wave T-matrix have formal expansions, and the locations of poles of the Padé approximants of these are proved to determine the couplings for bound states of given energy of the system.

The rigorous results of this section can be extended to the case where the potential $V(r)$ does not have a single sign. The potential is decomposed as $V(r) = V_1(r) + V_2(r)$, where the whole interaction with $V_l(r)$ can be treated analytically (e.g., the square-well or Coulomb potential) and where $V_2(r)$ has a single sign. The eigenstates of the Hamiltonian including $V_1(r)$ [but not $V_2(r)$] are used as basis states, and then $V_2(r)$ is treated as a perturbation. Using this method, a distorted-wave formalism allows rigorous results about convergence of Padé approximants to the scattering amplitude to be established. We refer to Michalík [1970], Alder, Trautman, and Viollier [1973], Giraud, Khalil, and Moussa [1976], and Khalil [1977] for details.

So far in this section, we have considered cases in which the direct Padé methods for the Bethe–Salpeter and Lippmann-Schwinger equations are proved to converge. Numerical experience has shown that the direct Padé method is extremely successful when applied to the Bethe–Salpeter equation for nucleon–nucleon scattering, although no convergence proofs exist in this case. Plausible assumptions are made about the nature of the internucleon exchange forces, and the results of these calculations are most interesting as tests of the underlying theories [Gersten et al., 1976; Fleischer and Tjon, 1975].

9.8 Variational Padé approximants

In Section 9.5, variational quantities such as $[E]$ in the Rayleigh–Ritz principle are minimized with respect to all possible variations of the trial functions. Indeed, we saw that the Padé method and the variational method give identical results using a particular basis for the trial functions. Naturally, this suggests that the Padé method may be improved by varying suitable parameters, such as the analogues of $\{\beta_{ij}\}$ described following (5.14) [Alabiso, Butera, and Prosperi, 1970, 1971, 1972a].

In the bound-state and scattering problems of Sections 9.4, 9.5 the full partial-wave Green's function $g(r, r', k^2)$ of (5.25) and its Padé approximant were expressed as functions of k^2, r, and r'. It is also convenient to use the momentum-space representation

$$\bar{g}(k', k'', k^2) = \int_0^\infty \int_0^\infty u_l(k'r) u_l(k''r') g(r, r', k^2) \, dr \, dr'. \tag{8.1}$$

The point is that the T-matrix elements and their Padé approximants are expressed as functions of r, r' (or k', k'') as well as k^2. The poles of the exact Green's functions and the exact T-matrices occur at values of k which do not depend on r, r' (or k', k''), as follows from the theory of Section 9.2. Thus it is natural to evaluate the Padé approximants at values of r, r' (or k', k'') which are stationary points for $[I]$ or $[T]$ in (5.20) or (5.28). For example, the poles of the $[M - 1/M]$ Padé approximant to $[T]$ depend on the off-shell momenta k' and k''. For scattering, the easiest method is to use on-shell values $k' = k'' = k$, but the best solution is to use values of k', k'' such that

$$\frac{\partial [T]}{\partial k'} = \frac{\partial [T]}{\partial k''} = 0. \tag{8.2}$$

Figures 9.8.1 and 9.8.2 show that this leads to a marked improvement in the accuracy of low-order approximants. However, Alabiso et al. [1971] indicate that the proportional improvement is worse with higher-order approximants. There are, of course, ambiguities about what to do if the turning point is not unique, but it is usually possible to discriminate among a few discrete alternatives. This variational method should be seen as an alternative to the vector methods of Section 8.4. However, they can be made equivalent by the use of Dirac weights in the vector method.

For a further insight into the techniques likely to be successful in quantum field theories, we return to equation (8.2). That use of the off-shell momentum as a variational parameter is an extraordinarily successful technique in scattering problems is an empirical fact. Nevertheless, the optimal value of this parameter is not given by the on-shell value of the momentum, even approximately. This deviation of empirical fact

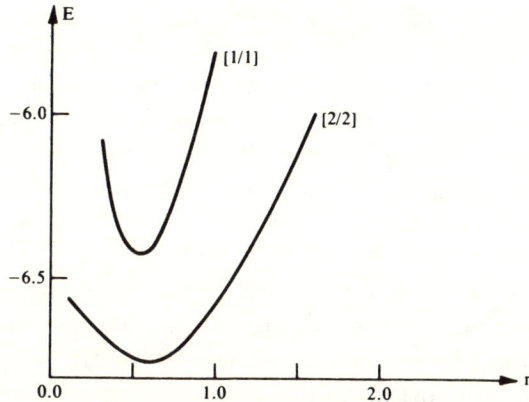

Figure 9.8.1. The energy of the first S-wave bound state, calculated from the $[1/1]$ and $[2/2]$ Padé approximants of the Green's function $g(r, r, k^2)$ in (5.25), against r, for the potential $V(r) = -29 \exp(-r)$. [C. Alabiso et al., 1970]

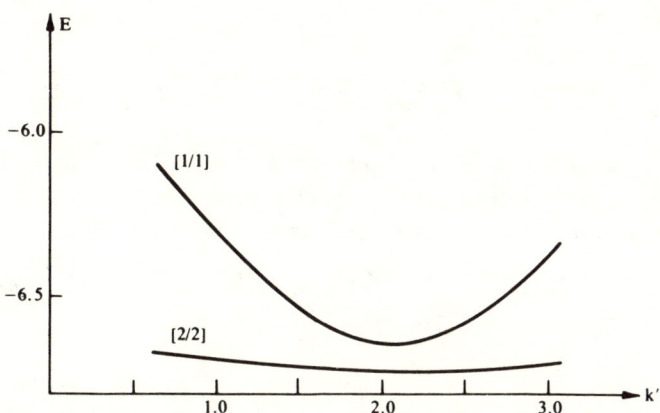

Figure 9.8.2. The energy of the first S-wave bound state, calculated from the $[1/1]$ and $[2/2]$ Padé approximants of the Green's function $\tilde{g}(k', k', k^2)$ in (7.1), against k', for the potential $V(r) = -29 \exp(-r)$. [C. Alabiso et al., 1970]

from physical intuition indicates a deficiency in the analysis, and it is the use of matrix Padé approximants (see Section 8.2) for the T-matrix that appears to offer the most natural solution to certain scattering problems.

The simplest matrix Padé method of calculating a scattering amplitude $\langle k|T|k \rangle$ consists of forming the power-series expansion of

$$\begin{bmatrix} \langle k|T|k \rangle & \langle k|T|k' \rangle \\ \langle k'|T|k \rangle & \langle k'|T|k' \rangle \end{bmatrix} = \sum_{i=1}^{\infty} \lambda^i t_i. \tag{8.3}$$

Here each t_i is a 2×2 matrix, k is the on-shell momentum, and k' is the off-shell momentum, which may be used as the variational parameter. Equation (8.3) may be understood as an expression of four Born series of a particular partial-wave amplitude, expanded in the coupling strength λ. Other interpretations using a similar formalism are possible, such as that of a forward-scattering amplitude $\langle \mathbf{k}|T|\mathbf{k}\rangle$, etc. The formalism of matrix Padé approximants of the series (8.3) enables the scattering amplitudes to be reconstructed; subsequently k' is varied until the reconstructed approximation of $\langle k|T|k\rangle$ is stationary.

We will show that this method unifies the techniques of variation of parameters of the trial wave function with the method of summation of the Born series, using the ideas of Section 9.5. Let $U(r)$ be the potential of unit strength, so that $V(r) = \lambda U(r)$ is the given potential. Our basic result is summarized by the result of Nuttall [1973]:

Theorem 9.8.1. *Let $g_1, g_2, \ldots, g_N$ and $h_1, h_2, \ldots, h_N$ be scattering-state wave functions, denoted by vectors $|\mathbf{g}\rangle$ and $\langle \mathbf{h}|$. The matrix Padé method requires formation of an $[M - 1/M]$ approximant of the matrix series*

$$T/\lambda = \sum_{i=0}^{\infty} \lambda^i \langle \mathbf{h}|(UG)^i U|\mathbf{g}\rangle.$$

If $1 \le \alpha \le N$, $1 \le \beta \le N$, the value of $T_{\alpha\beta} = \langle h_\alpha | T | g_\beta \rangle$ derived by this method is the same as that derived from the Schwinger principle (5.28) for $T_{\alpha\beta}$ using trial functions which are linear combinations of

$$g_1, g_2, \ldots, g_N, GUg_1, \ldots, GUg_N, (GU)^2 g_1, \ldots, (GU)^2 g_N, \ldots,$$
$$(GU)^{M-1} g_1, \ldots, (GU)^{M-1} g_N$$

and

$$h_1^*, h_2^*, \ldots, h_N^*, GUh_1^*, \ldots, GUh_N^*, (GU)^2 h_1^*, \ldots, (GU)^2 h_N^*, \ldots,$$
$$(GU)^{M-1} h_1^*, \ldots, (GU)^{M-1} h_N^*.$$

Proof. The method of proof is the same for arbitrary N as for $N = 2$; for ease of presentation of the matrices we take $N = 2$. Thus our trial wave functions are

$$|\psi_t\rangle = \sum_{i=0}^{M-1} d_i^{(1)} (GU)^i |g_1\rangle + \sum_{i=0}^{M-1} d_i^{(2)} (GU)^i |g_2\rangle$$

and

$$\langle \psi_t' | = \sum_{i=0}^{M-1} \tilde{d}_i^{(1)} \langle h_1 | (UG)^i + \sum_{i=0}^{M-1} \tilde{d}_i^{(2)} \langle h_2 | (UG)^i. \tag{8.4}$$

9.8 Variational Padé approximants

We consider the Schwinger bivariational functional

$$[T_{\alpha\beta}] = \langle \psi_t'|V|g_\beta\rangle + \langle h_\alpha|V|\psi_t\rangle - \langle \psi_t'|V - VGV|\psi_t\rangle, \qquad (8.5)$$

where $\alpha, \beta = 1, 2$. The functional is stationary when

$$\frac{\partial [T_{\alpha\beta}]}{\partial d_i^{(1)}} = \frac{\partial [T_{\alpha\beta}]}{\partial d_i^{(2)}} = \frac{\partial [T]_{\alpha\beta}}{\partial \tilde{d}_i^{(1)}} = \frac{\partial [T]_{\alpha\beta}}{\partial \tilde{d}_i^{(2)}} = 0$$

for $i = 0, 1, \ldots, M - 1$. These lead to the equations

$$\langle h_\alpha|U(GU)^i|g_1\rangle - \langle \psi_t'|(U - \lambda UGU)(GU)^i|g_1\rangle = 0, \qquad (8.6)$$

$$\langle h_\alpha|U(GU)^i|g_2\rangle - \langle \psi_t'|(U - \lambda UGU)(GU)^i|g_2\rangle = 0, \qquad (8.7)$$

$$\langle h_1|(UG)^iU|g_\beta\rangle - \langle h_1|(UG)^i(U - \lambda UGU)|\psi_t\rangle = 0, \qquad (8.8)$$

$$\langle h_2|(UG)^iU|g_\beta\rangle - \langle h_2|(UG)^i(U - \lambda UGU)|\psi_t\rangle = 0, \qquad (8.9)$$

which hold for $i = 0, 1, \ldots, M - 1$. Equations (8.6), (8.7) simplify to

$$T_{\alpha 1}^{(i)} - \sum_{j=0}^{M-1} \tilde{d}_j^{(1)}\{T_{11}^{(i+j)} - \lambda T_{11}^{(i+j+1)}\}$$

$$- \sum_{j=0}^{M-1} \tilde{d}_j^{(2)}\{T_{21}^{(i+j)} - \lambda T_{21}^{(i+j+1)}\} = 0,$$

$$T_{\alpha 2}^{(i)} - \sum_{j=0}^{M-1} \tilde{d}_j^{(1)}\{T_{12}^{(i+j)} - \lambda T_{12}^{(i+j+1)}\}$$

$$- \sum_{j=0}^{M-1} \tilde{d}_j^{(2)}\{T_{22}^{(i+j)} - \lambda T_{22}^{(i+j+1)}\} = 0,$$

which are $2M$ equations for $\tilde{d}_j^{(1)}, \tilde{d}_j^{(2)}$, $j = 0, 1, \ldots, M - 1$. They can be written conveniently in block matrix form as

$$\begin{bmatrix} T_{11} - \lambda T_{11}^{(+)} & T_{21} - \lambda T_{21}^{(+)} \\ T_{12} - \lambda T_{12}^{(+)} & T_{22} - \lambda T_{22}^{(+)} \end{bmatrix} \begin{bmatrix} \tilde{\mathbf{d}}^{(1)} \\ \tilde{\mathbf{d}}^{(2)} \end{bmatrix} = \begin{bmatrix} T_{\alpha 1} \\ T_{\alpha 2} \end{bmatrix},$$

where the elements of the $M \times M$ block matrices are

$$(T_{\alpha\beta})_{ij} = T_{\alpha\beta}^{(i+j)} \text{ and } (T_{\alpha\beta}^{(+)})_{ij} = T_{\alpha\beta}^{(i+j+1)}, \quad i, j = 0, \ldots, M - 1.$$

Hence

$$[\tilde{\mathbf{d}}^{(1)} \tilde{\mathbf{d}}^{(2)}] = [T_{\alpha 1} \quad T_{\alpha 2}] \begin{bmatrix} T_{11} - \lambda T_{11}^{(+)} & T_{12} - \lambda T_{12}^{(+)} \\ T_{21} - \lambda T_{21}^{(+)} & T_{22} - \lambda T_{22}^{(+)} \end{bmatrix}^{-1}. \qquad (8.10)$$

Similarly we may derive from (8.8), (8.9) that

$$\begin{bmatrix} \mathbf{d}^{(1)} \\ \mathbf{d}^{(2)} \end{bmatrix} = \begin{bmatrix} T_{11} - \lambda T_{11}^{(+)} & T_{12} - \lambda T_{12}^{(+)} \\ T_{21} - \lambda T_{21}^{(+)} & T_{22} - \lambda T_{22}^{(+)} \end{bmatrix}^{-1} \begin{bmatrix} T_{1\beta} \\ T_{2\beta} \end{bmatrix}. \quad (8.11)$$

Substituting (8.10), (8.11) in (8.4), (8.5) gives the stationary value of $[T_{\alpha\beta}]$ as

$$[T_{\alpha\beta}]_{\text{st}} = \begin{bmatrix} T_{\alpha 1} & T_{\alpha 2} \end{bmatrix} \begin{bmatrix} T_{11} - \lambda T_{11}^{(+)} & T_{12} - \lambda T_{12}^{(+)} \\ T_{21} - \lambda T_{21}^{(+)} & T_{22} - \lambda T_{22}^{(+)} \end{bmatrix}^{-1} \begin{bmatrix} T_{1\beta} \\ T_{2\beta} \end{bmatrix}. \quad (8.12)$$

This is the Nuttall compact form of the 2×2 matrix Padé approximant. The $N \times N$ generalized form of this formula is entirely obvious, and the result for general N follows by this proof.

Nuttall's theorem establishes the N-dimensional $[M - 1/M]$ matrix Padé method as a variational method in a $2MN$-dimensional space, thereby accounting for the remarkable accuracy of the method.

It is interesting to note that the $[0/1]$-two-dimensional variational matrix Padé method is exact for scattering from a square-well potential [Fratamico, Ortolani, and Turchetti, 1976; Graves-Morris, 1978b]:

$$g_1 = h_1^* \propto j_l(kr)$$

for the on-shell wave function [see (4.31a)], and

$$g_2 = h_2^* \propto j_l(qr)$$

for the off-shell wave function, depending on the variational parameter q. Let $V(r) = \lambda \theta(b - r)$ be a square-well potential of strength λ and range b. With units in which $\hbar = 2m = 1$, the momentum q inside the well is given by

$$q^2 = k^2 - \lambda. \quad (8.13)$$

With this value of q, $g_2(q)$ is the exact wave function within the well. However, (8.5) is independent of the wave function outside the well, and consequently g_2 is effectively the exact wave function if q is given by (8.13). The Schwinger principle asserts that $[T_{\alpha\beta}]$ is the exact scattering amplitude and that $|\psi_t\rangle$ is the exact wave function if the exact wave function is a linear combination of the basis wave functions. Therefore $[T_{\alpha\beta}]$ has a turning point and takes the exact value of the scattering amplitude when q is given by (8.13). Note that (8.13) does not necessarily give the only turning point of $[T_{\alpha\beta}]$, and that the order of Padé approximation is immaterial: the lowest-order $[0/1]$ approximant suffices for an exact result.

The moral of this analyis is that a variational 2×2 matrix Padé method is probably as efficient a method as any for scattering from a single Yukawa-like potential; what is best for the realistic, sign-changing nuclear potentials is not yet known.

As a prelude to the use of variational Padé approximants in quantum field theory, the Bethe–Salpeter equation was tackled [Alabiso et al., 1972b]. Equation (7.15) shows that ω', p', ω'', and p'' are the four available variational parameters. This cornucopia is grasped and controlled by the empirical observation that $\omega' = \omega'' = 0$ is usually the optimal choice. In short, on-shell energy variables should be used. The authors choose to work with a single symmetric variational parameter by taking $p' = p''$ in the Wick-rotated, once-subtracted Bethe–Salpeter R-matrix equation. Their results confirm the view that for good accuracy in low-order Padé approximation, a variational technique is essential. We show in Figure 9.8.3 their results for the first S-wave bound state, at $s = 1$, corresponding to $\omega = -\frac{1}{2}$ with $m = \mu = 1$ [see (7.5) and (7.6)]. The figure shows the values of λ needed for this bound state versus the variational momentum p'. There is a clear turning point near $p' = 1.25$, and the importance of using the variational principle to select this value is very evident.

In practice, variational methods have been very successful in conjunction with the solution of the Bethe-Salpeter equation for nucleon–nucleon scattering. A variety of tricks have been used to reduce the problems to manageable proportions, and we refer to Fleischer, Gammel, and Menzel [1973], Fleischer and Tjon [1980], and Bessis and Turchetti [1977] for details.

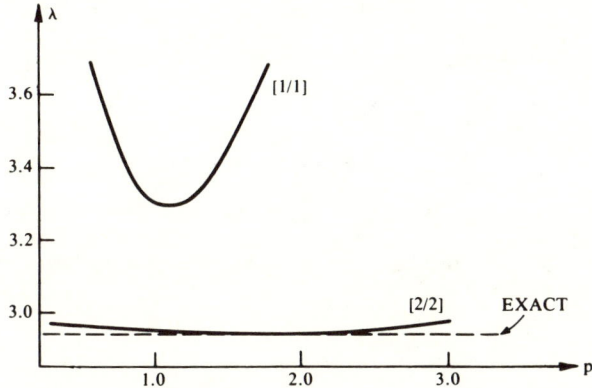

Figure 9.8.3. The coupling constant λ required to give a bound state at $s = 1$, against the off-shell momentum p' for the [1/1] and [2/2] Padé approximants to the S-wave Bethe–Salpeter R-matrix.

For further details on matrix Padé approximants and variational techniques, we refer to Turchetti [1976], Benofy and Gammel [1977], Benofy et al. [1976], Bessis et al. [1977], Mery [1977], and Pindor [1979b].

9.9 Singular potentials

A remarkable advantage of Padé approximants is their ability to treat the problems of singular potentials. A potential such as $V(r) = \lambda r^{-4}$ is so strong at short range that any negative value of λ causes bound states of infinite binding energy. This may be deduced from the Rayleigh–Ritz principle (5.2). Normally, only positive values of λ are considered to be physical. From a mathematical point of view, a Frobenius expansion of the radial Schrödinger equation

$$\psi_l(r) = r^\alpha(a_0 + a_1 r + a_2 r^2 + \cdots)$$

leads to an indicial equation for α which cannot be satisfied. Potentials more divergent at the origin than $V(r) = \lambda r^{-2}$ are called singular potentials. Schrödinger's equation still has scattering solutions corresponding to scattering from a repulsive singular potential, and it turns out that $\psi_l(r)$ has an essential singularity at $r = 0$ in this case. The review of Frank, Land, and Spector [1971] contains an exhaustive account of the best-known examples of singular potentials and their properties. The Padé method consists of regularizing the potentials with a cutoff [Garibotti, Pellicoro, and Villani, 1970]. The general method is quite clear from a specific example, and so we consider the potential

$$V(r) = \lambda r^{-4} \tag{9.1}$$

and define a cut-off potential

$$V_\varepsilon(r) = \lambda r^{-4} \theta(r - \varepsilon) \tag{9.2a}$$

$$= \lambda r^{-4}, \quad r > \varepsilon,$$

$$= 0, \quad r \leq \varepsilon. \tag{9.2b}$$

We are going to use the Jost method, and so the equation for the Jost functions, (4.5.1),

$$f(k, r) = e^{ikr} - \frac{1}{k}\int_r^\infty \sin k(r - r') V_\varepsilon(r') f(k, r') \, dr', \tag{9.3}$$

is a Volterra integral equation with an $\mathscr{L}^2$ kernel. If the potential were not cut off, the kernel would not be $\mathscr{L}^2$. We will use an approximation valid

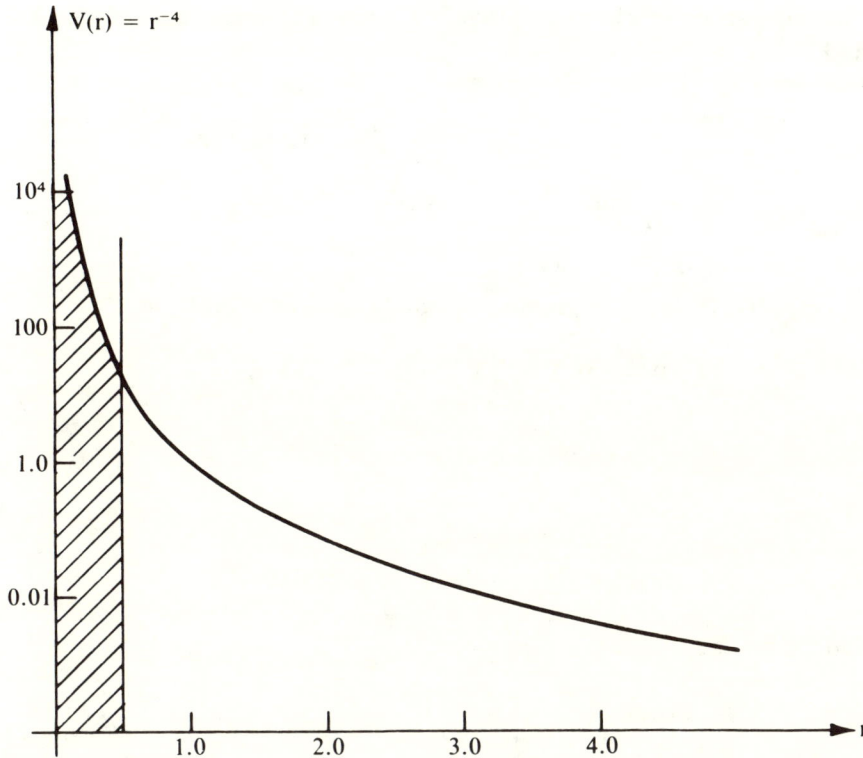

Figure 9.9.1. The potential $V(r) = r^{-4}$ plotted *logarithmically* against r. A cutoff at $r = 0.5$ is shown.

for small k. It follows that

$$f(0, r) = 1 + \int_r^\infty (r' - r) V_\varepsilon(r') f(0, r') \, dr', \tag{9.4a}$$

$$\frac{\partial f}{\partial k}(0, r) = ir + \int_r^\infty (r' - r) V_\varepsilon(r') \frac{\partial f}{\partial k}(0, r') \, dr'. \tag{9.4b}$$

Define

$$f_A(k, r) = f(0, r) + k \frac{\partial f}{\partial k}(0, r). \tag{9.5}$$

Then

$$f_A(k, r) = 1 + ikr + \int_r^\infty (r' - r) V_\varepsilon(r') f_A(k, r') \, dr'.$$

We solve this equation iteratively for $r > \varepsilon$. (Of course, this calculation is easy for our particular example, but even in more complicated situations it is quite straightforward to provide a numerical iterative solution.) We find

$$f_A(k, r) = 1 + ikr + \lambda\left(\frac{r^{-2}}{3!} + \frac{r^{-1}}{2!}ik\right) + \lambda^2\left(\frac{r^{-4}}{5!} + \frac{r^{-3}}{4!}ik\right) + \cdots$$

$$= \frac{r}{\sqrt{\lambda}}\sinh\frac{\sqrt{\lambda}}{r} + ikr\cosh\frac{\sqrt{\lambda}}{r} \quad \text{for } r > \varepsilon.$$

We need to take $r = 0$ for the Jost function; the easiest method of doing this is to use (9.4a, b) to provide a formula for $f_A(k, r)$ valid at $r = 0$:

$$f_A(k, 0) = 1 + \int_\varepsilon^\infty r^{-2}\left(\sqrt{\lambda}\sinh\frac{\sqrt{\lambda}}{r} + ik\lambda\cosh\frac{\sqrt{\lambda}}{r}\right) dr$$

$$= \cosh\frac{\sqrt{\lambda}}{\varepsilon} + ik\sqrt{\lambda}\sinh\frac{\sqrt{\lambda}}{\varepsilon}.$$

Hence

$$f_A(-k, 0) = \cosh\frac{\sqrt{\lambda}}{\varepsilon} - ik\sqrt{\lambda}\sinh\frac{\sqrt{\lambda}}{\varepsilon},$$

and from (4.52),

$$\exp(2i\delta) = \frac{1 - ik\sqrt{\lambda}\tanh(\sqrt{\lambda}/\varepsilon)}{1 + ik\sqrt{\lambda}\tanh(\sqrt{\lambda}/\varepsilon)} + O(k^2).$$

Our convention is that $\lambda > 0$ gives a positive, repulsive potential and normally a negative phase shift. But, for notational convenience in this section only, we use the nuclear-physics convention that the scattering length a is positive in this situation. Therefore the scattering length, which depends on the coupling strength λ and the cutoff ε, is given by

$$a(\varepsilon, \lambda) = \sqrt{\lambda}\tanh\frac{\sqrt{\lambda}}{\varepsilon}. \tag{9.6}$$

At this point we must assume that the scattering length $a(\varepsilon, \lambda)$ caused by the potential $V(r)$ is the limit as $\varepsilon \to 0$ of the cut-off potentials $V_\varepsilon(r)$. Then we deduce that $a(\lambda) = \lambda^{1/2}$. Let us note the following important features of the cutoff solution, which all follow (9.6):

(i) $$\frac{a(\varepsilon, \lambda)}{\lambda} = \frac{1}{\varepsilon} - \frac{1}{3}\frac{\lambda}{\varepsilon^3} + \frac{2}{15}\frac{\lambda^2}{\varepsilon^5} - \cdots = \frac{1}{\sqrt{\lambda}}\tanh\frac{\sqrt{\lambda}}{\varepsilon}.$$

This is a Stieltjes series which is regular at $\lambda = 0$, and has poles at $\lambda = -(\varepsilon n\pi/2)^2$ for n odd and integral.

(ii) $$\lim_{\varepsilon \downarrow 0}\left(\frac{\partial}{\partial\varepsilon}\right)^n a(\varepsilon, \lambda) = 0 \text{ for all } n > 0.$$

(iii) $\qquad a(\varepsilon, \lambda) < a(\lambda)$ for $\varepsilon > 0$.

(iv) $\qquad [N - 1/N]_{\lambda^{-1}a(\varepsilon,\lambda)} = \dfrac{1}{\varepsilon} \dfrac{p_{N-1}(\lambda/\varepsilon^2)}{q_N(\lambda/\varepsilon^2)},$

where p_{N-1} and q_N are polynomials of degrees $N - 1$ and N, respectively.

Let us define

$$a_N(\varepsilon) = [N/N]_{a(\varepsilon,\lambda)} = \lambda [N - 1/N]_{\lambda^{-1}a(\varepsilon,\lambda)}.$$

Then we find the surprising result that when the cutoff is zero or infinity,

$$a_N(0) = a_N(\infty) = 0.$$

These properties are shown in Figure 9.9.2, which is a plot of $a_N(\varepsilon, \lambda)/\lambda^{1/2}$ against $\varepsilon/\lambda^{1/2}$. The value of the scattering length with no cutoff, $\lambda^{1/2}$, is shown as a straight line with ordinate one, corresponding to the limiting values of $a_N(\varepsilon, \lambda)/\lambda^{1/2}$.

We are now able to understand what happens in the general case of an arbitrary singular repulsive potential at zero energy. The arguments of

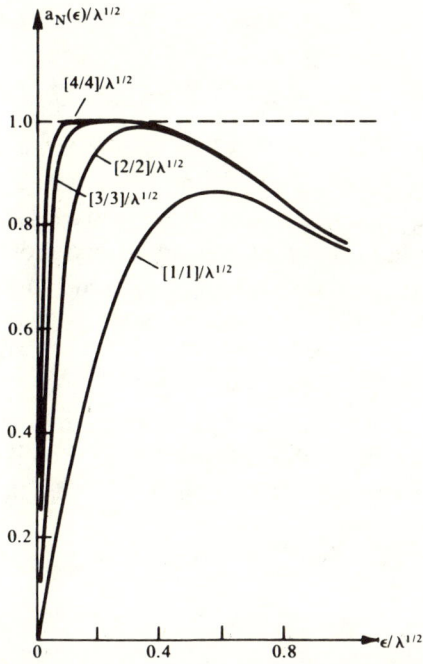

Figure 9.9.2. $[N/N]$ Padé approximants to the scattering length for the regularized potential $V(r) = \lambda r^{-4} \theta(r - \varepsilon)$. From Graves-Morris [1973a].

Section 9.7 show that $\lambda^{-1}a(\varepsilon, \lambda)$ is a Stieltjes series. Using either Jost solutions or the Lippmann–Schwinger equation, we can construct $a(\varepsilon, \lambda)$ as a power series in λ. We can then form Padé approximants to the series, and use the theorems of Chapter 5 to prove that

$$[N/N]_{\lambda^{-1}a(\varepsilon,\lambda)} \geq \frac{a(\varepsilon, \lambda)}{\lambda} \geq [N - 1/N]_{\lambda^{-1}a(\varepsilon,\lambda)}.$$

Hence

$$a_N(\varepsilon, \lambda) \equiv [N/N]_{a(\varepsilon,\lambda)} \leq a(\varepsilon, \lambda)$$

and

$$\lim_{N \to \infty} a_N(\varepsilon, \lambda) = a(\varepsilon, \lambda).$$

The whole purpose of our work is to find

$$a(\lambda) = \lim_{\varepsilon \to \infty} a(\varepsilon, \lambda) = \lim_{\varepsilon \to 0} \lim_{N \to \infty} a_N(\varepsilon, \lambda). \quad (9.7)$$

The example $V(r) = \lambda r^{-4}$ shows that the limits cannot be interchanged, because $a_N(0, \lambda) = 0$. In fact, $a_N(0, \lambda) = 0$ is a general result for all singular potentials, and the combined result

$$a_N(0, \lambda) = a_N(\infty, \lambda) = 0 \quad (9.8)$$

follows by power counting. To understand the limits of (9.7) we exhibit the entries in Table 9.9.1. We suppress the variable λ for conciseness. Table 9.9.1 shows $a_N(\varepsilon_i)$, the $[N/N]$ Padé approximant to the scattering length for a singular potential cut off at $r = \varepsilon_i$. Each column has a limit $a(\varepsilon_i)$ as $N \to \infty$, and each row has the limit zero as $\varepsilon_i \to 0$.

The mathematics indicates that we should extrapolate each column to its limit $a(\varepsilon_i)$ and deduce the scattering length $a(0)$ as $\varepsilon_i \to 0$. Even if this is feasible, it is an inefficient procedure, and there is a much better method.

Because all the entries in Table 9.9.1 are positive, (9.8) shows that the maximum entry of each row occurs at a finite, nonzero value of ε. Thus $a_N(\varepsilon, \lambda)$ has a maximum value at $\varepsilon = \bar{\varepsilon}(N)$, and if there are several maxima, $\bar{\varepsilon}(N)$ is the largest of these. Then we find that

$$a(0) = \lim_{N \to \infty} a_N(\bar{\varepsilon}(N)), \quad (9.9)$$

which is a stable procedure.

To prove this correct, we take $\eta > 0$ and prove that, for all $N > N_0(\eta)$,

$$|a_N(\bar{\varepsilon}(N)) - a(0)| < \eta.$$

Table 9.9.1. *[N/N] approximants of scattering lengths of cut-off potentials*

$a_1(\varepsilon_1)$	$a_1(\varepsilon_2)$	$a_1(\varepsilon_3)$	$\cdots$	$\rightarrow$	0
$a_2(\varepsilon_1)$	$a_2(\varepsilon_2)$	$a_2(\varepsilon_3)$	$\cdots$	$\rightarrow$	0
$a_3(\varepsilon_1)$	$a_3(\varepsilon_2)$	$a_3(\varepsilon_3)$	$\cdots$	$\rightarrow$	0
$\vdots$	$\vdots$	$\vdots$			
$\downarrow$	$\downarrow$	$\downarrow$			
$a(\varepsilon_1)$	$a(\varepsilon_2)$	$a(\varepsilon_3)$	$\cdots$	$\rightarrow$	$a(0)$

The proof begins by noting the existence of a cutoff ε, depending on η, for which, by our continuity hypothesis,

$$|a(\varepsilon) - a(0)| < \eta/2$$

and then noting the order $N_0(\eta)$ for which

$$|a_N(\varepsilon) - a(\varepsilon)| < \eta/2$$

for all $N > N_0(\eta)$. Since we choose $\bar{\varepsilon}(N)$ so that

$$a_N(\bar{\varepsilon}(N)) \geq a_N(\varepsilon_i)$$

for all ε_i in row N of the table, it follows that

$$|a_N(\bar{\varepsilon}(N)) - a(0)| \leq |a_N(\varepsilon) - a(0)|$$
$$\leq |a_N(\varepsilon) - a(\varepsilon)| + |a(\varepsilon) - a(0)|$$
$$< \eta$$

provided $N > N_0(\eta)$. Hence (9.9) is proved.

This argument formally proves the convergence of the method, as summarized by (9.9). The fact that we use the maximum entry in each row shows that the process is efficient. Figure 9.9.2 demonstrates the efficiency of the method for our particular example $V(r) = \lambda r^{-4}$.

The possibilities of application of this technique in divergent theories would seem to be immense. Some applications are briefly reviewed by Graves-Morris [1973a], and a full discussion of convergence is given by Bessis et al. [1974].

10

Connection with numerical analysis

10.1 Acceleration of convergence

In this section, some of the principles governing acceleration of convergence of series and their corresponding sequences are briefly introduced. As explained in Section 3.3, Padé approximation can be regarded as arising from the epsilon algorithm for sequence acceleration in the following way. We start with the sequence of real or complex numbers

$$S = \{S_0, S_1, S_2, \ldots\} \tag{1.1}$$

and form the elements of the corresponding series by $c_0 = S_0$ and

$$c_i = \Delta S_{i-1} = S_i - S_{i-1}, \quad i = 1, 2, 3, \ldots. \tag{1.2}$$

Then the generating function

$$f(z) = \sum_{i=0}^{\infty} c_i z^i \tag{1.3}$$

is defined. One may view the variable z as being introduced via (1.3) by multiplying each ΔS_i by z^{i+1} prior to summation. By introducing the variable z in this way, every method of acceleration of S generates a corresponding function $\mathcal{A}(z)$, and $\mathcal{A}(z)$ should approximate $f(z)$ accurately on the interval $[0, 1]$.

We have noted that the epsilon algorithm generates Padé approximants of $f(z)$. It may happen that the poles of the Padé approximants in the chosen sequence lie too close to $z = 1$ for the epsilon algorithm to be of much use. In the event that there is prior knowledge of where the poles of $\mathcal{A}(z)$ should be, for example, at $z_1, z_2, \ldots, z_M$, one can form

$$q_{L/M}(z) = \prod_{i=1}^{M}(z - z_i) \tag{1.4}$$

as the desired denominator of $\mathcal{A}(z)$. The corresponding numerator is

defined by
$$p_{L/M}(z) = [f(z)q_{L/M}(z)]_0^L \qquad (1.5)$$

and the corresponding approximant

$$\mathcal{A}_{L/M}(z) = p_{L/M}(z)/q_{L/M}(z) \qquad (1.6)$$

is called a Padé-type approximant [Brezinski, 1979, 1980a]; see also Baumel, Gammel, and Nuttall [1981]. The general principle of placing the poles is implemented in (1.4), the data are entered and used in (1.5), and the accelerated sequence is $\{\mathcal{A}_{L/M}(1)\}_{L=0}^\infty$. Theoretical justification for this approach is based on Eiermann's [1984] theorem, and a simpler proof of it is given by Cala Rodriguez and Wallin [1992].

The general theme of this book is that an appropriate sequence of Padé approximants is usually an excellent accelerator for a series (1.3) at a point z of analyticity of $f(z)$ and when the coefficients c_i are smoothly varying with i. Lubinsky [1985a, 1987, 1988] and Levin and Lubinsky [1986, 1990] exploit a useful definition of smoothness. Corresponding remarks apply to the sequence S after taking $z = 1$. However, when $z = 1$ is not a point of analyticity of $f(z)$, it is usually a limit point of poles of the chosen sequence of approximants [see Section 6.1, e.g., Fabry, 1896]. In this case, other methods, such as those described in this section, must be used to compute $f(1)$.

The E-algorithm of Håvie [1979] and Brezinski [1980b] for convergence acceleration is the most general known algorithm, in the sense that most algorithms for convergence acceleration turn out to be special cases of it. More importantly, it exemplifies the general principle of convergence acceleration, namely, that the error can be expressed as

$$S_n - S_\infty \approx \alpha_1 g_1(n) + \alpha_2 g_2(n) + \cdots + \alpha_k g_k(n), \quad n = 0, 1, 2, \ldots, \qquad (1.7)$$

where $\alpha_1, \alpha_2, \ldots, \alpha_k$ are constants, $g_1(n), g_2(n), \ldots, g_k(n)$ are prescribed functions that are often usefully chosen to depend on the elements of S, and $S_\infty = \lim_{n\to\infty} S_n$. After application of the E-algorithm, the elements of the new sequence $T_k = \{T_{0k}, T_{1k}, T_{2k}, \ldots\}$ are given by

$$T_{nk} = \frac{\begin{vmatrix} S_n & S_{n+1} & \cdots & S_{n+k} \\ g_1(n) & g_1(n+1) & \cdots & g_1(n+k) \\ \vdots & \vdots & & \vdots \\ g_k(n) & g_k(n+1) & \cdots & g_k(n+k) \end{vmatrix}}{\begin{vmatrix} 1 & 1 & \cdots & 1 \\ g_1(n) & g_1(n+1) & \cdots & g_1(n+k) \\ \vdots & \vdots & & \vdots \\ g_k(n) & g_k(n+1) & \cdots & g_k(n+k) \end{vmatrix}}. \qquad (1.8)$$

It is clear from (1.8) that the error terms shown explicitly in the right-hand side of (1.7) have been implicitly eliminated, by subtracting α_j times row $(j+1)$ from the first row of the numerator determinant, for each $j = 1, 2, \ldots, k$. It is also clear that (1.8) can be written as

$$T_{n,k} = \sum_{j=0}^{k} \beta_j S_{n+j} \tag{1.9}$$

and this represents a weighted average of $k+1$ of the elements of S, because $\sum_{j=0}^{k} \beta_j = 1$.

From (1.3.7)–(1.3.8), we see that the entries $\varepsilon_{2M}^{(J)}$ of the epsilon table that correspond to the $[M + J/M]$ Padé approximant are generated by taking $n = J$ and

$$g_1(n) = \Delta S_n, \quad g_2(n) = \Delta S_{n+1}, \ldots, g_k(n) = \Delta S_{n+k}. \tag{1.10}$$

Alternatively, (1.3.8) shows that the same result is given by taking

$$g_1(n) = \Delta S_n, \quad g_2(n) = \Delta^2 S_n, \ldots, g_k(n) = \Delta^k S_{n+k} \tag{1.11}$$

instead of (1.10). We refer to Brezinski [1980b] and Håvie [1979] for the important recursive implementation of this algorithm, to Schneider [1975] historically for its development from the Neville process and to Ford and Sidi [1987] for useful details.

Levin [1973] proposed three particular methods of sequence acceleration which exemplify a powerful general principle. It is based on the requirement that a function $R(n)$ can be found that mathematically models the remainders in the following way. A fundamental Levin model takes the form

$$S_n \approx S_\infty - R(n)\omega(n), \tag{1.12}$$

where $R(n)$ is a function which is chosen to model the error and

$$\omega(n) = \gamma_0 + \gamma_1/n + \cdots + \gamma_{k-1}/n^{k-1} \tag{1.13}$$

is smoothly varying. The parameters $\gamma_0, \gamma_1, \ldots, \gamma_{k-1}$ of $\omega(n)$ are calculated by requiring that

$$S_n = S_\infty - R(n)\omega(n), \quad n = N, N+1, \ldots, N+k, \tag{1.14}$$

for some chosen positive integers N and k. In fact, (1.14) represents $(k+1)$ linear equations for S_∞ and $\gamma_0, \gamma_1, \ldots, \gamma_{k-1}$; only the result S_∞ is ultimately of interest, but in principle the γ_i determine the background function $\omega(n)$, which is smooth for large n. The functional form, i.e., the shape, of $R(n)$ has to be known or to be modeled in advance. Obviously, (1.14) only makes sense if $R(n) \to 0$ as $n \to \infty$, and $R(n)$ need only be known up to a scale factor, because $\{\gamma_i\}$ determine the scale of the error.

Put crudely, $R(n)$ is chosen to represent the most rapidly varying n dependence of $\{S_n\}$, whereas $\omega(n)$ models a slowly varying correction.

Example 1. Sum the series

$$\sum_{i=1}^{\infty} c_i = \sum_{i=1}^{\infty} \frac{1}{i^2}. \tag{1.15}$$

Solution. The Padé approach suggests that we consider the associated function

$$f(z) = \sum_{i=1}^{\infty} c_i z^i = -\int_0^z \frac{1}{y} \ln(1-y)\, dy$$

which is a dilogarithm. The integrand has a logarithmic branch point at $y = 1$, and so the integral has an endpoint singularity of logarithmic type at $z = 1$.

Using contour-integral methods, one finds that

$$\int_{-\infty}^{\infty} \frac{\ln|w|}{1-w^2}\, dw = \tfrac{1}{2} \int_0^{\infty} \frac{\ln|w|}{1-w^2}\, dw = -\frac{\pi^2}{2}$$

and hence that

$$f(1) = -\int_0^1 \frac{1}{y} \ln(1-y)\, dy = \frac{\pi^2}{6} = 1.64493407 \ldots,$$

which is the familiar exact solution of (1.15), and so provides a useful test of numerical methods.

To formulate the method of (1.12)–(1.14), an integral estimator

$$R(n) = \int_n^{\infty} \frac{dv}{v^2} = \frac{1}{n} \tag{1.16}$$

is used to estimate the remainders [which actually are $\sum_{i=n+1}^{\infty} i^{-2}$]. Choosing $N = k = 1$, for example, (1.12) then takes the form

$$1 = S_{\infty} + \gamma_0 + \gamma_1,$$
$$1.25 = S_{\infty} + \gamma_0/2 + \gamma_1/3,$$
$$1.36\ldots = S_{\infty} + \gamma_0/3 + \gamma_1/9,$$

from which the estimate $S_{\infty} \approx 1.625$ follows. Other estimates of S_{∞} are given in Table 10.1.1, together with comparative estimates using the same data and other methods. The ε-algorithm is the $[l/l]$ Padé-approximant method, with $n = 2l + 1$, and rational interpolation refers to an $[l/l]$-type interpolant in the variable $w = n^{-1}$.

Table 10.1 *Estimates of the sum of the series* (1.15).

n	S_n	Method 1 ε-algorithm ($[l/l]$ Padé)	Method 2 Iterated ε-algorithm	Method 3 Rational interpolation (n-point Padé)	Method 4 Levin's interpolation
1	1.0	1.0	1.0	1.0	1.0
3	1.361 111	1.45	–	1.65	1.625
5	1.463 611	1.551 617	1.581 258	1.644 895	1.644 965
7	1.511 797	1.590 305	–	1.644 934	1.644 935
9	1.539 768	1.609 087	1.629 670	1.644 934	1.644 934

This example is rather particular, because the method given happens also to coincide with Richardson extrapolation (which is polynomial extrapolation to $n = \infty$ in terms of the variable n^{-1}). Other interesting comparisons of a similar kind have been made for the Madelung constant by Borwein, Borwein, and Taylor [1985], Sarkar and Bhattacharyya [1988], and Bhowmick, Roy, and Bhattacharyya [1988]. See also Bhattacharyya [1989a, b, c].

Of course, if a sequence (or its series) is given numerically rather than analytically, a formula for $R(n)$ like (1.16) cannot be constructed, nor can it for the important class of applications to divergent series. The standard suggestions [Levin, 1973; Weniger, 1989] are:

(i) For alternating series, the t-algorithm is defined with

$$R(n) = c_n. \tag{1.17}$$

(ii) For monotone, logarithmically convergent series, the u-algorithm is defined with

$$R(n) = (\zeta + n)c_n, \tag{1.18}$$

where ζ is a parameter to be chosen.

(iii) An Aitken-like model, called the v-algorithm, is defined with

$$R(n) = \frac{c_n c_{n+1}}{c_n - c_{n+1}}. \tag{1.19}$$

Notice that there is considerable scope for adaptation of these formulas (1.17)–(1.19) to particular problems. For each of these algorithms, the corresponding t, u, and v transformations are derived by solving the Vandermonde system (1.14), with the result that

$$T_{k,n} = \frac{\sum_{j=0}^{k}(-1)^j \binom{k}{j}\left(\frac{n+j}{n+k}\right)^{k-1}\frac{S_{n+j}}{R(n+j)}}{\sum_{j=0}^{k}(-1)^j \binom{k}{j}\left(\frac{n+j}{n+k}\right)^{k-1}\frac{1}{R(n+j)}}$$

[Levin, 1973].

The G-algorithm of Gray, Atchison, and McWilliams [1971] preceded this approach. More recent work on this theme has focused on the D-transformation for acceleration of convergence of infinite integrals in which the integrand satisfies a linear differential equation, and the corresponding d-transformation for series whose terms satisfy a linear difference equation [Levin and Sidi, 1981]. Levin's u-transformation has found particular application to the calculation of ground-state energies of various anharmonic oscillators [Weniger, Čížek, and Vinette, 1993; Homeier, 1993]. See also Čížek, Vinette and Weniger [1991] and Čížek and Vrscay [1982], and the extensive bibliography by Weniger [1994]. The topic of sequence acceleration is worth a whole book and that by Brezinski and Redivo-Zaglia [1991] is clear, comprehensive, and algorithmic. The theoretical side has recently been covered by Delahaye [1988]; there are many other texts of which the most relevant are listed in the bibliography, and the review by Guttmann [1989] is comprehensive.

10.2 Tchebycheff's inequalities for the density function

We resume the theory of Section 5.5, where only the first $k+1$ moments $\mu_0, \mu_1, \ldots, \mu_k$ are given, and these quantities are known to be moments of an unknown Stieltjes density function $\phi(u)$, which means that

$$\mu_j = \int_{-\infty}^{\infty} u^j \, d\phi(u), \quad j = 0, 1, 2, \ldots. \tag{2.1}$$

The information contained in the first $k+1$ moments in (2.1) is sufficient to bound the density function. This function $\phi(u)$ is an increasing and piecewise-continuous function, conventionally normalized by $\phi(-\infty) = 0$. If $u = u_0$ is a point of discontinuity, then the magnitude of the discontinuity is given by

$$\phi(u_0+) - \phi(u_0-) = \Delta\phi(u_0).$$

For a Stieltjes density, $\Delta\phi(u_0) \geq 0$, and there is an additive contribution of $\Delta\phi(u_0)\delta(u - u_0)$ in $\phi'(u)$.

The purpose of this section is to derive a piecewise-continuous density $\psi(u)$ associated with an arbitrary real point w. It will turn out that w is a point of increase of $\psi(u)$, $\psi(u)$ is nondecreasing and piecewise continuous, and the formula

$$\psi(w-) \leq \phi(w) \leq \psi(w+) \tag{2.2}$$

provides bounds for $\phi(u)$ at $u = w$.

The general method of approach to this type of problem, begun in Section 5.5, is to define coefficients

$$c_j = (-)^j \mu_j = (-)^j f_j, \quad j = 0, 1, 2, \ldots,$$

and a function

$$f(z) = \int_{-\infty}^{\infty} \frac{d\phi(u)}{1 + zu} \tag{2.3}$$

with the formal expansion $f(z) = \sum_{j=0}^{\infty} c_j z^j = \sum_{j=0}^{\infty} f_j(-z)^j$. The next step is the formation of Padé approximants of $f(z)$, from the known coefficients of its expansion. We will need

$$[M/M+1] = \frac{A^{[M/M+1]}(z)}{B^{[M/M+1]}(z)} = \sum_{i=1}^{M+1} \frac{\lambda_i}{1 + zu_i}. \tag{2.4}$$

This Padé approximant is associated with the distribution given by

$$d\psi_{M+1}(u) = \sum_{i=1}^{M+1} \lambda_i \delta(u - u_i), \tag{2.5a}$$

corresponding to a density function

$$\psi_{M+1}(u) = \sum_{i=1}^{M+1} \lambda_i \theta(u - u_i), \tag{2.5b}$$

because these justify the equality

$$[M/M+1] = \int_{-\infty}^{\infty} \frac{d\psi_{M+1}(u)}{1 + zu} = \int_{-\infty}^{\infty} \frac{d\phi(u)}{1 + zu} + O(z^{2M+2}).$$

Had we chosen the $[M/M]$ Padé approximant, we would have found

$$[M/M] = \frac{A^{[M/M]}(z)}{B^{[M/M]}(z)} = \lambda_0' + \sum_{i=1}^{M} \frac{\lambda_i'}{1 + zu_i'}, \tag{2.6}$$

which is associated with the distribution given by

$$d\tilde{\psi}_{M+1}(u) = \lambda_0' \delta(u) + \sum_{i=1}^{M} \lambda_i' \delta(u - u_i'), \tag{2.7a}$$

corresponding to the density function

$$\tilde{\psi}_{M+1}(u) = \lambda_0' \theta(u) + \sum_{i=1}^{M} \lambda_i' \theta(u - u_i'). \tag{2.7b}$$

Notice that these expressions (2.5), (2.7) for $\psi_{M+1}(u)$ and $\tilde{\psi}_{M+1}(u)$ are approximations to $\phi(u)$.

To construct the bounding distribution (2.2), first construct the polynomials

$$g(z) = G(z, w) = B^{[M/M]}(z) B^{[M/M+1]}(w) - B^{[M/M]}(w) B^{[M/M+1]}(z), \tag{2.8}$$

$$h(z) = H(z, w) = A^{[M/M]}(z) B^{[M/M+1]}(w) - B^{[M/M]}(w) A^{[M/M+1]}(z). \tag{2.9}$$

$g(z) = G(z, w)$ is related to the Christoffel–Darboux kernel, and vanishes at $z = w$. Its companion $h(z)$ is chosen so that

$$f(z)g(z) - h(z) = O(z^{2M+1}), \tag{2.10}$$

which follows from the Padé equations. $g(z)$ and $h(z)$ are polynomials of order $M + 1$ in z, dependent on the parameter w. Thus we may write

$$\frac{h(z)}{g(z)} = \sum_{j=1}^{M+1} \frac{\rho_j}{1 + z\zeta_j}, \tag{2.11}$$

which is a w-dependent expression, matching $f(z)$ and $M + 1$ of its derivatives at the origin. Notice that $\zeta_k = -w^{-1}$ for some k, $1 \leq k \leq M + 1$.

Lemma 1. *Equation (2.11) defines an $(M + 1)$-point quadrature formula which integrates polynomials of order $2M$ exactly.*

Proof. From (2.10) and (2.11),

$$\int_{-\infty}^{\infty} \frac{d\phi(u)}{1 + zu} = \sum_{j=1}^{M+1} \frac{\rho_j}{1 + z\zeta_j} + O(z^{2M+1}). \tag{2.12}$$

Equation (2.12) defines a weighted quadrature formula with points $\zeta_1, \zeta_2, \ldots, \zeta_{M+1}$ and weights $\rho_1, \rho_2, \ldots, \rho_{M+1}$. The accuracy-through-order conditions contained in (2.12) are

$$\int_{-\infty}^{\infty} u^i d\phi(u) = \sum_{j=1}^{M+1} \rho_j(\zeta_j)^i, \quad i = 1, 2, \ldots, 2M,$$

and so polynomials of order no greater than $2M$ are integrated exactly over the measure $d\phi(u)$.

Lemma 2. *The zeros of $B^{[M/M]}(z)$ and $B^{[M/M+1]}(z)$ interlace.*

Proof. Using (2.4), (2.6), and the

$$\begin{bmatrix} * \\ * \end{bmatrix}$$

identity,

$$\sum_{i=1}^{M+1} \frac{\lambda_i}{1 + zu_i} - \sum_{i=1}^{M} \frac{\lambda_i'}{1 + zu_i'} - \lambda_0' = \frac{C(M + 1/M)}{C(M/M)} \frac{z^{2M+1}}{B^{[M/M+1]}(z) B^{[M/M]}(z)}. \tag{2.13}$$

Because $f(z)$ is a Hamburger series,

$$\frac{C(M + 1/M + 1)}{C(M/M)} z^{2M+1} \leq 0 \quad \text{for } z \geq 0,$$

and the zeros of $B^{[M/M+1]}(z)$ and $B^{[M/M]}(z)$ are real and simple. Let z_i, $i = 1, 2, \ldots, M + 1$ be the zeros of $B^{[M/M+1]}(z)$, and z_i', $i = 1, 2, \ldots, M$ be the zeros of $B^{[M/M]}(z)$. Because $B^{[M/M+1]}(z) B^{[M/M]}(z)$ is a polynomial, the residues of $\{B^{[M/M+1]}(z) B^{[M/M]}(z)\}^{-1}$ alternate in sign. Because $f(z)$ is a Hamburger series, λ_i, λ_i' in (2.13) are positive. The interlacing property is a direct consequence of these two observations.

Lemma 3. *$g(z)$ has real roots.*

Proof. At a root $z = z_i$ of $B^{[M/M+1]}(z)$,

$$\text{sign}\{g(z)\} = \text{sign}\{B^{[M/M+1]}(w)\} \text{sign}\{B^{[M/M]}(z_i)\}.$$

Thus $g(z)$ has $M + 1$ sign changes, and therefore M real roots. Consequently the remaining root is real.

Theorem 10.2.1. *With the definitions (2.8), (2.9), and (2.11), let*

$$\frac{h(z)}{g(z)} = \frac{H(z, w)}{G(z, w)} = \sum_{i=1}^{M+1} \frac{\rho_i}{1 + z\zeta_i}. \tag{2.14}$$

The lemmas show that (2.14) defines a w-dependent distribution

$$d\psi(u) = \sum_{i=1}^{M+1} \rho_i \delta(u - \zeta_i) \tag{2.15}$$

and a density function

$$\psi(u) = \sum_{i=1}^{M+1} \rho_i \theta(u - \zeta_i). \tag{2.16}$$

10.2 Tchebycheff's inequalities for the density function

$\psi(u)$ has a point of increase at $u = \zeta_k = -w^{-1}$, and

$$\psi(\zeta_k+) \geq \phi(\zeta_k+) \quad \text{and} \quad \psi(\zeta_k-) \leq \phi(\zeta_k-). \tag{2.17}$$

Proof. Define a polynomial $r^{(k)}(u)$ by

$$r^{(k)}(\zeta_j) = 1, \quad 1 \leq j \leq k, \tag{2.18a}$$

$$r^{(k)}(\zeta_j) = 0, \quad k < j \leq M+1, \tag{2.18b}$$

$$\frac{d}{du} r^{(k)}(u) \bigg|_{u=\zeta_j} = 0, \quad 1 \leq j < k \text{ and } k < j \leq M+1, \tag{2.19}$$

where the zeros of $g(z)$ have been ordered so that

$$\zeta_1 < \zeta_2 < \zeta_3 < \cdots < \zeta_{M+1}.$$

The polynomial $r^{(k)}(u)$ may be chosen to satisfy the conditions (2.18) and (2.19) with degree at most $2M$. Because $d\psi(u)$ has $M+1$ points of increase, using (2.15), (2.18), we find

$$\psi(\zeta_k+) = \int_{-\infty}^{\infty} r^{(k)}(u) \, d\psi(u).$$

Because the quadrature formula has precision $2M$,

$$\int_{-\infty}^{\infty} r^{(k)}(u) \, d\psi(u) = \int_{-\infty}^{\infty} r^{(k)}(u) \, d\phi(u).$$

Because $\phi'(u) > 0$ and $r^{(k)}(u) \geq 0$,

$$\int_{-\infty}^{\infty} r^{(k)}(u) \, d\phi(u) \geq \int_{-\infty}^{\zeta_k+} r^{(k)}(u) \, d\phi(u).$$

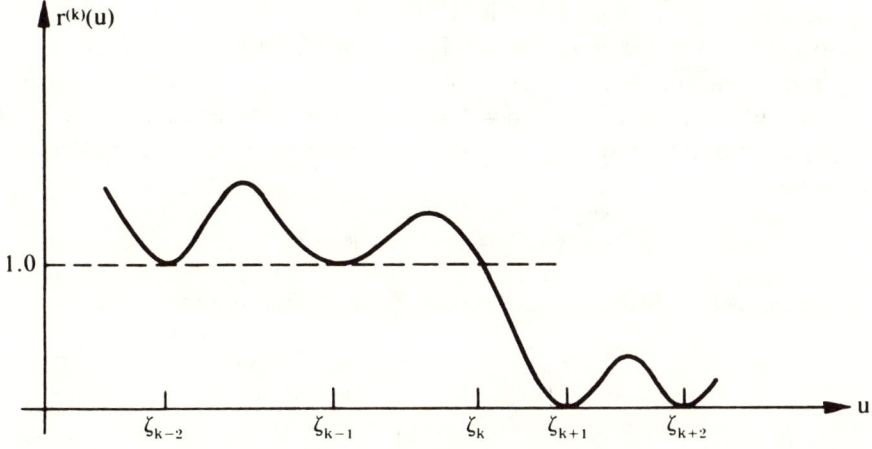

Figure 10.2.1. The polynomial $r^k(u)$.

Because $r^{(k)}(u) \geq 1$ on $-\infty < u < \zeta_k +$,

$$\int_{-\infty}^{\zeta_k+} r^{(k)}(u)\, d\phi(u) \geq \int_{-\infty}^{\zeta_k+} d\phi(u) = \phi(\zeta_k+).$$

We conclude that $\psi(\zeta_k+) \geq \phi(\zeta_k+)$. By the use of a different polynomial $t^{(k)}(u)$ for which

$$t^{(k)}(\zeta_j) = 0, \quad 1 \leq j < k,$$

$$t^{(k)}(\zeta_j) = 1, \quad k \leq j \leq M+1,$$

the other result that $\psi(\zeta_k-) \leq \phi(\zeta_k-)$ follows directly, and the theorem is proved.

There are manifold applications for these techniques in physics and chemistry, where densities are apt to be positive.

As a postscript to this section, we mention first the Markov problem. Given only the first $k+1$ moments $\mu_0, \mu_1, \ldots, \mu_k$ of an unknown Stieltjes density function $\phi(u)$, so that in principle all the moments are finite and given by

$$\mu_j = \int_{-1}^{1} u^j\, d\phi(u), \quad j = 0, 1, 2, \ldots,$$

$\phi(u)$ is restricted but not determined by these data. For a given function $h(u)$ and an x satisfying $-1 < x < 1$, the Markov problem is the determination of

$$\inf_{\phi} \int_{-1}^{x} h(u)\, d\phi(u) \quad \text{and} \quad \sup_{\phi} \int_{-1}^{x} h(u)\, d\phi(u).$$

We refer to Shohat and Tamarkin [1963, Chapter 3] for the solution of this problem for various conditions on $h(u)$. The method of solution uses techniques similar to those of this section.

Another interesting problem arises in the theory of equilibrium classical or quantum statistical mechanics, in which systems are described by a distribution over energy E of the form $\exp(-\beta E)\, d\psi(E)$. We are interested in the case where $\beta = 1/kT$, k is the Boltzmann constant, T is the temperature, and so $\beta > 0$. The problem consists of bounding the Stieltjes integral

$$Q(\beta) = \int_0^{\infty} e^{-\beta E}\, d\psi(E),$$

given the values of the first $2M$ moments of the distribution

$$\mu_i = \int_0^{\infty} E^i\, d\psi(E), \quad i = 0, 1, 2, \ldots, 2M-1.$$

This problem was solved by Gordon [1968] using the techniques of this section, but the resulting bounds are not extensive in the system size, and so are only useful for finite-sized systems. Another solution is possible using the methods of Section 7.2. As historical references, we cite Tchebycheff [1874], Stieltjes [1884], and Markov [1884]. Later, these techniques have been applied to bounding thermodynamic quantities by Isenberg [1963], Gordon [1968], and Corcoran and Langhoff [1977].

More recently interest has settled on evaluation of the Zolotarev constant

$$Z_{l,m}(E, F) = \inf_r \frac{\sup \{|r(z)|, z \in E\}}{\inf \{|r(z)|, z \in F\}},$$

where E, F are disjoint compact subsets of the complex plane and $r(z)$ is a rational function of type $[l/m]$. We refer to Istace and Thiran [1995] for details of its calculation, and to Levin and Saff [1994] for the existence and interpretation of $F(\tau)$ in the ray sequence limit

$$\lim_{m \to \infty} (Z_{l,m})^{1/(l+m)} = e^{-F(\tau)},$$

where $\tau = \lim_{l,m \to \infty} l/(l + m)$.

10.3 Collocation and the τ-method

The exponential function, $\exp(x)$, satisfies the differential equation

$$\frac{dy}{dx} - y = 0 \tag{3.1}$$

with the boundary condition

$$y(0) = 1. \tag{3.2}$$

The series solution is $y(x) = \sum_{k=0}^{\infty} c_k x^k$, and c_k are determined by

$$(k + 1)c_{k+1} - c_k = 0, \quad k = 0, 1, 2, \ldots, \tag{3.3}$$

from (3.1), and $c_0 = 1$ from (3.2). The conventional Lth-order polynomial approximation to $y(x)$, namely, $y(x) = \sum_{k=0}^{L} c_k x^k$, is an exact solution of a modification of (3.1), namely,

$$\frac{dy}{dx} - y = \tau x^L, \tag{3.4}$$

with the same boundary condition (3.2). Explicit solution reveals that $\tau = 1/L!$. To obtain a more accurate polynomial approximation to $y(x)$ for use at $x = z$, (3.1) and (3.4) are replaced by

$$\frac{dy}{dx} - y = \tau \pi(x), \tag{3.5}$$

where $\pi(x)$ is an Lth-order polynomial and $\tau\pi(x)$ is small in some sense. Specifically, the τ-method of Lanczos [1957, p. 438] requires that

$$\pi(x) = T_M^*\left(\frac{x}{z}\right) = T_M\left(\frac{2x}{z} - 1\right), \qquad (3.6)$$

which is a shifted Tchebycheff polynomial with the famous minimax property (see Section 6.6) in the range $0 \leq x \leq z$. We have assumed here that x, z are real and positive, because the analysis is simpler. If z is to be negative or complex, the range in question is $x = \alpha z$, where $0 \leq \alpha \leq 1$. The introduction of the τ-term into the right-hand side of (3.1) allows solutions which are polynomials in x to be found which are expected to be accurate over the range $0 \leq x \leq z$.

If $\pi(x)$ is a shifted Legendre polynomial,

$$\pi(x) = P_M^*\left(\frac{x}{z}\right) = P_M\left(\frac{2x}{z} - 1\right), \qquad (3.7)$$

then the polynomial solution of (3.5) at the particular point $x = z$ turns out to be the $[M/M]$ Padé approximant of the precise solution of (3.1) at $x = z$ [Luke, 1958, 1960]. We will justify (3.7) before proving the latter result.

One motive for choosing (3.7) in preference to (3.6) is that the explicit algebraic solution is easier to find. We know (Section 6.6) that the Tchebycheff polynomials $T_m(x)$ satisfy the minimax property for the class $\mathcal{P}_n$ of polynomials of order n, $n \geq 1$, with leading coefficient unity:

$$\inf_{p_n(x)\in\mathcal{P}_n} \sup_{-1\leq x\leq 1} |p_n(x)| = 2^{-(n-1)} \qquad (3.8)$$

is achieved when $p_n(x) \propto T_n(x)$. In this sense, the Legendre polynomials are remarkably small. With their conventional normalization

$$P_n(x) = \frac{1}{2^n n!}\left(\frac{d}{dx}\right)^n (x^2 - 1)^n, \qquad (3.9)$$

it is well known that [Isaacson and Keller, 1966, p. 219]

$$\sup_{-1\leq x\leq 1} |P_n(x)| = 1 \qquad (3.10)$$

and the leading coefficient is derived from (3.9) as

$$\frac{(2m)!}{2^m(m!)^2} = \frac{2^m}{\sqrt{m\pi}}[1 + O(m^{-1})]. \qquad (3.11)$$

Equations (3.9), (3.10), and (3.11) may be 'renormalized' to be comparable with (3.8). The Legendre polynomials have all their zeros in the interior of $[-1, 1]$; furthermore, the optimal polynomial for the right-hand side of (3.5) is unlikely to be independent of the specific structure of the equation, and thus selection of $\pi(x) = P_M^*(x/z)$ is to be regarded as a very satisfactory choice.

To be a little more general than choosing $\pi(x) = P_M^*(x/z)$, we choose

$$\pi(x) = x^J G_M(J+1, J+1, x/z). \tag{3.12}$$

For brevity we write $G_M(J+1, J+1, x/z) = R_M^J(x/z)$, which is a Jacobi polynomial normalized to have leading coefficient unity. The Jacobi polynomials $G_M(p, q; x)$ are orthogonal on $(0, 1)$ over a weight $w(x) = (1-x)^{p-q} x^{q-1}$, and so our choice of $R_M^J(x/z)$ selects polynomials orthogonal on $(0, z)$ over the weight x^J.

The special case of $J = 0$, with $R_M^0(x/z) \propto P_M^*(x/z)$, is of special interest, since it corresponds to (3.7).

From Abramowitz and Stegun [1964, p. 775], we find that

$$R_M^J\left(\frac{x}{z}\right) = \frac{(J+M)!}{(J+2M)!} \sum_{m=0}^M (-)^m \, {}^M C_m \frac{(J+2M-m)!}{(J+M-m)!} \left(\frac{x}{z}\right)^{M-m}$$

$$= \sum_{k=0}^M c_k^{(M,J)} \left(\frac{x}{z}\right)^k \tag{3.13}$$

and we must solve (3.5) and (3.12) as

$$\frac{dy}{dx} - y = \tau x^J R_M^J\left(\frac{x}{z}\right) \tag{3.14}$$

for a polynomial solution $y(x) = \sum_{k=0}^L d_k x^k$, subject to consistency. Equating coefficients of x^k on each side of (3.14), we find

$$(k+1)d_{k+1} - d_k = \tau c_{k-J}^{(M,J)} z^{-k+J} \quad \text{for } k = J, J+1, \ldots, L, \tag{3.15a}$$

$$(k+1)d_{k+1} - d_k = 0 \quad \text{for } k = 0, 1, \ldots, J-1. \tag{3.15b}$$

From (3.15b),

$$d_k = \frac{J!}{k!} d_J \quad \text{for } k = 0, 1, \ldots, J-1, \tag{3.16a}$$

and from (3.15a), because $y(x)$ is a polynomial and $d_{L+1} = 0$, it follows that

$$d_k = -\tau \sum_{r=k-J}^M c_r^{(M,J)} \frac{(r+J)!}{k!} z^{-r} \quad \text{for } k = J, J+1, \ldots, L. \tag{3.16b}$$

Substitute for $c_r^{(M,J)}$ from (3.13); then (3.16) shows that

$$d_0 = -\tau z^{-M} L!\, {}_1F_1(-M, -(L+M); -z). \tag{3.17}$$

Since $d_0 = 1$ is the boundary condition (3.2), τ is determined and is given by

$$\tau = -z^M[L!\,_1F_1(-M, -(L + M); -z)]^{-1}. \tag{3.18}$$

At this stage it is clear that the solution of (3.14) is a polynomial in x at the point $x = z$, the solution is a rational function of z, and the degree of its denominator is M.

Calculation of the full solution proceeds by using (3.16b) to prove

$$d_J = -\frac{\tau z^{-M} L!}{J!(L + M)!} \sum_{j=0}^{M} (-)^j \frac{M!(L + M - j)!}{(M - j)!j!} z^j. \tag{3.19}$$

It then follows from (3.16) and (3.19) that

$$y(z) = -\frac{\tau z^{-M} M! L!}{(L + M)!} \sum_{j=0}^{M} \sum_{k=0}^{L-j} \frac{z^{j+k}}{k!} \frac{(-)^j (L + M - j)!}{(M - j)!j!}$$

$$= (-\tau)z^{-M} L!\,_1F_1(-L, -(L + M); z). \tag{3.20}$$

Eliminating τ from (3.18) and (3.20) yields

$$y(z) = \frac{_1F_1(-L, -(L + M); z)}{_1F_1(-M, -(L + M); -z)}, \tag{3.21}$$

which is an $[L/M]$-type rational function of z. We can show that it is, in fact, the $[L/M]$ Padé approximant of $y(z)$, either by referring to (1.2.12) or by proving that

$$y(z) - \exp(z) = O(z^{L+m+1}) \tag{3.22}$$

by utilizing the present derivation.

Equation (3.14) is simplified, using an integrating factor, as

$$\frac{d}{dx}(e^{-x}y) = e^{-x}\tau x^J R_M^J\left(\frac{x}{z}\right). \tag{3.23}$$

Define $\varepsilon(x) = \exp(x) - y(x)$. Integration of (3.23) yields

$$\varepsilon(z) = -e^z \int_0^z e^{-x} \tau x^J R_M^J\left(\frac{x}{z}\right) dx$$

$$= -\tau e^z z^{J+1} \int_0^1 e^{-zt} R_M^J(t) t^J \, dt. \tag{3.24}$$

With Rodrigues's formula

$$R_M^J(t) = \frac{(M + J)!}{(2M + J)!} t^{-J} \left(\frac{d}{dt}\right)^M \{t^{M+J}(t - 1)^M\},$$

M-fold integration by parts of (3.24) leads to

$$\exp(z) - [L/M](z) = \frac{e^z z^{L+M+1} \int_0^1 e^{-zt} t^L (t-1)^M \, dt}{(L+M)! \, _1F_1(-M, -L-M; -z)}. \quad (3.25)$$

This is a well-known error expression [Perron, 1957, p. 436], and (3.22) is thereby validated. Its easier proof, assuming the integral representations, is also given by Braess [1986].

In summary, (3.14) with $J = L - M$ defines an $[L/M]$ Padé approximant for $\exp(z)$ for $L \geq M$, and the explicit solution is given by (3.21). Apparently, the optimal solution is given by $J = 0$. The choice $J = L$, $M = 0$ gives the ordinary Maclaurin polynomial solution.

Let us consider exactly what has been achieved in finding an approximate solution of (3.1) by the method given, which was introduced as the Lanczos τ-method, as described earlier

The method is also a collocation method [Wright, 1970]. Points x_i, $i = 1, 2, \ldots, L$, not necessarily distinct, which are in fact the zeros of $\pi(x)$ are selected in the interval $[0, z]$. The approximate solution is the polynomial solution $y_L(x)$ of order L which satisfies

$$\frac{dy_L}{dx} - y_L = 0 \quad \text{at } x = x_i, \, i = 1, 2, \ldots, L. \quad (3.26)$$

These hypotheses imply directly that

$$\frac{dy_L}{dx} - y_L = \tau \pi(x),$$

returning us to our starting point (3.5), and showing that the method is a collocation method. Implementation of the collocation method consists primarily of determining the coefficients d_j of the monomials x^j which constitute the polynomial $y_L(x) = \sum_{j=0}^{L} d_j x^j$.

The method may be viewed as an implicit Runge–Kutta method, using Gauss points $\{x_i\}$ in the interval $[0, z]$. The method returns a quadrature solution of

$$y(z) = 1 + \int_0^z y(x) \, dx$$

taking the form, with $x_j = z\xi_j$,

$$\sum_{r=0}^{L} c_r (z\xi_j)^r = 1 + \sum_{r=0}^{L} \frac{c_r (z\xi_j)^{r+1}}{r+1}, \quad j = 1, 2, \ldots, n.$$

It follows that we have L implicit equations for L unknown coefficients c_r, $r = 1, \ldots, n$. The function and its derivative are implicitly determined point-wise at the nodes.

In summary, the collocation method and the implicit Runge–Kutta method are equivalent in the present context; either leads to a solution which is a rational function of z. If this solution can be shown to satisfy the accuracy-through-order condition (3.22), and if z is identified as the step-length (commonly called h), then the solution is a Padé approximant. Furthermore, these methods are equivalent to certain Galerkin methods in which the inner products are evaluated by Gauss–Legendre quadrature: this is the collocation method in a different guise. The connection between implicit methods and Padé approximation appears again in the next section.

If we apply the method to the hypergeometric equation,

$$\left[(x - x^2)\frac{d^2}{dx^2} + \gamma\frac{d}{dx} - x(1 + \alpha + \beta)\frac{d}{dx} - \alpha\beta\right]y = 0, \quad (3.27)$$

we encounter immediately the problem that (3.27) is a second-order equation. But by writing it as

$$\frac{d}{dx}\left[(x - x^2)\frac{dy}{dx} + \{\gamma - 1 - x(\alpha + \beta - 1)\}y\right] + (\alpha + \beta - 1 - \alpha\beta)y = 0,$$

we see that it can be reduced to a first-order differential equation if $(\alpha - 1)(\beta - 1) = 0$. We take $\beta = 1$, and consider the solution of

$$(x - x^2)\frac{dy}{dx} + (\gamma - 1 - \alpha x)y = (\gamma - 1)y(0). \quad (3.28)$$

The τ-method is implemented by choosing

$$(1 - x)\frac{dy}{dx} + \left(\frac{\gamma - 1}{x} - \alpha\right)y = \tau P\left(\frac{x}{z}\right)x^J + \frac{(\gamma - 1)y(0)}{x}, \quad (3.29)$$

where $P(x)$ is a polynomial of order M to be specified. Equation (3.29) corresponds to modifying (3.27) to

$$(x - x^2)\frac{d^2y}{dx^2} + \gamma\frac{dy}{dx} - x(2 + \alpha)\frac{dy}{dx} - \alpha y = \tau\frac{d}{dx}\left\{x^{J+1}P\left(\frac{x}{z}\right)\right\},$$

but this is not directly useful.

To integrate (3.29), rewrite it as

$$\frac{dy}{dx} + \left[(\gamma - 1)\left(\frac{1}{x} + \frac{1}{1 - x}\right) - \frac{\alpha}{1 - x}\right]y = \frac{\tau x^J}{1 - x}P\left(\frac{x}{z}\right) + \frac{(\gamma - 1)y(0)}{x(1 - x)}. \quad (3.30)$$

The integrating factor is $x^{\gamma-1}(1 - x)^{1-\gamma+\alpha}$, and so (3.30) becomes

$$\frac{d}{dx}\{x^{\gamma-1}(1 - x)^{1-\gamma+\alpha}y\} = \tau x^{J+\gamma-1}(1 - x)^{\alpha-\gamma}P\left(\frac{x}{z}\right)$$

$$+ (\gamma - 1)x^{\gamma-2}(1 - x)^{\alpha-\gamma}y(0). \quad (3.31)$$

Integrating (3.31), and checking consistency at $x = 0$ when $\gamma > 1$, leads to

$$y(x) = x^{1-\gamma}(1 - x)^{\gamma-1-\alpha}$$

$$\times \int_0^x \left\{ \tau t^{J+\gamma-1}(1 - t)^{\alpha-\gamma} P\left(\frac{t}{z}\right) + (\gamma - 1)t^{\gamma-2}(1 - t)^{\alpha-\gamma}y(0) \right\} dt$$

$$(3.32)$$

Naturally, $F(x) = {}_2F_1(\alpha, 1, \gamma; x)$ satisfies a similar integral equation with $\tau = 0$, and hence $\varepsilon(x) = y(x) - F(x)$ obeys

$$\varepsilon(x) = x^{1-\gamma}(1 - x)^{\gamma-1-\alpha} \int_0^x \tau t^{J+\gamma-1}(1 - t)^{\alpha-\gamma} P\left(\frac{t}{z}\right) dt.$$

Hence

$$\varepsilon(z) = z^{J+1}\tau(1 - z)^{\gamma-1-\alpha} \int_0^1 x^{J+\gamma-1}(1 - xz)^{\alpha-\gamma} P(x) \, dx, \quad (3.33)$$

and the correct choice of $P(x)$ is that of a Jacobi polynomial orthogonal over weight $x^{J+\gamma-1}$, and we assume that $J + \gamma > 0$. From Abramowitz and Stegun [1964, p. 774] we find

$$P(x) = \frac{\Gamma(\gamma + J)}{\Gamma(\gamma + J + M)} \sum_{m=0}^{M} (-)^m \, {}^MC_m \frac{\Gamma(\gamma + J + 2M - m)}{\Gamma(\gamma + J + M - m)} t^{M-m},$$

which reduces to (3.13) when $\gamma = 1$. An analysis almost identical to that following (3.13) reveals that

$$\tau^{-1} = z^{-M} {}_2F_1(-(\alpha + L), -M, -(\gamma + L + M - 1); z), \quad (3.34)$$

and we deduce that

$$Q^{[L/M]}(z) = {}_2F_1(-(\alpha + L), -M, -(\gamma + L + M - 1); z). \quad (3.35)$$

The numerator $P^{[L/M]}(z)$ does not have a simple compact form in general. That these formulas are the numerators and denominators of Padé approximants follows from (3.33) and (3.34), since $\varepsilon(z) = O(z^{J+1+M+M}) = O(z^{L+M+1})$. It is interesting to compare this derivation of (3.35), the Padé denominator of ${}_2F_1(\alpha, 1; \gamma; z)$, with the methods of Section 1.2, and the continued-fraction method based on (4.6.14).

For a complete review of the explicit formulas obtainable using the techniques of this section, we refer to Luke [1964, 1969]. For further details of the connection between the τ-method, collocation, and implicit Runge–Kutta methods, we refer to Butcher [1963, 1964], Ehle [1968], Mason [1967], Axelsson [1969, 1972], and Chipman [1971].

10.4 Crank–Nicholson and related methods for the diffusion equation

The initial-value problem associated with the diffusion (or heat) equation is the problem of finding a continuous function $u(x, t)$ which satisfies

$$\frac{1}{\kappa}\frac{\partial u}{\partial t} - \frac{\partial^2 u}{\partial x^2} = 0, \quad t > 0, \tag{4.1a}$$

subject to the boundary condition

$$u(x, 0) = f(x), \quad -\infty < x < \infty. \tag{4.1b}$$

These equations define a mathematical problem (a pure initial-value problem) which describes many idealized diffusion processes, among them being the process of heat transfer along a bar. The temperature $u(x, 0) = f(x)$ is specified at the initial time $t = 0$, and the problem is to find the temperature at later times $t > 0$.

From the Green's function associated with (4.1), the solution to (4.1) is well known to be

$$u(x, t) = \frac{1}{\sqrt{4\pi\kappa t}} \int_{-\infty}^{\infty} \exp\left\{\frac{-(\xi - x)^2}{4\kappa t}\right\} f(\xi)\, d\xi.$$

Thus the solution of (4.1) is reduced to numerical integration for all times $t > 0$.

There is, of course, considerable interest in solving (4.1a) by other methods which can be used when the Green's-function method is inapplicable, for instance, when the equation is slightly different or the boundary conditions are different. From reactor physics, we might wish to consider the following partial differential equation for the neutron density $u(\mathbf{x}, t)$:

$$\underbrace{\frac{1}{\kappa}\frac{\partial u}{\partial t}}_{\text{time variation}} - \underbrace{\nabla^2 u}_{\text{diffusion}} = \underbrace{s(\mathbf{x}, t)u}_{\text{source term}}. \tag{4.2}$$

We might consider heat transfer along a finite bar, with the ends maintained at given temperatures and with various heat sources; the temperature $u(x, t)$ obeys an equation such as

$$\frac{1}{\kappa}\frac{\partial u}{\partial t} - \frac{\partial^2 u}{\partial x^2} = s(x, t), \quad 0 < x < 1, \tag{4.3a}$$

with the boundary conditions at the ends of the bar

$$\begin{aligned} u(0, t) &= T_0(t), \\ u(L, t) &= T_L(t), \end{aligned} \tag{4.3b}$$

10.4 Crank–Nicholson and related methods for the diffusion equation

and the specification of the initial temperature

$$u(x, 0) = f(x).$$

It is obvious enough from the physics of the heat problem that (4.3b) and (4.3c) are sufficient boundary conditions for the solution of (4.3a). The boundary conditions necessary with a mathematical formulation are not always quite so obvious.

To obtain numerical solutions of equations such as (4.2) and (4.3) and others of that type, various methods are available, e.g., Galerkin methods, but the simplest is to replace the derivatives by differences, and initially we discretize the x-variable in (4.3). We choose N interior points in $0 < x < L$, so that the mesh spacing Δx is given by $\Delta x = L/(N+1)$ and the points used are given by

$$x_i = i\Delta x, \quad i = 1, 2, \ldots, N.$$

The approximation scheme (method of lines) is defined by

$$u(x_i, t) \to U_j(t)$$

and

$$\left.\frac{\partial^2 u}{\partial x^2}\right|_{x=x_i} \to \frac{U_{i+1}(t) - 2U_i(t) + U_{i-1}(t)}{(\Delta x)^2}.$$

We solve the 'space-discretized' equations

$$\frac{1}{\kappa}\frac{\partial U_i}{\partial t} - \frac{U_{i+1} - 2U_i + U_{i-1}}{(\Delta x)^2} = s(x_i, t) \tag{4.4}$$

for the functions $U_1(t), U_2(t), \ldots, U_N(t)$. It is understood that $U_0(t) = T_0(t)$ and $U_{N+1}(t) = T_L(t)$. It is important to be able to prove that this system of equations is convergent; this statement means that as $\Delta x \to 0$ and $N \to \infty$, the approximation scheme implemented with exact arithmetic tends to the exact result, and we refer to Iserles and Nørsett [1991, Chapter 7] for the proofs.

Equation (4.4) can be written in matrix form:

$$\frac{\partial U_i}{\partial t} = \sum_{j=1}^{N} A_{ij} U_j(t) + S_i(t), \quad i = 1, 2, \ldots, N, \tag{4.5}$$

where $S_i(t) = \kappa S(x_i, t) + \kappa\{\delta_{i1} U_0(t) + \delta_{iN} U_{N+1}(t)\}/(\Delta x)^2$.

If we had been treating higher-dimensional systems, such as the three-dimensional reactor problem of (4.2), we would have been led to an equation similar to (4.5). However, the source terms would be different, and, more significantly, the matrix A would be a large sparse matrix. A sparse matrix has most of its entries zero, which means that special techniques may be used to avoid redundant arithmetic operations. The practical problem we have in mind is that of treating the case where A is a not necessarily symmetric but large matrix of dimension $N^3 \times N^3$.

Equation (4.5) typifies diffusion equations after space discretization, and the all too obvious method of solution is by time discretization. We use a sequence of time points $t_0 = 0$, t_1, t_2, t_3, ... and let $\Delta t_k = t_{k+1} - t_k$. The approximation consists of

$$U_i(t_k) \to U_i^{(k)}, \quad k = 1, 2, \ldots,$$

$$\left. \frac{\partial U_i}{\partial t} \right|_{t=t_k} \to \frac{U_i^{(k+1)} - U_i^{(k)}}{\Delta t_k}.$$

Then (4.5) is replaced by the equation

$$U_i^{(k+1)} = U_i^{(k)} = \Delta t_k \left\{ \sum_{j=1}^{N} A_{ij} U_j^{(k)} + S_i(t_k) \right\}, \tag{4.6}$$

which is represented schematically in Figure 10.4.1.

Let us now consider the form of the equations with all boundary and source terms set to zero. The major problems we will discover and solve are independent of these terms, and their presence would be largely immaterial. Then (4.6) is temporarily replaced by

$$\mathbf{U}^{(k+1)} = (I + \Delta t A)\mathbf{U}^{(k)}, \tag{4.7}$$

where $\mathbf{U}^{(k)} = (U_1^{(k)}, U_2^{(k)}, U_3^{(k)}, \ldots, U_N^{(k)})$ is a vector of values at time t_k. To discover the effect of the matrix operator $(I + \Delta t A)$, let us take the case of $N = 3$, using (4.4) we write

$$I + \Delta t A = I - \frac{\kappa \Delta t}{(\Delta x)^2} B_3,$$

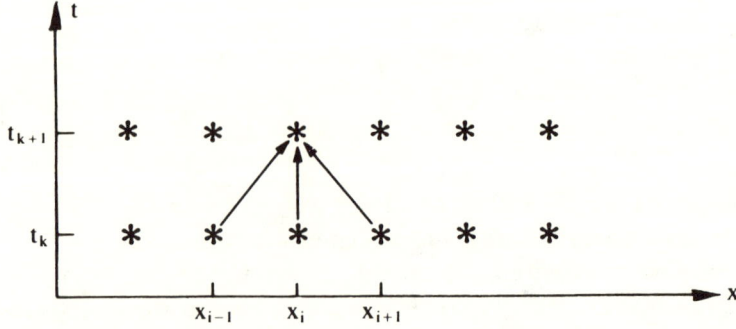

Figure 10.4.1. Diagrammatic representation of space and time points relevant to (4.6).

10.4 Crank–Nicholson and related methods for the diffusion equation

where

$$B_3 = \begin{bmatrix} 2 & -1 & 0 \\ -1 & 2 & -1 \\ 0 & -1 & 2 \end{bmatrix}.$$

For arbitrary N, we would define

$$(B_N)_{ij} = 2\delta_{ij} - \delta_{i-1,j} - \delta_{i,j-1}.$$

The eigenvalues of the matrix B_3 are easily found to be 2 and $2 \pm \sqrt{2}$. In general, the eigenvalues of B_N are positive and are given by

$$\lambda_i = 4\sin^2\left(\frac{\pi}{2} \cdot \frac{i}{N+1}\right), \quad i = 1, 2, \ldots, N. \tag{4.8}$$

It follows from Section 4.3 using a similar recursion relation to (4.3.4) for tridiagonal determinants, that the polynomial given by $\det(B_N - \lambda I)$ is simply related to Nth-order shifted Tchebycheff polynomials, yielding (4.8).

The essential point is that the eigenvalues of $(I + \Delta t A)$ are given by

$$\Lambda_i = 1 - \frac{\kappa \Delta t}{(\Delta x)^2} \lambda_i.$$

Since the operator $(I + \Delta t A)$ in (4.7) is to be used iteratively, an essential requirement for stability is that $|\Lambda_i| < 1$. Otherwise, rounding errors will grow exponentially during the iteration. This stability condition implies that

$$0 < \frac{\kappa \Delta t}{(\Delta x)^2} \lambda_i < 2, \tag{4.9}$$

and from this inequality we see that a necessary condition for stability is that the time step is limited by

$$\Delta t < 2(\Delta x)^2/\kappa. \tag{4.10}$$

Consequently, the explicit method defined by (4.7) is called conditionally stable. The upper limit on the time step given by (4.10) and imposed by the requirement of stability is unsatisfactory, and we will find a way of avoiding it.

We know, from the physics of the heat problem (4.3), that the initial behavior is transient heat flow in the bar and that then the system 'settles down.' This suggests that the final time steps may be much larger than the initial ones with a satisfactory method. In fact, the nature of the true solution of later times is dominated by the effect of the smallest eigenvalue λ_1 of (4.8) corresponding to a time variation of $\exp(-\lambda_1 \Delta t)$

between steps. Systems in which the numerical solution is controlled by a large eigenvalue and the true solution is controlled by a much smaller eigenvalue are called *stiff systems*. Clearly, our present explicit method of solution of such systems will not do.

Let us review our first attempts to see the solution to the dilemma, by looking at (4.5), with no source terms, as our idealized problem. It is

$$\frac{\partial U_i}{\partial t} = \sum_{j=1}^{N} A_{ij} U_j, \qquad (4.11)$$

where A is a known constant matrix with negative eigenvalues. Equation (4.11) has the solution

$$\mathbf{U}^{(k+1)} = \exp(A\Delta t)\mathbf{U}^{(k)}, \qquad (4.12)$$

and so the question becomes that of how the matrix $\exp(A\Delta t)$ is to be calculated.

Let $p(z)/q(z)$ denote a Padé approximant of type $[L/M]$ for $\exp z$, so that

$$\exp(A\Delta t) = p(A\Delta t)[q(A\Delta t)]^{-1} + O(\Delta t^{L+M+1}). \qquad (4.13)$$

Then (4.12) is approximated by

$$q(A\Delta t)\mathbf{u}^{(k+1)} = p(A\Delta t)\mathbf{u}^{(k)} \qquad (4.14)$$

with local truncation error (based on taking $\mathbf{U}^{(k)} = \mathbf{u}^{(k)}$)

$$|\mathbf{U}^{(k+1)} - \mathbf{u}^{(k+1)}| = O(\Delta t^{L+M+1}).$$

For example, by using the [1/1] Padé approximant, we have for (4.14)

$$(1 - \tfrac{1}{2}A\Delta t)\mathbf{u}^{(k+1)} = (1 + \tfrac{1}{2}A\Delta t)\mathbf{u}^{(k)} \qquad (4.15)$$

which is expressed concretely as

$$-\frac{\mu}{2}u_{l-1}^{(k+1)} + (1 + \mu)u_l^{(k+1)} - \frac{\mu}{2}u_{l+1}^{(k+1)} = \frac{\mu}{2}u_{l-1}^{(k)} + (1 - \mu)u_l^{(k)} + \frac{\mu}{2}u_{l+1}^{(k)}$$

$$(4.16)$$

and has error of $O(\Delta t^3)$. Equation (4.16) is the familiar Crank–Nicholson method. This method can equally well be motivated by making more accurate estimates of $\partial U_i/\partial t$ in (4.11) by using differences centered on $t_k + \tfrac{1}{2}\Delta t$ rather than t_k, which amounts to the trapezoidal rule for the integrated form of (4.9). It is represented by the heat-flow pattern of Figure 10.4.2.

If ξ is an eigenvalue of $A\Delta t$, then $\xi < 0$, and we notice that $\Xi(\xi) = (1 + \tfrac{1}{2}\xi)/(1 - \tfrac{1}{2}\xi)$ satisfies $|\Xi(\xi)| < 1$. This characteristic is obviously necessary for the stability of (4.15) as an iterative process. In fact,

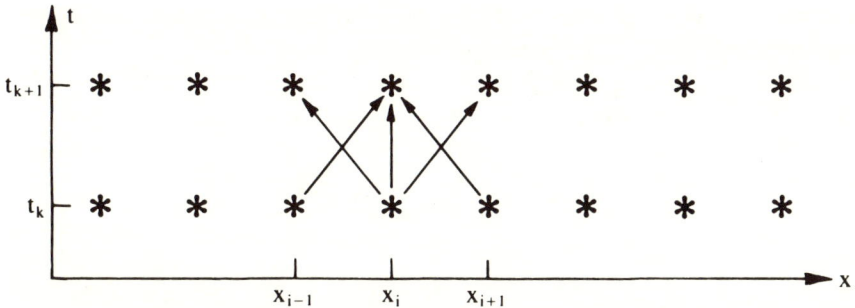

Figure 10.4.2. Diagrammatic representation of heat flow at the point x_i at sequential times.

rather stronger conditions are usually recommended, and we refer to Gear [1971] and Lambert [1974] for a discussion of these. Most often, A-stable methods are recommended.

Definition. Suppose that the values y_n, $n = 0, 1, 2, \ldots$, are obtained as the numerical solution of the test equation $y' = \lambda y$, so that $y_n \propto y(nh)$. The numerical method is A-*stable* if $y_n \to 0$ as $n \to \infty$ for every fixed λ, with $\operatorname{Re} \lambda < 0$ and $h > 0$.

A rational approximant $r(z)$ is associated with the method as in (4.13), (4.14), motivating a corresponding definition for the rational functions associated with A-stable methods.

Definition. A rational function $r(z)$ is A-*acceptable* if

$$|r(z)| < 1 \quad \text{for } \operatorname{Re} z < 0. \tag{4.17}$$

If large step lengths are required, $p(z)/q(z)$ should approximate $\exp(z)$ over a larger range. Approximants for which $r(-\infty) = 0$ are called L-acceptable, and Padé approximants with this property are a trivial form of two-point Padé approximation. These considerations have led to a number of methods based on polynomial and rational approximation, and interpolants for $\exp(z)$ on $-\infty < z \leq 0$, and we refer to Saff, Varga, and Ni [1976] for details and references.

An alternative approach to the analysis of the diffusion equation was introduced by Iserles [1985] and is reviewed by Iserles and Nørsett [1991]. To explain it, let us return to the model equation in one dimension and without source terms. The discretized numerical solution for it ultimately takes the form

$$\sum_{j=-\tau}^{\tau} \beta_j u_{l-j}^{(k+1)} = \sum_{j=-\sigma}^{\sigma} \alpha_j u_{l-j}^{(k)} \tag{4.18}$$

and we use a mesh $h = \Delta x$ and time step Δt. Let $\hat{u}(\theta)$ denote the Fourier transform of $\{u_i\}_{i=-\infty}^{\infty}$: the standard formulas are

$$\hat{u}(\theta) = \sum_{l=-\infty}^{\infty} u_l e^{il\theta h}, \quad -\pi/h < \theta \leq \pi/h, \tag{4.19}$$

$$u_n = \frac{h}{2\pi} \int_{-\pi/h}^{\pi/h} \hat{u}(\theta) e^{in\theta h} \, d\theta. \tag{4.20}$$

Using (4.19), we make the Fourier transform

$$F\left\{\sum_{j=-\tau}^{\tau} \beta_j u_{l-j}^{(k+1)}\right\} = \sum_{l=-\infty}^{\infty} \sum_{j=-\tau}^{\tau} \beta_j u_{l-j}^{(k+1)} e^{il\theta h}$$

$$= \sum_{j=-\tau}^{\tau} \beta_j e^{ij\theta h} \sum_{l=-\infty}^{\infty} u_l^{(k+1)} e^{il\theta h}$$

$$= \sum_{j=-\tau}^{\tau} \beta_j z^j \cdot \hat{u}^{(k+1)}(\theta),$$

where $z = e^{i\theta h}$. Thus (4.18) is transformed to

$$\hat{u}^{(k+1)}(\theta) = r(z)\hat{u}^{(k)}(\theta), \tag{4.21}$$

where

$$r(z) = \sum_{j=-\sigma}^{\sigma} \alpha_j z^j \bigg/ \sum_{j=-\tau}^{\tau} \beta_j z^j = \alpha(z)/\beta(z). \tag{4.22}$$

The system we seek to solve for is $u(x, t)$, and this is sampled at $x = \{lh\}_{l=-\infty}^{\infty}$. Corresponding to (4.20), we have

$$u(nh, t) = \frac{h}{2\pi} \int_{-\pi/h}^{\pi/h} \hat{u}(\theta, t) e^{-in\theta h} \, d\theta. \tag{4.23}$$

By using (4.23) as an interpolatory formula between integer values of n, we obtain

$$\frac{\partial^2 u(nh, t)}{\partial x^2} = \frac{-h}{2\pi} \int_{-\pi/h}^{\pi/h} \theta^2 \hat{u}(\theta, t) e^{-in\theta} \, d\theta$$

and hence the model equation Fourier transforms to

$$\frac{\partial}{\partial t} \hat{u}(\theta, t) = -\theta^2 \kappa \hat{u}(\theta, t)$$

with solution

$$\hat{u}(\theta, t + \Delta t) = e^{-\kappa \theta^2 \Delta t} \hat{u}(\theta, t). \tag{4.24}$$

10.4 Crank–Nicholson and related methods for the diffusion equation

By comparing (4.24) and (4.21), we see that we require

$$r(z) \approx e^{-\kappa\theta^2 \Delta t} = e^{-\mu(\theta \Delta x)^2} = e^{\mu(\ln z)^2}. \tag{4.25}$$

Suppose that approximation of order $\Delta x^p = h^p$ is required, which is equivalent to $O((z-1)^p)$. From (4.22), (4.25) we see that the optimal numerical method of the form (4.18) arises from osculatory rational approximants:

$$r(z) = \frac{\sum_{j=-\sigma}^{\sigma} \alpha_j z^j}{\sum_{j=-\tau}^{\tau} \beta_j z^j} = e^{\mu(\ln z)^2} + O((z-1)^{p+1}). \tag{4.26}$$

The parameters α_j, β_j are derived from the coefficients of the Padé approximant of type $[2\sigma/2\tau]$ to $z^{\sigma-\tau} \exp(\mu(\ln(z))^2)$ about $z = 1$, and hence $p \geq 2\sigma + 2\tau$ (assuming the normal case). Because $\exp(\mu(\ln z)^2)$ is invariant under $z \to z' = z^{-1}$ and $z = 1$ is a fixed point,

$$\frac{\sum_{j=-\sigma}^{\sigma} \alpha_j z^j}{\sum_{j=-\tau}^{\tau} \beta_j z^j} = \frac{\sum_{j=-\sigma}^{\sigma} \alpha_j z^{-j}}{\sum_{j=-\tau}^{\tau} \beta_j z^{-j}} + O((z-1)^{2\sigma+2\tau+1}). \tag{4.27}$$

Assuming normality yields that the fractions in (4.27) are irreducible, and hence $\alpha_j = \alpha_{-j}$, $\beta_j = \beta_{-j}$ in (4.18), which guarantees invariance under parity.

For the case of $\sigma = \tau = 1$, the [2/2]-type Padé approximant of $\exp(\mu(\ln z)^2)$ about $z = 1$ is needed, and this leads to

$$r(z) = \frac{1 + \left(\dfrac{1}{12} + \dfrac{\mu}{2}\right)\left(z - \dfrac{1}{z}\right)^2}{1 + \left(\dfrac{1}{12} - \dfrac{\mu}{2}\right)\left(z - \dfrac{1}{z}\right)^2}.$$

Iserles [1985] gives a comprehensive analysis of the remarkable properties of the diagonal Padé approximants of $\exp(\mu[\ln(w-1)]^2)$ using order-star theory: he finds that $p \geq 4\sigma + 1$, confirming that Crandall's method is of order $p = 5$ [Mitchell and Griffiths, 1980] and proving unconditional stability.

We refer to the books of Crank [1975], Carslaw and Jaeger [1959], and Bell and Glasstone [1970] for accounts of the origins of the parabolic equations discussed in this section. We refer to the books of Varga [1962], Gear [1971], Richtmyer and Morton [1967], and Lambert [1975], and the review of Argyris, Vaz, and William [1977] for accounts of the methods of solution of the equations. We refer to Baker [1960], Baker and Oliphant [1960], Ehle [1968, 1973, 1976], Cavendish, Culham, and Varga [1972], Fairweather [1971], Nørsett [1974], Siemieniuch [1976], Saff et al. [1976], Smith, Siemieniuch, and Gladwell [1977], and Saff, Schönhagen, and Varga [1978] as a selected set of references on the topics of this section.

10.5 Inversion of the Laplace transform

The problem of inversion of the Laplace transform is the determination of $f(t)$, given $\bar{f}(p)$. In principle, the functions are related by

$$f(t) = \frac{1}{2\pi i} \int_{\omega - i\infty}^{\omega + i\infty} \bar{f}(p) e^{pt}\, dp \qquad (5.1)$$

and

$$\bar{f}(p) = \int_0^\infty e^{-pt} f(t)\, dt. \qquad (5.2)$$

Equation (5.1) determines $f(t)$ from $\bar{f}(p)$ directly, provided a suitable contour is chosen on which the integrand is not unduly oscillatory. The trouble with this problem is that small oscillations in $\bar{f}(p)$ result from larger changes in $f(t)$ for large values of t. Here we discuss only the Padé method, which avoids the task of selecting a suitable contour and involves an implicit smoothness assumption. To motivate the method, we follow the original problem first solved this way, which has smoothness properties built in by the underlying physics.

Waves moving in an imperfectly elastic medium are conjectured to satisfy the equation for the displacement $y(x, t)$

$$\frac{\partial^2 y}{\partial t^2} = c^2 \left(1 + \tau \frac{\partial}{\partial t}\right) \frac{\partial^2 y}{\partial x^2}. \qquad (5.3)$$

Under initial conditions representing propagation of a transverse displacement caused by a unit-step-function displacement at time $t = 0$ at the end of the bar, $x = 0$, we use the Heaviside function $\theta(t)$ to define

$$y(x = 0, t) = y_0(t) = \theta(t). \qquad (5.4)$$

Using the transform

$$\bar{y}(x, p) = \int_0^\infty e^{-pt} y(x, t)\, dt,$$

equation (5.3) is simplified to become

$$\frac{\partial^2 \bar{y}(x, p)}{\partial x^2} = \frac{p^2}{c^2(1 + \tau p)} \bar{y}(x, p) \quad \text{for } x > 0,$$

10.5 Inversion of the Laplace transform

with the solution

$$\bar{y}(x, p) = \bar{y}_0(p) \exp\left\{\frac{-px}{c\sqrt{1 + \tau p}}\right\}$$

$$= \frac{1}{p} \exp\left\{\frac{-px}{c\sqrt{1 + \tau p}}\right\}. \quad (5.5)$$

The solution for $y(x, t)$ requires the inverse transform of (5.5). $\bar{y}(x, p)$ is meromorphic in p for $\operatorname{Re} p > -\tau^{-1}$, and analytic in $\operatorname{Re} p > 0$. Hence (5.1) is well defined in this application if $\omega > 0$. We omit the changes of variable which are, in practice, desirable for putting (5.5) in canonical form, because they are inessential to the description. The Padé method requires three simple steps.

Step 1. Write (5.5) as a Maclaurin series

$$p\bar{y}(x, p) = c_0 + c_1 p + c_2 p^2 + \cdots,$$

and form its diagonal Padé approximants $[M/M](p)$. The coefficients depend on x and the parameters c, τ.

Step 2. Re-express this Padé approximant as

$$\frac{1}{p}[M/M](p) = \sum_{i=0}^{M} \frac{\gamma_i^{(M)}(x)}{p - p_i^{(M)}(x)}, \quad \text{where } p_0^{(M)}(x) = 0. \quad (5.6)$$

If $\operatorname{Re} p_i^{(M')}(x) > 0$ for any pair (i, M'), then the $[M'/M']$ Padé approximant is an unacceptable approximation because it has the wrong analytic structure. The poles of the Padé approximants ought to be close to the essential singularity and the natural locations of the cuts of (5.5), which are in the left half plane. Furthermore, a pole in the right half plane would lead to an unphysical solution which increases indefinitely with time.

Step 3. The approximate inverse transform is given by the transform of (5.6), and is

$$y(x, t) \approx \sum_{i=0}^{M} \gamma_i^{(M)}(x) e^{p_i^{(M)}(x)t}.$$

This is a sum of damped exponentials; it is interesting to compare this method with that of the Baker–Gammel approximants (section 7.2) with an exponential kernel.

We make no claim that the method described in this section is a 'best method' of inversion of the Laplace transform. Improvements based on development of this method incorporate a best or near-best rational approximation of $\bar{f}(p)$ in (5.1) [Longman, 1974; Brezinski, 1979]. We refer to Talbot [1979] and Evans [1993] for a different method and a list of methods which achieve good accuracy in general cases, and to Luke [1978].

10.6 Connection with rational approximation

We have occasionally emphasized in earlier sections that Padé approximants are not usually best rational approximants. It is notoriously difficult, in general, to find best rational approximants either in principle or numerically.

For a function $f(z)$ whose singularities are confined to a finite number of regions, Runge's theorem, very crudely stated, shows that rational approximants exist having prescribed poles in the singularity regions only, and which converge to $f(z)$ in the complement regions where $f(z)$ is holomorphic. To explain what is easily attainable using rational approximation, we prove two introductory lemmas, and then we carefully state and prove Runge's theorem. This presentation is followed by a simple example showing the futility of polynomial approximation in a particular case for which rational approximation is exact. Finally we present Walsh's theorem and some associated results showing that Padé approximants are the best local rational approximants.

Lemma 1. *Let $\mathcal{D}$ be an open, connected, and bounded domain, with a boundary curve C consisting of an outer boundary C_0 and inner boundary curves $C_1, C_2, \ldots, C_n$. Let $f(z)$ be holomorphic in $\mathcal{D}$ and continuous on C. Then a sequence $\{P_k(z)\}$ of rational functions exists whose poles lie on C, and $P_k(z) \to f(z)$ uniformly on any compact subset of $\mathcal{D}$.*

Proof. Using Cauchy's theorem, we have

$$f(z) = \frac{1}{2\pi i} \int_C \frac{f(t)}{t - z} dt, \qquad (6.1)$$

where $C = C_0 - C_1 - C_2 - \cdots - C_m$, meaning that the contour C_0 is taken counterclockwise and the other contours are taken clockwise, as shown in Figure 10.6.1.

We prove convergence on compact subsets of $\mathcal{D}$; let these be at least a distance δ from the boundary C. Clearly, $|t - z| \geq \delta > 0$ in (6.1).

To extract a rational approximant from the representation (6.1), we choose $\eta > 0$, we divide C into p subarcs $\Gamma_1, \Gamma_2, \ldots, \Gamma_p$, each of length at

10.6 Connection with rational approximation

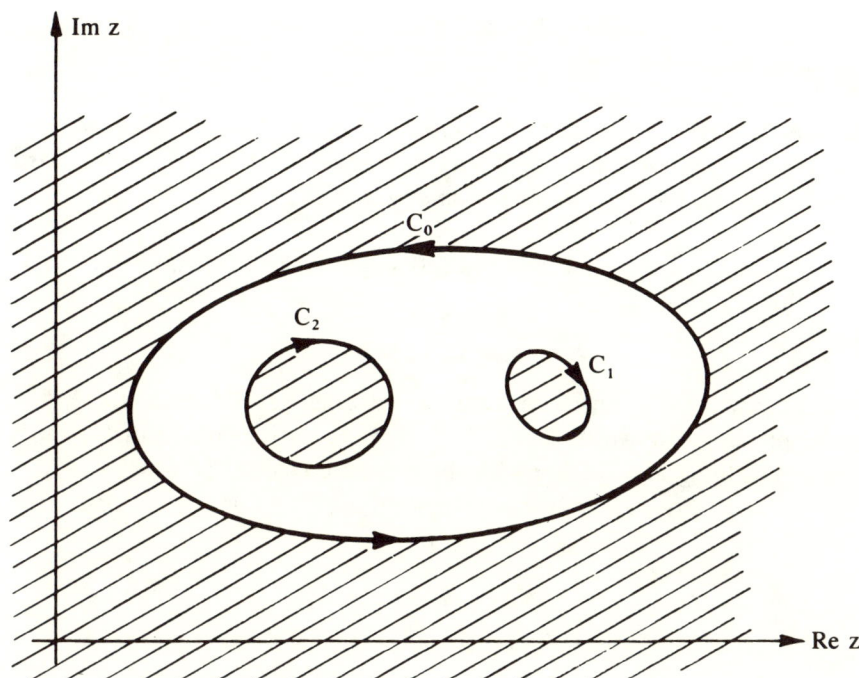

Figure 10.6.1. The domain of holomorphy of $f(z)$ is shown unshaded; the boundary curves C_0, C_1, and C_2 are also shown.

most η, and we choose arbitrary points $t_k \in \Gamma_k$. Define

$$\gamma_k = \frac{1}{2\pi i} \int_{\Gamma_k} f(t)\, dt$$

and

$$R(z, \eta) = \sum_{k=1}^{p} \frac{\gamma_k}{t_k - z}. \tag{6.2}$$

Equation (6.2) defines the rational approximant. Let $l(C)$ denote the length of C, and $M = \max |f(z)|$ over the compact subset of $\mathcal{D}$ in question. Then we have an easy error bound

$$|f(z) - R(z, m)| \leq \frac{Ml(C)}{2\pi\delta^2} \eta. \tag{6.3}$$

We may choose η to be arbitrarily small in (6.3); in particular, we choose $\eta \ll \delta^2 [Ml(C)]^{-1}$, and the theorem is proved.

Notice that (6.2) provides an explicit construction of the rational approximants. The next lemma generalizes Lemma 1 by allowing an infinite number of inner boundary curves which enclose nonanalytic points of $f(z)$.

Lemma 2. *Let $\mathcal{D}$ be an arbitrary connected, open, bounded domain, and let $f(z)$ be holomorphic in $\mathcal{D}$. Then a sequence of rational functions exists which converges to $f(z)$ uniformly on any compact subset of $\mathcal{D}$.*

Proof. We can exhaust $\mathcal{D}$ by a nested sequence of open domains $\{\mathcal{D}_n\}$, such that for all n

$$\overline{\mathcal{D}}_{n-1} \subset \mathcal{D}_n \subset \mathcal{D}. \tag{6.4}$$

A simple way to do this is to set up a Cartesian grid of mesh size 2^{-m} in the z-plane, as shown in Figure 10.6.2. Then let $\mathcal{E}_m$ consist of the open squares of the grid, together with their common sides, for which $\overline{\mathcal{E}}_m \subset \mathcal{D}$. We take m larger than some minimum m_0, so that $\mathcal{E}_m$ is open and

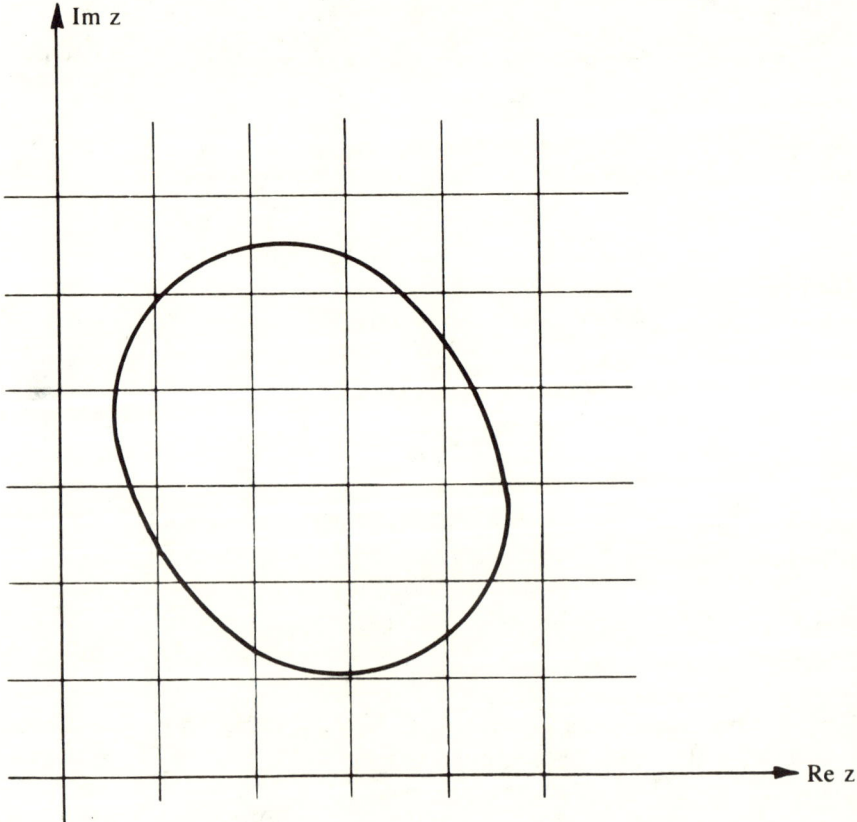

Figure 10.6.2. A Cartesian grid and the domain $\mathcal{D}$.

connected for $m > m_0$. Then a subsequence of $\{\mathscr{E}_m, m > m_0\}$, called $\{\mathscr{D}_n\}$, exists satisfying (6.4); for each $z \in \mathscr{D}$, n_z may be found such that $z \in \mathscr{D}_{n_z}$. Further, the boundary of $\mathscr{D}_n$, called C_n, consists of simple closed polygons. Let the minimum distance δ_n of $\bar{\mathscr{D}}_{n-1}$ from C_n be δ_n; we have arranged by (6.4) that $\delta_n > 0$. Using the previous lemma, we obtain rational functions $S_n(z)$ with poles on C_n such that for any $\eta_n > 0$,

$$|f(z) - S_n(z)| \leq \frac{M_n \lambda_n}{2\pi \delta_n^2} \eta_n,$$

where λ_n is the length of C_n and $M_n = \max |f(z)|$ for $z \in \mathscr{D}_n$. Hence λ_n, M_n are bounded. Whatever value δ_n may have, η_n may be chosen sufficiently small so that the sequence $S_n(z) \to f(z)$ uniformly on compact subsets of $\mathscr{D}$, and we may arrange that

$$|f(z) - S_n(z)| < 2^{-n}. \tag{6.5}$$

The construction of $S_n(z)$ in Lemma 1 consists of assigning poles on the boundary of $\mathscr{D}$, the domain of holomorphy of $f(z)$. These poles are not normally singularities of $f(z)$. The construction of Lemma 2 consists of assigning poles to $S_n(z)$ within the domain of holomorphy but outside the compact region of convergence. Runge's theorem removes this unappealing feature of the approximants by placing the poles of $S_n(z)$ at prescribed points in the excluded regions. For details, see Roth [1938] and Gauthier [1977].

***Theorem* 10.6.1. [Runge, 1885].** *Let $f(z)$ be holomorphic in $\mathscr{D}$, where $\mathscr{D}$ is a bounded open domain with a complement consisting of precisely $m + 1$ components, $\mathscr{F}_0, \mathscr{F}_1, \ldots, \mathscr{F}_m$. Let $\mathscr{F}_0$ contain the point $z = \infty$; then the other components $F_1, F_2, \ldots, F_m$ are compact. Let $z_k \in F_k$, $k = 0, 1, 2, \ldots, m$, be a chosen set of finite points. Then a sequence $\{R_n(z)\}$ of rational approximants exists for which $R_n(z) \to f(z)$ uniformly on compact subsets of $\mathscr{D}$, and all the poles of $R_n(z)$ are located at the preassigned points $\{z_k\}$. If $m = 0$, the same approximation is possible taking $\{R_n(z)\}$ to be polynomials instead of rational functions.*

Proof. We shall prove the case with $m \geq 1$. Using the method of Lemma 2, suppose that n is large enough that the boundary of the nth polygonal curve may be expressed as

$$C_n = C_{0n} - C_{1n} - C_{2n} - \cdots - C_{mn},$$

where the exterior of C_{0n} contains $\mathscr{F}_0$ and the interiors of $C_{1n}, C_{2n}, \ldots, C_{mn}$ contain $\mathscr{F}_1, \mathscr{F}_2, \ldots, \mathscr{F}_m$, respectively. $S_n(z)$ is a rational approximant constructed to have many simple poles on each of the $m + 1$ curves

C_{kn}. These poles are to be replaced by precisely $m + 1$ multipoles at z_0, $z_1, \ldots, z_m$ without appreciably altering the accuracy of the approximation.

To 'move' a pole from $z = a$ on C_k to $z = b$ which is closer to z_k, we use the identity

$$\frac{1}{z - a} = \frac{1}{z - b} + \frac{a - b}{(z - b)^2} + \cdots + \frac{(a - b)^{p-1}}{(z - b)^p} + \frac{(a - b)^p}{(z - a)(z - b)^p},$$

so that

$$\left| \frac{1}{z - a} - \sum_{j=1}^{p} \frac{(a - b)^{j-1}}{(z - b)^j} \right| = \frac{|a - b|^p}{|z - a||z - b|^p}.$$

For any $z \in \mathcal{D}_{n-1}$ and b such that

$$2|b - a| < |z - b|, \tag{6.6}$$

we have

$$\left| \frac{1}{z - a} - \sum_{j=1}^{p} \frac{(a - b)^{j-1}}{(z - b)^j} \right| < \frac{2^{-p}}{\delta_n}. \tag{6.7}$$

The expression (6.7) is the key to the theorem, because it shows that a pole at $z = a$ may be 'replaced' by a multipole of high order at $z = b$. Furthermore, any multipole located at $z = a$ [i.e., a polynomial in $(z - a)^{-1}$] may be similarly approximated. If we could take $b = z_k$, the theorem would now be proved, but in general (6.6) does not hold with $b = z_k$. Instead, we must find a sequence of points on a polygonal line L joining a to z_k, given by

$$b_0 = a, b_1, b_1, \ldots, b_N = z_k,$$

such that $|b_j - b_{j-1}| < \frac{1}{2}\delta$, $j = 1, 2, \ldots, N$, where δ is the distance from L to $\mathcal{D}_{n-1}$. This condition maintains the inequality

$$|b_j - b_{j-1}| < \tfrac{1}{2}|z - b_j|.$$

Now, by expansion and reexpansion, but with a finite total number of steps, we may find $R_n(z)$ with poles at $z_0, z_1, \ldots, z_k$ such that

$$|S_n(z) - R_n(z)| < 2^{1-n}. \tag{6.8}$$

Equations (6.5), (6.8) together show that $|f(z) - R_n(z)| < 2^{1-n}$ for $z \in \mathcal{D}_{n-1}$, completing the proof for $m > 0$. We leave the proof for $m = 0$ as a straightforward exercise [Hille, 1962, p. 299].

It is interesting to contrast Runge's theorem with Laurent's theorem about convergent rational approximation within an annulus of holomorphy. In this case, there are two multipoles, one in the annulus and one at infinity.

Thus far, we have established the possibility of obtaining uniform convergence with rational approximants to a function with a domain of holomorphy such as Figure 10.6.1 or any Hepworth sculpture. We are a long way from being able to prove that even a subsequence of Padé approximants converges in such a domain. However, we can easily show that polynomial approximation in certain cases is futile, and why Runge's theorem is so important.

We will show that $p_n(z) = 0$ is a best polynomial approximation of order n to $f(z) = z^{-1}$ on the annulus $1 \leq |z| \leq 2$. For suppose that $\tilde{p}_n(z)$ is a better polynomial approximation; then

$$\max_{1 \leq |z| \leq 2} |\tilde{p}_n(z) - z^{-1}| = r < 1. \tag{6.9}$$

By Cauchy's theorem

$$\int_{|z|=1} \{\tilde{p}_n(z) - z^{-1}\} \, dz = -2\pi i.$$

Hence

$$\int_{|z|=1} |\tilde{p}_n(z) - z^{-1}| \, |dz| \geq 2\pi,$$

which contradicts (6.9). Thus there is no better polynomial approximation (in the variable z) than $p_n = 0$ to z^{-1} on the annulus; of course, a rather simple rational approximant is exact.

The following two theorems show the role of Padé approximants in the theory of best rational approximation in the Tchebycheff sense. The first theorem we state without proof. It shows the similarity between the convergence properties of Padé approximants in the ideal conditions under which de Montessus's theorem holds, and best rational approximations under the same conditions.

Theorem 10.6.2 [Walsh, 1964a]. *Let $f(z)$ be analytic in the disk $|z| < r$ and be meromorphic with precisely M poles, counting multiplicities, in the larger disk $|z| < \rho$. If $0 < \varepsilon < r$, then the rational functions $R^{(L/M)}(\varepsilon, z)$ of type (L/M) of best approximation to $f(z)$ on the closed disk $|z| \leq \varepsilon$ converge as $L \to \infty$ uniformly to $f(z)$ on any compact subset of $|z| < \rho$ containing no poles of $f(z)$. Also, the poles of $R^{(L/M)}(\varepsilon, z)$ converge to the M poles of $f(z)$ lying in $r < |z| < \rho$.*

We omit the proof, because the theorem is only indirectly connected with Padé approximants. Next, we both state and prove another theorem which shows that Padé approximants are the limits of best rational approximants in a vanishingly small neighborhood of the origin.

Theorem 10.6.3 [Walsh, 1964b]. *Let $f(z)$ be analytic at the origin, and let $R^{(L/M)}(\varepsilon, z)$ be the best rational approximant of type (L/M) in $|z| \leq \varepsilon$. Provided $C(L/M) \neq 0$, then as $\varepsilon \to 0$, $R^{(L/M)}(\varepsilon, z) \to [L/M]_f(z)$ uniformly on any compact set containing no pole of $[L/M]_f(z)$.*

Proof. Because $R^{(L/M)}(\varepsilon, z)$ is the best rational approximant of $f(z)$, it has no pole in $|z| \leq \varepsilon$ for some preassigned positive ε, and therefore

$$R^{(L/M)}(\varepsilon, z) = \sum_{k=0}^{\infty} r_k(\varepsilon) z^k \quad \text{for } |z| < \varepsilon.$$

Using the contour representation, we further know that

$$r_k(\varepsilon) - c_k = \frac{1}{2\pi i} \int_{|z|=\varepsilon} \frac{R^{(L/M)}(\varepsilon, z) - f(z)}{z^{k+1}} dz.$$

Again, because $R^{(L/M)}(\varepsilon, z)$ is the best rational approximant on $|z| \leq \varepsilon$,

$$|R^{(L/M)}(\varepsilon, z) - f(z)| < |[L/M]_f(z) - f(z)| \quad \text{for } |z| = \varepsilon,$$

and so

$$|r_k(\varepsilon) - c_k| \leq \frac{1}{2\pi} \int_{|z|=\varepsilon} \frac{|[L/M]_f(z) - f(z)|}{|z|^{k+1}} |dz|$$
$$= O(\varepsilon^{L+M-k+1}).$$

Consequently $r_k(\varepsilon) \to c_k$ as $\varepsilon \to 0$ for $k = 0, 1, \ldots, L + M$. The coefficients c_k, $k = 0, 1, \ldots, L + M$ uniquely determine $[L/M]$ if $C(L/M) \neq 0$. Provided $|r_k(\varepsilon) - c_k|$ is sufficiently small for $k = 0, 1, \ldots, L + M$, $R^{(L/M)}(\varepsilon, z)$ is uniquely determined by $r_k(\varepsilon)$ for these values of k, and $R^{(L/M)}(\varepsilon, z) \to [L/M]_f(z)$ uniformly if the respective denominators do not vanish; these requirements are met by the conditions of the theorem.

This theorem of Walsh is important because it may be interpreted as the statement that Padé approximants are the best local rational approximants. There are two interesting related results, which we present as corollaries without proof.

Corollary 1. *Theorem 10.6.3 still holds when $R^{(L/M)}(\varepsilon, z)$ is defined as the best rational approximation on the real closed interval $0 \leq z \leq \varepsilon$ (instead of the disk $|z| \leq \varepsilon$).*

Corollary 2. Let $f(z)$ be a real-valued function with $m + n + 1$ continuous derivatives on $[0, \delta]$ for some $\delta > 0$. For each ε, $0 < \varepsilon < \delta$, let $R^{(L/M)}(\varepsilon, z)$ be the best (L/M)-type approximant to $f(z)$ on $[0, \varepsilon]$. Then $R^{(L/M)}(\varepsilon, z) \to [L/M]_f(z)$ as $\varepsilon \to 0$ on some closed interval $[0, \varepsilon_0]$, $0 < \varepsilon_0 \leq \delta$, using the classical definition, not the Baker definition, of $[L/M]_f(z)$.

Corollary 1 is simply a result different from but similar to Walsh's theorem. Corollary 2 shows that Padé approximants with the classical definition emerge as the limit of best local rational approximation [Chui, Shisha, and Smith, 1974].

For further details of recent progress with Walsh-type theorems, we refer to Chui, Shisha, and Smith, [1974, 1975], Chui, Smith, and Ward [1978], Walsh [1965a, b, 1970], Trefethen [1984], Trefethen and Gutknecht [1985], and Gutknecht [1990]. Examples which show the non-uniqueness of best rational complex-valued approximations to real functions on real intervals [Saff and Varga, 1978b] and of complex best rational approximations on a disk [Gutkneckt and Trefethen, 1983a, b] are cautionary and interesting. A good general reference on rational approximation per se is Hille [1962].

10.6.1 The Carathéodory–Fejér method

The Carathéodory–Fejér method, or CF method, is a method of finding a near-best rational approximation $r_{L/M}^{CF}(z)$ on the unit disk $D = \{z: |z| \leq 1\}$ to a function $f(z)$ which is analytic in the interior of D and continuous on the boundary S of D. The approximating function $r_{L/M}^{CF}(z)$ is required to be of type $[L/M]$. The function $f(z)$ may be given as

$$f(z) = \sum_{i=0}^{\infty} c_i z^i \qquad (6.10)$$

or else by its values on $S = \{z: |z| = 1\}$, from which the coefficients c_j are calculated using

$$c_j = \frac{1}{2\pi i} \int_S \frac{f(z)}{z^{j+1}} dz;$$

for this type of integral the standard numerical procedure is the use of the fast Fourier transform.

The CF method is based on the principle of arranging for the error function defined by

$$e_{L/M}^{CF}(z) = f(z) - r_{L/M}^{CF}(z) \qquad (6.11)$$

to have winding number at least $L + M + 1$ and then for $\|e_{L/M}^{CF}\|$ to be as small as possible. Here, the winding number $W\{e\}$ is (generically) defined to be the number of cycles of phase change that $e(z)$ undergoes as z encircles S once. The norm $\|\cdot\|$ is defined by

$$\|e\| = \sup_{z \in S} |e(z)| = \sup_{z \in D} |e(z)|, \qquad (6.12)$$

where the maximum-modulus theorem guarantees the last equality.

For example [Trefethen, 1981], Figure 10.6.3 shows the curve formed by the values of $e_{1/1}^{PA}(z)$ and $e_{1/1}^{CF}(z)$ on S, where

$$e_{1/1}^{PA}(z) = \exp(z) - \frac{1 + \frac{1}{2}z}{1 - \frac{1}{2}z} \qquad (6.13)$$

is the error of the Padé approximant of type [1/1] for the exponential function, and $e_{1/1}^{CF}(z)$ is the corresponding error for CF approximation. Both curves in Figure 10.6.3 have winding number equal to three. It is usually the case that greater accuracy is achieved on D by best and near-best approximations by sacrificing the precision of the equivalent Padé approximants at the origin, and this fact is also borne out in Figure 10.6.3.

The CF method is best thought of as using a full specification of $f(z)$ on S, or equivalently many values $\{c_i\}_{i=0}^{K}$ with K large. These data can then be used (as follows) to construct a comparatively low-order approximant $r_{L/M}^{CF}(z)$ with $L, M \ll K$. Then, in the limiting case of K large, the CF approximants are directly related to $f(z)$ rather than to a Maclaurin section of it.

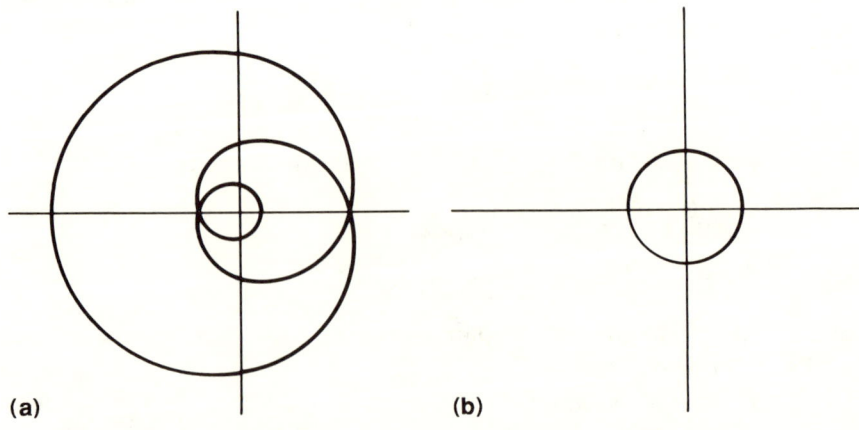

Figure 10.6.3. Error curves (a) for the [1/1] Padé approximant of $\exp(z)$ and (b) for the [1/1] near-best CF approximant of $\exp(z)$.

10.6 Connection with rational approximation

The CF construction process begins with the construction of an intermediate rational approximant,

$$\tilde{r}^{CF}(z) = \frac{\tilde{p}^{CF}(z)}{\tilde{q}^{CF}(z)}, \tag{6.14}$$

according to the formula

$$\sum_{i=0}^{K} c_i z^i - \frac{\tilde{p}^{CF}(z)}{\tilde{q}^{CF}(z)} = \tilde{e}^{CF}(z). \tag{6.15}$$

With an eye on the end result that $r_{L/M}^{CF}(z)$ should have at most M poles, all lying in $\mathbb{C}\backslash D$, we will arrange for $\tilde{q}^{CF}(z)$ to have at most M zeros in $\mathbb{C}\backslash D$ and (6.15) shows that we should seek an error function $\tilde{e}^{CF}(z)$ which has at most M poles in $\mathbb{C}\backslash D$.

We will arrange it [see (6.30)–(6.32)] so that $\tilde{r}^{CF}(z)$ is expressed by

$$\tilde{r}^{CF}(z) = \frac{\tilde{s}^{CF}(z)}{q_{L/M}^{CF}(z)}, \tag{6.16}$$

where its numerator

$$\tilde{s}^{CF}(z) = \frac{\tilde{p}^{CF}(z)}{\tilde{q}^{CF}(z)/q_{L/M}^{CF}(z)} \tag{6.17}$$

is a rational function which is analytic in $|z| \geq 1$ but may have poles in $|z| < 1$. As such, it possesses the Laurent expansion

$$\tilde{s}^{CF}(z) = \sum_{i=-\infty}^{L} s_i^{CF} z^i \tag{6.18}$$

which converges in $|z| \geq 1$. Thus we must obtain $\tilde{r}^{CF}(z)$ in (6.16) from the class $\tilde{R}_{L/M}$ of rational functions defined by

$$\tilde{R}_{L/M} = \left\{ \tilde{r}(z) \colon \tilde{r}(z) = \frac{\tilde{p}(z)}{\tilde{q}(z)} \text{ with } \tilde{p}(z) = \sum_{i=-\infty}^{L} \tilde{p}_i z^i \text{ and } \tilde{q}(z) = \sum_{i=0}^{M} \tilde{q}_i z^i \right\} \tag{6.19}$$

with the understanding that $\tilde{p}(z)$ is analytic in $|z| \geq 1$.

Error functions with the smallest norms are circular or nearly circular in the sense of Figure 10.6.3(b). The following theorem of de la Vallée-Poussin-type shows why this is so.

Theorem 10.6.4 [Trefethen, 1981]. *Let $f(z)$ be analytic in D and continuous in S, and let $\hat{r}(z) \in \tilde{R}_{L/M}$ be a best approximation to $f(z)$ on D. Let $\tilde{r}(z)$ be an approximation to $f(z)$ whose error function*

$$\tilde{e}(z) = f(z) - \tilde{r}(z)$$

is finite and nonzero for $z \in D$ and has winding number $W\{\tilde{e}\} \geq L + M + 1$. Then

$$\inf_{z \in S} |f(z) - \tilde{r}(z)| \leq \|f - \hat{r}\| \leq \|f - \tilde{r}\|. \tag{6.20}$$

Proof. The second inequality of (6.20) follows from the definition of $\hat{r}(z)$. For the first inequality, suppose to the contrary that

$$\|f - \hat{r}\| < \inf_{z \in S} |f(z) - \tilde{r}(z)|.$$

Then a geometrical argument justifies

$$|\arg(f(z) - \tilde{r}(z)) - \arg(\tilde{r}(z) - \hat{r}(z))| < \frac{\pi}{2}$$

for all $z \in S$. Hence the winding numbers of $f(z) - \tilde{r}(z)$ and $\tilde{r}(z) - \hat{r}(z)$ are the same. Therefore

$$W\{\tilde{r} - \hat{r}\} \geq L + M + 1. \tag{6.21}$$

However, $\tilde{r}(z) - \hat{r}(z) \in \tilde{R}_{L+M/2M}$, it has at most $2M$ poles in $\mathbb{C} \setminus D$, and at $z = \infty$ it is meromorphic and of order z^{L-M}. Therefore $W\{\tilde{r} - \hat{r}\} \leq 2M + (L - M)$, which contradicts (6.21). Therefore (6.20) is established.

There follows immediately the

Corollary. *If the error function is circular, i.e., if*

$$|e(z)| = \|e\|, \quad z \in S, \tag{6.22}$$

and if $W\{e\} \geq L + M + 1$, then the corresponding approximant $\tilde{r}(z)$ is a best approximant to $f(z)$ in D from $\tilde{R}_{L/M}$.

The remarkable result of Carathéodory and Fejér is the construction of an error function satisfying (6.22): we will show concretely how $\tilde{r}^{CF}(z)$ in (6.14) is constructed so that (6.22) is satisfied. To begin, let $K \geq J = L - M$ and consider the $K - J$ equations

$$\begin{bmatrix} c_{J+1} & c_{J+2} & \cdots & c_K \\ c_{J+2} & c_{J+3} & \cdots & 0 \\ \vdots & \vdots & \ddots & \vdots \\ c_K & 0 & \cdots & 0 \end{bmatrix} \begin{bmatrix} u_0^{(j)*} \\ u_1^{(j)*} \\ \vdots \\ u_{K-J-1}^{(j)*} \end{bmatrix} = \sigma_j \begin{bmatrix} u_0^{(j)} \\ u_1^{(j)} \\ \vdots \\ u_{K-J-1}^{(j)} \end{bmatrix} \tag{6.23}$$

which are denoted concisely by

$$C\mathbf{u}^{(j)*} = \mathbf{u}^{(j)} \sigma_j.$$

10.6 Connection with rational approximation

Equation (6.23) represents the jth column of the singular-value decomposition of C:

$$CU^* = U\Sigma,$$

where $\Sigma = \text{diag}(\sigma_1, \sigma_2, \ldots, \sigma_{K-J})$ consists of singular values of C, ordered with $\sigma_1 \geq \sigma_2 \geq \cdots \geq \sigma_{K-J}$ [Golub and Van Loan, 1989, p. 71]. The matrix C is symmetric, but not Hermitian if its elements c_i are complex rather than real. The singular-value decomposition stably provides the solutions $\{(\sigma_j, \mathbf{u}^{(j)})\}_{j=1}^{K-J}$ for (6.23); the appropriate value of j will be given shortly.

For each j, a polynomial $p^{(j)}(z)$ of degree $2K - J - 1$ at most is defined by

$$\left(\sum_{i=0}^{K} c_i z^i\right)(u_0^{(j)*} z^{K-J-1} + u_1^{(j)*} z^{K-J-2} + \cdots + u_{K-J-1}^{(j)*}) - p^{(j)}(z)$$

$$= \sigma_j z^K (u_0^{(j)} + u_1^{(j)} z + \cdots + u_{K-J-1}^{(j)*} z^{K-J-1}). \quad (6.24)$$

Notice that (6.23) implies that the coefficients of z^K, z^{K+1}, ..., z^{2K-J-1} on the right-hand side of (6.24) equal those of the first term on the left-hand side, and hence $p^{(j)}(z)$ is actually a polynomial of degree $K - 1$ at most. In this respect, the equations (6.23) resemble the Padé equations. We now have the representation

$$\sum_{i=0}^{K} c_i z^i - \frac{p^{(j)}(z)}{q^{(j)}(z)} = e^{(j)}(z), \quad (6.25)$$

where

$$q^{(j)}(z) = u_0^{(j)*} z^{K-J-1} + u_1^{(j)*} z^{K-J-2} + \cdots + u_{K-J-1}^{(j)*}, \quad (6.26)$$

$$r^{(j)}(z) = \frac{p^{(j)}(z)}{q^{(j)}(z)} = \frac{\left[\left(\sum_{i=0}^{K} c_i z^i\right) q^{(j)}(z)\right]_0^{K-1}}{q^{(j)}(z)}, \quad (6.27)$$

and

$$e^{(j)}(z) = \sigma_j z^K \frac{u_0^{(j)} + u_1^{(j)} z + \cdots + u_{K-J-1}^{(j)*} z^{K-J-1}}{u_0^{(j)*} z^{K-J-1} + u_1^{(j)*} z^{K-J-2} + \cdots + u_{K-J-1}^{(j)*}}. \quad (6.28)$$

In fact, (6.28) shows that $e^{(j)}(z)$ is a perfectly circular error function. To see this, and to prepare for the next step, suppose that μ zeros $z_1, z_2, \ldots, z_\mu$ of $q^{(j)}(z)$ lie in $|z| > 1$ and that λ zeros $z_{\mu+1}, z_{\mu+2}, \ldots, z_{\mu+\lambda}$ lie in $|z| < 1$. Then (6.28) becomes

$$e^{(j)}(z) = \sigma_j z^{J+1} \prod_{i=1}^{\lambda+\mu} \frac{1 - z_i^* z}{1 - z_i/z}. \quad (6.29)$$

For simplicity later, we assume that $q^{(j)}(z)$ has no zeros on S, even though the factors associated with such zeros could cancel in the product in (6.29). This product is a Blaschke product: because z_i, z_i^{*-1} are inverse points with respect to S, it follows that

$$|e^{(j)}(z)| = 1 \quad \text{for } z \in S,$$

and therefore that $e^{(j)}(z)$ is an ideal, circular error function indicative of best approximation.

Provided that the singular values $\{\sigma_j\}$ are distinct, they form a monotonically decreasing sequence of values of $\{\|e^{(j)}\|\}$. Each of $e^{(1)}(z)$, $e^{(2)}(z), \ldots, e^{(K-J)}(z)$ is an ideal error function, which must correspond respectively to the cases of best approximation with no poles, one pole, two poles, etc. allowed in $|z| > 1$. Consequently we take $j = M + 1$. With this assignment, the singular value and singular vector of (6.23) are determined uniquely once M is given. It is then the case that $q^{(M+1)}(z)$ has precisely M zeros in $|z| > 1$, and $\mu = M$ in (6.29). To summarize, we have

$$\tilde{q}^{\text{CF}}(z) = q^{(M+1)}(z), \quad \tilde{e}^{\text{CF}}(z) = e^{(M+1)}(z), \tag{6.30}$$

$$\tilde{p}^{\text{CF}}(z) = p^{(M+1)}(z) = \left[\left(\sum_{i=0}^{K} c_i z^i\right) \tilde{q}^{\text{CF}}(z)\right]_0^{K-1} \tag{6.31}$$

as formulas for the denominator, error, and numerator of the intermediate rational approximant $\tilde{r}^{\text{CF}}(z)$ defined in (6.14). If we assume that $u_0^{(M+1)} \neq 0$, $u_{K-J-1}^{(M+1)} \neq 0$, then $\tilde{r}^{\text{CF}}(z)$ is of type $[K-1/K-J-1]$. Because $\tilde{q}^{\text{CF}}(z)$ has precisely M zeros in $|z| > 1$, it follows from (6.29), (6.30) that $\tilde{e}^{\text{CF}}(z)$ has $(K-J-1) - M = K-L-1$ zeros in $|z| > 1$ and a pole of order K at infinity. Hence

$$W\{\tilde{e}^{\text{CF}}\} = K + M - (K - L - 1) = L + M + 1.$$

Therefore $\tilde{r}^{\text{CF}}(z)$ is a best approximation from $\tilde{R}_{L/M}$ for $f(z)$ on D, assuming that $u_0^{(M+1)} \neq 0$, $u_{K-J-1}^{(M+1)} \neq 0$ and $q^{(M+1)}(z)$ has no zeros in S, and that all the singular values of C are distinct.

We have noted that $\tilde{r}^{\text{CF}}(z)$ is of type $[K-1/K-J-1]$ which is not the type $[L/M]$ required for $r_{L/M}^{\text{CF}}(z)$. Nevertheless, we can use the values of the M zeros $\{z_i\}_{i=1}^{M}$ lying in $|z| > 1$ of $\tilde{q}^{\text{CF}}(z)$ to define

$$q_{L/M}^{\text{CF}}(z) = \prod_{i=1}^{M}(1 - z/z_i). \tag{6.32}$$

The CF method of Gutknecht and Trefethen consists of truncating (6.18) at $i = 0$ and defining

$$p_{L/M}^{\text{CF}}(z) = \sum_{i=0}^{L} s_i^{\text{CF}} z^i \tag{6.33}$$

in terms of the coefficients s_i^{CF} obtained from (6.17). This truncation is justified by the empirical observation that $\{s_i\}_{i=-\infty}^{-1}$ are small in practice for problems in which the coefficients c_i vary smoothly with i. Then $r_{L/M}^{CF}(z) = p_{L/M}^{CF}(z)/q_{L/M}^{CF}(z)$ is the CF near-best approximant of type $[L/M]$ to $f(z)$ on D.

We refer to Trefethen [1981, 1983] for useful examples. Suffice it to say that in using the CF method to find $r_{1/1}^{CF}(z)$ for $\exp(z)$ on D, (6.32) takes the form

$$q_{1/1}^{CF}(z) = 1 - 0.43417z,$$

(6.18) takes the form

$$\tilde{s}^{CF}(z) = \cdots + 0.00001z^{-2} + 0.00025z^{-1} + 0.99613 + 0.58955z,$$

and hence the CF approximant is

$$r_{1/1}^{CF}(z) = \frac{0.99613 + 0.58955z}{1 - 0.43417z}.$$

The CF method evolved from the fundamental papers of Carathéodory and Fejér [1911] (see Akhiezer [1956, Appendix D]) and Adamian, Arov, and Krein [1971]; its history is given by Trefethen [1983] and Gutknecht [1984]. The latter paper includes the important extension to CF approximation on an interval, as well as on a circle. An early important result which used this method was a proof by Magnus [1988] of the '$\frac{1}{9}$th' conjecture. He showed that

$$\lim_{n \to \infty} (e_{n/n})^{1/n} = 1/9.289 \ldots$$

for the rate of error decrease of best rational approximation to e^{-x} on the positive real axis, where $e_{n/n}$ is defined by

$$e_{n/n} = \inf_{r \in R_{n/n}} \sup_{0 \leq x < \infty} |e^{-x} - r_{n/n}(x)|^{1/n}$$

where $R_{n/n}$ is the class of rational approximants of type $[n/n]$. Corresponding results for $\exp(-x^2)$ and $\exp(-x^3)$ are also given by Magnus [1988]. The same approach also led to the proof of Meinardus's conjecture [Braess, 1984, 1986].

We notice that the matrix C in (6.23) has a Hankel structure similar to that of the matrix in (1.1.5) for Padé approximation. It is therefore to be expected that there will be as many convergence problems with CF approximants as there are for Padé approximants when the coefficients c_n do not vary smoothly with n. This expectation is borne out by the theorem of Saff and Totik [1989].

10.7 Padé approximants for the Riccati equation

A Riccati equation is a nonlinear ordinary differential equation of the form

$$A(x)\frac{du}{dx} + B(x)u + C(x)u^2 = D(x). \tag{7.1}$$

Riccati equations arise naturally in the analysis of homogeneous, second-order, ordinary differential equations. For example, we consider the one-dimensional Schrödinger equation, which may be expressed as

$$\phi'' - D(x)\phi = 0. \tag{7.2}$$

By making the substitution $u(x) = \phi'(x)/\phi(x)$, we get for (7.2)

$$u' + u^2 = D(x),$$

which is a speical case of (7.1). In fact, the general Riccati equation of the form (7.1) can always be reduced to a second-order, linear homogeneous ordinary differential equation [Ince, 1944, p. 24]. If the functions $A(x)$, $B(x)$, $C(x)$, and $D(x)$ occurring in (7.1) are polynomials, it may be possible to solve (7.1) using a continued-fraction variant of the Frobenius method. Notice that the substitution

$$u(x) = \frac{1}{f(x)}$$

in (7.1) leads to the equation

$$A(x)f' - B(x)f + D(x)f^2 = C(x), \tag{7.3}$$

which is also a Riccati equation. In this sense, the Riccati equation is 'form invariant under reciprocation.'[†] We explain a method of solution by means of an illustrative example. Consider the Riccati equation

$$(ax + b)y' + (cx + d)y + (ex + g)y^2 = h, \tag{7.4}$$

where a, b, c, d, e, g, and h are constants. Substitute

$$y(x) = \frac{h}{cx + \alpha - u(x)}, \tag{7.5}$$

which leads to the equation

$$(ax + b)u' + (cx + \alpha - u)(d - \alpha + u) + h(ex + g) - c(ax - b) = 0. \tag{7.6}$$

[†] J. Zinn-Justin, private communication.

Equations (7.5), (7.6) are consistent with the hypothesis that

$$u(x) = O\left(\frac{1}{x}\right), \tag{7.7}$$

as $|x| \to \infty$, provided that the value of α is taken to be

$$\alpha = d - a + eh/c. \tag{7.8}$$

Then (7.6) reduces

$$(ax + b)u' + (cx - d + 2\alpha)u - u^2 = \alpha(\alpha - d) - gh + cb. \tag{7.9}$$

Comparison of (7.4) and (7.9) shows that this Riccati equation is form invariant under the substitution (7.5). The constants a, b and c are unchanged, and

$$d \to 2\alpha - d, \quad e \to 0, \quad g \to -1, \quad h \to \alpha(\alpha - d) - gh + cb. \tag{7.10}$$

A solution of an arbitrary Riccati equation of the type (7.4) can therefore be expressed as a continued fraction by making repeated substitutions of the form (7.5). A systematic investigation of this procedure leads to the value-invariance theorem [Baker, 1984a; Baker, Benofy, and Enting, 1986]. As an example of a Riccati equation of this kind, we consider

$$xy' - (x + n - 1)y = -1. \tag{7.11}$$

Continued-fraction solutions of differential equations of this type were first found by Laguerre [1879, 1885], who used the equation (7.11) in the case $n = 1$ as an explicit example. Equation (7.11) has the solution

$$y(x) = e^x E_n(x), \tag{7.12}$$

where

$$E_n(x) = \int_1^\infty e^{-xt} t^{-n} \, dt = x^{n-1} \int_x^\infty e^{-u} u^{-n} \, du \tag{7.13}$$

for $\operatorname{Re} x > 0$. Following (7.5), (7.8), we substitute

$$y(x) = \frac{-1}{-x - n - u(x)}, \tag{7.14}$$

and from (7.10) we find that $u(x)$ satisfies

$$xu' - (x + n + 1)u - u^2 = n. \tag{7.15}$$

Now we consider the inductive hypothesis

$$xu_r' - (x + n + 2r - 1)u_r - u_r^2 = r(n + r - 1), \tag{7.16}$$

and note that (7.15) is precisely (7.16) in the case $r = 1$. Using (7.5), we substitute

$$u_r(x) = \frac{r(n + r - 1)}{-x + \alpha_r - u_{r+1}(x)}. \tag{7.17}$$

Using (7.8), we find that

$$\alpha_r = -n - 2r \tag{7.18}$$

and from (7.10) we find that $u_{r+1}(x)$ satisfies

$$xu'_{r+1} - (x + n + 2r + 1)u_{r+1} - u_{r+1}^2 = (r + 1)(n + r). \tag{7.19}$$

Equation (7.19) establishes the inductive hypothesis (7.16) for all positive integers r. We find the solution of (7.11) from (7.14), (7.17), and (7.18) to be

$$y(x) = \frac{-1}{-x - n -} \frac{n}{-x - n - 2 -} \frac{2n + 2}{-x - n - 4 -} \cdots$$
$$\frac{(r + 1)(n + r)}{-\; -x - n - 2(r + 1)\; -} \cdots$$
$$= \frac{1}{x + n -} \frac{n}{x + n + 2 -} \frac{2n + 2}{x + n + 4 -} \cdots$$
$$\frac{(r + 1)(n + r)}{-\; x + n + 2(r + 1)\; -} \cdots . \tag{7.20}$$

In terms of the variable $z = x^{-1}$, the Mth convergent of (7.20) is the $[M/M]$ Padé approximant of $y(1/z)$, by virtue of the accuracy-through-order conditions imposed by (7.5), (7.7), and (7.8). In fact the fraction (7.20) is precisely that of (4.6.8), although the method of derivation is quite different. The foregoing analysis makes it clear that neat, closed-form expressions for the solution of Riccati equations can only be obtained in exceptional cases. In general, the technique exemplified by (7.4), (7.5), (7.8), and (7.10) leads to a continued-fraction representation of $y(x)$, the solution of the Riccati equation, provided the accuracy-through-order conditions can be imposed consistently. A necessary condition for this consistency in the case that the coefficients $A(x)$, $B(x)$, $C(x)$, and $D(x)$ occurring in (7.1) are polynomials is that the degrees of these polynomials satisfy

$$\deg\{D(x)\} < \deg\{B(x)\},$$
$$\deg\{A(x)\} \leq \deg\{B(x)\}, \tag{7.21}$$
$$\deg\{C(x)\} \leq \deg\{B(x)\}.$$

10.7 Padé approximants for the Riccati equation

The books of Luke [1969, Chap 10] and Khovanskii [1963] describe the origins of Padé–Riccati approximation. For further details of the theory of Riccati equations, we refer to Reid [1972]; for their connection with the Lanczos τ-method, we refer to Fair [1964]; and for an application to Laguerre's method for the expansion of the Hamburger function

$$G_\beta(x) = \int_{-\infty}^{\infty} \exp\left(\frac{-t^2/2 - \beta t^4}{x - t}\right),$$

we refer to Bessis [1979]. A full specification of the degree conditions which lead to value- and argument-invariance properties is given by Baker [1984a]. Baker, Benofy, and Enting [1986] show that Padé–Riccati approximants have the capacity to represent confluent singularities, which are important for Ising-model calculations, and Magnus [1984] gives their application to one-dimensional disordered structures. Matrix Riccati equations have been investigated by Common and Roberts [1986] and applied to the Toda lattice [Common and Hafez, 1990].

11

Connection with quantum field theory

11.1 Perturbed harmonic oscillators

This chapter consists of three sections. In the first section, we discuss the arguments which lead us to think that Padé approximants are an important tool for the summation of the perturbation expansions of quantum field theories. In the second section, we discuss two exemplary models of pion–pion scattering in which Padé approximants have revitalized the prospects of perturbation theory. In the last section we give a flavor of the application of Padé methods to critical phenomena and indicate how this subject relates to field theory.

We cannot adequately summarize here the many books on quantum theory. The relevant background of quantum mechanics used in this section is given in *The Principles of Quantum Mechanics* [Dirac, 1958]. Elementary field theory and quantum electrodynamics are described by Schweber [1961]; we refer especially to the *Cargèse Lectures in Physics* [Bessis, 1972] for an advanced treatment of the relevant background of quantum field theory.

The world of quantum electrodynamics (QED) is the world of photons and electrons. The vacuum is postulated to be the lowest energy state. A basic feature of QED is the possibility of the formation of virtual electron–positron pairs. This phenomenon is the origin of vacuum polarization, which has a measurable and accurately verified effect on the energy levels of atomic hydrogen. Exploding black holes may be a more dramatic realization of this effect. Dyson [1952] considered a configuration of many nearby electrons and, some distance away, as many nearby positrons. We show an example in Figure 11.1. The self-interaction energy of such a virtual system is dominated by the short-range repulsion energy of each group. Thus the configuration has a large positive self-energy, corresponding to a very small temporal duration and little effect on any calculation.

11.1 Perturbed harmonic oscillators

Figure 11.1.1. A configuration of virtual electrons and positrons.

Suppose that any quantum-electrodynamic quantity is calculated from a perturbation series in $\lambda = e^2$, and that this series converges. Then its radius of convergence is at least e^2, and so it also converges for $\lambda = -\frac{1}{2}e^2$. This corresponds physically to reversing the sign of the electron–positron interaction. The configuration of Figure 11.1.1 now has a large negative energy and becomes exceedingly probable. This physical effect is imaginatively called the exploding vacuum. It should be represented by a divergence of any perturbation calculation in QED. If any series expansion of QED is to be associated with a function analytic in $|\lambda| < \varepsilon$, $|\arg(\lambda)| < \pi$, we can only conclude that such a series expansion should have zero radius of convergence.

We can illustrate this physical argument by two models which support the contention that Padé approximants or similar techniques will have to be used to interpret the perturbation series. These are quantum-mechanical harmonic-oscillator models which are analogues of a zero-dimensional field theory.

11.1.1 The Peres model

Peres considered two independent one-dimensional quantum-mechanical harmonic oscillators. The mth excited state of one is described by the wave function

$$|m\rangle_1 = \psi_m(x) = c_m e^{-(1/2)x^2} H_m(x), \tag{1.1a}$$

and the nth excited state of the other by

$$|n\rangle_2 = \psi_n(y) = c_n e^{-(1/2)y^2} H_n(y). \tag{1.1b}$$

where c_m, $m = 0, 1, 2, \ldots$, are normalization constants. The combined wave function of the oscillators is

$$|m, n\rangle = \psi_{m,n}(x, y) = c_m c_n e^{-(1/2)(x^2+y^2)} H_m(x) H_n(y), \tag{1.2}$$

which is an eigenfunction of the Hamiltonian

$$\mathcal{H}_0 = \tfrac{1}{2}(p_x^2 + p_y^2 + x^2 + y^2). \tag{1.3}$$

The time-independent solution (1.2) is perturbed by the instantaneous interaction

$$\mathcal{H}_1 = gx^2 y \delta(t). \tag{1.4}$$

Its effect is determined by the time-dependent Schrödinger equation

$$\{\mathcal{H}_0 + \mathcal{H}_1\}\Psi(x, y, t) = i\frac{\partial \Psi(x, y, t)}{\partial t}.$$

For $t < 0$, the solutions are the unperturbed solutions

$$\Psi_{m,n}(x, y, t) = \psi_{m,n}(x, y)e^{-i(E_m + E_n)t}$$

and so for $t = 0+$, the perturbed wave function is

$$\Psi_{m,n}(x, y, 0+) = e^{-igx^2 y}\Psi_{m,n}(x, y, 0-).$$

The S-matrix for the interaction is therefore given by the overlap integral

$$S(m'n', mn) = \int_{-\infty}^{\infty}\int_{-\infty}^{\infty} \psi_{m',n'}^*(x, y) e^{-igx^2 y} \psi_{m,n}(x, y)\, dx\, dy. \tag{1.5}$$

The ground-state wave function, given by (1.1) and normalized, is

$$\psi_{0,0}(x, y) = \frac{1}{\sqrt{\pi}} e^{-(1/2)(x^2 + y^2)}.$$

Upon substitution in (1.5), the ground-state-to-ground-state transition amplitude is

$$S_{00}(g^2) = \frac{1}{\pi}\int_{-\infty}^{\infty}\int_{-\infty}^{\infty} e^{-x^2 - y^2 - igx^2 y}\, dx\, dy \tag{1.6}$$

$$= \frac{2}{\sqrt{\pi}}\int_0^{\infty} e^{-x^2 - (1/4)g^2 x^4}\, dx \tag{1.7}$$

$$= \pi^{-1/2} g^{-1} e^{1/(2g^2)} K_{1/4}\left(\frac{1}{2g^2}\right). \tag{1.8}$$

$S_{00}(g^2)$ is well defined by (1.6), (1.7), or (1.8) for all real values of g. Further, (1.7) shows that $S_{00}(g^2) \to 1$ as $g \to 0$ for real g. However, (1.7) is undefined for Re $g^2 < 0$, and it is plain that $S_{00}(g^2)$ has a singularity at $g^2 = 0$. While (1.8) is an explicit expression for $S_{00}(g^2)$, it is more convenient to integrate over x first in (1.6) and obtain

$$S_{00}(g^2) = \frac{1}{\sqrt{\pi}}\int_{-\infty}^{\infty} e^{-y^2} \frac{1}{\sqrt{1 + igy}}\, dy$$

$$= \frac{1}{\sqrt{\pi}}\int_{-\infty}^{\infty} e^{-y^2}\left\{1 + \frac{(-\frac{1}{2})(-\frac{3}{2})}{2!}(-g^2 y^2)\right.$$

11.1 Perturbed harmonic oscillators

$$= \frac{1}{\sqrt{\pi}} \int_{-\infty}^{\infty} e^{-y^2} \left\{ 1 + \frac{(-\frac{1}{2})(-\frac{3}{2})(-\frac{5}{2})(-\frac{7}{2})}{4!}(-g^2 y^2)^2 + \cdots \right\} dy$$

$$= \frac{1}{\sqrt{\pi}} \int_{-\infty}^{\infty} e^{-y^2} \left\{ 1 + \frac{(\frac{1}{4})(\frac{3}{4})}{(\frac{1}{2})1!}(-g^2 y^2) + \frac{(\frac{1}{4})(\frac{3}{4})(\frac{5}{4})(\frac{7}{4})}{(\frac{1}{2})(\frac{3}{2})2!}(-g^2 y^2)^2 + \cdots \right\} dy.$$

Using the recursion relation for the gamma function and

$$\int_0^{\infty} e^{-y^2} y^{2r} \, dy = \Gamma(r + \tfrac{1}{2}),$$

we obtain

$$S_{00} = 1 + \frac{(\frac{1}{4})(\frac{3}{4})(-g^2)}{1!} + \frac{(\frac{1}{4})(\frac{3}{4})(\frac{5}{4})(\frac{7}{4})}{2!}(-g^2)^2 + \cdots \tag{1.9}$$

$$= {}_2F_0(\tfrac{1}{4}, \tfrac{3}{4}, -g^2). \tag{1.10}$$

These formal manipulations with divergent series are justified by the use of Carleman's theorem. Also we see that $S_{00}(g^2)$ is a Stieltjes series with zero radius of convergence. The asymptotic series (1.9) shows that finite-order perturbation theory with sufficiently weak coupling can give reasonably accurate results. However, arbitrarily accurate results for any specified coupling cannot be obtained by simple series summation. Diagonal Padé approximants to (1.9) are known to converge, and numerical calculations confirm this impressively [Baker and Chisholm, 1966].

We review this analysis using the Heisenberg representation, in which the space variable x and the momentum p_x are time-dependent operators. They satisfy the equal-time commutation relation

$$[x(t), p_x(t)] = i.$$

The solution is given by

$$x(t) = x_0 \cos t + p_{x0} \sin t,$$

$$p_x(t) = p_{x0} \cos t - x_0 \sin t,$$

and it follows that

$$[x(t), x(t')] = -i \sin(t - t'). \tag{1.11}$$

We consider next the operators

$$\eta = \frac{p + ix}{\sqrt{2}}$$

and

$$\eta^* = \frac{p - ix}{\sqrt{2}},$$

which play the roles of creation and annihilation operators. Following Dirac, we find that the Heisenberg energy eigenstates are given by $\eta^n|0\rangle$ with eigenvalues $E_n = n + \frac{1}{2}$. Then $x = (\eta - \eta^*)/\sqrt{2}$ corresponds to the quantum field, and its propagator is given by (1.11). Hence we see that the interaction Hamiltonian (1.4) is analogous to a $g\bar{\psi}\psi\phi$ interaction in a field theory with no space dimensions, and similarly to the electron–photon interaction of QED. We are led to expect a singularity at zero coupling strength in the QED perturbation expansions.

11.1.2 The anharmonic oscillator

The quantum-mechanical one-dimensional anharmonic oscillator considered here is derived from the Hamiltonian

$$\mathcal{H} = p^2 + x^2 + \beta x^4, \quad \beta > 0. \tag{1.12}$$

The quantum-mechanical problem consists of finding the energy eigenvalues and eigenstates of the Schrödinger equation

$$-\frac{d^2\psi}{dx^2} + (x^2 + \beta x^4)\psi = E\psi. \tag{1.13}$$

If $\beta = 0$, the solution is given by (1.1a) and perturbation theory provides a method for calculating E and $\psi(x)$ as power series in β.

The leading term of the asymptotic solution of (1.13) is

$$\psi(x) \sim e^{-\sqrt{\beta} x^3/3} \quad \text{as } x \to +\infty.$$

We see that the character of the wave function changes drastically at $\beta = 0$. It is better to consider the alternative problem with

$$H = p^2 + \alpha x^2 + x^4$$

and solve the Schrödinger equation

$$-\frac{d^2\psi}{dx^2} + (\alpha x^2 + x^4)\psi = E\psi.$$

For this, the perturbative solutions can be proved to be analytic in α at $\alpha = 0$, and the expansion converges for small $|\alpha|$. More generally, we consider the Hamiltonian

$$H = p^2 + \alpha x^2 + \beta x^4, \tag{1.14}$$

which has energy levels $E = E_n(\alpha, \beta)$. We have stated that $E_n(\alpha, 1)$ is analytic at $\alpha = 0$.

Consider the scale change in (1.13) of

$$x \to x' = \lambda x.$$

Any one solution of (1.13) leads to a whole class of solutions of (1.14), and we find

$$E_n(\alpha, \beta) = \lambda^{-2} E_n(\alpha \lambda^4, \beta \lambda^6).$$

By taking $\alpha = 1$, $\beta \lambda^6 = 1$, this becomes

$$E_n(1, \beta) = \beta^{1/3} E_n(\beta^{-2/3}, 1), \qquad (1.15)$$

which shows that $E_n(1, \beta)$ has a three-sheeted branch cut ending at ∞ in the β-plane. Further, it suggests strongly (and correctly) that the three-sheeted branch point is at $\beta = 0$. Since $E_n(1, \beta)$ is real symmetric, the cut is taken along $-\infty < \beta \leq 0$, and we emphasize the consequence that the perturbation series of $E_n(1, \beta)$ has zero radius of convergence. Again, it can be proved that $E_n(1, \beta)$ is a Stieltjes series in β with a convergent sequence of diagonal Padé approximants. For further details of the theory of the anharmonic oscillator and its relation to field theory, we refer to Bender and Wu [1968] and Simon [1970]. The paper of Loeffel et al. [1969] is important for proving the convergence of the Padé method for the βx^4 anharmonic oscillator. For the βx^8 anharmonic oscillator, the perturbation series diverge so rapidly that the moment problem is indeterminate [Graffi and Grecchi, 1978]. For further details, we refer to Killingbeck [1980], Graffi, Grecchi, and Simon [1971], Graffi, Grecchi, and Turchetti [1970], and Ruijgrok [1972].

In the Heisenberg picture, the βx^4 perturbation in (1.12) corresponds to ϕ^4 field theory. The inference is that the perturbation expansion of field theory is always divergent and that Padé approximants should be used. Indeed, recent explicit estimates show that the perturbation-series coefficients in four dimensions diverge like

$$c_k \sim \left(\frac{-1}{16\pi^2}\right)^k (k + \tfrac{7}{2})! \quad \text{as } k \to \infty$$

[Brezin, Le Guillou, and Zinn-Justin, 1977].

11.2 Pion–pion scattering

In this section we discuss some applications of Padé approximants by which fundamental theories of elementary particles are tested. We concentrate on how the properties of the Padé approximants are useful in these applications rather than on the details of the dynamics. No longer are we in a situation of having arbitrarily many terms of a power series available to arbitrary accuracy. With pion dynamics, at most five terms are available at present, and laborious calculation is required to obtain even a few decimal places of accuracy of the highest-order term. Space does not permit a full discussion of the technicalities of elementary-particle theory in this section. Again we refer to the *Cargèse Lectures in Physics* [Bessis, 1972] for details.

A variety of calculations have been made using Padé approximants on nucleon–nucleon and meson–nucleon systems. The prototypical Padé approximant calculation of the hadronic spectrum is that of the ϕ^4 model of pion–pion scattering. The perturbation series is derived from the Lagrangian

$$L = \tfrac{1}{2}(\partial_\mu \phi_\alpha)^2 + \tfrac{1}{2}m^2 \phi_\alpha^2 + \tfrac{1}{4}\lambda(\phi_\alpha \phi_\alpha)^2. \tag{2.1}$$

The T-matrix elements are derived from the Feynman graphs of Figure 11.2.1, and formally

$$T(s, t, u) = \sum_{i=0}^{\infty} T_i(s, t, u).$$

Each independent closed loop of each graph represents a four-dimensional space-time integral, and each internal line represents a propagator.

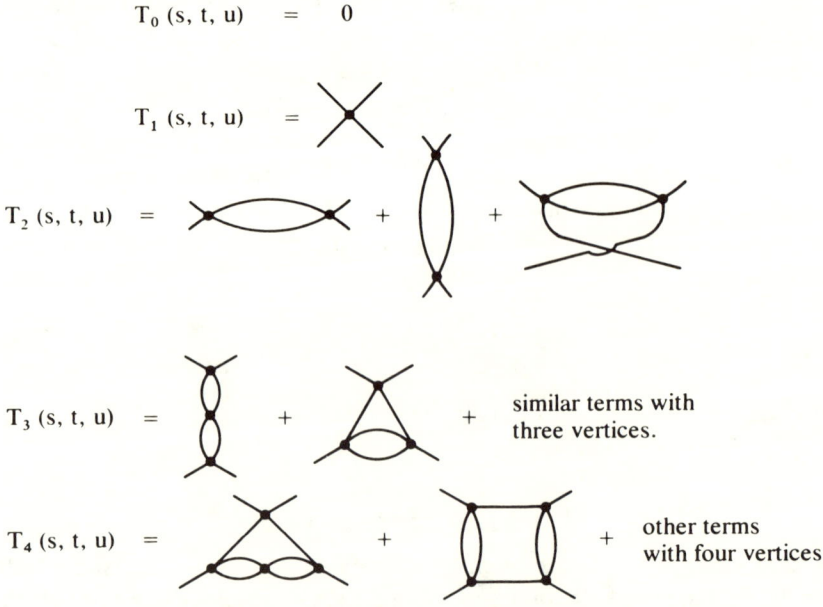

Figure 11.2.1. Some Feynman graphs.

A factor of λ is associated with each four-line vertex. If the graph represents a scattering process in which the intermediate particles can satisfy classical energy and momentum conservation, as if the pions were colliding billiard balls, then the graphs are singular and necessarily require multidimensional singular quadratures. Furthermore, such graphs are always numerically significant, so accurate evaluation of them is important.

From an analysis of the theory of Feynman graphs, it is known that the T-matrix is crossing symmetric, which means that $T(s, t, u)$ is a symmetric function of all three variables s, t, and u. The variable s corresponds to the energy available for the scattering process, and t to the momentum transfer. s, t, and u are relativistic invariants formed from the pions' four-momenta, and $s + t + u = 4$; consequently, u is a dependent variable. Ignoring the complication of isospin, the perturbation series of the T-matrix is crossing symmetric order by order:

$$T_i(s, t, u) = T_i(t, u, s) = T_i(u, t, s). \tag{2.2}$$

Consequently formation of any $[L/M]$ Padé approximant to the T-matrix preserves crossing symmetry.

Another important characteristic of the T-matrix is its unitary property. We form partial waves by making the Legendre decomposition

$$t_l(k^2) = \rho(k^2) \int_{-1}^{1} T(4 + 4k^2, -2k^2(1-x), -2k^2(1+x)) P_l(x)\, dx,$$

where $\rho(k^2)$ is a kinematical factor. k is the modulus of momentum of any of the pions in the center-of-mass frame, and $x = \cos\theta$ is the cosine of the scattering angle (θ). The unitary property of the T-matrix is expressed by

$$[1 + it_l(k^2)][1 + it_l^*(k^2)] = 1. \tag{2.3}$$

Just as in potential theory [see (9.4.12)], any finite order of perturbation theory is not exactly unitary. However, diagonal approximants of the S-matrix and $[L/M]$ approximants of the T-matrix perturbation series are unitary, provided $M \geq L$ (see Section 1.5: Theorem 1.5.5 and Example 1).

We summarize the situation by noting that all Padé approximants of the full amplitude are crossing symmetric but not unitary, whereas diagonal approximants of the partial-wave amplitudes are unitary but are not components of a crossing-symmetric amplitude. We are faced with a dilemma about which is the best method to choose. In fact, it is best to use both methods and compare them for stability and consistency.

So far, it has been tacitly assumed that Padé approximants are used to accelerate convergence of the perturbation series, but there is also a strong additional reason for forming the approximants, namely, their suitability for representing resonances. Many resonances occur in the various channels of pion elastic scattering, such as the ρ-meson in the $I = J = 1$ channel, where I denotes the isotopic spin and J denotes the angular momentum of the pions. If the $I = J = 1$ channel is open, the ρ-meson is the dominant feature of the pion–pion interaction at center-of-mass energy near 760 MeV, corresponding to $k^2 \approx 6.4$ in pion mass units.

It is widely thought that the existence of this meson is a consequence of an effective pion–pion Lagrangian, such as (2.1). Hence the ρ-meson would correspond to a singularity of the amplitude computed from the Lagrangian, and in fact the ρ-meson would be a pole of the partial-wave amplitude. More precisely, this means that $t_l(k^2)$ with g constant has a pole near $k^2 = 6.4$. However, it is natural to think of the Padé approximant as generating a pole in the coupling strength λ^2 at fixed k^2. In fact the pole of t_l in λ^2 at fixed k^2 corresponds to two second-sheet poles of $t_l(k^2)$ at fixed λ^2 if unitarity and analyticity are properly satisfied.

We are now in a position to assess qualitatively the achievements of theories based on a $\lambda \phi^4$ Lagrangian. The $I = J = 1$ amplitude should contain the ρ pole, and we assume that the coupling strength is adjusted so that a particular Padé approximant of the partial-wave perturbation series has a pole at the right energy. The choice of which approximant to use is rather restricted because the constant and first-order terms of the λ^2 expansion of $t_1^1(k^2)$ are zero. Bessis and Pusterla [1967] used the [2/2] approximant, calculated λ^2 from the ρ mass, and thereby fixed the only free parameter of the theory. The predictions of the theory are the partial-wave phase shifts, the resonances, and their widths. These quantities should satisfy stability tests, such as comparison with the results from [2/1] and [1/2] approximants, and the whole model should not violate crossing symmetry too much. In short, the model predictions are that the ρ-meson is clearly present and stable, but much too narrow; the $f^0(I = 0, J = 2)$ is predicted but much too narrow; the S-waves are very wrong; and an exotic $(I = 2, J = 2)$ meson is incorrectly predicted. Crossing symmetry is tested by the Martin inequalities and is well satisfied.

The alternative approach, using Padé approximants of the full amplitude, requires isolation of the 'pole term.' Rather imprecisely, we expect

$$[2/2]_T(k^2, \cos\theta) = \frac{P_1(\cos\theta)}{k^2 - 6.4 - i\Gamma k} \cdot \text{constant} + \text{background}.$$

The denominator of the Padé approximant is a function of (s, t) or equivalently $(k^2, \cos\theta)$. The zero of the denominator defines the poloid, as it is called, and this should occur at a value of k^2 which is independent of $\cos\theta$. The flatness of the poloid in the physical region, $-1 \leq \cos\theta \leq 1$, is an important consistency test of this approach, which is well satisfied in the $\lambda\phi^4$ fourth-order model. With this approach, the partial-wave amplitudes are defined by

$$t_l(k^2) = \rho(k^2)\int_{-1}^{1} P_l(x)[2/2]_T(4 + 4k^2, -2k^2(1-x), -2k^2(1+x))\,dx.$$

These should be unitary in the sense that (2.3) is approximately satisfied. Again, violation by the $\lambda\phi^4$ model is only a few percent.

We may summarize the results of the $\lambda\phi^4$ theory using low-order Padé approximants of either the full amplitude or the partial-wave amplitudes by noting that the internal-consistency tests are well satisfied and that only some of the predictions of the model are correct. These results were convincing enough to stimulate considerable further work, but also indicated that the $\lambda\phi^4$ Lagrangian is probably incomplete. We do not consider here the deeper question of whether the results of the renormalized perturbation series for $\lambda\phi^4$ theory in four dimensions in fact represent the true results of the field theory. In two and three dimensions, the perturbation series is Borel summable to the Schwinger functions [Eckmann, Magnen, and Sénéor, 1974; Dimock, 1974; Magnen and Sénéor, 1976; Feldman and Osterwalder, 1976].

The best physical predictions of pion–pion scattering have come from the σ-model. The method is basically the same as the one described, except that the Lagrangian is

$$\mathcal{L} = \tfrac{1}{2}[(\partial_\mu\sigma)^2 + (\partial_\mu\pi_\alpha)^2] - \tfrac{1}{2}\mu^2[\sigma^2 + \pi_\alpha^2] - \tfrac{1}{4}\lambda[\sigma^2 + \pi_\alpha^2]^2 + c\sigma. \quad (2.4)$$

We have introduced a scalar field σ which is a two-pion ($I = 0$, $J = 0$) resonance. The Lagrangian has chiral symmetry, only broken by the $c\sigma$ term. There are two expansion parameters, c and λ; their values are fixed by the ρ mass and the pion decay rate (through partial conservation of the axial-vector current). The order of the graphs in the perturbation expansion is determined by the number of loops, enabling ordinary one-variable Padé approximants to be used. Regrettably, the model was only calculated in the one-loop approximation, permitting formation of the [1/1] Padé approximant. However, even in this low order, the $I = 0$ and $I = 2$ low-energy S-wave phase shifts are in good agreement with experiment, which is remarkable. Furthermore, the good features of $\lambda\phi^4$ model, namely, the existence of ρ and f^0 mesons, are retained, and the consistency tests are well satisfied. On the debit side, there are insufficient terms of the series for stability tests, the $I = 2$ exotic meson is still predicted, and the ρ and f^0 mesons are too narrow. Even so, the most exciting thing about the σ-model is that the best features of the $\lambda\phi^4$ theory are retained and the worst feature of the $\lambda\phi^4$ theory—the bad S-waves—have been corrected.

Just which ingredients of the complete correct theory are lacking in (2.4) is not at all clear. Undoubtedly, the status of the σ-model could be clarified using more modern techniques, such as using the off-shell momentum as a variational parameter [Alabiso et al., 1972a, b], and two-variable approximants [Graves-Morris and Samwell, 1975]. This penultimate remark reflects the theme of this work—how to make the most of numerical power series.

For further details of the status of the model described, we refer to the reviews of Basdevant [1969, 1972a,b, 1973] and Zinn-Justin [1970], and to the original papers: Bessis and Pusterla [1967, 1968], Copley and Masson [1967], Copley, Elias, and Masson [1968], Basdevant, Bessis, and Zinn-Justin [1968, 1969], Basdevant and Lee [1969a], and Bessis and Turchetti [1972].

11.3 Lattice-cutoff $\lambda\phi_n^4$ Euclidean field theory, or the continuous-spin Ising model

In this section we address some of the ways in which Padé approximants can be used to study the deeper questions of the existence of a field theory and its relation to the renormalized perturbation theory. For a fuller treatment (up to the date it was written) see the review article of Baker [1984c]. Our approach here is to relate field theory to a corresponding problem in statistical mechanics. An important step is Nelson's theorem [1973], which showed that boson field theory could be completely studied in Euclidean space and that the corresponding Minkowski-space theory could always be constructed from the Euclidean theory. In Euclidean space the noncommuting operators of the usual field theory are replaced by correlated random fields. Once in Euclidean space, a lattice-type ultraviolet cutoff is used. The prototypical Lagrangian, $\mathcal{L}$ [similar to (4.2.1), except with just one field component], and the action, $\mathcal{A}$, are related by

$$\mathcal{A} = \int d\mathbf{x}\mathcal{L}(\phi(\mathbf{x})) = \tfrac{1}{2}\int_{-\infty}^{+\infty} d\mathbf{x}\{[\nabla\phi(\mathbf{x})]^2 + m_0^2{:}\phi^2{:} + \tfrac{1}{12}\lambda_0{:}\phi^4{:}\}, \quad (3.1)$$

where m_0 is the bare mass, λ_0 is the bare coupling constant, and $:\phi^{2p}:$ is the normal-ordered product. For the present purpose, $:\phi^{2p}:$ can be thought of as an even monic polynomial in ϕ of degree $2p$ whose coefficients are numerical functions of m_0^2, the spatial dimension, and the

11.3 Lattice-cutoff $\lambda\phi_n^4$ Euclidean field theory, or the continuous-spin Ising model

ultraviolet cutoff. These coefficients in a space of two or more dimensions become infinite when the cutoff is removed. The lattice-cutoff fundamental function is the partition function

$$Z(H) = M^{-1} \int_{-\infty}^{+\infty} \cdots \int \prod_{i=1}^{N} d\phi_i$$

$$\times \exp\left[-\frac{v}{2}\sum_i \left\{\frac{2d}{q}\sum_{\{\delta\}} \frac{(\phi_i - \phi_{i+\delta})^2}{a^2} + m_0^2{:}\phi_i^2{:} + \tfrac{1}{12}\lambda_0{:}\phi_i^4{:}\right\}\right.$$

$$\left. + \sum_i H_i \phi_i\right], \tag{3.2}$$

where $v \propto a^d$ is the specific volume per lattice site, d is the spatial dimension, a is the lattice spacing, q is the lattice coordination number, $\{\delta\}$ is the set of one-half the nearest-neighbor sites, H_i is the magnetic field at site $\mathbf{i}$, and M is a normalizing constant.

The field theory is defined by the Schwinger functions

$$\langle \phi_a \phi_b \cdots \phi_v \rangle = \left.\frac{\partial^n \ln Z}{\partial H_a \cdots \partial H_v}\right|_{H_i=0}, \tag{3.3}$$

scaled for appropriate (possibly lattice-spacing dependent) choices of m_0 and λ_0 in the limit as the lattice spacing goes to zero.

By a suitable rescaling of the ϕ field, we can re-express (3.2) as

$$Z(\widetilde{H}) = M^{-1}\int_{-\infty}^{+\infty}\cdots\int \prod_i d\sigma_i \exp\left[\sum_i\left\{K\sum_{\{\delta\}} \sigma_i\sigma_{i+\delta} - \tilde{\lambda}_0 \sigma_i^4 - \widetilde{A}\sigma_i^2 + H_i\sigma_i\right\}\right], \tag{3.4}$$

which is immediately of the form of the continuous-spin Ising model on a d-dimensional spatial lattice. The infinite-volume limit of the field theory is now just the long-studied thermodynamic limit in statistical mechanics. The limit as the lattice spacing goes to zero can be thought of as that in which the ratio of the limit between the lattice spacing and the other length in the problem (i.e., the correlation length) goes to zero. This situation corresponds precisely to the approach to the critical point, where the fluctuations are large and correlations are long range. An example is the highest temperature at which a ferromagnet, a bar of iron, for example, can support a spontaneous magnetization (in the absence of a magnetic field). The illumination of the deep questions of field theory is seen in this light as equivalent to the study in statistical mechanics of the approach to the critical point. Baker and Kincaid [1979, 1981] have begun such a study, and obtained remarkable results by Padé methods and the methods of Section 7.3.

To see how this investigation is aided by Padé methods, we will look at a few of the applications to statistical mechanics of critical phenomena. This area is one of the best developed of the applications of Padé methods to practical problems. This application is treated in detail in a recent book by Baker [1990]. From (3.4) we may define the usual free energy per spin as

$$f = -(\beta N)^{-1} \ln Z, \qquad (3.5)$$

where $\beta = 1/kT$, k is Boltzmann's constant, and T is the temperature. The typical thermodynamic quantities of interest can be derived directly from (3.5). At the critical point the various thermodynamics quantities have branch-point singularities, whose nature is of considerable physical interest. One can develop power-series expansions in $K \propto 1/kT$, for example, of all of these quantities. It is then the idea to use Padé methods to find the critical point and to determine the nature of the branch-point singularities. Typical quantities of interest would be

$$C_H \propto \left.\frac{\partial^2 f}{\partial \beta^2}\right|_H \propto |K - K_c|^{-\alpha},$$

$$M \propto \left.\frac{\partial f}{\partial H}\right|_\beta \propto |K - K_c|^\beta, \qquad (3.6)$$

$$\chi \propto \left.\frac{\partial^2 f}{\partial H^2}\right|_\beta \propto |K - K_c|^{-\gamma},$$

where C_H is the specific heat at constant magnetic field, M is the spontaneous magnetization, χ is the magnetic susceptibility, and K_c is the critical value of K. We remark that

$$M = 0 \quad \text{for } K \leq K_c$$

and

$$M > 0 \quad \text{for } K > K_c. \qquad (3.7)$$

In the case of the spin-$\frac{1}{2}$ Ising model (limit as $\tilde{\lambda}_0 \to \infty$, with $\langle \sigma_i^2 \rangle = 1$ for $K = 0$ fixed) in two dimensions, a number of the critical properties are known exactly [McCoy and Wu, 1973]; for example, $\gamma = 1.75$, $\beta = \frac{1}{8}$, and $C_H \propto \ln|K - K_c|$, which corresponds to $\alpha = 0$. For higher dimensions ($d = 3, 4$, etc.), no exact solution is known. For statistical mechanics, $d = 3$ is of great physical interest, and for field theory $d = 4$ is.

Now, from (3.6), the problem for analysis is to design efficient procedures to utilize series-expansion information under the hypothesis

11.3 Lattice-cutoff $\lambda\phi_n^4$ Euclidean field theory, or the continuous-spin Ising model

that the leading behavior near the critical temperature is

$$g(x) \sim A(x)\left(1 - \frac{x}{x_0}\right)^{-\phi} + B(x), \tag{3.8}$$

with A and B regular near x_0. We seek estimates for x_0, ϕ, $A(x_0)$, etc. The procedures (i)–(vi) described in Section 2.2 are designed to give just these estimates, and estimates of the apparent errors are given by the methods of (2.2.25), (2.2.26). Using these techniques, various workers [Domb, 1974] have generally been able to compute the critical-point locations to 4 or 5 decimal places and the critical indices to 3 or 4.

Another way in which Padé techniques have been used for the spin-$\frac{1}{2}$ Ising model is in connection with the high-field expansions. If in (3.4) we set all $H_i = H$, we can expand the free energy about an ordered (all spins parallel) state in powers of $\mu = e^{-2H}$. By use of a remarkable theorem [Lee and Yang, 1952] which states that for real K, $\tilde{\lambda}_0$, $\tilde{A}$ all the singularities in the μ plane lie on the unit circle $|\mu| = 1$ and come from zeros of Z, we can show [Bessis, Drouffe, and Moussa, 1976], in terms of the variable

$$v = \frac{4\mu}{(1 + \mu)^2}, \tag{3.9}$$

that the magnetization is

$$M(v) = 2(\tanh H)\int_0^{(1+\cos\theta_0)/2} \frac{d\phi(x)}{1 - xv}, \tag{3.10}$$

where, for $K < K_c$, $\theta_0 > 0$ defines a singularity free sector, $-\theta_0 < \arg\mu < \theta_0$, and for $K \geqslant K_c$, $\theta_0 = 0$. Since $M(v)/\tanh H$ is a Stieltjes series, the results of Chapter 5 apply and lead to bounds on M, θ_0, etc.

In addition to the high-temperature and high-field expansions already discussed, the spin-$\frac{1}{2}$ Ising model also has low-temperature expansions. Here there is an added complication. Instead of just one nearest, physically interesting singularity, there is a ring of close singularities. By use of the very long series available and the G^3J approximants of (7.3.9) which result from the solution of a linear, first-order, inhomogeneous integral equation, even this complicated structure has been successfully approximated [Hunter and Baker, 1979].

Of these various methods, the high-temperature series (K expansion) has been applied to the analysis of what is called the dimensionless, renormalized coupling constant

$$\lambda \propto \frac{\partial^2\chi/\partial H^2}{\chi^2 \xi^d}, \tag{3.11}$$

where ξ is the correlation length in units of the lattice spacing. The methods of Padé approximants and first-order integral approximants (Section 8.5) have been applied to the computation of this quantity as a function of both temperature and bare coupling constant λ_0 to produce contour maps of the renormalized coupling constant λ [Baker and Kincaid, 1981]. Insofar as the field-theory limit is concerned, it can be constructed by this method to give $\lambda(\lambda_0)$. Strong numerical evidence for $d = 1$, 2, and 3 has thereby been obtained that the field-theory results given by the asymptotic series for λ in powers of λ_0 are correct. These results are in accord with, and extend, the rigorous results of the constructive field theorists (see Baker [1984c] for references). They also agree with the Padé–Borel summation (Section 7.2) of the λ_0 series [Baker et al., 1976; Baker, Nickel, and Meiron, 1978]. In four dimensions the theory is found to be trivial, that is to say, there is no scattering for any $0 \le \lambda_0 \le \infty$. This conclusion is a consequence of the computation of the field-theory limit $\lambda = 0$, for all λ_0. This result leaves as a deep unanswered question the reason for the excellent results described in Section 11.2, based on pre-renormalized field theory.

In the case of three dimensions, an interesting feature was found. In the 'middle' of the contour plot of λ a saddle point was found. The value of λ at that saddle point was, within numerical accuracy, equal to the field-theory value λ^* of λ for the fixed-point attractor given by the zero of the Callan–Symanzik β-function. The occurrence of this saddle point destroys the theoretical basis [Schrader, 1976] for the claim that the fixed point is a universal attractor, and the hoped for necessary universality in this class of models. Thus Padé analysis has found in this well-studied physical problem a richer structure than had been dreamed of. Recently, exact results, confirming part of this richness, have been obtained for the one dimensional case [Baker, 1988a]. It has been found that, while the field-theory limit for this case is independent of $\tilde{\lambda}_0$, the rate of approach to the limit is not. Thus the statistical mechanics, which depends on the rate, is different in the spin-$\frac{1}{2}$ Ising case from that for $0 < \tilde{\lambda}_0 < \infty$. For instance, the behavior of the correlation length is given asymptotically by $\log \xi \sim 2K$ for the spin-$\frac{1}{2}$ Ising case, and asymptotically by $\log \xi \propto K^2/\tilde{\lambda}_0$ for $0 < \tilde{\lambda}_0 < \infty$.

Another example of the use of Padé-type methods to the benefit of the field-theory problems addressed in this section is the study of the method of 'phantom fields' by Baker and Johnson [1985, 1987] and Baker [1986]. In this method a term $\vartheta_0{:}\phi^6{:}$ is added to the integrand of (3.1) for the case of four dimensions. The effects of this term are so strong (actually divergent if ϑ_0 is finite) that they still cause a finite effect on the theory when $\vartheta_0 \to 0$ and the lattice-type ultraviolet cutoff is removed (the lattice

11.3 Lattice-cutoff $\lambda\phi_n^4$ Euclidean field theory, or the continuous-spin Ising model

spacing goes to zero). They have found that the resultant field theory exists and satisfies all the necessary axioms (rotational invariance is not yet rigorously demonstrated). By means of Padé methods, this theory is found to be nontrivial [Baker and Johnson, 1987]. So far as we know, this theory is the very best candidate for a nontrivial, boson field theory in four spacetime dimensions.

APPENDIX: A FORTRAN FUNCTION

1. *Specification*. PADE(C,IC,A,IA,B,IB,L,M,X,W,W1,W2,IK,IW) is a FORTRAN IV FUNCTION for calculating Padé approximants.

2. *Method*. The method is based on the solution of the linear equations (1.1.6), (1.1.7). A successful FUNCTION call causes the numerator and denominator coefficients of the Padé approximant to be calculated and returns the value of the approximant at some prespecified point X as the value of PADE. If the FUNCTION call is not successful, the approximant has failed a simple empirical degeneracy test which is built into the Gauss–Jordan elimination. The method is designed for data coefficients with full single-precision accuracy. Higher-precision data coefficients are often warranted: in this case, the code must be modified correspondingly.

3. *Parameters*.

    C   a REAL ARRAY of dimension IC. On entry, C contains the given series coefficients and is unchanged on exit.

    A   a REAL ARRAY of dimension IA. On exit, A(1), A(2), ..., A(L + 1) contain the numerator parameters a_0, a_1, ..., a_L, respectively.

    B   a REAL ARRAY of dimension IB. On exit, B(1), B(2), ..., B(M + 1) contain the numerator parameters b_0, b_1, ..., b_M, respectively.

    L   an INTEGER specifying the degree of the numerator.

    M   an INTEGER specifying the degree of the denominator.

    W   a two-dimensional DOUBLE PRECISION REAL ARRAY of dimensions (IW,IW) which must be provided for workspace.

  W1,W2 two DOUBLE PRECISION REAL ARRAYs of dimension IW which must be provided for workspace.

    IK an INTEGER ARRAY of dimension IW which must be provided for workspace.

Appendix: A FORTRAN FUNCTION   691

IC,IA,IB,IW   INTEGERs specifying various array dimensions, to be set on entry and unchanged on exit.

X   the REAL variable of the given power series which must be set with some value on entry. X is unchanged on exit.

4. *Error indicators.* If the integers IC, IA, IB, and IW are inconsistently chosen (we require that $L \geq 0$, $M \geq 0$, $IW \geq M$, $IB > M$, $IA > L$, and $IC > L + M$), or the approximant is degenerate, an error message is printed, and control is returned to the calling program. If the approximant is evaluated at a pole, an error message is printed. The arrays A, B are set correctly, and the artificial value PADE( ) = 0 is returned.

5. *Example.* The following program calculates the [4/4] Padé approximant of $\exp(x)$ to estimate e. ($e = 2.71828182\ldots$). Channel 7 is used for input and channel 2 is used for output.

```
        PROGRAM BOOKPROG
        DIMENSION CC(10),AA(5),BB(5),WW(10,10),WX(10),WY(10),IK(10)
        DOUBLE PRECISION WW,WX,WY
        CC(1)=1.0
        DO 1 I=1,8
 1      CC(I+1)=CC(I)/I
 2      FORMAT(1X,33HTHE GIVEN SERIES COEFFICIENTS ARE/9F13.8)
        WRITE(2,2) (CC(I),I=1,9)
 3      FORMAT (2I3,F13.5)
        READ (7,3) L,M,X
        LP1=L+1
        MP1=M+1
 4      FORMAT(1X,//11H FORM THE [,I4,1H/,I4,13H] PADE AT X=,F13.8//)
        WRITE (2,4) L,M,X
        Y=PADE(CC,10,AA,5,BB,5,L,M,X,WW,WX,WY,IK,10)
 5      FORMAT(1X,30HTHE NUMERATOR COEFFICIENTS ARE//9F13.8)
        WRITE(2,5) (AA(I),I=1,LP1)
 6      FORMAT(1X/33H THE DENOMINATOR COEFFICIENTS ARE//9F13.8)
        WRITE(2,6) (BB(I),I=1,MP1)
 7      FORMAT(1X/42H THE VALUE OF THE PADE APPROXIMANT AT X IS,F13.8)
        WRITE(2,7) Y
        STOP
        END

        FUNCTION PADE (C,IC,A,IA,B,IB,L,M,X,W,W1,W2,IK,IW)
C
C       THE REAL FUNCTION PADE RETURNS THE VALUE OF THE [L/M] PADE
C       APPROXIMANT EVALUATED AT X. THE PADE NUMERATOR COEFFICIENTS
C       ARE STORED IN THE FIRST L+1 LOCATIONS OF THE ARRAY A.
C       THE DENOMINATOR COEFFICIENTS ARE STORED IN THE FIRST M+1
C       LOCATIONS OF THE ARRAY B.
C       DOUBLE PRECISION ARRAYS W(IW,IW), W1(IW), W2(IW) AND AN
C       INTEGER ARRAY IK(IW) ARE REQUIRED FOR WORKSPACE. THE STRICT
```

C   INEQUALITIES IW>0 AND IW .GE. M, IB>M, IA>L, IC>L+M ARE
C   NECESSARY TO FULFIL THE STORAGE REQUIREMENTS AT ENTRY.
C
      DIMENSION C(IC),B(IB),A(IA),W(IW,IW),W1(IW),W2(IW),IK(IW)
      DOUBLE PRECISION W,W1,W2
      DOUBLE PRECISION ONE,OUGHT,DET,T
      DATA ONE,OUGHT/0.1D 01,0.0D 00/
C   NEXT STATEMENT APPLIES TO COMPUTERS WITH 12 FIGURE PRECISION
      EPS=0.1E-10
      LP1=L+1
      B(1)=1.0
      IF(L) 40,1,1
    1 IF(M) 40,6,2
    2 IF(IW-M) 40,3,3
    3 IF(IB-M) 40,40,4
    4 IF(IA-L) 40,40,5
    5 IF(IC-L-M) 40,40,8
    6 DO 7 I=1,LP1
    7 A(I)=C(I)
      D=1.0
      GO TO 35
    8 DO 13 I=1,M
      DO 12 J=1,M
    9 IF (L+I-J) 11,10,10
   10 I1=L+I-J+1
      W(I,J)=C(I1)
      GO TO 12
   11 W(I,J)=OUGHT
   12 CONTINUE
      I1=L+I+1
      W1(I)=-C(I1)
   13 CONTINUE
C   SOLVE DENOMINATOR EQUATIONS USING GAUSS JORDAN ELIMINATION
C   WITH FULL PIVOTING AND DOUBLE PRECISION.
      DET=ONE
      DO 14 J=1,M
   14 IK(J)=0
      DO 30 I=1,M
      T=OUGHT
      DO 19 J=1,M
      IF(IK(J)-1) 15,19,15
   15 DO 18 K=1,M
      IF(IK(K)-1) 16,18,30
   16 IF(DABS(T)-DABS(W(J,K))) 17,17,18
   17 IROW=J
      ICOL=K
      T=W(J,K)
   18 CONTINUE
   19 CONTINUE
      IK(ICOL)=IK(ICOL)+1
```
```

```
        IF(IROW-ICOL) 20,22,20
20      DET=-DET
        DO 21 N=1,M
        T=W(IROW,N)
        W(IROW,N)=W(ICOL,N)
21      W(ICOL,N)=T
        T=W1(IROW)
        W1(IROW)=W1(ICOL)
        W1(ICOL)=T
22      W2(I)=W(ICOL,ICOL)
        IM1=I-1
        IF(I.EQ.1) GO TO 25
        T=DEXP(DLOG(DABS(DET))/DBLE(FLOAT(IM1)))
        IF(DABS(W2(I))-T*DBLE(FLOAT(I))*EPS) 23,25,25
23      DET=OUGHT
24      FORMAT(1X,5HTHE [,I4,1H/,I4,
       *45H] PADE APPROXIMANT APPARENTLY IS REDUCIBLE OR/
       *49H DOES NOT EXIST. IF THIS IS SO, TRY USING A LOWER/
       *50H ORDER APPROXIMANT. OTHERWISE TRY HIGHER PRECISION/
       *41H THROUGHOUT OR THE NAG LIBRARY ALGORITHM.)
        WRITE (2,24) L,M
        RETURN
25      DET=DET*W2(I)
        W(ICOL,ICOL)=ONE
        DO 26 N=1,M
26      W(ICOL,N)=W(ICOL,N)/W2(I)
        W1(ICOL)=W1(ICOL)/W2(I)
        DO 29 LI=1,M
        IF(LI-ICOL) 27,29,27
27      T=W(LI,ICOL)
        W(LI,ICOL)=OUGHT
        DO 28 N=1,M
28      W(LI,N)=W(LI,N)-W(ICOL,N)*T
        W1(LI)=W1(LI)-W1(ICOL)*T
29      CONTINUE
30      CONTINUE
C       EVALUATE DENOMINATOR
        DO 31 I=1,M
31      B(I+1)=W1(I)
        D=B(M+1)
        DO 32 I=1,M
        I1=M+1-I
32      D=X*D+B(I1)
C       EVALUATE NUMERATOR
        DO 34 I=1,LP1
        K=M+1
        IF(K.GT.I)K=I
        Y=0.0
        DO 33 J=1,K
        I1=I-J+1
```

```
33   Y=Y+B(J)*C(I1)
     A(I)=Y
34   CONTINUE
     Y=ABS(D)/M
     IF(Y.LT.EPS) GO TO 42
35   IF(L) 40,36,37
36   PADE=C(1)/D
     RETURN
37   Y=A(L+1)
     DO 38 I=1,L
     I1=L-I+1
38   Y=Y*X+A(I1)
     PADE=Y/D
     RETURN
39   FORMAT(4H IC=,I4,4H IA=,I4,4H IB=,I4,3H L=,I4,3H M=,I4,4H IW=,
    *I4/ 43H AND THERE IS AN ERROR AMONG THESE INTEGERS)
40   WRITE(2,39) IC,IA,IB,L,M,IW
     PADE=0.0
     RETURN
41   FORMAT(1X,E16.8,40H APPARENTLY IS A POLE OF THE APPROXIMANT)
42   WRITE(2,41) X
     PADE=0.0
     RETURN
     END
```

The output from the program BOOKPROG is:

```
THE GIVEN SERIES COEFFICIENTS ARE
    1.00000000   1.00000000   0.50000000   0.16666667   0.04166667 ...
FORM THE [ 4/ 4] PADE AT X =    1.00000000
THE NUMERATOR COEFFICIENTS ARE
    1.00000000   0.50000000   0.10714286   0.01190476   0.00059524
THE DENOMINATOR COEFFICIENTS ARE
    1.00000000  −0.50000000   0.10714286  −0.01190476   0.00059524
THE VALUE OF THE PADE APPROXIMANT AT X IS    2.71828172
```

BIBLIOGRAPHY

The chapter and section numbers [included in brackets] following most of the entries state where these references are cited. Some references to material of interest in the area of Padé approximation, but not covered in the text, are included in this list without such back-referencing.

Abd-Elall, L. F.; Delves, L. M.; and Reid, J. K. 1970. A numerical method for locating the zeroes and poles of a meromorphic function. In *Numerical Methods for Nonlinear Algebraic Equations*, P. Rabinowitz (ed.), pp. 47–59. Gordon and Breach, London. [2§2]

Abramowitz, M., and Stegun, I. 1964. *Handbook of Mathematical Functions*. Dover, New York. [5§5, 10§3]

Adamian, V. M.; Arov, D. Z.; and Krein, M. G. 1971. Analytic properties of Schmidt pairs for a Hankel operator and the Schur–Takagi problem. *Mat. Sbornik* **86** (128); *Math. USSR Sbornik* **15**, 31–73. [10§6.1]

Adler, J.; Moshe, M.; and Privman, V. 1983. Corrections to scaling for percolation. Chap. 17 in *Percolation Structures and Processes*, G. Deutscher, R. Zallen, and J. Adler (eds.). Annals of Israel Physical Society, vol. 5. Hilger, Bristol.

Aitken, A. C. 1926. On Bernoulli's numerical solution of algebraic equations. *Proc. Roy. Soc. Edin.* **46**, 289–305. [3§1]

Aitken, A. C. 1964. *Determinants and Matrices*. Oliver and Boyd, London. [6§2]

Akhiezer, N. I. 1956. *Theory of Approximation*. Ungar, New York. [10§6.1]

Akhiezer, N. I. 1965. *The Classical Moment Problem*. Oliver and Boyd, London. [5§6, 7§2]

Alabiso, C., and Butera, P. 1975. N-variable rational approximants and method of moments. *J. Math. Phys.* **16**, 840–41. [7§6]

Alabiso, C.; Butera, P.; and Prosperi, G. M. 1970. Resolvent operator, Padé approximation and bound states in potential scattering. *Lett. Nuovo. Cim.* **3**, 831–9. [9§8]

Alabiso, C.; Butera, P.; and Prosperi, G. M. 1971. Variational principles and Padé approximants. Bound states in potential theory. *Nucl. Phys.* B **31**, 141–62. [9§8]

Alabiso, C.; Butera, P.; and Prosperi, G. M. 1972a. Variational principles and

Padé approximants. Resonances and phase shifts in potential scattering. *Nucl. Phys. B* **42**, 493–517. [9§8, 11§2]

Alabiso, C.; Butera, P.; and Prosperi, G. M. 1972b. Variational Padé solution of the Bethe–Salpeter equation. *Nucl. Phys. B* **46**, 593–614. [9§8, 11§2]

Alder, K.; Trautmann, D.; and Viollier, R. D. 1973. Anwendung der Padé Approximation in der Kernphysik. *Z. für Naturforschung* **28**, 321–31. [9§7]

Allen, G. D.; Chui, C. K.; Madych, W. R.; Narcowich, F. J.; and Smith, P. W. 1974. Padé approximation and Gaussian quadrature. *Bull. Austr. Math. Soc.* **10**, 263–70. [5§4]

Allen, G. D.; Chui, C. K.; Madych, W. R.; Narcowich, F. J.; and Smith, P. W. 1975. Padé approximants of Stieltjes series. *J. Approx. Theory* **14**, 302–16. [5§3]

Allen, G. D., and Narcowich, F. J. 1975. On representation and approximation of a class of operator-valued analytic functions. *Bull. Am. Math. Soc.* **81**, 410–12. [5§6]

Ammar, G. S., and Gragg, W. B. 1986. Implementation and use of the generalised Schur algorithm. In *Computational and Combinatorial Methods in Systems Theory*, C. I. Byrnes and A. Lindquist (eds.), pp. 265–79. Elsevier, New York. [3§6]

Amos, A. T. 1978. Padé approximants and Rayleigh Schrödinger perturbation theory. *J. Phys. B* **11**, 2053–60. [9§5]

Anderson, B. M.; Brasch, F. M., Jr.; and Lopresti, P. V. 1975. The sequential construction of minimal partial realizations from finite input–output data. *SIAM J. Control* **13** (3), 552–71. [8§3]

Anderson, G. 1965. 1740 Letter from Anderson in Leyden to Jones. In *Correspondence of Scientific Men of the Seventeenth Century*, S. J. Rigaud (ed.), vol. 1, p. 361. George Olms, Hildesheim. [Preface to the first edition]

Antoulas, A. C. 1986. On recursiveness and related topics in linear systems. *IEEE Transactions* **AC-31**, 1121–35. [4§4, 8§3]

Antoulas, A. C. 1988. Rational interpolation and the Euclidean algorithm. *Lin. Alg. Applcns.* **108**, 157–71. [7§1]

Antoulas, A. C., and Anderson, B. D. O. 1986. On the scalar rational interpolation problem. *IMA J. Math. Control Info.* **3**, 61–88. [7§1]

Antoulas, A. C., and Anderson, B. D. O. 1989. On the problem of stable rational interpolation. *Lin. Alg. Applcns.* **122–4**, 301–29. [8§3]

Aptekharev, A. I. 1985. Analytic properties of two dimensional P-fraction expansions with constant coefficients and their simultaneous Padé–Hermite approximants. Keldysh Institute of Applied Mathematics, USSR Academy of Sciences, preprint 79. [8§1]

Argyris, J. H.; Vaz, L. E.; and William, K. J. 1977. Higher order methods for transient diffusion analysis. *Computer Methods in Appl. Mech. and Engineering* **12**, 243–78. [10§4]

Arms, R. J., and Edrei, A. 1970. The Padé tables and continued fractions generated by totally positive sequences. In *Mathematical Essays*, H. Shankar (ed.), pp. 1–21. Ohio University Press, Athens, Ohio. [5§7]

Arndt, H. 1980. Ein verallgemeinerter Kettenbruch–Algorithmus zur rationalen Hermite-interpolation. *Num. Math.* **36**, 99–107. [4§2]

Axelsson, O. 1969. A class of A-stable methods. *BIT* **9**, 185–9. [10§3]

Axelsson, O. 1972. A note on a class of strongly A-stable methods. *BIT* **12**, 1–4. [10§3]

Baker, G. A., Jr. 1960. An implicit, numerical method for solving the n-dimensional heat equation. *Quarterly of Appl. Math.* **17**, 440–3. [10§4]

Baker, G. A., Jr. 1961. Application of the Padé approximant method to the investigation of some magnetic properties of the Ising model. *Phys. Rev.* **124**, 768–74. [2§2, 7§3]

Baker, G. A., Jr. 1965. The theory and application of the Padé approximant method. *Adv. in Theoretical Phys.* **1**, 1–58. [6§4]

Baker, G. A., Jr. 1967. Convergent, bounding approximation procedures to the ferromagnetic Ising model. *Phys. Rev.* **161**, 434–45. [2§2, 7§2]

Baker, G. A., Jr. 1969. Best error bounds for Padé approximants to convergent series of Stieltjes. *J. Math. Phys.* **10**, 814–20. [5§2, 7§1]

Baker, G. A., Jr. 1970. The Padé approximant and some related generalizations. In *The Padé Approximant in Theoretical Physics*, G. A. Baker, Jr., and J. L. Gammel (eds.), pp. 1–39. Academic Press, New York. [6§4, 7§2]

Baker, G. A., Jr. 1972. Converging bounds for the free energy in certain statistical mechanical problems. *J. Math. Phys.* **13**, 1862–4.

Baker, G. A., Jr. 1973a. Recursive calculation of Padé approximants. In *Padé Approximants and Their Applications*, P. R. Graves-Morris (ed.), pp. 83–92. Academic Press, London. [2§1]

Baker, G. A., Jr. 1973b. Existence and convergence of subsequences of Padé approximants. *J. Math. Anal. Appl.* **43**, 498–528. [1§4, 6§7]

Baker, G. A., Jr. 1973c. Generalized Padé approximant bounds for critical phenomena. In *Padé Approximants and Their Applications*, P. R. Graves-Morris (ed.), pp. 147–58. Academic Press, London.

Baker, G. A., Jr. 1974. Certain invariance properties of the Padé approximant. *Rocky Mtn. J. Math.* **4**, 141–9. [6§4]

Baker, G. A., Jr. 1975a. *The Essentials of Padé Approximants*. Academic Press, New York. [8§6.6]

Baker, G. A., Jr. 1975b. Convergence of Padé approximants using the solution of linear functional equations. *J. Math. Phys.* **16**, 813–22. [9§3, 9§7]

Baker, G. A., Jr. 1975c. Self-interacting boson quantum field theory and the thermodynamic limit in d dimensions. *J. Math. Phys.* **16**, 1324–46.

Baker, G. A., Jr. 1976. A theorem on convergence of Padé approximants. *Studies in Applied Math.* **55**, 107–17. [6§4]

Baker, G. A., Jr. 1977a. The application of Padé approximants to critical phenomena. In *Padé and Rational Approximation*, E. B. Saff and R. H. Varga (eds.), pp. 323–37. Academic Press, New York. [2§2]

Baker, G. A., Jr. 1977b. Analysis of hyperscaling in the Ising model by the high temperature series method. *Phys. Rev. B* **15**, 1552–9.

Baker, G. A., Jr. 1979. The continuous-spin Ising model of field theory and the renormalization group. In *Bifurcation Phenomena in Mathematical Physics and Related Topics*, C. Bardos and D. Bessis (eds.), pp. 91–114. Reidel, Dordrecht.

Baker, G. A., Jr. 1984a. Invariance properties in Hermite–Padé approximation theory. *J. Comp. & Appl. Math.* **11**, 49–55. [8§6.3, 8§6.4, 10§7]

Baker, G. A., Jr. 1984b. Approximate analytic continuation beyond the first Riemann sheet. In *Rational Approximation and Interpolation* Springer Lecture Notes in Mathematics 1105, P. R. Graves-Morris, E. B. Saff, and R. S. Varga (eds.), pp. 285–94. Springer, Berlin. [8§6.6]

Baker, G. A., Jr. 1984c. Critical point statistical mechanics and quantum field theory. In *Phase Transitions and Critical Phenomena*, C. Domb and J. L. Lebowitz (eds.), vol. 9, pp. 233–311. Academic Press, New York. [11§3]

Baker, G. A., Jr. 1986. Four-dimensional boson field theory. II. Existence. *J. Math. Phys.* **27**, 2379–93. [11§3]

Baker, G. A., Jr. 1988a. Pluralism in the critical phenomena of the one-dimensional continuous-spin Ising model. *Phys. Rev. Letts.* **60**, 1844–7. [11§3]

Baker, G. A., Jr. 1988b. Integral approximants for functions of higher monodromic dimension. In *Nonlinear Numerical Methods and Rational Approximation*, A. A. M. Cuyt (ed.), pp. 3–22. Reidel, Dordrecht. [8§6.6]

Baker, G. A., Jr. 1990. *Quantitative Theory of Critical Phenomena*. Academic Press, Boston. [8§6.1, 8§6.6, 8§6.7]

Baker, G. A., Jr. 1994. The accuracy through order principle and the equivalence properties in the algebraic approximant. In *Nonlinear Numerical Methods and Rational Approximation II*, A. A. M. Cuyt (ed.), pp. 283–90. Kluwer, Dordrecht. [8§6.2, 8§6.3]

Baker, G. A., Jr.; Benofy, L. P.; and Enting, J. G. 1986. Yang–Lee edge for the two-dimensional Ising model. *Phys. Rev. B* **33**, 3187–208. [8§6.4, 10§7]

Baker, G. A., Jr.; Bessis, D.; and Moussa, P. 1992. Asymptotic behavior of some Hankel–Toeplitz determinants. *Reviews in Math. Phys.* **4**, 65–94. [2§3]

Baker, G. A., Jr., and Chisholm, J. S. R. 1966. The validity of perturbation series with zero radius of convergence. *J. Math. Phys.* **7**, 1900–2. [11§1]

Baker, G. A., Jr., and Gammel, J. L. 1961. The Padé approximant. *J. Math. Anal. Appl.* **2**, 21–30. [9§2]

Baker, G. A., Jr., and Gammel, J. L. (eds.). 1970. *The Padé Approximant in Theoretical Physics*. Academic Press, London.

Baker, G. A., Jr., and Gammel, J. L. 1971. Application of the principle of the minimum maximum modulus to generalized moment problems and some remarks on quantum field theory. *J. Math. Anal. Appl.* **33**, 197–211. [5§6]

Baker, G. A., Jr.; Gammel, J. L.; and Wills, J. G. 1961. An investigation of the applicability of the Padé approximant method. *J. Math. Anal. Appl.* **2**, 405–18. [1§5, 6§7, 9§2]

Baker, G. A., Jr.; Gilbert, H. E.; Eve, J.; and Rushbrooke, G. S. 1967. High temperature expansions for the spin-$\frac{1}{2}$ Heisenberg model. *Phys. Rev.* **164**, 800–17.

Baker, G. A., Jr., and Graves-Morris, P. R. 1977. Convergence of rows of the Padé table. *J. Math. Anal. Appl.* **57**, 323–39. [6§1]

Baker, G. A., Jr., and Graves-Morris, P. R. 1982. The convergence of sequences of Padé approximants. *J. Math. Anal. Applcns.* **87**, 382–94. [6§4]

Baker, G. A., Jr., and Graves-Morris, P. R. 1990. Definition and uniqueness of integral approximants. *J. Comp. & Appl. Math.* **31**, 357–72. [8§5, 8§5.1, 8§5.2, 8§6.2, 8§6.3, 11§3]

Baker, G. A., Jr., and Graves-Morris, P. R. 1991. Convergence for rows of Hermite–Padé integral approximants. *Rocky Mtn. J. Math.* **21**, 41–69. [8§6.6]

Baker, G. A., Jr., and Graves-Morris, P. R. 1993a. Determinants and the structure of the integral approximant table. *J. Comp. & Appl. Math.* **47**, 67–81. [8§5.2, 8§5.3]

Baker, G. A., Jr., and Graves-Morris, P. R. 1993b. Results on sequences of Hermite–Padé, integral approximants. *J. Comp. & Appl. Math.* **47**, 241–52. [8§6.6]

Baker, G. A., Jr., and Graves-Morris, P. R. 1995a. Existence of certain sequences of Hermite–Padé approximants. *J. Comp. & Appl. Math.*, in press. [8§5.4]

Baker, G. A., Jr., and Graves-Morris, P. R. 1995b. Some convergence and divergence theorems for Hermite–Padé approximants. *J. Comp. Appl. Math.*, in press. [8§6.6]

Baker, G. A., Jr., and Gubernatis, J. E. 1981. An asymptotic Padé approximant method for Legendre series. In *Proc. 1979 Int. Christoffel Symposium*, P. L. Butzer (ed.), pp. 232–42. Birkhäuser, Basel. [7§2]

Baker, G. A., Jr., and Hunter, D. L. 1973. Methods of series analysis II: Generalized and extended methods with application to the Ising model. *Phys. Rev. B* **7**, 3377–92. [2§2]

Baker, G. A., Jr., and Johnson, J. D. 1985. Four-dimensional boson field theory. *J. Phys. A* **18**, L261–6. [11§3]

Baker, G. A., Jr., and Johnson, J. D. 1987. Four-dimensional boson field theory. III. Nontriviality. *J. Math. Phys.* **28**, 1146–58. [11§3]

Baker, G. A., Jr., and Kincaid, J. M. 1979. The continuous spin Ising model and $\lambda{:}\phi^4{:}_d$ field theory. *Phys. Rev. Lett.* **42**, 1431–4; *Errata* **44**, 434 (1980). [11§3]

Baker, G. A., Jr., and Kincaid, J. M. 1981. The continuous-spin Ising model, $g_0{:}\phi^4{:}_d$ field theory, and the renormalization group. *J. Stat. Phys.* **24**, 469–528. [11§3]

Baker, G. A., Jr., and Lubinsky, D. S. 1987. Convergence theorems for rows of differential and algebraic Hermite–Padé approximations. *J. Comp. & Appl. Math.* **18**, 29–52. [8§6.6]

Baker, G. A., Jr., and Moussa, P. 1978. Relation in the Ising model of the Lee–Yang branch point and critical behaviour. *J. Appl. Phys.* **49**, 1360–2. [7§6]

Baker, G. A., Jr.; Nickel, B. G.; Green, M. S.; and Meiron, D. I. 1976. Ising model critical indices in three dimensions from the Callan–Symanzik equation. *Phys. Rev. Lett.* **36**, 1351–4. [11§3]

Baker, G. A., Jr.; Nickel, B. G.; and Meiron, D. I. 1978. Critical indices from perturbation analysis of Callan–Symanzik equation. *Phys. Rev. B* **17**, 1365–74. [11§3]

Baker, G. A., Jr.; Oitmaa, J.; and Velgakis, M. J. 1988. Series analysis of multivalued functions. *Phys. Rev. A* **38**, 5316–31. [8§6.4, 8§6.5, 8§6.6, 8§6.7]

Baker, G. A., Jr., and Oliphant, T. A. 1960. An implicit, numerical method for solving the two-dimensional heat equation. *Quarterly of Appl. Math.* **17**, 361–73. [10§4]

Baker, G. A., Jr.; Rushbrooke, G. S.; and Gilbert, H. E. 1964. High temperature series expansions for the spin-$\frac{1}{2}$ Heisenberg model by the method of irreducible representations of the symmetric group. *Phys. Rev. A* **135**, 1272–7.

Balk, M. F. 1960. The Padé interpolation process for certain analytic functions. In *Issledovaniya Po Sovremennym Problemam Teorii Funkcii Kompleksnogo Peremennogo*, pp. 234–57. Gosudarstv. Izdat. Fiz.-Mat. Lit., Moscow. (Abstract: see *Zentralblatt für Mathematik* **173**(1969), 73; *MR* **22**(1961) No. 9780.) [4§3]

Bareiss, E. H. 1969. Numerical solution of linear equations with Toeplitz matrices. *Num. Math.* **13**, 404–24. [7§5]

Barnsley, M. F. 1973. The bounding properties of the multipoint Padé approximant to a series of Stieltjes on the real line. *J. Math. Phys.* **14**, 299–313. [7§1]

Barnsley, M. F. 1975. Bounds on dynamical polarizabilities at high frequencies. *Molecular Physics* **29**, 1377–86.

Barnsley, M. F. 1976. Padé approximant bounds for the difference of two series of Stieltjes. *J. Math. Phys.* **17**, 559–65. [5§6]

Barnsley, M. F., and Baker, G. A., Jr. 1976. Bivariational bounds in a complex Hilbert space, and correction terms. *J. Math. Phys.* **17**, 1019–27. [9§6]

Barnsley, M. F., and Bessis, D. 1979. Constructive methods based on analytic characterizations and their applications to nonlinear elliptic and parabolic differential equations. *J. Math. Phys.* **20**, 1135–45.

Barnsley, M. F.; Bessis, D.; and Moussa, P. 1979. The Diophantine moment problem and the analytic structure in the activity of the ferromagnetic Ising model. *J. Math. Phys.* **20**, 535–46. [5§4.2]

Barnsley, M. F., and Robinson, P. D. 1974a. Padé approximant bounds and approximate solutions for Kirkwood–Risemann integral equations. *J. Inst. Math. Appl.* **14**, 251–85. [9§6]

Barnsley, M. F., and Robinson, P. D. 1974b. Bivariational bounds. *Proc. Roy. Soc. Lond. A* **338**, 527–33. [9§6]

Barnsley, M. F., and Robinson, P. D. 1974c. Dual variational principles and Padé-type approximants. *J. Inst. Math. Appl.* **14**, 229–49. [9§6]

Barnsley, M. F., and Robinson, P. D. 1978. Rational approximant bounds for a class of two variable Stieltjes functions. *SIAM J. Math. Anal.* **9**, 272–90. [7§6]

Basdevant, J.-L. 1969. Padé approximants. In *Methods in Subnuclear Physics*, M. Nikolic (ed.) pp. 1–40. Gordon and Breach, New York. [11§2]

Basdevant, J.-L. 1972a. Padé approximants in strong interaction physics. In *Cargèse Lectures in Physics*, D. Bessis (ed.), vol. 5, pp. 431–59. Gordon and Breach, New York. [11§2]

Basdevant, J.-L. 1972b. The Padé approximant and its physical applications. *Fortschr. Phys.* **20**, 282–331. [11§2]

Basdevant, J.-L. 1973. Strong interaction physics and the Padé approximant in quantum field theory. In *Padé approximants*, P. R. Graves-Morris (ed.), pp. 77–100. Inst. of Physics, Bristol. [11§2]

Basdevant, J.-L.; Bessis, D; and Zinn-Justin, J. 1968. Padé approximants in strong interactions. Two body pion and Kaon systems. *Phys. Lett.* **27B**, 230–3. [11§2]

Basdevant, J.-L.; Bessis, D; and Zinn-Justin, J. 1969. Padé approximants in strong interactions. Two body pion and Kaon systems. *Nuovo Cim.* **60A**, 185–238. [11§2]

Basdevant, J.-L., and Lee, B. W. 1969a. Padé approximants in the σ-model with unitary π-π amplitudes with the current algebra constraints. *Phys. Lett.* **29B**, 437–41. [9§5, 11§2]

Basdevant, J.-L., and Lee, B. W. 1969b. Padé approximants and bound states: exponential potential. *Nucl. Phys. B* **13**, 182–8. [9§5]

Basu, S., and Bose, N. K. 1983. Matrix Stieltjes series and network models. *SIAM J. Math. Anal.* **14**, 209–22. [8§2]

Bauer, F. L. 1959. The quotient-difference and epsilon algorithm. In *Numerical Approximation*, R. Langer (ed.), pp. 361–70. Univ. of Wisconsin Press, Madison, WI. [3§3]

Bauer, F. L.; Rutishauser, H.; and Stiefel, E. 1963. New aspects in numerical quadrature. In *Proc. of 15th Symposium in Applied Mathematics*, N. C. Metropolis, A. H. Traub, J. Todd, and C. B. Tompkins (eds.), vol. 15, pp. 199–218. Amer. Math. Soc. Press, Providence, RI. [3§3]

Bauhoff, W. 1977. Padé approximants in the Wick–Cutkosky model. *J. Phys. A* **10**, 129–34.

Baumel, R. T.; Gammel, J. L.; and Nuttall, J. 1981. Placement of cuts in Padé-like approximation. *J. Comp. Appl. Math.* **7**, 135–40. [10§1]

Bausset, M. (ed.). 1973. *Accélération des Convergences*. Centre Universitaire de Toulon, Toulon.

Baxter, G., and Hirschman, I. I. 1964. An explicit inversion formula for finite section Wiener–Hopf operators. *Bull. Am. Math. Soc.* **70**, 820–3. [3§6]

Beardon, A. F. 1968a. The convergence of Padé approximants. *J. Math. Anal. Appl.* **21**, 344–6. [6§4]

Beardon, A. F. 1968b. On the location of poles of Padé approximants. *J. Math. Anal. Appl.* **21**, 469–74. [6§1]

Beckenbach, E. F., and Bellman, R. 1965. *Inequalities*. Springer, New York. [8§2]

Beckermann, B. 1990. The structure of the singular solution table of the M-Padé approximation problem. *J. Comp. & Appl. Math.* **32**, 3–15. [8§5.1, 8§5.2, 8§5.3]

Beckermann, B. 1993. Recursiveness in matrix rational interpolation problems and vector M-Padé approximation with infinity. Hannover preprint.

Beckermann, B., and Carstensen, C. 1992. A reliable modification of the cross rule for rational Hermite interpolation. *Num. Algs.* **3**, 29–44. [7§1]

Beckermann, B., and Carstensen, C. 1993. QD-type algorithms for the non-normal Newton–Padé approximation table. Hannover preprint. [7§1]

Beckermann, B., and Labahn, G. 1992. A uniform approach for Hermite–Padé and simultaneous Padé approximants and their matrix type generalizations. *Num. Algs.* **3**, 45–54. [8§1]

Beckermann, B., and Labahn, G. 1994. A uniform approach for the fast, reliable computation of matrix-type Padé approximants. *SIAM J. Matrix Analy. Applcns.* **15**, 804–23. [8§1]

Beckermann, B.; Neuber, A.; and Muhlbach, G. 1992. Shanks' transformation revisited. *Lin. Alg. Applcns.* **173**, 191–219. [3§7]

Bell, G. I., and Glasstone, S. 1970. *Nuclear Reactor Theory*. Van Nostrand-Reinhold, New York. [10§4]

Bellman, R. 1970. *Introduction to Matrix Analysis*. McGraw-Hill, New York. [7§5]

Bender, C. M., and Orszag, S. A. 1978. *Advanced Mathematical Methods for Scientists and Engineers*. McGraw-Hill, New York. [3§3, 5§5, 5§6, 7§2]

Bender, C. M., and Wu, T. T. 1968. Analytic structure of energy levels in a field theory model. *Phys. Rev. Lett.* **21**, 406–9. [11§1]

Benofy, L. P., and Gammel, J. L. 1977. Variational principles and matrix Padé approximants. In *Padé and Rational Approximation*, E. B. Saff and R. S. Varga (eds.), pp. 339–56. Academic Press, New York. [9§5, 9§8]

Benofy, L. P.; Gammel, J. L.; and Mery, P. 1976. The off-shell momentum as a variational parameter in calculations of matrix Padé approximants in potential scattering. *Phys. Rev. D* **13**, 3111–4. [9§8]

Berlekamp, E. R. 1968. *Algebraic Coding Theory*. McGraw-Hill, New York. [4§4]

Bernstein, S. 1928. Sur les fonctions absolument monotones. *Acta Math.* **52**, 1–66. [5§4.1]

Bessis, D. (ed.). 1972. *Cargèse Lectures in Physics*. Vol. 5. Gordon and Breach, New York. [11§1, 11§2]

Bessis, D. 1973. Topics in the Theory of Padé Approximants, in *Padé Approximants*, P. R. Graves-Morris (ed.), pp. 19–44. Inst. of Phys., Bristol. [8§2]

Bessis, D. 1976. Construction of variational bounds for the N-body eigenstate problem by the method of Padé approximants. In *Padé Approximants Method and Its Applications to Mechanics*, H. Cabannes (ed.), pp. 17–31. Springer Lecture Notes in Phys., no. 37. Springer, Berlin. [9§5]

Bessis, D. 1979. A new method for the combinatorics of the topological expansion. *Comm. Math. Phys.* **69**, 147–63. [5§1, 10§9]

Bessis, D.; Drouffe, J. M.; and Moussa, P. 1976. Positivity constraints for the Ising ferromagnetic model. *J. Phys. A* **9**, 2105–24. [11§3]

Bessis, D.; Epele, L.; and Villani, M. 1974. Summation of regularized perturbation expansions for singular interactions. *J. Math. Phys.* **15**, 2071–8. [9§9]

Bessis, D.; Gilewicz, J.; and Mery, P. (eds.). 1975. *Proceedings of a Workshop on Padé Approximants*. C.N.R.S., Marseille.

Bessis, D.; Mery, P.; and Turchetti, G. 1974. Angular momentum analysis of four-nucleon Green's function. *Phys. Rev. D* **10**, 1992–2009.

Bessis, D.; Mery, P.; and Turchetti, G. 1977. Variational bounds from matrix Padé approximants in potential scattering. *Phys. Rev. D* **15**, 2345–53. [9§8]

Bessis, D.; Moussa, P.; and Villani, M. 1975. Monotonic converging variational approximants to the functional integrals in quantum statistical mechanics. *J. Math. Phys.* **16**, 2318–25.

Bessis, D., and Pusterla, M. 1967. Unitary Padé approximants for the S-matrix in strong coupling field theory and application to the calculation of ρ and f^0 Regge trajectories. *Phys. Lett.* **25B**, 279–81. [11§2]

Bessis, D., and Pusterla, M. 1968. Unitary Padé approximants in strong coupling field theory and application to the ρ and f^0 meson trajectories. *Nuovo Cim.* **54A**, 243–94. [11§2]

Bessis, D., and Turchetti, G. 1972. Renormalization of the σ-model through Ward identities. In *Cargèse Lectures in Physics*, D. Bessis (ed.), vol. 5, pp. 119–78. Gordon and Breach, New York. [11§2]

Bessis, D., and Turchetti, G. 1977. Variational matrix Padé approximants in potential scattering and low energy Lagrangian field theory. *Nucl. Phys. B* **123**, 173–88. [9§8]

Bessis, D., and Villani, M. 1975. Perturbative-variational approximations to the spectral properties of semi-bounded Hilbert space operators, based on the moment problem with finite or diverging moments. Application to quantum mechanical systems. *J. Math. Phys.* **16**, 462–74. [9§5]

Bhattacharyya, K. 1989a. Asymptotic response of observables from divergent power-series expansions. *Phys. Rev. A* **39**(12), 6124–8. [10§1]

Bhattacharyya, K. 1989b. On the temperature dependence of the partition function: Molecular free internal rotation. *Chem. Phys. Lett.* **159**(1), 40–4. [10§1]

Bhattacharyya, K. 1989c. On the maximum entropy method of handling divergent perturbation series. *Chem. Phys. Lett.* **161**(3), 259–64. [10§1]

Bhowmick, S.; Roy, D,; and Bhattacharyya, K. 1988. Evaluation of lattice sums by convergence acceleration transforms. *Chem. Phys. Lett.* **148**, 317–22. [10§1]

Birkhoff, G., and Varga, R. S. 1965. Discretisation errors for well set Cauchy problems. *Int. J. Math. Phys.* **44**, 1–23. [5§7]

Biswas, S. N., and Vidhani, T. 1973. Solution of Schrödinger equation with continued fractions. *J. Phys. A* **6**, 468–77.

Blanch, G. 1964. The numerical evaluation of continued fractions. *SIAM Rev.* **6**, 383–421. [4§1]

Boas, R. P. 1954. *Entire Functions*. Academic Press, New York.

Bombelli, R. 1572. *L'Algebra*. Venezia. [4§1]

Borwein, D.; Borwein, J. M.; and Taylor, K. E. 1985. Convergence of lattice sums and Madelung's constant. *J. Math. Phys.* **26**, 2999–3009. [10§1]

Borwein, P. B. 1986. Quadratic Hermite–Padé approximation to the exponential function. *Constr. Approx.* **2**, 291–302.

Borwein, P. B. 1988. Padé approximants for the q-elementary functions. *Constr. Approx.* **4**, 391–402. [2§3]

Bose, N. K. 1982. *Applied Multidimensional Systems Theory*. Van Nostrand-Reinhold, New York. [7§6]

Bose, N. K., and Basu, S. 1980. Two-dimensional matrix Padé approximants: Existence, nonuniqueness, and recursive computation. *IEEE Trans. Auto. Control* **AC-25**, 509–14. [7§6]

Bosgra, O. H., and van der Weiden, A. J. J. 1980. Input output invariants for linear multivariable systems. *IEEE Trans. Auto. Control* **AC-25**, 20–36.

Braess, D. 1984. On the conjecture of Meinardus on rational interpolation of e^x II. *J. Approx Theory* **40** 375–379. [10§6.1]

Braess, D. 1986. Non-linear Approximation Theory, Springer, Berlin. [10§6.1]

Brent, R. P.; Gustavson, F. G.; and Yun, D. Y. Y. 1980. Fast solution of Toeplitz systems of equations and computation of Padé approximants. *J. Algorithms* **1**, 259–95. [3§6]

Brézin, E.; Le Guillou, J. C.; and Zinn-Justin, J. 1977. Perturbation theory at large order. I. The ϕ^{2N} interaction. *Phys. Rev. D* **15**, 1544–58. [11§1]

Brezinski, C. 1971. Accélération de suites à convergence logarithmique. *Comptes Rendus* **273A**, 727–30.

Brezinski, C. 1972. Conditions d'application et de convergence de procédés d'extrapolation. *Num. Math.* **20**, 64–79.

Brezinski, C. 1974. Review of acceleration methods. *Rendiconti di Mat.* **7**, 303–16.

Brezinski, C. 1975. Généralisations de la transformation de Shanks, de la table de Padé, et de l'ε-algorithme. *Calcolo* **12**, 317–60.

Brezinski, C. 1976. Computation of Padé approximants and continued fractions. *J. Comp. Appl. Math.* **2**, 113–23.

Brezinski, C. 1977. *Accélération de la Convergence en Analyse Numérique*. Springer Lecture Notes in Mathematics, no. 584. Springer, Berlin. [3§1, 3§3, 5§2]

Brezinski, C. 1978a. Convergence acceleration of some sequences by the ε-algorithm. *Num. Math.* **29**, 173–7. [5§4.1]

Brezinski, C. 1978b. Padé type approximants for double power series. *J. Indian Math. Soc.* **42**, 267–82. [7§6]

Brezinski, C. 1979. Rational approximation to formal power series. *J. Approx. Theory* **25**, 295–317. [10§1, 10§5]

Brezinski, C. 1980a. *Padé-Type Approximation and General Orthogonal Polynomials*. Birkhäuser, Basel. [10§1]

Brezinski, C. 1980b. A general extrapolation algorithm. *Num. Math.* **35**, 175–87. [10§1]

Brezinski, C. 1988. Partial Padé approximation. *J. Approx. Theory* **54**, 210–33.
Brezinski, C. 1990. *History of Continued Fractions and Padé Approximants*. Springer, Berlin. [4§1]
Brezinski, C., and Redivo-Zaglia, M. 1991. *Extrapolation Methods: Theory and Practice*. North-Holland, Amsterdam. [3§1, 10§1]
Brezinski, C., and Rieu, A. C. 1974. Solution of systems of equations using the ε-algorithm, and application to boundary value problems. *Math. Comp.* **28**, 731–4.
Brezinski, C., and Sadok, H. 1993. Lanczos-type algorithms for solving systems of linear equations. *Applied Numerical Mathematics* **11**, 443–73. [9§3]
Brezinski, C., and Van Iseghem. 1994. *Padé Approximations*. Vol. 3 of *Handbook of Numerical Analysis*, P. G. Ciarlet and J. L. Lions (eds.). North-Holland, Amsterdam. [8§1]
Brophy, F., and Salazar, A. C. 1973. Considerations of the Padé approximant technique in the synthesis of recursive digital filters. *IEEE Trans. Audio Electroacoust.* **Au-21**, 500–5.
de Bruin, M. G. 1974. Generalized continued fractions and a multidimensional Padé table. Ph.D. Thesis, University of Amsterdam. [8§1]
de Bruin, M. G. 1978. Convergence of generalized C-fractions. *J. Approx. Theory* **24**, 177–204.
de Bruin, M. G. 1982. Generalised Padé tables and some algorithms therein. In *Padé Approximation and Convergence Acceleration Techniques*, J. Gilewicz (ed.), pp. 9–16. C.N.R.S., Marseille.
de Bruin, M. G. 1984. Some convergence results in simultaneous rational approximation to the set of hypergeometric functions $\{{}_1F_1(1; c_i; z)\}_{i=1}^n$. In *Padé Approximation and Its Applications*, H. Werner and H. J. Bünger (eds.), pp. 12–33. Springer LNIM 1071. Springer, Berlin.
de Bruin, M. G. 1988. Simultaneous rational approximation to some q-hypergeometric functions. In *Nonlinear Numerical Methods and Rational Approximation*, A. M. Cuyt (ed.), pp. 135–42. Reidel, Dordrecht. [8§1]
de Bruin, M. G.; Driver, K. A.; and Lubinsky, D. S. 1992. Convergence of simultaneous Hermite-Padé approximants to the n-tuple of q-hypergeometric series $\{{}_1\phi_1((1, 1); (c, \gamma_i); z)\}_{i=1}^n$. *Num. Algs.* **3**, 185–92. [8§1]
Bulirsch, R., and Stoer, J. 1964. Fehler Abschätzungen und Extrapolation mit rationalen Funktionen bei Verfahren vom Richardson-Typus. *Num. Math.* **6**, 413–27. [7§1]
Bultheel, A. 1979. Recursive algorithms for the Padé table: two approaches. In *Padé Approximation and Its Applications*, L. Wuytack (ed.), pp. 211–30. Springer Lecture Notes in Mathematics, no. 765. Springer, Berlin.
Bultheel, A. 1980a. Recursive algorithms for the matrix Padé problem. *Math. Comp.* **35**, 875–92. [4§4]
Bultheel, A. 1980b. Division algorithms for continued fractions and the Padé table. *J. Comp. Appld. Math.* **6**, 259–66. [4§2, 4§4]
Bultheel, A. 1980c. On the relation between Padé algorithms and Levinson/Schur recursive methods. In *Signal Processing and Applications*, M. Kunt and F. Coulon (eds.), pp. 517–23. North-Holland, Amsterdam. [7§5]
Bultheel, A. 1987. *Laurent Series and Their Padé Approximations*. OT27. Birkhäuser, Basel. [4§2, 7§5]
Bultheel, A., and Van Barel, M. 1986. Padé techniques for model reduction in linear system theory: A survey. *J. Comp. Appld. Math.* **14**, 401–38. [4§4]

Bultheel, A., and Van Barel, M. 1990a. Minimal vector Padé approximation. *J. Comp. Appld. Math.* **32**, 27-37.

Bultheel, A., and Van Barel, M. 1990b. A matrix Euclidean algorithm and the matrix minimal Padé approximation problem. In *Continued Fractions and Padé Approximants*, C. Brezinski (ed.), pp. 11-51. Elsevier, New York. [4§4, 8§3]

Bultheel, A., and Wuytack, L. 1981. Stability of numerical methods for computing Padé approximants. In *Approximation Theory III*, E. W. Cheney (ed.), pp. 267-72. Academic Press, New York.

Bunch, J. R. 1985. Stability of methods for solving Toeplitz systems of equations. *SIAM J. Sci. Stat. Comp.* **6**, 349-64. [3§6]

Bus, J. C. P., and Dekker, T. J. 1975. Two efficient algorithms with guaranteed convergence for finding a zero of a function. *ACM Trans. on Math. Software* **1**, 330-45. [7§1]

Buslaev, V. I.; Gončar, A. A.; and Suetin, S. P. 1983. On the convergence of subsequences of the mth row of the Padé table. *Mat. Sbornik* **120**(162); *Math. USSR Sbornik.* **48**, 535-40. [6§1]

Busonnais, D. 1978. Tous les algorithmes de calcul par récurrence des approximants de Padé d'une série. Construction des fractions continues correspondantes. Séminaire d'Analyse Numérique 293, Université de Grenoble. [4§2]

Butcher, J. C. 1963. Coefficients for the study of Runge-Kutta integration processes. *J. Australian Math. Soc.* **3**, 185-201. [10§3]

Butcher, J. C. 1964. Implicit Runge-Kutta processes. *Math. Comp.* **18**, 50-64. [10§3]

Cabannes, H. (ed.). 1976. *Padé Approximant Method and Its Application to Mechanics. Proceedings of the Euromech Colloquium, Toulon, 1975*. Springer Lecture Notes in Physics, no. 47. Springer, Berlin.

Cabay, S., and Jackson, L. W. 1976. A polynomial extrapolation method for finding limits and antilimits of vector sequences. *SIAM J. Numer. Anal.* **13**, 734-52. [8§4]

Cabay, S., and Labahn, G. 1992. A superfast algorithm for multidimensional Padé systems. *Num. Algs.* **2**, 201-24. [8§1]

Cabay, S.; Labahn, G.; and Beckermann, B. 1992. On the theory of non-perfect Padé-Hermite approximants. *J. Comp. Appld. Math.* **39**, 295-313. [8§1]

Cabay, S., and Meleshko, R. 1993. A weakly stable algorithm for Padé approximants and the inversion of Hankel matrices. *SIAM J. Matrix Anal. Applcns.* [3§6]

Cala Rodriguez, F., and Wallin, H. 1992. Padé-type approximants and a summability theorem by Eiermann. *J. Comp. Appld. Math.* **39**, 15-21. [10§1]

Carathéodory, C., and Fejér, L. 1911. Über den Zusammenhang der Extremen von harmonische Funktionen mit ihre Koeffizienten und über den Picard-Landauschen Satz. *Rend. Circ. Mat. Palermo* **32**, 218-39. [10§6.1]

Carleman, T. 1926. *Les Fonctions Quasi-Analytiques*. Gauthier Villars, Paris; translated by J. L. Gammel, Los Alamos Tech. report 4702 (1971). [5§5]

Carroll, A.; Baker, G. A., Jr.; and Gammel, J. L. 1977. A comment on a paper of Hamer applying Padé approximants to a 1 + 1 dimensional lattice model of QCD. *Nucl. Phys.* B **129**, 361-4.

Carslaw, H. S., and Jaeger, J. C. 1959. *Conduction of Heat in Solids*. Oxford University Press, Oxford. [10§4]

Cartan, H. 1928. Sur les systèmes de fonctions holomorphes à variétés linéaires et leurs applications. *Ann. Sci. Ecole. Norm. Sup (3)* **45**, 255–346. [6§6]

Caser, S.; Piquet, C.; and Vermeulen, J. L. 1969. Padé approximants for a Yukawa potential. *Nucl. Phys.* B **14**, 119–32. [9§4]

Cauchy, A.-L. 1821. *Cours d'analyse de l'Ecole Royale Polytechnique*, première partie. L'Imprimerie Royale, Paris. [7§1]

Cavendish, J. C.; Culham, W. E.; and Varga, R. S. 1972. A comparison of Crank–Nicholson and Chebychev rational methods for numerically solving linear parabolic equations. *J. Comp. Phys.* **10**, 354–68. [10§4]

Char, B. W. 1980. On Stieltjes continued fraction for the Gamma function. *Math. Comp.* **34**, 547–51. [5§5]

Cheney, E. W. 1966. *Introduction to Approximation Theory*. McGraw-Hill, New York. [6§6, 7§4]

Chipman, F. H. 1971. A-stable Runge–Kutta processes. *BIT* **11**, 384–8. [10§3]

Chisholm, J. S. R. 1963. Solution of linear integral equations using Padé approximants. *J. Math. Phys.* **4**, 1506–10. [9§1]

Chisholm, J. S. R. 1966. Approximation by sequences of Padé approximants in regions of meromorphy. *J. Math. Phys.* **7**, 39–46. [6§4]

Chisholm, J. S. R. 1973. Rational approximants defined from double power series. *Math. Comp.* **27**, 841–8. [7§6]

Chisholm, J. S. R. 1977a. N-variable rational approximants. In *Padé and Rational Approximation*, E. B. Saff and R. H. Varga (eds.), pp. 23–42. Academic Press, New York. [7§6]

Chisholm, J. S. R. 1977b. Multivariate approximants with branch points. I. Diagonal approximants. *Proc. Roy. Soc. Lond.* A **358**, 351–66. [7§6]

Chisholm, J. S. R. 1978a. Multivariate approximants with branch cuts. In *Multivariate Approximation*, D. C. Handscomb (ed.), pp. 31–45. Academic Press, London. [7§6]

Chisholm, J. S. R. 1978b. Multivariate approximants with branch points. II. Off-diagonal approximants. *Proc. Roy. Soc. Lond.* A **362**, 43–56. [7§6]

Chisholm, J. S. R., and Common, A. K. 1981. Generalizations of Padé approximation for Chebyshev and Fourier series. In *Proc. 1979 Int. Christoffel Symposium*, P. L. Butzer and F. Fehér (eds.), pp. 212–31. Birkhäuser, Basel. [7§4]

Chisholm, J. S. R.; Genz, A. C.; and Rowlands, G. E. 1972. Accelerated convergence of sequences of quadrature approximations. *J. Comp. Phys.* **10**, 284–307.

Chisholm, J. S. R., and Graves-Morris, P. R. 1975. Generalizations of the theorem of de Montessus to two-variable approximants. *Proc. Roy. Soc. Lond.* A **342**, 341–72. [6§2, 7§6]

Chisholm, J. S. R., and Hughes Jones, R. 1975. Relative scale covariance of N-variable approximants. *Proc. Roy. Soc. Lond.* A **344**, 465–70. [7§6]

Chisholm, J. S. R., and McEwan, J. 1974. Rational approximants defined from power series in N-variables. *Proc. Roy. Soc. Lond.* A **336**, 421–52. [7§6]

Chisholm, J. S. R., and Roberts, D. E. 1976. Rotationally covariant approximants derived from double power series. *Proc. Roy. Soc. Lond.* A **351**, 585–91. [7§6]

Christoffel, E. B. 1877. Sur une classe particulière de fonctions entières et de fractions continues. *Ann. di Mat.* **8**, 1–10.

Chudnovsky, G. V. 1980. Padé approximation and the Riemann monodromy problem. In *Bifurcation Problems in Mathematical Physics and Related Topics*, C. Bardos and D. Bessis (eds.), pp. 449–510. Reidel, Boston.

Chui, C. K. 1976. Recent results on Padé approximants and related problems. In *Approximation Theory*, G. G. Lorentz, C. K. Chui, and L. L. Schumaker (eds.), pp. 79–116. Academic Press, New York.

Chui, C. K., and Chan, A. K. 1979. A two sided rational approximation method for recursive digital filtering. *IEEE Trans. Acous. Speech and Signal Proc.* **ASSP-27**, 141–5.

Chui, C. K.; Shisha, O.; and Smith, P. W. 1974. Padé approximants as limits of best rational approximants. *J. Approx. Theory* **12**, 201–4. [10§6]

Chui, C. K.; Shisha, O.,; and Smith, P. W. 1975. Best local approximation. *J. Approx. Theory* **15**, 371–81. [10§6]

Chui, C. K.; Smith, P. W.; and Su, L. Y. 1977. A minimization problem related to Padé synthesis of recursive digital filters. In *Padé and Rational Approximation*, E. B. Saff and R. S. Varga (eds.), pp. 247–56. Academic Press, New York.

Chui, C. K.; Smith, P. W.; and Ward, J. D. 1978. Best L_2 local approximation. *J. Approx. Theory* **22**, 254–61. [10§6]

Čížek, J.; Vinette, F.; and Weniger, E. J. 1991. Symbolic computation in physics and chemistry: Applications of the inner projection technique and of a new summation method for divergent series. In *Proceedings of the International Symposium on Quantum Chemistry, Solid State Theory, and Molecular Dynamics*, P.-O. Lowdin (ed.), Int. J. Quantum Chemistry: Quantum Chemistry Symposium 25, pp. 209–23. Wiley, New York. [10§1]

Čížek, J., and Vrscay, E. R. 1982. Large order perturbation theory in the context of atomic and molecular physics—Interdisciplinary aspects. *Int. J. Quantum Chemistry* **21**, 27–68. [10§1]

Claessens, G. 1976a. Some aspects of the rational Hermite interpolation table and its applications. Ph.D. Thesis, University of Antwerp. [7§1]

Claessens, G. 1976b. A new algorithm for osculatory rational interpolation. *Num. Math.* **27**, 77–83. [7§1]

Claessens, G. 1978a. On the Newton–Padé approximation problem. *J. Approx. Theory* **22**, 150–60. [7§1]

Claessens, G. 1978b. On the structure of the Newton–Padé table. *J. Approx. Theory* **22**, 304–19. [7§1]

Claessens, G. 1978c. Generalized ε-algorithm for rational interpolation. *Num. Math.* **29**, 227–31. [7§1]

Claessens, G., and Wuytack, L. 1979. On the computation of nonnormal Padé approximants. *J. Comp. Appl. Math.* **5**, 283–9. [4§3]

Clenshaw, C. W., and Lord, K. 1974. Rational approximations from Chebyshev series. In *Studies in Numerical Analysis*, B. K. P. Scaife (ed.), pp. 95–113. Academic Press, London. [7§4]

Cody, W. J.; Meinardus, G.; and Varga, R. S. 1969. Chebyshev rational approximations to e^{-x} in $[0, \infty)$ and applications in heat conduction equations. *J. Approx. Theory* **2**, 50–65.

Common, A. K. 1968. Padé approximation and bounds to series of Stieltjes. *J. Math. Phys.* **9**, 32–8. [5§2]

Common, A. K. 1969a. Properties of Legendre expansions related to Stieltjes series and applications to π–π scattering. *Nuovo Cim.* **63A**, 863–91. [7§2]

Common, A. K. 1969b. Properties of generalisations to Padé approximants. *J. Math. Phys.* **10**, 1875–80. [7§2]

Common, A. K. 1970. Some consequences of relations of π–π partial wave amplitudes to Stieltjes series. *Nuovo Cim.* **65A**, 581–96.

Common, A. K. 1979. Calculation of Yukawa scattering amplitude and impact parameter amplitudes using Legendre–Padé approximants. *J. Phys. A* **12**, 2563–72.

Common, A. K., and Graves-Morris, P. R. 1974. Some properties of Chisholm approximants. *J. Inst. Math. Appl.* **13**, 229–32. [7§6]

Common, A. K., and Hafez, S. T. 1990. Continued fraction solution to the Riccati equation and nonlinear lattice systems. *J. Phys. A* **23**, 455–66. [10§9]

Common, A. K., and Roberts, D. E. 1986. Solutions of the Riccati equation and their application to the Toda lattice. *J. Phys. A* **19**, 1889–98. [10§9]

Common, A. K. and Stacey, T. W. 1979a. The convergence of Legendre–Padé approximants to the Coulomb and other scattering amplitudes. *J. Phys. A* **11**, 275–89. [7§2]

Common, A. K., and Stacey, T. W. 1979b. Legendre–Padé approximants and their application in potential scattering. *J. Phys. A* **11**, 259–73. [7§2]

Common, A. K., and Stacey, T. W. 1979c. Convergent series of Legendre–Padé approximants to the real and imaginary parts of the scattering amplitudes. *J. Phys. A* **12**, 1399–417. [7§2]

Conn, R. W. 1974. Higher order variational principles and Padé approximants for linear functionals. *Nucl. Sci. Eng.* **55**, 468–70. [9§5]

Conte, S. D., and de Boor, C. 1980. *Elementary Numerical Analysis*. McGraw-Hill, New York.

Coope, I. D., and Graves-Morris, P. R. 1993. The rise and fall of the vector epsilon algorithm. *Num. Algs.* **5**, 275–86. [8§4]

Cooper, S.; Magnus, A.; and McCabe, J. H. 1986. On the non-normal Padé table. *J. Comp. Appld. Math.* **16**, 371–80. [7§1]

Copley, L. A.; Elias, D. K.; and Masson, D. 1968. Broken symmetry model of meson interactions. *Phys. Rev.* **173**, 1552–63. [11§2]

Copley, L. A., and Masson, D. 1967. Padé approximant calculation of π–π scattering. *Phys. Rev.* **164**, 2059–62. [11§2]

Copson, E. T. 1948. *An Introduction to the Theory of a Complex Variable*. Oxford University Press, Oxford. [4§7, 6§6, 7§4]

Corcoran, C. T., and Langhoff, P. W. 1977. Moment theory approximations for nonnegative spectral densities. *J. Math. Phys.* **18**, 651–7. [10§2]

Cordellier, F. 1974. L'ε-algorithme vectoriel: interpretation géometrique et règles singulières. *Exposé au Colloque d'Analyse de Gourette*. [3§7, 8§4]

Cordellier, F. 1979a. Sur la régularité des procédés δ^2 d'Aitken et W de Lubkin. In *Padé Approximation and Its Applications*, L. Wuytack (ed.), pp. 20–35. Springer Lecture Notes in Mathematics, no. 765. Springer, Berlin. [3§1]

Cordellier, F. 1979b. Demonstration algébrique de l'extension de l'identité de Wynn aux tables de Padé non-normales. In *Padé Approximation and Its Applications*, L. Wuytack (ed.), pp. 36–60. Springer Lecture Notes in Mathematics, no. 765. Springer, Berlin. [3§7]

Cordellier, F. 1989. Thesis, University of Lille. [3§4, 3§7, 8§4]

Courant, R., and Hilbert, D. 1953. *Methods of Mathematical Physics*. Vol. 1. Interscience, New York. [5§2]

Crank, J. 1975. *The Mathematics of Diffusion*. Oxford University Press, Oxford. [10§4] da Cunha, R. D., and Hopkins, T. R. 1994. A comparison of acceleration techniques applied to the SOR method. In *Nonlinear Numerical Methods and Rational Approximation II*, A. A. M. Cuyt (ed.), pp. 247–60. Kluwer, Dordrecht. [8§4]

Cuyt, A. A. M. 1982. Padé approximants in operator theory for the solution of nonlinear integral and differential equations. *Comp. & Maths. with Applcns.* **8**, 445–66. [8§2]

Cuyt, A. A. M. 1983. The QD algorithm and multivariate Padé approximants. *Num. Math.* **42**, 259–69. [7§6]

Cuyt, A. A. M. 1984. *Padé Approximants for Operators: Theory and Applications*. Springer Lecture Notes in Mathematics no. 1065. Springer, Berlin. [8§2]

Cuyt, A. A. M. 1988. A multivariate QD-like algorithm. *BIT* **28**, 98–112.

Cuyt, A. A. M. 1990. A multivariate convergence theorem of the "de Montessus de Ballore" type. *J. Comp. Appld. Math.* **32**, 47–57. [7§6]

Cuyt, A. A. M. 1992. Extension of "A multivariate convergence theorem of the 'de Montessus de Ballore' type" to multipoles. *J. Comp. Appld. Math.* **41**, 323–30. [7§6]

Cuyt, A. A. M. 1993. Where do columns of the multivariate QD algorithm go? *Acta Appl. Math.* **33**, 273–302. [7§6]

Cuyt, A. A. M.; Jones, W. B.; and Verdonk, B. M. 1991. Model reduction and stability of two-dimensional recursive systems. *Rocky Mtn. J. Math.* **21**, 187–208.

Cuyt, A. A. M., and Verdonk, B. M. 1988a. Multivariate reciprocal differences for branched Thiele continued fraction expansions. *J. Comp. Appld. Math.* **21**, 145–60. [7§4]

Cuyt, A. A. M., and Verdonk, B. M. 1988b. A review of branched continued fraction theory for the construction of multivariate rational approximants. *Appld. Num. Math.* **4**, 263–71.

Cuyt, A. A. M., and Verdonk, B. M. 1994. Different techniques for the construction of multivariate rational interpolants. In *Nonlinear Numerical Methods and Rational Approximation II*, A. A. M. Cuyt (ed.) pp. 167–90, Kluwer, Dordrecht. [7§6]

Dancis, J. 1991a. The optimal ω is not the best for the SOR iteration method. *Lin. Alg. Applcns.* **154–6**, 819–45. [8§4]

Dancis, J. 1991b. Diagonalizing the adaptive SOR iteration method. *SIAM J. Matrix Anal. Applcns.* **12**, 661–78. [8§4]

Darboux, G. 1878. Sur l'approximation des fonctions de très-grands nombres et sur une classe étendue de développements en série. *J. de Math. (3)* **4**, 377–416. [8§6.6]

Delahaye, J.-P. 1988. *Sequence Transformations*. Springer, Berlin. [3§1, 10§1]

Della Dora, J. 1980. Contribution à l'approximation de fonctions de la variable complexe au sens de Hermite–Padé et de Hardy. Thèse, Université de Grenoble. [8§5.2, 8§5.3]

Della Dora, J., and Di Crescenzo, C. 1980. Approximation de Padé–Hermite. In *Padé Approximation and Its Applications*, L. Wuytack (ed.), pp. 88–115. Springer Lecture Notes in Mathematics, no. 765. Springer, Berlin. [8§5.2, 8§5.3]

Delves, L. M., and Phillips, A. C. 1969. Present status of the nuclear three body problem. *Rev. Mod. Phys.* **41**, 497–530. [9§5]

Dennis, J. J., and Wall, H. S. 1945. The limit-circle case for a positive definite J-fraction. *Duke Math. J.* **12**, 255–73. [4§7]

Dickinson, B. W.; Morf, M.; and Kailath, T. 1974. A minimal realisation algorithm for matrix sequences. *IEEE Trans. Auto. Control* **AC-19**, 31–8. [8§3]

Dienes, P. 1957. *The Taylor Series*. Dover, New York.

Dijkstra, D. 1977. A continued fraction expansion for a generalization of Dawson's integral. *Math. Comp.* **31**, 503–10. [7§1]

Dimock, J. 1974. Asymptotic perturbation expansion in the $P(\phi)_2$ quantum field theory. *Comm. Math. Phys.* **35**, 347–56. [11§2]

Dirac, P. A. M. 1958. *Principles of Quantum Mechanics*. 4th ed. Oxford University Press, Oxford. [11§1]

Domany, E.; Mukamel, D.; and Fisher, M. E. 1976. Destruction of first order transitions by symmetry breaking fields. *Phys. Rev. B* **15**, 5432–41.

Domb, C. 1974. Ising model. In *Phase Transitions and Critical Phenomena*, vol. 3, C. Domb and M. S. Green (eds.), pp. 357–485. Academic Press, London. [11§3]

Domb, C., and Sykes, M. F. 1961. Use of series expansions for the Ising model susceptibility and excluded value problem. *J. Math. Phys.* **2**, 63–7. [7§3]

Donnelly, J. D. P. 1966. The Padé table. In *Methods of Numerical Approximation*, D. C. Handscomb (ed.), pp. 125–30. Pergamon Press, Oxford.

Draux, A. 1983. *Polynômes Orthogonaux Formels — Applications*. Springer Lecture Notes in Mathematics no. 974. Springer, Berlin. [7§1]

Draux, A. 1987. Two point Padé type and Padé approximants in a noncommutative algebra. In *Rational Approximation and Its Applications in Mathematics and Physics*, J. Gilewicz, M. Pindor, and W. Siemaszko (eds.), Springer Lecture Notes in Mathematics no. 1237, pp. 51–62. Springer, Berlin. [7§1]

Drew, D., and Murphy, J. A. 1977. Branch points, M-fractions, and rational approximants generated by linear equations. *J. Inst. Math. Appl.* **19**, 169–85. [7§1]

Driver, K. A., and Lubinsky, D. S. 1990. Convergence of Padé approximants for Wynn's power series II. *Colloquia Mathematica Societatis Janos Bolyai* **58**, 221–39. [2§3]

Driver, K. A., and Lubinsky, D. S. 1991. Convergence of Padé approximants for Wynn's power series I. *Aequationes Mathematicae* **42**, 85–106. [2§3]

Driver, K. A., and Lubinsky, D. S. 1993. Convergence of Padé approximants for a q-hypergeometric series (Wynn's power series III). *Aequationes Mathematicae* **45**, 1–23. [2§3]

Driver, K. A., and Stahl, H. 1994. Normality and error formulae for simultaneous rational approximants to Nikishin systems. In *Nonlinear Numerical Methods and Rational Approximation II*, A. A. M. Cuyt (ed.), pp. 97–106. Kluwer, Dordrecht. [8§1]

Dunn, A. G.; Essam, J. W.; and Ritchie, D. S. 1975. Series expansion study of the pair connectedness in bond percolation models. *J. Phys. C* **8**, 4219–35.

Dyson, F. J. 1952. Divergence of perturbation theory in quantum electro-dynamics. *Phys. Rev.* **85**, 631–2. [11§1]

Eckmann, J.-P.; Magnen, J.; and Sénéor, R. 1974. Decay properties and Borel summability for the Schwinger functions in $P(\phi)_2$ theories. *Comm. Math. Phys.* **39**, 251–71.

Eddy, R. P. 1968. Causes and countermeasures for ping-pong oscillations in the output of the DX-DXG FFLS DD07 computer program (Appendix A. Mathematical theory of iteration). Applied Mathematics Laboratory Technical Note AML-35-68, Naval Ship Research and Development Center, Washington, DC.

Eddy, R. P. 1979. Extrapolating to the limit of a vector sequence. In *Information Linkage between Applied Mathematics and Industry*, P. C. C. Wang (ed.), pp. 387–96. Academic Press, New York. [8§4]

Edrei, A. 1939. Sur les determinants récurrents et les singularités d'une fonction données par son développement de Taylor. *Compositio Math.* **7**, 20–88. [1§5]

Edrei, A. 1953. Proof of a conjecture of Schoenberg on the generating function of a totally positive sequence. *Can. J. Math.* **5**, 86–94. [5§7]

Edrei, A. 1975a. Convergence of complete Padé tables of trigonometric functions. *J. Approx. Theory* **15**, 278–93. [5§7]

Edrei, A. 1975b. The Padé table of functions having a finite number of essential singularities. *Pacific J. Math.* **56**, 429–53. [6§5]

Edrei, A. 1975c. The complete Padé tables of certain series of simple fractions. *Rocky Mtn. J. Math.* **5**, 559–84.

Ehle, B. L. 1968. High order A-stable methods for numerical solution of systems of differential equations. *BIT* **8**, 276–8. [10§3, 10§4]

Ehle, B. L. 1973. A-stable methods and Padé approximations to the exponential. *SIAM J. Math. Anal.* **4**, 671–80. [5§7, 10§4]

Ehle, B. L. 1976. On certain order constrained Chebyshev rational approximations. *J. Approx. Theory* **17**, 297–306. [10§4]

Eiermann, M. 1984. On the convergence of Padé-type approximants to analytic functions. *J. Comp. Appld. Math.* **10**, 219–27. [10§1]

Erdélyi, A. 1956. *Asymptotic Expansions*. Dover, New York. [5§5]

Essam, J. W., and Fisher, M. E. 1963. Padé approximant studies of the lattice gas and Ising ferromagnet below the critical point. *J. Chem. Phys.* **38**, 802–12.

Euler, L. 1737. De fractionibus continuis. *Comm. Acad. Sci. Imper. Petropol* **9**, 98–137. [4§1]

Evans, G. A. 1993. Numerical inversion of Laplace transforms. *Int. J. Comp. Math* **50**, 144–62. [10§5]

Fabry, E. 1896. Sur les points singuliers d'une fonction donnée par son développement en série et sur l'impossibilité du prolongation analytique dans les cas très généraux. *Ann. Ecole Normale (3)* **13**, 107–14. [6§1, 10§1]

Fahmy, M. F. 1980. The use of Padé approximants in the derivation of low-pass filters with simultaneous flat amplitude and decay characteristics. *Circuit Theory* **8**, 197–204.

Fair, W. G. 1964. Padé approximants to the solution of the Ricatti equation. *Math. Comp.* **18**, 627–334. [10§9]

Fair, W., and Luke, Y. L. 1970. Padé approximants to the operator exponential. *Num. Math.* **14**, 379–82.

Fairweather, G. 1971. A survey of discrete Galerkin methods for parabolic equations in one space variable. *Math. Colloq. U.C.T.* **7**, 43–77. [10§4]

Feldman, J. S., and Osterwalder, K. 1976. The Wightman axioms and the mass gap for weakly coupled $(\phi^4)_3$ quantum field theories. *Ann. Phys.* **97**, 80–135. [11§2]

Ferer, M.; Moore, M. A.; and Wortis, M. 1971. Some critical properties of the nearest-neighbor, classical Heisenberg model for the f.c.c. lattice in finite field for temperatures greater than T_c. *Phys. Rev. B* **4**, 3954–63.

Ferrar, W. L. 1938. *Convergence*. Oxford University Press, Oxford.

Fike, C. T. 1968. *Computer Evaluation of Mathematical Functions*. Prentice-Hall, Englewood Cliffs, NJ.

Fisher, M. E. 1967. The theory of equilibrium critical phenomena. *Rept. Prog. Phys.* **30**, 615–731.

Fisher, M. E. 1977. Series expansion approximants for singular functions of many variables. In *Statistical Mechanics and Statistical Methods in Theory and Application*, Uzi Landman (ed.), pp. 3–31. Plenum Press. [7§6]

Fisher, M. E., and Au-Yang, H. 1979. Inhomogeneous differential approximants for power series. *J. Phys. A* **12**, 1677–92. [8§6.2]

Fisher, M. E., and Burford, R. J. 1967. Theory of critical point scattering and correlations I. The Ising model. *Phys. Rev.* **156**, 583–622.

Fisher, M. E., and Kerr, R. M. 1977. Partial differential approximants for multicritical singularities. *Phys. Rev. Lett.* **39**, 667–70. [7§6]

Fisher, M. E., and Styer, D. F. 1982. Partial differential approximants for multivariable power series I. Definitions and faithfulness. *Proc. Roy. Soc. Lond. A* **384**, 259–87. [7§6]

Fisher, M. E., and Styer, D. F. 1983. Partial differential approximants for multivariable power series II. Invariance properties. *Proc. Roy. Soc. Lond. A* **388**, 75–102. [7§6]

Fleischer, J. 1972. Analytic continuation of scattering amplitudes and Padé approximants. *Nucl. Phys. B* **37**, 59–76.

Fleischer, J. 1973a. Nonlinear Padé approximants for Legendre series. *J. Math. Phys.* **14**, 246–8. [7§4]

Fleischer, J. 1973b. Generalizations of Padé approximants. In *Padé Approximants*, P. R. Graves-Morris (ed.), pp. 126–31. Inst. of Phys., Bristol. [7§4]

Fleischer, J.; Gammel, J. L.; and Menzel, M. T. 1973. Matrix Padé approximants for the $^1S_0$ and $^3P_0$ partial waves in nucleon–nucleon scattering. *Phys. Rev. D* **8**, 1545–52. [9§8]

Fleischer, J., and Tjon, J. A. 1975. Bethe–Salpeter equation for $J = 0$ nucleon–nucleon scattering with one boson exchange. *Nucl. Phys. B* **84**, 375–96. [9§7]

Fleischer, J., and Tjon, J. A. 1980. Bethe–Salpeter equation for elastic nucleon–nucleon scattering. *Phys. Rev. D* **21**, 87–94. [9§7, 9§8]

Fogli, G. L.; Pellicoro, M. F.; and Villani, M. 1971. A summation method for a class of series with divergent terms. *Nuovo Cim.* **6A**, 79–97.

Fogli, G. L.; Pellicoro, M. F.; and Villani, M. 1972. An approach to the radiative corrections in QED in the framework of the Padé method. *Nuovo Cim.* **11A**, 153–77.

Ford, W. B. 1960. *Asymptotic Series and Divergent Series*. Chelsea, New York. [5§5]

Ford, W. F., and Sidi, A. 1987. An algorithm for a generalization of the Richardson extrapolation process. *SIAM J. Numer. Anal.* **24**, 1212–32. [10§1]

Frank, W. M.; Land, D. J.; and Spector, R. M. 1971. Singular potentials. *Rev. Mod. Phys.* **43**, 36–98. [9§9]

Fratamico, G.; Ortolani, F.; and Turchetti, G. 1976. Exact solutions from the variational [1/1] matrix Padé approximant in potential scattering. *Lett. Nuovo Cim.* **17**, 582–4. [9§8]

Freud, G. 1976. *Orthogonal Polynomials*. Pergamon, Oxford. [5§4]

Freund, R. W.; Golub, Gene; and Nachtigal, N. 1992. Iterative solution of linear systems. In *Acta Numerica*, A. Iserles (ed.), pp. 57–100. Cambridge University Press, Cambridge. [7§5]

Frobenius, G. 1881. Ueber relationen zwischen den Näherungsbrüchen von Potenzreihen. *J. für reine und angewandte Math.* **90**, 1–17. [Preface, 1§4, 3§5]

Froissart, M. 1969. Private communication to J. L. Basdevant. In "Padé Approximants," in *Methods of Subnuclear Physics*, M. Nikolic (ed.), pp. 1-60. Gordon and Breach, London.

Gallucci, M. A., and Jones, W. B. 1976. Rational approximations corresponding to Newton series (Newton–Padé approximants). *J. Approx. Theory* **17**, 366-92. [7§1]

Gammel, J. L. 1973a. Effect of random errors (noise) in the terms of a power series on the convergence of the Padé approximants. In *Padé Approximants*, P. R. Graves-Morris (ed.), pp. 132-3. Inst. of Phys., Bristol.

Gammel, J. L. 1973b. Review of two recent generalizations of the Padé approximant. In *Padé Approximants and Their Applications*, P. R. Graves-Morris (ed.), pp. 3-9. Academic Press, London. [8§6.2]

Gammel, J. L. 1974. Continuation of functions beyond natural boundaries. *Rocky Mtn. J. Math.* **4**, 203-6.

Gammel, J. L., and McDonald, F. A. 1966. Applications of the Padé approximant to scattering theory. *Phys. Rev.* **142**, 1245-54. [1§5, 8§2]

Gammel, J. L., and Nuttall, J. 1973. Convergence of Padé approximants to quasi-analytic functions beyond natural boundaries. *J. Math. Anal. Appl.* **43**, 694-6.

Gammel, J. L.; Rousseau, C. C.; and Saylor, D. P. 1967. A generalization of the Padé approximant. *J. Math. Anal. Appl.* **20**, 416-20. [7§2]

Gander, W.; Golub, Gene H.; and Gruntz, D. 1989. Solving linear equations by extrapolation. In *Supercomputing, Trondheim, 1989*, NATO Adv. Sci. Inst. Ser. F, pp. 279-93. Springer, Berlin. [8§4]

Gargantini, J., and Henrici, P. 1967. A continued fraction algorithm for the computation of higher transcendental functions in the complex plane. *Math. Comp.* **21**, 18-29. [5§2]

Garibotti, C. R. 1972. Schwinger variational principle and Padé approximants. *Ann. Phys.* **71**, 486-96. [9§5]

Garibotti, C. R., and Grinstein, F. F. 1978a. Summation of partial waves for long range potentials I. *J. Math. Phys.* **19**, 821-9.

Garibotti, C. R., and Grinstein, F. F. 1978b. Punctual Padé approximants as a regularization procedure for divergent and oscillatory partial wave expansion of the scattering amplitude. *J. Math. Phys.* **19**, 2405-9.

Garibotti, C. R., and Grinstein, F. F. 1979. Summation of partial waves for long range potentials II. *J. Math. Phys.* **20**, 141-7.

Garibotti, C. R.; Pellicoro, M. F.; and Villani, M. 1970. Padé method in singular potentials. *Nuovo Cim.* **66A**, 749-66. [9§9]

Garibotti, C. R., and Villani, M. 1969a. Continuation in the coupling constant for the total T and K matrices. *Nuovo Cim.* **59A**, 107-23. [9§3, 9§7]

Garibotti, C. R., and Villani, M. 1969b. Padé approximant and the Jost function. *Nuovo Cim.* **61A**, 747-54. [9§7]

Garnett, J. 1973. *Analytic Capacity and Measure*. Springer Lecture Notes in Mathematics, no. 297. Springer, New York.

Garside, G. R.; Jarratt, P.; and Mack, C. 1968. A new method for solving polynomial equations. *Comp. J.* **11**, 87-90. [7§1]

Gasper, G., and Rahman, M. 1990. *Basic Hypergeometric Series*. Cambridge University Press, New York. [2§3]

Gaunt, D. S., and Guttmann, A. J. 1974. Asymptotic analysis of coefficients. In *Phase Transitions and Critical Phenomena*, vol. 3, C. Domb and M. S. Green (eds.), pp. 181-245. Academic Press, New York.

Gauss, C. F. 1813. Disquisitiones generales circa seriam infinitam

$$1 + \frac{\alpha\beta}{1\cdot\gamma}x + \frac{\alpha(\alpha+1)\beta(\beta+1)}{1\cdot 2\gamma(\gamma+1)}x\cdot x + \cdots.$$

Commentationes Societatis Regiae Scientiorum Goettingensis Recentiores **2**. [4§6]

Gauthier, P. M. 1977. On the possibility of rational approximation. In *Padé and Rational Approximation*, E. B. Saff and R. S. Varga (eds.), pp. 261–4. Academic Press, New York. [10§6]

Gautschi, W. 1967. Computational aspects of three term recurrence relations. *SIAM Review* **9**, 24–82. [4§1]

Gautschi, W. 1975. Computational methods in special functions—A survey. In *Theory and Application of Special Functions*, R. A. Askey (ed.), pp. 1–98. Academic Press, New York.

Gear, C. W. 1971. *Numerical Initial Value Problems in Ordinary Differential Equations*. Prentice-Hall, Englewood Cliffs, NJ. [10§4]

Geddes, K. O. 1979. Symbolic computation of Padé approximants. *A.C.M. Trans. Math. Software* **5**, 218–33. [2§1]

Geddes, K. O. 1981. Block structure in the Chebyshev–Padé table. *SIAM J. Num. Analy.* **18**, 844–61. [7§4]

Gekeler, E. 1972. On the solution of systems of equations by the epsilon algorithm of Wynn. *Math. Comp.* **26**, 427–36. [3§3]

Genz, A. C. 1972. An adaptive multidimensional quadrature procedure. *Comp. Phys. Comm.* **4**, 11–15.

Genz, A. C. 1973. The application of the ε-algorithm to quadrature problems. In *Padé Approximants*, P. R. Graves-Morris (ed.), pp. 105–16. Inst. of Physics, Bristol.

Genz, A. C. 1974. Some extrapolation methods for the numerical calculation of multidimensional integrals. In *Software for Numerical Mathematics*, D. J. Evans (ed.), pp. 159–72. Academic Press, New York.

Genz, A. C. 1977. A nonlinear method for the convergence of multidimensional sequences. *J. Comp. Appl. Math.* **3**, 181–4. [7§6]

Germain-Bonne, B. 1979. Ensembles des suites et des procédés liés pour l'accélération de la convergence. In *Padé Approximation and Its Applications*, L. Wuytack (ed.), pp. 116–34. Springer Lecture Notes in Mathematics, no. 765, Springer, Berlin. [3§1]

Gersten, A.; Owen, D. A.; Gammel, J. L.; Mery, P.; and Turchetti, G. 1976. Matrix Padé approximants and the Bethe–Salpeter equation of the nucleon–nucleon interaction. *Phys. Rev. D* **13**, 1140–3. [9§7]

Gilewicz, J. 1978. *Approximants de Padé*. Springer Lecture Notes in Mathematics, no. 667. Springer, Berlin.

Gilewicz, J., and Magnus, A. P. 1991. Sharp inequalities for the Padé approximant errors in the Stieltjes case. *Rocky Mtn. J. Math.* **21**, 227–34. [5§4]

Giraud, B. G. 1978. Lower bounds to bound state eigenvalues. *Phys. Rev. C* **17**, 800–9. [9§5]

Giraud, B. G.; Khalil, A. B.; and Moussa, P. 1976. Padé approximants for distorted waves. *Phys. Rev. C* **14**, 1679–87. [9§7]

Giraud, B. G., and Turchetti, G. 1978. Rigorous lower bounds to the energy levels with finite summations. *Lett Nuovo Cim.* **21**, 605–8. [9§5]

Glasser, M. L., and Zucker, I. J. 1980. Lattice sums. In *Theoretical Chemistry: Advances and Perspectives*, vol. 5, pp. 67–139. Academic Press, New York.

Gohberg, I. C., and Heinig, G. 1974. Inversion of finite Toeplitz matrices made of elements of a noncommutative algebra (in Russian). *Rev. Roumaine. Math. Pures Appl.* **19**, 623–63. [3§6]

Gohberg, I. C., and Semencul, A. A. 1972. On the inversion of finite Toeplitz matrices and their continuous analogs (in Russian). *Mat. Issled* **2**, 201–33. [3§6]

Goldberger, M. L., and Watson, K. M. 1964. *Collision Theory*. Wiley, New York. [9§4]

Goldhammer, P., and Feenberg, E. 1956. Refinement of the Brillouin Wigner perturbation method. *Phys. Rev.* **101**, 1233–4. [9§5]

Golub, Gene H., and Van Loan, C. F. 1989. *Matrix Computations*. Johns Hopkins University Press, Baltimore. [2§1, 3§6, 9§3, 10§6.1]

Gončar, A. A. 1973. On the convergence of Padé approximants. *Mat. Sbornik* **92**(134); *Math. USSR Sbornik* **21**, 155–6.

Gončar, A. A. 1981. Poles of rows of the Padé table and meromorphic continuation of functions. *Mat. Sbornik* **115**(157); *Math. USSR Sbornik* **43**, 527–46. [6§1]

Gončar, A. A. 1982. On uniform convergence of diagonal Padé approximants. *Mat. Sbornik* **118**(160); *Math. USSR Sbornik* **46**, 539–59. [6§4]

Gončar, A. A., and Guillermo-Lopez, L. 1978. On Markov's theorem for multipoint Padé approximants. *Mat. Sbornik* **105**(147); *Math. USSR Sbornik* **34**, 449–59. [7§1]

Gončar, A. A., and Rakhmanov, E. A. 1981. On the convergence of simultaneous Padé approximants for systems of functions of Markov type. *Trudi Mat. Inst. Steklov* **157**; *Proc. Steklov Inst. Maths.* **3**, 31–50. [8§1]

Gončar, A. A.; Rakhmanov, E. A.; and Suetin, S. P. 1992. On the rate of convergence of Padé approximants of orthogonal expansions. In *Progress in Approximation Theory*, A. A. Gončar and E. B. Saff (eds.), pp. 169–90. Springer Verlag, New York. [7§4]

Gordon, R. G. 1968. Error bounds in equilibrium statistical mechanics. *J. Math. Phys.* **9**, 655–63. [10§2]

Götz, U.; Rösel, F.; Trautmann, D.; and Jochin, H. 1976. Influence of the Mott–Schwinger interaction on the elastic scattering of protons. *Z. Phys. A* **278**, 139–43.

Graffi, S., and Grecchi, V. 1978. Borel summability and indeterminacy of the Stieltjes moment problem: application to the anharmonic oscillators. *J. Math. Phys.* **19**, 1002–6. [11§1]

Graffi, S.; Grecchi, V.; and Simon, B. 1970. Borel summability: application to the anharmonic oscillator. *Phys. Lett.* **32B**, 631–4. [7§2, 11§1]

Graffi, S.; Grecchi, V.; and Turchetti, G. 1971. Summation methods for the perturbation series of the generalized harmonic oscillator. *Nuovo Cim.* **4B**, 313–40. [11§1]

Gragg, W. B. 1965. On extrapolation algorithms for ordinary initial value problems. *SIAM J. Num. Anal.* **2**, 384–403. [7§1]

Gragg, W. B. 1968. Truncation error bounds for g-fractions. *Num. Math.* **11**, 370–9. [5§4.1]

Gragg, W. B. 1970. Truncation error bounds for π-fractions. *Bull. Am. Math. Soc.* **76**, 1091–4.

Gragg, W. B. 1972. The Padé table and its relation to certain algorithms of numerical analysis. *SIAM Review* **14**, 1–62. [3§6, 3§8, 6§2]

Gragg, W. B. 1974. Matrix interpretations and applications of the continued fraction algorithm. *Rocky Mtn. J. Math.* **4**, 213–25. [7§4]

Gragg, W. B. 1977. Laurent, Fourier, and Chebychev Padé tables. In *Padé and Rational Approximation*, E. B. Saff and R. H. Varga (eds.), pp. 61–70. Academic Press, New York. [7§4]

Gragg, W. B.; Gustavson, F. G.; Warner, D. D.; and Yun, D. Y. Y. 1982. On fast computation of superdiagonal Padé fractions. *Math. Programming Study* **18**, 39–42.

Gragg, W. B., and Johnson, G. D. 1974. The Laurent Padé table. *Info. Proc. 74, North-Holland, Amsterdam* **3**, 632–7. [7§4]

Gragg, W. B., and Lindquist, A. 1983. On the partial realisation problem. *Lin. Alg. Applcns.* **50**, 277–319. [4§4]

Grant, J. A. and Hitchens, G. D. 1971. An always convergent minimisation technique for the solution of polynomial equations. *J. Inst. Math. Applcns.* **8**, 122–129. [7§1]

Graves-Morris, P. R. 1973a. Padé approximants and potential scattering. In *Padé Approximants*, P. R. Graves-Morris (ed.), pp. 64–76. Inst. of Phys., Bristol. [9§2, 9§3, 9§9]

Graves-Morris, P. R. 1973b. *Padé Approximants and Their Applications*. Academic Press, New York.

Graves-Morris, P. R. 1975. Convergence of rows of the Padé table. In *Padé Approximation and Its Application to Mechanics*, H. Cabannes (ed.), pp. 55–68. Springer Lecture Notes in Physics, no. 47. Springer, Berlin.

Graves-Morris, P. R. 1977. Generalisations of the theorem of de Montessus using Canterbury approximants. In *Padé and Rational Approximation*, E. B. Saff and R. S. Varga (eds.), pp. 73–82. Academic Press, New York. [6§2, 7§6]

Graves-Morris, P. R. 1978a. Padé approximants for integral equations? *J. Inst. Math. Appl.* **21**, 375–8. [9§2]

Graves-Morris, P. R. 1978b. Applications of matrix Padé approximants in potential theory. *Ann. Phys.* **114**, 290–5. [9§8]

Graves-Morris, P. R. 1979. The numerical calculation of Padé approximants. In *Padé Approximation and Its Applications*, L. Wuytack (ed.), pp. 231–45. Springer Lecture Notes in Mathematics, no. 765, Springer, Berlin. [2§1, 7§1]

Graves-Morris, P. R. 1980a. Practical, reliable, rational interpolation. *J. Inst. Math. Appl.* **25**, 267–86. [7§1]

Graves-Morris, P. R. 1980b. A generalized Q.D. algorithm. *J. Comp. Appl. Math.* **6**, 247–9. [7§1]

Graves-Morris, P. R. 1981a. The convergence of ray sequences of Padé approximants of Stieltjes functions. *J. Comp. Appld. Math.* **7**, 191–201. [5§4]

Graves-Morris, P. R. 1981b. Efficient, reliable, rational interpolation. In *Padé Approximation and Its Applications*, H. Van Rossum and M. G. de Bruin (eds.), pp. 28–63. Springer Lecture Notes in Mathematics no. 888. Springer, Amsterdam. [7§1]

Graves-Morris, P. R. 1982. Toeplitz equations and Kronecker's algorithm. *CSSP Journal* **1**, 289–304.

Graves-Morris, P. R. 1983. Vector-valued rational interpolants I. *Numer. Math.* **42**, 331–48. [8§4]

Graves-Morris, P. R. 1984a. Vector-valued rational interpolants II. *IMA J. Num. Anal.* **4**, 209–24.

Graves-Morris, P. R. 1984b. Symmetrical formulas for rational interpolants. *J. Comp. Appld. Math.* **10**, 107–11. [7§1]

Graves-Morris, P. R. 1988. The critical index of the magnetic susceptibility of 3D Ising models. *J. Phys. A* **21**, 1867–81.

Graves-Morris, P. R. 1990. Solution of integral equations using generalised inverse, function-valued Padé approximants I. *J. Comp. Appld. Math.* **32**, 117–24. [8§4.1]

Graves-Morris, P. R. 1992. Extrapolation methods for vector sequences. *Num. Math.* **61**, 475–87. [8§4]

Graves-Morris, P. R. 1994. A review of Padé methods for the acceleration of convergence of a sequence of vectors. *Applied Numerical Mathematics*, **15**, 153–174 [8§4]

Graves-Morris, P. R., and Hopkins, T. R. 1981. Reliable rational interpolation. *Num. Math.* **36**, 111–28. [7§1]

Graves-Morris, P. R., and Hughes Jones, R. 1976. An analysis of two-variable rational approximants. *J. Comp. Appld. Math.* **2**, 41. [7§7]

Graves-Morris, P. R.; Hughes Jones, R.; and Makinson, G. J. 1974. The calculation of some rational approximants in two variables. *J. Inst. Math. Appl.* **13**, 311–20. [7§6]

Graves-Morris, P. R., and Jenkins, C. D. 1986. Vector-valued rational interpolants III. *Constr. Approx.* **2**, 263–89. [3§7, 8§4]

Graves-Morris, P. R., and Jenkins, C. D. 1989. Degeneracies of generalized inverse, vector-valued Padé approximants. *Constr. Approx.* **5**, 463–85. [8§4]

Graves-Morris, P. R., and Johnson, C. R. 1990. Confluent Hermitian Cauchy matrices are diagonally signed. *Lin. Alg. Applcns.* **142**, 129–40. [6§2]

Graves-Morris, P. R., and Johnson, C. R. 1995. Determinantal inequalities for diagonally signed matrices and an application to Gram–Cauchy matrices. *Lin Alg. Applcns.*, in press. [6§2]

Graves-Morris, P. R., and Rennison, J. F. 1974. Analyticity in the coupling strength. *J. Math. Phys.* **15**, 230–3. [9§4]

Graves-Morris, P. R., and Roberts, D. E. 1975. Calculation of Canterbury approximants. *Comp. Phys. Comm.* **10**, 234–44; *Erratum* **13** (1977), 72. [7§6]

Graves-Morris, P. R., and Roberts, D. E. 1994. From matrix to vector Padé approximants. *J. Comp. Appld. Math.*, **51**, 205–36. [3§7, 8§4]

Graves-Morris, P. R., and Saff, E. B. 1984. A de Montessus theorem for vector-valued rational interpolants. In *Rational Approximation and Its Applications*, P. R. Graves-Morris, E. B. Saff, and R. S. Varga (eds.), pp. 227–42. Springer Lecture Notes in Mathematics no. 1105. Springer, Berlin. [6§3, 8§1, 8§4]

Graves-Morris, P. R., and Saff, E. B. 1988. Row convergence theorems for generalised inverse vector-valued Padé approximants. *J. Comput. Appl. Math.* **23**, 63–85.

Graves-Morris, P. R., and Saff, E. B. 1991a. An extension of a row convergence theorem for vector Padé approximants. *J. Comp. Appld. Math.* **34**, 315–24.

Graves-Morris, P. R., and Saff, E. B. 1991b. Divergence of vector-valued rational interpolants to meromorphic functions. *Rocky Mtn. J. Math.* **21**, 245–61. [8§1]

Graves-Morris, P. R., and Samwell, C. J. 1975. Canterbury approximants in potential scattering. *J. Phys. G* **1**, 805–14. [11§2]

Graves-Morris, P. R., and Thukral, R. 1992. Solution of integral equations using function-valued Padé approximants II. *Num. Algs.* **3**, 223–34. [8§4.1]

Graves-Morris, P. R., and Wilkins, J. M. 1987. A fast algorithm to solve Kalman's partial realisation problem for single input, multi-output systems. In *Approximation Theory*, E. B. Saff (ed.), pp. 1–20. Springer Lecture Notes in Mathematics no. 1287. Springer Verlag, New York. [8§1, 8§3]

Gray, H. L.; Atchison, T. A.; and McWilliams, G. V. 1971. Higher order G-transformations. *SIAM J. Num. Anal.* **8**, 365–81. [10§1]

Grenander, U., and Szegö, G. 1958. *Toeplitz Forms and Their Applications*. Univ. of California Press, Berkeley, CA.

Griffiths, R. B. 1972. Rigorous results and theorems. In *Phase Transitions and Critical Phenomena*, vol. 1, C. Domb and M. S. Green (eds.), pp. 7–109. Academic Press, London.

Grinstein, F. F. 1980. Summation of partial wave expansions in the scattering by short range potentials. *J. Math. Phys.* **21**, 112–19.

Gupta, D. P.; Ismail, M. E. H.; and Masson, D. P. 1992. Contiguous relations, basic hypergeometric functions, and orthogonal polynomials. II Associated big q-Jacobi polynomials. *J. Math. Anal. Applcns.* **171**, 477–97. [2§3]

Gustavson, F. G., and Yun, D. Y. Y. 1979. Fast algorithms for rational Hermite approximation and solution of Toeplitz systems. *IEEE Trans. Circuits and Systems CAS* **26**, 750–5.

Gutknecht, M. H. 1984. Rational Carathéodory–Fejér approximation on a disk, a circle, and an interval. *J. Approx. Theory* **41**, 257–78. [10§6.1]

Gutknecht, M. H. 1989. Continued fractions associated with the Newton Padé table. *Num. Math.* **56**, 547–89. [7§1]

Gutknecht, M. H. 1990. In what sense is the rational interpolation problem well posed? *Constr. Approx.* **6**, 437–50. [10§6]

Gutknecht, M. H. 1993. Stable row recurrences for the Padé table and generically superfast look-ahead solvers for non-Hermitian Toeplitz systems. *Lin. Alg. Applcns.* **188–9**, 351–422. [3§6, 4§5, 7§5]

Gutknecht, M. H., and Trefethen, L. N. 1983a. Real and complex Chebyshev approximations in the unit disk and interval. *Bull. Am. Math. Soc.* **8**, 455–7. [10§6]

Gutknecht, M. H., and Trefethen, L. N. 1983b. Nonuniqueness of rational Chebyshev approximations on the unit disk. *J. Approx. Theory* **39**, 275–88. [10§6]

Guttman, A. J. 1969. Determination of critical behaviour in lattice statistics from series expansions III. *J. Phys. C* **2**, 1900–7.

Guttman, A. J. 1975a. On the recurrence relation method of series analysis. *J. Phys. A* **8**, 1081–8.

Guttman, A. J. 1975b. Derivation of "mimic expansions" from regular perturbation expansions in fluid mechanics. *J. Inst. Math. Appl.* **15**, 307–15.

Guttman, A. J. 1989. Asymptotic analysis of power-series expansions. In *Phase Transitions and Critical Phenomena*, vol. 13, C. Domb and J. L. Lebowitz (eds.), pp. 1–234. Academic Press, New York. [7§3, 8§6.2, 10§1]

Guttman, A. J., and Joyce, G. S. 1972. On a new method of series analysis in lattice statistics. *J. Phys. A* **5**, L81–4. [7§3, 8§6.2]

Guttman, A. J., and Joyce, G. S. 1973. A new method of series analysis. In *Padé*

Approximants and Their Applications, P. R. Graves-Morris (ed.), pp. 163–7. Academic Press, New York. [8§6.2]
Guttman, A. J.; Thompson, C. J.; and Ninham, B. W. 1970. Determination of critical behaviour from series expansions in lattice statistics IV. *J. Phys. C* **3**, 1641–51.
Hadamard, J. 1892. Essai sur l'étude des fonctions données par leur developpement de Taylor, 2ième partie. *J. de Math.* **4**, 101–86. [6§2]
Hadamard, J. 1968. *Oeuvres*. Vol. 2, pp. 24–60. C.N.R.S., Paris. [6§2]
Hamburger, H. 1920. Ueber eine Erweiterung des Stieltjesschen Momenten Problems. *Math. Annalen* **81**, 235–319. [5§6]
Hamburger, H. 1921. Ueber eine Erweiterung des Stieltjesschen Momenten Problems. *Math. Annalen* **82**, 120–64, 168–87. [5§6]
Hardy, G. H. 1956. *Divergent Series*. Oxford University Press, Oxford. [5§5]
Hastings-James, R., and Mehra, S. K. 1977. Extensions of the Padé approximant technique for the design of recursive digital filters. *IEEE Trans. ASSP* **25**, 501–9.
Håvie, T. 1979. Generalized Neville type extrapolation schemes. *BIT* **19**, 204–13. [10§1]
Haymaker, R. W. 1968. Application of analyticity properties to the numerical solution of the Bethe–Salpeter equation. *Phys. Rev.* **165**, 1790–802. [9§4]
Haymaker, R. W., and Schlessinger, L. 1970. Padé approximant as a computational tool for solving the Schrödinger and Bethe–Salpeter equations. In *The Padé Approximant in Theoretical Physics*, G. A. Baker, Jr., and J. L. Gammel (eds.), pp. 257–88. Academic Press, New York. [9§4]
Heine, H. E. 1878. *Handbuch der Kugelfunktionen*. Riemer, Berlin. [4§3]
Heinig, G. 1983. Inversion of Toeplitz and Hankel matrices with singular sections. *Wiss. Z. d. Techn. Hochsch. Karl Marx Stadt* **25**, 326–33.
Heinig, G., and Rost, K. 1984. *Algebraic Methods for Toeplitz-like Matrices and Operators*. Akademic-Verlag, Berlin and Birkhaüser, Basel/Stuttgart. [3§6]
Hendriksen, E., and Van Rossum, H. 1982. Moment problems in Padé approximation. *J. Approx. Theory* **35**, 250–63. [4§3]
Henrici, P. 1958. The quotient difference algorithm. *Nat. Bur. Stand.—Appl. Math. Series* **49**, 23–46. [3§8]
Henrici, P. 1964. *Elements of Numerical Analysis*. Wiley, New York. [3§1]
Henrici, P. 1974. *Applied and Computational Complex Analysis*. Vol. I. Wiley, New York. [3§8]
Henrici, P., and Pfluger, P. 1966. Truncation error estimates for Stieltjes fractions. *Num. Math.* **9**, 120–38. [5§2, 5§5]
Hermite, C. 1893. Sur la généralisation des fractions continues algébriques. *Annali di Mathematica Pura e Applicata, Sér. 2* **21**, 289–308. [8§5]
Hermite, C. 1917. Sur quelques approximations algébriques. *Oeuvres*, t. III, 146–9; Sur la fonction exponentielle. *Oeuvres*, t. III, 150–81; Sur la généralisation des fractions continues algébriques. *Oeuvres*, t. IV, 357–77. Gauthier-Villars, Paris.
Higham, N. J. 1991. Algorithm 694; A collection of test matrices in MATLAB. *ACM Trans. Math. Soft.* **17**, 289–305. [2§1]
Hildebrand, F. B. 1956. *Introduction to Numerical Analysis*. McGraw-Hill, New York. [7§1]
Hille, E. 1959. *Analytic Function Theory*. Vol. 1. Ginn, Boston. [6§4]
Hille, E. 1962. *Analytic Function Theory*. vol. 2. Ginn, Boston. [6§4, 6§6, 10§6]

Hillion, P. 1977a. Remarks on rational approximation of multiple power series. *J. Inst. Math. Appl.* **19**, 281–93. [7§6]

Hillion, P. 1977b. Approximating functions with a given singularity. *J. Math. Phys.* **18**, 465–70.

Hofman, H. M.; Starkand, Y.; Kirson, M. W. 1976. Perturbation theory and Padé approximants in realistic large-matrix models of the nuclear effective interaction. *Nucl. Phys. A.* **266**, 138–62.

Holdeman, J. T., Jr. 1969. A method for the approximation of functions defined by formal series expansion in orthogonal polynomials. *Math. Comp.* **23**, 275–87. [7§4]

Homeier, H. H. H. 1993. Some applications of nonlinear convergence accelerators. *Int. J. Quantum Chem.* **45**, 545–62. [10§1]

Hopkins, T. R. 1982. On the sensitivity of the coefficients of Padé approximants with respect to their defining power series coefficients. *J. Comp. Appl. Math* **8**, (2), 105–9. [2§1]

Householder, A. S. 1970. *The Numerical Treatment of a Single Nonlinear Equation*. McGraw-Hill, New York. [2§4]

Householder, A. S. 1971. The Padé table, the Frobenius identities, and the Q.D. algorithm. *Lin. Algebra Applcns.* **4**, 161–74. [2§4]

Householder, A. S., and Stewart, G. W., III. 1969. Bigradients, Hankel determinants, and the Padé table. In *Constructive Aspects of the Fundamental Theorem of Algebra*, B. Dejon and P. Henrici (eds.), pp. 131–50. Academic Press, New York. [2§4, 7§1]

Hughes Jones, R. 1976. General rational approximants in N-variables. *J. Approx. Theory* **16**, 201–33. [7§6]

Hughes Jones, R., and Makinson, G. J. 1974. The generation of Chisholm rational approximants to power series in two variables. *J. Inst. Math. Appl.* **13**, 299–310. [7§6]

Hunter, D. L., and Baker, G. A., Jr. 1973. Methods of series analysis I: Comparison of methods used in critical phenomena. *Phys. Rev. B* **7**, 3346–76. [2§2, 8§6.2]

Hunter, D. L., and Baker, G. A., Jr. 1979. Methods of series analysis III: Integral approximant methods. *Phys. Rev. B* **19**, 3808–21. [7§3, 8§6.4, 8§6.7, 11§3]

Ince, E. L. 1944. *Ordinary Differential Equations*. Dover, New York. [8§6.2, 10§7]

Isaacson, E., and Keller, H. B. 1966. *Analysis of Numerical Methods*. Wiley, New York. [10§3]

Isenberg, C. 1963. Moment calculations in lattice dynamics I; F.C.C. lattice with nearest neighbor interactions. *Phys. Rev.* **132**, 2427–33. [10§2]

Iserles, A. 1979a. A note on Padé approximations and generalized hypergeometric functions. *BIT* **19**, 543–5.

Iserles, A. 1979b. On the generalized Padé approximations to the exponential function. *SIAM J. Num. Anal.* **16**, 631–6. [5§7]

Iserles, A. 1985. Order stars, approximations, and finite differences III. Finite differences for $u_t = \omega u_{xx}$. *SIAM J. Math. Anal.* **16**, 1020–33. [10§4]

Iserles, A., and Nørsett, S. P. 1988. On the theory of biorthogonal polynomials. *Trans. Amer. Math. Soc.* **306**, 455–74. [10§4]

Iserles, A., and Nørsett, S. P. 1991. *Order Stars*. Chapman and Hall, London. [1§2, 8§1]

Iserles, A., and Powell, M. J. D. 1981. On the A-acceptability of rational

approximations that interpolate the exponential function. *IMA J. Numer. Anal.* **1**, 241–51. [5§7]
Iserles, A., and Saff, E. B. 1987. Biorthogonality in rational approximation. *J. Comp. Appld. Math.* **19**, 47–54. [8§1]
Ismail, M. E. H., and Rahman, M. 1991. Associated Askey–Wilson polynomials. *Trans. Am. Math. Soc.* **328**, 201–39. [2§3]
Istace, M.-P., and Thiran, J.-P. 1995. On the third and fourth Zolotarev problems in the complex plane. *SIAM J. Num. Anal.*, **32**, 249–59. [10§2]
Jackson, L. B. 1986. *Digital Filters and Signal Processing*. Kluwer Academic, Norwell, MA. [7§5]
Jacobi, C. G. J. 1846. Über die Darstellung einer reihe gegebner Werthe durch eine gebrochne rationale Function. *J. für Reine und Angewandte Math.* **30**, 127–56. [Preface, 1§1, 7§1]
Jacobsen, L. (ed.). 1989. *Analytic Theory of Continued Fractions III*. Springer Lecture Notes in Mathematics no. 1406. Springer, Berlin. [7§1]
Jarratt, P. 1970. A review of methods for solving nonlinear algebraic equations in one variable. In *Numerical Methods for Nonlinear Algebraic Equations*, P. Rabinowitz (ed.), pp. 1–26. Gordon and Breach, London. [7§1]
Jones, W. B. 1977. Multiple point Padé tables. In *Padé and Rational Approximation*, E. B. Saff and R. H. Varga (eds.), pp. 163–72. Academic Press, New York.
Jones, W. B.; Njåstad, O.; and Thron, W. J. 1986. Continued fractions associated with the trigonometric and other strong moment problems. *Constr. Approx.* **2**, 197–211. [4§5, 7§1]
Jones, W. B., and Thron, W. J. 1966. Further properties of T-fractions. *Math. Annalen.* **166**, 106–18.
Jones, W. B., and Thron, W. J. 1971. A posteriori bounds for the truncation error of continued fractions. *SIAM J. Num. Anal.* **8**, 693–705.
Jones, W. B., and Thron, W. J. 1974a. Numerical stability in evaluating continued fractions. *Math. Comp.* **28**, 795–810. [4§1]
Jones, W. B., and Thron, W. J. 1974b. Proceedings of the international conference on Padé approximants, continued fractions, and related topics. *Rocky Mtn. J. Math.* **4**, 135–397
Jones, W. B., and Thron, W. J. 1975. On the convergence of Padé approximants. *SIAM J. Math. Anal.* **6**, 9–16. [6§4]
Jones, W. B., and Thron, W. J. 1977. Two point Padé tables and T-fractions. *Bull. Am. Math. Soc.* **83**, 388–90.
Jones, W. B., and Thron, W. J. 1979. Sequences of meromorphic functions corresponding to formal Laurent series. *SIAM J. Math. Anal.* **10**, 1–17. [6§4, 7§1]
Jones, W. B., and Thron, W. J. 1980. *Continued fractions: Analytic theory and applications*. Vol. 11 of *Encyclopedia of Mathematics and Its Applications*, G.-C. Rota (ed.). Addison-Wesley, Reading, MA. [4§1, 4§5]
Jones, W. B., and Thron, W. J. 1988. Continued fractions in numerical analysis. *Appld. Num. Math.* **4**, 143–230. [4§5, 7§1]
Jones, W. B.; Thron, W. J.; and Waadeland, H. 1980. A strong Stieltjes moment problem. *Trans. Am. Math. Soc.* **261**, 503–28. [4§5, 7§1]
Jones, W. B.; Thron, W. J.; and Waadeland, H. (eds.). 1982. *Analytic Theory of Continued Fractions*. Springer Lecture Notes in Mathematics no. 932. Springer, Berlin. [7§1]

Joyce, D. C. 1971. Survey of extrapolation processes in numerical analysis. *SIAM Rev.* **13**, 435–90.

Joyce, G. S., and Guttman, A. J. 1973. A new method of series analysis. In *Padé Approximants and Their Applications*, P. R. Graves-Morris (ed.), pp. 163–7. Academic Press, London.

Kahaner, D. K. 1972. Numerical quadrature by the ε-algorithm. *Math. Comp.* **26**, 689–94.

Kailath, T. 1980. *Linear Systems*. Prentice-Hall, Englewood Cliffs, N.J. [4§4, 8§3]

Kailath, T.; Vieira, A.; and Morf, M. 1978. Inverses of Toeplitz operators, innovations and orthogonal polynomials. *SIAM Review* **20**, 106–19. [7§5]

Kalyagin, V. A. 1979. On a class of polynomials defined by two orthogonality relations. *Mat. Sbornik* **110**(152), 609–27; *Math. USSR Sbornik* **38**, 563–80 (1981). [8§1]

Kamke, E. 1951. Differentialgleichungen Lösungsmethoden und Lösungen. Vol. 1. Akademic Verlagsgesellschaft, Leipzig. [8§6.6]

Karlin, S. 1968. *Total Positivity I*. Stanford Univ. Press, Stanford, CA. [5§7]

Karlsson, J. 1976. Rational interpolation and best rational interpolation. *J. Math. Anal. Appl.* **53**, 38–52. [6§6, 7§1]

Karlsson, J., and von Sydow, B. 1976. The convergence of Padé approximants to series of Stieltjes. *Arkiv for Matematik* **14**, 43–53. [5§3]

Karlsson, J., and Wallin, H. 1977. Rational approximation by an interpolation procedure in several variables. In *Padé and Rational Approximation*, E. B. Saff and R. H. Varga (eds.), pp. 83–100. Academic Press, New York. [7§6]

Khalil, A. B. 1977. The K-matrix in distorted wave theory. *Nuovo Cim.* **36A**, 354–66. [9§7]

Khovanskii, A. N. 1963. *Application of Continued Fractions and Their Generalizations to Problems in Approximation Theory*. Noordhoff, Leiden. [10§9]

Killingbeck, J. 1977. Quantum mechanical perturbation theory. *Rep. Prog. Phys.* **40**, 963–1031. [9§5]

Killingbeck, J. 1980. The harmonic oscillator with λx^m perturbation. *J. Phys. A* **13**, 49–56. [11§1]

Koenig, J. 1884. Über eine Eigenschaft der Potenzreihen. *Math. Ann.* **23**, 447–9.

Kogbetlianz, E. G. 1960. Generation of elementary functions. In *Mathematical Methods for Digital Computers*, A. Ralston and H. S. Wilf (eds.), pp. 7–35. Wiley, New York.

Kraichnan, R. H. 1970. Turbulent diffusion: evaluation of primitive and renormalised perturbation series by Padé approximants and by expansion of Stieltjes transforms into contributions from continuous orthogonal functions. In *The Padé Approximant in Theoretical Physics*, G. A. Baker, Jr., and J. L. Gammel (eds.), pp. 129–70. Academic Press, New York.

Kronecker, L. 1881. Zur Theorie der Elimination einer Variabeln aus zwei algebraischen Gleichungen. *Monatsberichte der Königlisch Preussichen Akademie der Wissenschaften zu Berlin* 535–600. [2§1, 7§1]

Kuchminskaya, K. I. 1978. Corresponding and associated branching continued fractions for the double power series. *Doklady Academy Nauk Ukraine A* **7**, 613–17. [7§6]

Kuchminskaya, K. I. 1989. On the convergence of two-dimensional continued fractions. *Trudi Matem. Inst. Steklov* **180**; *Proc. Steklov Inst. Math.* (1989) 171–2. [7§6]

Kuchminskaya, K. I., and Siemasko, W. 1987. Rational approximation and interpolation of functions by branched continued fractions. In *Rational Approximation and Its Applications in Mathematics and Physics*, J. Gilewicz, M. Pindor, and W. Siemaszko (eds.), Springer Lecture Notes in Mathematics no. 1237, pp. 24–40. Springer, Berlin. [7§6]

Kung, S. 1977. Multivariable and multidimensional systems: analysis and design. Ph.D. Thesis, Stanford University, California. [8§3]

Labahn, G. 1992. Inversion components of block Hankel-like matrices. *Lin. Alg. Applcns.* **177**, 7–48. [8§1]

Labahn, G., and Cabay, S. 1989. Matrix Padé fractions. *SIAM J. Computing* **18**, 639–57. [8§2]

Labahn, G.; Choi, D. K.; and Cabay, S. 1990. The inverses of block Hankel and block Toeplitz matrices. *SIAM J. Comput.* **19**, 98–123. [3§6, 8§2]

Laguerre, E. 1879. Sur la réduction en fractions continues d'une fonction qui satisfait à une équation linéaire du premier ordre à coefficients rationnels. *Bull. Soc. Math. de France* **8**, 21–7. [10§7]

Laguerre, E. 1885. Sur la réduction en fractions continues d'une fonction qui satisfait à une équation différentielle linéaire du premier ordre dont les coefficients sont rationnels. *J. de Math, Pures et Appliqués* **1**, 135–65. [10§7]

Lambert, F. 1980. Padé approximants and closed form solutions of the KdV and MKdV equations. *Zeit. für Physik (Particles and Fields)* **5**, 147–50.

Lambert, F., and Musette, M. 1984. Solitary waves, Padéons, and solitons. In *Padé Approximations and Its Applications, Bad Honnef, 1983*, H. Werner and H. J. Bünger (eds.), Springer Lecture Notes in Mathematics no. 1071, pp. 197–212. Springer, Berlin.

Lambert, J. D. 1974. Two unconventional classes of methods for stiff systems. In *Stiff Differential Equations*, R. A. Willoughby (ed.), pp. 171–86. Plenum, New York. [10§4]

Lambert, J. D. 1975. *Computational Methods in Ordinary Differential Equations*. Wiley, New York. [10§4]

Lambert, J. D., and Shaw, B. 1965. On the numerical solution of $y' = f(x, y)$ by a class of formulae based on rational approximation. *Math. Comp.* **19**, 456–62. [7§1]

Lambert, J. D., and Shaw, B. 1966. A generalization of multistep methods for ordinary differential equations. *Num. Math.* **8**, 250–63. [7§1]

Lanczos, C. 1952. Solution of systems of linear equations by minimized iterations. *J. Res. Nat. Bur. Stand.* **49**, 33–53. [9§3]

Lanczos, C. 1957. *Applied Analysis*. Pitman Press, London. [10§3]

Landkof, N. S. 1972. *Foundations of Modern Potential Theory*. Springer, Berlin. [6§6, 8§1, 8§6.6]

Langhoff, P. W., and Karplus, M. 1970. Application of Padé approximants to dispersion force and optical polarizability computations. In *The Padé Approximant in Theoretical Physics*, G. A. Baker, Jr., and J. L. Gammel (eds.), pp. 41–97. Academic Press, New York.

Larkin, F. M. 1967. Some techniques for rational interpolation. *Comp. J.* **10**, 178–87. [7§1]

Larkin, F. M. 1981. Root finding by divided differences. *Num. Math.* **37**, 93–104. [7§1]

Lee, T. D., and Yang, C. N. 1952. Statistical theory of equations of state and phase transitions II. Lattice gas and Ising model. *Phys. Rev.* **87**, 410–19. [11§3]

Leighton, W., and Scott, W. T. 1939. A general continued fraction expansion. *Bull. Am. Math. Soc.* **45**, 596–605. [4§2]

Levin, A. L., and Lubinsky, D. S. 1986. Best rational approximation of entire functions whose Maclaurin series decrease rapidly and smoothly. *Trans. Am. Math. Soc.* **293**, 533–46. [10§1]

Levin, A. L., and Lubinsky, D. S. 1990. Rows and diagonals of the Walsh array for entire functions with smooth Maclaurin series coefficients. *Constr. Approx.* **6**, 257–86. [10§1]

Levin, A. L., and Saff, E. B. 1994. Optimal ray sequences of rational functions connected with the Zolotarev problem. *Constr. Approx.* **10**, 235–74. [10§2]

Levin, D. 1973. Development of nonlinear transformations for improving convergence of sequences. *Int. J. Comp. Math.* **B3**, 371–88. [10§1]

Levin, D. 1976. General order Padé type rational approximants defined from double power series. *J. Inst. Math. Appl.* **18**, 1–8. [7§6]

Levin, D., and Sidi, A. 1981. Two new classes of nonlinear transformations for accelerating the convergence of infinite integrals and series. *Appld. Math. and Computation* **9**, 175–215. [10§1]

Levrie, P., and Bultheel, A. 1993. A note on Thiele n-fractions. *Num. Algs.* **4**, 225–39. [7§1]

Liu, A., and Fisher, M. E. 1989. The three dimensional Ising model revisited numerically. *Physica A* **156**, 35–76. [7§6]

Loeffel, J. J.; Martin, A.; Simon, B.; and Wightman, A. S. 1969. Padé approximants and the anharmonic oscillator. *Phys. Lett.* **30B**, 656–8. [11§1]

Longman, I. M. 1966. The application of rational approximants to the solution of problems in theoretical seismology. *Bull. Seismol. Soc. America* **56**, 1045–65.

Longman, I. M. 1972. Numerical Laplace transform inversion of a function arising in viscoelasticity. *J. Comp. Phys.* **10**, 224–31.

Longman, I. M. 1973. Use of Padé table for approximate Laplace transform inversion. In *Padé Approximants and Their Applications*, P. R. Graves-Morris (ed.), pp. 131–4. Academic Press, London.

Longman, I. M. 1974. Best rational function approximation for Laplace transform inversion. *SIAM J. Math. Anal* **5**, 574–80. [10§5]

Lopez, C., and Yndurain, F. J. 1973. Model independent calculation of Kp dispersion relations; extrapolation with Padé techniques. *Nucl. Phys. B* **64**, 315–33.

Lorentzen, L., and Waadeland, H. 1992. *Continued Fractions with Applications*. North-Holland, Amsterdam. [4§5, 7§1]

Lovelace, C. 1964. Practical theory of three-particle states I. Nonrelativistic. *Phys. Rev.* **135B**, 1225–9. [9§4]

Lovelace, C., and Masson, D. 1962. Calculation of Regge poles by continued fractions. *Nuovo Cim.* **26**, 472–84.

Lubinsky, D. S. 1980a. Exceptional sets of Padé approximants. Ph.D. thesis, Univ. of Witwatersrand. [6§6]

Lubinsky, D. S. 1980b. Diagonal Padé approximants and capacity. *J. Math. Anal. Applcns.* **78**, 58–67. [6§6]

Lubinsky, D. S. 1985a. Padé tables of entire functions of very slow and smooth growth. *Constr. Approx.* **1**, 349–58. [6§7, 10§1]

Lubinsky, D. S. 1985b. Padé tables of a class of entire functions. *Proc. Am. Math. Soc.* **94**, 399–406. [6§7]

Lubinsky, D. S. 1987. Uniform convergence of rows of the Padé table for functions with smooth Maclaurin series coefficients. *Constr. Approx.* **3**, 307–30. [10§1]

Lubinsky, D. S. 1988. Padé tables of entire functions of very slow and smooth growth. II. *Constr. Approx.* **4**, 321–39. [10§1]

Lubinsky, D. S. 1991. Distribution of poles of diagonal rational approximants to functions of fast rational approximability. *Constr. Approx.* **7**, 501–19. [6§7]

Lubinsky, D. S. 1992. Spurious poles in diagonal rational approximation. In *Progress in Approximation Theory*, A. A. Gončar and E. B. Saff (eds.), pp. 191–213. Springer, Berlin. [6§7]

Lubinsky, D. S., and Saff, E. B. 1987. Convergence of Padé approximants of partial theta functions and Rogers–Szegö polynomials. *Constr. Approx.* **3**, 331–61. [4§3, 6§1]

Lubkin, S. 1952. A method of summing infinite series. *J. Res. Nat. Bur. Stand.* **48**, 228–54. [3§1]

Luke, Y. L. 1958. The Padé table and the τ-method. *J. Math. and Phys.* **37**, 110–27. [10§3]

Luke, Y. L. 1960. On the economic representation of transcendental functions. *J. Math. and Phys.* **38**, 279–94. [10§3]

Luke, Y. L. 1962. On the approximate inversion of some Laplace transforms. In *Proc. 4th U.S. Nat. Congr. Appl. Mech.*, R. M. Rosenberg (ed.), pp. 269–76. American Society of Mechanical Engineers, New York. [10§5]

Luke, Y. L. 1964. *The Special Functions and Their Approximations*, Vol. 1. Academic Press, New York. [10§3]

Luke, Y. L. 1969. *The Special Functions and Their Approximations*, Vol. 2. Academic Press, New York. [10§3, 10§7]

Luke, Y. L. 1970. Evaluation of the gamma function by means of Padé approximations. *SIAM J. Num. Anal.* **1**, 266–81.

Luke, Y. L. 1975. On the error in the Padé approximants for a form of incomplete gamma function including the exponential. *SIAM J. Math. Anal* **6**, 829–39. [4§6]

Luke, Y. L. 1977. On the error in Padé approximations for functions defined by Stieltjes integrals. *Comp. and Maths. with Appls.* **3**, 307–14.

Luke, Y. L. 1978. Error estimation in numerical inversion of Laplace transform using Padé approximation. *J. Franklin Inst.* **305**, 259–73. [10§5]

Luke, Y. L. 1980. Computations of coefficients in the polynomials of Padé approximations by solving systems of linear equations. *J. Comp. and Appl. Math.* **6**, 213–18.

Luke, Y. L. 1982. A note on the evaluation of coefficients in the polynomials of Padé approximants by solving systems of linear equations. *J. Comp. Appld. Math.* **8**, 93–9. [2§1]

Luke, Y. L.; Fair, W.; and Wimp, J. 1975. Predictor-corrector formulas based on rational interpolants. *Int. J. Computers and Math. with Appl.* **1**, 3–12. [7§1]

Lutterodt, C. L. 1974. A two-dimensional analogue of Padé approximant theory. *J. Phys. A* **7**, 1027–37. [7§6]

Lyness, J., and Ninham, B. W. 1967. Numerical quadrature and asymptotic expansions. *Math. Comp.* **21**, 162–78.

McCabe, J. H. 1974. A continued fraction expansion, with a truncation estimate, for Dawson's integral. *Math. Comp.* **28**, 811–16. [4§5, 7§1]

McCabe, J. H. 1975. A formal extension of the Padé table to include two point Padé quotients. *J. Inst. Math. Appl.* **15**, 363–72. [4§5, 7§1]

McCabe, J. H., and Murphy, J. A. 1976. Continued fractions which correspond to power series at two points. *J. Inst. Math. Appl.* **17**, 233–47. [4§5, 7§1]

McCoy, B. M., and Wu, T. T. 1973. *The Two Dimensional Ising Model*. Harvard Univ. Press, Cambridge, MA. [11§3]

McEliece, R. J., and Shearer, J. B. 1978. A property of Euclid's algorithm and its application to Padé approximation. *SIAM J. Appl. Math.* **34**, 611–15. [3§7]

McInnes, A. W. 1992. Existence and uniqueness of algebraic function approximants. *Constr. Approx.* **8**, 1–21. [8§6.2]

McLeod, J. B. 1971. A note on the ε-algorithm. *Computing* **7**, 17–24. [8§4]

Maehly, H. J. 1956. October monthly progress report. Institute for Advanced Study, Princeton, NJ. [7§4]

Maehly, H. J. 1960. Rational approximants for transcendental functions. In *Proceedings of the International Conference on Information Processing*, pp. 57–62. Butterworth, London.

Maehly, H. J., and Witzgall, C. 1960. Tschebyscheff—Approximationen in kleinen Intervallen II. Stetigkeitsätze für gebrochen rationale Approximationen. *Num. Math.* **2**, 293–307. [7§1]

Magnen, J., and Sénéor, R. 1976. The infinite volume limit of the ϕ_3^4 model. *Ann. d'Inst. H. Poincaré A* **24**, 95–159. [11§2]

Magnus, A. 1962a. Certain continued fractions associated with the Padé table. *Math. Zeit.* **78**, 361–74.

Magnus, A. 1962b. Expansion of power series into P-fractions. *Math. Zeit.* **80**, 209–16. [4§2]

Magnus, A. P. 1984. Riccati acceleration of Jacobi continued fractions and Laguerre–Hahn orthogonal polynomials. In *Padé Approximation and Its Applications*, H. Werner and H. J. Bünger (eds.), pp. 213–30. Springer Lecture Notes in Mathematics no. 1071, Springer, Berlin. [10§7]

Magnus, A. P. 1988. On the use of the Carathéodory–Fejér method for investigating "1/9" and similar constants. In *Nonlinear Numerical Methods and Rational Approximation*, A. A. M. Cuyt (ed.), pp. 105–32. Reidel, Dordrecht. [10§6.1]

Mahler, K. 1968. Perfect systems. *Compositio Math.* **19**, 95–166. [8§1]

Mall, J. 1934. Grundlagen für eine Theorie der mehrdimensionalen Padéschen Tafel. Thesis, Munich University. [8§1]

Markov, A. 1884. Démonstration de certaines inégalités de Tchebychef. *Math. Ann.* **24**, 172–80. [5§1, 5§4, 10§2]

Marx, I. 1963. Remark concerning a nonlinear sequence to sequence transformation. *J. Math. and Phys.* **42**, 334–35. [3§1]

Mason, J. C. 1967. Chebychev polynomial approximations for the L-membrane eigenvalue problem. *SIAM J. Appl. Math.* **15**, 172–81. [10§3]

Massey, J. L. 1969. Shift-register synthesis and BCH decoding. *IEEE Trans. Info. Theory IT* **15**, 122–7. [4§4]

Masson, D. 1967a. Padé approximant and the partial-wave integral equation. *J. Math. Phys.* **8**, 512–14. [1§5, 9§6]

Masson, D. 1967b. Analyticity in the potential strength. *J. Math. Phys.* **8**, 2308–14. [1§5, 9§6]

Meinguet, J. 1970. On the solubility of the Cauchy interpolation problem. In *Approximation Theory*, A. Talbot (ed.), pp. 137–63. Academic Press, London. [7§1]

Meleshko, R., and Cabay, S. 1991. On computing Padé approximants quickly and accurately. *Congressus Numerantium* **80**, 245–55. [3§6]

Mery, P. 1977. A variational approach to operator and matrix Padé approximation. Applications to potential scattering and field theory. In *Padé and Rational Approximation*, E. B. Saff and R. S. Varga (eds.), pp. 375–88. Academic Press, New York. [9§8]

Merz, G. 1968. Padésche Näherungsbrüche und Iterationsverfahren höherer Ordnung. *Computing* 3, 165–83. [7§1]

Mesina, M. 1977. Convergence acceleration for the iterative solution of the equations $X = AX + f$. *Comput. Methods Appl. Mech. Engrg.* 10, 165–73.

Michalik, B. 1970. Distortion operator method and Padé approximants. *Ann. Phys.* 57, 201–13. [9§7]

Mills, W. H. 1975. Continued fraction and linear recurrences. *Math. Comp.* 29, 173–80. [3§3]

Milne-Thomson, L. M. 1960. *Calculus of Finite Differences*. Macmillan, London.

Mitchell, A. R., and Griffiths, D. F. 1980. *The Finite Difference Method in Partial Differential Equations*. Wiley, Chichester. [10§4]

Moler, C., and Van Loan, C. 1978. Nineteen dubious ways to compute the exponential of a matrix. *SIAM Review* 20, 801–36.

de Montessus de Ballore, R. 1902. Sur les fractions continues algébriques. *Bull. Soc. Math. de France* 30, 28–36. [6§2, 8§1]

de Montessus de Ballore, R. 1905. Sur les fractions continues algébriques. *Rend. di Palermo* 19, 1–73.

Moore, M. A.; Jasnow, D.; Wortis, M. 1969. Spin–spin correlation function of the three-dimensional Ising ferromagnet above the Curie temperature. *Phys. Rev. Lett.* 22, 940–3.

Muir, T. 1960. *A Treatise on the Theory of Determinants*. Revised and enlarged by W. H. Metzler. Dover, New York.

Murphy, J. A. 1971. Certain rational fraction approximations to $(1 + x^2)^{-1/2}$. *J. Inst. Math. Appl.* 7, 138–50. [7§1]

Murphy, J. A., and O'Donohoe, M. R. 1977. A class of algorithms for rational approximation of functions formally defined by power series. *J. Appl. Math. and Phys. (ZAMP)* 28, 1121–31.

Murphy, J. A., and O'Donohoe, M. R. 1978. A two variable generalization of the Stieltjes type continued fraction. *J. Comp. Appld. Math.* 4, 181–90. [7§6]

Muskhelishvili, N. I. 1977. *Singular Integral Equations*. Noordhoff, Leiden. [6§6]

Nagle, J. F., and Bonner, J. C. 1970. Numerical study of Ising chain with long range ferromagnetic interactions. *J. Phys. C.* 3, 352–65.

Narcowich, F. J., and Allen, G. D. 1975. Convergence of diagonal operator-valued Padé approximants to the Dyson expansion. *Comm. Math. Phys.* 45, 153–7. [5§6, 8§2]

Nelson, E. 1973. Construction of quantum fields from Markoff fields. *J. Funct. Anal.* 12, 97–112. [11§3]

Newton, R. G. 1966. *Scattering Theory of Waves and Particles*. McGraw-Hill, New York. [9§4]

Nikisin, E. M. 1980. On simultaneous Padé approximants. *Mat. Sbornik* 113(155); *Math. USSR Sbornik* 41, 409–25 (1982).

Ninham, B. W., and Lyness, J. 1969. Asymptotic expansions for the error functional. *Math. Comp.* 23, 71–83.

Nørsett, S. P. 1974. One step methods of Hermite type for numerical integration of stiff systems. *BIT* 14, 63–77. [10§4]

Nuttall, J. 1966. Padé approximants and bounds on the Bethe–Salpeter amplitude. *Phys. Lett.* 23, 492. [9§5]

Nuttall, J. 1967. Convergence of Padé approximants for the Bethe–Salpeter amplitude. *Phys. Rev.* **157**, 1312–16. [9§5, 9§7]
Nuttall, J. 1970a. Connection of Padé approximants with stationary variational principles and the convergence of certain Padé approximants. In *The Padé Approximant in Theoretical Physics*, G. A. Baker, Jr., and J. L. Gammel (eds.), pp. 219–30. Academic Press, New York. [9§5]
Nuttall, J. 1970b. Convergence of Padé approximants of meromorphic functions. *J. Math. Anal. Appl.* **31**, 147–53. [6§5, 6§6, 9§2]
Nuttall, J. 1973. Variational principles and Padé approximants. In *Padé Approximants and Their Applications*, P. R. Graves-Morris (ed.), pp. 29–40. Academic Press, New York. [9§8]
Nuttall, J. 1977. The convergence of Padé approximants to functions with branch points. In *Padé and Rational Approximation*, E. B. Saff and R. S. Varga (eds.), pp. 101–9. Academic Press, New York. [6§6, 9§5]
Nuttall, J. 1980. Sets of minimum capacity, Padé approximants, and the bubble problem. In *Bifurcation Problems in Mathematical Physics and Related Topics*, C. Bardos and D. Bessis (eds.), pp. 185–201. Reidel, Boston.
Nuttall, J. 1990. Padé polynomials asymptotics from a singular integral equation. *Constr. Approx.* **6**, 157–66. [6§6]
Nuttall, J., and Singh, S. R. 1977. Orthogonal polynomials and Padé approximants associated with a system of arcs. *J. Approx. Theory* **21**, 1–42. [6§6]
Ostrowski, A. 1925. Über Folgen analytischer Funktionen und einige verschärfungen des Picarden Satzes. *Math. Zeit.* **24**, 215–58. [6§4]
Padé, H. 1892. Sur la répresentation approchée d'une fonction par des fractions rationelles. *Ann. de l'Ecole Normale Sup. 3iéme Série* **9**, Suppl., 1–93. [Preface, 1§1, 1§2, 8§5.1, 8§6.2]
Padé, H. 1894a. Sur les séries entières convergents ou divergents, et les fractions continues rationelles. *Acta Math.* **18**, 97–112. [8§5]
Padé, H. 1894b. Sur la généralisation des fractions continues algébriques. *J. de Math. Pures et Appl.* **10**, 291–329. [8§5]
Padé, H. 1899. Memoire sur les développements en fractions continues de la fonction exponentielle pouvant servir d'introduction à la théorie des fractions continues algébriques. *Ann. Sci. Ecole Norm. Sup.* **16**, 395–426.
Padé, H. 1900. Sur la distribution des réduites anormales d'une fonction. *Comptes Rendus* **130**, 102–4. [2§4, 8§5.1]
Padé, H. 1901. Sur l'expression générale de la fonction rationelle approchée de $(1+x)^m$. *Comptes Rendus* **132**, 754–6.
Padé, H. 1907. Recherches sur la convergence de développements en fractions continues d'une certaine catégorie de fonctions. *Ann. Sci. Ecole Norm. Sup.* **24**, 341–400.
Paszkowski, S. 1982. Quelques algorithmes de l'approximation de Padé–Hermite. Université de Sciences et Techniques de Lille, publication ANO 89, 1–67. [8§5.2, 8§5.3]
Paszkowski, S. 1987. Recurrence relations in Padé–Hermite approximation. *J. Comp. Appl. Math.* **19**, 99–107. [8§5.3]
Paszkowski, S. 1990. Hermite–Padé approximation (basic notions and theorems). *J. Comp. & Appl. Math.* **32**, 229–36. [8§5.1]
Paydon, J. F., and Wall, H. S. 1942. The continued fraction as a sequence of linear transformations. *Duke Math. J.* **9**, 360–72. [4§7]
Pekeris, C. L. 1958. Ground state of two electron atoms. *Phys. Rev.* **112**, 1649–58. [9§5]

Pekeris, C. L. 1959. Binding energy of helium. *Phys. Rev.* **115**, 1216–21. [9§5]

Peres, A. 1963. Mechanical model for quantum field theory. *J. Math. Phys.* **4**, 332–3. [11§1]

Perron, O. 1957. Die Lehre von der Kettenbrüchen. Vol. 2. Teubner, Stuttgart. [5§1, 5§5, 5§6, 6§1, 8§1, 10§3]

Pfeuty, P.; Jasnow, D.; and Fisher, M. E. 1974. Crossover scaling functions for exchange anisotropy. *Phys. Rev. B* **10**, 2088–112. [7§6]

Pindor, M. 1976. A simplified algorithm for calculating the Padé table derived from the Baker and Longman schemes. *J. Comp. Appl. Math.* **2**, 255–8.

Pindor, M. 1979a. Padé approximants and rational functions as tools for finding poles and zeros of analytical functions measured experimentally. In *Padé Approximation and Its Applications*, L. Wuytack (ed.), pp. 338–47. Springer Lecture Notes in Mathematics, no. 765, Springer, Berlin.

Pindor, M. 1979b. Matrix variational Padé approximants and multistep square well potentials. *Lett. Math. Phys.* **3**, 223–8. [8§2, 9§8]

Pindor, M. 1980. Unambiguous results from variational matrix Padé approximants. Centre of Theoretical Physics Report, Trieste.

Pindor, M. 1984. Operator continued fractions and bound states. *Nuovo Cim.* **B84**, 105–20. [8§2]

Pindor, M. 1987. Operator rational functions and variational methods for the model operator. In *Rational Approximation and its Applications in Mathematics and Physics*, J. Gilewicz, M. Pindor, and W. Siemaszko (eds.), pp. 305–14. Springer LNIM 1237, Springer, Berlin.

Pindor, M., and Turchetti, G. 1982. Padé approximants and variational methods for operator series. *Nuovo Cim.* **71A**, 171–86. [8§2]

Pommerenke, C. 1973. Padé approximants and convergence in capacity. *J. Math. Anal. Appl.* **41**, 775–80. [6§5, 9§2]

Powell, M. J. D. 1967. On the maximum errors of polynomial approximations defined by interpolation and by least squares criteria. *Comput. J.* **9**, 404–7. [7§4]

Pringsheim, A. 1910. Über Konvergenz und Funktionen-theoretischen Charakter gewisser limitärperiodischer Kettenbrüche. *Sitzungsberichte der Math.-Physikalische Klasse der Kgl. Bayerischen Akademie der Wissenschaften zu München* **6**, 1–52.

Reed, M., and Simon, B. 1979. *Scattering Theory.* Vol. 3 of *Methods of Modern Mathematical Physics.* Academic Press, New York. [9§4]

Rehr, J. J.; Joyce, G. S.; and Guttman, A. J. 1980. A recurrence technique for confluent singularity analysis of power series. *J. Phys. A* **13**, 1587–602. [8§6.7]

Reid, W. T. 1972. *Riccati Equations.* Academic Press, London. [10§9]

Richardson, L. F. 1927. The deferred approach to the limit, I. Single lattice. *Phil. Trans. Roy. Soc. Lond.* **A226**, 299–349.

Richtmyer, R. D., and Morton, K. W. 1967. *Difference Methods for Initial Value Problems.* Wiley, New York. [10§4]

Riemann, B. 1953. *Collected Works of Bernhard Riemann.* H. Weber (ed.), pp. 379–90. Dover, New York. [8§6.1]

Riesz, F., and Szökefalvi-Nagy, B. 1955. *Functional Analysis.* Ungar, New York. [5§1, 5§6, 9§2]

Riesz, M. 1992a. Sur le problème des moments. *Arkiv för Matematik, Astronomi och Fysik* **16**(12), 1–23. [7§2]

Riesz, M. 1922b. Sur le problème des moments. *Arkiv för Matematik, Astronomi och Fysik* **16**(19), 1–21. [7§2]

Riesz, M. 1923. Sur le problème des moments. *Arkiv för Matematik, Astronomi och Fysik* **17**(16), 1–52. [7§2]

Riordan, J. 1966. *Combinatorial Identities*. Wiley, New York. [6§2]

Rissanen, J. 1972. Recursive evaluation of Padé approximants for matrix sequences. *IBM J. Res. Develop.* 401–6.

Rissanen, J. 1973. Algorithms for triangular decomposition of block Hankel and Toeplitz matrices with applications to factoring positive matrix polynomials. *Math. Comp.* **27**, 147–55.

Rivlin, T. J. 1969. *An Introduction to the Theory of Functions*. Blaisdell, New York. [6§6]

Roberts, D. E. 1977. An analysis of double power series using rotationally covariant approximants. *J. Comp. Appl. Math.* **3**, 257–62. [7§6]

Roberts, D. E. 1990. Clifford algebras and vector-valued rational forms I. *Proc. Roy. Soc. A* **431**, 285–300. [8§4]

Roberts, D. E. 1992. Clifford algebras and vector-valued rational forms II. *Numerical Algorithms* **3**, 371–82. [8§4]

Roberts, D. E. 1995. On the algebraic foundations of vector epsilon algorithm. In *Clifford Algebras and Spinor Structures*, P. Lounesto and R. Ablamowicz (eds.), *Mathematics and Its Applications* **321**, 343–61. Kluwer, Dordrecht [6§3]

Roberts, D. E.; Griffiths, H. P.; and Wood, D. W. 1975. The analysis of double power series using Canterbury approximants. *J. Phys. A* **8**, 1365–72. [7§6]

Roberts, D. E. 1996. Vector Padé approximants. In *Clifford Algebras with CLICAL and other Computer Algebra Systems* R. Ablamowicz, P. Lounesto, and J. M. Parra (eds.), Kluwer, Dordrecht, in press. [8§4]

Robinson, P. D., and Barnsley, M. F. 1979. Pointwise bivariational bounds on solutions of Fredholm integral equations. *SIAM J. Num. Anal.* **16**, 135–44. [9§6]

Rogers, L. J. 1907. On the representation of certain asymptotic series as convergent continued fractions. *Proc. Lond. Math. Soc.* **4**, 72–89. [4§6]

Roman, S. M., and Rota, G.-C. 1978. The umbral calculus. *Adv. Math.* **27**, 95–188. [4§3]

Roth, A. 1938. Approximationseigenschaften und Strahlgrenzwerte unendlich vieler linearer Gleichungen. *Comm. Math. Helv.* **11**, 77–125. [10§6]

Rudin, W. 1976. *Principles of Mathematical Analysis*. 3rd ed. McGraw-Hill, New York. [5§1]

Ruijgrok, T. W. 1968. A new formulation of the theory of strong coupling. *Nucl. Phys. B* **8**, 591–608. [9§4]

Ruijgrok, T. W. 1972. A numerical study of analyticity in the coupling constant. *Nucl. Phys. B* **39**, 616–42. [11§1]

Runge, C. 1885. Zur Theorie der eindeutigen analytischen Funktionen. *Acta Math.* **6**, 229–44. [10§6]

Rutishauser, H. 1954. Der Quotienten-Differenzen-Algorithmus. *Zeit. für Angewandte Math. Physik* **5**, 233–51. [3§8, 4§3]

Rutishauser, H. 1957. *Der Quotienten-Differenzen-Algorithmus*. Birkhaüser, Basel. [4§3]

Saff, E. B. 1972. An extension of Montessus de Ballore's theorem on the convergence of interpolating rational functions. *J. Approx. Theory* **6**, 63–7. [6§3, 7§1]

Saff, E. B. 1983. Incomplete and orthogonal polynomials. In *Approximation Theory IV*, C. K. Chui, L. L. Schumaker, and J. D. Ward (eds.), pp. 219–56. Academic Press, New York. [5§4]

Saff, E. B.; Schönhage, A.; and Varga, R. S. 1978. Geometrical convergence to e^{-z} by rational functions with real poles. *Num. Math.* **25**, 307–22.

Saff, E. B., and Totik, V. 1989. Limitations of the Carathéodory–Fejér method for polynomial approximation. *J. Approx. Theory* **58**, 284–96. [10§6.1]

Saff, E. B., and Varga, R. S. 1975. On the zeros and poles of Padé approximants to e^z. *Num. Math.* **25**, 1–14. [5§7]

Saff, E. B., and Varga, R. S. 1976a. Zero-free parabolic regions for sequences of polynomials. *SIAM J. Math. Anal.* **7**, 344–57.

Saff, E. B., and Varga, R. S. 1976b. On the sharpness of theorems concerning zero free regions for certain sequences of polynomials. *Num. Math.* **26**, 345–54. [5§7]

Saff, E. B., and Varga, R. S. (eds.). 1977a. *Padé and Rational Interpolation*. Academic Press, New York.

Saff, E. B., and Varga, R. S. 1977b. On the zeros and poles of Padé approximants to e^z, II. In *Padé and Rational Approximation*, E. B. Saff and R. S. Varga (eds.), pp. 195–214. Academic Press, New York. [5§7]

Saff, E. B., and Varga, R. S. 1978a. On the zeros and poles of Padé approximants to e^z, III. *Num. Math.* **30**, 241–66. [5§7, 10§4]

Saff, E. B., and Varga, R. S. 1978b. Nonuniqueness of best rational approximations to real functions on real intervals. *J. Approx. Theory* **23**, 78–85. [10§6]

Saff, E. B.; Varga, R. S.; and Ni, W.-C. 1976. Geometric convergence of rational approximations to e^{-z} in infinite sectors. *Num. Math.* **26**, 211–25. [5§7, 10§4]

Salzer, H. E. 1962. Note on osculatory rational interpolation. *Math. Comp.* **16**, 486–91. [4§2]

Sarkar, B., and Bhattacharyya, K. 1988. On the Padé method for sequence acceleration and calculation of Madelung constants. *Chem. Phys. Lett.* **150**, 419–23. [10§1]

Sarkar, B., and Bhattacharyya, K. 1989. Forcing of convergence in pathological self-consistent field calculations: A Padé-(MC)SCF strategy. *Chem. Phys. Lett.* **162**, 61–6.

Sarkar, B., and Bhattacharyya, K. 1993. Accurate calculation of Coulomb sums: efficacy of Padé like methods. *Phys. Rev. B* **48**, 6913–19.

Scanlan, J. O. 1973. Circuit theory. In *Padé Approximants*, P. R. Graves-Morris (ed.), pp. 101–111. Institute of Physics, Bristol.

Schlessinger, L., and Schwartz, C. 1966. Analyticity as a computational tool. *Phys. Rev. Lett.* **16**, 1173–4. [9§4]

Schneider, C., and Werner, W. 1991. Hermite interpolation: The barycentric approach. *Computing* **46**, 35–51. [7§1]

Schoenberg, I. J. 1951. On the Pólya frequency functions I: The totally positive functions and their Laplace transforms. *J. d'Anal. Math.* **1**, 331–74. [5§7]

Schofield, D. F. 1972. Continued fraction method for perturbation theory. *Phys. Rev. Lett.* **29**, 811–14. [9§5]

Schrader, R. 1976. A possible constructive approach to ϕ_4^4. *Commun. Math. Phys.* **49**, 131–53; A possible constructive approach to ϕ_4^4 III. *Commun. Math. Phys.* **50**, 97–102. [11§3]

Schwartz, C. 1966. Information content of the Born series. *J. Comp. Phys.* **1**, 21–8. [9§4]

Schwartz, C., and Zemach, C. 1966. Theory and calculation with the Bethe–Salpeter equation. *Phys. Rev.* **141**, 1454–67. [9§7]

Schweber, S. S. 1961. *Introduction to Relativistic Quantum Field Theory*. Harper and Row, New York. [11§1]

Scott, R. E. 1977. Low order rational approximations to the delay operator. *IEEE Trans. Ind. Electron. and Control Instrum.* **IECI-24**, 61–4.

Scott, W. T., and Wall, H. S. 1940a. Continued fraction expansions for arbitrary power series. *Ann. Math.* **41**, 328–49. [4§2]

Scott, W. T., and Wall, H. S. 1940b. A convergence theorem for continued fractions. *Trans. Am. Math. Soc.* **47**, 155–72. [4§2, 4§7]

Shafer, R. E. 1974. On quadratic approximation. *SIAM J. Num. Anal.* **11**, 447–60. [7§3, 8§6.2]

Shamash, Y. 1975. Linear system reduction using Padé approximation to allow retention of dominant modes. *Int. J. of Control* **21**, 257–72.

Shanks, D. 1955. Nonlinear transformations of divergent and slowly convergent sequences. *J. Math. Phys.* **34**, 1–42. [1§3, 3§3, 3§4]

Sheludyak, Y. E., and Rabinovich, V. A. 1979. On the values of the critical indices of the three dimensional Ising model. *High Temp.* **17**, 40–3. [2§2]

Shohat, J. A., and Tamarkin, J. D. 1963. *The Problem of Moments*. American Math. Soc. Publ., vol. 1. AMS, Providence, RI. [10§2]

Short, L. 1978. The practical evaluation of multivariate approximants with branch points. *Proc. Roy. Soc. Lond. A* **362**, 57–69. [7§6]

Sidi, A. 1979. Convergence properties of some nonlinear sequence transformations. *Math. Comp.* **33**, 315–26.

Sidi, A. 1980a. Some aspects of two-point Padé approximants. *J. Comp. Appld. Math.* **6**, 9–17.

Sidi, A. 1980b. Analysis of the convergence of the T-transformation for power series. *Math. Comp.* **35**, 833–50.

Sidi, A. 1986. Convergence and stability properties of minimal polynomial and reduced rank algorithms. *SIAM J. Num. Anal.* **23**, 197–209. [8§4]

Sidi, A. 1990. Quantitative and constructive aspects of the generalized Koenig's and de Montessus's theorems for Padé approximants. *J. Comp. Appld. Math.* **29**, 257–91. [8§4]

Sidi, A. 1994. Rational approximations from power series of vector-valued meromorphic functions. *J. Approx. Theory* **77**, 89–111. [8§4]

Siemieniuch, J. L. 1976. Properties of certain rational approximations to e^{-z}. *BIT* **16**, 172–91.

Silver, D. M., and Wilson, S. 1982. Special invariance properties of $[N + 1/N]$ Padé approximants in Rayleigh–Schrödinger perturbation theory II. Molecular interaction energies. *Proc. Roy. Soc. Lond. A.* **383**, 477–83.

Simon, B. 1970. Coupling constant analyticity for the anharmonic oscillator. *Ann. Phys.* **58**, 76–136. [11§1]

Simon, B. 1972. The anharmonic oscillator — a singular perturbation theory. In *Cargèse Lecture Notes in Physics*, D. Bessis (ed.), vol. 5, pp. 383–414. Gordon and Breach, London.

Simon, B. 1982. Large orders and summability of eigenvalue perturbation theory: A mathematic overview. *Int. J. Quantum. Chem* **21**, 3–25. [7§2]

Skelboe, S. 1980. Computation of the periodic steady-state response of nonlinear networks by extrapolation methods. *IEEE Trans. Circuits and Systems* **27**, 161–75.

Skorobogatko, V. 1983. *Branched Continued Fractions*. Nauka, Moscow. [7§6]
Smith, D. A., and Ford, W. F. 1979. Acceleration of linear and logarithmic convergence. *SIAM J. Num. Anal.* **16**, 223–40.
Smith, D. A.; Ford, W. F.; and Sidi, A. 1987. Extrapolation methods for vector sequences. *SIAM Rev.* **29**, 199–233; Erratum **30**, 623–4.
Smith, I. M.; Siemieniuch, J. L.; and Gladwell, I. 1977. Evaluating Nørsett methods for integrating differential equations in time. *Int. J. Num. Anal. Meth. in Geomechanics* **1**, 57–74. [10§4]
Smithies, F. 1958. *Integral Equations*. Cambridge University Press, Cambridge. [9§2]
Sobhy, M. I. 1973. Applications of Padé approximants in electrical network problems. In *Padé Approximants and Their Applications*, P. R. Graves-Morris (ed.), pp. 321–36. Academic Press, London.
Springer, G. 1957. *Introduction to Riemann Surfaces*. Addison Wesley, Reading, MA. [8§6]
Stahl, H. 1976. Beiträge zum Problem der Konvergenz von Padé Approximierenden. Thesis, Tech. Univ., Berlin. [5§4]
Stahl, H. 1980. On the convergence of Padé approximants to Hamburger functions. *Constr. Function Theory, 1977, Sofia* 497–500. [5§4]
Stahl, H. 1987. Three different approaches to a proof of convergence for Padé approximants. In *Rational Approximation and Its Applications in Mathematics and Physics*, J. Gilewicz, M. Pindor, and W. Siemaszko (eds.), pp. 79–124. Lecture Notes in Mathematics no. 1237. Springer, Berlin. [6§6]
Stahl, H. 1988. Asymptotics of Hermite–Padé polynomials and related convergence results. In *Nonlinear Numerical Methods and Rational Approximation*, A. A. M. Cuyt (ed.), pp. 23–54. Reidel, Dordrecht. [8§6.6]
Stahl, H. 1989. On the convergence of generalised Padé approximants. *Constr. Approx.* **5**, 221–40. [8§1]
Starkand, Y. 1976. Subroutine for calculation of matrix Padé approximants. *Comm. Comp. Phys.* **11**, 325–30.
Stern, M. S., and Warburton, A. E. A. 1972. Finite difference solution of the partial wave Schrödinger equation. *J. Phys. A* **5**, 112–24. [9§4]
Stewart, G. W. 1984. Rank degeneracy. *SIAM J. Stat. Comput.* **5**(2), 403–13. [2§1]
Stewart, G. W. 1993. Determining rank in the presence of error. NATO Workshop on Large Scale Linear Algebra, Leuven. [7§5]
Stieltjes, T. J. 1884. Quelques recherches sur les quadratures dites méchaniques. *Ann. Sci. Ecole. Norm. Sup.* **1**, 409–26. [10§2]
Stieltjes, T. J. 1889. Sur la réduction en fraction continue d'une série précédent suivant les puissances descendents d'une variable. *Ann. Fac. Sci. Toulouse* **3H**, 1–17. [5§1]
Stieltjes, T. J. 1894. Recherches sur les fractions continues. *Ann. Fac. Sci. Toulouse* **8J**, 1–122; **9A**, 1–47. [5§1]
Stoer, J. 1961. Über zwei Algorithmen zur Interpolation mit rationalen Funktionen. *Num. Math.* **3**, 285–304. [7§1]
Stoer, J., and Bulirsch, R. 1979. *Introduction to Numerical Analysis*. Springer, New York.
Stone, M. H. 1932. *Linear Transformations in Hilbert Space*. Am. Math. Soc., Providence, RI. [5§6]
Styer, D. F. 1983. Partial differential approximants for multivariable power series

III. Enumeration of invariance transformations. *Proc. Roy. Soc. Lond. A* **390**, 321–39. [7§6]

Suetin, S. P. 1983. On the poles of the mth row of a Padé table. *Mat. Sbornik* **120** (162); *Math. USSR Sbornik* **48**, 493–7 (1984). [6§1]

Suetin, S. P. 1984. On the inverse problem for the mth row of the Padé table. *Mat. Sbornik* **124** (166); *Math. USSR Sbornik* **52**, 231–44 (1985). [6§1]

Szegö, G. 1967. *Orthogonal Polynomials*. American Math. Soc., Providence, RI. [5§4, 7§5]

Takehasi, H., and Mori, M. 1971. Estimation of errors in the numerical quadrature of analytic functions. *Applicable Analysis* **1**, 201–29.

Talbot, A. 1979. The accurate numerical inversion of Laplace transforms. *J. Inst. Math. Appl.* **23**, 97–120. [10§5]

Tani, S. 1965. Padé approximants in potential scattering. *Phys. Rev.* **139**, B1011–20.

Tani, S. 1966a. Complete continuity of the kernel in generalized potential scattering I. Short range interaction without strong singularity. *Ann. Phys.* **37**, 411–50. [9§7]

Tani, S. 1966b. Complete continuity of the kernel in generalized potential scattering II. Generalized Fourier series expansion. *Ann. Phys.* **37**, 451–86. [9§7]

Taylor, J. M. 1978. The condition of Gram matrices and related problems. *Proc. Roy. Soc. Edin.* **80A**, 45–56. [2§1]

Tchebycheff, P. 1858. Sur les fractions continues. *J. de Math.* **8**, 289–323. [5§1]

Tchebycheff, P. 1874. Sur les valeur limites des intégrales. *J. de Math. Pures et Appl.* **19**, 157–60. [10§2]

Thacher, H. C. 1974. Numerical application of the generalized Euler transformation. In *Information Processing 74*, J. L. Rosenfeld (ed.), pp. 627–31. North-Holland, Amsterdam. [1§5]

Thacher, H. C., and Tukey, J. 1960. Rational interpolation made easy by a recursive algorithm. Unpublished manuscript, Argonne National Laboratory. [7§1]

Thiele, T. N. 1909. Interpolationsrechnung. Teubner, Stuttgart. [7§1]

Thompson, C. J.; Guttmann, A. J.; and Ninham, B. W. 1969. Determination of critical behaviour in lattice statistics from series expansions, II. *J. Phys. C* **2**, 1889–99.

Thron, W. J. 1948. Some properties of continued fractions $1 + d_0 z + K(z/(1 + d_n z))$. *Bull. Am. Math. Soc.* **54**, 206–18. [4§5, 7§1]

Thron, W. J. 1961. Convergence regions for continued fractions and other infinite processes. *Am. Math. Monthly* **68**, 734–50.

Thron, W. J. 1974. A survey of recent convergence results for continued fractions. *Rocky Mtn. J. Math.* **4**, 273–82. [4§7]

Thron, W. J. 1977. Two point Padé tables, T-fractions, and sequences of Schur. In *Padé and Rational Approximation*, E. B. Saff and R. S. Varga (eds.), pp. 215–25. Academic Press, New York.

Thron, W. J. (ed.). 1986. *Analytic Theory of Continued Fractions II*. Springer Lecture Notes in Mathematics no. 1199. Springer, Berlin. [7§1]

Titchmarsh, E. C. 1939. *Theory of Functions*. Oxford University Press, Oxford. [2§2, 4§7, 5§1, 5§2, 5§5, 5§7, 7§4]

Tjon, J. A. 1970. The Padé approximation in three body calculations. *Phys. Rev. D.* **1**, 2109–12. [9§4]

Tjon, J. A. 1973. Application of Padé approximants in the three body problem. In *Padé Approximants*, P. R. Graves-Morris (ed.), pp. 241-52. Inst. of Phys., Bristol. [9§4]

Tjon, J. A. 1977. Operator Padé approximants and three body scattering. In *Padé and Rational Approximation*, E. B. Saff and R. S. Varga (eds.), pp. 389-96. Academic Press, New York. [9§4]

Traub, J. F. 1964. *Iterative Methods for the Solution of Equations*. Prentice-Hall, Englewood Cliffs, NJ. [3§1]

Trefethen, L. N. 1981. Rational Chebyshev approximation on the unit disk. *Num. Math.* **37**, 297-320. [10§6.1]

Trefethen, L. N. 1983. Chebyshev approximation on the unit disk. In *Computational Aspects of Complex Analysis*, H. Werner et al. (eds.), pp. 309-23. Reidel, Dordrecht. [10§6.1]

Trefethen, L. N. 1984. Square blocks and equioscillation in the Padé, Walsh, and CF tables. In *Rational Approximation and Interpolation*, P. R. Graves-Morris, E. B. Saff, and R. S. Varga (eds.), Springer Lecture Notes in Mathematics no. 1105. Springer, Berlin. [10§6]

Trefethen, L. N., and Gutknecht, M. H. 1985. On convergence and degeneracy in rational Padé and Chebyshev approximation. *SIAM J. Math. Anal.* **16**, 198-210. [7§4, 10§6]

Trefethen, L. N., and Gutknecht, M. H. 1987. Padé, stable Padé, and Chebyshev-Padé approximation. In *Algorithms for Approximation*, J. C. Mason and M. G. Cox (eds.), pp. 227-64. Oxford University Press, Oxford. [7§4]

Trench, W. 1964. An algorithm for the inversion of finite Toeplitz matrices. *SIAM J. Appl. Math.* **12**, 515-22.

Trench, W. 1965. An algorithm for the inversion of finite Hankel matrices. *SIAM J. Appl. Math.* **13**, 1102-7.

Tricomi, F. G. 1957. *Integral Equations*. Interscience, New York. [9§2]

Trudi, N. 1862. *Teoria de Determinanti e loro Applicazioni*. Libreria Scientifica e Industriale de B. Pellerano, Napoli. [2§4]

Turchetti, G. 1976. Variational principles and matrix approximants in potential theory. *Lett. Nuovo Cim.* **15**, 129-33. [9§8]

Turchetti, G. 1978. Variational matrix Padé approximants in two body scattering. *Fortsch. Physik* **26**, 1-28.

Turchetti, G. 1980. Padé approximants and soliton solutions of the KdV equation. *Lett. Nuovo Cim.* **27**, 107-10.

Van Barel, M. 1989. Nested minimal partial realisations and related matrix rational approximants. Thesis, Katholieke Universiteit Leuven, Belgium. [8§3]

Van Barel, M., and Bultheel, A. 1987a. A minimal partial realization algorithm for MIMO systems. Report TW79, Department of Computer Science, Katholieke Universiteit Leuven, Belgium. [8§3]

Van Barel, M., and Bultheel, A. 1987b. A minimal partial realization algorithm for MIMO systems II. A matrix continued fraction. Report TW91, Department of Computer Science, Katholieke Universiteit Leuven, Belgium. [8§3]

Van Barel, M., and Bultheel, A. 1987c. A minimal partial realization algorithm for MIMO systems III. A parametrization of all solutions. Report TW93, Department of Computer Science, Katholieke Universiteit Leuven, Belgium. [8§3]

Van Barel, M., and Bultheel, A. 1989a. A canonical matrix continued fraction solution of the minimal (partial) realization problem. *Lin. Alg. Applcns.* **122-4**, 973-1002. [8§3]
Van Barel, M., and Bultheel, A. 1989b. A matrix Euclidean algorithm for minimal partial realization. *Lin. Alg. Applcns.* **121**, 674-82.
Van Barel, M., and Bultheel, A. 1991. The computation of non-perfect Padé Hermite approximants. *Num. Algs.* **1**, 285-304. [8§1]
Van Barel, M., and Bultheel, A. 1992a. A general module theoretic framework for vector M-Padé and matrix rational interpolation. *Num. Algs.* **3**, 451-61. [8§1]
Van Barel, M., and Bultheel, A. 1992b. A new formal approach to the rational interpolation problem. *Num. Math.* **62**, 87-122. [7§1]
Van Dyke, M. 1958. The supersonic blunt body problem — review and extension. *J. of Aerospace Sciences* **25**, 485-96.
Van Iseghem, J. 1985. Vector Padé approximants. In *Numerical Mathematics and Applications*, R. Vichnevetsky and J. Vignes (eds.), pp. 73-7. North-Holland, Amsterdam. [8§1]
Van Iseghem, J. 1986. An extended cross rule for vector Padé approximants. *Applied Num. Math.* **2**, 143-55. [8§1]
Van Iseghem, J. 1987a. Vector orthogonal relations. Vector QD algorithm. *J. Comp. Appld. Math.* **19**, 141-50. [8§1]
Van Iseghem, J. 1987b. Approximants de Padé Véctoriels. Thesis, University of Lille. [8§1]
Van Iseghem, J. 1989. Convergence of the vector QD algorithm. Zeros of vector orthogonal polynomials. *J. Comp. Appld. Math.* **25**, 33-46. [8§1]
Van Rossum, H. 1953. A theory of orthogonal polynomials based on the Padé table. Thesis, University of Utrecht, Van Gorcum, Assen. [4§3]
Van Rossum, H. 1971. On the poles of Padé approximants to e^z. *Nieuw Archief voor Wiskunde* **29**, 37-45.
Van Rossum, H. 1979. Orthogonal expansions in indefinite inner product spaces. In *Padé Approximation and Its Applications*, L. Wuytack (ed.), pp. 172-83. Springer Lecture Notes in Mathematics, no. 765. Springer, Berlin.
Van Tuyl, A. H. 1971. Use of Padé fractions in the calculation of blunt body flows. *AIAA J.* **9**, 1431-3.
Van Tuyl, A. H. 1973. Calculation of nozzle flows using Padé fractions. *AIAA J.* **11**, 537-41.
Van Vleck, E. B. 1903. On an extension of the 1894 memoir of Stieltjes. *Trans. Am. Math. Soc.* **4**, 297-332. [5§1]
Van Vleck, E. B. 1904. On the convergence of algebraic continued fractions whose coefficients have limiting values. *Trans. Am. Math. Soc.* **5**, 253-62. [4§7]
Varga, R. S. 1961. On higher order stable implicit methods for solving parabolic partial differential equations. *J. Math. Phys.* **40**, 220-31.
Varga, R. S. 1962. *Matrix Iterative Analysis*. Prentice-Hall, Englewood Cliffs, NJ. [8§4, 10§4]
Vavilov, V. V.; Prokhorov, V. A.; and Suetin, S. P. 1983. The poles of the mth row of the Padé table and the singular points of a function. *Mat. Sbornik* **122** (164), 475-80; *Math. USSR Sbornik* **50**, 457-63 (1985). [6§1]
Villani, M. 1972. A summation method for perturbative series with divergent terms. In *Cargèse Lecture Notes in Physics*, D. Bessis (ed.), vol. 5, pp. 461-74. Gordon and Breach, New York.

Viskovatov, B. 1803–6. De la méthode générale pour réduire toutes sortes des quantités en fractions continues. *Memoires de L'Academie Impériale des Sciences de St. Petersburg* **1**, 226–47. [4§2]

Vorobyev, Y. V. 1965. *Method of Moments in Applied Mathematics*. Gordon and Breach, New York. [9§2]

Waadeland, H. 1979. General T-fractions corresponding to functions satisfying certain boundedness conditions. *J. Approx. Theory* **26**, 317–28.

Wall, H. S. 1929. On the Padé approximants associated with the continued fraction and series of Stieltjes. *Trans. Am. Math. Soc.* **31**, 91–115.

Wall, H. S. 1931. On the Padé approximants associated with a positive definite power series. *Trans. Am. Math. Soc.* **33**, 511–32. [4§5]

Wall, H. S. 1932a. On the relationship among the diagonal files of a Padé table. *Bull. Am. Math. Soc.* **38**, 752–60. [4§5]

Wall, H. S. 1932b. General theorems on the convergence of sequences of Padé approximants. *Trans. Am. Math. Soc.* **34**, 409–16. [4§5]

Wall, H. S. 1945. Note on a certain continued fraction. *Bull. Am. Math. Soc.* **51**, 930–4. [4§7]

Wall, H. S. 1948. *The Analytic Theory of Continued Fractions*. Van Nostrand, Princeton, NJ. [4§5, 4§6, 5§1, 5§4.1, 5§5, 5§6]

Wallin, H. 1972. On the convergence theory of Padé approximants. In *Linear Operators and Approximation (Proc. Conf. at Oberwolfach)*, P. L. Butzer, J.-P. Kahane, and B. Sz.-Nagy (eds.), pp. 461–9. Birkhäuser, Basel.

Wallin, H. 1974. The convergence of Padé approximants and the size of the power series coefficients. *Appl. Anal.* **4**, 235–51. [6§6]

Wallin, H. 1984. Convergence and divergence of multipoint Padé approximants of meromorphic functions. In *Rational Approximation and Interpolation*, P. R. Graves-Morris, E. B. Saff, and R. S. Varga (eds.), pp. 272–84. Springer Lecture Notes in Mathematics no. 1105. Springer, Berlin. [6§3, 7§1]

Walsh, J. L. 1964a. Convergence of sequences of rational functions of best approximation I. *Math. Ann.* **155**, 252–64. [7§1, 10§6]

Walsh, J. L. 1964b. Padé approximants as limits of rational functions of best approximation. *J. Math. and Mech.* **13**, 305–12. [7§1, 10§6]

Walsh, J. L. 1965a. Convergence of sequences of rational functions of best approximation II. *Trans. Am. Math. Soc.* **116**, 227–37. [7§1, 10§6]

Walsh, J. L. 1965b. The convergence of sequences of rational functions of best approximation with some free poles. In *Approximation of Functions*, H. L. Garabedian (ed.), pp. 1–16. Elsevier, Amsterdam. [7§1, 10§6]

Walsh, J. L. 1967. On the convergence of sequences of rational functions. *SIAM J. Num. Anal.* **4**, 211–21.

Walsh, J. L. 1969. *Interpolation and Approximation by Rational Functions in the Complex Domain*. Am. Math. Soc. Colloq. Pub., vol. 20. AMS, Providence, RI. [6§6]

Walsh, J. L. 1970. Approximation by rational functions: open problems. *J. Approx. Theory* **3**, 236–42. [6§6, 6§7, 10§6]

Wanner, G.; Hairer, E.; and Nørsett, S. P. 1978. A note on unconditionally stable linear multistep methods. *BIT* **7**, 65–70. [5§7]

Warburton, A. E. A., and Stern, M. S. 1968. Two body off-shell potential scattering. *Nuovo Cim.* **60A**, 131–59. [9§4]

Warner, D. D. 1974. Hermite interpolation with rational functions. Ph.D. Thesis, University of California, San Diego. [7§1]

Warner, D. D. 1976. An extension of Saff's theorem on the convergence of interpolating rational functions. *J. Approx. Theory* **18**, 108–18. [7§1]

Watson, P. J. S. 1973. Algorithms for differentiation and integration. In *Padé Approximants and Their Applications*, P. R. Graves-Morris (ed.), pp. 93–8. Academic Press, New York.

Watson, P. J. S. 1974. Two variable rational approximants—a new method. *J. Phys. A* **7**, L167–70. [7§6]

Weniger, E. J. 1989. Nonlinear sequence transformations for the acceleration of convergence and the summation of divergent series. *Comput. Phys. Rep.* **10**, 189–371. [10§1]

Weniger, E. J. 1994. Verallgemeinerte Summationsprozesse als Numerische Hilfsmittel für Quantenmechanische und Quantenchemische Rechnung. Thesis, University of Regensburg, Germany.

Weniger, E. J.; Čížek, J.; and Vinette, F. 1993. The summation of the ordinary and renormalized perturbation series for the ground state energy of the quartic, sextic, and octic anharmonic oscillators using nonlinear sequence transformations. *J. Math. Phys.* **34**, 571–609. [10§1]

Werner, H. 1979. A reliable method for rational interpolation. In *Padé Approximation and Its Applications*, L. Wuytack (ed.), pp. 257–77. Springer Lecture Notes in Mathematics, no. 765, Springer, Berlin. [4§2, 7§1]

Werner, H. 1980. Ein Algorithmus zur rationalen Interpolation. In *Numerische Methoden der Approximationstheorie 5*, L. Collatz, G. Meinardus, and H. Werner (eds.), vol. 5, pp. 319–37. Birkhaüser, Basel. [4§2]

Werner, H., and Schaback, R. 1972. *Praktische Mathematik II*. Springer, Berlin. [7§1]

Werner, H., and Wuytack, L. 1978. Nonlinear quadrature rules in the presence of a singularity. *Comp. & Math. with Applcns.* **4**, 237–45.

Wetterling, W. 1963. Ein Interpolationsverfahren zur Lösung der linearen Gleichungssysteme, die bei der rationalen Tchebyscheff-Approximation auftreten. *Archiv. Rat. Mech. und Anal.* **12**, 403–8. [7§1]

Wheeler, J. C., and Gordon, R. G. 1970. Bounds for averages using moment constraints. In *The Padé Approximant in Theoretical Physics*, G. A. Baker, Jr., and J. L. Gammel (eds.), pp. 99–128. Academic Press, New York.

Widder, D. V. 1972. *The Laplace Transform*. Princeton University Press, Princeton, NJ. [5§4.1, 5§5]

Wigert, S. 1923. Sur les polynomes orthogonaux et l'approximation des fractions continues. *Ark. Mat. Astr. Fys.* **17**, 18. [4§3]

Wilf, H. S. 1970. *Finite Sections of Some Classical Inequalities*, Springer, Berlin. [2§1]

Wilkinson, J. H. 1963. *Rounding Error in Algebraic Processes*. Notes in Applied Science, No. 32, H.M.S.O., London. [2§1]

Wilkinson, J. H. 1965. *The Algebraic Eigenvalue Problem*. Oxford University Press, Oxford. [2§1]

Wilson, R. 1927. Divergent continued fractions and polar singularities. *Proc. Lond. Math. Soc.* **26**, 159–68. [6§1]

Wilson, R. 1928a. Divergent continued fractions and polar singularities II. Boundary pole multiple. *Proc. Lond. Math. Soc.* **27**, 497–512. [6§2]

Wilson, R. 1928b. Divergent continued fractions and polar singularities III. Several boundary poles. *Proc. Lond. Math. Soc.* **28**, 128–44. [6§2]

Wilson, R. 1930. Divergent continued fractions and nonpolar singularities. *Proc. Lond. Math. Soc.* **30**, 38–57. [6§2]
Wimp, J. 1981. *Sequence Transformations and Their Applications*. Academic Press, New York.
Wimp, J. 1987. Explicit formulas for the associated Jacobi polynomials and some applications. *Can. J. Math.* **39**, 983–1000. [2§3]
Wood, D. W., and Griffiths, H. P. 1974. Chisholm approximants and critical phenomena. *J. Phys. A* **7**, L101–4.
Woodcock, C. F., and Graves-Morris, P. R. 1993. Solution to Problem 92–20. *SIAM Review* **35**, 649–651
Woodcock, C. F., and Graves-Morris, P. R. 1995. Generalised discriminants. *Bull. Australian Math. Soc.*, in press. [8§4]
Wragg, A., and Davies, C. 1975. Computation of the exponential of a matrix II. Practical considerations. *J. Inst. Math. Appl.* **15**, 273–8.
Wright, K. 1970. Some relationships between implicit Runge–Kutta, collocation and Lanczos τ-methods and their stability properties. *BIT* **10**, 217–27. [10§3]
Wuytack, L. 1973. An algorithm for rational interpolation similar to the Q.D.-algorithm, *Num. Math.* **20**, 418–24. [7§1]
Wuytack, L. 1974a. Numerical integration by using nonlinear techniques. *J. Comp. Appl. Math.* **1**, 261–72.
Wuytack, L. 1974b. On some aspects of the rational interpolation problem. *SIAM J. Num. Anal.* **11**, 52–60.
Wuytack, L. 1975. On the osculatory rational interpolation problem. *Math. Comp.* **29**, 837–43.
Wuytack, L. 1976. Applications of Padé approximants in numerical analysis. In *Approximation Theory*, R. Schaback and K. Schere (eds.), pp. 453–61. Springer Lecture Notes in Mathematics, no. 556. Springer, Berlin.
Wuytack, L. (ed.). 1979a. *Padé Approximation and Its Applications*. Springer Lecture Notes in Mathematics, no. 765. Springer, Berlin.
Wuytack, L. 1979b. Commented bibliography. In *Padé Approximation and Its Applications*, L. Wuytack (ed.), pp. 375–92. Springer Lecture Notes in Mathematics, no. 765. Springer, Berlin.
Wynn, P. 1956. On a device for calculating the $e_m(S_n)$ transformation. *Math. Tables and Auto. Comp.* **10**, 91–6. [3§3]
Wynn, P. 1960. The rational approximation of functions which are formally defined by power series expansions. *Math. Comp.* **14**, 147–86. [3§3]
Wynn, P. 1961a. The epsilon algorithm and operational formulas of numerical analysis. *Math. Comp.* **15**, 151–8. [3§3, 3§4]
Wynn, P. 1961b. L'ε-algoritmo e la tavola di Padé. *Rendi. Mat. Roma* **20**, 403–8.
Wynn, P. 1962a. The numerical efficiency of certain continued fraction expansions. *Proc. Kon. Ned. Akad. Wetensch A* **65**, 127–54. [7§1]
Wynn, P. 1962b. Acceleration techniques for iterated vector and matrix problems. *Math. Comp.* **16**, 301–22. [3§3, 8§2, 8§4]
Wynn, P. 1963. Continued fractions whose coefficients obey a noncommutative law of multiplication. *Arch. Rat. Mech. Anal.* **12**, 273–312. [8§2]
Wynn, P. 1964. General purpose vector epsilon algorithm procedures. *Num. Math.* **6**, 22–36. [3§3, 8§4]
Wynn, P. 1966a. Upon systems of recursions which obtain among quotients of the Padé table. *Num. Math.* **8**, 264–9.

Wynn, P. 1966b. On the convergence and stability of the ε-algorithm. *SIAM J. Num. Anal.* **3**, 91–122. [5§4.1]

Wynn, P. 1967. A general system of orthogonal polynomials. *Quart. J. Math.* **18**, 81–96. [4§3]

Wynn, P. 1968a. Vector continued fractions. *Lin. Alg. Applcns.* **1**, 357–95. [8§4]

Wynn, P. 1968b. Upon the Padé table derived from a Stieltjes series. *SIAM J. Num. Anal.* **5**, 805–34. [5§2]

Wynn, P. 1973. On the zeros of certain confluent hypergeometric functions. *Proc. Am. Math. Soc.* **40**, 173–82.

Wynn, P. 1974. Some recent developments in the theories of continued fractions and the Padé table. *Rocky Mtn. J. Math.* **4**, 297–323.

Xu, G. L., and Bultheel, A. 1990. Matrix Padé approximation: definitions and properties. *Lin. Alg. Applcns.* **137–8**, 67–136. [8§3]

Young, R. C.; Biedenharn, L. C.; and Feenberg, E. 1957. Continued fraction approximants to the Brillouin–Wigner perturbation series. *Phys. Rev.* **106**, 1151–5. [9§5]

Zinn-Justin, J. 1970. Strong interaction dynamics with Padé approximants. *Phys. Repts.* **1C**, 56–102. [2§2, 5§4, 7§1]

Zinn-Justin, J. 1971. Convergence of Padé approximants in the general case. In *Colloquium on Advanced Computing Methods in Theoretical Physics*, A. Visconti (ed.), pp. 88–102. C.N.R.S., Marseille. [6§4, 6§5]

Zinn-Justin, J. 1973. Recent developments in the theory of Padé approximants. In *International Colloquium on Advanced Computing Methods in Theoretical Physics*, A. Visconti (ed.), vol. 2, pp. C-XIII-1. C.N.R.S. Marseille. [11§2]

Zohar, S. 1974. The solution of a Toeplitz set of linear equations. *J. Assoc. Comp. Mech.* **21**, 272–6. [7§5]

INDEX

acceleration of convergence, 17, 67–85, 490, 628–33
accuracy, numerical, 38–44, 54–6
accuracy-through-order, 1
   for algebraic approximants, 494, 531–8
   for integral approximants, 532
   for simultaneous Padé approximants, 416
   for vector Padé approximants, 468, 490
Aitken's Δ^2 method, 67–70, 74
algebraic approximants, 524–69
   convergence of, 552
   definition of, 531
   equivalence theorem for, 539
   invariance under argument transformations, 539
   invariance under value transformations of, 540
   pointwise convergence of, 546
   uniqueness of convergence of, 544
algebraic polynomials, 531
   existence of essentially unique, 552
ε-algorithm, 73–85, 632
   for vector sequences, 466
   generalized, 348
η-algorithm, 77
algorithms for rational interpolation, 341–50
anharmonic oscillator, 678
annulus theorem of Saff and Varga, 270
Arzela's theorem, 210
associated continued fraction, 166
A-acceptability, 267, 651
A-stability, 651
asymptotic convergence, 238
asymptotic expansion, 258

backward recursion method, 127
Baker algorithm, 43
Baker definition, 21, 432
Baker–Gammel approximants, 362–72, 385
   asymptotic, 371

Baker, Gammel, and Wills
   theorem of, 33
   conjecture of, 332–4
Baker–Hunter method, 52
Beardon's theorems, 277
Beckermann minimal region, 500
Beckermann minimality, 500
Beckermann structures, 503
Berlekamp–Massey algorithm, 153–65
Bethe–Salpeter equation, 610
bigradients, 62–6, 95, 439
   polynomial, 62–6, 95
binomial function, 174, 179
biorthogonal algorithm, 581–4
blocks, 25–31, 108–15, 138–41, 157–63
bounds for Stieltjes functions, 223
Boutroux–Cartan lemma, 325
Brezinski's topological epsilon algorithm, 419

calculation of Padé approximants
   algebraic, 8–13, 43, 56–62, 148–53
   numerical, 38–44
Canterbury approximants, 403–8
capacity, 319
Carathéodory–Fejér method, 663–9
cardioid theorem, 186
Carleman's criterion, 238–43
   application of, 677
   for Hamburger series, 260
   for Stieltjes series, 240
Cartan's lemma, 316
Cauchy–Binet formula, 281
Cauchy–Jacobi problem, 335
C-fraction, 165
Chebychev. *See* Tchebycheff
Chisholm approximants, 403–8
chordal metric, 297–9
Clenshaw–Lord algorithm, 387
Clifford algebra, 473–7

C(L/M), definition of, 22
coefficient problem, 43, 341
collocation, 639–45
complementary error function, 175, 179
complete analytic function, 525
condition number, 41, 95–100
confluent singularities, 53
continued fractions, 122–92
    associated, 166
    contraction of, 166, 182
    convergents of. *See* convergents
    corresponding, 165
    definition of, 123, 126
    divergence condition for, 183
    elements of, 123, 143
    equivalence transformation of, 124
    recursion relations for, 125, 130, 144
    regular corresponding, 165
    summation formula, 127
    terminating, 123
convergence
    in capacity, 316–30, 564
    in measure, 305–23
convergence of
    diagonal sequences, 304–34
    equicontinuous sequences, 210
    general ray sequences, 312–30
    general sequences, 305–30
    Padé approximants for Hamburger series, 260
    Padé approximants for Stieltjes series, 201–45
    ray sequences for exponential function, 273
    ray sequences for Stieltjes series, 230
    real J-fractions, 261
    row sequences, 276–308
    staircase sequences, 185–92
    vector Padé approximants, 488
convergents, 123, 132, 141–6
Cordellier's identity, 113
counterexamples of
    Buslaev, Gončar, and Suetin, 278
    Gammel, 331
    Graves-Morris, 575
    Labahn and Cabay, 435
    Lubinsky and Saff, 279
    Perron, 278
coupling constant renormalization, 688
Crandall's method, 653
Crank–Nicholson method, 646–54
critical point, methods for, 51–6, 372–8, 564–9, 687
C-table, 23, 30

Dawson integral, 361
defect, 48–50, 54
deficiency index, 21
deficiency of algebraic polynomials, 534
degree of contact bound for algebraic polynomials, 536

de Montessus' theorem, 282–96
    generalization of, 350
de Montessus type theorems, 407–8, 421, 488
density function
    bounds for, 633
    construction of, 216
determinacy of moment problem, 213, 232, 236
determinantal formulas
    for denominator of a vector Padé approximant, 470
    for Hermite–Padé polynomials, 495
    for N-point Padé polynomials, 339
    for Padé polynomials, 4–8, 142
    for simultaneous Padé denominators, 417–19
determinantal identities for Stieltjes series coefficients, 200
determinantal inequalities
    for Hamburger series coefficients, 246, 264
    for Polyá series coefficients, 275
    for Stieltjes series coefficients, 197, 245
diagonal sequences: 18, 33, 276, 304–34
    See also paradiagonal sequences
diffusion processes, 646–54
disk monodromy theorem, 543
disk problems, 304, 306
divergence condition, 183, 191
divided differences, 336
D-log Padé approximants, 51, 373
$D(m, n)$, definition of, 197
duality, 32, 265, 276

E-algorithm, 629
energy integral, 425
epsilon algorithm. *See* ε-algorithm
equicontinuous sequences, 209, 248, 301
equivalence transformation, 124
error formula for Padé approximants of
    Hamburger series, 227
    Stieltjes series, 220, 222
error formula for Padé approximation, 6, 291
    from variational principles, 606–8
error function, 175, 361
essential singularity, 45
Euclidean algorithm, 106
Euclidean field theory, 684–9
Euler's corresponding fraction, 167
Euler's function and series, 18, 238
Euler's recurrence theorem, 125
Euler's summation formula, 127
existence of Padé approximants, 28–9, 31
exploding vacuum, 675
exponential approximants, 368
exponential function
    continued fractions for, 173
    Padé approximants for, 8–15, 643, 664
    rational approximation of, 669

Saff–Varga theorems for, 268–73
exponential integral
  continued fraction for, 175, 179, 672
  extended Stieltjes series. *See* Hamburger series
eye theorem of Saff and Varga, 271

factorization theorem, 405
factorization condition, 468
Fisher approximants, 411
forward recursion method, 126
Frobenius definition, 20, 22
Frobenius' formula, 23
  generalized, 507
Frobenius identities, 82, 85–9, 394
functional Padé approximants, 492–4

gamma function, Binet's formula for, 240
Gammel–Guttman–Gaunt–Joyce approximants, 372–8
Gammel's counterexample, 331
general C-fraction, 135
generalized hypergeometric function, 60–2, 150–2
generalized inverse, 466
gnomic theory, 580
Green's functions, 326, 646
  bound-state, 615
  partial wave
    free, 593
    Jost solution, 594
    momentum space, 616
    standing wave, free, 593
  S-wave, exponential potential, 601
  three-dimensional
    free, 586
    relativistic, free, 611
    relativistic, standing wave, free, 611
    standing wave, free, 589
Green's potentials, 562
Gregory's series, 75, 78

Hadamard's determinant theorem, 65, 284
Hamburger functions as real J-fractions, 261–2
Hamburger function, definition of, 245
Hamburger moment, definition of, 246
Hamburger series
  definition of, 246
  Padé approximation of, 102–5, 246–62
Hamburger's theorem, 258
Hankel determinant, 7
Hankel matrix, 93–105
Hankel matrix, condition number of, 41, 95–100
Hardy's puzzle, 76
Hausdorff measure, α-dimensional, 324
Hausdorff moment problem, 233–4
Herglotz functions, 262
Hermite–Padé polynomials, 427, 494–569
  axis sequence of, 524

basis set of, 504
definition of, 495
definition of minimal, 497
determinantal identities for, 506–12
diagonal identity for, 518
existence of subsequences of, 523
linear chain identity for, 520–4
modified minimality for, 521
near diagonal sequence of, 524
near partially diagonal sequence of, 524
nearest neighbor sequence of, 523
optimal, 499
table of, 501–15
Hermite's formula, 290, 336
high field expansions, 687
Hilbert space methods, 409–10, 573–627
homographic invariance, 32–5, 404, 406, 440
Hughes Jones approximants, 405
hybrid vector Padé approximants, 489–91
hyperbolic tangent, 174, 179
hypergeometric functions, 176–7
  $_2F_1(.)$, Padé approximants for, 57–61, 176, 178, 181, 190, 644
  $_1F_1(.)$, Padé approximants for, 57–61, 177, 180
  $_2F_0(.)$, Padé approximants for, 57–61, 176, 177, 180, 236
  generalized, 60–2, 150–2

identities for neighboring approximants, 85–92
inclusion regions, 206, 243
incomplete Gamma function, 176, 179, 361
inequalities for density function, 633–9
integer moment problem, 234
integral approximants, 524–69
  convergence of, 553
  definition of, 532
  equivalence theorem for, 538
  invariance under argument transformations for, 541
  invariance under value transformations for, 542
  pointwise convergence of, 546
  separation property for, 543
  uniqueness of convergence of, 544
integral equations, 492, 570–627
integral polynomials
  definition of, 531
  existence of essentially unique, 552
  first order theorem for, 555
interlacing, 202–4, 247, 264
invariance, homographic, 32–5, 404, 440
invariance properties of algebraic approximants, 539–42
invariance properties of integral approximants, 542
inverse hyperbolic tangent, 174
inverse problems, 280
inverse tangent, 174, 179, 376

Index

Ising model
  continuous spin, 684
  one dimensional, 688

J-fraction, 166
Jost method, 622

kernels
  compact, 575
  completely continuous, 575
  finite rank, 571, 580
K-matrix, 589, 609, 612
Kronecker indices, 448–53
Kronecker's algorithm, 43, 106–12, 341–3, 477
Kronecker's theorem, 453
Kung–Kailath method, 445

lacunary series, 5, 46
Laguerre's method, 671
Lanczos biorthogonal algorithm, 581–4
Lanczos τ-method, 639
Laplace transform, inversion of, 654–6
Laurent–Padé approximation, 389–402
Laurent's theorem, 44
lattice-cutoff field theory, 684–9
Legendre polynomials, 576, 591, 640
lemniscates, 316–25
Le Roy function, 372
Levin algorithms, 630
Levinson algorithm, 396
Lippman–Schwinger equation, 587, 614
local monodromy dimension, 543
logarithmic capacity, 324
L-acceptability, 651

McLeod's theorem, 484
Magnus' P-fraction, 135
Mahler's theorem, 427–9
Markov parameters, 164, 442, 463–4
Markov problem, 638
Markov's theorem, 228
matrix Padé approximants, 429–66
maximum modulus principle, 318, 329
measure, convergence in, 305–323
meromorphic functions, convergence of sequences of, 300
τ-method, 639
minimal domain theorem, 563
minimal element of a block, 28, 108, 157
minimal Hermite–Padé polynomials, 497
minimal partial realization problem, 442–66
minimality, definition of, 497
  Beckermann, 500
Mittag-Leffler star, 46
moment
  definition of Hamburger, 246
  definition of Hausdorff, 233
  definition of Stieltjes, 193
moment method, 409–10

moment problems, 213–64
moments, bounds using, 638
monodromy dimension, local, 543
monodromy matrices, 526
monodromy theory, 525
Moore–Penrose inverse, 466
multi-index, 281
multi-niqueness theorem, 496
multiple input, multiple output systems, 442
multipoint Padé approximation, 335–62
multipole, 45
multiseries approximants, 415
multivalued (multiform) functions, 524
multivariable approximants, 402–14

natural logarithm, 174, 179
Newton interpolating polynomials, 337
Newton–Padé approximants, 335, 350
N-point Padé approximants, 335
nullity lemma, 504
numerical accuracy, 38–44, 54–6, 127
Nuttall's compact form, 16–17
Nuttall's notation, 137
Nuttall's theorem, 310

O_+, O_- notation, 352
operator Padé approximants, 429–42
optimal Hermite–Padé approximants, 499
order star, 14, 266–8, 653
orthogonal polynomials, 217–22, 227, 232, 247
osculatory rational interpolation, 335

Padé approximant
  Baker definition of, 21
  Frobenius definition of, 20
Padé–Borel approximation, 369
Padé denominator
  Baker definition of, 21
  determinantal definition of, 4
Padé equations, 2
Padé–Fourier approximants, 382–3
Padé–Frobenius definition, 20, 22
Padé–Laurent approximants, 378–83
Padé–Legendre series, 362, 369
Padé polynomials, determinantal definition of, 4–6
Padé table, 7, 28
Padé–Tchebycheff approximants, 383–9
Padé type approximant, 629
parabola theorem for continued fractions, 185
parabola theorem of Saff and Varga, 270
paradiagonal sequences, 31, 97, 276, 438
partial ordering, 500
partial theta function, 151, 279
Peres model, 675
Perron's counterexample, 278
perturbed harmonic oscillator, 674–9
P-fraction, 135

PC fractions, 172
phantom fields, 688
pion–pion scattering, 679–84
poles and zeros of Padé approximants, 14–15, 44–50, 115–21, 153, 268, 305–34
    for Hamburger series, 246, 264
    for Stieltjes series, 201–4, 220, 221, 229–32
polewise independence, 421
poloids, 683
Pólya frequency series, 264–75
polynomially independent, 496
Pommerenke's theorem, 312
potential scattering, 584–627
potentials of polynomials, 425
progression theorems, 93, 398–402, 437
projection techniques, 578–84
prong method, 406
Prym's function, 175

q-binomial coefficient, 280
q-hypergeometric functions, 60–2, 150–2
Q.D. algorithm, 43, 115–21, 146–53
    generalized, 347
    for T-fractions, 359
Q.D. table, 119, 148–50
quadratic approximants, 376, 413
quantum theory, 584–627
quasi-analytic functions, 47

ratio method, 373
rational approximation, 656–69
rational interpolation, 335–62
ray sequences, 225, 312–30
real J-fractions, 167, 261
real symmetric functions, 195
rectangular matrix Padé approximants, 442–66
recursion relations
    backward, 127
    forward, 126
recursion relations for
    continued fractions, 125, 130
    Hermite–Padé polynomials, 515–21
    Padé polynomials, 85–90, 144, 154–65
recursive calculation of Padé approximants, 81–5, 92–115
regular C-fraction, 165
reliability, 39, 95–103, 110, 139–41, 341
remainder series, 93
renormalized perturbation theory, 684–9
reverse automorphism, 476
rhombus rule, 74–8
Riccati equation
    generalized, 542
    Padé approximants for, 670–3
Riemann's monodromy theorem, 528
Riemann–Papperitz equation, 542
Riemann sphere, 297–9

Rogers–Szegö polynomials, 280
root problem, 115–21
Rouché's theorem, 320
Runge's theorem, 656–61

Saff's theorem, 350
Saff's theorem, 350
Saff–Varga theorems, 268–74
scattering theory, quantum mechanical, 584–627
Schur's algorithm, 395
Schwarz's lemma, 223, 226
Schwein's identity, 507
sectorial theorem of Saff and Varga, 268
Seidel's theorem, 183
separation property, 543
    convergence theorem, 554
    divergence theorem, 557
sequence
    acceleration of convergence of, 17, 67–85
    column, 31
    diagonal, 31, 33
    paradiagonal, 31, 97
    row, 31
series, accleration of convergence of, 17, 67–85
series analysis, 51–56, 372–8
S-fraction, 166, 199, 244
Shafer approximants, 375–7
signal processing, 161–5, 455–64
simplex, isolated, 510
    rule of, 509
    vertex solution for isolated, 512
simplices, intersecting, 513
simultaneous Padé approximants, 415–29
single input, single output, 163
single-sign potentials, 608
singular index computations, 51–6, 372–8, 564–9
singular potentials, 622–7
singularities, decipherment of, 44–56
S-matrix, partial wave, 593
SOR iteration, 486
spherical convergence, 297–316
stability, 40
Stahl's extremal domain theorem, 326
Stahl's Padé convergence theorem, 326
staircase sequences, 132, 154–63
star identity, 23
    generalized, 508
Stieltjes function, 166, 199, 244
    bounds for, 223
    definition of, 193
Stieltjes inversion formula, 257–8
Stieltjes series
    definition of, 194
    inequalities for Padé approximants for, 204, 223
    example of, 196
    simultaneous approximation for, 422–6
    S-fraction for, 199

Sturm sequence, 201
superfast algorithms, 104
Sylvester resultant, 530, 545
Sylvester's theorem, 24

tangent function, 174
tau-method, 639
Tchebycheff inequalities, density function for, 633–8
Tchebycheff polynomials, 317–18, 640
T-fractions, 170–3, 352–62
Thacher–Tukey algorithm, 345
Thiele's reciprocal difference method, 343
T-matrix, 586, 602, 680
   partial wave, 593
Toeplitz determinants, 275, 392
Toeplitz matrix, 390, 393–4
topological epsilon algorithm, 419
totally monotone sequence, 234
totally positive series, 264
transfer function, 164, 455
Trudi's theorem, 66
truncation theorem, 35
two-point Padé approximants, 167, 400

uniform boundedness convergence criterion, 303
unimodular matrix, 164
uniqueness of
   minimal Hermite–Padé polynomials, 498
   operator Padé approximants, 432
   optimal Hermite–Padé approximants, 499
   Padé approximants, 28–9
   vector Padé approximants, 469

unitarity, 36, 405, 440, 681
universality, 688

value problem, 43, 341
Vandermonde determinants, 285
Van Vleck's theorem, 187
variational Padé approximants, 616–22
variational principle, 440
   Bessis, 604
   Kohn, 600, 606
   Rayleigh–Ritz, 596
   Schwinger, 600
vector epsilon-algorithm, 466
vector epsilon-table, 468
vector Padé approximants, 466–94
vector-valued continued fractions, 472
vector-valued power series, 468
Viskovatov's method, 133–4
   generalized, 135–41, 472
V-matrix, 164, 461–3

Walsh theorems, 661–2
Weierstrass's (approximation) theorem, 574
Weierstrass's (double series) theorem, 211
Weierstrass's (essential singularity) theorem, 47
Werner's algorithm, 135
Wronskian, 591
Wynn's identity, 81–5, 482

zeros of functions, numerical methods for, 351
Zolotarev constant, 639